DEPARTMENT OF THE ARMY
U.S. Army Corps of Engineers
Washington, DC 20314-1000

EM 1110-2-1100
Change 3

CECW-CE

Manual
No. 1110-2-1100

28 September 2011

Engineering and Design
COASTAL ENGINEERING MANUAL

1. This Change 3 to EM 1110-2-1100, 30 April 2002, adds Part VI, the second of two engineering-based parts oriented toward design of coastal projects. Part VI contains specific information and procedures for designing the structural elements used in coastal navigation and shore protection projects.

2. The added Part VI has 8 chapters.

Chapter 1 – Introduction to Coastal Project Element Design.
Chapter 2 – Types and Functions of Coastal Structures.
Chapter 3 – Site Specific Design Conditions.
Chapter 4 – Materials and Construction Aspects.
Chapter 5 – Fundamentals of Design.
Chapter 6 – Reliability Based Design of Coastal Structures.
Chapter 7 – Example Problems.
Chapter 8 – Monitoring, Maintenance, and Repair of Coastal Projects.

DIONYSIOS ANNINOS
Colonel, Corps of Engineers
Chief of Staff

Outline – EM 1110-2-1100 Part VI
757 pages total

Chapter 5 – Fundamentals of Design – 364 pages

Chapter 6 – Reliability Based Design of Coastal Structures – 56 pages

Chapter 7 – Example Problems, 93 pages

Chapter 8 – Monitoring, Maintenance, and Repair of Coastal Projects – 74 pages

CHAPTER VI-1

Introduction to Coastal Project Element Design

TABLE OF CONTENTS

List of Tables

CHAPTER VI-1

Introduction to Coastal Project Element Design

VI-1-1. Introduction to Part VI. Part VI is the second of two engineering-based parts oriented toward design of coastal projects. Part V (Coastal Project Planning and Design) focuses on aspects related to overall coastal project design and development, whereas Part VI contains specific information and procedures for designing the structural elements used in coastal navigation and shore protection projects.

a. Part VI Overview. Part VI contains eight chapters as described below.

(1) Chapter VI-1 briefly overviews and summarizes the contents of each Part VI chapter and provides overall guidance on using the design procedures.

(2) Chapter VI-2, "Types and Functions of Coastal Structures," begins with brief descriptions of the various types and functions of coastal structures most often used in coastal projects. Typical cross sections and layouts are illustrated in the second section along with descriptions of the usual construction for each structure type. The third section introduces the types of concrete armor units used on coastal structures. The chapter ends with a comprehensive overview of failure modes associated with each structure type. Understanding potential failure modes is critical for design, and this understanding is aided by many figures in the chapter illustrating the various failure modes.

(3) Chapter VI-3, "Site-Specific Design Conditions" focuses on site-specific design information that may need to be considered during preliminary and final design. Whereas several project sites might experience similar design wave and water-level conditions, there may be unique conditions at each site that could significantly influence final design. Site specific design conditions and considerations discussed in this chapter include foundation and geotechnical requirements, seasonal profile variation, structure flanking, seismic activity, ice, environmental considerations, aspects related to construction, maintenance, aesthetics, etc.

(4) Chapter VI-4, "Materials and Construction Aspects," begins with a section discussing general requirements of materials used in coastal projects such as material properties, durability, adaptability, cost, availability, handling, and maintenance requirements. More detail and specific design guidance and considerations are given in individual sections for earth and sand, stone, concrete, steel and other metals, wood, geotextiles, and plastics. Each section overviews typical use of the material in coastal projects, lists the physical and mechanical properties of the material, and discusses placement considerations and potential environmental impacts.

(5) Chapter VI-5, "Fundamentals of Design," is the heart of Part VI, and contains most of the guidance and methodology needed for designing coastal structure cross sections (with the exception of hydrodynamic criteria, which are presented in Part II, "Coastal Hydrodynamics"). A brief introduction is followed by a section giving guidance related to the hydraulic response of the waves at the structure (i.e., wave runup, overtopping, transmission, and reflection). Next are two sections containing methodology for determining loading and response for sloping-front

rubble-mound structures and vertical-front structures, respectively. Foundation loads are also reviewed in this chapter. The intention is to generate awareness of when geotechnical aspects should be considered in the design process. Estimation of scour for different situations is covered next along with procedures for designing scour protection. Methods for estimating wave forces on slender cylindrical piles are followed by the final section, which summarizes briefly impact forces and ice forces that might occur at structures.

(6) Chapter VI-6 introduces reliability-based design of coastal structures. Familiar deterministic design equations are cast into the form of "failure functions," which can be evaluated in terms of random loading and response variables. Sophisticated probability methods are overviewed, but application requires the distribution of the random variables be reasonably well established. The partial safety factor system is presented that permits a less rigorous, but still useful, reliability analysis based on tabulated coefficients for specific design formulas. Overall system reliability can be analyzed based on failure probability of system components.

(7) Chapter VI-7, "Example Problems," illustrates application of design guidance and formulations given in previous chapters through case studies and example problems. The examples cover the most common applications such as wave runup, wave overtopping, armor layer stability, and stability of vertical-walled bulkheads and caissons, and forces on cylindrical piles.

(8) Chapter VI-8 begins with an overview of coastal project maintenance followed by a section on project monitoring and inspection. Condition monitoring occurs over the life of the structure whereas performance monitoring is usually short-term monitoring to access project performance. Each type of monitoring has different elements, but often shares the same instrumentation or techniques. Coverage includes evaluation of structure condition, types of instruments, and monitoring plan considerations. Repair and rehabilitation of rubble-mound structures after damage or deterioration is needed to assure continued functionality. Techniques for different levels of repair are discussed.

b. Comparison between Part VI and the *Shore Protection Manual*. Part VI corresponds mainly to the topics covered in Chapter 7 of the *Shore Protection Manual* ("Structural Design: Physical Factors"). Part VI is substantially larger than the equivalent portions of the older document, with most topics receiving expanded coverage. Several topics not included in the *Shore Protection Manual* have been added to Part VI. Table VI-1-1 shows the main topics of Part VI and how coverage compares with similar content in the *Shore Protection Manual*.

(1) The two main organizational differences between the *Coastal Engineering Manual* and *Shore Protection Manual* relate to presentation of design guidance and example calculations. Much of the design guidance in the CEM is presented in the form of tables. Each table applies to a specific design formula or procedure. Typically the table includes all the information needed for applying the formula or procedure to a specific structure type including necessary coefficients, variables, range of applicability, and when necessary, a cross-section or planview sketch illustrating the specific structure for which the methodology applies. The tables contain only minimal discussion and background information about the methodology so that Part VI

could be kept to a manageable length. If additional information is needed, see the provided reference to the original source material.

Table VI-1-1		
Topic Coverage in Part VI Compared to *Shore Protection Manual* (1984)		
Section	Topic	Changes from *Shore Protection Manual* (1984)
VI-2-1	Structure types	Similar coverage
VI-2-4	Failure modes	New topic
VI-3	Site-specific design criteria	Revised/updated
VI-4	Construction materials and procedures	New topic
VI-5-2	Wave runup and overtopping	Revised/updated
VI-5-2	Wave transmission and reflection	Revised/updated
VI-5-3	Rubble-mound structures	Revised/expanded
VI-5-4	Vertical-front structures	Revised
VI-5-5	Marine foundations	New topic
VI-5-6	Scour and scour protection	New topic
VI-5-7	Forces on piles	Same/revised
VI-5-8	Other forces on structures	Revised
VI-6	Reliability in design	New topic
VI-7	Design of specific project elements	Same/new examples using new guidance
VI-8	Monitoring, maintenance and repair	New topic

(2) In the *Shore Protection Manual*, some of the design guidance was immediately illustrated by example problems embedded in the text at that location, whereas in the *Coastal Engineering Manual* example applications of various design formulas are collected separately in Part VI-7. Separation of examples from the tables of design guidance is part of the modular design of Part VI. Many of the *Coastal Engineering Manual* example problems use the same conditions given for the corresponding problems in the *Shore Protection Manual*. This allows side-by-side comparison to show how the new guidance of the *Coastal Engineering Manual* differs from the older *Shore Protection Manual* results.

(3) Even with the increased number of pages, there are several instances where specific guidance contained in the *Shore Protection Manual* has not been included in the *Coastal Engineering Manual*. Decisions on whether or not to include certain material were based on considerations of guidance reliability and usefulness, with priority given to more common design needs. For those few situations where no guidance is included in the *Coastal Engineering Manual*, it should be acceptable to use the older guidance found in the *Shore Protection Manual* (or other sources) subject to the restrictions and caveats noted and provided improved procedures are not available in other design manuals. For example, the *Shore Protection Manual* contains provisional guidance for estimating slamming forces on vertical walls caused by breaking waves (Minikin method). This procedure produces high forces, and it is viewed with skepticism by some. Because no commonly accepted procedure was available at the time of writing to replace Minikin's method, it was decided to not include this topic. In other cases, the guidance was

deemed sound, but the application was considered fairly rare compared to the included procedures. These were tough decisions, but the important point to remember is that the *Coastal Engineering Manual* is not all-inclusive; and just because a particular design formula or procedure is not given in the *Coastal Engineering Manual*, this does not imply that it is not valid. That judgment must be made separately based on available information. However, where design guidance provided in the *Shore Protection Manual* differs from newer design methods contained in the *Coastal Engineering Manual*, strong justification is needed before choosing the *Shore Protection Manual* guidance over the preferred *Coastal Engineering Manual* recommendation.

c. Logical Connections to Other *Coastal Engineering Manual* Parts. Part VI is most closely linked with Part V, "Coastal Project Planning and Design" and Part II, "Coastal Hydrodynamics." Part V guides the engineer in selecting and evaluating alternatives to solve coastal problems, and Part VI tells how to design the specific elements (structures) that comprise each alternative. For each alternative, functional project design results in planform dimensions and multiple design objectives related to the purpose and intended function of the structure types used in the project. Contained in these design objectives are specific structure performance criteria such as whether to allow overtopping, permissible wave transmission, allowable damage levels, etc. Some information on general design criteria related to individual structure types can also be found in Part V.

(1) Hydrodynamic loading (i.e., wave parameters, design water levels, currents, etc.) related to specific structure geographic location and exposure are estimated using information provided in *Coastal Engineering Manual* Part II. These estimates form the hydrodynamic design criteria for each structure that is part of the coastal project. Hydrodynamic input requirements are specified for each design formula, and some parameters may vary between formulas. For example, design waves may be specified as significant wave height (H_s or $H_{1/3}$), zeroth-moment wave height (H_{mo}), or some other statistical representative wave height such as $H_{10\%}$ or $H_{1\%}$. Therefore, it is necessary to determine from Part VI what form the hydrodynamic criteria take when estimating the criteria from Part II.

(2) **Important note:** Part II was authored several years before Part VI, and in Part II-7, "Harbor Hydrodynamic," older guidance related to wave reflection and transmission was included that has since been superseded by the updated design guidance provided in Part VI. Although the older guidance provides reasonable estimates, using the formulas for wave reflection and transmission provided in Part VI-5-2 is recommended.

d. Using Part VI in the Design Process. In some cases more than one procedure is available for estimating the result of wave loading on a specific structure type, and separate tables are given for each methodology. Usually, similar results are produced by the different methods, but the range of applicability may differ. Therefore, it is important to first verify the appropriateness of a selected procedure for the specific design problem. If more than one set of guidelines is appropriate, it is wise to perform the calculation by both methods and compare results. Results that are nearly the same indicate either method can be used with confidence. In cases where answers from two methodologies diverge, the engineer must investigate further the background of the competing methods with primary focus on extent and range of experimental measurements on which the method is based, and the general acceptance of the method in the coastal community. For example, one method might be based primarily on shallow water conditions, whereas the other method examined deeper water conditions. An older formula might be based on monochromatic wave experiments, but cover a wide range of incident wave conditions, whereas a newer formula might have been developed with a small number of irregular wave observations. At the end of the day, engineering judgment based on past experience may be needed, and that is acceptable provided concerns and caveats are made known to those with a stake in the project.

VI-1-2. References.

Shore protection manual. (1984). 4th ed., 2 Vol, U.S. Army Engineer Waterways Experiment Station, U.S. Government Printing Office, Washington, DC.

VI-1-3. Acknowledgements.

Author: Steven A. Hughes, PhD, Coastal and Hydraulics Laboratory (CHL), Engineer Research and Development Center, Vicksburg, Mississippi.

Reviewer: Frank A. Santangelo, U.S. Army Engineer District, New York.

VI-1-4. Symbols.

H_{mo}	Zeroth-moment wave height
H_s or $H_{1/3}$	Significant wave height (length)
$H_{10\%}$ or $H_{1\%}$	Statistical representative wave heights

CHAPTER 2

Types and Functions of Coastal Structures

TABLE OF CONTENTS

List of Figures

List of Tables

CHAPTER VI-2

Types and Functions of Coastal Structures

VI-2-1. Applications. Coastal structures are used in coastal defense schemes with the objective of preventing shoreline erosion and flooding of the hinterland. Other objectives include sheltering of harbor basins and harbor entrances against waves, stabilization of navigation channels at inlets, and protection of water intakes and outfalls. An overview of the various types of coastal structures and their application is given in Table VI-2-1. Overall planning and development of coastal projects is covered in Part V.

a. Sea dikes. Sea dikes are onshore structures with the principal function of protecting low-lying areas against flooding. Sea dikes are usually built as a mound of fine materials like sand and clay with a gentle seaward slope in order to reduce the wave runup and the erodible effect of the waves. The surface of the dike is armored with grass, asphalt, stones, or concrete slabs.

b. Seawalls.

(1) Seawalls are onshore structures with the principal function of preventing or alleviating overtopping and flooding of the land and the structures behind due to storm surges and waves. Seawalls are built parallel to the shoreline as a reinforcement of a part of the coastal profile. Quite often seawalls are used to protect promenades, roads, and houses placed seaward of the crest edge of the natural beach profile. In these cases a seawall structure protruding from the natural beach profile must be built. Seawalls range from vertical face structures such as massive gravity concrete walls, tied walls using steel or concrete piling, and stone-filled cribwork to sloping structures with typical surfaces being reinforced concrete slabs, concrete armor units, or stone rubble.

(2) Erosion of the beach profile landward of a seawall might be stopped or at least reduced. However, erosion of the seabed immediately in front of the structure will in most cases be enhanced due to increased wave reflection caused by the seawall. This results in a steeper seabed profile, which subsequently allows larger waves to reach the structure. As a consequence, seawalls are in danger of instability caused by erosion of the seabed at the toe of the structure, and by an increase in wave slamming, runup, and overtopping. Because of their potential vulnerability to toe scour, seawalls are often used together with some system of beach control such as groins and beach nourishment. Exceptions include cases of stable rock foreshores and cases where the potential for future erosion is limited and can be accommodated in the design of the seawall.

c. Revetments. Revetments are onshore structures with the principal function of protecting the shoreline from erosion. Revetment structures typically consist of a cladding of stone, concrete, or asphalt to armor sloping natural shoreline profiles. In the Corps of Engineers, the functional distinction is made between seawalls and revetments for the purpose of assigning project benefits; however, in the technical literature there is often no distinction between seawalls and revetments.

d. Bulkheads. Bulkhead is the term for structures primarily intended to retain or prevent sliding of the land, whereas protecting the hinterland against flooding and wave action is of secondary importance. Bulkheads are built as soil retaining structures, and in most cases as a vertical wall anchored with tie rods. The most common application of bulkheads is in the construction of mooring facilities in harbors and marinas where exposure to wave action is minimized. Some reference literature may not make a distinction between bulkheads and seawalls.

Table VI-2-1
Types and Functions of Coastal Structures

Type of Structure	Objective	Principal Function
Sea dike	Prevent or alleviate flooding by the sea of low-lying land areas	Separation of shoreline from hinterland by a high impermeable structure
Seawall	Protect land and structures from flooding and overtopping	Reinforcement of some part of the beach profile
Revetment	Protect the shoreline against erosion	Reinforcement of some part of the beach profile
Bulkhead	Retain soil and prevent sliding of the land behind	Reinforcement of the soil bank
Groin	Prevent beach erosion	Reduction of longshore transport of sediment
Detached breakwater	Prevent beach erosion	Reduction of wave heights in the lee of the structure and reduction of longshore transport of sediment
Reef breakwater	Prevent beach erosion	Reduction of wave heights at the shore
Submerged sill	Prevent beach erosion	Retard offshore movement of sediment
Beach drain	Prevent beach erosion	Accumulation of beach material on the drained portion of beach
Beach nourishment and dune construction	Prevent beach erosion and protect against flooding	Artificial infill of beach and dune material to be eroded by waves and currents in lieu of natural supply
Breakwater	Shelter harbor basins, harbor entrances, and water intakes against waves and currents	Dissipation of wave energy and/or reflection of wave energy back into the sea
Floating breakwater	Shelter harbor basins and mooring areas against short-period waves	Reduction of wave heights by reflection and attenuation
Jetty	Stabilize navigation channels at river mouths and tidal inlets	Confine streams and tidal flow. Protect against storm water and crosscurrents

Training walls	Prevent unwanted sedimentation or erosion and protect moorings against currents	Direct natural or man-made current flow by forcing water movement along the structure
Storm surge barrier	Protect estuaries against storm surges	Separation of estuary from the sea by movable locks or gates
Pipeline outfall	Transport of fluids	Gravity-based stability
Pile structure	Provide deck space for traffic, pipelines, etc., and provide mooring facilities	Transfer of deck load forces to the seabed
Scour protection	Protect coastal structures against instability caused by seabed scour	Provide resistance to erosion caused by waves and current

e.   Groins.

(1) Groins are built to stabilize a stretch of natural or artificially nourished beach against erosion that is due primarily to a net longshore loss of beach material. Groins function only when longshore transport occurs. Groins are narrow structures, usually straight and perpendicular to the preproject shoreline. The effect of a single groin is accretion of beach material on the updrift side and erosion on the downdrift side; both effects extend some distance from the structure. Consequently, a groin system (series of groins) results in a saw-tooth-shaped shoreline within the groin field and a differential in beach level on either side of the groins.

(2) Groins create very complex current and wave patterns. However, a well-designed groin system can arrest or slow down the rate of longshore transport and, by building up of material in the groin bays, provide some protection of the coastline against erosion. Groins are also used to hold artificially nourished beach material, and to prevent sedimentation or accretion in a downdrift area (e.g., at an inlet) by acting as a barrier to longshore transport. Deflecting strong tidal currents away from the shoreline might be another purpose of groins.

(3) The orientation, length, height, permeability, and spacing of the groins determine, under given natural conditions, the actual change in the shoreline and the beach level. Because of the potential for erosion of the beach downdrift of the last groin in the field, a transition section of progressively shorter groins may be provided to prevent the formation of a severe erosion area. Even so, it might be necessary to protect some part of the downdrift beach with a seawall or to nourish a portion of the eroded area with beach material from an alternative source.

(4) Groins are occasionally constructed non-perpendicular to the shoreline, can be curved, have fishtails, or have a shore-parallel T-head at their seaward end. Also, shore-parallel spurs are provided to shelter a stretch of beach or to reduce the possibility of offshore sand transport by rip currents. However, such refinements, compared to the simple shape of straight perpendicular groins, are generally not deemed effective in improving the performance of the groins.

(5) In most cases, groins are sheet-pile or rubble-mound constructions. The latter is preferably used at exposed sites because of a rubble-mound structure's ability to withstand severe wave loads and to decrease wave reflection. Moreover, the risk of scouring and formation of strong rip currents along rubble groins is reduced.

(6) The landward end of the groins must extend to a point above the high-water line in order to stay beyond the normal zone of beach movement and thereby avoid outflanking by back scour. The groins must, for the same reason, reach seawalls when present or connect into stable back beach features. The position of the seaward end is determined such that the groin retains some proportion of the longshore transport during more severe wave conditions. This means that the groin must protrude some distance into the zone of littoral transport, the extent of which is largely determined by surf zone width. Groins can be classified as either *long* or *short*, depending on how far across the surf zone they extend. Groins that transverse the entire surf zone are considered *long*, whereas those that extend only part way across the surf zone are considered *short*. These terms are relative, since the width of the surf zone varies with water level, wave height, and beach profile. Most groins are designed to act as *short* structures during severe sea states and as *long* structures under normal conditions. Groins might also be classified as *high* or *low*, depending on the possibility of sediment transport across the crest. Significant cost savings can be achieved by constructing groins with a variable crest elevation that follows the beach profile rather than maintaining a constant crest elevation. These groins would maintain a constant cross section and allow increasing amounts of sand to bypass as water depth increases. At some point the crest of the groin becomes submerged. *Terminal groins* extend far enough seaward to block all littoral transport, and these types of groins should never be used except in rare situations, such as where longshore transported sand would be otherwise lost into a submarine canyon.

(7) Some cross-groin transport is beneficial for obtaining a well-distributed retaining effect along the coast. For the same reason *permeable groins,* which allow sediment to be transported through the structure, may be advantageous. Examples of permeable groins include rubble-mound structures built of rock and concrete armor units without fine material cores, and structures made of piles with some spacing. Most sheet-pile structures are impermeable. Low and permeable groins have the benefit of reduced wave reflection and less rip current formation compared with high and impermeable groins.

f.   Detached breakwaters.

(1) Detached breakwaters are small, relatively short, nonshore-connected nearshore breakwaters with the principal function of reducing beach erosion. They are built parallel to the shore just seaward of the shoreline in shallow water depths. Multiple detached breakwaters spaced along the shoreline can provide protection to substantial shoreline frontages. The gaps between the breakwaters are in most cases on the same order of magnitude as the length of one individual structure.

(2) Each breakwater reflects and dissipates some of the incoming wave energy, thus reducing wave heights in the lee of the structure and reducing shore erosion. Beach material transported along the beach moves into the sheltered area behind the breakwater where it is deposited in the lower wave energy region. The nearshore wave pattern, which is strongly

influenced by diffraction at the heads of the structures, will cause *salients* and sometimes *tombolos* to be formed, thus making the coastline similar to a series of pocket beaches. Once formed, the pockets will cause wave refraction, which helps to stabilize the pocket-shaped coastline.

(3) Like groins, a series of detached breakwaters can be used to control the distribution of beach material along a coastline. Just downdrift of the last breakwater in the series there is an increased risk of shoreline erosion. Consequently, it might be necessary to introduce a transition section where the breakwaters gradually are made smaller and placed closer to the shoreline. In addition, seawall protection of the downdrift stretch of beach might be necessary.

(4) Detached breakwaters are normally built as rubble-mound structures with fairly low crest levels that allow significant overtopping during storms at high water. The low-crested structures are less visible and help promote a more even distribution of littoral material along the coastline. Submerged detached breakwaters are used in some cases because they do not spoil the view, but they do represent a serious nonvisible hazard to boats and swimmers.

(5) Properly designed detached breakwaters are very effective in reducing erosion and in building up beaches using natural littoral drift. Moreover, they are effective in holding artificially nourished beach material.

(6) Optimizing detached breakwater designs is difficult when large water level variations are present, as is the case on coastlines with a large tidal range or in portions of the Great Lakes, which may experience long-term water level fluctuations.

g. Reef breakwaters. Reef breakwaters are coast-parallel, long or short submerged structures built with the objective of reducing the wave action on the beach by forcing wave breaking over the *reef*. Reef breakwaters are normally rubble-mound structures constructed as a homogeneous pile of stone or concrete armor units. The breakwater can be designed to be stable or it may be allowed to reshape under wave action. Reef breakwaters might be *narrow crested* like detached breakwaters in shallow water or, in deeper water, they might be *wide crested* with lower crest elevation like most natural reefs that cover a fairly wide rim parallel to the coastline. Besides triggering wave breaking and subsequent energy dissipation, reef breakwaters can be used to regulate wave action by refraction and diffraction. Reef breakwaters represent a nonvisible hazard to swimmers and boats.

h. Submerged sills. A submerged sill is a special version of a reef breakwater built nearshore and used to retard offshore sand movements by introducing a structural barrier at one point on the beach profile. However, the sill may also interrupt the onshore sand movement. The sill introduces a discontinuity into the beach profile so that the beach behind it becomes a *perched* beach as it is at higher elevation and thus wider than adjacent beaches. Submerged sills are also used to retain beach material artificially placed on the beach profile behind the sill. Submerged sills are usually built as rock-armored, rubble-mound structures or commercially available prefabricated units. Submerged sills represent a nonvisible hazard to swimmers and boats.

i. Beach drains. Beach drains are installed for the purpose of enhancing accumulation of beach material in the drained part of the beach. In principal, the drains are arranged at an elevation just beneath the lowest seasonal elevation of the beach profile in the swash zone. Pumping water from the drains causes local lowering of the groundwater table, which helps reduce the backwash speed and the groundwater outflow in the beach zone. This allows more beach material to settle out on the foreshore slope. Beach drains are built like normal surface drain systems consisting of a stable granular filter, with grain sizes ranging from that of the beach material to coarse materials like pebbles, arranged around closely spaced perforated pipes. The drain pipes are connected to few shore-normal pipelines leading to a pump sump in the upper part of the beach profile. Replacing the granular filter with geotextiles is not recommended because of the increased tendency to clog the drainage system.

j. Beach nourishment and dune construction.

(1) Beach nourishment is a *soft structure* solution used for prevention of shoreline erosion. Material of preferably the same, or larger, grain size and density as the natural beach material is artificially placed on the eroded part of the beach to compensate for the lack of natural supply of beach material. The beachfill might protect not only the beach where it is placed, but also downdrift stretches by providing an updrift point source of sand.

(2) Dune construction is the piling up of beach quality sand to form protective dune fields to replace those washed away during severe storms. An essential component of dune reconstruction is planting of dune vegetation and placement of netting or snow fencing to help retain wind-blown sand normally trapped by mature dune vegetation. Storm overwash fans may be a viable source of material for dune construction.

k. Breakwaters.

(1) Breakwaters are built to reduce wave action in an area in the lee of the structure. Wave action is reduced through a combination of reflection and dissipation of incoming wave energy. When used for harbors, breakwaters are constructed to create sufficiently calm waters for safe mooring and loading operations, handling of ships, and protection of harbor facilities. Breakwaters are also built to improve maneuvering conditions at harbor entrances and to help regulate sedimentation by directing currents and by creating areas with differing levels of wave disturbance. Protection of water intakes for power stations and protection of coastlines against tsunami waves are other applications of breakwaters.

(2) When used for shore protection, breakwaters are built in nearshore waters and usually oriented parallel to the shore like *detached breakwaters*. The layout of breakwaters used to protect harbors is determined by the size and shape of the area to be protected as well as by the prevailing directions of storm waves, net direction of currents and littoral drift, and the maneuverability of the vessels using the harbor. Breakwaters protecting harbors and channel entrances can be either detached or shore-connected.

(3) The cost of breakwaters increases dramatically with water depth and wave climate severity. Also poor foundation conditions significantly increase costs. These three environmental factors heavily influence the design and positioning of the breakwaters and the harbor layout.

(4) Breakwaters can be classified into two main types: *sloping-front* and *vertical-front* structures. Sloping-front structures are in most cases rubble-mound structures armored with rock or concrete armor units, with or without wavewall superstructures. Vertical-front structures are in most cases constructed of either sandfilled concrete caissons or stacked massive concrete blocks placed on a rubble stone bedding layer. In deep water, concrete caissons are often placed on a high mound of quarry rock for economical reasons. These breakwaters are called *composite structures*. The upper part of the concrete structure might be constructed with a sloping front to reduce the wave forces. For the same reason the front wall might be perforated with a wave chamber behind to dissipate wave energy. Smaller vertical structures might be constructed of steel sheetpiling backfilled with soil, or built as a rock-filled timber cribwork or wire cages. In milder wave climates sloping reinforced concrete slabs supported by batter piles is another possibility.

l. Floating breakwaters. Floating breakwaters are used in protected regions that experience mild wave climates with very short-period waves. For example, box-shaped reinforced concrete pontoons are used to protect marinas in sheltered areas. Floating docks affixed to piles are also used in marinas.

m. Jetties. Jetties are used for stabilization of navigation channels at river mouths and tidal inlets. Jetties are shore-connected structures generally built on either one or both sides of the navigation channel perpendicular to the shore and extending into the ocean. By confining the stream or tidal flow, it is possible to reduce channel shoaling and decrease dredging requirements. Moreover, on coastlines with longshore currents and littoral drift, another function of the jetties is also to arrest the crosscurrent and direct it across the entrance in deeper water where it represents less hazard to navigation. When extended offshore of the breaker zone, jetties improve the maneuvering of ships by providing shelter against storm waves. Jetties are constructed similar to breakwaters.

n. Training walls. Training walls are structures built to direct flow. Typical training wall objectives might be to improve mooring conditions in an estuary or to direct littoral drift away from an area of potential deposition. Most training walls are constructed using sheet piles.

o. Storm surge barriers. Storm surge barriers protect estuaries against storm surge flooding and related wave attack. These barriers also prevent excessive intrusion of salt-water wedges during high-water episodes. In most cases the barrier consists of a series of movable gates that normally stay open to let the flow pass but will be closed when storm surges exceed a certain level. The gates are sliding or rotating steel constructions supported in most cases by concrete structures on pile foundations. Scour protection on either side of the barrier sill is an important part of the structure because of high flow velocities over the sill.

p. Pipelines. Pipelines in the coastal zone are typically used for outlet of treated sewage, transport of oil and gas from offshore fields, and water supply between islands/mainlands and across inlets. Typical types of pipelines are small-diameter flexible PVC pipes used for water supply and small sewage outfalls, large low-pressure sewage outfalls constructed of stiff reinforced concrete pipe elements up to several meters in diameter, and semi-flexible concrete-covered steel pipes used for high-pressure transport of oil and gas. Diffusers at the offshore terminal of sewage outlets are in most cases concrete structures placed on or in the seabed.

Pipelines might be buried or placed on the seabed with or without surface protection, depending on the risk of damage caused by scour and flow-induced instability, or damage by surface loads from collision with ships, anchors, and fishing gear. Where significant changes in the seabed are expected, e.g., surf zones and eroding beaches, it is common to bury the pipelines to depths below the expected maximum eroded profile. In some cases it is prudent to provide scour protection due to uncertainty in predicting the eroded beach profile.

q. Pile structures. The most common pile structures in coastal engineering are bridge piers extending from the shore into the water where they are exposed to loads from waves, currents, and in cold regions, ice loads. The purpose of pile structures might be to provide open coast moorings for vessels, in which case the deck and the piles must carry loads from traffic, cranes, goods, and pipeline installations. Piers are also used for recreational purposes by providing space for fishing, outlook platforms, restaurants, shops, etc. The supporting pile structure might consist of slender wood, steel or reinforced concrete piles driven into the sea-bed, or of large diameter piles or pillars placed directly on the seabed or on pilework, depending on the bearing capacity and settlement characteristics of the seabed. Large diameter piles would commonly be constructed of concrete or be steel pipes filled with mass concrete. Pillars would most commonly be constructed as concrete caissons, concrete blockwork, or backfilled steel sheet piling.

r. Scour protection. The function of scour protection of the seabed is to prevent instability of coastal structures with foundations that rely on stable seabed or beach levels. Both granular material and clay can be eroded by the action of waves and currents. Scour potential is especially enhanced by a combination of waves and currents. In most cases scour protection consists of a rock bed on stone or geotextile filter; however, several specially designed concrete block and mattress systems exist. Scour protection is commonly used at the toe of seawalls and dikes; and in some instances scour protection is needed around piles and pillars, at the toe of vertical-front breakwaters, and at groin heads. Scour protection might also be needed along structures that cause concentration of currents, such as training walls and breakwaters extending from the shoreline. Highly reflective structures like impermeable vertical walls are much more susceptible to near-structure scour than sloping rubble-mound structures.

VI-2-2. Typical Cross Sections and Layouts.

a. Sea dikes. Sea dikes are low-permeability (watertight) structures protecting low-lying areas against flooding. As a consequence fine materials such as sand, silty sand, and clay are used for the construction. The seaside slope is usually very gentle in order to reduce wave runup and wave impact. The risks of slip failures and erosion by piping determine the steepness of the rear slope. The seaward slope is armored against damage from direct wave action. Steeper slopes require stronger armoring. Figure VI-2-1 shows asphalt armoring on slopes of 1:5 and 1:3, while Figure VI-2-2 shows grass armoring on a slope of 1:10. When risk of lowering of the foreshore is present, it is important either to design an embedded toe or a flexible toe that can sink and still protect the slope when the foreshore is eroded as illustrated in Figure VI-2-1.

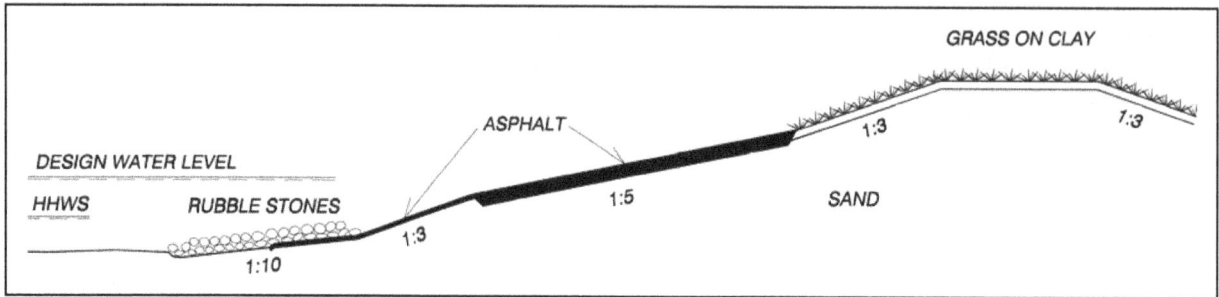

Figure VI-2-1. Example of asphalt-armored sea dike

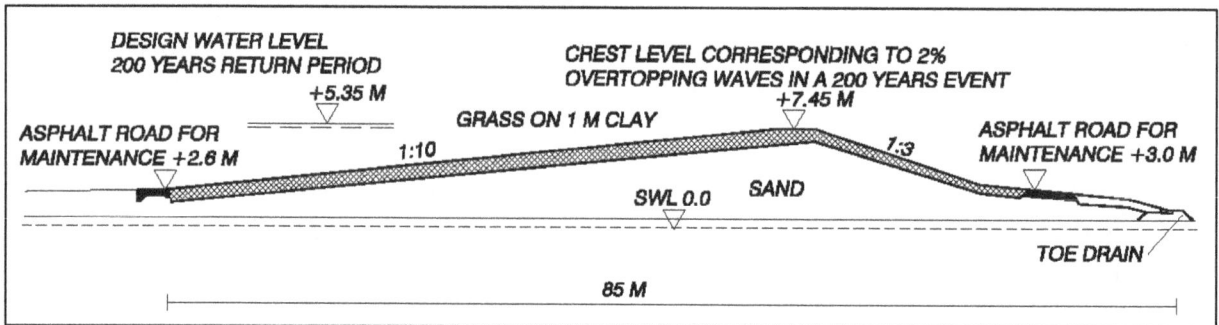

Figure VI-2-2. Example of grass-armored sea dike design from the North Sea coast of Denmark

b. Seawalls and revetments.

(1) Although seawalls and revetments differ by function, they often are similar in construction detail. Seawalls and revetments can be classified as sloping-front and vertical-front structures. Sloping-front structures might be constructed as flexible rubble-mound structures which are able to adjust to some toe and crest erosion. Figure VI-2-3 shows three examples with randomly placed armor. Figure VI-2-4 shows sloping-front structures with pattern-placed concrete armor units. In the United States pattern-placed block slopes are more commonly found on revetments. The stability of the slope is very dependent on an intact toe support. In other words, loss of toe support will likely result in significant armor layer damage, if not complete failure of the armored slope.

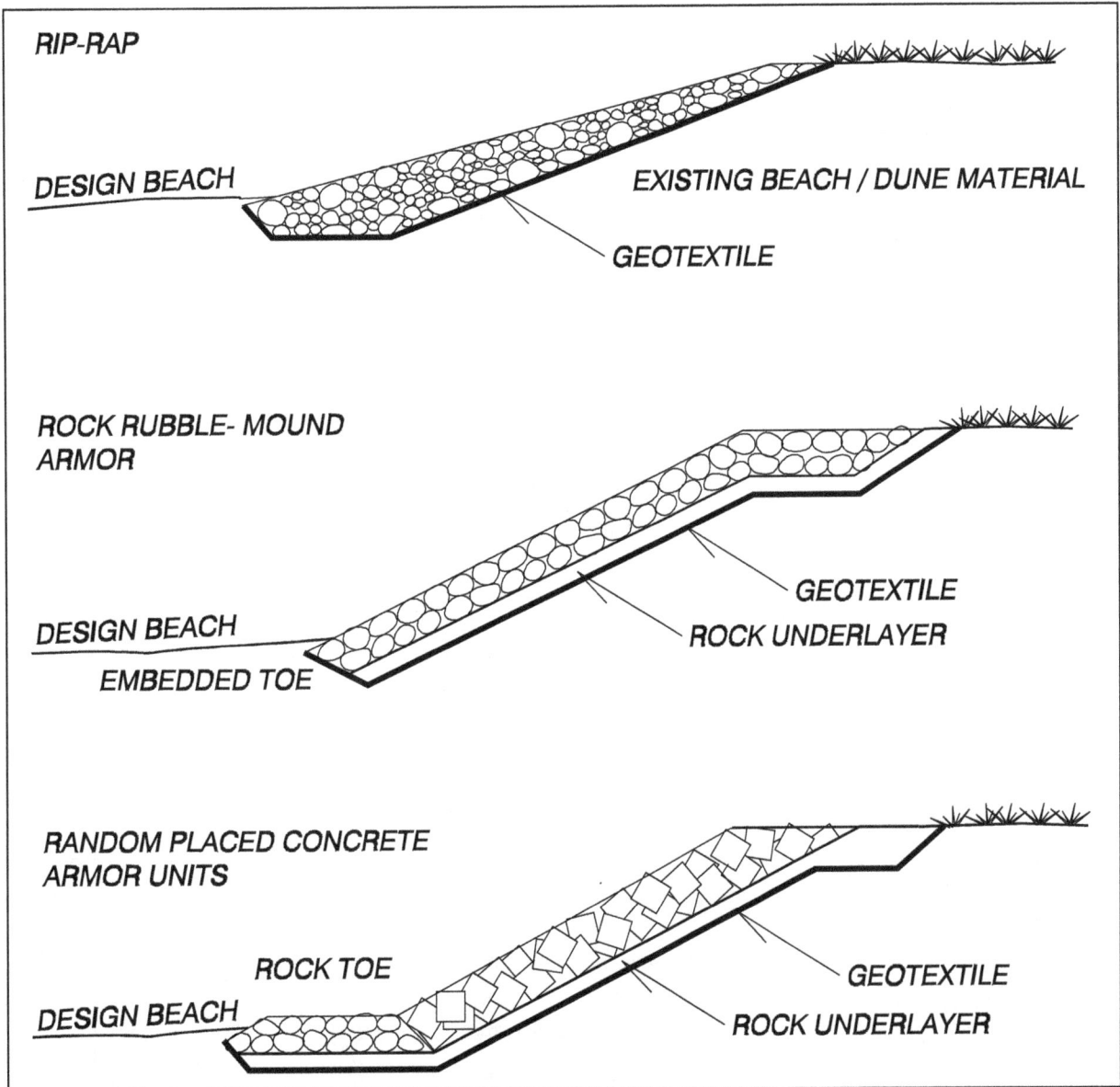

Figure VI-2-3. Examples of sloping front rubble-mound seawall/revetment structures

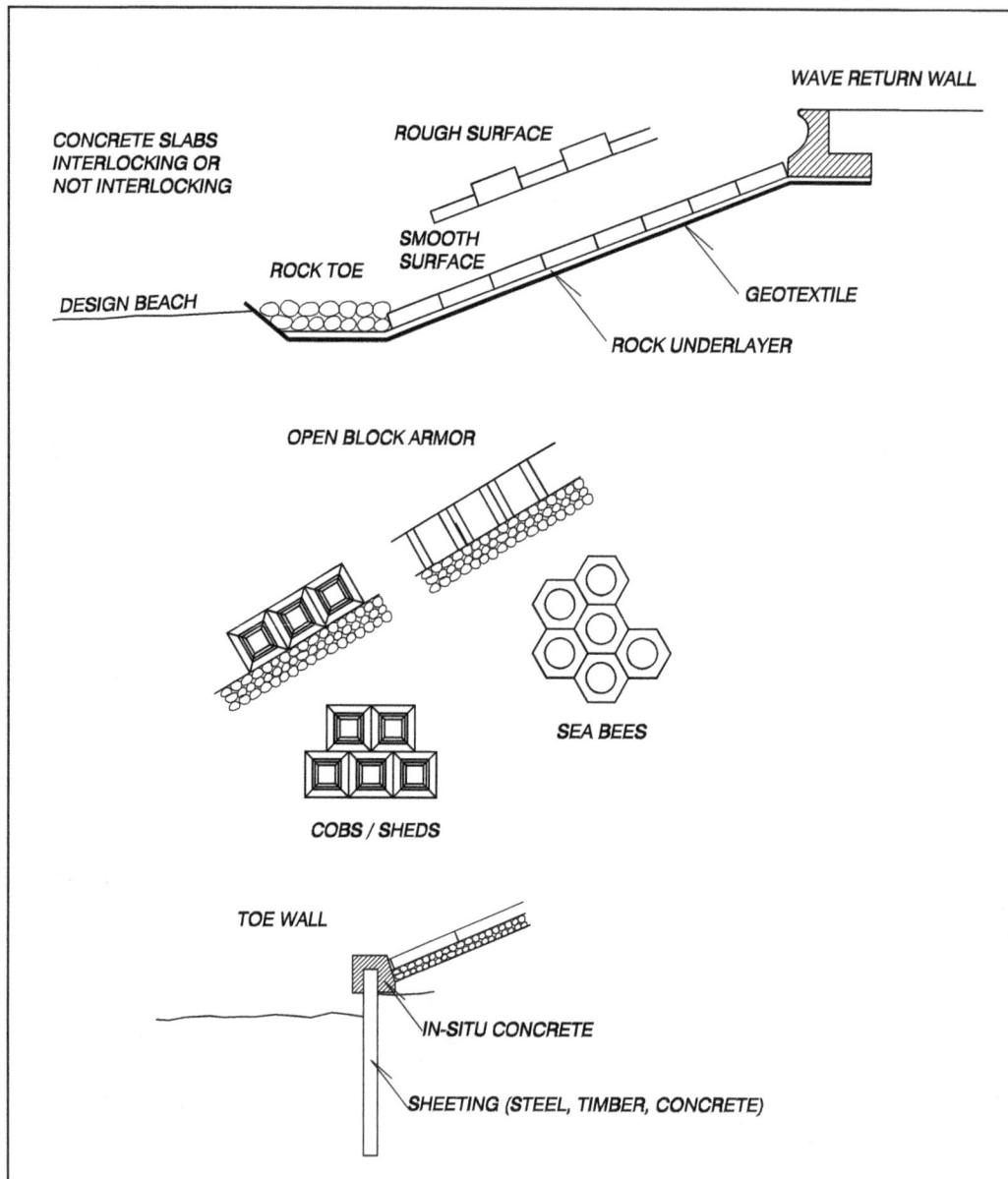

Figure VI-2-4. Examples of sloping-front seawalls/revetments with pattern-placed
concrete armor units

(2) The top portion of Figure VI-2-5 shows an example of a sloping front revetment with
fixed asphalt layer surface and the bottom portion shows a seawall with a sloping front of in situ
cast concrete. Asphalt structures with either rubble toes or thin asphalt carpet toes are flexible
structures able to survive rather substantial beach erosion, whereas the rigid concrete structures
are vulnerable to any form of undermining. Figure VI-2-6 shows examples of sloping front
revetment designs from the Danish North Sea coast. Vertical-front seawalls can be constructed
as tied walls using steel, concrete, and timber piling; as stone-filled cribwork; and as massive
gravity concrete walls. Figure VI-2-7 is an example of a gravity wall structure.

c. Groins. Groins are in most cases constructed as sloping-front structures or as piled vertical face structures. Figure VI-2-8 shows a typical beach configuration with groins. Figure VI-2-9 shows examples of groin structure designs. Timber groins are used for smaller and less exposed applications, whereas rubble-mound groins are used for all conditions. On very exposed coastlines the armor is often concrete armor units. The timber planking groin is weaker and much more vulnerable to scour failure than the timber pile groin.

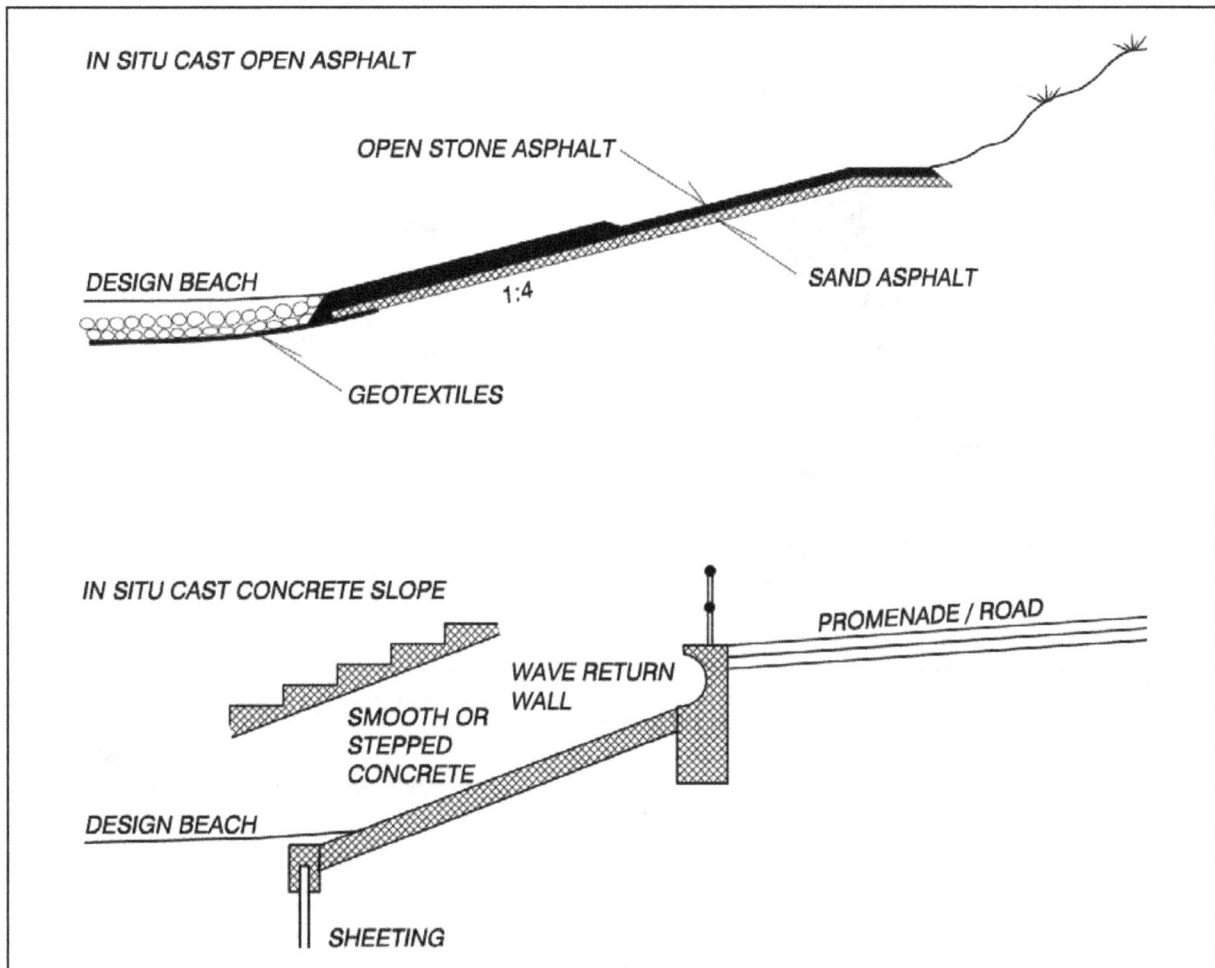

Figure VI-2-5. Examples of sloping front seawalls/revetments with fixed surfaces of asphalt and in situ cast concrete

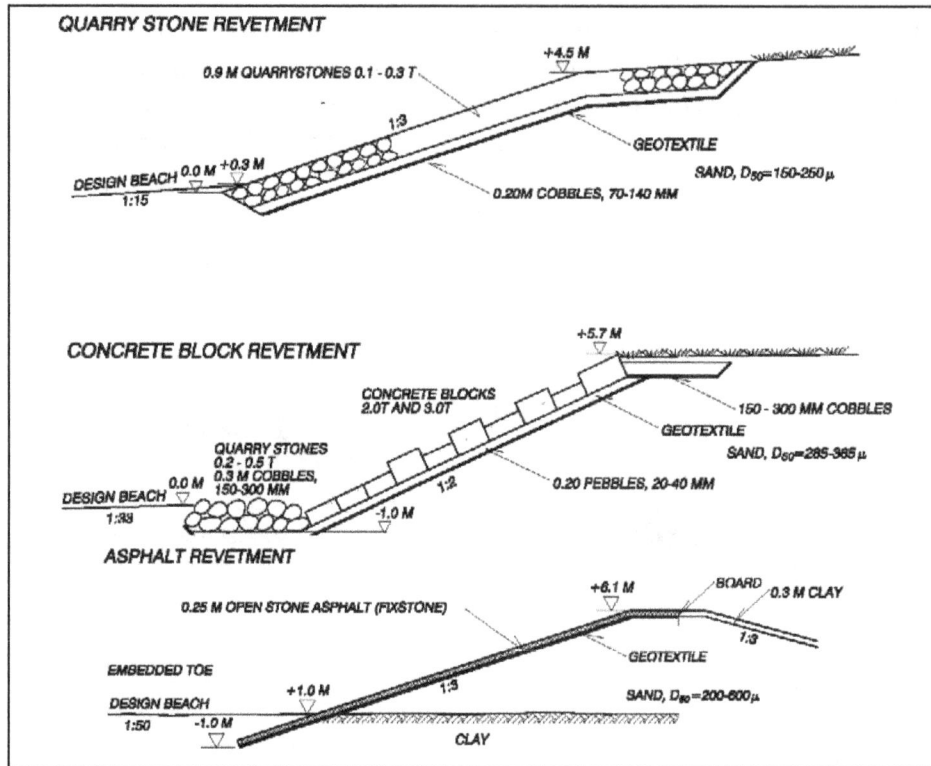

Figure VI-2-6. Examples of sloping front revetment designs from the
Danish North Sea coast (Danish Coast Authority)

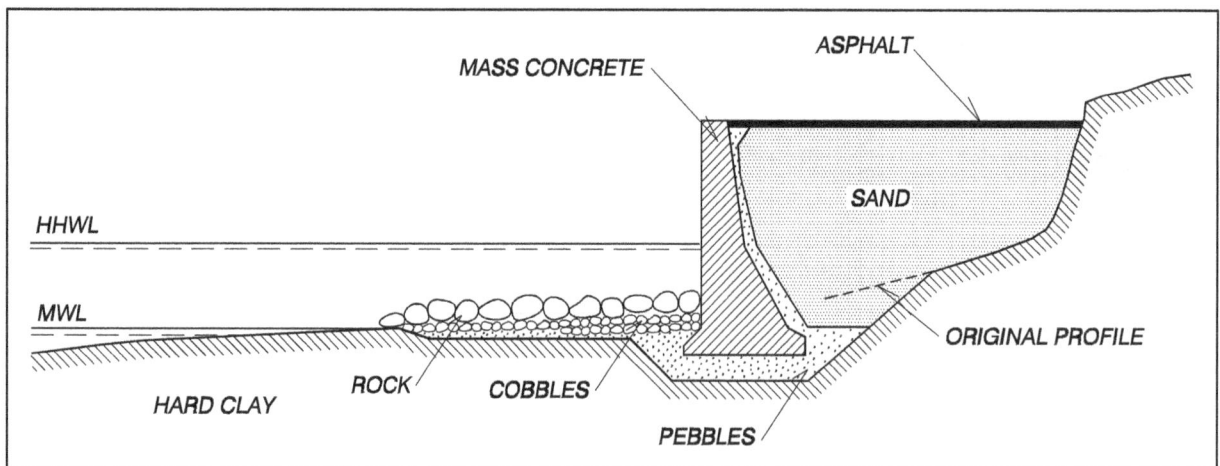

Figure VI-2-7. Example of a vertical front seawall

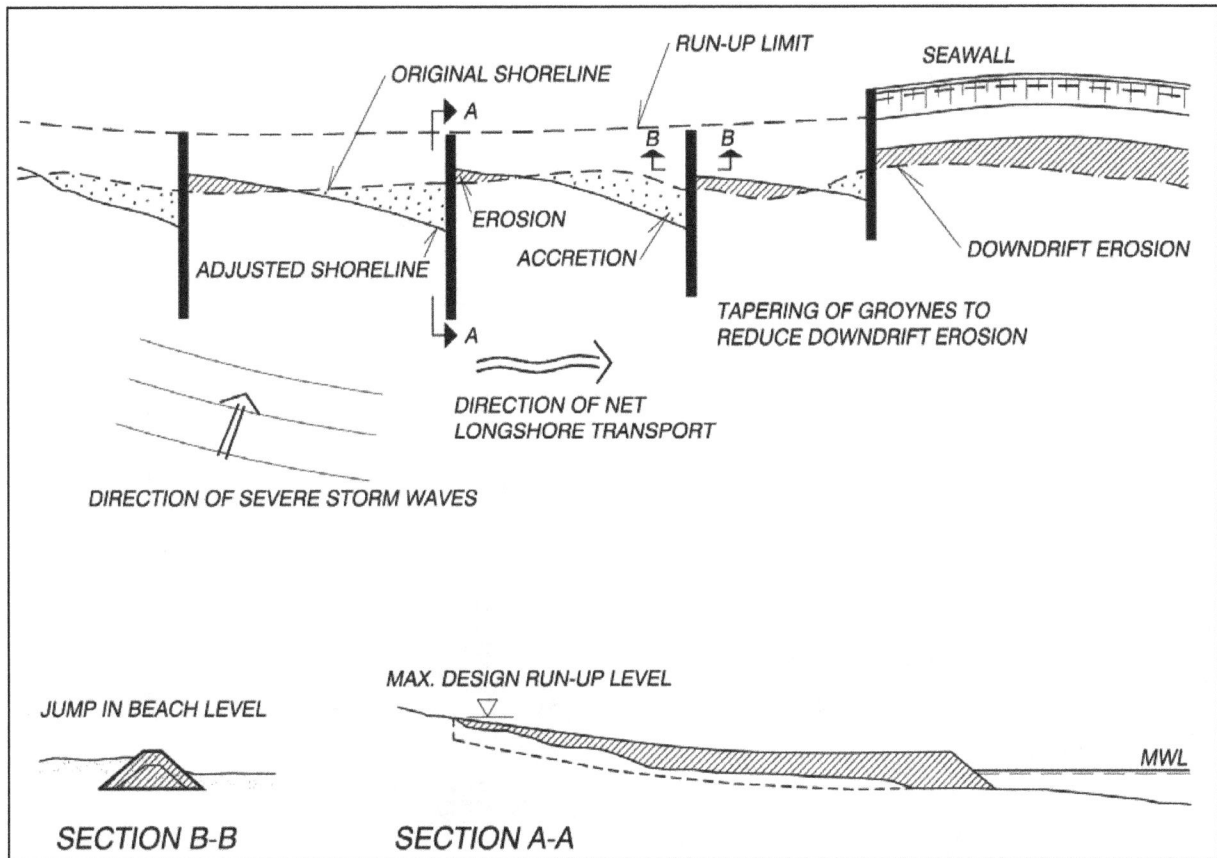

Figure VI-2-8. Typical beach configuration with groins

d. Detached breakwaters. Detached breakwaters are almost always built as rubble-mound structures. Typical cross sections are as shown for the rubble-mound groin in Figure VI-2-9. Typical beach configurations with detached nearshore breakwaters are shown in Figure VI-2-10. Whether or not the detached breakwaters become attached to shore is a function of placement distance offshore. Tombolos are more likely to form when breakwaters are constructed within the surf zone. The two examples of detached breakwaters shown in Figure VI-2-10 serve different functions. See Part V-4 for functional design guidance on detached breakwaters.

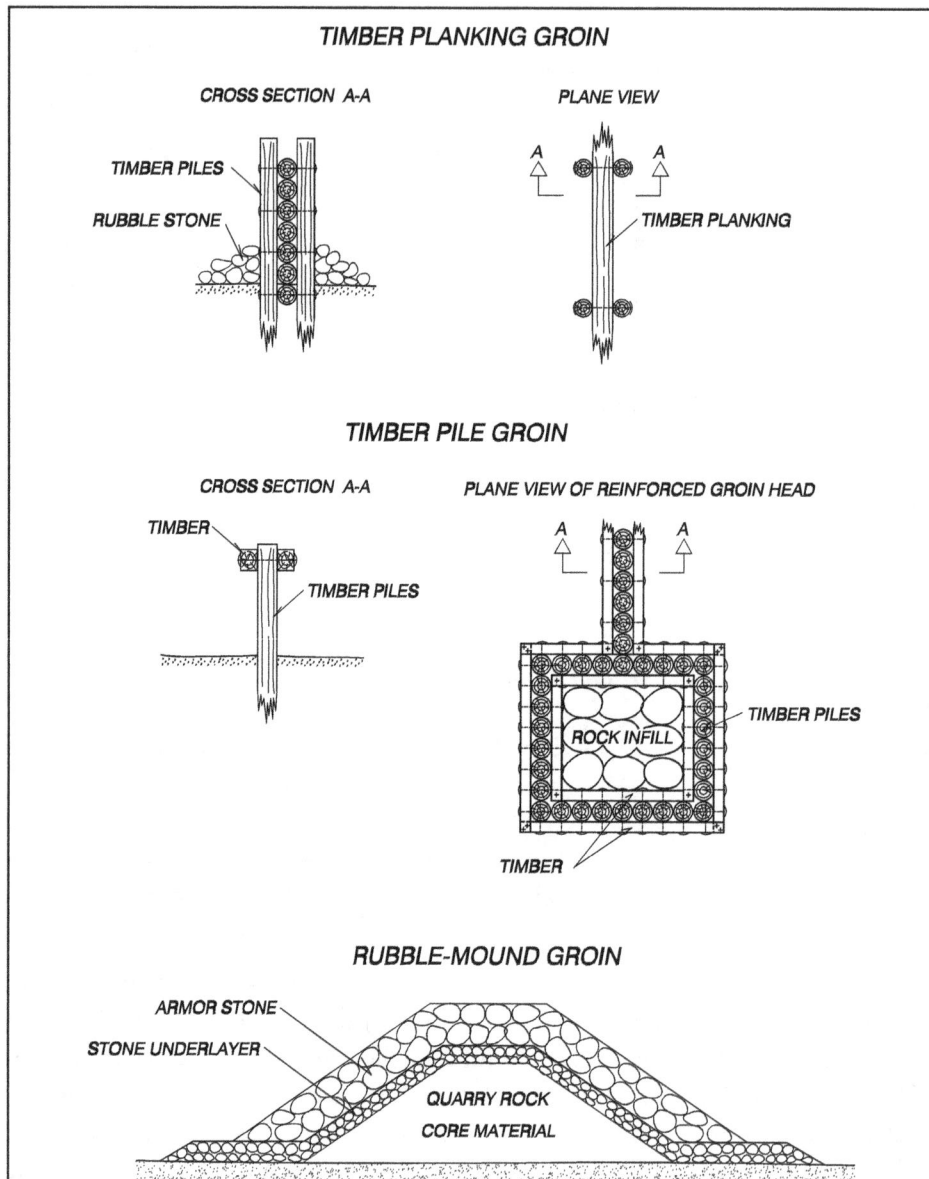

Figure VI-2-9. Examples of groin structures

e. Rubble-mound breakwaters.

(1) Rubble-mound breakwaters are the most commonly applied type of breakwater. In its most simple shape it is a mound of stones. However, a homogeneous structure of stones large enough to resist displacements due to wave forces is very permeable and might cause too much penetration not only of waves, but also of sediments if present in the area. Moreover, large stones are expensive because most quarries yield mainly finer material (quarry run) and only relatively few large stones. As a consequence the conventional rubble-mound structures consist of a core of finer material covered by big blocks forming the so-called armor layer. To prevent finer material being washed out through the armor layer, filter layers must be provided. The filter layer just beneath the armor layer is also called the underlayer. Structures consisting of armor layer, filter layer(s), and core are referred to as multilayer structures. The lower part of the armor

layer is usually supported by a toe berm except in cases of shallow- water structures. Figure VI-2-11 shows a conventional type of rubble-mound breakwater.

(2) Concrete armor units are used as armor blocks in areas with rough wave climates or at sites where a sufficient amount of large quarrystones is not available. Main types of armor units are discussed in Part VI-2-3.

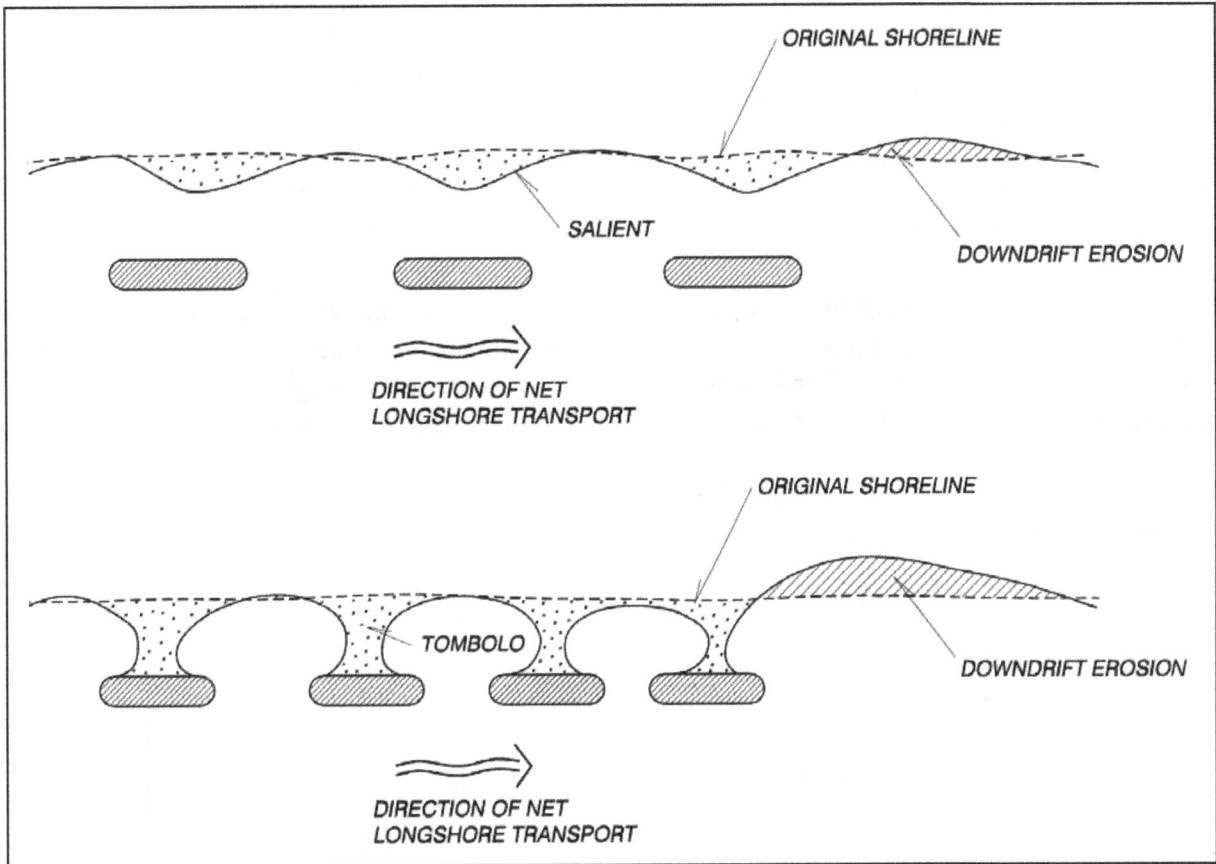

Figure VI-2-10. Typical beach configurations with detached nearshore breakwaters

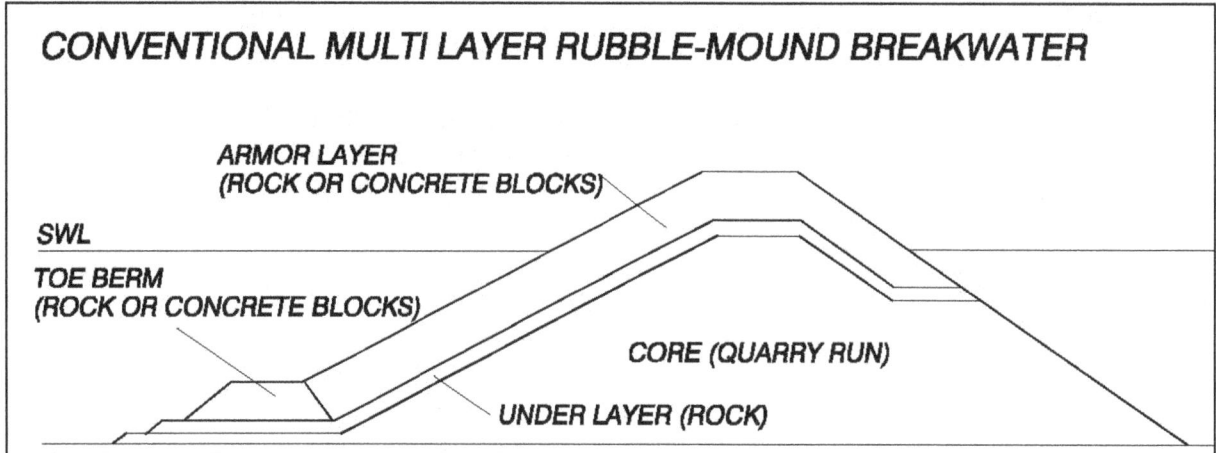

Figure VI-2-11. Conventional multilayer rubble-mound breakwater

(3) The front slope of the armor layer is in most cases straight. However, an S-shaped front or a front with a horizontal berm might be used to increase the armor stability and reduce overtopping. For these types of structures, optimization of the profiles might be difficult if there are large water level variations. Figure VI-2-12 illustrates these types of front profiles.

(4) Overtopping can be reduced by a wave-wall superstructure as shown in Figure VI-2-13.

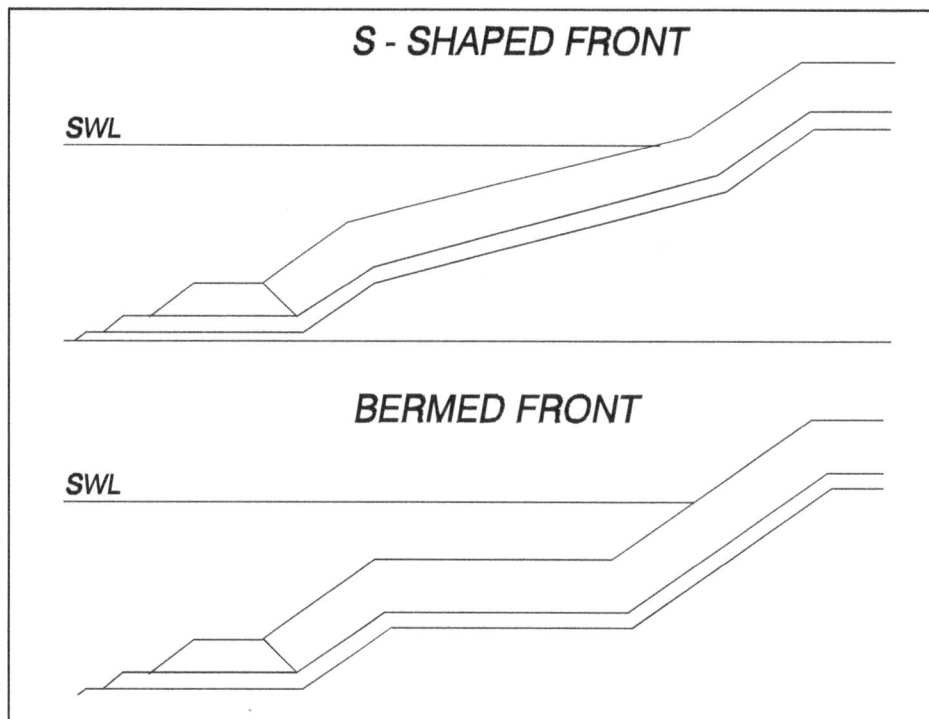

Figure VI-2-12. Rubble-mound structures with S-shaped and bermed fronts

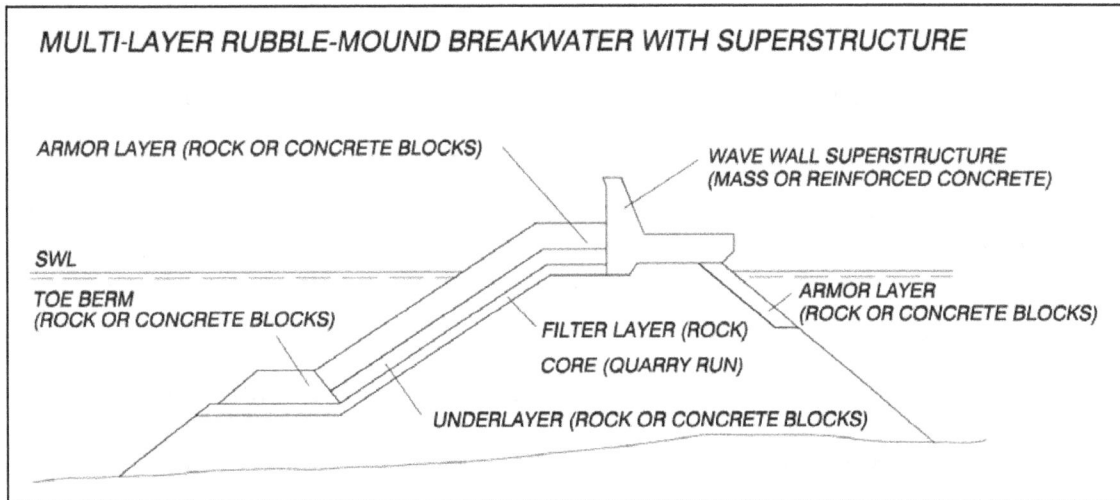

Figure VI-2-13. Example of rubble-mound breakwater with concrete superstructure

(5) Superstructures can serve several purposes, e.g., providing access for vehicles, including cranes for maintenance and repair, and accommodation of installations such as pipelines.

(6) The armor units in conventional multilayer structures are designed to stay in place as built, i.e., the profile remains unchanged with displacement of only a minor part of the armor units.

f. Reshaping rubble-mound breakwaters.

(1) Reshaping rubble-mound breakwaters is based on the principle of natural adjustment of the seaward profile to the actual wave action, as illustrated by Figure VI-2-14. In this way the most efficient profile in terms of armor stability (and possibly minimum overtopping) is obtained for the given size and quantity of armor stone.

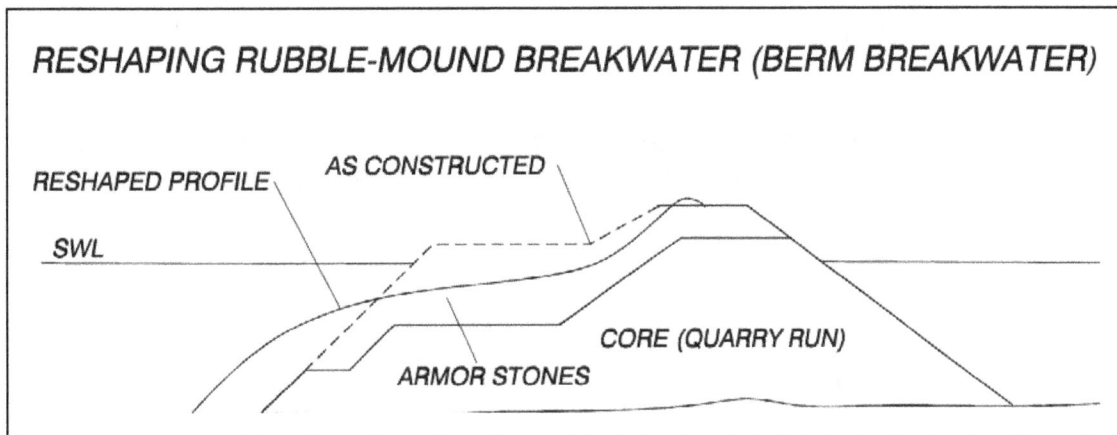

Figure VI-2-14. Reshaping rubble-mound breakwater

(2) Because of natural reshaping, the structure can be built in a very simple way by first dumping the core material consisting of quarry run, and then dumping the armor stones in a berm profile with seaward slope equal to the natural angle of repose for the stone material. Due to the initial berm profile, this type of structure is also known as a berm breakwater.

(3) The natural adjusted S-profile allows smaller armor stones to be used compared to the armor stones in conventional rubble structures. The smaller the armor stones, the flatter the S-profile will be. The minimum size of the armor stones is often selected to limit transport of stones along the structure under oblique wave attack.

g. Reef breakwaters. Reef breakwaters are in principle designed as a rubble-mound structure with submerged crests, as shown in Figure VI-2-15. Both homogeneous and multilayer structures are used. This example shows a mound of smaller stones protected by an armor layer of larger stones.

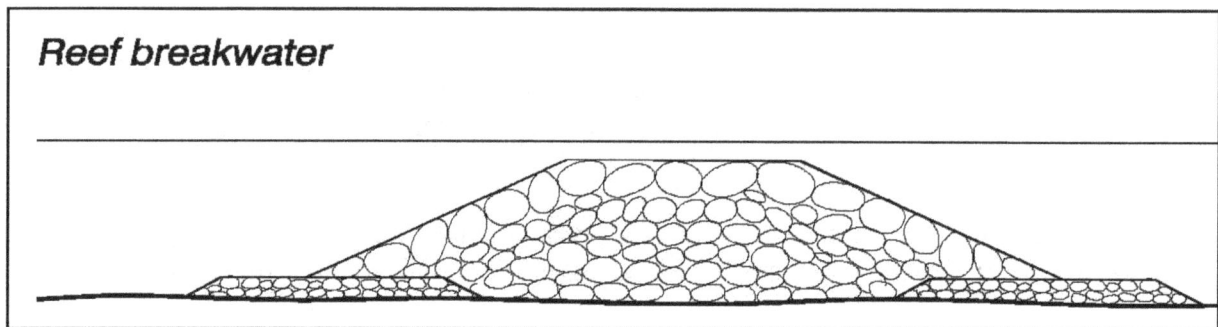

Figure VI-2-15. Example of a reef breakwater

h. Vertical-front breakwaters. Vertical-front breakwaters are another major class of breakwater structures. The basic structure element is usually a sandfilled caisson made of reinforced concrete, but blockwork types made of stacked precast concrete blocks are also used. Caisson breakwaters might be divided into the following types:

(1) Conventional, i.e., the caisson is placed on a relatively thin stone bedding layer (Figure VI-2-16).

(2) Vertical composite, i.e., the caisson is placed on a high rubble-mound foundation (Figure VI-2-17). This type is economical in deep waters. Concrete caps may be placed on shore-connected caissons.

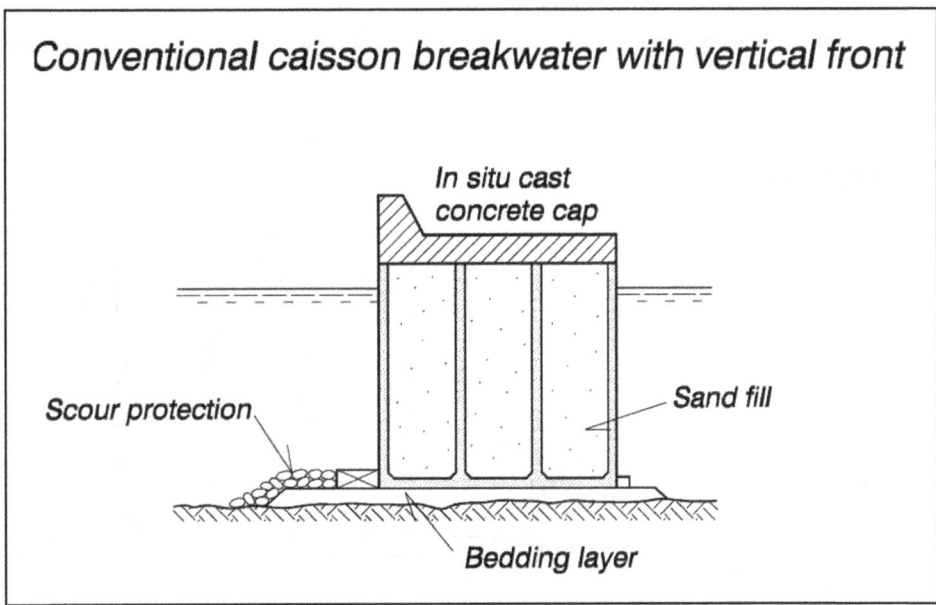

Figure VI-2-16. Conventional caisson breakwater with vertical front

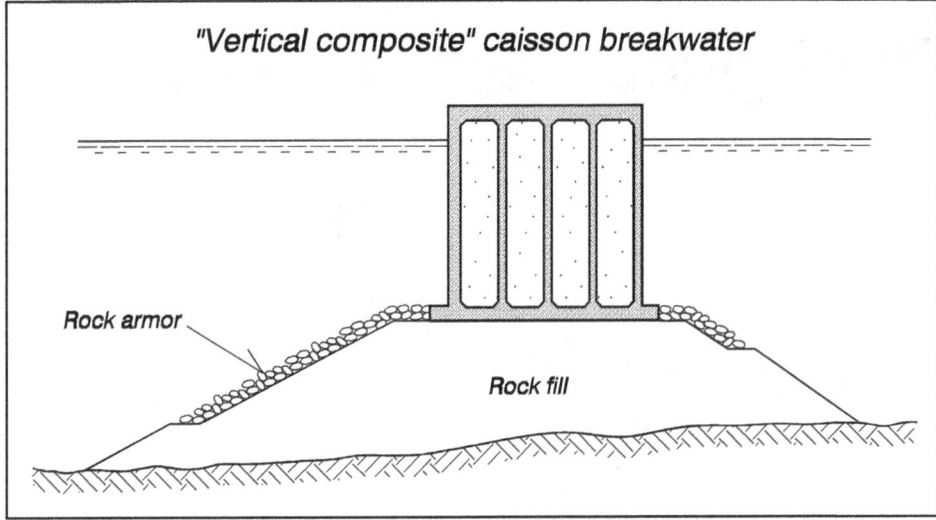

Figure VI-2-17. Vertical composite caisson breakwater

(3) Horizontal composite, i.e., the front of the caisson is covered by armor units or a rubble-mound structure (multilayered or homogeneous) (Figure VI-2-18). This type is typically used in shallow water; however, there have been applications in deeper water where impulsive wave pressures are likely to occur. The effects of the mound are reduction of wave reflection, wave impact, and wave overtopping. Depending on bottom conditions, a filter layer may be needed beneath the rubble-mound portion.

(4) Sloping top, i.e., the upper part of the front wall above still-water level is given a slope with the effect of a reduction of the wave forces and a much more favorable direction of the wave forces on the sloping front (Figure VI-2-19). However, overtopping is larger than for a vertical wall of equal crest level.

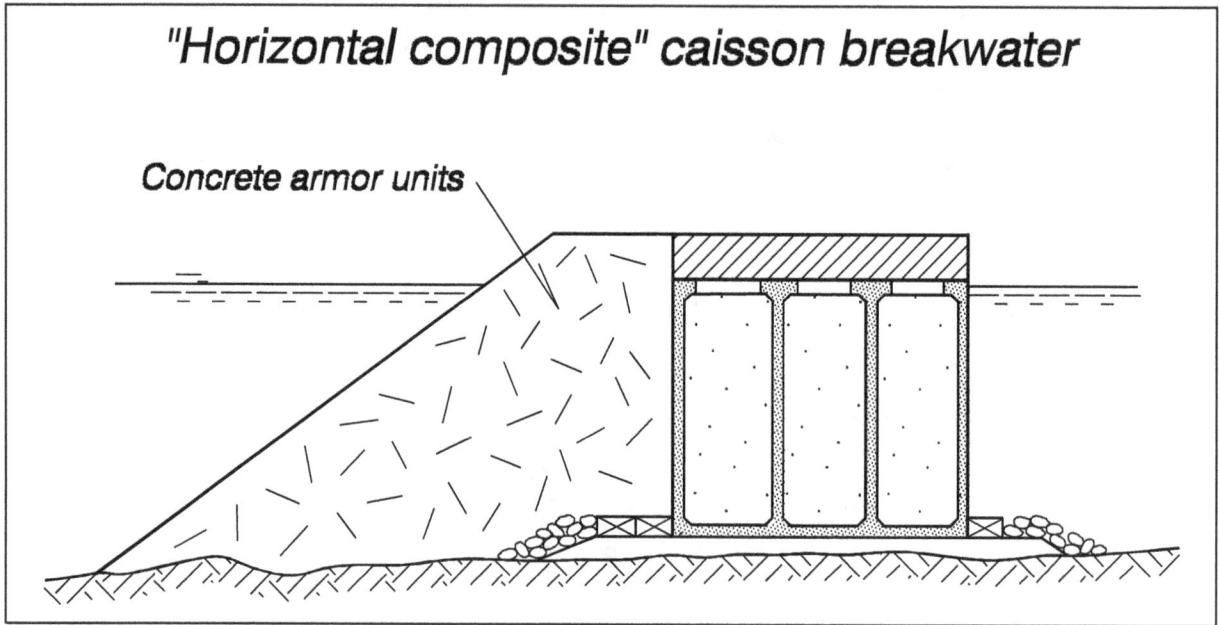

Figure VI-2-18. Horizontal composite caisson breakwater

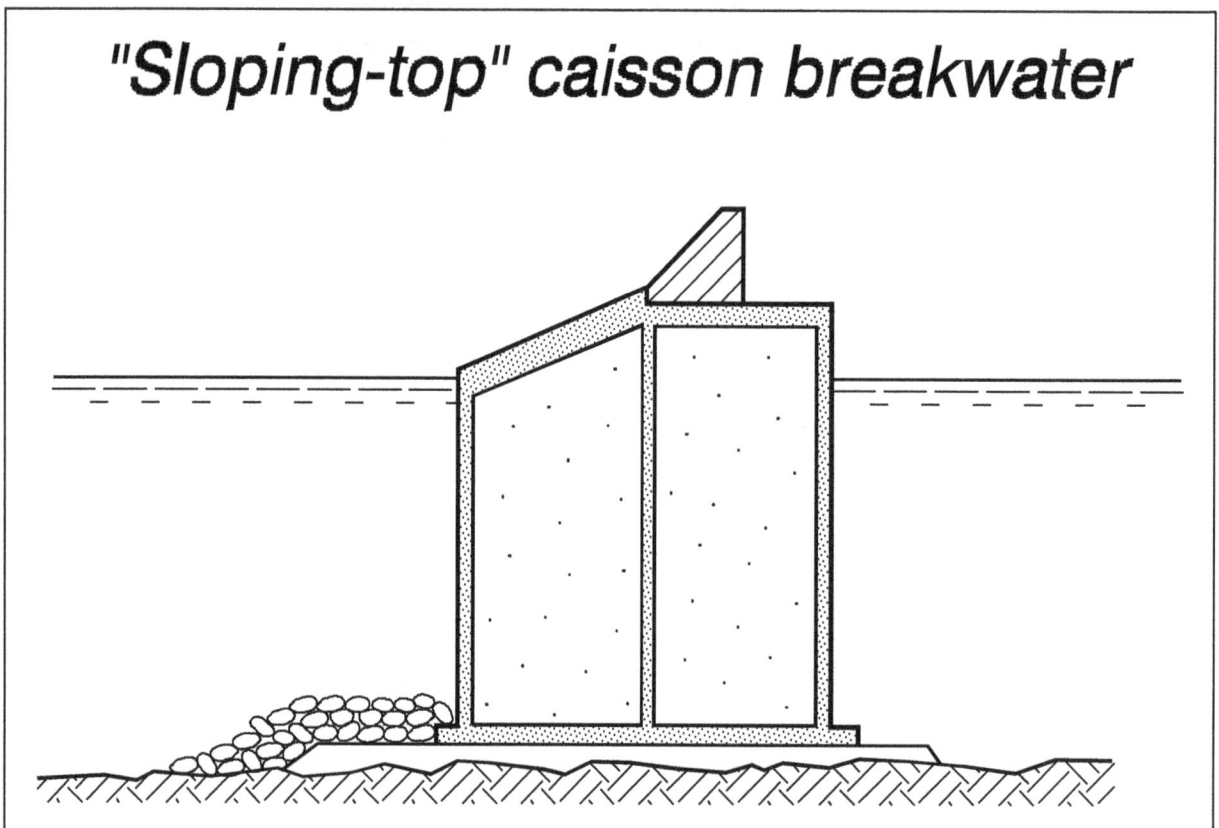

Figure VI-2-19. Sloping-top caisson breakwater

(5) Perforated front wall, i.e., the front wall is perforated by holes or slots with a wave chamber behind (Figure VI-2-20). Dissipation of energy reduces both wave forces on the caisson and wave reflection. Caisson breakwaters are generally less economical than rubble-mound structures in shallow water. Moreover, they demand stronger seabed soils than rubble structures. In particular, the blockwork type needs to be placed on rock seabeds or on very strong soils due to very high foundation loads and sensitivity to differential settlements (Figure VI-2-21).

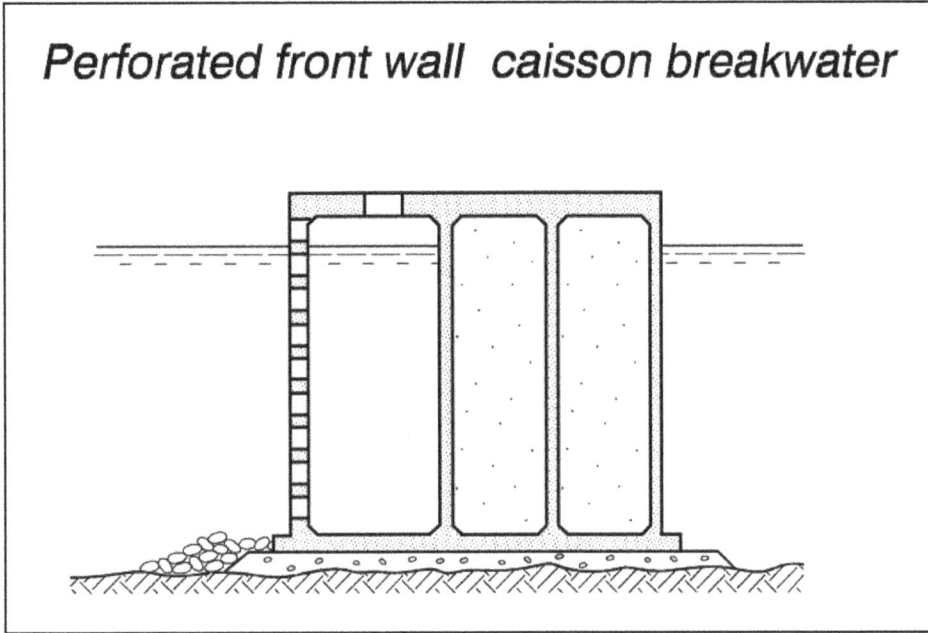

Figure VI-2-20. Perforated front wall caisson breakwater

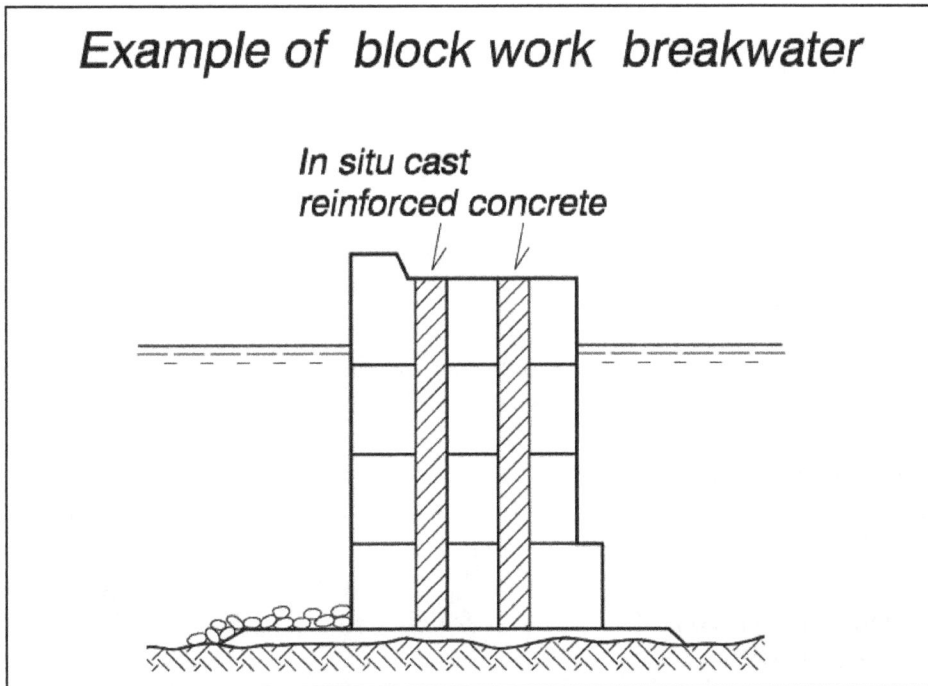

Figure VI-2-21. Example of blockwork breakwater

i.   Piled breakwaters. Piled breakwaters consist of an inclined or vertical curtain wall mounted on pile work (Figure VI-2-22). This type of breakwater is applicable in less severe wave climates on sites with weak and soft subsoils.

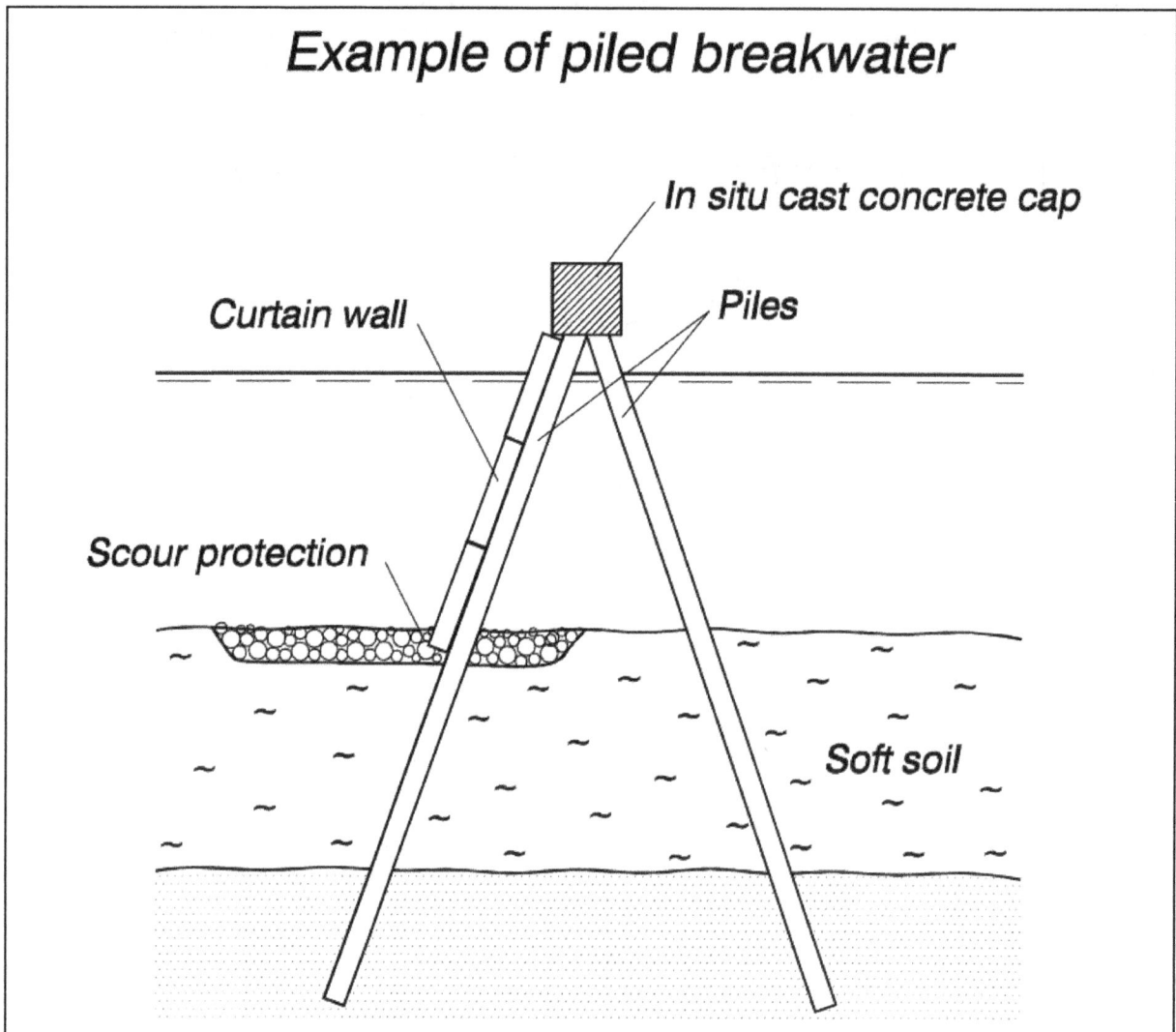

Example of piled breakwater

In situ cast concrete cap

Curtain wall

Piles

Scour protection

Soft soil

Figure VI-2-22. Example of piled breakwater

j.   Jetties. Jetties are in most cases designed as rubble-mound structures (breakwaters and groins) except that the outer part must be armored on both sides.

k.   Storm surge barriers. Storm surge barriers are generally designed as movable segmented gates made of steel. The segments might span between caisson structures, either hinged to be the caisson sidewalls or hanging in a hoist arrangement. A solution with no visible structures was proposed for the protection of the Venice Lagoon where the segments are hinged to a concrete foundation placed in the seabed as sketched in Figure VI-2-23. This structure had not been built as of this writing.

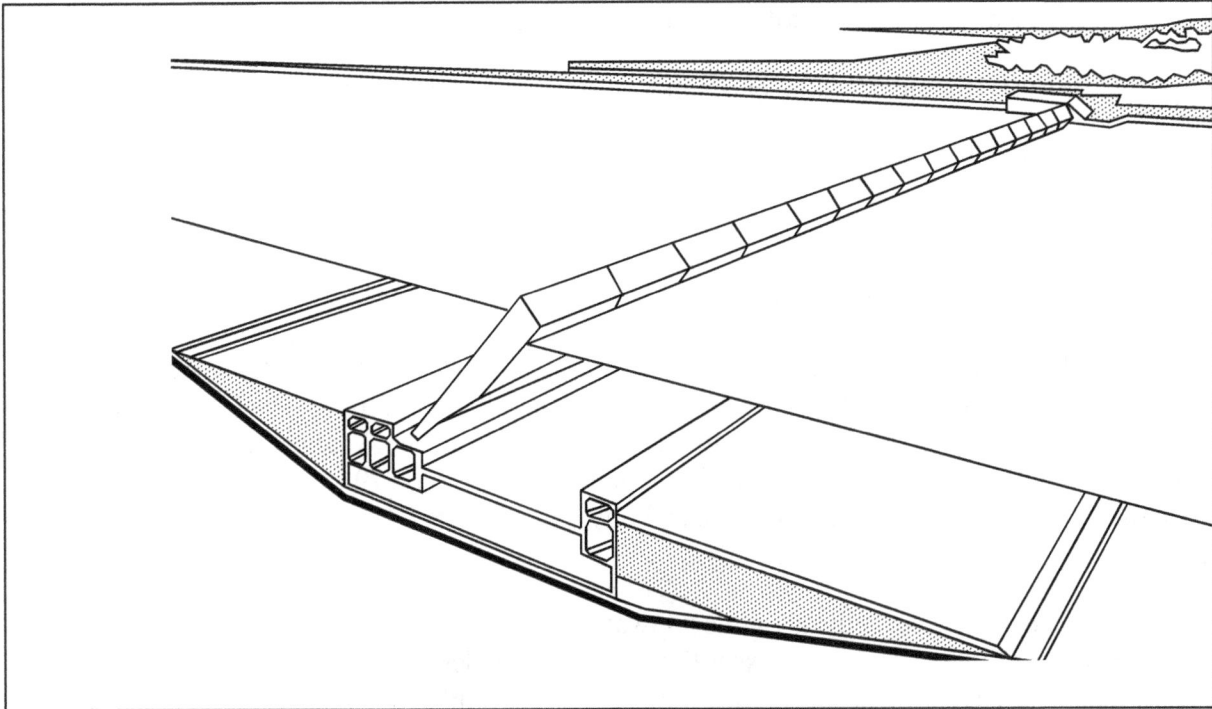

Figure VI-2-23. Storm surge barrier proposed for the Venice Lagoon

VI-2-3. <u>Main Types of Concrete Armor Units</u>.

    a. Figure VI-2-24 shows examples of the many existing types of concrete armor units.

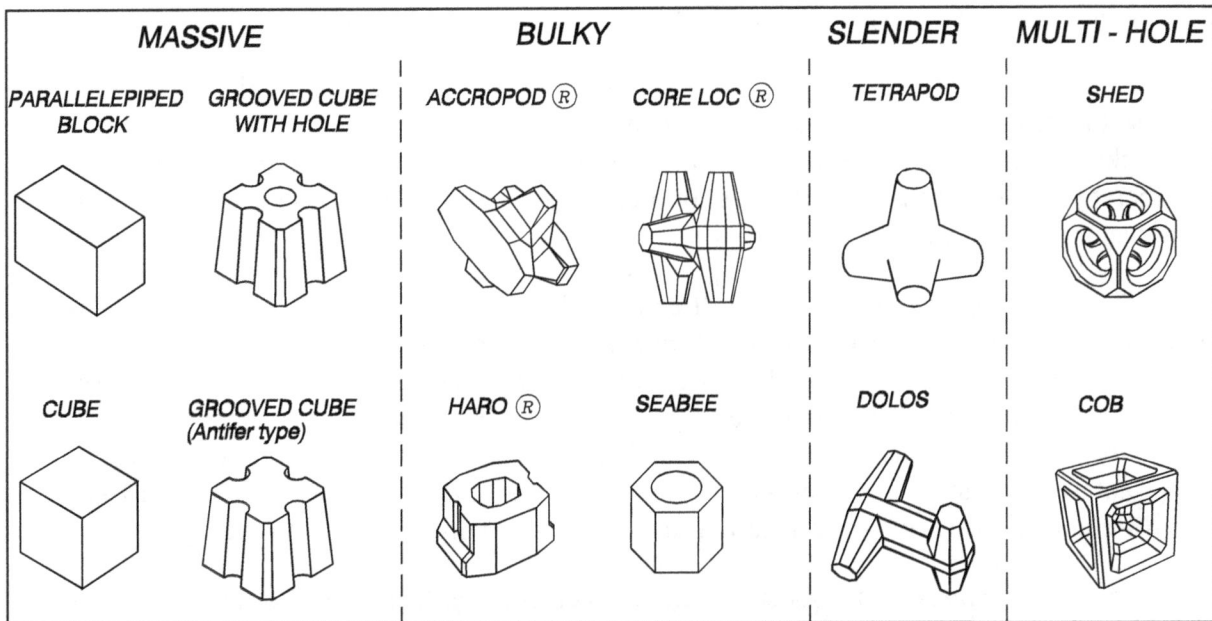

Figure VI-2-24. Examples of concrete armor units

The units can be divided into the following categories related to the structural strength:

(1) Massive or blocky (e.g., cubes incl. grooved types, parallelepiped block)

(2) Bulky (e.g., Accropode7, Core Loc7, Haro7, Seabee)

(3) Slender (e.g., Tetrapod, Dolos)

(4) Multi-hole cubes (e.g., Shed, Cob)

b. The hydraulic efficiency might be expressed in terms of the resistance against movements per volume of concrete required to armor a unit area of the slope. The hydraulic efficiency increases from massive units to slender units to multi-hole cubes. Because the porosity of randomly placed armor also increases in the same way (Price 1979), there appears to be an explainable correlation between hydraulic stability and porosity (Burcharth and Thompson 1983).

c. Concrete armor units are almost always placed randomly on the slope in a layer that has a thickness of two armor units. Exceptions are Accropodes[7] and Core Locs[7], which are placed in a layer having thickness of one armor unit, and multi-hole cubes which are placed orderly in a regular pattern where each unit rests against adjacent units.

d. Generally, concrete armor units are made of conventional unreinforced concrete except for some of the multi-hole cubes where fiber reinforcement is used. For slender units, such as Dolos with small waist ratios, various types of high-strength concrete and reinforcement (conventional bars, prestressing, fibers, scrap iron, steel profiles) have been considered. But these solutions are generally less cost-effective, and they are seldom used.

e. The hydraulic stability of armor layers is decreased if the armor units disintegrate because this reduces the stabilizing gravitational force acting on the unit, and possibly decreases interlocking effects. Moreover, broken armor unit pieces can be thrown around by wave action and thereby trigger accelerated breakage. In order to prevent breakage it is necessary to ensure the structural integrity of the armor units.

f. Unreinforced concrete is a brittle material with a low *tensile strength* S_T (on the order of 2-5 MPa) and a *compressive strength* S_C that is one order of magnitude larger than S_T. Consequently, crack formation and breakage are nearly always caused by load-induced *tensile stresses* σ_T exceeding S_T. Therefore, the magnitude of S_T is more critical to concrete armor unit design than S_C, and this fact should be reflected in specifications for armor unit concretes. It is important to note that S_T decreases with repeated load due to *fatigue* effects (Burcharth 1984). The different categories of concrete armor units are not equally sensitive to breakage.

g. *Slender units* are the most vulnerable to cracking and breaking because the limited cross-sectional areas give rise to relatively large tensile stresses. Many failures of breakwaters armored with Tetrapods and Dolosse were caused by breakage of the units before the hydraulic stability of the unbroken units was exceeded. Much of the damage could have been avoided if design diagrams for concrete armor unit structural integrity had been available during design.

h. These failures caused a decline in the use of slender armor units and a return to the use of massive blocks, especially the Antifer Cube types. This also led to the development of bulky units like the Haro[7], the Accropode[7], and the Core Loc[7]. The tendency toward massive blocks will not change until reliable design strength diagrams exist for the slender units. Presently, structural integrity diagrams are available only for Dolos (Burcharth 1993, Melby 1993) and Tetrapod armor (Burcharth et al. 1995).

i. *Massive units* generally will have the smallest tensile stresses due to the large cross-sectional areas. However, breakage can take place if the units experience impacts due to application of less restrictive hydraulic stability criteria and if the concrete quality is poor in terms of a low tensile strength. Cracking can also occur in larger units where temperature differences during the hardening process can create tensile stresses that exceed the strength of the weak young concrete, resulting in microcracking of the material, also known as *thermal stress cracking* (Burcharth 1983). If massive units are made of good quality concrete, units are not damaged during handling, and if the units are designed for marginal displacements, there will be no breakage problems. With the same precautions the *bulky units* are also not expected to have breakage problems. No structural integrity design diagrams exist for the massive concrete armor units.

j. *Multi-hole cubes* will experience very small solid impact loads provided they are placed correctly in patterns that exclude significant relative movements of the blocks. Due to the slender structural members with rather tiny cross sections, the limiting factors (excluding impacts) for long-term durability are material deterioration, abrasion on sandy coasts, and fatigue due to wave loads.

VI-2-4. Failure Modes of Typical Structure Types.

a. Failure.

(1) For many people, the word "failure" implies a total or partial collapse of a structure, but this definition is limited and not accurate when discussing design and performance of coastal structures. In the context of design reliability, it is preferable to define failure as:

> FAILURE: Damage that results in structure performance and
> functionality below the minimum anticipated by design.

Thus, partial collapse of a structure may be classified as "damage" provided the structure still serves its original purpose at or above the minimum expected level. For example, subsidence of a breakwater protecting a harbor would be considered a failure if it resulted in wave heights within the harbor that exceed operational criteria. Conversely, partial collapse of a rubble-mound jetty head might be classified as damage if resulting impacts to navigation and dredging requirements are minimal or within acceptable limits.

(2) Coastal project elements fail for one or more of the following reasons:

(a) Design failure occurs when either the structure as a whole, including its foundation, or individual structure components cannot withstand load conditions within the design criteria. Design failure also occurs when the structure does not perform as anticipated.

(b) Load exceedance failure occurs because anticipated design load conditions were exceeded.

(c) Construction failure arises due to incorrect or bad construction or construction materials.

(d) Deterioration failure is the result of structure deterioration and lack of project maintenance.

(3) New or innovative coastal project design concepts are more susceptible to design failure due to lack of previous experience with similar designs. In these situations, allowances should be made for unknown design effects, and critical project elements should be extensively tested using laboratory and/or numerical model techniques before finalizing the design.

(4) Practically all projects accept some level of failure probability associated with exceedance of design load conditions, but failure probability increases at project sites where little prototype data exist on which to base the design. These cases may require a conservative factor of safety (for information on probabilistic design see Part V-1-3, "Risk Analysis and Project Optimization," and Part VI-6, "Reliability in Design").

(5) In the design process all possible failure modes must be identified and evaluated in order to obtain a balanced design. An overview of the most important and common failure modes for the main types of fixed coastal structures is given in this chapter. Some failure modes are common to several types of structures. Examples include displacement of armor stones and toe erosion which are relevant to most rubble structures such as seawalls, groins, and breakwaters. It should be noted that in this chapter the common failure modes are shown only for one of the relevant structures. The most comprehensive sets of failure modes are related to breakwaters, and for this reason they are discussed first.

b. Sloping-front structures.

(1) Breakwaters. Figure VI-2-25 provides an overview of the failure modes relevant to rubble-mound breakwaters. The individual failure modes are explained in more detail in Figures VI-2-26 to VI-2-42.

(2) Seawalls/revetments. Typical failure modes for seawalls and revetments are shown in Figures VI-2-43 to VI-2-46.

(3) Dikes. Figures VI-2-47 through VI-2-51 illustrate dike failure modes.

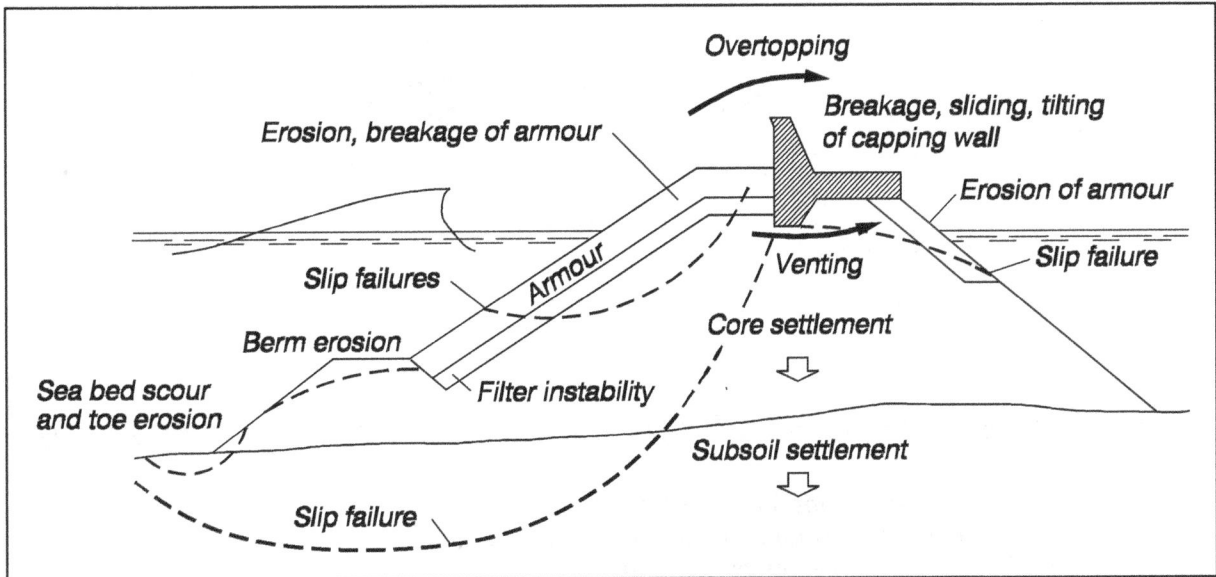

Figure VI-2-25. Overview of rubble-mound breakwater failure modes

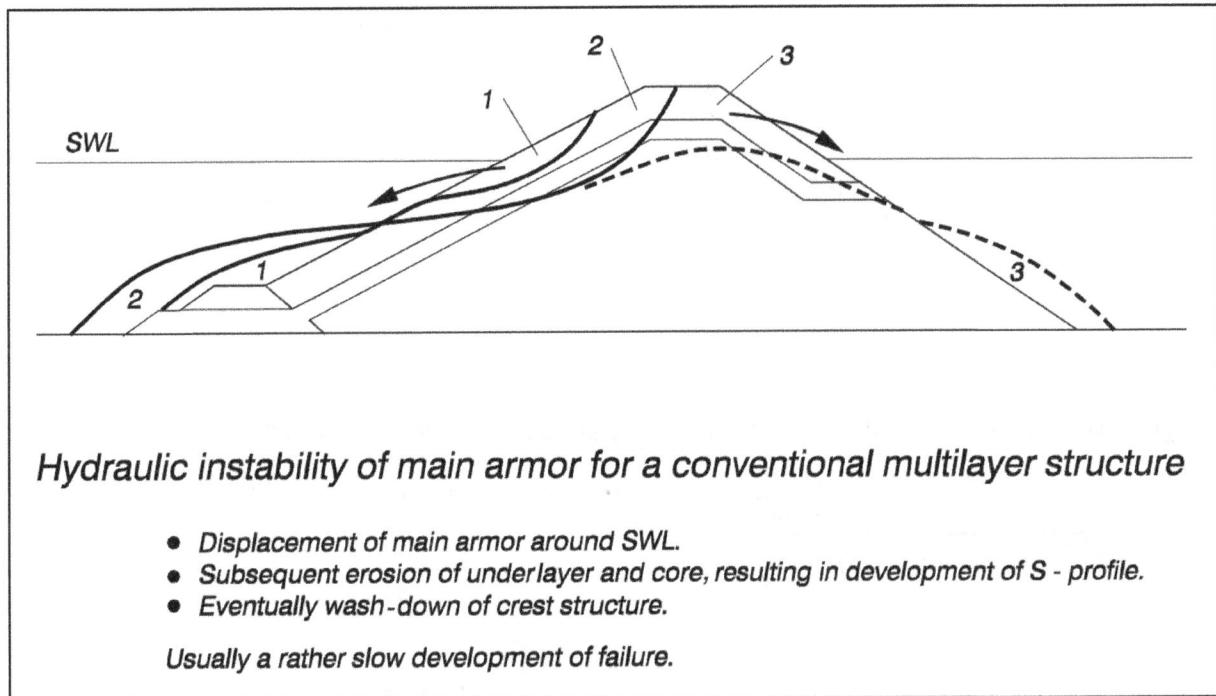

Hydraulic instability of main armor for a conventional multilayer structure

- Displacement of main armor around SWL.
- Subsequent erosion of underlayer and core, resulting in development of S - profile.
- Eventually wash-down of crest structure.

Usually a rather slow development of failure.

Figure VI-2-26. Main armor layer instability

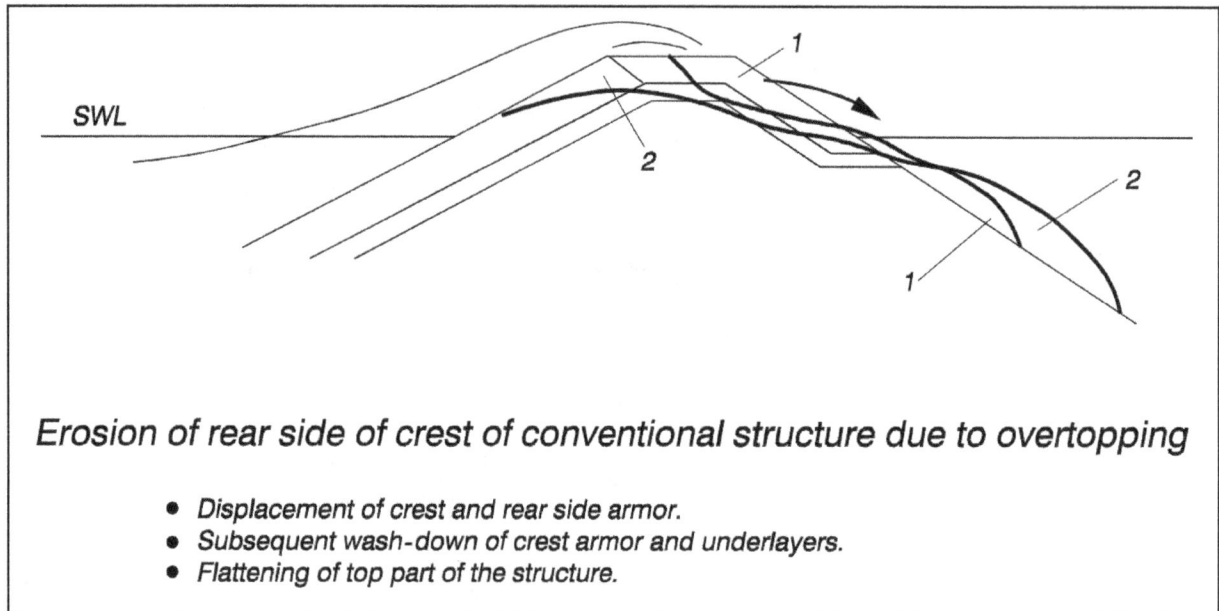

Erosion of rear side of crest of conventional structure due to overtopping

- *Displacement of crest and rear side armor.*
- *Subsequent wash-down of crest armor and underlayers.*
- *Flattening of top part of the structure.*

Figure VI-2-27. Rear side erosion of crest

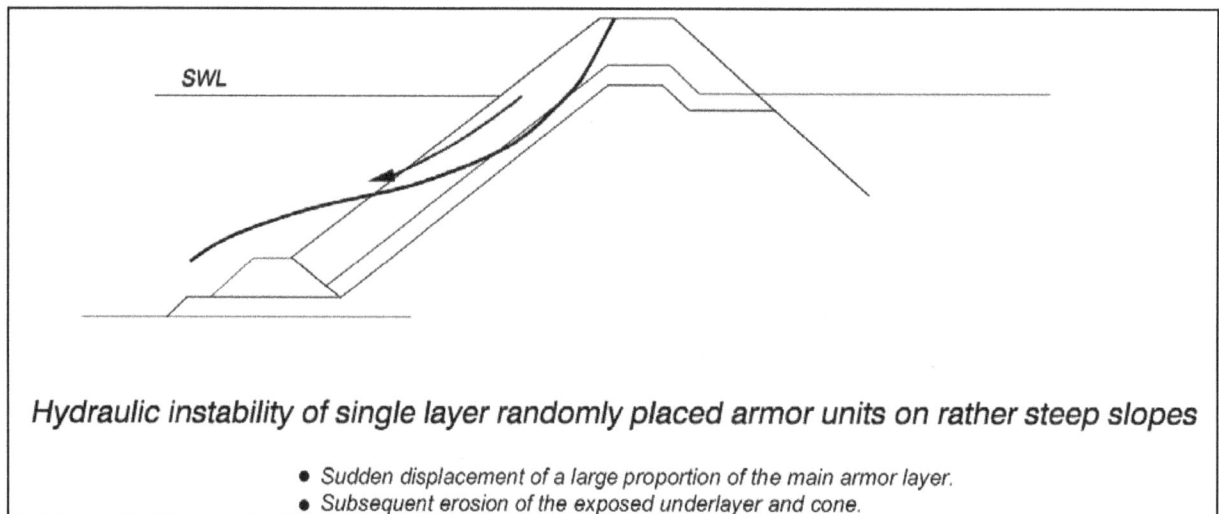

Hydraulic instability of single layer randomly placed armor units on rather steep slopes

- *Sudden displacement of a large proportion of the main armor layer.*
- *Subsequent erosion of the exposed underlayer and cone.*

Figure VI-2-28. Hydraulic instability on steep slopes

Breakage of armor units

- The armor units break when the stresses caused by gravity and wave-induced forces exceed the strength of the concrete.
- Breakage of complex types of armor units causes collapse of the armor layer if a substantial proportion of the units (say more than 15%) break.

Figure VI-2-29. Armor unit breakage

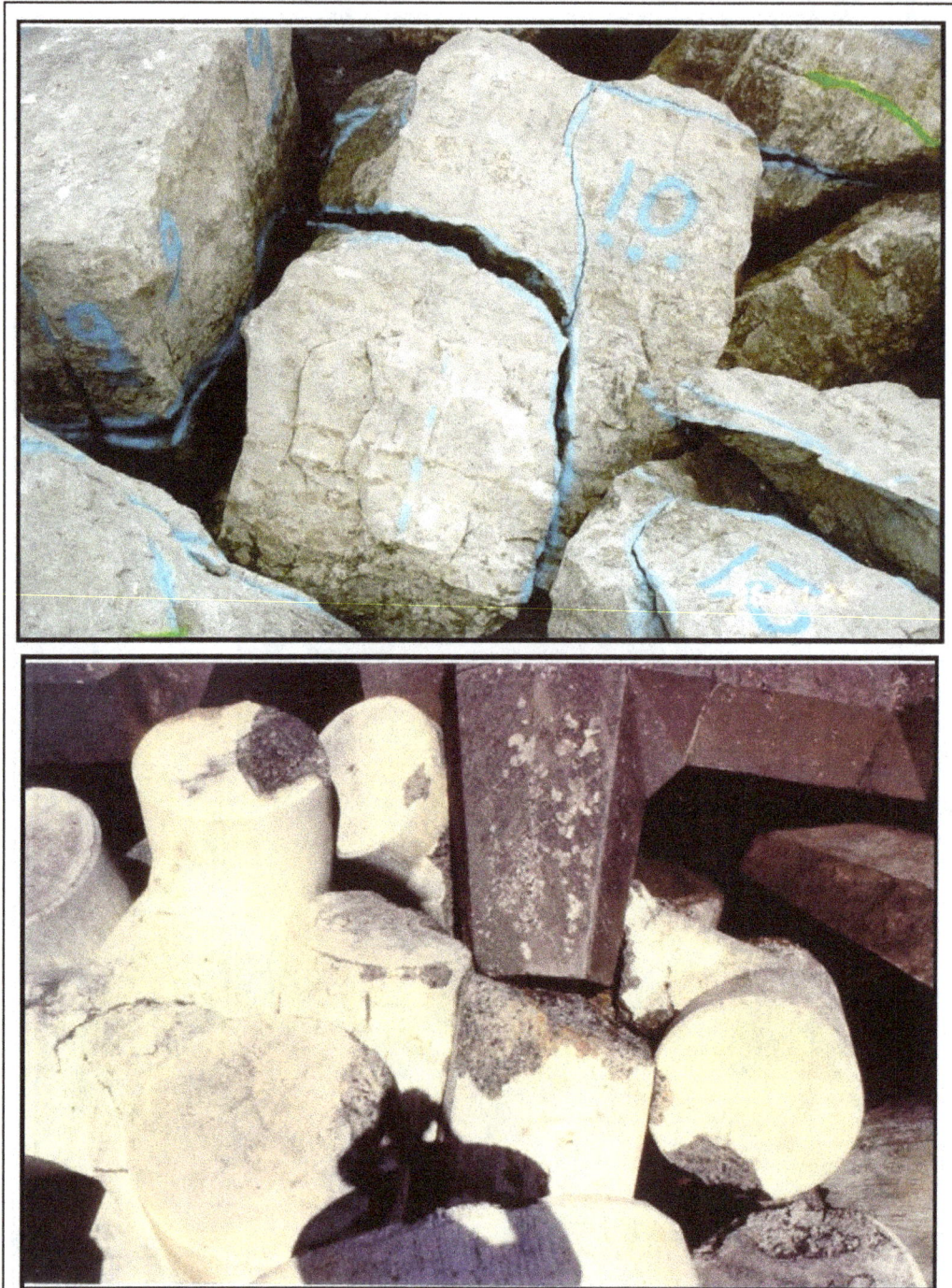

Deterioration of rock and concrete armor blocks

- Weathering caused by chemical reactions, temperature variations (including freeze - thaw) and abrasion results in loss of strength and mass.
- Subsequently there is a higher risk of breakage and in reduced hydraulic stability.

Figure VI-2-30. Armor unit deterioration

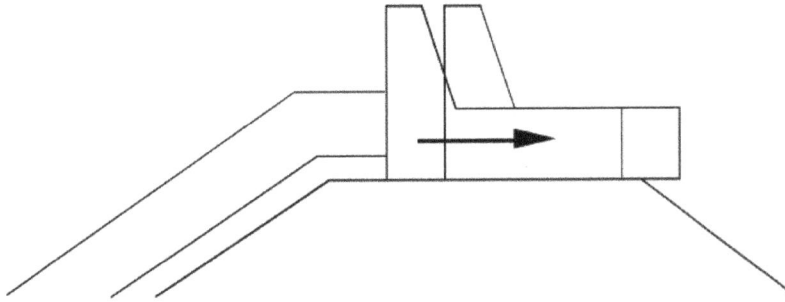

Sliding of wave wall superstructure

The superstructure slides backwards when the horizontal wave force exceeds the friction forces underneath the base plate.

Large displacements lead to tilting and wash-down of the superstructure.

Figure VI-2-31. Sliding of superstructure

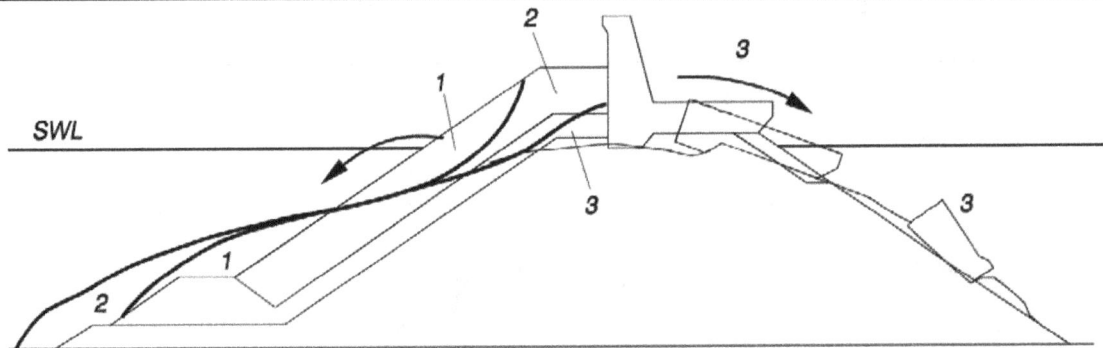

Hydraulic instability and/or breakage of complex types of armor units leading to failure of wave wall superstructure

- *Displacement of intact or broken units.*
- *Subsequent exposure of the wave wall to large wave impacts and uplift pressures.*
- *Eventual breakage and wash-down of superstructure.*

The failure can proceed rather fast if the armor unit breakage is extensive.

Figure VI-2-32. Failure due to armor unit breakage

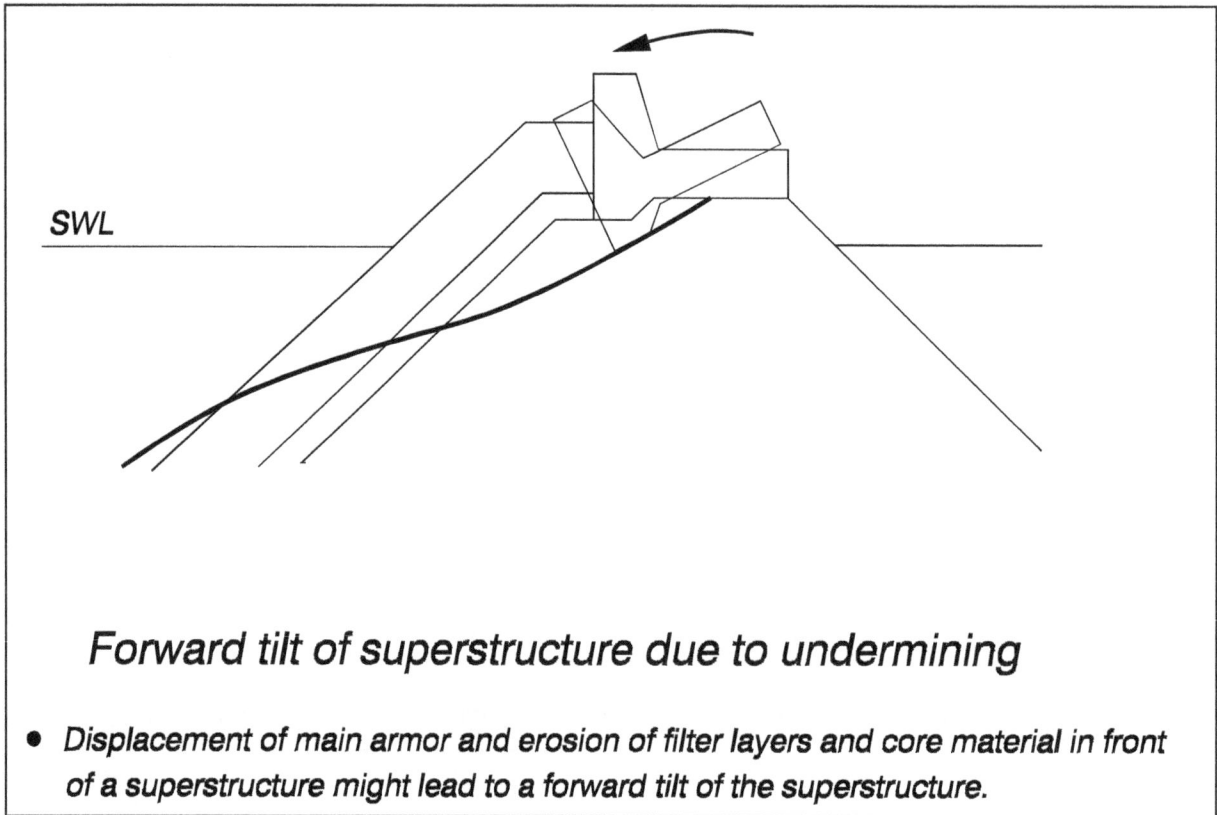

Forward tilt of superstructure due to undermining

- *Displacement of main armor and erosion of filter layers and core material in front of a superstructure might lead to a forward tilt of the superstructure.*

Figure VI-2-33. Forward tilting of superstructure

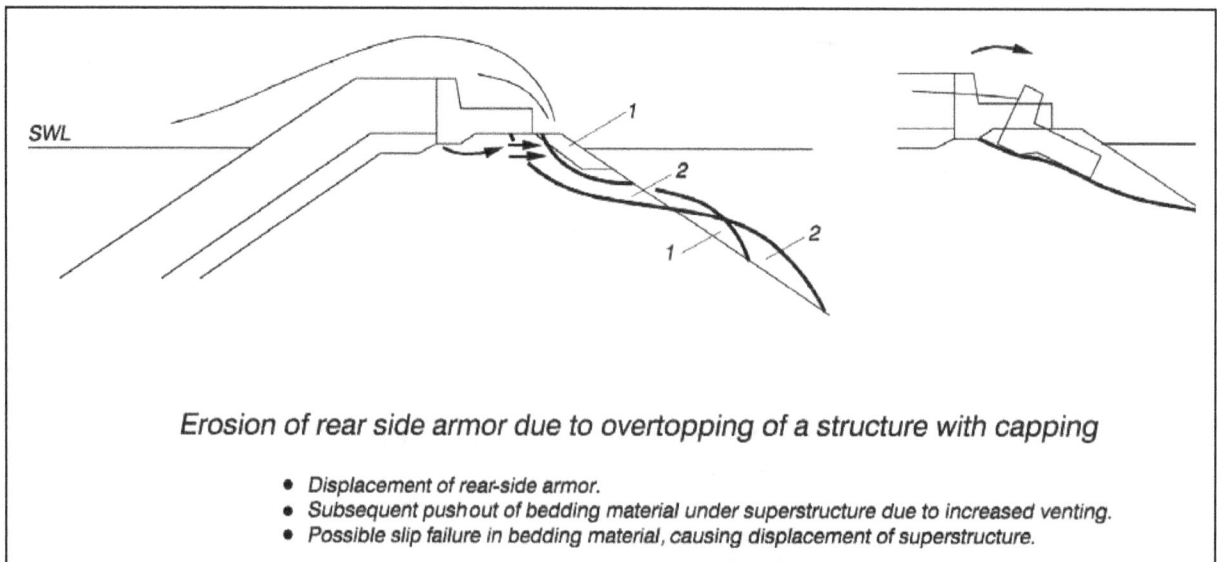

Erosion of rear side armor due to overtopping of a structure with capping

- *Displacement of rear-side armor.*
- *Subsequent pushout of bedding material under superstructure due to increased venting.*
- *Possible slip failure in bedding material, causing displacement of superstructure.*

Figure VI-2-34. Rear-side erosion due to overtopping

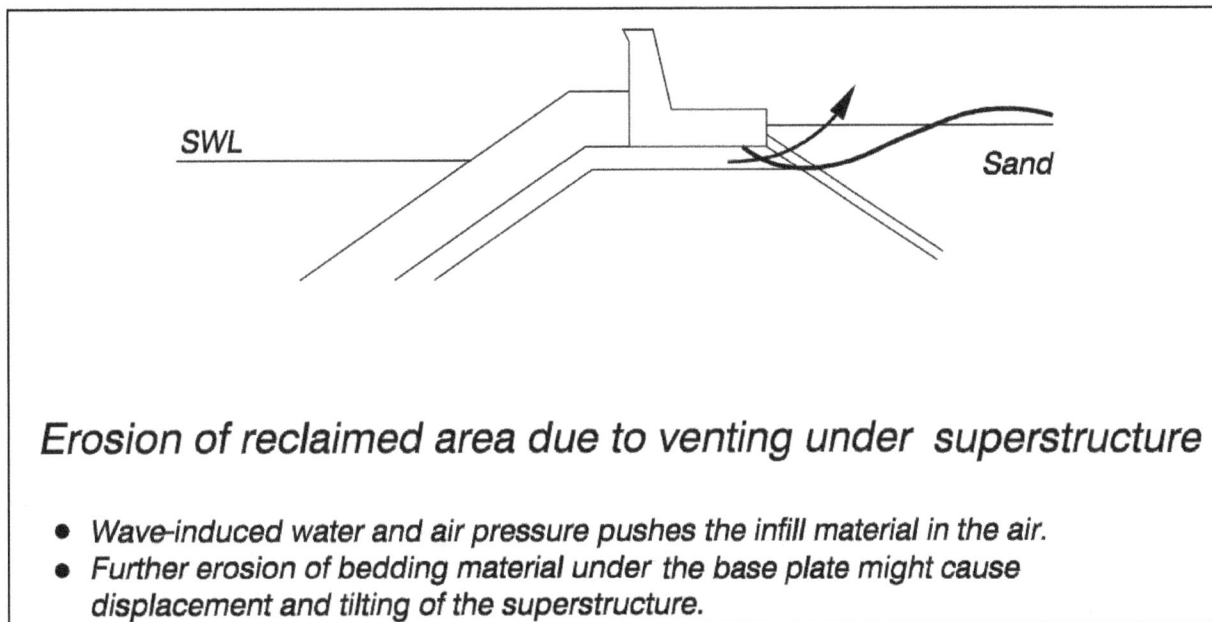

Erosion of reclaimed area due to venting under superstructure

- *Wave-induced water and air pressure pushes the infill material in the air.*
- *Further erosion of bedding material under the base plate might cause displacement and tilting of the superstructure.*

Figure VI-2-35. Erosion due to venting

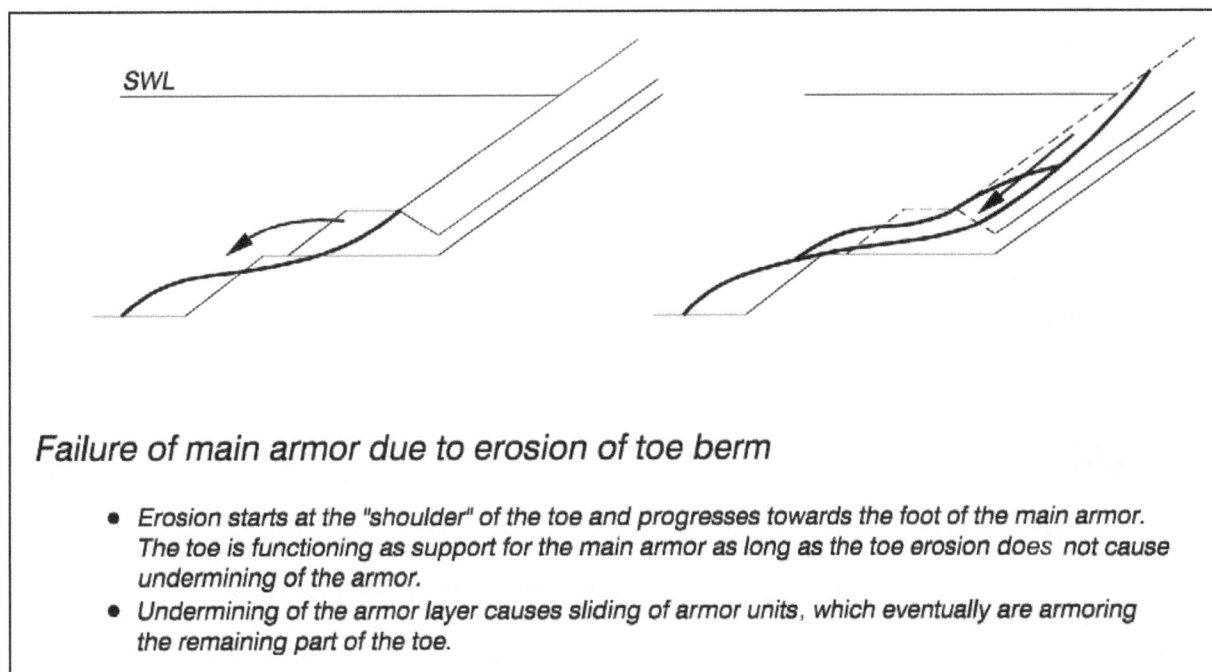

Failure of main armor due to erosion of toe berm

- *Erosion starts at the "shoulder" of the toe and progresses towards the foot of the main armor. The toe is functioning as support for the main armor as long as the toe erosion does not cause undermining of the armor.*
- *Undermining of the armor layer causes sliding of armor units, which eventually are armoring the remaining part of the toe.*

Figure VI-2-36. Failure due to toe berm erosion

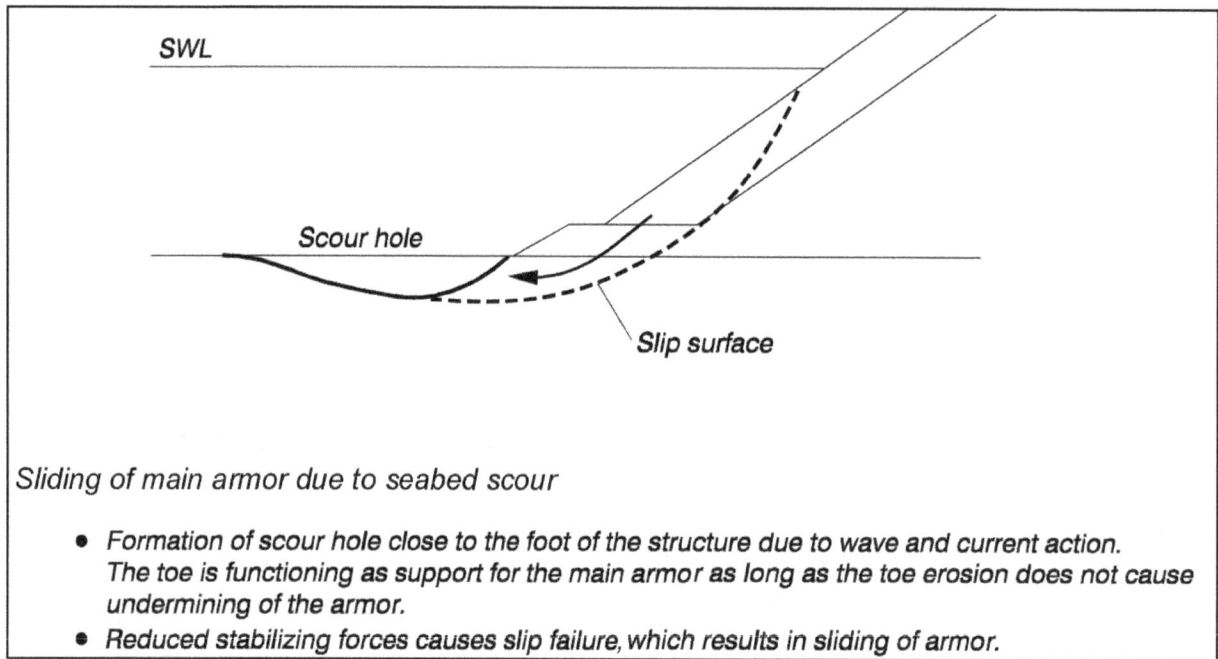

SWL

Scour hole

Slip surface

Sliding of main armor due to seabed scour

- *Formation of scour hole close to the foot of the structure due to wave and current action. The toe is functioning as support for the main armor as long as the toe erosion does not cause undermining of the armor.*
- *Reduced stabilizing forces causes slip failure, which results in sliding of armor.*

Figure VI-2-37. Scour-induced armor displacement

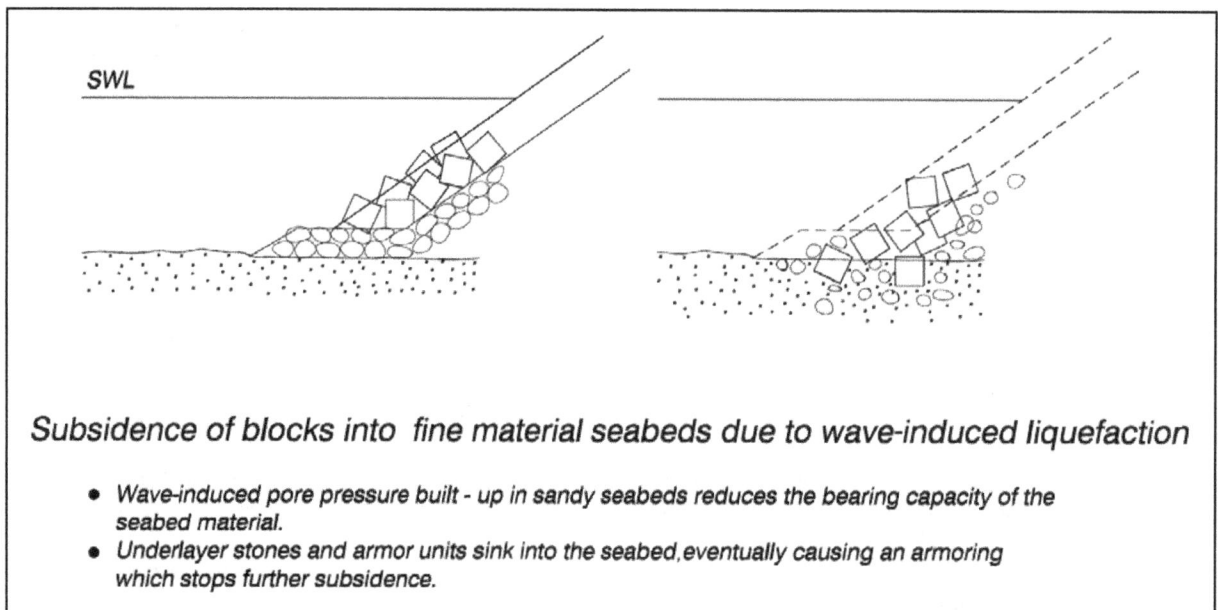

SWL

Subsidence of blocks into fine material seabeds due to wave-induced liquefaction

- *Wave-induced pore pressure built - up in sandy seabeds reduces the bearing capacity of the seabed material.*
- *Underlayer stones and armor units sink into the seabed, eventually causing an armoring which stops further subsidence.*

Figure VI-2-38. Block subsidence due to liquefaction

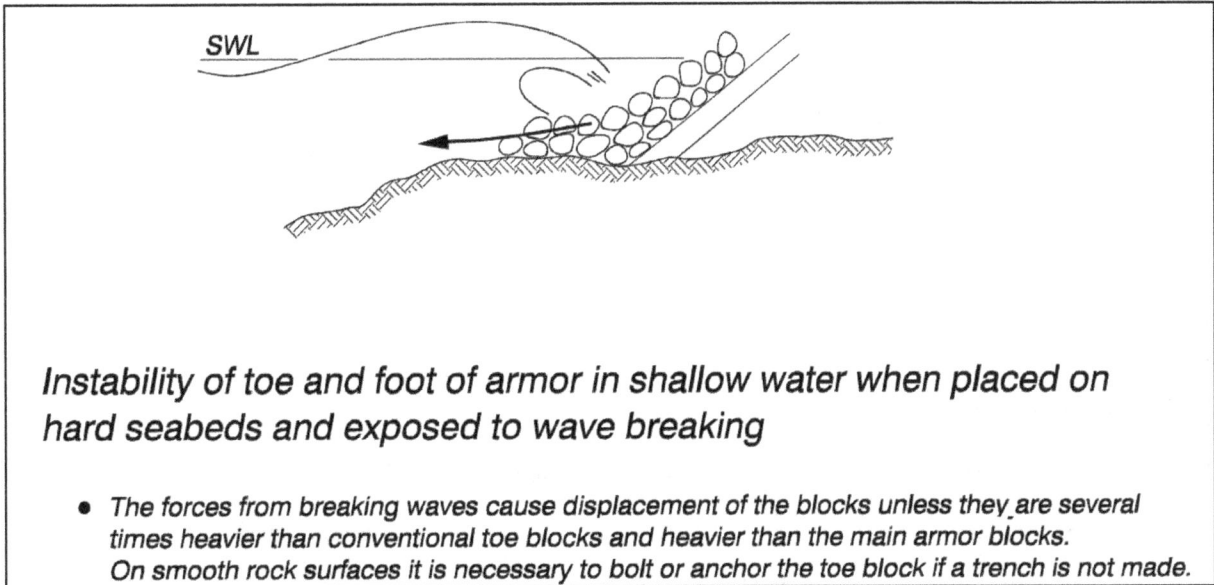

Instability of toe and foot of armor in shallow water when placed on hard seabeds and exposed to wave breaking

- *The forces from breaking waves cause displacement of the blocks unless they are several times heavier than conventional toe blocks and heavier than the main armor blocks.*
  On smooth rock surfaces it is necessary to bolt or anchor the toe block if a trench is not made.

Figure VI-2-39. Toe instability on hard bottoms

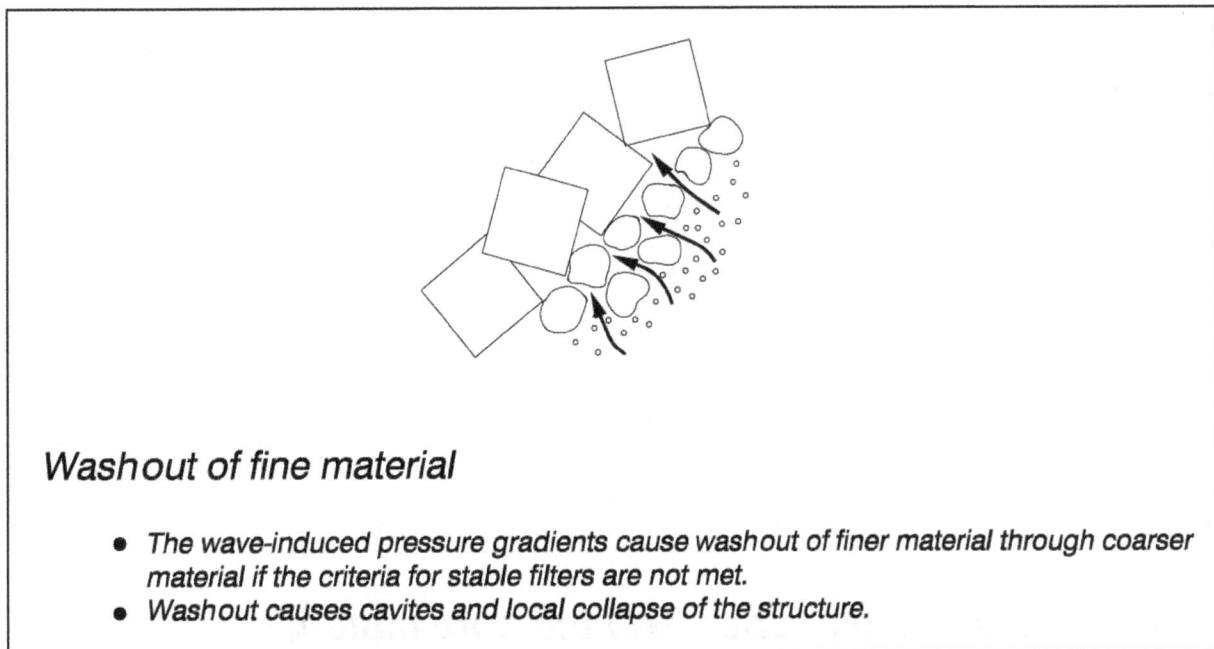

Washout of fine material

- *The wave-induced pressure gradients cause washout of finer material through coarser material if the criteria for stable filters are not met.*
- *Washout causes cavites and local collapse of the structure.*

Figure VI-2-40. Washout of underlayer material

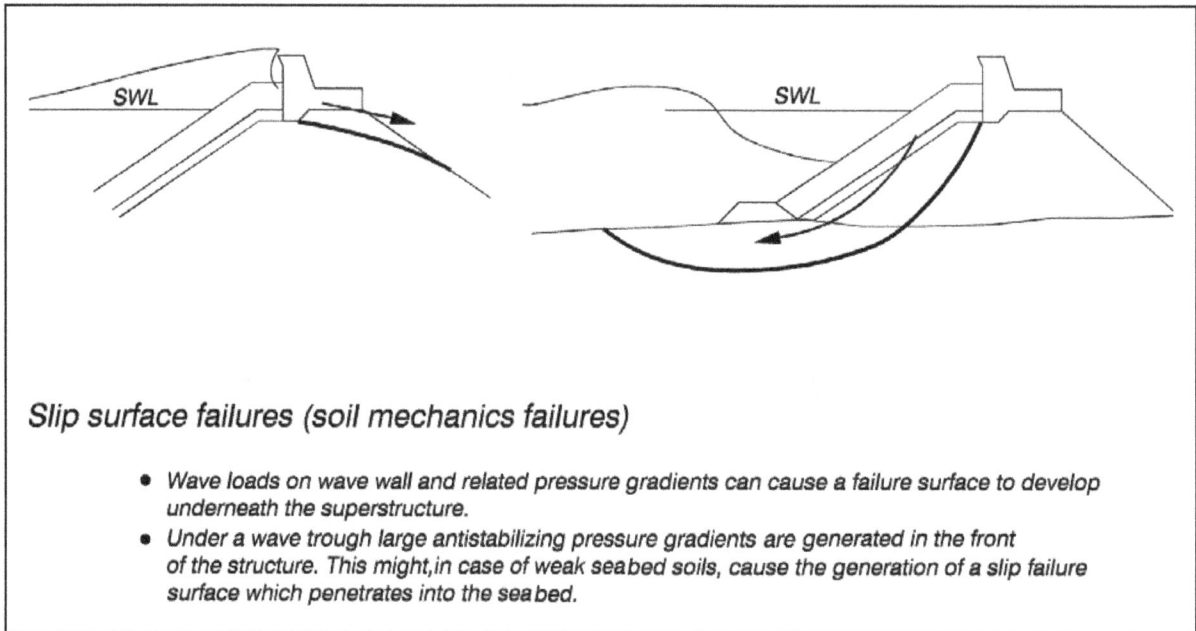

Slip surface failures (soil mechanics failures)

- *Wave loads on wave wall and related pressure gradients can cause a failure surface to develop underneath the superstructure.*
- *Under a wave trough large antistabilizing pressure gradients are generated in the front of the structure. This might, in case of weak seabed soils, cause the generation of a slip failure surface which penetrates into the seabed.*

Figure VI-2-41. Slip surface failure

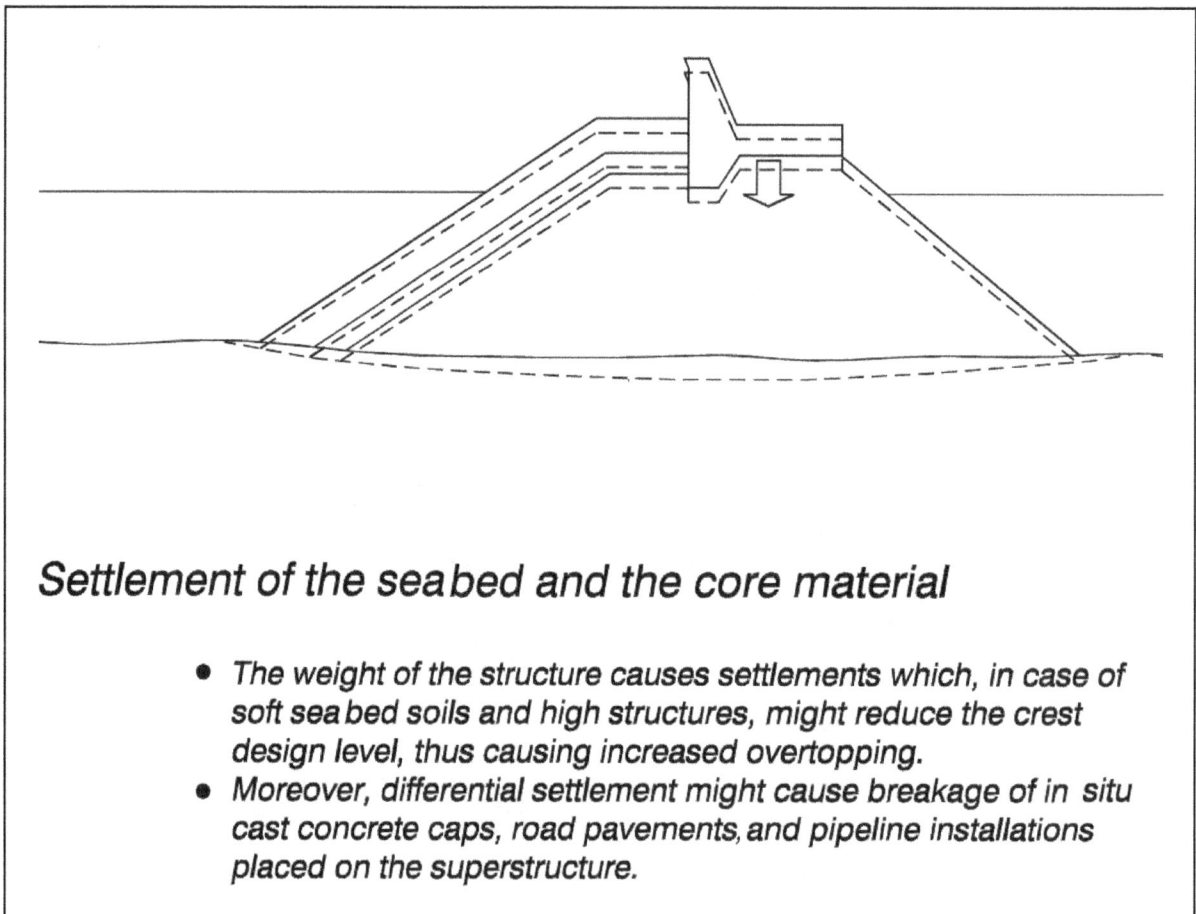

Settlement of the seabed and the core material

- *The weight of the structure causes settlements which, in case of soft seabed soils and high structures, might reduce the crest design level, thus causing increased overtopping.*
- *Moreover, differential settlement might cause breakage of in situ cast concrete caps, road pavements, and pipeline installations placed on the superstructure.*

Figure VI-2-42. Structure settlement failure

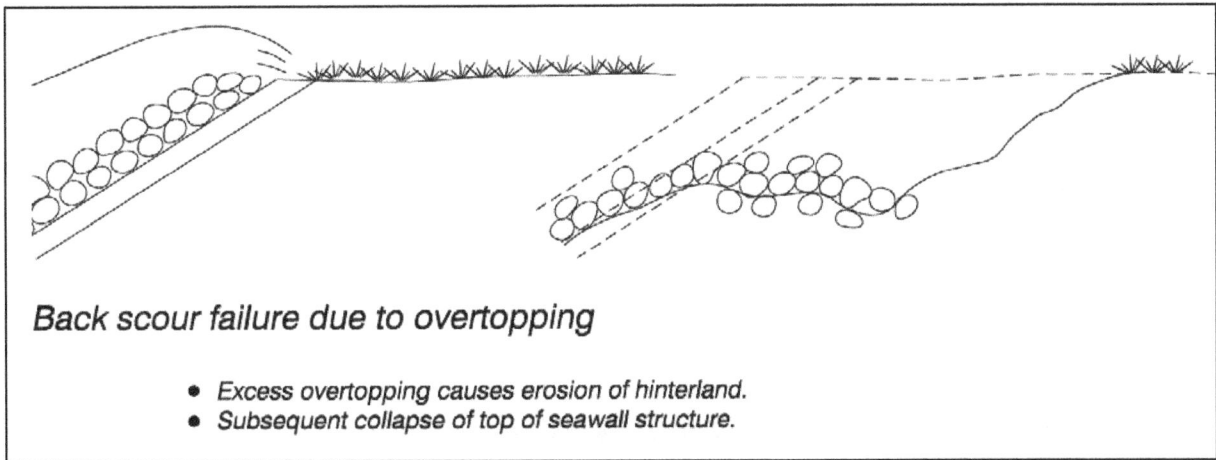

Back scour failure due to overtopping

- *Excess overtopping causes erosion of hinterland.*
- *Subsequent collapse of top of seawall structure.*

Figure VI-2-43. Scour due to overtopping

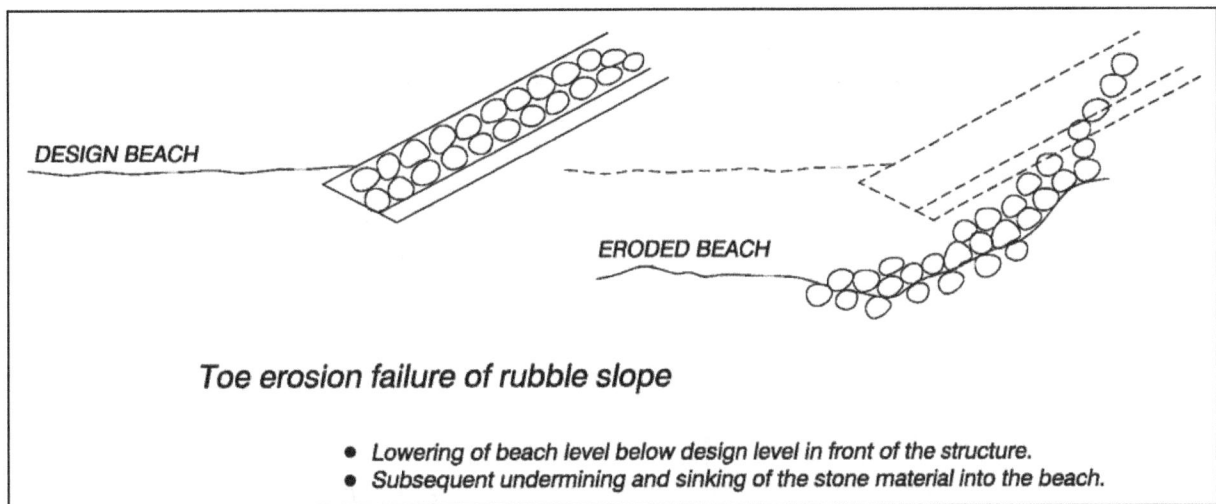

DESIGN BEACH

ERODED BEACH

Toe erosion failure of rubble slope

- *Lowering of beach level below design level in front of the structure.*
- *Subsequent undermining and sinking of the stone material into the beach.*

Figure VI-2-44. Toe erosion failure of rubble slope

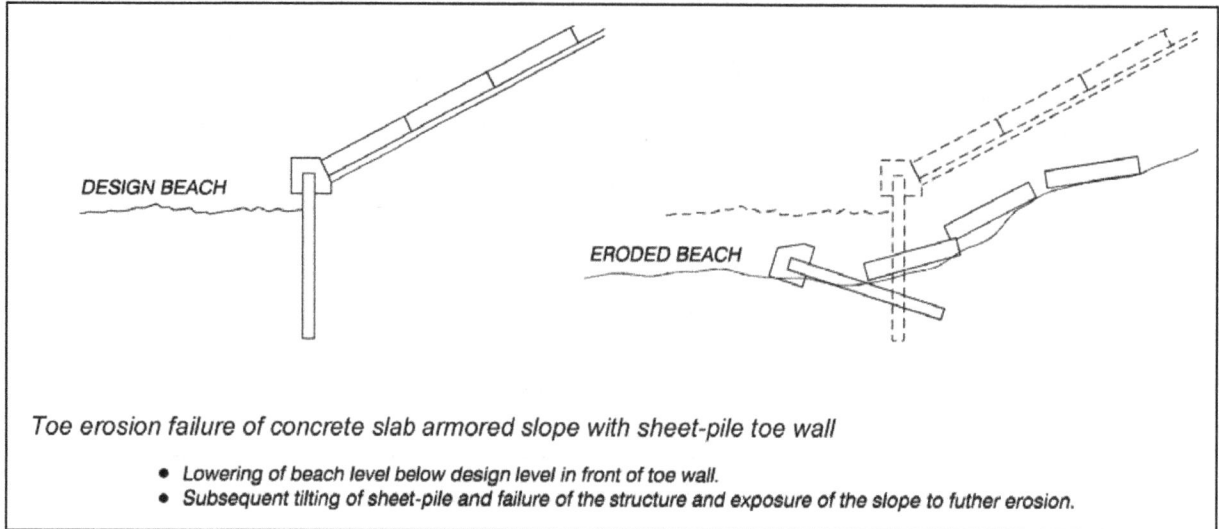

Toe erosion failure of concrete slab armored slope with sheet-pile toe wall

- Lowering of beach level below design level in front of toe wall.
- Subsequent tilting of sheet-pile and failure of the structure and exposure of the slope to futher erosion.

Figure VI-2-45. Failure of sheet-pile toe wall

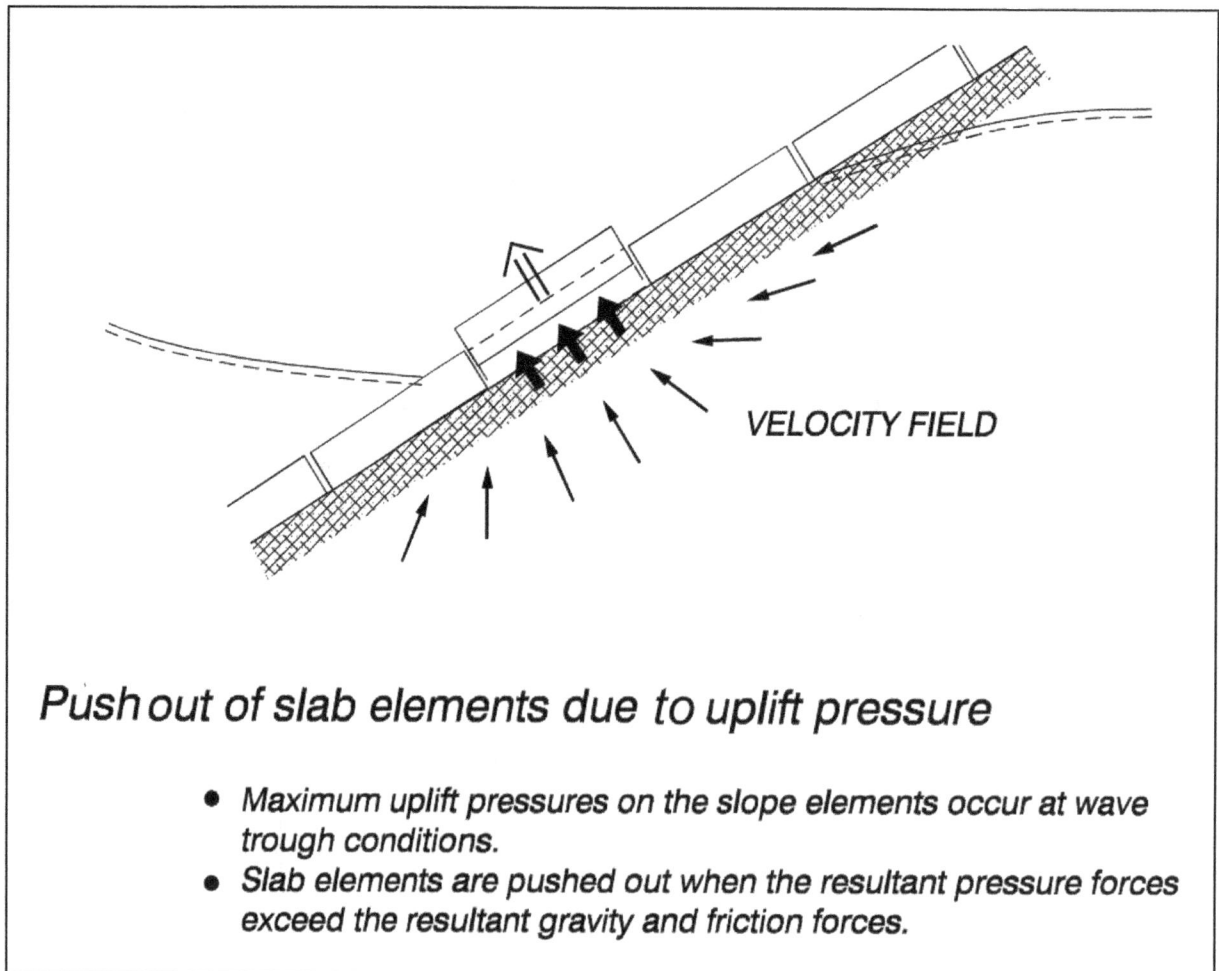

Push out of slab elements due to uplift pressure

- Maximum uplift pressures on the slope elements occur at wave trough conditions.
- Slab elements are pushed out when the resultant pressure forces exceed the resultant gravity and friction forces.

Figure VI-2-46. Pressure blowout of slab elements

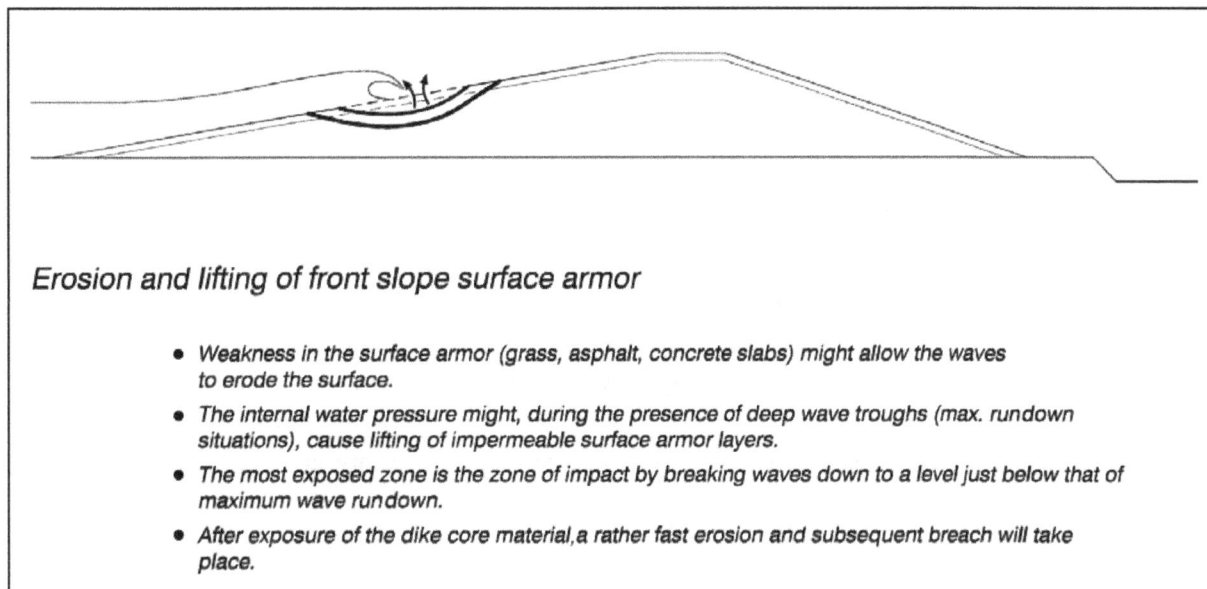

Erosion and lifting of front slope surface armor

- *Weakness in the surface armor (grass, asphalt, concrete slabs) might allow the waves to erode the surface.*
- *The internal water pressure might, during the presence of deep wave troughs (max. rundown situations), cause lifting of impermeable surface armor layers.*
- *The most exposed zone is the zone of impact by breaking waves down to a level just below that of maximum wave rundown.*
- *After exposure of the dike core material, a rather fast erosion and subsequent breach will take place.*

Figure VI-2-47. Erosion of dike slope protection

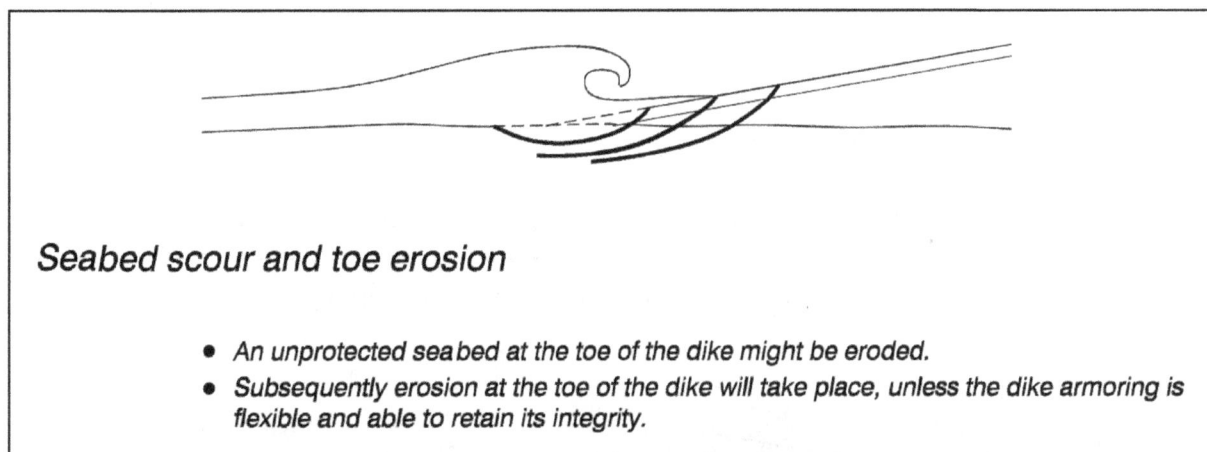

Seabed scour and toe erosion

- *An unprotected seabed at the toe of the dike might be eroded.*
- *Subsequently erosion at the toe of the dike will take place, unless the dike armoring is flexible and able to retain its integrity.*

Figure VI-2-48. Toe scour erosion of dike

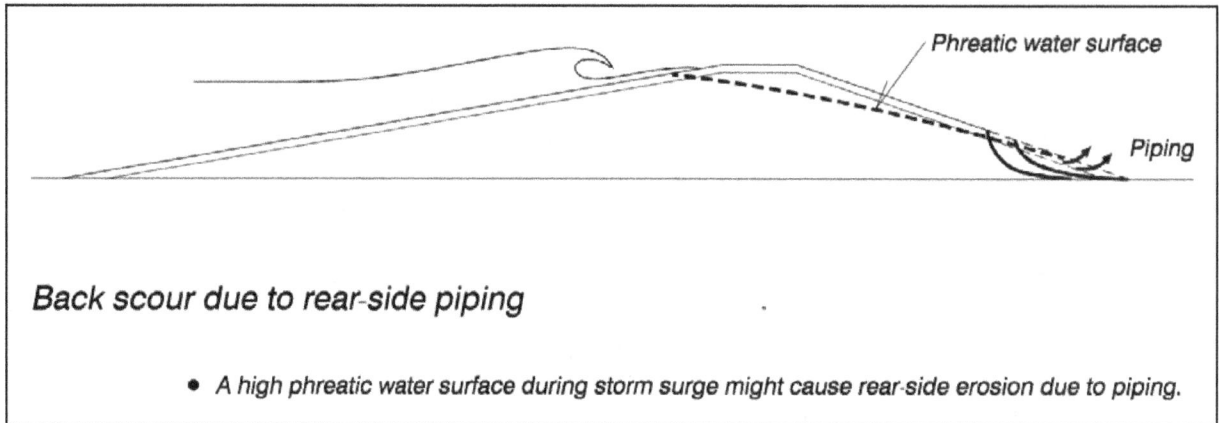

Figure VI-2-49. Dike crest erosion by overtopping

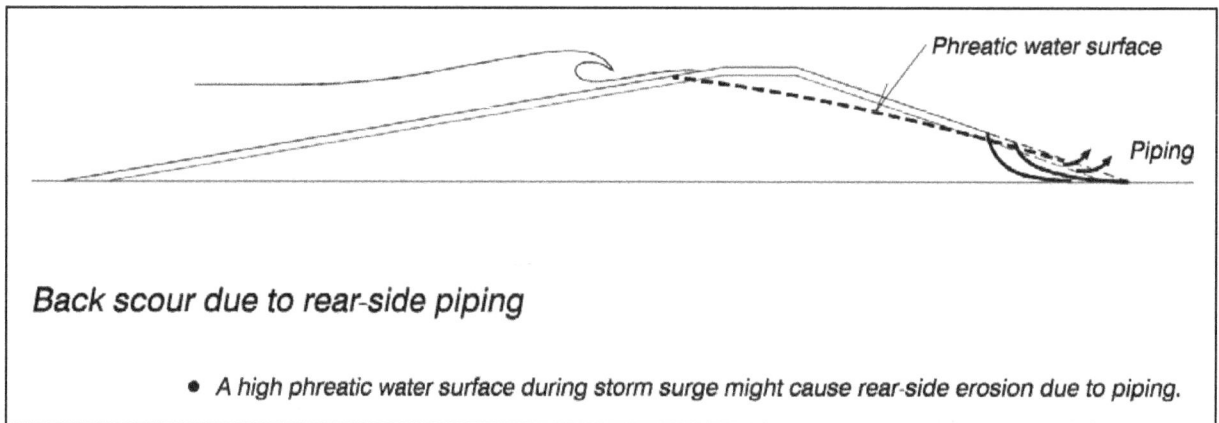

Figure VI-2-50. Dike backscouring due to piping

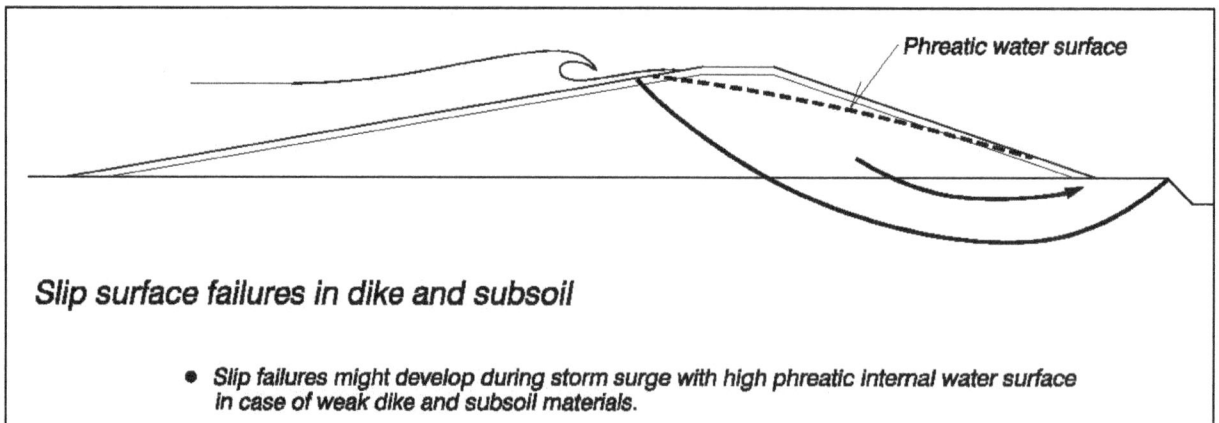

Figure VI-2-51. Dike slip surface failure

c. Vertical-front structures.

(1) Caisson and Blockwork Breakwaters. Failure modes can be classified based on Table VI-2-2.

Table VI-2-2 Failure Modes of Caisson and Blockwork Breakwaters	

Overall (global) instability of monoliths	Foundation failure modes:
	    Slip surface failures
	    Excess settlement
	Overturning
	Lateral displacement or sliding on foundation

Local instability	Hydraulic instability of rubble foundation Hydraulic instability of rubble-mound slope protection in front of caissons and breakage of blocks Seabed scour in front of the structure Breakage and displacement of structural elements

The local stability failure modes can trigger the overall stability failure modes. Figures VI-2-52 to VI-2-62 illustrate the failure modes for caisson and blockwork breakwaters.

(2) Seawalls/revetments.

(a) Gravity walls. Figures VI-2-63 to VI-2-68 illustrate common failure modes for gravity-type seawalls/revetments.

(b) Tied walls. Failure modes of thin walls supported by tie-backs are shown in Figures VI-2-69 through VI-2-73.

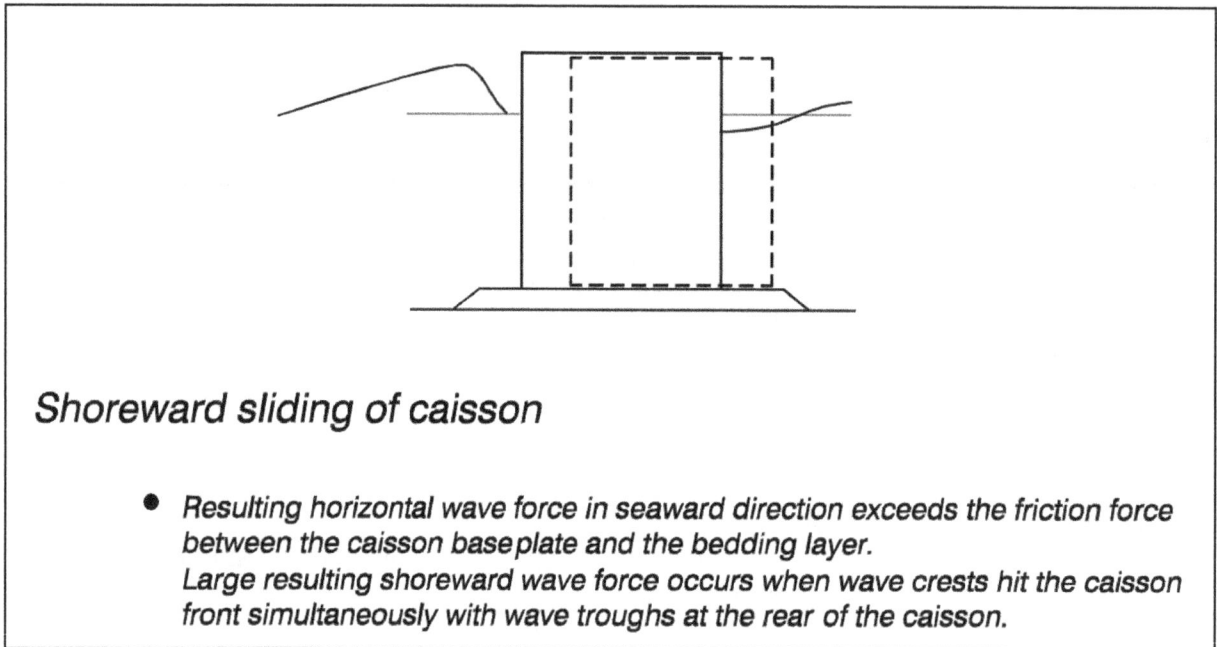

Shoreward sliding of caisson

- *Resulting horizontal wave force in seaward direction exceeds the friction force between the caisson baseplate and the bedding layer.*
  Large resulting shoreward wave force occurs when wave crests hit the caisson front simultaneously with wave troughs at the rear of the caisson.

Figure VI-2-52. Sliding of caisson on foundation

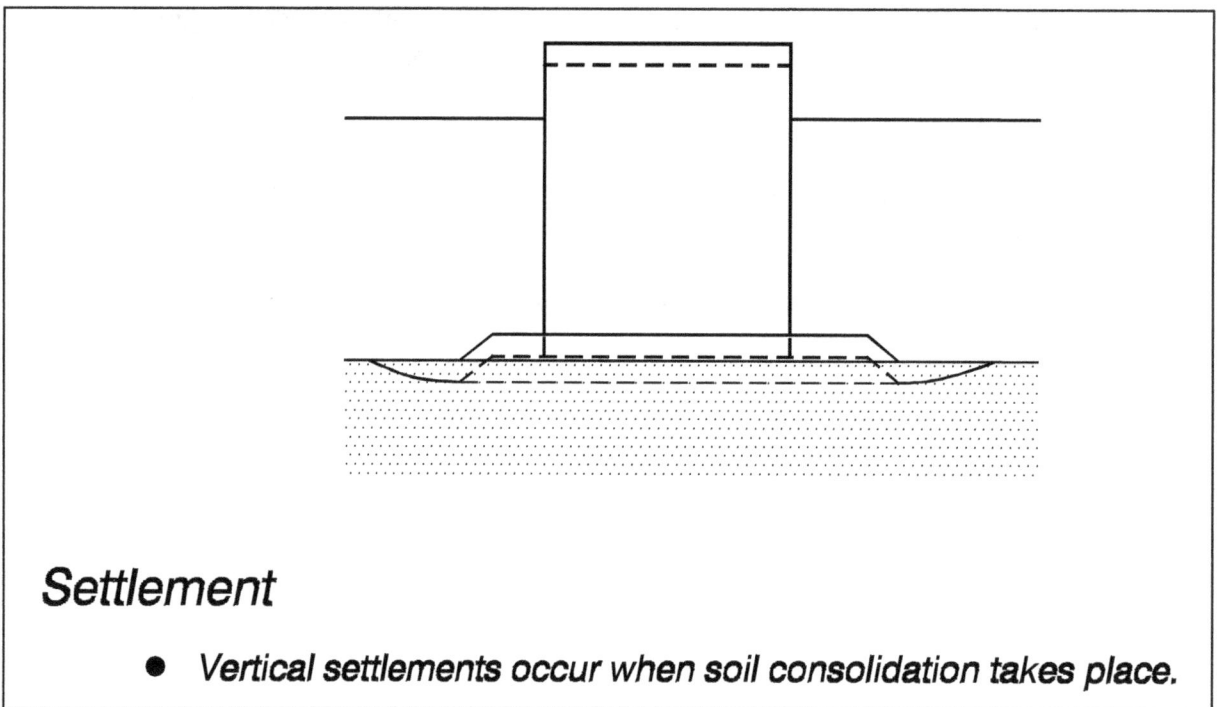

Settlement

- *Vertical settlements occur when soil consolidation takes place.*

Figure VI-2-53. Caisson settlement

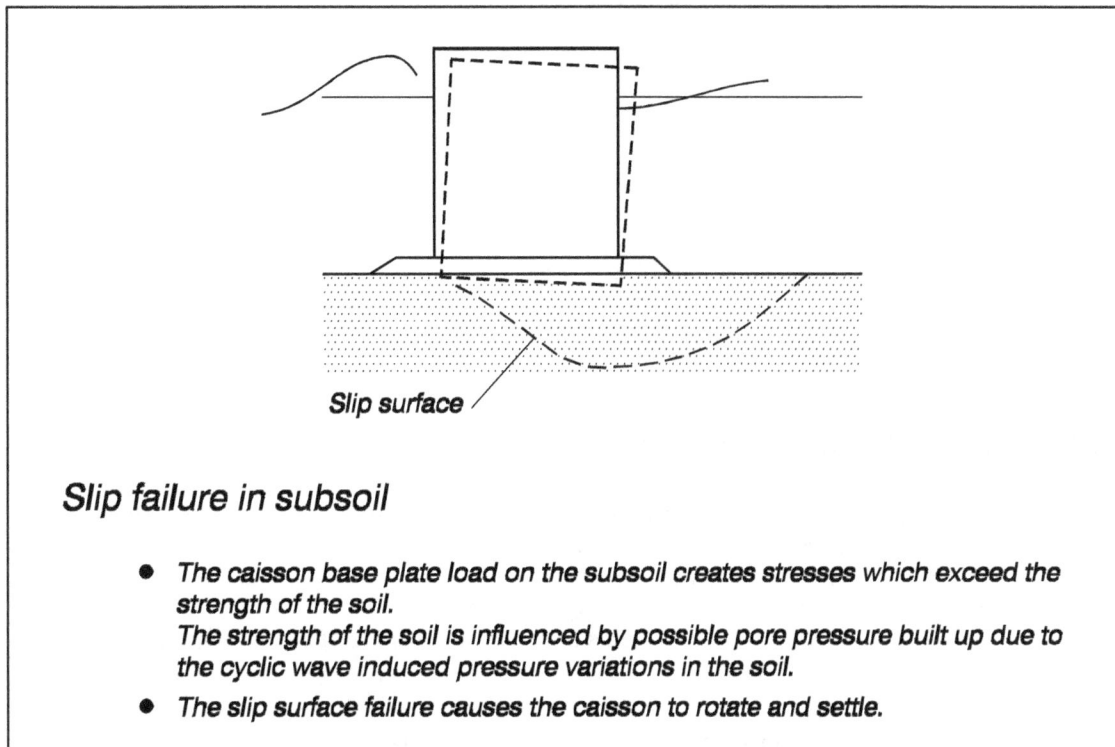

Slip surface

Slip failure in subsoil

- *The caisson base plate load on the subsoil creates stresses which exceed the strength of the soil.*
  The strength of the soil is influenced by possible pore pressure built up due to the cyclic wave induced pressure variations in the soil.
- *The slip surface failure causes the caisson to rotate and settle.*

Figure VI-2-54. Soil foundation slip surface failure

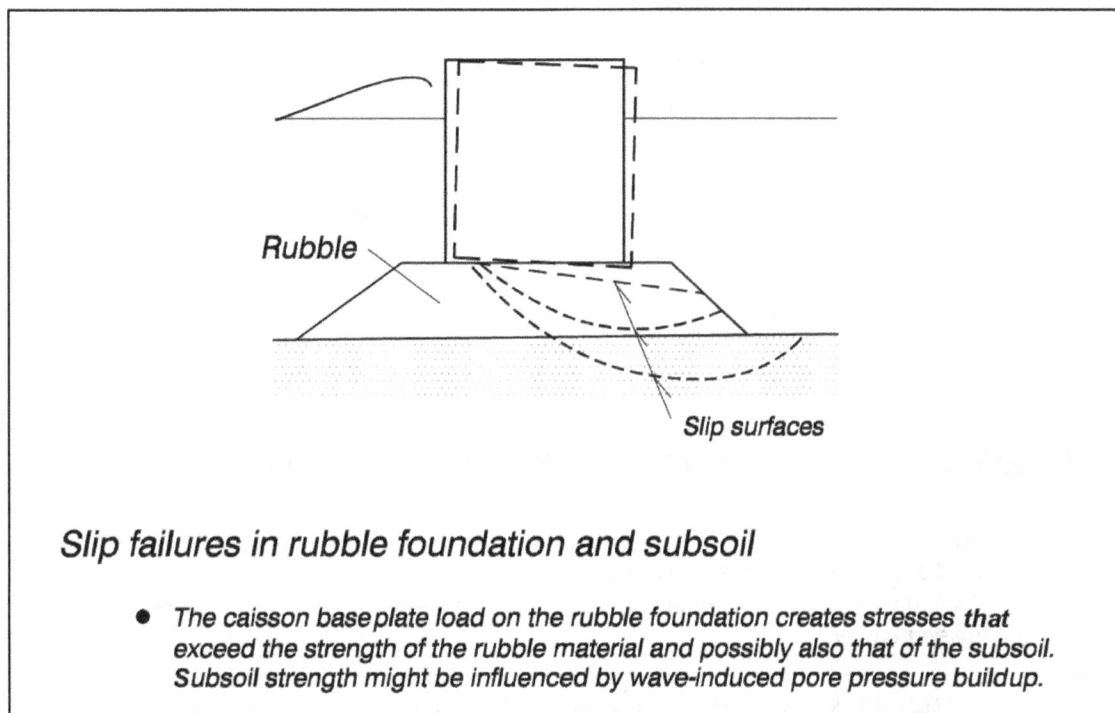

Rubble

Slip surfaces

Slip failures in rubble foundation and subsoil

- *The caisson base plate load on the rubble foundation creates stresses that exceed the strength of the rubble material and possibly also that of the subsoil. Subsoil strength might be influenced by wave-induced pore pressure buildup.*

Figure VI-2-55. Slip surface failure of rubble foundation

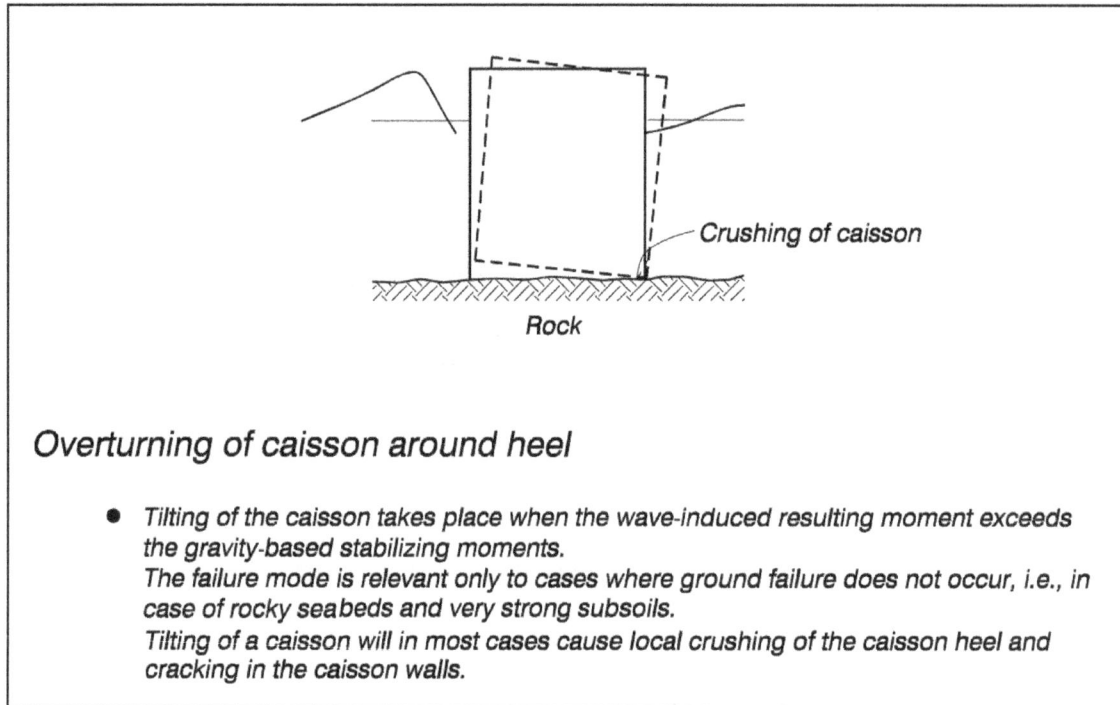

Overturning of caisson around heel

● Tilting of the caisson takes place when the wave-induced resulting moment exceeds the gravity-based stabilizing moments.
The failure mode is relevant only to cases where ground failure does not occur, i.e., in case of rocky seabeds and very strong subsoils.
Tilting of a caisson will in most cases cause local crushing of the caisson heel and cracking in the caisson walls.

Figure VI-2-56. Caisson overturning

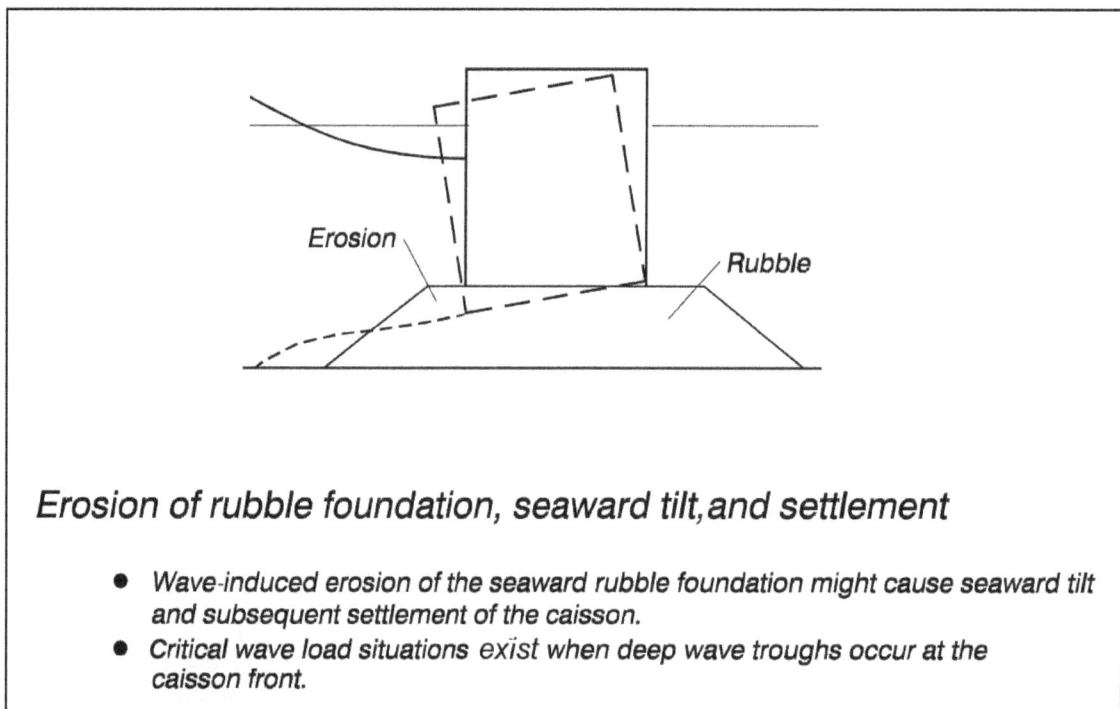

Erosion of rubble foundation, seaward tilt, and settlement

● Wave-induced erosion of the seaward rubble foundation might cause seaward tilt and subsequent settlement of the caisson.
● Critical wave load situations exist when deep wave troughs occur at the caisson front.

Figure VI-2-57. Seaward tilting and settlement due to erosion of rubble base

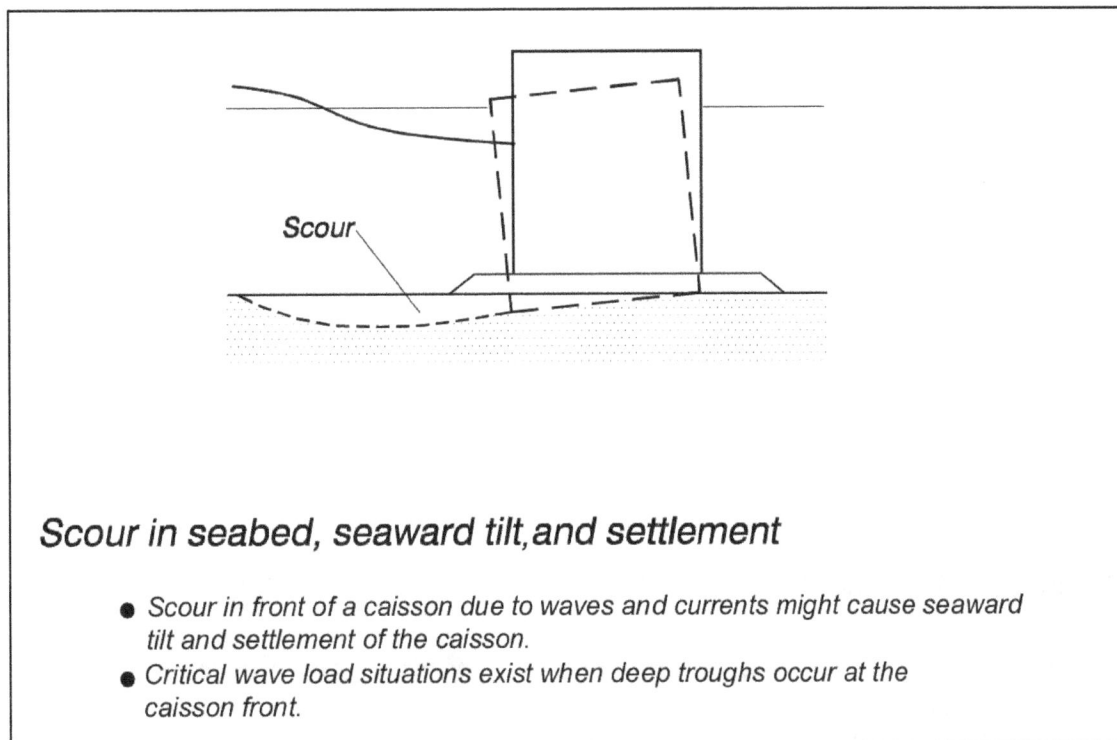

Scour in seabed, seaward tilt, and settlement

- Scour in front of a caisson due to waves and currents might cause seaward tilt and settlement of the caisson.
- Critical wave load situations exist when deep troughs occur at the caisson front.

Figure VI-2-58. Seaward tilting and settlement due to scour

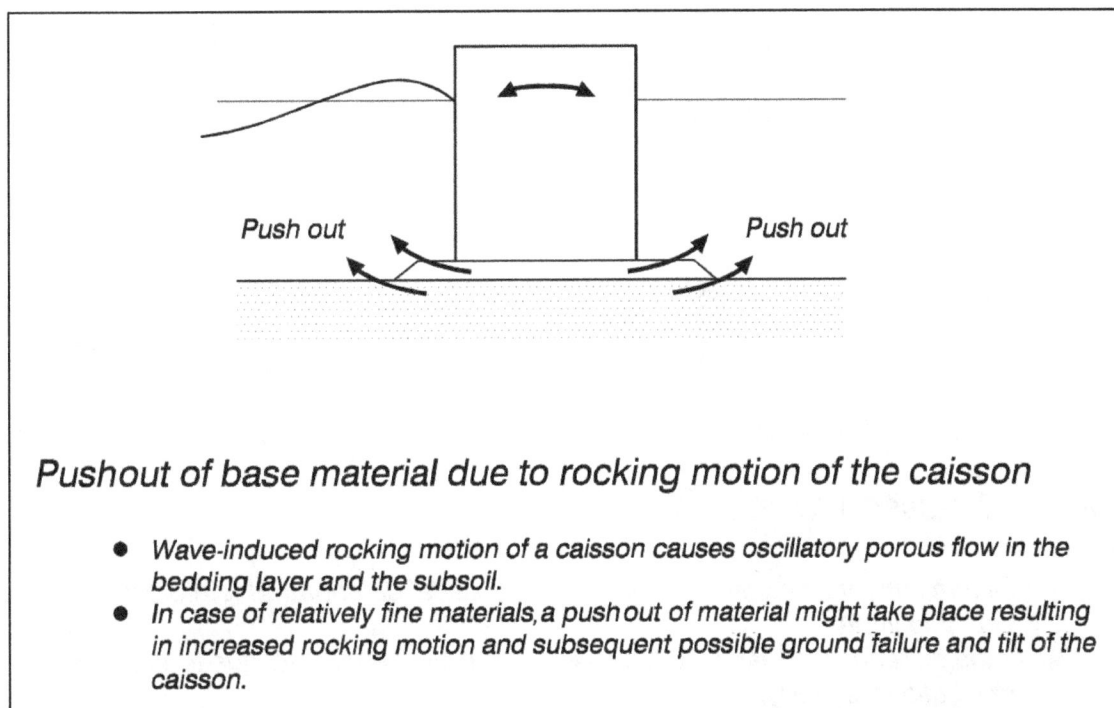

Pushout of base material due to rocking motion of the caisson

- Wave-induced rocking motion of a caisson causes oscillatory porous flow in the bedding layer and the subsoil.
- In case of relatively fine materials, a push out of material might take place resulting in increased rocking motion and subsequent possible ground failure and tilt of the caisson.

Figure VI-2-59. Loss of foundation material due to caisson motion

Armour units

Breakage and displacement of armor units in front of caisson

- *Wave action might lead to breakage and/or displacement of the armor units.*
- *The subsequent increase in the wave forces on the caisson front wall might cause the caisson to slide.*
- *Damage to the armor protection might increase the overtopping.*

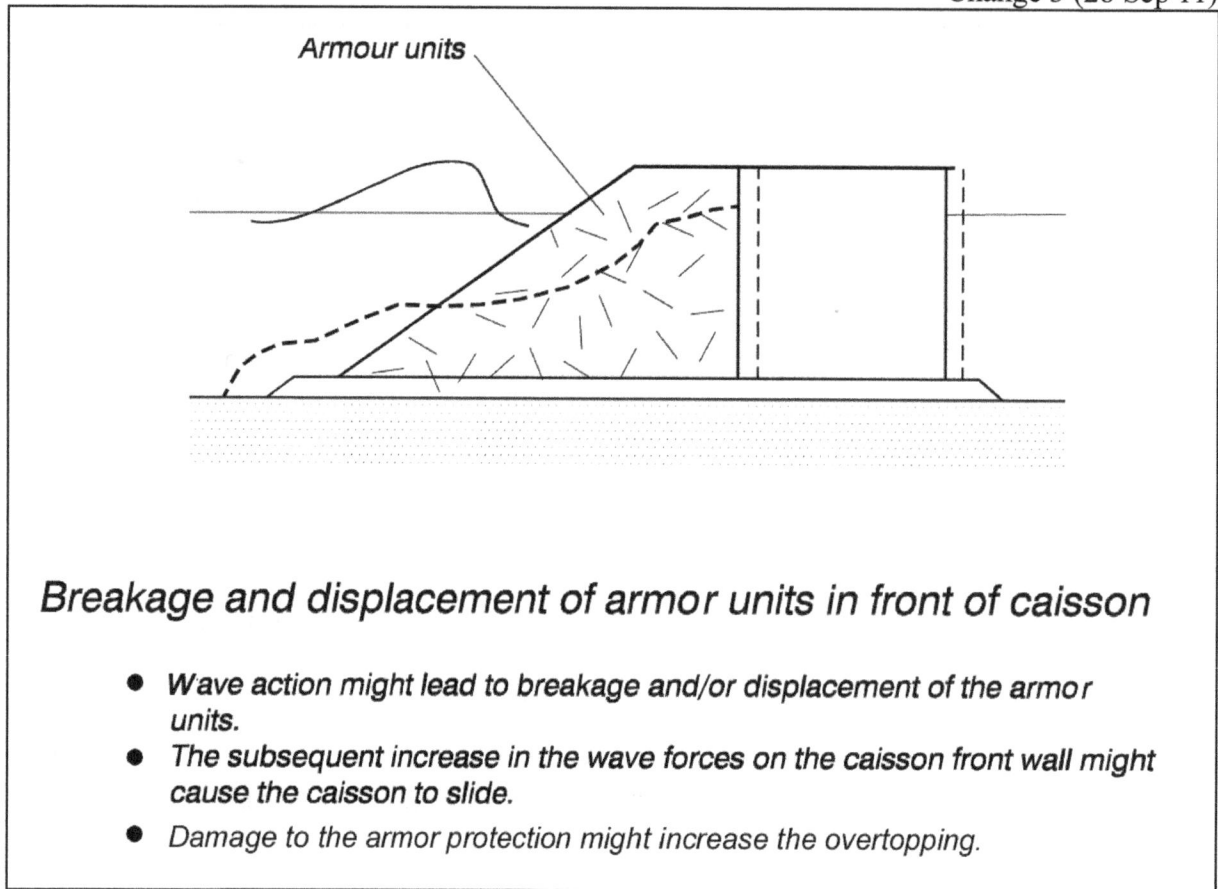

Figure VI-2-60. Failure of fronting armor units

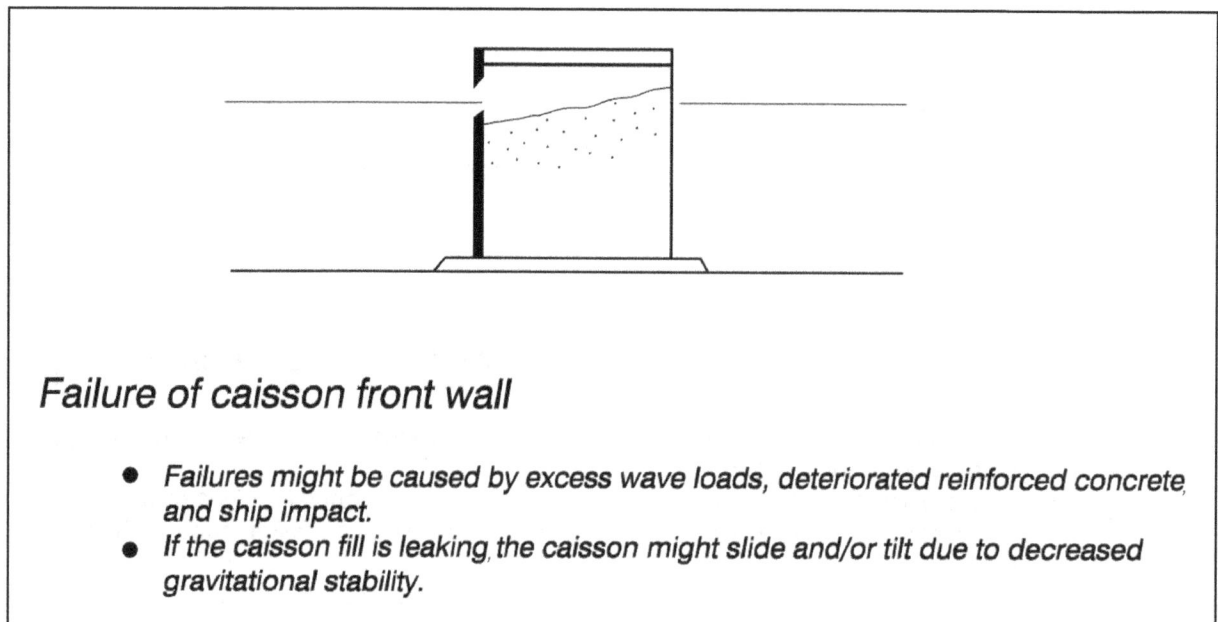

Failure of caisson front wall

- *Failures might be caused by excess wave loads, deteriorated reinforced concrete, and ship impact.*
- *If the caisson fill is leaking, the caisson might slide and/or tilt due to decreased gravitational stability.*

Figure VI-2-61. Caisson front wall failure

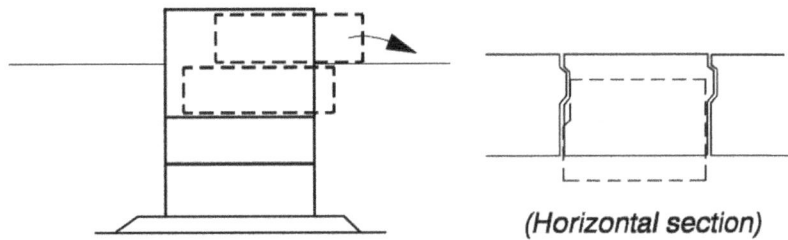

Pushout of elements in blockwork structures and failure of shear keys between blocks and caissons

- *Wave-induced loads might cause elements in blockwork structures to be pushed out of position with subsequent failure of the wall.*
- *Failure of shear keys between blocks and caissons due to large differential wave loads or due to differential settlements in case of caisson will lead to displacements of blocks and caissons.*

Figure VI-2-62. Displacement of individual blocks

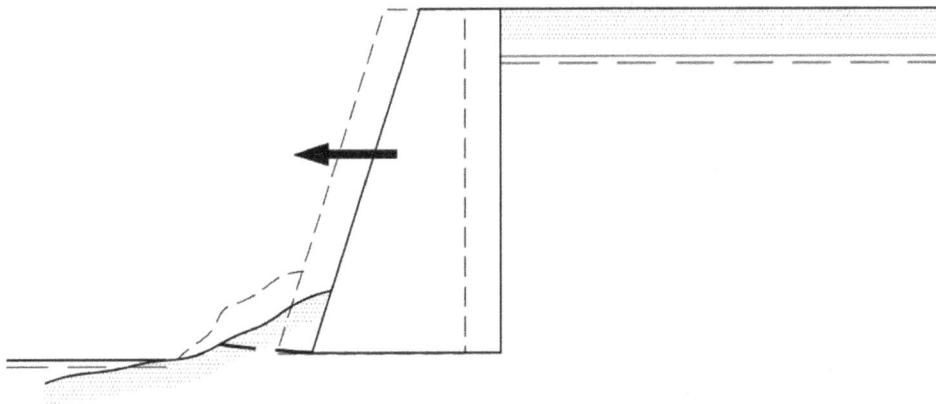

Sliding of gravity wall

- *The wall slides when the resulting pressure on the rear of the wall from active soil pressure and the groundwater exceeds the sum of the frictional resistance over the base of the wall and the passive resistance at the toe.*

Figure VI-2-63. Seaward sliding of gravity wall

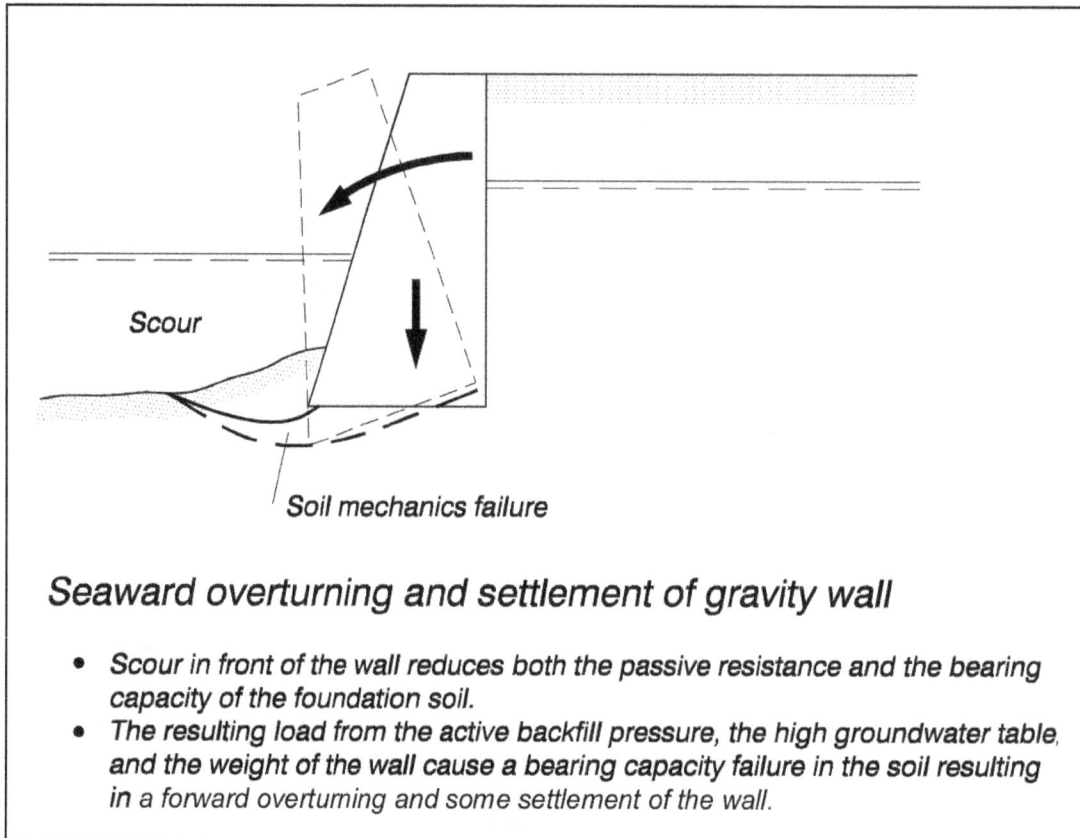

Seaward overturning and settlement of gravity wall

- *Scour in front of the wall reduces both the passive resistance and the bearing capacity of the foundation soil.*
- *The resulting load from the active backfill pressure, the high groundwater table, and the weight of the wall cause a bearing capacity failure in the soil resulting in a forward overturning and some settlement of the wall.*

Figure VI-2-64. Seaward overturning of gravity wall

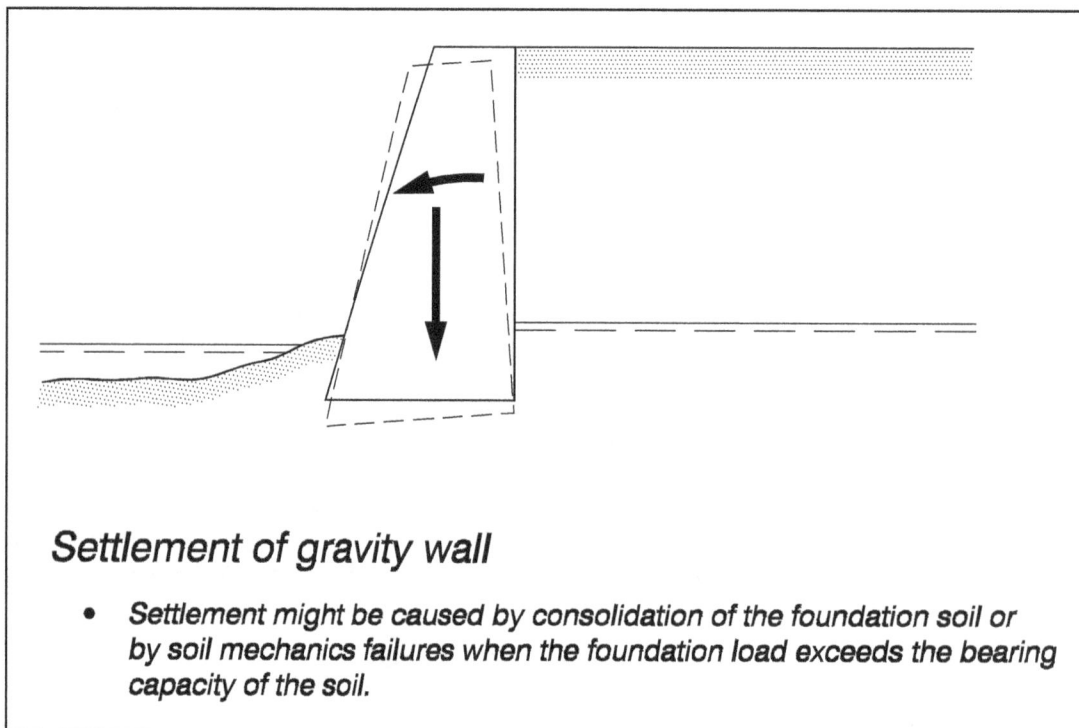

Settlement of gravity wall

- *Settlement might be caused by consolidation of the foundation soil or by soil mechanics failures when the foundation load exceeds the bearing capacity of the soil.*

Figure VI-2-65. Gravity wall settlement

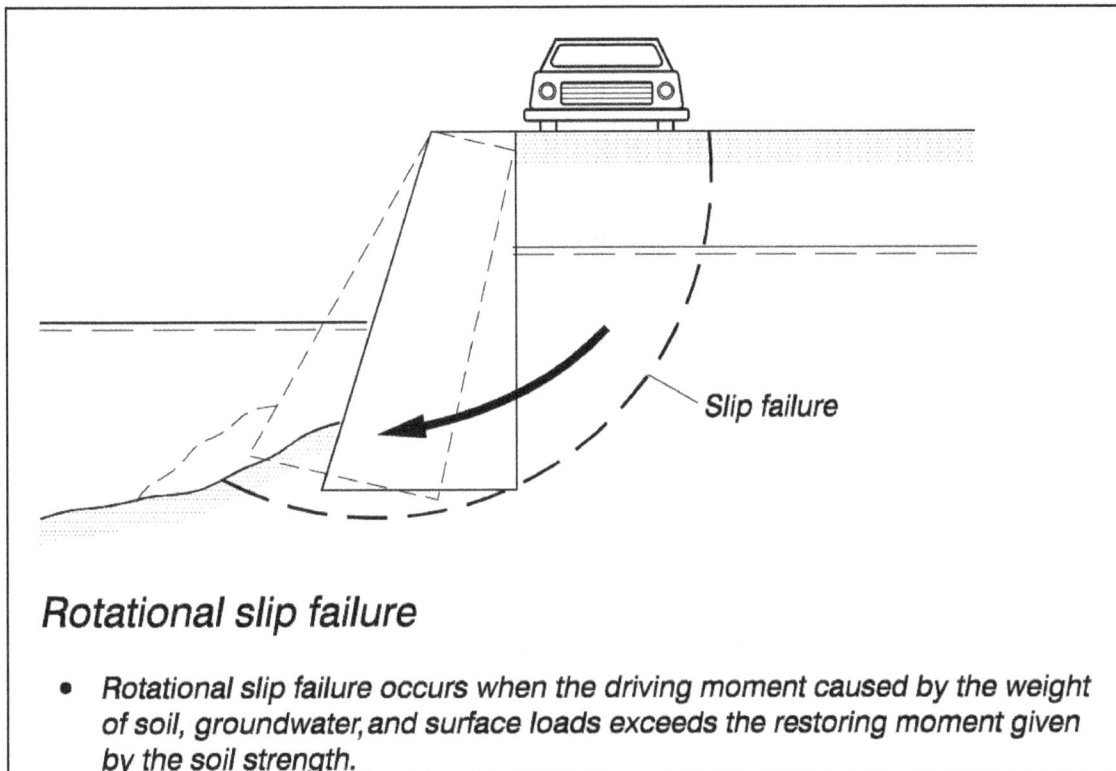

Rotational slip failure

- *Rotational slip failure occurs when the driving moment caused by the weight of soil, groundwater, and surface loads exceeds the restoring moment given by the soil strength.*

Figure VI-2-66. Rotational slip failure of gravity wall

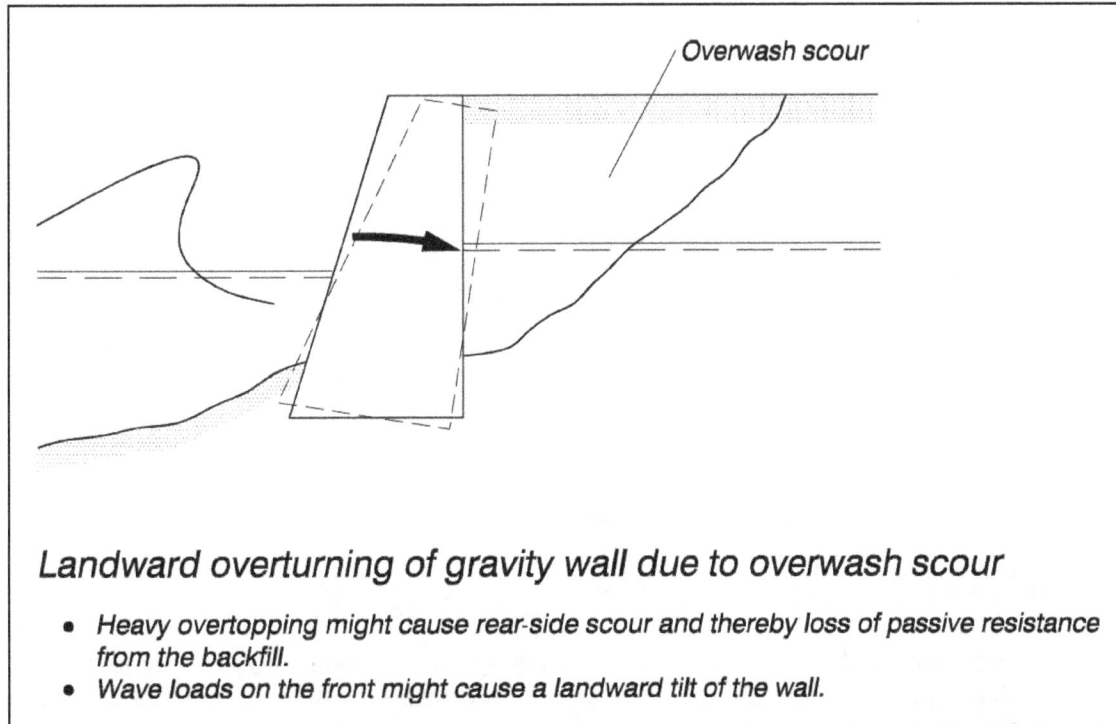

Landward overturning of gravity wall due to overwash scour

- *Heavy overtopping might cause rear-side scour and thereby loss of passive resistance from the backfill.*
- *Wave loads on the front might cause a landward tilt of the wall.*

Figure VI-2-67. Landward overturning of gravity wall

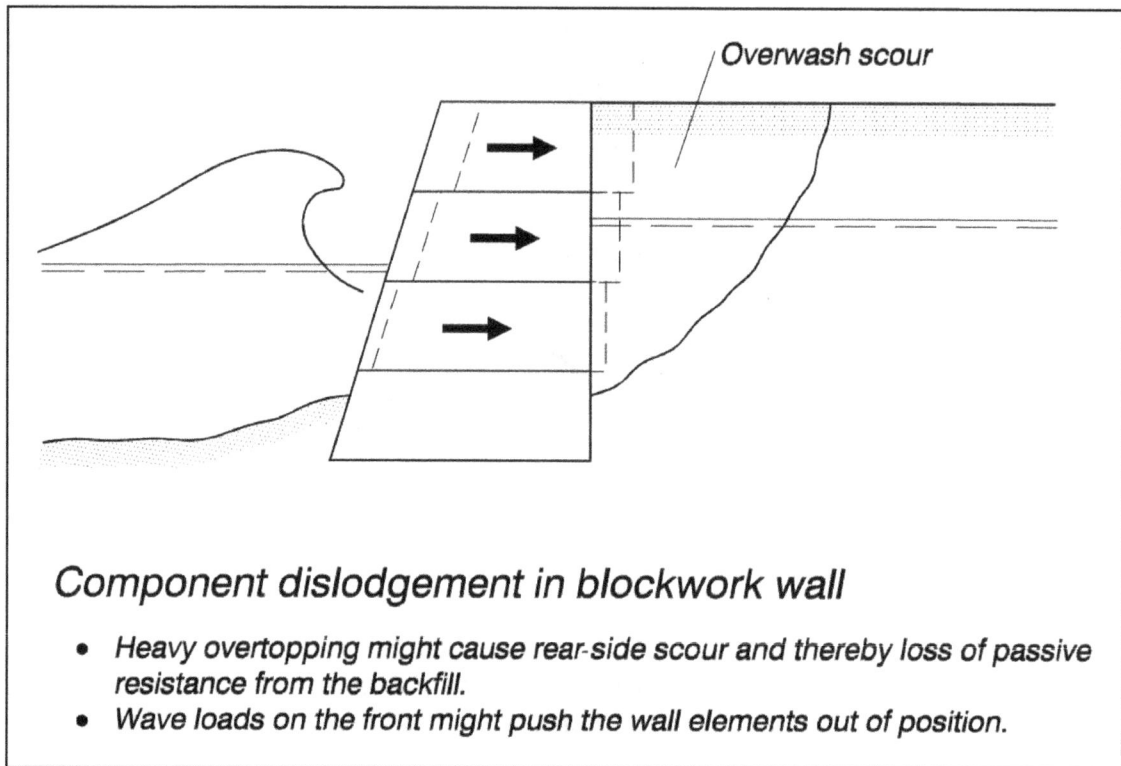

Component dislodgement in blockwork wall

- *Heavy overtopping might cause rear-side scour and thereby loss of passive resistance from the backfill.*
- *Wave loads on the front might push the wall elements out of position.*

Figure VI-2-68. Displacement of individual gravity wall components

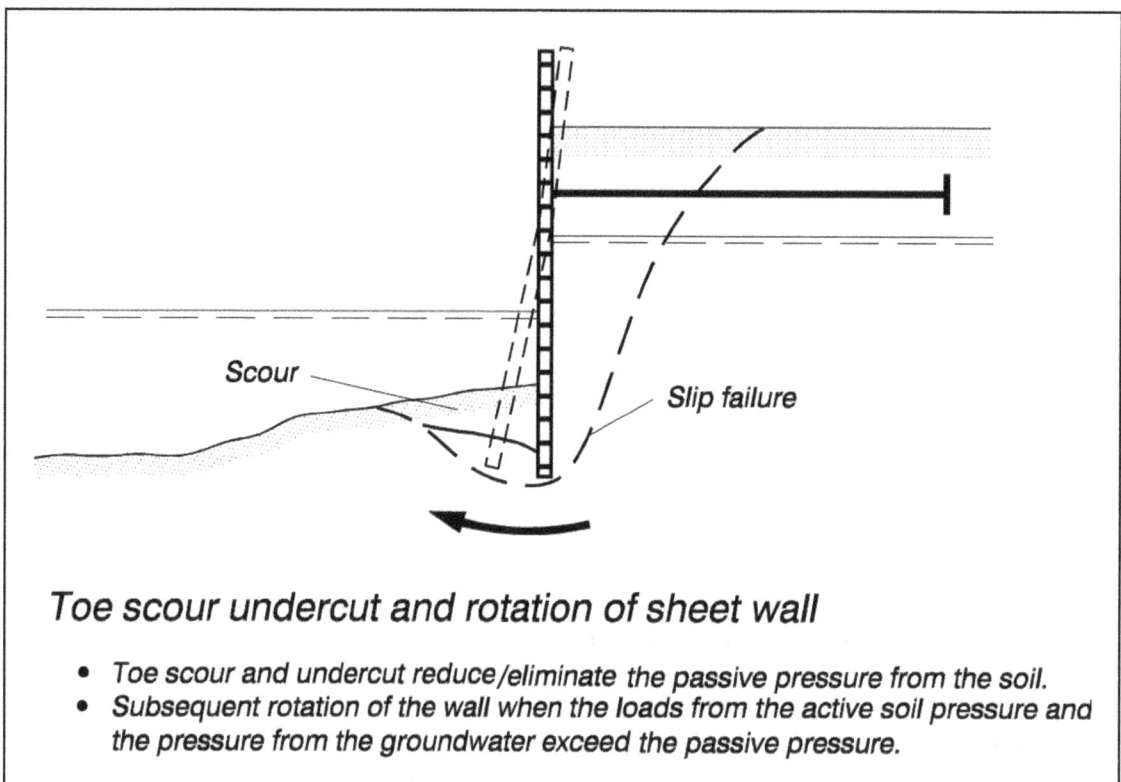

Toe scour undercut and rotation of sheet wall

- *Toe scour and undercut reduce/eliminate the passive pressure from the soil.*
- *Subsequent rotation of the wall when the loads from the active soil pressure and the pressure from the groundwater exceed the passive pressure.*

Figure VI-2-69. Failure due to toe scour

Rotational slip failure

- *Rotational slip failure occurs when the driving moments from the weight of the soil and the surface loads exceed the restoring moment given by the soil strength.*

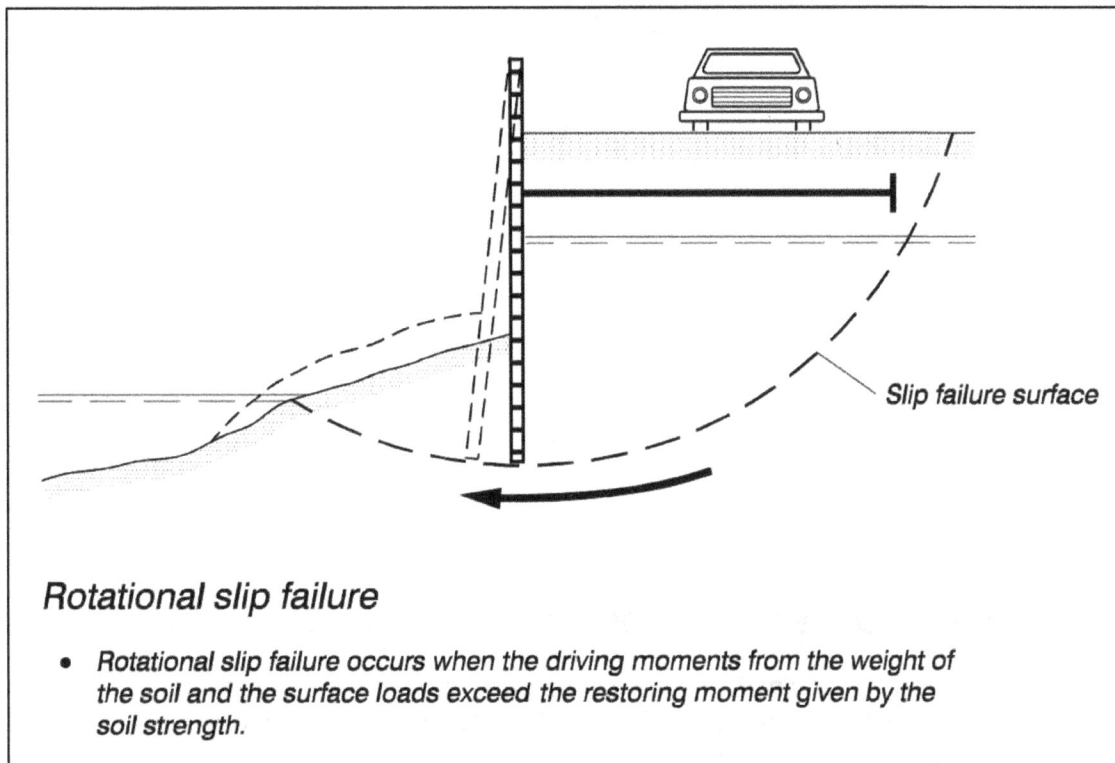

Figure VI-2-70. Rotational slip surface failure

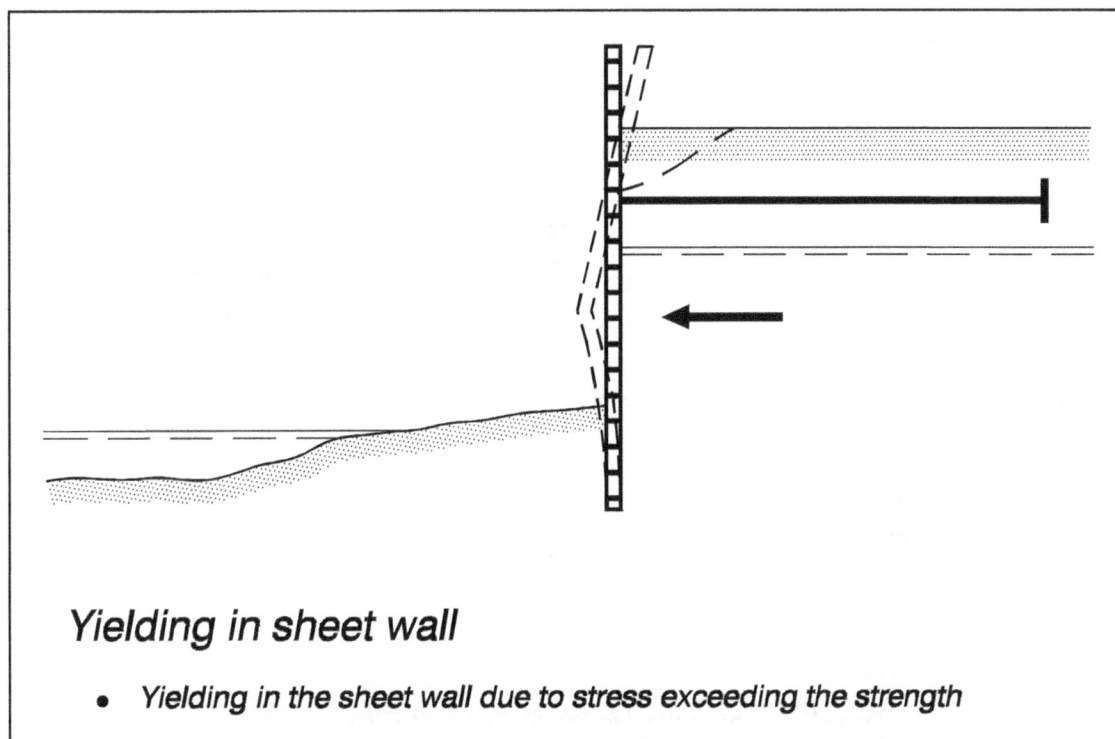

Yielding in sheet wall

- *Yielding in the sheet wall due to stress exceeding the strength*

Figure VI-2-71. Failure of thin wall construction material

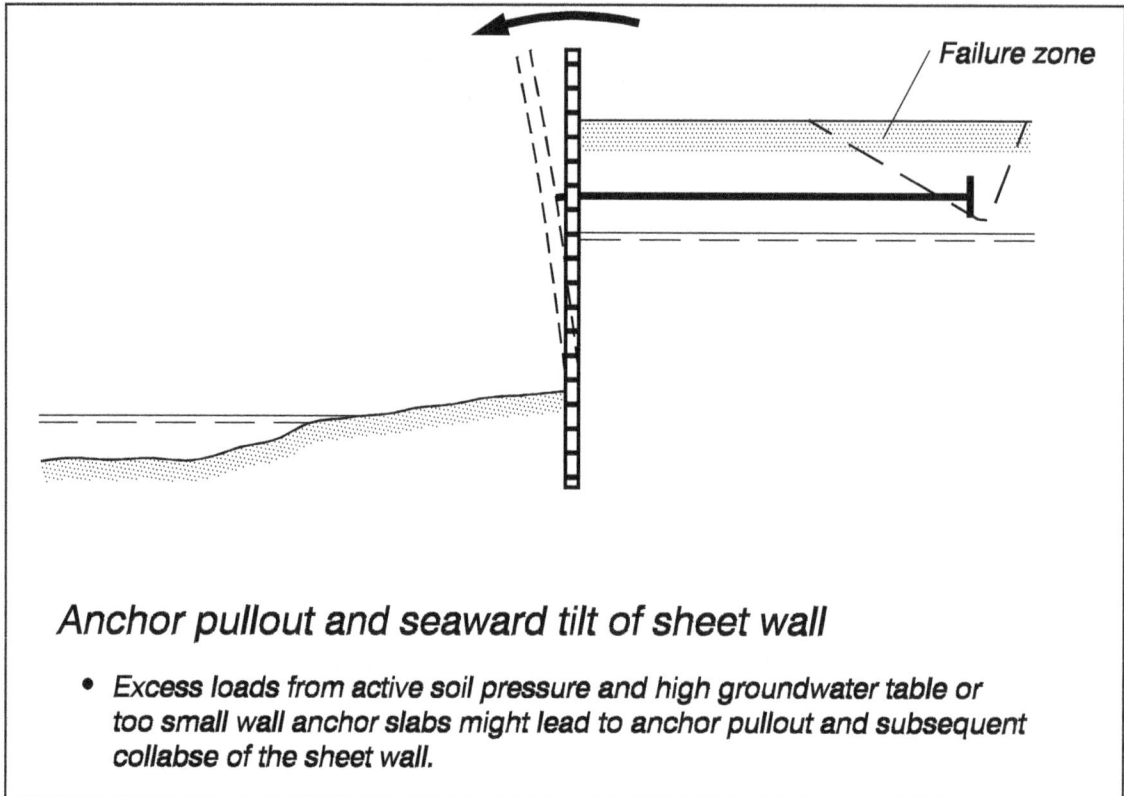

Anchor pullout and seaward tilt of sheet wall

- *Excess loads from active soil pressure and high groundwater table or too small wall anchor slabs might lead to anchor pullout and subsequent collabse of the sheet wall.*

Figure VI-2-72. Failure due to anchor pullout

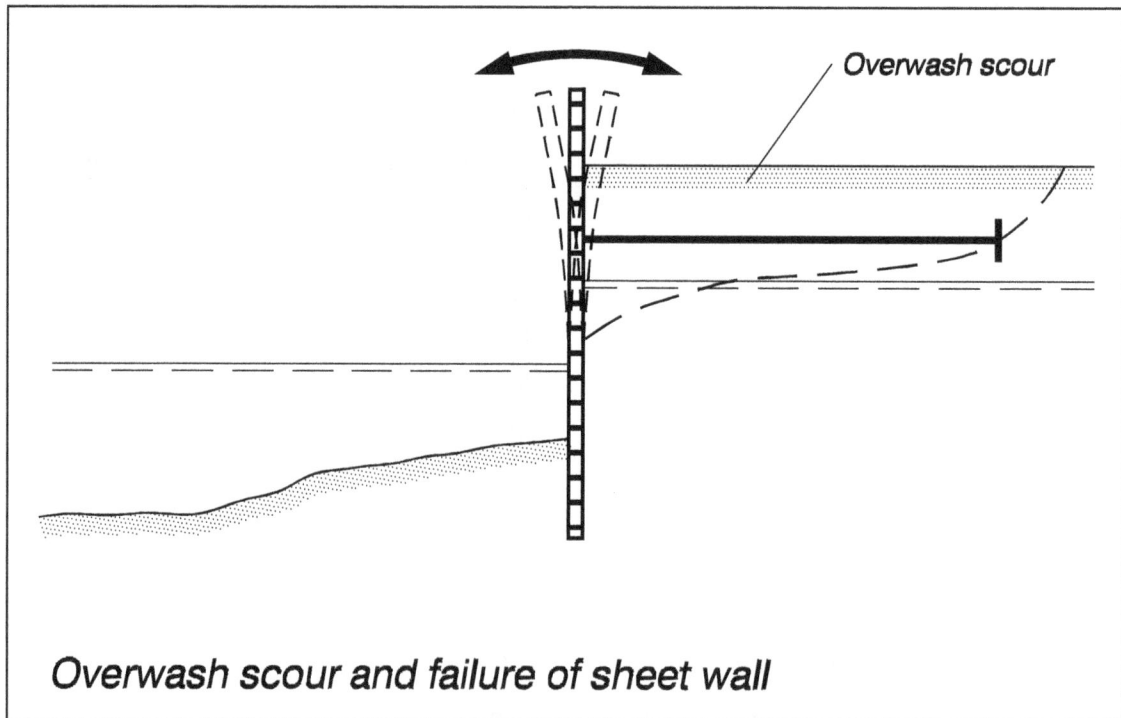

Overwash scour and failure of sheet wall

Figure VI-2-73. Back scour and thin wall failure

d. Floating structures.

(1) Floating structures used in coastal engineering applications have four different failure modes associated with structural design:

(a) Failure of floating sections.

(b) Failure of floating section connectors.

(c) Failure of anchor system or pile supports.

(d) Flooding and sinking.

(2) Material used to fabricate individual floating sections must be able to withstand wave slamming forces, vessel impacts, marine corrosion, and concentrated loading at connection points and mooring points. Floating sections fabricated with concrete must withstand deterioration. Damage to the floating section construction material may lead to flooding of the unit, displacement of the section from its moored location, or rendering of the section unsafe for traffic.

(3) Many floating structures consist of several units joined with flexible connections. These connectors must be able to resist transmitted force and moment loads induced by environmental loading and differential movements between floating sections. Failure of the connectors results in a substantial decrease in the structure=s functionality.

(4) The mooring/anchoring system of a floating structure is critical to a successful design. Mooring systems fail when the mooring lines break or separate from the connection points, or when the anchoring method does not resist the mooring loads. Dead weight anchors rely primarily on anchor mass with additional help from frictional resistance between the anchor and soil. Penetration anchors rely on soil shear strength to resist Apullout@ under load conditions. Extreme loads may cause anchors to drag or pullout, but the floating structure may survive intact at a displaced location. Failure occurs if the structure displacement results in the floating structure breaking up, suffering impact damage, or becoming unretrievable. Similar damage can occur to pile-supported floating structures when the piles fail or the floating structure breaks free of the pilings.

(5) Many floating structures have compartments of entrapped air that provide necessary buoyancy. If these compartments flood, the structure may sink. Flooding can occur because of impact damage, construction material failure, excessive structure tilting, or wave overtopping.

(6) Finally, floating structures may fail because they do not perform at the expected level. For example, a floating breakwater that does not reduce wave heights sufficiently and a floating boat slip with excessive motion would both be considered failures.

e. Beach fills.

(1) Judging the success or failure of beach fill projects is somewhat subjective and arises more from a political position rather than an assessment in terms of the beach fill project's original design goals. Beach fill projects are usually built in areas that experience shoreline erosion, and it is expected that the beach fill will gradually disappear in time as erosion continues. Part of the design process is estimating how long the beach fill will serve its function under typical wave conditions. Such estimates are difficult, at best, because of wave climate uncertainty and the complexity of beach fill response to storm conditions. A new project may suffer a severe storm immediately upon completion, resulting in massive fill losses, or the beach fill may serve for many years without ever being exposed to design storm conditions.

(2) Despite the controversy about failure or success of beach fill projects, there are several recognized failure modes for beach fills:

(a) Failure to protect upland property or structures during storm events.

(b) Movement of fill material to undesired locations, such as into inlets or harbors.

(c) Loss of fill material at a rate greater than anticipated for some reason other than design wave exceedance.

(3) When used to protect upland property, beach fills are sacrificial soft structures, somewhat analogous to automobile fenders that are designed to crumple on impact to absorb the energy.

f. Scour potential and toe failure. Any coastal structure resting on, driven into, or otherwise founded on soil or sand is susceptible to scour and possible toe failure when exposed to waves and currents. Generally, scour potential around impermeable structures is enhanced in regions of flow concentration due to directed currents, high wave reflection, etc. Scour potential is decreased around sloping permeable structures. Failure modes due to scour for specific structure types are illustrated in the figures of this section.

VI-2-5. <u>References</u>.

Burcharth 1983
Burcharth, H. F. 1983. "Materials, Structural Design of Armour Units," *Proceedings of Seminar on Rubble Mound Breakwaters*, Royal Institute of Technology, Stockholm, Sweden, Bulletin No. TRITA-VBI-120.

Burcharth and Thompson 1983
Burcharth, H. F., and Thompson, A. C. 1983. "Stability of Armour Units in Oscillatory Flow," *Proceedings of Coastal Structures '83*, American Society of Civil Engineers, pp 71-82.

Burcharth 1984
Burcharth, H. F. 1984. "Fatigue in Breakwater Concrete Armour Units," *Proceedings of the 19th International Conference on Coastal Engineering*, American Society of Civil Engineers, Vol. 3, pp 2592-2607.

Burcharth 1993
Burcharth, H. F. 1993. "Structural Integrity and Hydraulic Stability of Dolos Armour Layers," Doctoral Thesis (Series Paper 9), Department of Civil Engineering, Aalborg University, Aalborg, Denmark.

Burcharth, et al. 1995
Burcharth, H. F., Jensen, M. S., Liu, Z., Van der Meer, J. W., and D'Angremond, K. 1995. "Design Formula for Tetrapod Breakage," *Proceedings of the Final Workshop, Rubble Mound Breakwater Failure Modes*, Sorrento, Italy.

Melby 1993
Melby, J. A. 1993. "Dolos Design Procedure Based on Crescent City Prototype Data," Technical Report CERC-93-10, U.S. Army Engineer Waterways Experiment Station, Coastal Engineering Research Center, Vicksburg, MS.

Price 1979
Price, W. A. 1979. "Static Stability of Rubble Mound Breakwaters," Dock and Harbour Authority, Vol. LX (702).

VI-2-6. <u>Acknowledgments</u>.

Authors: Dr. Hans F. Burcharth, Department of Civil Engineering, Aalborg University, Aalborg, Denmark; Dr. Steven A. Hughes, Coastal and Hydraulics Laboratory (CHL), U.S. Army Engineer Research and Development Center, Vicksburg, MS.

Reviewers: Han Ligteringen, Delft University of Technology, The Netherlands; John H. Lockhart, Headquarters, U.S. Army Corps of Engineers, Washington, DC (retired); Joan Pope, U.S. Army Engineer Research and Development Center, Vicksburg, MS.

CHAPTER 3

Site Specific Design Conditions

TABLE OF CONTENTS

List of Figures

List of Tables

CHAPTER VI-3

Site-Specific Design Conditions

VI-3-1. <u>Foundation/Geotechnical Requirements</u>. This section presents an overview of site-specific design information that may be required to complete preliminary and final foundation design for coastal projects. Foundation failure modes are overviewed in Part VI-2-4, "Failure Modes of Typical Structure Types," and foundation design procedures are discussed in Part VI-5-5, "Foundation Loads."

    a. Introduction.

    (1) Most man-made coastal project elements are designed to be built or placed directly on top of the natural soil, sand, or other bottom material of the project site (the major exceptions are pile-supported structures). Depending on the particular functional requirements of the coastal project element, it may be subjected to environmental loadings that include waves, currents, fluctuating water levels, and seismic vibrations, along with specific loadings such as vessel impacts and ice surcharge.

    (2) The interaction between a coastal project and the soil upon which it is placed may be a critical aspect of a project=s performance. The underlying soil, or *foundation*, must be capable of resisting that portion of the loading that is transferred to the foundation in addition to supporting the weight of the structure. Resistance to the total imposed time-dependent loading and resultant changes in soil stresses within the soil layers must be achieved without undue structure deformation and with sufficient reserve soil strength to assure that the probability of soil instability is sufficiently low.

    (3) Foundation design for coastal structures requires the engineer to make reasonable estimates of the expected loading conditions, to determine the appropriate site-specific foundation soil engineering properties, and to understand reasonably the structure/soil interaction. Geotechnical investigations are conducted to gather necessary information about the soil layers beneath the project so the engineer can complete the foundation design to a level commensurate with each stage of project design.

    b. Foundation loads.

    (1) For typical coastal structures exposed to waves and currents, the underlying foundation soils must contend with static, dynamic, and impact force loads. Static loads are caused by the structure and foundation soil self-weight; and in most cases, these forces are relatively constant over the life of the project. It is important to remember that buoyancy effectively reduces the weight of that portion of the structure beneath the water surface. Consequently, the structure self-weight load on the foundation soil will vary with tide elevation.

    (2) A structure=s weight distribution and the differential loading applied to the foundation must be evaluated, particularly for gravity-type structures extending into greater depths or spanning different soil types. Lateral forces due to imbalanced hydrostatic pressure must also be considered.

(3) Waves, currents, tides, storm surges, and wind are the primary dynamic forces acting on coastal structures; however, in some regions of the world earthquake ground motions may also induce severe dynamic loads. Dynamic loads vary greatly in time, duration, and intensity, and the worst likely load combinations should be examined during foundation design.

(4) Impact loads on structures may arise from ship or ice collisions, partial failure of some portion of the structure, or even from breaking waves slamming into the structure. Importance of impacts depends on the type of structure and magnitude of impact. Structures such as rubble-mound breakwaters are somewhat flexible and can absorb a portion of the impact load, whereas monolithic structures are more likely to transmit a greater portion of the impact load to the foundation as lateral shear or overturning forces. However, the large mass and natural frequency of monolithic structures help to reduce the transmitted loads. Finally, the proposed project construction method should be examined to determine if any significant construction loads might adversely affect the foundation soil stability.

c. Foundation soil responses. Structure static loads applied to the foundation soil, along with dynamic and impulsive force loads transmitted by a coastal structure to the foundation, can evoke several soil responses that concern design engineers.

(1) Soil consolidation may occur due to the structure=s weight. Consolidation is a reduction in soil void space that occurs over time as compressive loads force water out of the voids. This results in a denser soil with increased soil strength properties. Densification of the soil may result in structure settlement or differential settlement that could impact the structure's functionality. Other factors that influence settlement include compression of softer subsoil layers, squeezing of very soft sublayers, or collapse of underground cavities (Construction Industry Research and Information Association/Centre for Civil Engineering Research and Codes (CIRIA/CUR) 1991). Estimates of potential settlement are used to assess the need for structure crest overbuild or to determine stability of structures sensitive to differential movement.

(2) Soil shear stresses are induced when lateral forces and overturning moments on the structure are transferred to the foundation. If soil strength is exceeded, foundation damage may occur which can be either localized or widespread, such as in the case of slip-surface failure of a soil slope. Rapid soil stress loading will cause excess pore pressures and a corresponding decrease in soil shear strength which may lead to soil liquefaction. Cyclic loading of noncohesive sand can also cause excess pore pressure buildup, and when combined with strong accelerations from earthquakes, liquefaction of the foundation and consequent failure of the foundation may be catastrophic.

(3) Finally, the design engineer needs to consider the possibility that foundation material will be eroded from beneath the structure or immediately adjacent to the structure (scour). Induced excess pore pressures or pressure gradients within the soil can contribute to this loss. Steps must be taken to provide adequate protection to keep the foundation material in place (See Part VI-5-6, "Scour and Scour Protection").

d. Geotechnical investigations.

(1) The wide variety of soil conditions encountered in the coastal regime, coupled with the range of coastal projects, precludes standardization of the study components comprising geotechnical investigations. A general guiding criterion is that the investigations should include sufficient subsurface investigations, lab testing, and analysis to assure the adequacy of project design and constructability. This may involve discovering answers for these questions:

(a) What are the soil types and strata at the site?

(b) What are the mechanical properties of the soil relative to its strength and deformation under loading?

(c) What is the range of conditions to which the soil might be exposed? (For example, flooding/drying or freeze/thaw cycles might be important for land-based structures.)

(d) Is the soil condition expected to degrade over the project life?

(e) Is the soil fissured or weathered?

(2) The three overlapping phases of a geotechnical investigation are site reconnaissance, preliminary exploration, and detailed design exploration. Determining what information is to be gathered in the latter phases depends, in part, on findings from previous investigation phases.

(a) The **Site Reconnaissance Phase** is primarily a desk study that seeks to assemble existing geological data to characterize the nature of the proposed project site. The results of this phase are important in helping to establish data collection requirements for subsequent phases. The goal of the site reconnaissance is to glean from available data an understanding of the geological stratification, formation, and history; the groundwater regime; and possibly the seismicity of the site (Pilarczyk 1990). Sources of information for this phase include: topographic and geologic maps, aerial photography, groundwater maps, past geotechnical studies at the proposed site and at adjacent locations, local observations and reports in the local media, recorded ground movements, published geotechnical and geological descriptions, and historical records of previous coastal projects in the vicinity. In particular, it is important to note partial removal of previous coastal structures because structure remnants in the foundation area may cause construction difficulty for new projects.

• Toward the end of the site reconnaissance phase it is essential for the engineer responsible for planning the geotechnical investigation to conduct a field trip to the project site. This trip allows the engineer to reconcile the assembled information with existing site conditions and to uncover any previously unrevealed factors (Eckert and Callender 1987). The focus of the trip should be on surface evidence of subsoil conditions such as existing cuts, landslides, surrounding geology, etc.

• The final step of the site reconnaissance phase is to develop a program for the preliminary exploration phase that will best fill remaining information gaps vital for site selection, planning, and preliminary design decisions. A major challenge is to optimize the data collection within the constraints of the project budget.

(b) The goals of the **Preliminary Exploration Phase** are the following: (1) to recognize potential geotechnical problems, (2) to obtain sufficient geotechnical information at alternative project sites so that the final site selection can be made, and (3) to determine adequately those geotechnical parameters necessary for preliminary project design. Results from the preliminary exploration generally form the basis of a Corps of Engineers survey report that is used to obtain project authorization. Therefore, the exploration should be sufficient to reveal any soil condition that might adversely influence project cost and constructability. Project size, cost, and importance all factor into the extent of the investigation. See Eckert and Callender (1987) for three useful examples that span the typical range of coastal projects.

- During the preliminary exploration it is necessary to collect site information to determine the following over the project area: (1) the approximate depth, thickness, and composition of the various soil strata, (2) depth to the soil-bedrock interface, (3) variations in the groundwater level, (4) estimates of critical soil parameters, and (5) potential sources of construction borrow materials.

- A variety of geophysical investigation methods can be employed to obtain these data over a wide area at a reasonable cost. Continuous seismic reflection surveys are commonly utilized for marine investigations. The seismic waves are reflected by the boundaries between soil strata, and the depth to each strata is determined by the arrival time of the reflected pulse. Different seismic frequencies provide varying depth coverage. Side-scan sonar images can provide information about the seafloor surface soil characteristics adjacent to the trackline of the survey vessel. Concurrent bathymetric soundings aid the interpretation of side-scan sonar images. Dry-land geophysical investigation methods include electro-resistivity, electromagnetic, and seismic refraction and reflection. Interpretation of most geophysical survey results requires an experienced professional.

- When feasible, geophysical survey results should be supplemented with a small number of in situ borings to aid in calibrating survey results. Ideally, the borings should be obtained at critical locations along the proposed project alignment, but often severe environmental conditions make coring at these locations dangerous or expensive. The core samples confirm the geophysical survey interpretation.

- Information collected during the preliminary exploration should be sufficient to make the final site selection and to develop a preliminary design. Once this has been achieved and the project receives final approval, planning for necessary detailed geophysical measurements can proceed.

(c) The purpose of the **Detailed Design Exploration Phase** is to collect and analyze specific soil data (beyond that gathered in the preliminary phases) to determine those geotechnical parameters necessary for completing the final foundation design. Critical to this phase is specifying which soil parameters need to be determined, at what locations and depths the determinations should be made, and how best to collect and analyze the data to achieve desired results within the time and monetary allowance. Secondary considerations include when to conduct the investigation, who will perform the work, and who will be responsible for the laboratory analyses and data interpretation. A well-planned exploration program that provides realistic soil parameters can often save more than it cost. Uncertainties about soil strength may result in unnecessary structure overdesign, or conversely, a detailed site investigation may reveal a soil weakness that could result in structure failure if adequate provisions are not enacted. The decision about which soil parameters must be determined depends on the anticipated foundation failure modes. Table VI-3-1 (from CIRIA/CUR (1991)) lists the main foundation failure modes and identifies those soil parameters that are useful in evaluating each mode.

Table VI-3-1
Soil Data Required to Evaluate Foundation Failure Modes (CIRIA/CUR 1991)

Macro-Instability			Macro-Failure	Micro-Instability	
Slip Failure	Liquefaction	Dynamic Failure	Settlements	Filter Erosion	Parameter
A	A	A	A	A	Soil profile
A	A	A	A	A	Classification/grain size
A	A	A	B	A	Piezometric pressure
B	B	B	A	A	Permeability
A	B	B	A	B	Dry/wet density
-	A	B	-	-	Relative density, porosity
A	B	B	-	C	Drained shear strength
A	-	-	-	C	Undrained shear strength
B	-	-	A	-	Compressibility
A	-	-	A	-	Rate of consolidation
B	B	A	A	-	Moduli of elasticity
B	A	A	A	-	In situ stress
-	A	B	A	-	Stress history
B	A	A	B	-	Stress/strain curve

NOTE: A - Very important.
B - Important.
C - Less important.

• Spacing, depth, and location of soil samples and borings are specified for each site based on the known geology and particular type of project. When soil conditions are relatively uniform, the number of borings can be decreased and the spacing can be increased. Conversely, areas where geotechnical problems have been identified will require denser boring spacing to delineate the problem area=s geotechnical parameters. Boring depth is a function of the estimated project surcharge loading and the in situ soil profile.

• Field tests conducted on the in situ soil include (Eckert and Callender 1987):

– Penetration and vane shear devices to estimate in situ soil strength.

    – Pressure meters and plate load equipment to estimate load-deformation characteristics.

    – Nuclear densimeters and sand cone devices to measure density.

    – Specialized equipment to measure permeability and pore pressure.

    – Test loading of piles.

    – Instrumentation of embankments and foundations.

    – Monitoring of soils during vibratory and impulse loading.

    • Modified versions of these tests (with an additional cost factor) are used for subaqueous bottom investigations.

    • Laboratory tests conducted on soil samples can include the use of triaxial or related apparatus to test for strength, compressibility, and dynamic response; consolidation test devices; and equipment to measure parameters such as water content, gradation, plasticity, permeability, and relative density (see Part VI-4-2, "Earth and Sand").

    • The information presented above is little more than a brief introduction to a coastal project's geotechnical design needs. Geotechnical engineers responsible for the foundation design will have the knowledge and information resources to guide the geotechnical design from the initial site reconnaissance phase through to the final design. Eckert and Callender (1987) and CIRIA/CUR (1991) provide additional guidance specifically related to coastal project geotechnical investigations, including details on soil collection and testing methods. Other more general information sources include Engineer Manual (EM) 1110-1-1802, EM 1110-1-1804, EM 1110-2-1906, EM 1110-1-1906; textbooks; and various soil testing standards.

    e.   Geotechnical design criteria for shallow foundations.

    (1) Foundations are classed as either shallow or deep depending on the depth of the soil strata at which the structure=s load is to be supported. Most coastal structures rely on shallow foundations for support, i.e., the load is supported by the soil just beneath the bottom. The exceptions are pile-supported structures and piers, which require deep foundations. Shallow foundations (e.g., spread footings, mats) effectively widen the bearing area over which the load is distributed so the underlying soil can safely carry the burden.

    (2) Environmental factors that may influence design of shallow foundations in coastal projects include currents, tides/storm surges, waves, and seismic activity. These loads (individually or in combination) may induce lateral or vertical forces, excess pore pressures, dynamic forces, or scour. Specific site considerations include soil type and strength, topography, water depth, and structure positioning. Other considerations related to foundation design include construction materials, construction techniques, and the anticipated foundation load distribution. Not all of the factors listed above will pertain to all foundation designs, so the geotechnical engineer must determine which factors are important for the particular site and foundation type.

(3) In shallow foundation design, ultimate bearing capacity and expected foundation settlement are calculated separately, and the in situ soil properties will largely determine whether the design is governed by bearing capacity or settlement criteria. Shallow foundation design on cohesionless soils is generally controlled by total and differential settlements because ultimate bearing capacity is very high for sand. Both settlement and bearing capacity must be considered when designing shallow foundations on cohesive soils, because either or both may be critical for the specific coastal project foundation.

(4) Allowable settlement depends on the rigidity and intended function of the supported project element. For example, a rubble-mound structure will tolerate more differential settlement without damage than a caisson-type structure. Internal structural stresses are less severe during uniform settlement; however, the overall decrease in crest elevation may impact the structure's functionality.

(5) Settlement in cohesionless soils is rapid with little time-delayed soil consolidation. In fact, much of the settlement occurs during construction, allowing onsite correction to achieve desired structure crest elevation. Total settlement in cohesive soils occurs in three steps. *Immediate settlement* is the soil distortion that occurs concurrently with soil loading. *Primary consolidation* occurs over time as water is pushed from the soil=s voids. Finally, *secondary compression* occurs as the soil structure adjusts to the effective applied load after consolidation.

(6) Coastal project elements situated on slopes or embankments may be susceptible to slip-surface failures passing through or under the structure. This applies mainly to retaining-type structures like bulkheads, seawalls, revetments, and earthen dikes. Design parameters related to the pore pressure distribution in the subsoil are needed to complete slip-failure analyses. Slip-surface failures are seldom a concern for subaqueous foundations such as those supporting navigation structures, breakwaters, and groins, unless these structures are built on weak soils. However, slip-surface damage could occur at channel structures if the navigation channel meanders too close to the structure toe.

(7) A key design criteria for retaining structures is determining the earth pressure caused by the backfill. The backfill pressure might be increased due to saturation by wave overtopping or rainfall runoff, or the pressure might be decreased by backfill erosion under certain conditions. Scour potential at the retaining structure toe should also be examined (see Part VI-5-6, "Scour and Scour Protection").

(8) The prospect of project damage by seismic activity should be evaluated for projects located in high seismic risk zones. Generally, seismic-related damage to rubble-mound structures does not result in catastrophic failure, and it is possible for earthquake-damaged rubble-mound structures to continue to fulfill their intended function. (The possible exception might be structures armored with nonreinforced slender concrete armor units that could experience significant breakage during an earthquake.) Potential repair costs resulting from seismic damage should be factored into the project selection criteria if conditions warrant. Seismic design of waterfront retaining structures is covered by Ebeling and Morrison (1993).

VI-3-2. <u>Seasonal Profile Variation</u>.

a. Many shore-normal cross-sectional beach profiles exhibit a distinct variation with season, characterized by a lowering of the profile at the shoreline during the stormy winter months and a building up of the profile at the shoreline during calmer summer months. The winter wave conditions contribute to movement of sand in a seaward direction and a general flattening of the profile. Winter profile erosion occurs fairly rapidly during storms, and some periods of beach recovery may occur between storm events. The calmer wave conditions typical of summer months contribute to the shoreward movement of sediment and a rebuilding of the beach at the shoreline with a steepening of the profile. Beach recovery occurs fairly continuously, but at a slow rate. Thus, complete recovery from a series of severe winter storms may not occur during the following summer season.

b. Seasonal weather patterns may also influence the direction and magnitude of net sediment transport at a project site, resulting in beach profile changes as longshore sediment supply increases or diminishes according to the site characteristics. Of course, pronounced profile changes beyond the seasonal variations can occur during any season in conjunction with strong storm wave conditions, and some profiles may be experiencing long-term erosional changes that are more subtle than seasonal variations.

c. The extent of seasonal cross-shore profile variations can be an important design consideration for shore protection projects, such as seawalls, revetments, beachfills, pipelines, and offshore breakwaters. For example, knowledge of the seasonal lowering of the beach profile fronting a seawall will be a factor when evaluating the type and extent of toe protection needed for seawall stability. Cross-shore profile seasonal variations are of lesser importance in the design of navigation structures such as breakwaters and jetties.

d. Seasonal variations in alongshore transport may influence design of coastal projects constructed normal to the shoreline, such as groins, jetties, piers, beachfills, and sand bypassing plants. For example, a groin at the boundary of a beach renourishment project might be needed to retain a beachfill during seasons of high longshore sediment movement.

e. As discussed in Part V, functional project design considers the potential impacts a coastal project may have on adjacent or fronting beaches. Conversely, the design of specific project components must anticipate whether or not the expected post-construction shoreline or profile changes will impact the component design. The type and location of a coastal project may cause substantial changes to the seasonal cross-shore and alongshore beach profile variations due to the influence of the project on the before-project coastal processes.

f. The seasonal extent of beach profile variations at a location can be approximated from historical profile data or with periodic site inspections over several years provided the yearly wave climatology during the observation period is typical for the site. Reliability of seasonal profile change estimated from measured beach profile data is a function of yearly profiling frequency, the number of years represented in the profile records, and the accuracy of the surveys. Judging seasonal variations using a few isolated profiles is not likely to produce a meaningful result. Always be aware that a single extraordinary storm could cause profile variation and beach recession several times that of a typical year, thus masking the true seasonal variation.

g. Profile variations due to seasonal changes in longshore sediment transport are more difficult to estimate than profile changes caused by storm activity. Generally, the coastal engineer needs to be aware of time periods when the supply of longshore sediment may be curtailed, which would result in a lowering of the profile.

VI-3-3. Flanking Possibility.

a. Some coastal projects, particularly shore protection projects located on or near the active shoreline, may be vulnerable to flanking damage due to continued beach erosion beyond the project boundaries. Flanking of a coastal structure, if left unchecked, will eventually lead to progressive damage of the project; and eventually periodic maintenance or rehabilitation will be required. Special attention should be given to designing suitable transitions between the project and adjacent non-project areas.

b. Ideally, shore protection projects should extend shoreward past the zone of active erosion to a stable portion of the beach or should be tied into a less erodible feature, such as a low bluff or dune. However, this is often not feasible due to increased costs, property boundaries, or other practical reasons. End transition sections for coastal structures should retreat landward and include enhanced toe protection in anticipation of increased erosion at the project terminus.

c. Possibility of flanking should also be considered when designing shore-connected structures such as jetties and groins. Navigation channel jetty structures are vulnerable to breaching on their landward end due to erosion on the seaward side and/or the interior bay side, as illustrated in Figure VI-3-1. This problem is usually associated with jetties stabilizing barrier island tidal inlets. The landward extent of the shore-connected structure must be sufficient to preclude the possibility of breaching due to shoreline recession, and it may be necessary to armor the bay-side shoreline with revetment (as shown on Figure VI-3-1) to stem potential erosion.

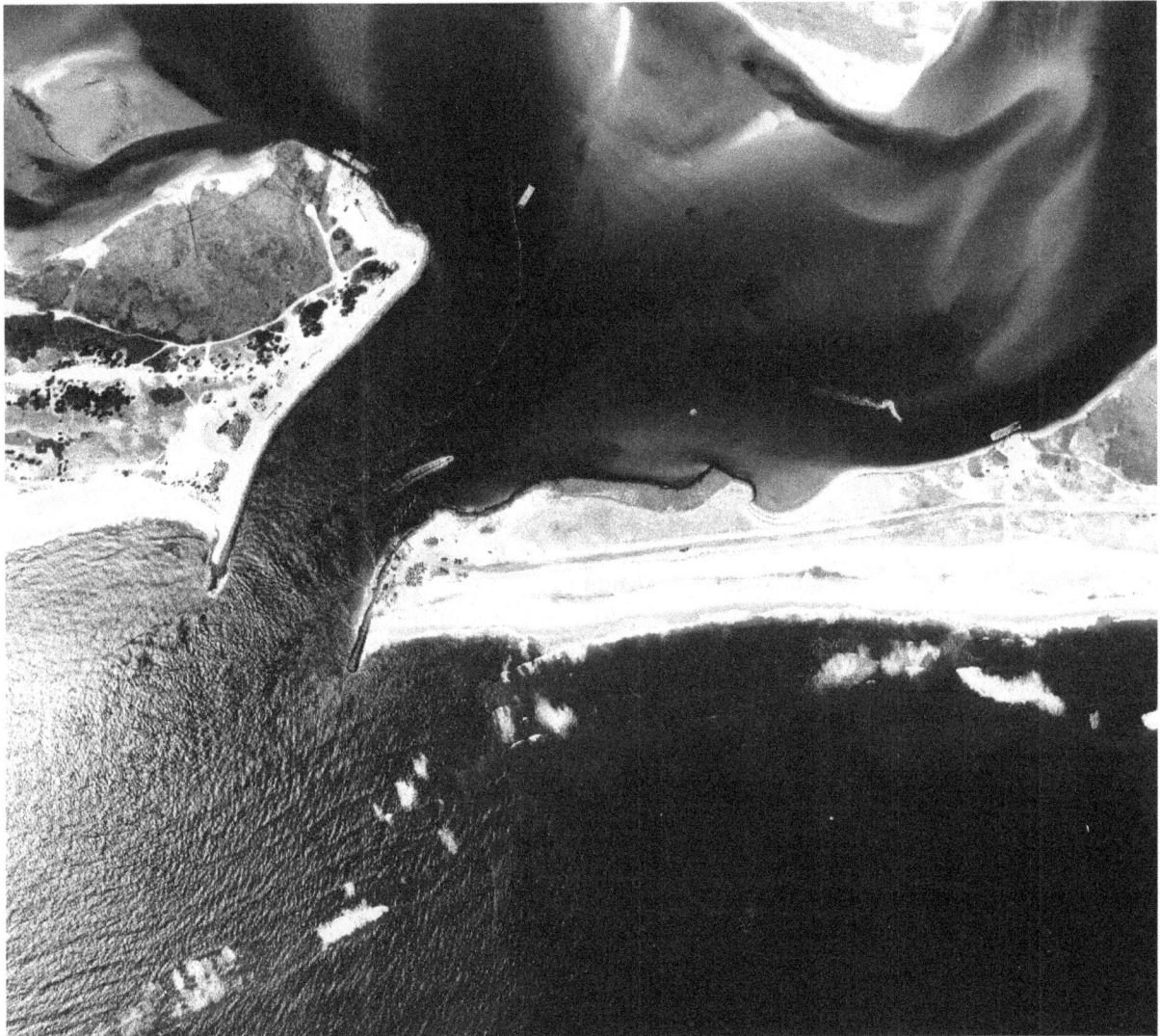

Figure VI-3-1. Bayside erosion and protective revetment at east jetty of Moriches Inlet

VI-3-4. Seismic Activity.

a. Coastal projects constructed in regions known to experience seismic activity may need to consider potential impacts related to ground deformation and severe liquefaction. Seismic loading may also be a concern in design of confined dredged material berms (subaerial) and caps (subaqueous) where liquefaction could release contaminated sediments.

b. Designing for seismic activity depends largely on the type and function of the project. For example, partial or complete failure of a breakwater or jetty during an earthquake probably will not result in catastrophic damage or loss of lives; therefore, these structures are usually not designed to withstand seismic loadings. Conversely, seismic failure of some coastal structures may carry substantial consequences if human life is at risk, repair costs are high, or vital services or commerce might be interrupted. Port and harbor facilities in particular fall into this category. The earthquake that struck Kobe, Japan in 1994 (magnitude about 7.0 on the Richter scale) resulted in ground motions and liquefaction so severe that quay walls sunk, gantry cranes were toppled, and 179 of 186 berths at the port had to be shut down (Matso 1995). In addition to direct damage of coastal project elements, engineers must consider potential damage to adjacent facilities that could result from failure or partial collapse of a coastal structure. In Kobe, an approach span to a harbor bridge collapsed when liquefaction resulted in a 2-m lateral movement of a seawall and highway column foundations being supported on the retained fill. Monolithic coastal structures in Japan are designed to resist earthquakes.

c. Until 1994, Corps of Engineers= experience with earthquake effects on coastal structures generally had indicated relatively minor damage to Corps-maintained rubble-mound breakwaters and jetties in Southern California. However, a 6.7-magnitude earthquake (Richter scale) at Northridge in 1994 was thought to have caused additional damage to the Channel Islands north jetty that had been previously damaged by a storm (Department of the Army 1995). The relative flexibility of rubble-mound structures makes them more suitable for withstanding earthquake loads with usually only minor settlement or damage to the armor layers. Monolithic-type structures are less likely to survive seismic loading unscathed.

d. Waterfront retaining structures typical of ports and harbors often have cohesionless soils beneath and behind them with relatively high water tables. During strong ground vibrations, there is a possibility of pore pressure buildup and associated liquefaction. Designing for such an occurrence is still an evolving art, with past experience and empirical results forming a substantial portion of the design guidance. Ebeling and Morrison (1993) provide a useful overview of specific design procedures applicable to the design of waterfront retaining structures.

e. The decision to allow for seismic loadings in coastal project design should be made on a case-by-case basis. When loss of life and interruption of vital services are not considerations, the decision to design for seismic loading may hinge on such factors as estimated repair costs versus replacement costs, or the risk of damage versus increased initial construction costs.

VI-3-5. Ice.

a. Ice loading.

(1) At some latitudes, freshwater lakes and coastal regions experience annual ice formation during portions of the year. Thus, in planning stages it is important to determine if the presence of ice adversely impacts the project=s functionality; and during design, it is important to consider the effect that ice loads and impacts might have on individual coastal project elements. (Also see Part V-3-13-d.)

(2) Most cases of ice action on coastal project elements fall into one of the below categories:

(a) **Dynamic ice forces** from floating ice sheets and floes driven by winds or currents are normally the most critical for coastal marine structures. At vertical structures the ice fails by crushing and/or splitting, which develops horizontal loads on the vertical face of the structure. At sloping structures the ice fails by bending and/or shear, which produces both vertical and horizontal loads on the sloping face of the structure.

(b) **Static ice forces** are developed when more or less intact ice sheets encompassing structures undergo thermal expansion and contraction, or when the ice mass exerts a steady pressure due to winds or water currents. The ice undergoes plastic deformation around the structure rather than failing outright.

(c) **Broken ice forces** occur when a mass of broken pack ice is driven against a structure much like a river ice jam or ice piling up along a lakeshore. This condition may be crucial in the design of small isolated structures. Because of flexibility within a broken ice field, loading pressures on structures usually are less than pressures developed by solid ice sheets.

(d) **Uplift and drawdown forces** are associated with ice that has frozen to a structure, such as a vertical pile, or with ice that rests atop a structure such as a breakwater. Changes in water level that suspend or submerge a portion of the ice create gravity or buoyancy loads, respectively.

(3) The above-listed forces pertain to the structural loading on the larger coastal project elements. Smaller additions to the project, such as railings, navigation aids, lights, or other relatively fragile structures, are easily damaged by ice riding over the structure.

(4) Table VI-3-2 summarizes the effects ice may have on the design of coastal project elements. Design guidance and appropriate references for use in estimating ice loads are given in Part VI-5-8-b, "Ice Forces." Additional information can be found in a monograph from the ASCE Technical Council on Cold Regions Engineering (Chen and Leidersdorf 1988) and in proceedings of the *International Offshore and Polar Engineering Conference* series sponsored by the International Society of Offshore and Polar Engineers.

b. Ice on sloping-sided (rubble-mound) structures.

(1) Sackinger (1985) distinguished several categories of ice action that could occur at rubble-mound structures:

(a) rideup of sheet ice on the structure slope,

(b) piling up of fractured ice fragments on the slope,

(c) ice sheets or fragments overriding the structure crest,

(d) dislocation of individual armor units by a moving ice sheet,

(e) damage to individual armor units by ice fragments,

(f) lateral forces on the entire structure by an ice sheet, and

(g) grounded ice rubble adjacent to the structure that could impede functionality.

(2) Massive rubble-mound structures, such as breakwaters and jetties built to protect harbors on open coasts are seldom affected to any great extent by ice loading. In these cases, the design wave loads are comparable in magnitude to the maximum possible pressure that could be developed in an ice sheet; and because maximum wave loads and ice thrust cannot occur at the same time, no special provision is made in the design. Smaller armor stones and concrete armor units may be frozen into the ice and displaced vertically with the ice during periods of water level increase (e.g., tides or seiche). However, small displacements of individual armor units should not adversely impact structure stability due to the random nature of rubble-mound armor layers. Increased breakage of 1,800-kg (2,000-lb) dolos on the Cleveland East Breakwater during winter months was attributed to a combination of increased wave action and ice forces (Pope, Bottin, and Rowen 1993).

Table VI-3-2
Ice Effects in Coastal Project Design (after Peyton (1968))

Direct Results of Ice Forces on Structures	
Horizontal forces on structures caused by:	Failure of laterally moving ice sheets by crushing.
	Failure of laterally moving ice sheets by bending.
	Impact by large floating ice masses.
	Plucking of individual armor units frozen to ice.
Vertical forces on structures caused by:	Weight of ice frozen to structure and suspended at low tide.
	Buoyancy of ice frozen to structure and submerged at high tide.
	Vertical component of ice sheet bending failure induced by ice breakers.
	Diaphragm bending forces during water level change of ice sheets frozen to structural elements.
	Weight of ice on superstructure elements caused by ice spray.
Second-order effects on structures caused by:	Movement during thawing of ice frozen to structure elements.
	Expansion during freezing of entrapped water.
	Jamming of ice rubble between structural framing members.
Indirect Results of Ice Forces on Structures	
Mooring loads caused by impingement of ice sheets on moored vessels.	
Ship impacts during mooring that are greater than normally expected.	
Abrasion and subsequent corrosion of structural elements.	
Low-Risk, But Catastrophic Considerations	
Collision by a ship caught in fast-moving, ice-covered waters.	
Collision by an extraordinarily large ice mass of very low probability of occurrence.	

c. Ice on vertical-wall structures.

(1) Vertical-wall structures must account for lateral ice loads caused by wind or currents acting on ice sheets. Heavy ice in the form of solid ice sheets or floating ice fields may exert sufficient lateral loads to dislodge monolithic structures off their base, and adequate precautions should be taken to secure the structure against sliding on its base. Generally, this should only be a concern for smaller structures designed for mild wave conditions, and in these cases it may be necessary to operate ice breakers to avoid potential buildup of large ice sheets. Lateral ice loads also could cause high overturning moment loads on the foundation.

(2) Uplift forces can occur with changes in water level when ice freezes to the structure, and additional ice surcharge needs to be included in the foundation design loading. Abrasion of the vertical face by ice rubble could lead to spalling of concrete or damage to timber wales. Large ice floes may ground on any submerged rubble berm structure, resulting in damage to the rubble mound.

d. Ice on the shoreline and shore protection structures. Occasionally, ice formations can cause damage to the shoreline and shoreline protective structures, but often the net ice effects are largely beneficial. Freezing spray on banks and structures covers them with a protective ice layer; however, thawing of frozen bluffs may contribute to bluff collapse. Ice driven ashore and piled up on beaches and structures generally does not cause serious damage, and in many instances the ice provides additional protection against winter storm waves. Ice formations may cause abrasion of timber or poorly fabricated concrete structures, and individual structural members may be bent or broken by the weight of the ice.

e. Ice on floating breakwaters. Floating breakwaters are particularly susceptible to ice impact, ice buoyancy lifting, and lateral ice loads. An additional concern is the transfer of ice loads to the floating structure=s mooring system, and the possibility of mooring line breakage or anchor dragging. Many floating structures are used seasonally and removed during winter months. Because most floating structures are not designed for severe wave loading conditions, ice loading may be the most critical design condition for those floating structures that serve through the winter.

f. Ice on piles and piers. Lightly loaded (tapered) piles can be lifted when ice that is frozen to the pile undergoes upward motion due to water level fluctuations caused by tides, or in some cases, passing vessels. Lifting of the pile is contingent on the ice sheet freezing to the pile in a relatively short time, and the force necessary to lift the pile is less than the force that would fracture the ice sheet. Lowering of the tide level does not return the pile to its original position because driving a pile takes more force. This problem can be alleviated by placing fiberglass, PVC, or plastic sleeves around piles to cover the region from high water to below the depth of freezing. When ice freezes to the sleeve, the ice sheet can oscillate freely without exerting vertical loads on the pile. An alternate method is to keep the region around the pile free of ice by using "bubble curtains" that continually circulate the warmer water at the bottom up to the surface. This is accomplished by forcing compressed air through perforated pipes placed on the bottom. Piles and pier structures are also subject to lateral ice loads, impacts, and abrasion by ice floes. For example, ships maneuvering in an ice field can induce lateral displacement of ice, resulting in lateral forces on nearby piles.

VI-3-6. Environmental Considerations.

a. Understanding and mitigating environmental impacts of coastal projects are key considerations throughout the planning, design, construction, and maintenance phases of all projects. Potential environmental impacts need to be identified early in the planning process and proactively addressed during subsequent functional design. Parts V-3-12, "Environmental Considerations," and V-4-1, "Project Assessment and Alternative Selection," provide details about environmental aspects that could influence the coastal project design.

b. Once environmental concerns have been identified and project alternatives have been developed to minimize environmental impacts, the engineer must design individual project elements to conform to the environmental guidelines established for each alternative. Each project site will have its own unique environmental considerations, so it is difficult to generalize what allowances will have to be made in project element design. Often design parameters that best fulfill the environmental requirements may not result in a project that is the most cost-effective or easiest to construct. For example, crest elevation for a seawall might be established so as not to block the view of the fronting beach and water; however, this could lead to unacceptable wave overtopping during storm events. In this case, the coastal engineer must consider structure alternatives, such as a milder structure slope or recurved seawall face, to compensate for the lowering of crest elevation.

c. Another environmental consideration that might influence the actual design of individual project elements relates to project construction. Concern for various species may constrain the time periods when construction can occur. For example, beach nourishment projects cannot be constructed during turtle nesting season, and dredging activities should avoid fish spawning periods. Construction during acceptable periods may expose the partially constructed project to adverse weather conditions, and the design should allow for these increased loads during construction. Construction methods that cause significant dust, noise, water turbidity, or disruption to local activities may need to be altered to comply with environmental standards. Some changes in construction procedure could result in changes to the project design.

d. In general most environmental design parameters are established during project functional design and carried over into design of individual project elements. The engineer must develop a viable design that meets the environmental design criteria or state compelling reasons why this is not feasible.

VI-3-7. Construction Considerations.

a. Fundamental to engineering design is the skillful combination of design elements necessary to resist the imposed loads along with practical elements related to project construction. This is particularly true of many coastal engineering projects where construction often involves massive quantities of material that must be accurately placed into the water when environmental conditions are less than ideal. Design optimization of coastal project elements without factoring in construction considerations will likely result in an elegant design that is expensive and difficult (if not impossible) to build.

b. Availability of construction material, equipment, and skilled labor determine, in part, the project construction procedure. Practical knowledge and/or experience about how construction will proceed helps the engineer to evaluate the possibilities and modify the design to best accommodate construction needs. Severe constraints in construction procedures will impact the design accordingly. Depending on the type of coastal project, construction may require land-based plant, floating plant, or some combination thereof. In cases where either option is viable, this becomes an important decision that should be weighed carefully. The following sections highlight some of the construction factors that influence or modify engineering design of coastal project elements.

(1) Availability of materials. The primary materials used in construction of coastal projects are stone, concrete, beach sand, steel, timber, and geotextiles (Part VI-4, "Materials and Construction Aspects," provides a more in-depth examination of materials and material properties). Large material quantities are required for many coastal projects, and considerable savings in transportation cost and future maintenance costs can be achieved if suitable materials can be obtained locally, or if the design can be adapted to use the locally available materials. For example, it may be less expensive to armor a coastal structure using concrete armor units if no local quarries can produce sufficient quantities of required stone sizes. Other considerations include methods of material transport and whether the required material quantities can be delivered when needed for construction.

(a) Rubble-mound structures depend on availability of large amounts of suitably sized stone at low cost. Source and availability of stone should be investigated during design, not after the design has been completed. If possible, the design should be tailored to the known output capability of the quarry expected to be used as the supplier. Quarry production records are helpful in assessing rock quality, density, durability, sizes, and gradation. Part V-3-15, "Availability of Materials," contains information and references on quarry inspection and stone quality.) If the quarry is unable to deliver the ideal stone size and gradation at a reasonable cost, it may be necessary to design a structure with milder slopes that can be protected with smaller stone. Alternately, a dynamic-slope structure could be specified with the initial slope being reshaped by the waves into an equilibrium profile.

(b) Designs specifying significant amounts of concrete require an affordable source of quality aggregates and sand. Beachfill projects rely on nearby sources of inexpensive, beach-quality sand. Beachfill construction techniques, and possibly the construction sequence, may depend on whether the selected sand source is inland or offshore of the project.

(c) When possible, construction of project components should use standard off-the-shelf items rather than custom manufactured components. Typical components might include sheet piles, piles, timbers, anchoring systems, steel members, prestressed concrete beams, etc. Substantial cost savings can be realized if minor design modifications result in the use of standard components.

(2) Availability of construction equipment. Coastal engineering construction can involve highly specialized equipment, such as heavy-duty cranes, barges, dredges, pipeline dredges, and large trucks. If a vital piece of equipment is unavailable, construction schedules and delays will add significantly to the costs. When the option exists for either land-based or floating construction, the availability and capacity of plants to handle the selected materials and construction procedure are key factors in the decision. Construction time can be decreased if the design permits more equipment to work in tandem.

(a) Additional equipment-related considerations are the time and costs associated with mobilization and demobilization of construction equipment. Land-based equipment mobilization time is generally shorter than the time needed to mobilize floating equipment. This is especially true if terminal facilities have to be constructed (CIRIA/CUR 1991). Barges and floating construction equipment and the skilled labor needed to operate the equipment are not as readily available as for land-based equipment.

(b) Placement of armor stones on rubble-mound structures is critically dependent on the capacity of the crane, which is determined by the maximum armor stone weight at the longest reach. Therefore, placement of toe stone and berm armor will impose the worst loads on the crane. For large rubble-mound structures the design engineer should consider the ramifications of decreasing structure slope and reducing the armor stone size accordingly, if the change facilitates use of a smaller crane. In situations where the stone size cannot be reduced and the capacity of existing cranes is inadequate, floating equipment can be employed to place the armor units beyond the safe reach of land-based cranes.

(c) Concrete armor units are used where stone of sufficient size is unavailable. Casting of the units requires a nearby concrete plant, a ready supply of materials, a casting yard large enough to stockpile enough units to keep abreast of construction, and a good supply of concrete forms. Economics may justify using existing forms for concrete armor units, even if the forms are larger than the size determined by the design analysis. The increased cost in materials is offset by not having to fabricate new forms. For information on availability of existing concrete armor unit forms used in previous construction, check with contractors and the Government agency or construction firm responsible for the project.

(d) Regardless of the type of coastal project being constructed, it is important for the design engineer to be aware of the types of equipment that will be required and to consider the entire construction sequence. Design modifications that avoid any obvious weak links related to availability of equipment may be crucial to project success.

(3) Constructability. Certain types of coastal projects can be constructed using either land-based or water-based construction techniques. The project design may need to be altered to facilitate one method over the other, and the best alternative might be a combination of both techniques. Land-based equipment is almost always preferred to floating equipment, and barge dumping is often more expensive. Therefore, when feasible, land-based construction should be used.

(a) Project construction with a floating plant depends primarily on water depth, tide range, currents, wave conditions, structure configuration, and equipment availability. Construction using floating equipment is possible for placing materials at levels deeper than 3 m below the low water level relative to vessel draft (CIRIA/CUR 1991). This allows rapid and efficient barge dumping of the core material. Long structures extending into deeper water are better suited to construction using floating equipment, and work can progress at several project locations simultaneously. Existing terminal facilities at the project site help to reduce costs when loading material and equipment onto barges.

(b) Cranes on floating platforms may have difficulty accurately placing heavy loads on the higher portion of structures like breakwaters and jetties because of the long reach. Likewise, underwater placement is also difficult. In areas with a large tide range, it may be possible to plan the construction procedure to take advantage of the differing water levels. Risk of damage to floating equipment is an important concern, and water-based construction has a greater probability of work stoppage during harsh wave and wind conditions.

(c) Land-based construction requires sufficient maneuvering space for the construction equipment. For shore protection structures located on the shoreline, access to the immediate area behind the structure is usually required. Construction of shore protection structures can proceed at more than one location, but care must be taken to avoid weakness where different sections join.

(d) Breakwaters and jetties can be built out seaward from dry land equipment located on a road built on the structure crest; generally construction can proceed on only one front. Crest elevation may need to be increased over that established to meet the overtopping criterion to assure the safety of construction equipment and personnel during heavy wave action, and sufficient crest width is needed for trucks and other equipment to pass or to accommodate special equipment. It may be necessary to add special turnaround areas to the structure. Some of the changes to accommodate land-based construction may increase the structure cross section beyond that required for stability and functionality. Risk of damage to land-based equipment is usually less than for water-based construction, and there will tend to be less work interruption due to storm wave conditions. Care must be taken to protect equipment from vandalism and theft.

(e) Accurate underwater placement of construction materials is a function of water depth, water clarity, wave conditions, and equipment. If accurate placement under water is expected to be difficult, design of that portion of the project will have to compensate for less than optimal construction. Placing geotextiles under water in a wave and current environment is also difficult, and the engineer should consider how the placement will be accomplished. Land-based construction of the underwater portion of rubble-mound structures and toes is difficult, and there may be a tendency to oversteepen the underwater slope.

(f) Construction of coastal projects requires experienced contractors, crane operators, and labor crews. Contractors should be given some leeway in fulfilling the essential aspects of constructing the project according to design specifications. Novel or unique projects will challenge even experienced contractors, and the engineer should be open to design modifications suggested by the winning bidder on the project. Experienced construction inspectors also may have good suggestions based on practices they have witnessed on previous projects.

(4) Design requirements during construction. Most completed coastal projects are expected to withstand severe environmental conditions with little or no damage (beachfills are a notable exception), but these same projects may be quite vulnerable to damage if exposed to high waves during the construction phase when not yet fully armored. Although large tide ranges can be beneficial to construction, there is also the possibility that storm waves could break on the partially completed structure during some stage of the tide.

(a) Land-based construction is concentrated around the crane position, so it is usually possible to build the structure to its full strength as construction progresses. Therefore, only a small portion of the unprotected project is exposed at any one time. Temporary stability of placed materials is necessary, and an approaching storm may necessitate temporary protection of incomplete construction in order to withstand the storm with minimum damage. Project construction may concentrate currents at the structure head and cause scour holes to develop. Infilling of the holes will add additional expense and delay construction.

(b) Water-based construction can proceed over a wider area and the risk of damage to uncompleted portions can be limited by not exposing the underlayers to breaking waves unless it can be immediately protected by the primary armor. Likewise, scour hole development can be curtailed by providing scour protection well in advance of the structure.

(c) Temporary roads, construction access, or construction supports on the project should be anticipated and allowed for as part of the design loads (also see Part V-3-16, "Accessibility"). In addition, removal of temporary engineering works must be undertaken so as not to damage or weaken the structure. Project construction may disrupt ongoing activities in the vicinity, such as navigation, dredging, beach recreation, etc. These construction impacts should be minimized if possible. Onsite facilities and storage areas for materials and equipment should be sufficiently large to supply the project at all times. Limited storage areas or supply lines may necessitate a construction procedure that reduces risk of damage to partially completed structures. Floating breakwaters require a means of transporting project components to the site and a safe method of connecting the modules into a continuous floating breakwater.

VI-3-8. Other Design Considerations.

a. Regulatory compliance. As discussed in Part V-2-1, "Planning and Design Process," and Part V-3-13, "Regional Considerations," coastal projects require regulatory approval from Federal, state, and local agencies. These approvals will likely be contingent upon the project meeting certain criteria. For example, local permitting agencies may be unwilling to grant construction approval for a seawall if the crest elevation blocks the view from a popular boardwalk or if adequate beach access is lacking. Likewise, construction of a project may require additional work to mitigate project impacts to an acceptable level. Fulfilling the regulatory requirements may impact the project design, the method of construction, the transportation of materials to the site, or even the choice of construction materials. Therefore, the design engineer must have a clear understanding of provisions expected to appear in various permits and approvals so that the design will meet all approval criteria. Failure to consider these important aspects will result in delays, added expenses, or possibly a nonviable project.

b. Project maintenance. The design engineer should be aware of maintenance requirements for each project element and assure that the design permits necessary maintenance to take place. Accommodating maintenance is particularly important for coastal projects, such as beachfills and rubble-mound structures, which are expected to suffer some degree of damage over the life of the project.

(1) Projects built using floating equipment will generally require floating equipment for maintenance activities. Projects constructed with land-based equipment may have adequate access for maintenance using land-based equipment, but this will depend on costs to mobilize the necessary equipment. For example, jetties with an installed concrete cap and road provide easy access and mobilization costs would be low, whereas jetties without a cap would require construction of a road over the structure crest before equipment could be moved into place. In the latter case it might be more economical to perform maintenance and repair using floating equipment.

(2) Monitoring and periodic inspections of coastal projects may be required to determine when maintenance should be performed (Part VI-8-2, "Inspecting and Monitoring Coastal Structures"). If the type of expected monitoring has been determined, it may be wise to include monitoring aids as part of the design. Such aids might include surveying targets, aerial photogrammetry targets, in situ monitoring instruments, etc.

(3) Maintenance considerations for floating structures include replacing connections and anchoring system components, removing marine growth which could affect the flotation height of the structure, replacing unsafe guardrails, and taking steps to prevent concrete deterioration. The designer should anticipate how the maintenance can be accomplished without subjecting the structure to additional risk. Design of super-structure, guardrails, walkways, etc., on coastal projects should strive for low maintenance requirements.

c. Disposal of dredged materials. Dredging may be required to gain access to the project site, for entrenching toe materials, for backfilling higher quality foundation material, or for other reasons. When dredging is to occur, dredging volumes should be estimated, and the method of dredged material transport and disposal should be determined. Beneficial uses of the dredged material should be considered, particularly if the displaced material consists primarily of beach-quality sediment. Guidance on dredging disposal and beneficial uses of dredged material can be found in Engineer Manuals 1110-2-5025 (Department of the Army 1983) and 1110-2-5026 (Department of Army 1987). Also, papers from technical specialty conferences, e.g., *Dredging '94* (American Society of Civil Engineers 1994), provide useful information.

d. Aesthetics. Coastal projects should be pleasing in form as well as functional. Good workmanship and close adherence to design contribute to project aesthetics. Repair sections should be geometrically similar to the original structure, and transitions between new and existing project elements should be made attractive, if possible. Public reaction to existing projects can serve as input to new designs and modifications. Examples of projects that require aesthetic consideration are low-cost shore protection devices, which may be viewed as unsightly, or high-crested structures, which may block a scenic ocean view.

e. Aids to navigation. Prior to construction of any coastal project that may impact navigation, or interrupt any existing aids to navigation, complete project information should be provided to local authorities (Coast Guard District Commander). This information should include details about project authorization, the proposed construction schedule, and a detailed drawing showing the project location relative to existing features. Local authorities may require a set of "as-built" plans after the project has been completed, and it may be necessary to include new aids to navigation as part of the project design.

f. Fishing platforms. Coastal structures normally provide excellent habitat for fish, which in turn attract recreational fishermen to the structures. Where safe and justified, project designs should include accommodations for recreational fishing. However, recognize that many coastal structures, such as low-crested rubble-mound breakwaters and jetties, are inherently unsafe during larger waves and higher water levels, and there is a substantial risk of fishermen being swept into the water. This risk, combined with the difficulty of providing guardrails on rubble-mound structures, may preclude fishing activities at the project, and provisions may be needed to prevent site access to unauthorized personnel.

g. Vandalism and theft. At some project sites it may be necessary to consider the potential consequences of vandalism and theft of materials. If vandalism and theft are potential threats to a project, construction materials must be chosen that cannot be easily cut, carried away, dismantled, or damaged. For example, sand-filled geotextile bags can be cut, small concrete blocks can be stolen, and wire gabions can be opened with wire cutters. Such damage could initiate considerable damage to the structure. On the other hand, there are no documented thefts of 30-ton armor stones.

VI-3-9. <u>References</u>.

EM 1110-1-1802
Geophysical Exploration for Engineering and Environmental Investigations

EM 1110-1-1804
Geotechnical Investigations

EM 1110-1-1906
Soil Sampling

EM 1110-2-1906
Laboratory Soils Testing

EM 1110-2-5025
Dredging & Dredged Material Disposal

EM 1110-2-5026
Beneficial Uses of Dredged Material

American Society of Civil Engineers 1994
American Society of Civil Engineers. 1994. *Proceedings of the Second International Conference on Dredging and Dredged Material Placement*, E. Clark McNair, ed., two volumes, American Society of Civil Engineers, New York.

Chen and Leidersdorf 1988
Chen, A. T., and Leidersdorf, C. B., eds. 1988. "Arctic Coastal Processes and Slope Protection Design," Technical Council on Cold Regions Engineering Monograph, American Society of Civil Engineers, New York.

CIRIA/CUR 1991
Construction Industry Research and Information Association (CIRIA) and Centre for Civil Engineering Research and Codes (CUR). 1991. "Manual on the Use of Rock in Coastal and Shoreline Engineering," CIRIA Special Publication 83/CUR Report 154, CIRIA, London and CUR, The Netherlands.

Department of the Army 1995
U.S. Army Engineer District, Los Angeles. 1995. "Channel Islands Harbor, Ventura County, CA: Basis for Design, North Jetty Repair," Los Angeles, CA.

Ebeling and Morrison 1993
Ebeling, R. M., and Morrison, E. E. 1993. "The Seismic Design of Waterfront Retaining Structures," NCEL Technical Report R-939, Naval Civil Engineering Laboratory, Port Hueneme, CA.

Eckert and Callender 1987
Eckert, J., and Callender, G. 1987. "Geotechnical Engineering in the Coastal Zone," Instruction Report CERC-87-1, U.S. Army Engineer Waterways Experiment Station, Coastal Engineering Research Center, Vicksburg, MS.

Matso 1995
Matso, K. 1995. "Lessons from Kobe," *Civil Engineering*, American Society of Civil Engineers, Vol. 65, No. 4, pp 42-47.

Peyton 1968
Peyton, H. R. 1968. "Ice and Marine Structure," *Ocean Industry Magazine*, Parts 1-3, March, September, and December.

Pilarczyk 1990
Pilarczyk, K. W. 1990. "Design of Seawalls and Dikes - Including Overview of Revetments," in *Coastal Protection*, K. Pilarczyk, ed., A. A. Balkema Publishers, Rotterdam, The Netherlands.

Pope, Bottin, and Rowen 1993
Pope, J., Bottin, R. R., Jr., and Rowen, D. 1993. "Monitoring of East Breakwater Rehabilitation at Cleveland Harbor, Ohio," Miscellaneous Paper CERC-93-5, U.S. Army Engineer Waterways Experiment Station, Coastal Engineering Research Center, Vicksburg, MS.

Sackinger 1985
Sackinger, W. 1985. "Ice Action Against Rock Mound Structure Slopes," *Design and Construction of Mounds for Breakwaters and Coastal Protection*, P. Bruun, ed., Elsevier, Amsterdam.

VI-3-10. Acknowledgments.

Author: Dr. Steven A. Hughes, Coastal and Hydraulics Laboratory (CHL), U.S. Army Engineer Research and Development Center, Vicksburg, MS.

Reviewers: Dr. Hans F. Burcharth, Department of Civil Engineering, Aalborg University, Aalborg, Denmark; Han Ligteringen, Delft University of Technology, The Netherlands; John H. Lockhart, Headquarters, U.S. Army Corps of Engineers, Washington, DC (retired); Charlie Johnson, U.S. Army Engineer District, Chicago, Chicago, IL (retired); Michael C. Mohr, U.S. Army Engineer District, Buffalo, Buffalo, NY; and Joan Pope, U.S. Army Engineer Research and Development Center, Vicksburg, MS.

CHAPTER 4

Materials and Construction Aspects

TABLE OF CONTENTS

List of Figures

List of Tables

CHAPTER VI-4

Materials and Construction Aspects

VI-4-1. <u>Material Requirements</u>. Materials used to construct coastal engineering projects are critically important to the success and longevity of the project. Selected construction materials often must withstand the rigors of relentless wave pounding in a corrosive environment that may undergo freeze-thaw cycles. Primary material selection criteria are physical properties and strength, durability, adaptability, cost, availability, handling requirements, maintenance requirements, and environmental impact. Knowledge of past material performance on similar coastal projects is an important consideration for the design engineer. Much of the information presented in the following sections was condensed from a comprehensive Special Report entitled Α*Construction Materials for Coastal Structures*@ by Moffatt and Nichol (1983).

a. Material properties and strength. In practically all cases, common materials having well- documented physical properties and strengths are used in construction of coastal project elements. Sections in this chapter, beginning with Part VI-4-2, give properties for widely used construction materials. General aspects of key material physical properties are listed below.

(1) Specific gravity. Specific gravity is a fundamental property for all coastal construction materials. Coastal structures, such as breakwaters, rely on self-weight of the structure to resist applied loads. Thus, materials with high specific gravity, like rock and concrete, are ideally suited for these types of applications, particularly for submerged portions where water buoyancy decreases effective structure weight. Specific gravity is also important for structures such as surge barriers and piers, which must be designed to support the weight of the component structural members. Materials with lower specific gravities, such as wood and plastics, also have uses in coastal construction. Beach renourishment projects function best if the placed beach fill material has a specific gravity the same as, or greater than, the native beach sand.

(2) Strength. Depending on the application, materials used in coastal construction may need to resist tension, compression, and flexure stresses. Material strength properties help determine the size, shape, and stability of component structural members. Structures built of stone, earth, concrete, and asphalt are capable of withstanding compression, shear, and impact loading; but they generally cannot resist tensile loads. Tensile loads in concrete structures can be tolerated provided there is sufficient steel reinforcing or prestressing of the member to carry the tensile stress. Geosynthetics add tensile strength to the soil mass.

(a) Steel, and most other metals, can accommodate high levels of tensile, compressive and torsional stresses and impact. Often steel structural members undergo considerable flexing or displacement when subjected to bending moments, and this displacement must be considered in the design. Metals also expand and contract with temperature change, which can introduce additional stress into the structure.

(b) Wood also exhibits good tensile and compressive strengths, but wood is not isotropic and its strength depends on orientation of the wood grain relative to the applied loads. Wood components can tolerate significant deflection and movement without failing.

(c) Geotextile fabrics are subjected mainly to tension, impacts, flexing, and fatigue. Synthetic structural components can resist compression, tension, shear, and torsion to varying degrees depending on the particular synthetic. Some plastics will undergo enormous deflection before yielding, whereas some plastics have very little elongation prior to failure. Strength characteristics of some synthetic materials will decrease in time due to ultraviolet radiation or other environmental factors, and precautions must be taken when using these materials. Also plastics can experience a slow, permanent deformation under constant load.

(3) Resistance to cyclic, impact, and seismic loads. Coastal engineering project elements are often exposed to continual cyclic wave loading, impact loading from waves or vessels, and occasionally accelerations due to seismic activity. Surviving these load conditions may require that portions of rigid structures be able to absorb the load without exceeding the elastic yield limit of the materials. Stone or earth structures resist these types of loads by providing stress relief through differential settlement, nesting of stone layers, or local areas of damage.

(4) Flexibility. Flexibility is the property of a material that allows it to bend without breaking. Materials with good flexibility will help absorb cyclic and impact loads, but continual flexing might eventually lead to fatigue failure, plastic deformation, and crack formation. Material flexibility is a relative term, and it depends on both the material and the shape of the structural member. For example, steel columns and beams can be designed for little deformation whereas steel rods and cables can be highly flexible. Generally, concrete and stone are considered to have little flexibility, followed by the more flexible steel and wood. Rubber and some synthetic materials are highly flexible. Flexibility can also be used to describe the response of coastal projects. For instance, the individual armor stones on a jetty have no flexibility, but the entire armor layer is capable of movement and settlement to a new position without undue loss of functionality, thus making it a "flexible" structure. Likewise, beach fills can be termed flexible structures even though the individual sand grains are rigid.

(5) Compatibility. Many projects combine different materials, and compatibility problems may arise due to differences in material physical or chemical properties. The constituent materials in composites such as concrete and asphalt must be compatible to attain adequate strength. Rigidly combining structural components of different flexibilities or different expansion coefficients may induce additional stresses or component failure. Different materials (or materials in which properties vary) undergo abrasion at different rates. For example, armor stones of different hardness may degrade at different rates, which may lead to weak spots in the armor layer. Contact between different types of metals in the marine environment can cause a galvanic reaction and rapid corrosion. Corrosion can also stem from contact with chemicals. For example, materials used to contain contaminated sediment must be able to withstand any chemical reactions that may result from direct contact with the contaminant.

b. Material durability. Durability is a relative term describing how well a material withstands the rigors of the environment into which it is placed. The durability of a particular coastal project element is a combination of the durability of the construction materials and the capability of the project to continue functioning at an acceptable level even after the construction material has begun to degrade. Therefore, material durability needs to be considered in terms of the project's design life, first costs, and projected maintenance expenses. Projects with short design lives can tolerate less durable materials at a reduced cost. Factors that affect a material's durability include its ability to resist abrasion, chemical attack and corrosion, marine biodegradation, wet/dry cycles, freeze/thaw cycles, and temperature extremes.

(1) Earth and sand. Earth is generally considered durable unless changes in water content or chemistry reduce grain size to the range of silts and clays. Quartz sand is very durable, but sand mixtures with high carbonate content from shell material will be more vulnerable to chemical attack if the water is acidic. Also shell particles are not as hard as quartz and are more susceptible to abrasion.

(2) Stone. Igneous rock is considered to be the most durable, but this depends partially on the geology of the rock. Sedimentary rock is usually stratified and subject to failure through shear stress, impact, chemical deterioration, or changes in water content. Sedimentary armor stones generally are more easily worn down by abrasion. Any armor stone that develops small cracks may eventually fracture due to freeze/thaw cycles, irrespective of the type of rock.

(3) Concrete and asphalt. Concrete is considered to be durable and is usually expected to last throughout the lifetime of most coastal projects, provided the concrete is not exposed to adverse chemicals or excessive abrasion, and loads are within design limits. Cracks in concrete may lead to spalling of the surface and exposure of steel reinforcement, which will immediately begin to rust. Rough handling of individual concrete armor units during placement may result in chipping or cracking of slender members. Asphalt it not considered to be a durable material because it has low strength in both compression and tension, it is subject to chemical reaction, its stiffness changes with temperature, and it is not resistant to impact or abrasion.

(4) Steel. Standard grade steel is considered very durable if properly protected from rust and corrosion throughout the project lifetime. Bare steel will rapidly deteriorate in the corrosive coastal environment. Sacrificial anodes should be provided to protect steel exposed to seawater. Abrasion of steel components by sand, particularly near the seabed, is also a problem. Stainless steel is more durable, but this advantage is often offset by increased cost.

(5) Wood. Although wood is considered less durable than concrete, lengthy service life can be obtained for wood components. Wood durability depends on the characteristics of the wood, its usage and exposure to the elements, and project maintenance. Wood is an organic material that can be attacked by plants and marine animals if precautions are not taken. Fasteners and connectors, such as bolts, nails, etc., must also be protected from corrosion to assure wood structure longevity. Dry wood is the least fire-resistant material commonly used in coastal projects.

(6) Geotextiles and plastics. Geotextiles and many plastics are generally resistant to chemical and biological attack, but will deteriorate when exposed to ultraviolet radiation. The rate of deterioration can be reduced by adding UV inhibitors, coatings, or by covering the geosynthetic with soil, sand, water, or even algal growth. Use of synthetic materials in coastal construction projects is relatively new, thus long-term durability of some synthetic materials in the coastal environment has yet to be determined. Some synthetic materials are vulnerable to fire and can generate toxic fumes when ignited. For these reasons and other functional requirements, geotextiles are generally covered with soil.

c. Material adaptability. Non-rigid mound-type stone and earthen structures can be constructed in a variety of shapes and sizes, and these structures can accommodate changes in foundation elevation and structure slope without losing functionality and structural stability. Stone and earth can be used in most weather conditions and temperature extremes without significant consequences.

(1) Concrete is very adaptable for use in coastal projects; however, cost often limits its usage to applications that cannot be effectively constructed using less expensive materials such as stone. For example, concrete vertical caisson breakwaters are used when water depths are too great for conventional rubble-mound structures or when mooring facilities are needed adjacent to the structure. Concrete is also viable for use as rubble-mound structure armor units, piles, and sheetpiling.

(2) Steel is very adaptable for complex structures, support frameworks, structures with movable parts, floating structures, and structure components. Except in the above cases, costs generally limit steel usage to piles, sheetpiling, and beams.

(3) Wood is considered to be fairly adaptable for use in smaller structures and as structure components, and it is easily stored and handled during construction. Synthetic materials usually have specific functions determined by their hydraulic and strength properties, such as geotechnical filter, separation of soil reinforcement. Geotextile tubes are finding a variety of uses due to their capability to retain fine-grained material.

d. Material costs. Because of the large quantities of material needed for most coastal projects, material cost is an important design consideration. Historically, coastal structures have been built using common, readily available materials that were obtained locally at low cost. When evaluating material costs, the cost of transporting the material to the job site must be included. If the material is not locally available, transportation costs could equal or exceed material costs per unit volume. Consequently, a more expensive local source may be preferable to a less expensive alternative located further away from the project site. Any material selection based on cost must include consideration of further maintenance expenses associated with the selection. For example, selecting a local source of lesser-quality stone for a breakwater may result in initial construction cost savings, but this choice may result in increased maintenance expense due to stone fracturing and stone abrasion. At every juncture of the design process, the coastal engineer should evaluate the costs associated with material specification. Significant cost savings can be realized for bulk materials because of the vast quantities required. However, practical choices are somewhat limited for most coastal projects. Any project design that requires fabricated components should attempt to specify common "off-the-shelf" items rather than custom-made parts. When feasible, this will result in both cost and time savings. Finally, consider the costs associated with any special material handling requirements (see below). These costs may more than offset any material cost savings.

e. Material availability. Availability of suitable materials for coastal project construction and future maintenance is an important design consideration. Lack of viable local sources for primary construction materials may limit design options or significantly increase construction costs and time of construction. For example, use of concrete in remote locations may not be feasible unless good quality sand and aggregate are locally available for onsite mixing. The projected rate of material usage must be matched with the rate that material can be supplied. It may be necessary to stockpile material onsite to compensate for an intermittent supply and to avoid slack work periods (Thomas and Hall 1992). If plans call for future project replacement, modification, or maintenance, sufficient sources of similar (or required) materials should be determined a priori. (See Part VI-3-7, "Construction Considerations," for site-specific design factors related to material availability.)

(1) Earth and sand. In most locations an adequate local source of earth exists for use in dikes, fills, and foundations. Exceptions include areas characterized by deltaic deposits of silts and clay and some rocky coastal regions. Less common are local sources of high-quality beach sand for use in placed beach fills.

(2) Stone. Stone is generally abundant in most regions of the continental United States. However, some locations, such as the coastline of the Gulf of Mexico and South Atlantic, can be as far as 250 km or more from stone sources. Other locations may have huge quantities of stone, but the quality may not be adequate for coastal projects because of low density or low strength. An example is volcanic rock on Pacific islands. Along high wave energy coasts, coastal projects may require huge stones that are difficult to produce from local quarries.

(3) Concrete and asphalt. Cement, stone aggregate, and sand suitable for use in concrete mixtures are available in all coastal regions in the United States. Concrete materials may have to be transported to some remote locations, such as some of the smaller Pacific islands. Also, difficult local access to material sources in remote regions may make importation of concrete materials economically feasible. Generally, asphalt is available at most project sites in the United States, but use of asphalt at other locations depends on availability of the asphalt components and handling equipment.

(4) Steel. Standard grades of steel in common cross sections and stock lengths are generally available for coastal projects. Special cross sections or less-common steel specifications (such as high strength steel or even stainless steel) are less likely to be available locally and may require substantial transportation costs between the mill and construction site. Availability of prefabricated steel components depends largely on the project's proximity to qualified steel fabrication yard.

(5) Wood. In the past, wood was one of the most available materials for construction of coastal projects. However, in recent years certain types and sizes of durable hardwoods have become more difficult to obtain. This has resulted in fewer coastal projects in the United States being constructed with wood as the primary construction material. Where available locally, hardwood often compares favorably in terms of cost and utility to other construction materials for projects such as bulkheads and piers.

(6) Geotextiles and plastics. It is unlikely that most coastal project sites will have a local source of manufactured geotextiles and plastics. However, these materials are economically transported to all regions of the United States. Availability of large quantities of synthetics may require special orders to the factory with plenty of lead time to assure ontime delivery.

f. Material handling requirements. A substantial portion of project construction cost involves material handling. Included in handling costs are transportation of materials to the construction site, onsite storage of materials, onsite material mixing and component fabrication, and placement of materials to build the project. Projects in isolated locations must consider site access and availability of equipment to handle materials. Conversely, projects in urbanized coastal regions must consider impacts of large-sized material transport vehicles on congested streets and space requirements for onsite material storage. Most materials can be transported by conventional methods such as rail, barge, truck, or ship. Special allowances are needed for oversized loads and loads exceeding usual United States highway load limits of 180-215 kN (20-24 tons) per truck. Another important transportation consideration is projected future site access for bringing in materials needed for long-term maintenance or rehabilitation. Just as important is the ease with which materials can be handled either by hand or with conventional equipment. Materials that are awkward to handle, require special handling techniques and equipment, or require particular labor skills and specialized training add to project costs.

(1) Earth and sand. Earth is easily handled with conventional earth-moving equipment and transportation methods. The availability of earth compaction equipment will determine how earth fills will be compacted, which in turn factors into design load bearing capacity. If earth handling results in formation of dust clouds, workers must wear some sort of breathing filters. Sand from land-based sources is handled similarly to earth. However, sand obtained from offshore sources must be dredged and pumped or transported to the project site. In these cases, material handling will be a substantial portion of the project cost. Cost of earth and sand will increase if sorting into acceptable grain size ranges is required.

(2) Stone. Stone handling limitations arise primarily with large armor stone sizes. Availability of adequate handling equipment at quarries is a critical factor, as well as the cost of quarrying and transporting large armor stones. Some quarries have equipment capable of handling stones larger than allowed on public highways. Road weight limitations not only influence armor layer design, but careful planning is also required to maximize usage of trucks or rail transportation. Equipment must be available for handling of large armor stones at the project site. Cranes must have sufficient lift capacity and must be able to reach outward sufficient distances to place armor stones accurately at the toe of the structure. Approach roads and staging areas must be able to support the heavy truck loads.

(3) Concrete and asphalt. Handling requirements for concrete and asphalt beyond normal batch processing, truck hauling, and truck placement are a function of the particular structure design. Some designs may require special handling equipment, such as cranes with buckets, pumps, or roller compaction equipment. Availability of this equipment may influence the structural design. Air or water temperature and underwater placement may have an impact on concrete and asphalt handling requirements. Forms are needed to cast concrete armor units, and special equipment is needed to fabricate reinforced or prestressed concrete piles. Consideration should be given to whether special equipment, such as concrete forms, is reusable. Time should be allowed for concrete armor units to cure before placing them on the structure.

(4) Steel. Conventional steel members and framework can be fabricated for easy transport and handling using conventional equipment; however, some site assembly may be

required. Unusual steel fabrications, or very heavy steel components, may require specially designed or modified handling equipment.

(5) Wood. Typical wooden structural components present no difficulty in transporting and handling. Application of chemical preservatives may require special equipment to assure sufficient wood penetration.

(6) Geotextiles and plastics. Most synthetic materials can be transported by conventional means. Special handling equipment and techniques may be required to place geotextile fabrics, particularly in underwater applications. If the geotextile has specific weight less than water, provisions must be made to hold the fabric in place until it is overlain with denser material. Similarly, in above-water applications wind can lift sections of geotextile fabric unless it is weighted.

g. Material maintenance requirements. Project maintenance requirements depend in part on how selected materials deteriorate over time due to physical and chemical processes.

(1) Earth and sand. It is not necessary to protect earth and sand used in coastal projects from physical or chemical deterioration, but it is necessary to prevent or retard removal of material by wind or water erosion. The only maintenance costs will be associated with replacing eroded material, and this cost will be affected by access to the earth and/or sand portions of the project.

(2) Stone. The main concern with stone is reduction in size through abrasion and splitting. Armor stones broken into smaller pieces can be removed from a structure by wave action. Maintenance consists of replacing damaged or missing stones, which can entail significant mobilization costs. Preservation of stone material is generally not feasible.

(3) Concrete and asphalt. Concrete quality is determined by the quality of its component materials and the method of mixing and placement. Like stone, the main maintenance requirement is periodically taking steps to prevent deterioration, or mending portions that have cracked, broken, spalled, etc. Protective coatings can be applied to exposed concrete surfaces to help prevent flaking and to seal cracks that might allow water to penetrate the surface and cause corrosion of steel reinforcement. Some concrete sealants may become less effective when exposed to certain chemicals that react with the sealant. Broken concrete armor units should be replaced with new units. Care must be taken to assure replacement armor units are interlocked into the armor layer rather than simply placed on top. During original construction, future maintenance costs can be reduced by casting a suitable number of replacement armor units and stockpiling them onsite. Maintenance of asphalt structures consists primarily of patching or replacing damaged areas. Underlying earth materials may shift and settle, opening large cracks in the asphalt cover layer. These cracks must be repaired before the fill material erodes. Also, repeated cycles of large temperature change may open significant cracks in the asphalt. Continuous maintenance of asphalt roadway surfaces is required to avoid damage to vehicles and equipment.

(4) Steel. Steel must be protected from chemical and galvanic corrosion, unless it is made of special alloys such as stainless steel. Exposed steel surfaces corrode very rapidly in coastal

settings, especially in the wet-dry regions and at the sandline where sand particles continually abrade the paint and protective rust. Most steel maintenance involves reapplying protective coatings like paint, replacing corroded structural members and fasteners, and servicing the cathodic protection system by replacing sacrificial anodes. Steel structural members damaged by vessel impacts or debris should be replaced as soon as possible if the damage is severe enough to threaten structural integrity. For example, a buckled steel strut could result in failure of adjacent members at loads considerably below design values. Cosmetic damage, such as dents, can be addressed during scheduled maintenance.

(5) Wood. Wood structure components are susceptible to biological attack at all places except below the mud line. Most wood deterioration occurs in the wet and dry tidal range. Wood maintenance consists of reapplying protective surface coatings such as paint and replacing deteriorated wood portions with new material. It is usually not practical to re-treat deteriorated pressure-treated, chemical-impregnated wood. These members should be replaced. Surface coatings consist of antifouling paints or coating materials that resist borer penetration, such as a 0.5-mm-thick coating of epoxy. Maintenance of wood structures also involves replacement of wood members damaged by vessel or wave impacts, fire, or exposure to harmful chemicals. Broken structural members should be immediately replaced to avoid additional damage to adjacent structure components. Pollutants in some harbors may be harmful to wood, but a side benefit is the almost complete absence of marine life harmful to wood structures.

(6) Geotextiles and plastics. Maintenance requirements of synthetic materials vary widely, depending on the material and its application. Maintenance of geotextiles may be warranted if the fabric is exposed for a period of time. For example, loss of armor stone and underlayer stone might expose the geotextile filter cloth, which could then be damaged by debris or sunlight. Geotextiles used in sand-filled bags can usually withstand ultraviolet radiation, but the bags can be torn or vandalized, requiring immediate repair. Repair can be accomplished by sewing, overlapping, or gluing a patch to the damaged geotextile. Plastics can withstand practically all naturally occurring chemicals found in coastal regions. However, pollutants or spilled fuels may react with some plastics, causing rapid deterioration or change in the material's characteristics. Plastics can be physically damaged by impacts and by fatigue brought about by cyclic loading. Determining whether or not broken plastic components is needed will depend upon the importance of the plastic component to overall structural integrity.

h. Material environmental impacts. Long-term project success relies on the ability of the selected construction materials to resist attacks from the surrounding environment by such diverse factors as force loadings, corrosive chemicals, marine organisms, abrasion, fire, wet/dry cycles, freeze/thaw cycles, etc. Equally important is minimizing effects that construction materials may have on the natural environment in which they are placed. Strong justification is needed to use any construction material that introduces adverse chemicals into the environment that might impact plant and animal life in the immediate project vicinity. Coastal construction can produce nonchemical adverse impacts such as high levels of turbidity from earth and sand placement or from foundation dredging. Impacts also arise from burying or displacing species during construction, although many mobile animal species simply migrate out of the area temporarily. Completed coastal projects often provide viable habitat, thus offsetting somewhat the negative environmental consequences of construction. An evaluation of potential environmental impacts of a project should consider future impacts that could arise from project deterioration, vandalism, and subsequent repair or maintenance. Environmental impacts may be reduced during repair and rehabilitation if materials from the original construction can be reused. Finally, present and future visual impacts of the project definitely should not be ignored.

VI-4-2. Earth and Sand.

a. Uses of earth and sand in coastal construction. Coastal projects tend to be fairly large and require a significant volume of construction materials. When feasible, structures are designed to use earth or sand as an economical filler material, and in many cases the mechanical strength properties of the soil are an integral part of the design. Below are some of the common uses of earth and sand in coastal construction:

(1) Rubble-mound breakwaters. Sand may be used as core material to provide a structure with a nearly impervious core, although sand-only cores are not common practice. The sand can contain clays, but cohesive clay-like materials alone are unsuitable for breakwater cores. Sand cores must be protected by geotextile or gravel filters and successively larger stone layers to prevent loss of sand due to piping under wave and current action.

(2) Caissons. Sand or soil is used to fill the compartments of concrete caissons and "cell-type" structures made of steel sheetpiles. Sand is preferred if the filler material is expected to support road works. Fill material must be protected from wave action that could wash away the soil.

(3) Bulkheads and vertical-front seawalls. Sand and soil are most often used as backfill or as foundation material for bulkheads and seawalls. The backfill usually is compacted to provide supportive soil pressure to resist wave loads and hydrostatic pressures. Soil may be needed to level the working area for foundations, or in weak soil conditions, to replace unsatisfactory in situ soil. Some circumstances may require coarser backfill material to promote rapid draining.

(4) Dikes. Earthen dikes constructed of sand, clay, or a combination of both, are used as dredged material containment structures and as storm protection structures. Dikes exposed to wave action need to be protected against erosion, i.e., armored like a revetment.

(5) Beach and dune restoration. Beach-quality sand from either land or offshore sources is the key ingredient for successful beach nourishment and dune restoration projects. Constructed sand dunes can be temporarily stabilized using snow fencing while dune vegetation is being established. Useful guidelines on stabilizing dunes with vegetation were given by Woodhouse (1978).

(6) Land reclamation. Construction of coastal facilities such as harbors and marinas often involves creation of new above-water land areas. Earth and sand used in these projects may come from dredging or from inland sources. Soils used in most reclamation projects are expected to have some degree of load- bearing capacity, depending on project requirements.

(7) Construction roads. Access to coastal projects may require construction of temporary or permanent roads using earth, sand, and gravel. Initial construction or major rehabilitation of shore-connected rubble-mound structures requires a roadway along the structure crest capable of supporting a crane and heavy trucks. If a permanent crest road is not part of the structure design, a temporary gravel road can be constructed that will eventually be washed away by storm waves.

(8) Concrete aggregate. Sand and gravel are essential ingredients in concrete and grouts used in coastal construction.

b.   Physical and mechanical properties of earth and sand. Part III-1 (Coastal Sediment Properties) provides a thorough overview of sand composition, properties, and engineering applications. The following sections cover a broader range of soils.

(1) General soil properties and classification. The terms "earth" and "soil" are used to describe mixtures of a large assortment of materials comprised of various size particles. Soils are classified according to grain size into groups that share similar engineering characteristics. One such system is the widely used Unified Soil Classification System (USCS) as presented in Table III-1-2 of Part III-1, "Coastal Sediment Properties." This classification system spans the particle size range that includes boulders, cobbles, gravels, sands, silts, and clays. Listed below are some general engineering characteristics of soils classified according to the USCS (Eckert and Callender 1987):

(a) **Boulders and cobbles**. "Boulders" and "cobbles" are rounded to angular, bulky, hard rock particles. Boulders have an average diameter greater than 300 mm, whereas cobbles have diameters spanning the range between 75 and 300 mm. Boulders and cobbles are very stable components for fill and for stabilizing slopes, particularly when the particles are angular. Including these larger particles as aggregates in finer grained soils helps improve the soil capacity to support foundation loads.

(b) **Gravels and sands**. Gravels and sands are rounded to angular bulky, hard, rock particles that can be naturally occurring or made by crushing larger stones. Gravels span the range of grain diameters from 4.76 to 75 mm, and sands cover grain sizes in the range from 0.074 to 4.75 mm. Within each category there are further divisions such as "coarse" and "fine." Gravel and sand have essentially the same engineering properties; they differ mainly in degree. They are easily compacted, little affected by moisture content, and unaffected by frost. Gravels are more permeable than sands, and they are generally more resistant to erosion and piping. Stability of sands and gravels generally decreases as the grain-size distribution becomes narrower.

(c) **Silts and clays**. Soil particles with diameters less than 0.074 mm are silts or clays, and the distinction between the two arises from its behavior under certain conditions. Silts are inherently unstable, particularly when moisture content is increased, and they may reach a "quick" state when saturated. Silts are difficult to compact, highly susceptible to frost heave, and are easily eroded. Clays exhibit plastic behavior and have cohesive strength, which increases as moisture content decreases. Clays have low permeability, are difficult to compact when wet, and are difficult to drain. Clays resist erosion and piping when compacted, and they are not susceptible to frost heave. However, clays do expand and contract with changes in moisture content. In general, highly expansive clays should not be used to backfill coastal structures. The most important engineering properties of soils are density, shear strength, compressibility, and permeability. These properties are used to estimate slope stability, bearing capacity, settlement, and erosion rate. Some of the basic soil properties can be determined using field and laboratory tests. For other properties it is necessary to correlate the soil parameters with results from previous experience. In the sections that follow, several key soil parameters are discussed. More detailed information on these and other soil properties such as water content and grain-size distribution are given in Eckert and Callender (1987) or in any geotechnical engineering textbook.

(2) Soil density. Soil is a multiphase mixture composed of solid particles and void spaces that are filled with water and/or gas. Consequently, in soil mechanics the term "*density*" describes the overall soil density as a function of particle density and the relative proportion of solids and voids in the sample. Table VI-4-1 shows a number of density-related parameters commonly used by geotechnical engineers. Note that specific gravity G is determined using the unit weight of fresh water. Typically, G ranges between 2.5 and 2.8 with preliminary calculation "default" values of 2.65 for sand and 2.70 for clays. Void ratio and porosity are indicators of soil compressibility and permeability. Geotechnical engineers prefer using void ratio because the volume of solids remains constant during any soil volume change. Void ratio can range from 0.15 for well-compacted soils having a wide grain-size distribution to 4.0 for very loose clays with high organic material content. Densely packed uniform spheres have a minimum void ratio of 0.35. Table III-1-4 in Part III-1 gives typical density values for common coastal sediments.

Table VI-4-1
Soil Density Parameters

Name	Symbol	Defining Equation
Basic Parameters		
Weight of solids	W_s	
Weight of water	W_w	
Volume of solids	V_s	
Volume of voids	V_v	
Total volume	V	$V_s + V_v$
Water unit weight	γ_w	
Derived Parameters		
Dry soil Unit weight	γ_d	$\dfrac{W_s}{V}$
Moist soil Unit weight	γ	$\dfrac{W_s + W_w}{V}$
Saturated soil Unit weight	γ_{sat}	$\dfrac{W_s + V_v \gamma_w}{V}$
Immersed soil Unit weight	γ_{sub}	$\dfrac{W_s - V_s \gamma_w}{V}$
Specific gravity	G	$\dfrac{\gamma_d}{\gamma_w}$

| Void ratio | e | $\dfrac{V_v}{V_s}$ or $\dfrac{n}{100-n}$ |
| Porosity | n | $\dfrac{V_v}{V}\times100\%$ or $\dfrac{e}{1+e}\times100\%$ |

(3) Soil relative density and relative compaction. These two parameters give a measure of a soil's in situ density relative to the range of possibilities for that particular soil.

(a) **Relative density** is used for noncohesive sands, and it is defined as the percentage given by the expression

$$D_r = \frac{e_{max} - e}{e_{max} - e_{min}} \times 100\% \qquad \text{(VI-4-1)}$$

where the numerator is "*the difference between the void ratio of a cohesionless soil in the loosest state (e_{max}) to any given void ratio, e,*" and the denominator is "*the difference between void ratios in the loosest and densest (e_{min}) states.*" Relative density provides a measure of the compactness of granular materials used in coastal projects such as sand backfill or dike cores. In the field, relative density is found using standard penetration tests or Dutch cone penetration tests. Actual estimation of relative density should follow the American Society for Testing and Materials (ASTM) Standards (ASTM D-4254 1994) or EM 1110-2-1906 (Department of the Army 1986). However, because of the difficulty in establishing the loosest and densest states of cohesionless soils, significant variations occur in determination of relative density, and correlations with other soil engineering properties should be avoided except for use in preliminary calculations.

(b) **Relative compaction** describes the relative density of *compacted* soils, and it is defined as "the ratio of the unit dry weight of an in situ material (γ_d) to the unit dry weight of the soil when compacted to its maximum density (γ_{dmax})," or

$$R_c = \frac{\gamma_d}{\gamma_{d\,max}} \times 100\% \qquad \text{(VI-4-2)}$$

The Standard Proctor Method, given in EM 1110-2-1906 (Department of the Army 1986) is recommended for determining maximum unit dry weight for coastal fills and embankment applications. Relative compaction is normally used to describe cohesive soils (placed or pre-existing) that have been stabilized or improved using compaction techniques.

(4) Soil shear strength. Soil fails when shear displacement occurs along a plane on which soil stress limit is exceeded. For all but preliminary design, soil strength should be determined using appropriate in situ or laboratory testing procedures as described in EM 1110-2-1906 (Department of the Army 1986) or ASTM Standards. Commonly performed tests are the Unconsolidated-Undrained triaxial test, Consolidated- Undrained triaxial test, and the Consolidated-Drained triaxial test. These tests produce stress-strain curves for the tested loading condition, and the shear strength is defined as the first maximum that occurs on the curve. The

tests also reveal conditions of failure for the soil. Soil strength is usually presented in terms of Mohr circles and Mohr failure envelopes. This allows shear strength to be expressed in terms of cohesion, maximum stress, and the angle of internal friction. Noncohesive, granular soils (i.e. sand) resist shearing through two mechanisms: (a) the frictional resistance between particles due to the normal force acting at the point of contact; and (b) the interlocking of particles as they attempt to shift past one another during strain. Frictional resistance is the principal source of soil strength, and it is a function of the soil confining stress. Soil shear strength increases with increases in confining stress. Highly compacted soils with low void ratios have increased strength due to particle interlocking. The shear strength of placed or backfilled cohesive soil will depend to a large extent on the moisture content (pore-pressure) and the compaction the soil receives. Tests should be conducted after compaction to verify that design strength levels have been achieved or surpassed. Shear strength of in situ cohesive soils depends on the method of original deposition and the past overburden history. Undisturbed clays may be over-consolidated, normally consolidated, or under-consolidated. Shear strength is determined using the triaxial tests mentioned above.

(5) Soil compressibility. Soil compressibility is an indication of settlement that will occur over time due to a given load condition or a change in groundwater level. Compressibility of noncohesive materials is governed by the relative density of the soil, and estimates of soil settlement are straightforward. Consolidation of cohesive soils is more complex and occurs in three stages. Immediate settlement is compression of the soil matrix without any dissipation of pore pressure or water expulsion. Some immediate settlement may be due to compression of trapped gases in the soil. Primary consolidation occurs over time as increased pore pressures force water from the soil voids. This process continues until all the excess pore pressure is relieved. The rate of consolidation depends on soil permeability and the drainage characteristics of the adjacent soil. After primary consolidation, Secondary compression can occur in soils having higher plasticity or significant organic content, such as soft marine or estuarine deposits. Consolidation tests are used to establish the coefficients necessary to estimate settlement of silts and clays. Eckert and Callender (1987) describe the test and analysis methods, and they provide an example application.

(6) Soil permeability. Permeability is a soil parameter related to laminar (viscous) flow of water through the soil under the influence of gravity. Coastal geotechnical problems affected by soil permeability include seepage through beach sand, consolidation of backfills and hydraulically placed fills, and settlement of foundations. Viscous flow through soils is calculated with an empirical relationship known as Darcy's Law, which is applicable for soils from clays and silts up to coarse sands. In its simplest form, Darcy=s equation for steady flow through uniform soil is

$$Q = K A \frac{\Delta h}{L} \tag{VI-4-3}$$

where

Q = discharge

A = flow cross-sectional area

L = length of flow path

Δh = head difference over the flow length

The empirical coefficient K in Darcy's equation is called the *coefficient of permeability*, and it is a function of both the soil and the pore fluid. Soil particle size and gradation have the largest influence on the coefficient of permeability. Soil permeability is best determined in the field using pumping tests (see Eckert and Callender (1987) for an overview and references). Less accurate permeability coefficients can be obtained with laboratory tests using falling- or constant-head permeameters as described in EM 1110-2-1906 (Department of the Army 1986) or ASTM Standards. Many empirical equations have been proposed to relate permeability to characteristics of the soil such as effective grain size. However, these equations are generally suited only for compacted, clean, coarse soils, whereas naturally occurring soils will exhibit significant variation. Table VI-4-2 gives typical coefficients of permeability for common soils. These values are suitable for use in preliminary design calculations.

Table VI-4-2
Typical Soil Permeability Coefficients (from Eckert and Callender (1987))

Soil Types	Particle Size Range, cm		"Effective" Size D_{10}, mm	Permeability Coefficient, k	
	D_{max}	D_{min}		(cm/sec)	(ft/yr)
Uniform, coarse sand	0.2	0.05	0.6	0.4	0.4×10^6
Uniform, medium sand	0.05	0.025	0.3	0.1	0.1×10^6
Clean, well-graded sand and gravel	1.0	0.0005	0.1	0.01	0.01×10^6
Uniform, fine sand	0.025	0.005	0.06	40×10^{-4}	4,000
Well-graded, silty sand and gravel	0.5	0.001	0.02	4×10^{-4}	400
Silty sand	0.2	0.0005	0.01	1×10^{-4}	100
Uniform silt	0.005	0.0005	0.006	0.5×10^{-4}	50
Sandy clay	0.10	0.0001	0.002	0.05×10^{-4}	5
Silty clay	0.005	0.0001	0.0015	0.01×10^{-4}	1
Clay (30 to 50 percent clay sizes)	0.005	0.00005	0.0008	0.001×10^{-4}	0.1

(7) Soil mixtures. Depending on the borrow source, backfill material may be composed of a mixture containing some fraction of gravel, sand, silt, or clay, along with a significant percentage of organic materials such as vegetable matter or shell fragments. Soil properties of soil mixtures containing a wide range of components will vary tremendously, and the soil should be tested to assure compliance with specified strength and density requirements. Soil mixtures containing organic materials are usually considered detrimental and should not be used because they tend to be more compressible and have lower shear strengths.

c. Placement considerations for earth and sand. The method chosen for earth placement depends on such factors as location of material borrow source (land or offshore), type of fill, availability of suitable equipment, environmental impacts of the method, and project economics.

(1) Dumped placement. Earth or sand obtained from upland sources or dredged from the sea bottom can be transported to the construction site and dumped into place. For land-based construction the mode of transport can be trucks, scrapers, conveyor belts, or other means, depending on the transport distance. Typical land-based projects include backfilling seawalls and bulkheads, placing foundation material, and placing the cores of shore-connected rubble-mound structures. Offshore earth and sand can be placed by dumping from barges or by using draglines and buckets for more precision. Dumped material that is not compacted will have low relative densities, and settlement should be expected to occur over time. Barge dumping at sea creates turbulence that will segregate material by grain size as it falls and increase turbidity as fine particles are suspended in the water column.

(2) Hydraulic placement. Soils dredged from the sea or lake bottom can be transported and placed hydraulically by moving the material as a slurry through a pipeline. The pipeline may extend directly from the dredge to the project site, as in the case of some beach nourishment projects; or barges that bring the material to the construction site can be emptied with hydraulic pumping. Hydraulic placement offers greater accuracy than dumping for offshore applications such as the cores of rubble-mound structures. Material placed underwater, either hydraulically or by dumping, may be moved by waves and currents before it can be adequately protected with overlying filters and armor layers. Land placement of earth and sand by hydraulic means involves a large amount of wash water runoff that can erode sediment along the drainage path or leave segregated pockets of fine-grained sediment that have engineering characteristics vastly different from the rest of the fill.

(3) Compaction. Above water, earth and sand fills can be compacted by a number of methods depending on the degree of compaction necessary to reach the specified soil parameters. Construction documents should specify the required density, moisture limits, and lift thickness. In situ testing is needed to verify that the compacted fill meets specifications. Mechanical compaction of soils placed underwater is not practical; however, in some situations cyclic wave loading will help compact placed sand.

d. Environmental effects on earth and sand.

(1) Effects of soils on the environment. Polluted soils should not be used in coastal projects because contaminants may be released in coastal waters either by leeching out of the placed fill material or through project damage and erosion of the fill material during storms. Potential soil contaminants include industrial wastes such as toxic heavy metals (mercury, cadmium, lead, and arsenic), chlorinated organic chemicals (DDT and PCB=s), and pathogens (bacteria, viruses, and parasites) (Eckert and Callender 1987). Use of dredged materials in coastal construction must be limited to good quality materials free of toxic wastes. See Engineer Manual 1110-2-1204 (Department of the Army 1989) and Engineer Manual 1110-2-5025 (Department of the Army 1998) for related design guidance. Also examine recent Federal and state environmental regulations pertaining to use of dredged material.

(2) Effects of the environment on soils. The particles comprising mixtures of earth and sand are generally unaffected by the natural environment over the project life span. However, structural components constructed using earth and sand are subject to natural forces that can degrade the performance and functionality of the project. Erosion of materials and subsequent decrease in fill volume can be caused by wind, rain, ice, currents, waves, burrowing animals, or human activities. This may reduce the capacity of the soil to resist applied loads and result in project damage. For example, vertical seawall designs often rely on the backfilled soil to help resist wave impacts and water pressures. Unconsolidated sands and silts are most susceptible to erosion. Gravel is more stable against erosion due to the size of the particles, and clays are more stable because of tractive forces between particles. Liquefaction of submerged loose fine sand and silts can occur in areas of high seismic activity or high wave action.

VI-4-3. Stone.

a. Use of stone in coastal construction. In the context of coastal engineering, "stone" refers to individual blocks, or to fragments that have been broken or quarried from bedrock exposures or obtained from boulders and cobbles in alluvium (Moffatt and Nichol 1983). Commercial-grade stone can be classified according to size, shape, size distribution, and various physical properties of the material. Stone is used extensively to construct coastal structures, and it is by far the most common material used in the United States for breakwaters, jetties, groins, revetments, and seawalls. Larger projects may contain more than a million tonnes of stone; 80 percent in the core and 20 percent in the armor layers (CIRIA/CUR 1991). Stone used as aggregate and riprap is crushed, broken, or alluvial stone in which the shape of individual stones has not been specified and the size distributions are fairly wide. Quarrystones are larger rock pieces that are "blocky" in shape rather than elongated or "slabby." A principal use of quarrystone is in the armor layer of rubble-mound structures. Below are listed the major uses of stone in coastal construction. Undoubtedly there are additional uses not mentioned. For example, quarrystones make great gifts for your geologist friends.

(1) **Rubble-mound structures**. Large quarrystone with specified weight, density, and durability are used for the primary armor layer of most rubble-mound structures. Underlayers are composed of progressively smaller stone sizes; and in many cases, the rubble-mound core material may be riprap or "*quarry-run*" stone. Quarry stone is also used to construct the base of "*composite structures*" where a monolithic, vertical-front structure is placed on a rubble-mound base.

(2) **Riprap structures**. Riprap is used more for shore and bank protection structures that are not exposed to high waves or strong currents. The wider size distribution of riprap provides a less uniform armor layer that is more susceptible to damage by strong waves and currents. Riprap is less expensive than uniform stone, and placement on the slope is usually less precise (e.g., dumping from trucks).

(3) **Toe protection**. Graded stone is used to protect the toes of sloping- and vertical-front structures from undermining by scour. Stable stone sizes are selected based on the anticipated maximum waves or currents.

(4) **Scour blankets**. Stone blankets are placed on the seafloor in areas subject to scour by waves and/or currents. Often the scour blanket is a remediation response to scour that was not anticipated in the original project design. Protection of bridge and pier pilings with scour blankets is a routine application.

(5) **Stone fill**. Stone is used as a filler material for coastal structures such as cribs, caissons, and gabions. (Gabions are steel wire cages filled with small stones that can be stacked to form steep revetments and bank protection.)

(6) **Filter layers**. Smaller stones are used for filter layers over the foundation soil or in drainage applications. Placement is usually by dumping. Selection of stone for a particular project depends on the purpose of the project, design loads, and local availability of suitable stone. In some cases, it may be necessary to evaluate the benefits of using inferior locally available stone as opposed to transporting higher-quality stone from a distant source.

b. Physical and mechanical properties of rock. The paragraphs below provide an overview of rock properties crucial for coastal engineering applications. These and other rock properties are covered in much greater detail in the *Manual on the Use of Rock in Coastal and Shoreline Engineering* (CIRIA/CUR 1991).

(1) Types of rock. Rock, as it occurs in nature, is classified into three distinct groups. **Igneous rocks** are formed by crystallization and solidification of molten silicate magma. **Sedimentary rocks** are formed by sedimentation (usually underwater) and subsequent lithification of mineral grains. **Metamorphic rocks** are transformed igneous or sedimentary rocks in which textures and minerals have been altered by heat and pressure over geological time periods (CIRIA/CUR 1991). Within each major rock category are additional subdivisions based mainly on composition and texture (Table VI-4-3). Some of the more common stone types are described below:

(a) **Granite** is a term applied to medium- and coarse-grained igneous rocks consisting mainly of feldspar and quartz. Mica may also exist in small quantities, but large amounts of mica may result in fracture planes within the rock. Most granites are dense, hard, strong, have low porosity, and are resistant to abrasion and impacts. These characteristics make granite a good choice for riprap and armor stone.

(b) **Basalt** is a term applied to various dense, fine-grained, volcanic rocks (dacite, andesite, trachyte, latite, basalt). Basaltic rock was formed by cooling lava, and it is composed primarily of feldspar and ferromagnesian minerals. Some basalts may not be suitable for concrete aggregates if they contain reactive substances in the pores. Basalts are generally very dense, hard, tough, and durable, and they are good choices for aggregates, riprap, and armor stone.

Table VI-4-3
Engineering Characteristics of Unweathered Common Rocks (from CIRIA/CUR (1991))

Rock Group Name	Rock Specific Weight (kN/m^3)	Unconfined Compressive Strength (MPa) x 10^8	Water Absorption (%)	Porosity (%)
Igneous				
Granite	24.5-27.5	160-260	0.2-2.0	0.4-2.4
Diorite	25.5-30.4	160-260	---	0.3-2.7
Gabbro	27.5-31.4	180-280	0.2-2.5	0.3-2.7
Rhyolite	22.6-27.5	100-260	0.2-5.0	0.4-6.0
Andesite	23.5-29.4	160-260	0.2-10	0.1-10
Basalt	24.5-30.4	160-280	0.1-1.0	0.1-1.0
Sedimentary				
Quartzite	25.5-27.5	220-260	0.1-0.5	0.1-0.5
Sandstone	22.6-27.5	15-220	1.0-15	5-20
Siltstone	22.6-27.5	60-100	1.0-10	5-10
Shale	22.6-26.5	15-60	1.0-10	5-30
Limestone	22.6-26.5	30-120	0.2-5.0	0.5-20
Chalks	14.7-22.6	5-30	2.0-30	20-30
Metamorphic				
Phyllite	22.6-26.5	60-90	0.5-6.0	5-10
Schist	26.5-31.4	70-120	0.4-5.0	5-10
Gneiss	25.5-27.5	150-260	0.5-1.5	0.5-1.5
Marble	26.5-27.5	130-240	0.5-2.0	0.5-2.0
Slate	26.5-27.5	70-120	0.5-5.0	0.5-5.0

(c) **Carbonate** is a broad term applied to limestone, dolomite, and marble. These rocks contain varying amounts of calcite and span the range from fine-grained to very coarse-grained. Often a high percentage of clays make some carbonate rock unsuitable for use as stone in coastal construction. Conversely, high sand or silica content may harden carbonates. Marble is limestone or dolomite transformed by metamorphic processes into a harder, more crystalline structure. Carbonate stone that is physically sound, dense, tough, and strong is suitable for concrete aggregate, riprap, and armor stone.

(d) **Sandstone** is sedimentary rock composed of small (0.25-6.0 mm) particles cemented together. Strength and durability of sandstone varies greatly depending on the cementing material. Rock cemented by silica or calcite is suitable for use as crushed and broken stone, whereas rock cemented with clay or iron oxide is inadequate for most applications. Sandstone is more porous than granite and basalt. Other less common rocks may be available for use in coastal construction, and many types have attributes necessary for use as armor stone and riprap. Moffatt and Nichol (1983) and CIRIA/CUR (1991) describe several additional rock types.

(2) Specific weight. Most coastal applications of stone require that the stones remain stable and stationary under all imposed wave and current forces. For structures in which the armor layer stones are not bound together by concrete or asphalt, stability is achieved through the relatively high specific weight of stone, assisted to some degree by the friction and mechanical interlocking that occurs between adjacent stones. Table VI-4-3 includes typical ranges of specific weight for common stone. Stones with high specific weight are best for primary layer armor units, but less dense stones can be used successfully. Specific weight is not as important for core material and underlayer stones, which are held in place by the primary armor layer. Design methods used to calculate stable armor stone weight depend on stone specific weight. Therefore, once the design is complete and stone specific weight has been specified, it is important to ensure stones used in the project meet or exceed the assumed specific weight used in design. Armor stones are usually purchased by weight, whereas core and secondary layer stones may be specified according to volume.

(3) Stone size and distribution. Quarries produce crushed and broken stone in sizes ranging from small gravel to huge blocks that cannot be handled and transported without special equipment. A rough estimate of stone size for a somewhat round stone is given as the diameter of an equivalent-volume sphere, i.e.,

$$D_s = 1.24 (\frac{W_s}{\gamma_s})^{1/3}$$

(VI-4-4)

where

W_s = stone weight

γ_s = stone specific weight in compatible units

Quarry output can be categorized according to median stone diameter and size distribution about the median. Categories of stone based on size and gradation include the following:

(a) **Armor stones** are selected by weight and density to resist wave loads. Ideally, all armor stones are blocky in shape and nearly uniform in size. The largest stone dimension on an individual stone should be no more than three times the shortest dimension.

(b) **Underlayer stones** are smaller stones randomly placed in a layer to support the primary armor layer. The size distribution of underlayer stone can be reasonably wide, provided the smallest stones in the distribution are still too large to pass through voids in the covering layer of larger stones.

(c) **Quarry-run** or quarry-waste materials are often used for cores of rubble-mound breakwaters and jetties. Generally the material should be sound and reasonably well-graded with no more than 10 percent fines. Smaller median sizes and wider distributions produce less porous structures.

(d) **Riprap** is comprised of heavy irregular stone fragments having a fairly wide size distribution. Riprap is used to protect slopes from erosion in less severe wave conditions. Riprap is also used in emergency repairs because sufficient quantities are usually readily available.

(e) **Bedding and filter layer stones** are typically smaller stones with narrow gradations. These layers prevent piping loss of underlying soils.

The above stone classifications are general. Specific guidance on median sizes and allowable size distributions for stone used in coastal structures is given in Part VI-5-2, "Wave/Structure Interactions" and Part VI-7, "Design of Specific Project Elements."

(4) Stone shape. Stone shape is an important factor in stability of armor stones. Angular, blocky stones are preferred for armor layers because they wedge and interlock well with adjacent stones when placed randomly, they can be placed on steeper slopes, and they provide a more porous armor layer that more effectively dissipates wave energy. Well-rounded armor stones are less stable, cannot be placed on steep slopes, and are more difficult to handle than angular stones. In addition, dislodged round stones will tend to roll downslope to the structure toe, whereas angular stones are more likely to find a new resting place on the armor slope. Quarry-produced stones are typically angular, whereas stones from glacial deposits and alluvial sources are usually rounded. Stones mined from older coastal structures could have become more rounded from years of service and weathering. Examples of stone shape and classification are given in CIRIA/CUR (1991). Many examples exist of coastal structures constructed of closely fitted blocky stones that resemble the work of stone masons. Gaps between stones can be grouted to provide a more impervious structure; however, sufficient openings must be left in the armor layer to relieve hydrostatic uplift pressures. Underlayers also should have sufficient angularity to be stable on the slope during construction. Underlayer stone angularity helps lessen the discontinuity between armor and underlayer. Highly angular stones placed directly on geotextile fabric are more likely to puncture the fabric during placement or subsequent movement.

(5) Durability. Stone durability is a qualitative measure of the stone's ability to retain its physical and mechanical properties throughout its service in an engineering project. Stone durability is related to properties of the basic rock from which the stones were produced (texture, structure, mineral composition, etc.), method of quarrying (blasting or cutting), handling of the stone prior to final placement, environmental conditions to which the stone is exposed, and loads applied to the stone (Magoon and Baird 1991). Generally, stone that is dense or fine-textured, hard, and tough is the most durable. Durability of stone placed in a coastal structure is a very important design consideration. However, stone durability is not well understood, and best durability estimates for stones from a particular quarry may come from past performance of stone from the same quarry that was placed in similar environments. Stone degradation by cracking or chipping reduces the average weight and angularity of armor stone resulting in a less stable armor layer. Stones that are expected to degrade rapidly lead to higher maintenance costs and may necessitate initial overdesign of armor stone size and placement on milder slopes. Economics may dictate using higher-quality stone from a distant site if local stone is not sufficiently durable. Useful information on stone durability experience in the United States was presented at the specialty conference *Durability of Stone for Rubble Mound Breakwaters*

(Magoon and Baird 1991). Papers in this conference covered theoretical and laboratory analysis of stone durability, engineering and design practices, quarry and construction topics, and case histories of stone durability. In one of the conference papers, Lutton (1991) gave the relative stone durability rankings shown on Table VI-4-4 for use in preliminary planning. Lutton also presented "approximate" criteria for evaluating stone durability shown on Table VI-4-5 (also given in Department of the Army (1990)). Descriptions of various tests used to quantify durability characteristics of stone are beyond the scope of the manual. See CIRIA/CUR (1991), Department of the Army (1990), Latham (1991), and Lienhart (1991) for information on these testing procedures. These sources also cite applicable testing standards of the American Society for Testing and Materials.

Table VI-4-4
Durability Ranking for Common Stone
Most Durable to Least Durable
1. Granite
2. Quartzite
3. Basalt
4. Limestone and Dolomite
5. Rhyolite and Dacite
6. Andesite
7. Sandstone
8. Breccia and Conglomerate

(6) Strength. Stone used in coastal projects is usually selected according to its specific weight, durability, and shape properties. Seldom are there any tensile or compressive strength requirements. Generally, stones must be sufficiently strong in compression to support the load of any overlying stone or structure without crushing. Table VI-4-3 gives compressive strength ranges for the listed stone. Generally, high density stone is also very strong in compression. Fittings such as ringbolts can be epoxied into holes drilled into stone, and usually the tensile stone strength is sufficient to withstand substantial loads on the fitting.

(7) Porosity and water absorption. **Stone porosity** is the volume of voids contained in a unit volume of stone. This term should not be confused with **bulk porosity** of a stone armor layer (which is related to the volume of voids between stones). **Water absorption** is the mass of water absorbed per unit of dry stone mass at atmospheric pressure, and it will be less than the absorption that would occur if all the voids of the stone were saturated. Values of stone porosity and water absorption are listed in Table VI-4-3. Stone water absorption is the single most important indicator of stone durability, particularly in applications where the stones undergo cyclic stresses caused by freeze/thaw cycles. Primary armor layer stones should have low values of water absorption to help ensure good weathering characteristics and less stone breakage. A limit of 1 percent absorption is considered reasonable (Department of the Army 1990).

Table VI-4-5
Approximate Criteria[1] for Evaluating Stone

Test	Approximate Criterion for Suitability
Petrography	Fresh, interlocking crystalline, with few pores, no clay minerals, and no soluble minerals
Bulk specific gravity (saturated, surface dry)	Greater than 2.60
Absorption	Less than 1.2 percent[2]
$MgSO_4$ soundness	Less than 2 percent loss in five cycles[1]
Glycol soundness	No deterioration except minor crumbs from surface
Abrasion	Less than 25 percent loss in 1,000 revolutions[2]
Freezing-thawing	Largely unaffected in 20 cycles
Wetting-drying	No major progressive cracking in 35 cycles
Field visual	Distinctions based on color, massiveness, and other visual characteristics
Field index	Distinctions based on scratch, ring, and other physical characteristics
Field drop test	No breakage or cracking
Field set-aside	No loss or cracking in 12-month exposure

[1] Criteria are broad generalizations useful for preliminary judgment only rather than being reflective of any official standard.
[2] Coarse aggregate sizes.

(8) Abrasion and soundness. Resistance to abrasion is an important stone property for materials handled in bulk such as core material, riprap, filter stone, etc. Weaker stones will break into smaller pieces as the materials are loaded into trucks, dumped, and rehandled onsite. This could result in changed size distributions by the time the stone is placed. Waterborne sand and cobbles can slowly wear away at weak armor stone, but this is not an overriding design concern. Dynamic armor layers that are reshaped by wave action should be constructed using abrasion-resistant stone. Stone soundness depends on the amount of fissures, fractures, laminations, and other discontinuities in the stone. Some stone fissures may be the result of blasting in the quarry, other weaknesses may develop with multiple handling and stockpiling of larger stones.

c. Quarrystone procurement and inspection guidelines. The following are suggested general guidelines for specifying and inspecting quarrystone for coastal projects. It will be necessary to supplement these guidelines on a case-by-case basis. Additional guidance is provided in EM 1110-2-2301 (Department of the Army 1994), CIRIA/CUR (1991), and Moffatt and Nichol (1983).

(1) Contractor bids should be reviewed to ensure bid items are not underpriced in anticipation of potential claims for extra payments.

(2) Any environmental, historic preservation, and biologic constraints on quarrying must be resolved by obtaining all relevant Federal, state, and local permits.

(3) Inspection visits to the quarry during production are needed to ensure adequate stone quality and gradation.

(4) Over-blasting, which may lead to unacceptable fracturing of armor stones, should be avoided.

(5) Well-trained inspectors familiar with blasting procedures, stone quality, and stone inspection techniques should be employed.

(6) A record of stone quality from known quarries should be maintained for reference. Quarries with records of producing unsatisfactory stone should be disqualified up front.

(7) Qualified personnel (e.g., a geologist) should identify specific areas of unacceptable in situ stone within the quarry and make the inspector aware of its location. This prevents the manufacture of potentially unsuitable stone.

(8) Stones representing the approved rock type in several different weights should be set aside and clearly marked for visual reference by the inspector and contractor.

(9) Stones should be spread out in the quarry for inspection prior to loading for transport. Armor stones should be rotated to inspect all sides.

(10) Weights of delivered stone should be checked periodically to ensure contract compliance, and an adequate supply of stone across the specified gradation should be maintained at the construction site.

d. Placement considerations for stone. The success of any coastal project built using stone depends critically on careful stone placement conforming to design specifications. Structures in which stones are carelessly placed will inevitably suffer damage at loads below design levels. The following stone placement guidelines (condensed from Moffatt and Nichol (1983)) are based on Corps of Engineers' experience in building rubble structures. These guidelines are intended to be general in nature with the recognition that Corps Districts and other entities may prefer their own specifications based on past experience and local knowledge.

(1) General placement considerations. On slopes, stone placement should begin at the toe and proceed upslope to produce a layer with maximum interlocking of stones and minimum voids. Larger stones that are individually placed should be oriented so the longest axis is approximately perpendicular to the structure slope. Armor stones should be "seated" on the underlayer stones to avoid slipping, rocking, or displacement under wave action or weight of overlying stones. Some settlement of the armor layer is expected, but ideally this will be a tightening of the matrix without significant lateral stone movements. Controlled stone placement provides improved armor layer stability, but it depends on skilled and experienced equipment operators and personnel. Typical extreme tolerances for rubble slopes are $\forall$30 cm (12 in.) from the design finished surface for underwater placement, and $\forall$15 cm (6 in.) for above-water portions. Underlayer and bedding layer tolerances may be as tight as $\forall$8 cm (3 in.), whereas up to $\forall$45 cm (18 in.) may be allowed for large armor stones. Rubble-mound structures exposed to wave action during construction should be completed and armored in short sections to minimize

damage risk from storms. Structures built through the surf zone may require stone blankets placed in advance of construction to reduce scour effects. Toe protection armor should be evenly distributed over the area with a minimum percentage of voids.

(2) Filters, bedding, and core materials. Stone used for rubble-mound cores, filter layers, and bedding layers should be handled and placed in a manner that minimizes segregation of the material size distribution. Material placed by clamshell, dragline, or similar equipment should not be dropped distances greater than 0.6 m (2 ft) above the bottom or previously placed stone. Self-unloading vessels like bottom dump scows (when permitted) should proceed along lines directly over the final dumping location and parallel to the structure center line. Placing bedding material over soft and organic bottom materials should force the soft material outward toward the edges of the bedding layer. When finished, filter and bedding layers should be free of mounds and windrows and coverage should be complete.

(3) Underlayer stone. Underlayer stone should be placed to full underlayer thickness in a manner that does not displace underlying materials or soil as construction progresses from the toe up the slope. The goal is to achieve an even distribution of the graded material with minimum voids in the underlayer. For smaller structures like revetments, unsegregated stone may be lowered in buckets and placed directly on the underlying material. Placing stone in any manner that results in stone segregation is not permitted. Drop heights for underlayer stone generally cannot exceed 0.6 m (2 ft).

(4) Armor layer stone. Armor layer stone can be placed uniformly, randomly, and by a special placement method.

(a) **Uniform placement** is used only for cut or dressed stones that are uniform in size and shape. Uniform stones are placed in an orderly pattern or arrangement in which the stones are closely spaced. Such arrangements make it more difficult for individual stones to be dislodged, but it also provides a less permeable structure with more runup and overtopping. This is the most expensive method of armor placement. Figure VI-4-1 illustrates uniform placement.

(b) **Random placement** covers a range of placement techniques from careful placement of individual angular quarrystones in a random pattern to underwater dumping of stones from barges. In the case of armor stones, significant variations in stability are likely to occur between underwater and above-water placement even when placement is by crane. Furthermore, the degree of armor interlocking achieved varies between crane operators, and even between structures constructed by the same crane operator. Figure VI-4-2 illustrates random placement. Placing individual armor stones should not displace underlayer stones and should not result in any armor damage other than minor chipping. Stone armor layers are at least two stones in thickness, and the layer should be constructed to this thickness as armoring progresses up the slope from the toe. This provides better interlocking than placing first one layer of stone and then covering it with a second layer. Placed armor stone should be stable, keyed, and interlocked with neighboring stones. "*Floater*" stones having minimal contact or not wedged against adjacent stones are more likely to be dislodged during storms. During construction, the crane operator should be able to select the best sized stone for a particular position from a number of armor stones stockpiled nearby. Smaller stones in the allowed size distribution should be used to fill

Figure VI-4-1. Uniform placement

Figure VI-4-2. Random placement

gaps between larger stones. In this way skilled operators are able to build "*tight*" armor layers. Equipment used for placing armor stones should be capable of positioning the stones to their final position before release (even at the toe), and the crane should be able to pick up and reposition stones after initial placement. Dropping stones more than 30 cm (1 ft) or pushing stones downslope should not be permitted. Final shaping of the armor layer slope to design grade should be achieved during stone placement.

(c) **Selective placement** is used by some Corps of Engineer field offices to increase structure stability. Selective placement is the careful selection and placement of individual armor stones to achieve a higher degree of interlocking. Although careful selective placement increases armor layer stability, the variation expected between projects does not warrant increasing the values of stability coefficients. In some respects selective placement is simply carefully constructed random placement. Figure VI-4-3 illustrates selective placement.

Figure VI-4-3. Selective placement

(d) **Special placement** applies only to parallelepiped-shaped stones, and this method of placement requires special efforts to align the longest axis of parallelepiped-shaped stones perpendicular to the structure slope. Special placement also requires careful supervision during construction with clear communication to the contractor about proper placement procedures. If feasible, construction supervisors with previous special placement experience should be employed. Special placement requires more time for selection, handling, and placement of the armor, along with increased costs of construction. Figure VI-4-4 illustrates special placement. Construction techniques for special placement have been suggested to supplement the recommendations given above for random placement. The lowest tier of armor stones should be keyed into the seafloor or bedding layer. Subsequent tiers should be placed in the saddle points of the next lower tier. Construction should proceed upslope and diagonally toward the crane operator. Spotters should be used to help direct placement and ensure grade line is maintained. Each stone should be oriented so the heavier end of the parallelepiped-shaped stone is closer to the underlayer, and stones should be keyed and fitted so there are at least three points of contact with adjacent stones. All capstone should be placed closely together. The top tier of armor stones on the seaward side should extend slightly above the level of the capstone to protect the cap from wave forces, whereas on the lee side the top tier should be slightly lower than the capstone. No stone should protrude out of the armor face more than one fifth of its major dimension. This is particularly important for single layer construction. Armor gradation should be fairly uniform, and stone on the landward face of breakwaters should not be reduced in size because wave transmission through permeable structures may dislodge leeside armor stones. In general, turbid water conditions do not allow special placement below the water level. Stones placed on underwater portions of the structure must be placed by "feel," and this results in a more random placement. Stones must be carefully fitted at the transition between random and special placement (around the low water level). In addition, care must be taken with special placement around the waterline because damage by breaking waves is more likely to occur in this region.

Figure VI-4-4. Special placement

(5) Riprap. Placement of riprap is less precise than armor stone, but the basic objectives are similar. Placement should not disturb underlying materials or damage geotextile fabric, and dumping should not segregate the riprap distribution. Dumping into chutes is likely to produce unacceptable segregation, and this practice should not be allowed. Riprap placement should be to full layer thickness in one operation; placing in multiple layers should not be permitted. After placement the riprap gradation should be similar throughout the structure with no obvious weak spots, with even distribution of larger stones, and with a minimum of voids. Rearrangement of individual stones with equipment or by hand may be needed to provide a reasonable gradation of stone sizes or to reinforce layer weaknesses. Pushing riprap up or down the slope is not allowed because it segregates the material and may damage the underlayer. Chink stones should be forced into voids in the riprap layer by rodding, spading, or similar methods.

e. Environmental effects on stone. Stone selected for use in coastal structures is very durable and is little affected by the natural environment.

(1) Wave action. Hydrodynamic forces caused by wave action on stone structures generally do not damage individual stones. However, waves which cause stone movement and

impacts between stones can lead to chipping and breakage. Waves also carry abrasive particles that can deteriorate weak stone over long time periods.

(2) Temperature and fire. Stone expands and contracts with temperature change, but most stone has reasonable tolerance to normal environmental temperature changes. Stone will be damaged to some degree by high temperatures caused by fire, and granite is particularly vulnerable to cracking and spalling caused by unequal expansion of differentially heated stone. This is due to its irregular crystalline structure and mineral composition. At temperatures greater than 100EC limestones start to decompose. Sandstones and other sedimentary stone will tend to crack along lamination planes after an extreme heating and cooling cycle.

(3) Freezing and thawing. Water that freezes in stone cracks produces stresses that may lead to stone breakage or spalling after a number of cycles. This problem increases with the porosity of the stone.

(4) Chemical attack. Calcareous stones are subject to decomposition by acids that may be formed by the combination of moisture and naturally occurring gases such as sulfur dioxide. This may cause disintegration of sandstones, which are cemented by calcium carbonate (Moffatt and Nichol 1983).

VI-4-4. Portland Cement Concrete and Bituminous Concrete. The sections below are intended to give a brief overview of portland cement concrete, and to a much lesser extent bituminous concrete, with emphasis on those characteristics important to coastal projects. Following common usage, the term "concrete" will be used to denote portland cement concrete, and "asphalt" will be used to denote bituminous concrete. Additional information is available in any of the literally hundreds of textbooks and design manuals that cover nearly all aspects of concrete and asphalt and their use as a construction material.

  a. Use of concrete and asphalt in coastal construction.

(1) Concrete. Concrete is one of the most common and adaptable materials used in coastal construction. Suitable aggregates and sand for mixing concrete are usually available near coastal project sites, and the widespread use of concrete in conventional land-based construction usually assures a nearby source for cement and steel reinforcement. Concrete components of coastal projects can consist of: (a) huge cast-in-place gravity structures, such as re-curved seawalls and roadways; (b) large components that are cast and then moved into position, such as caissons that are floated into position and sunk; (c) smaller components that are assembled into a larger coastal structure, such as armor layers constructed of concrete armor units or revetment blocks; and (e) prestressed beams, columns, and piles. Some of the more important coastal applications of concrete include the following:

(a) Seawalls, Revetments, Bulkheads. Massive cast-in-place concrete seawalls have survived many decades with need of only minor repair. Solid vertical-faced, recurved, or stepped concrete seawalls provide excellent protection of upland property from severe wave action. Specially shaped concrete blocks can be placed as an armor layer on sloping revetments. The interlocking block layer can tolerate minor movement without damage. Poured concrete cover layers can only be used for above-water revetments or when the slope has been dewatered.

Bulkheads can be constructed in numerous configurations using poured concrete or concrete sheet-piles.

(b) Jetties and Breakwaters. Concrete can be used as a grout in rubble-mound structures to reduce permeability or as a binder to hold stones together. Concrete is often used to construct rib caps for jetties. In milder wave climates, cellular jetties and breakwaters can be constructed of concrete, filled with earth or rocks, and capped with concrete. Weir sections in jetties can be constructed of prestressed concrete sheet piles.

(c) Groins. Groins can be constructed using prestressed concrete sheet piles or keyed kingpiles supporting concrete panels. A cast-in-place concrete cap ties the prestressed components together. Concrete-filled bags are also used as groins in low-wave climates.

(d) Caissons. In deeper water, concrete caissons are used as breakwaters and jetties. The caissons are placed either directly on the seafloor foundation or atop a rubble-mound base structure. Concrete is used to cap the filled caissons and to build additional structural features such as parapet walls or mooring and port facilities.

(e) Concrete Armor Units. Concrete is used to fabricate reinforced and unreinforced armor units of various sizes and shapes. Concrete armor units are used when suitably sized stone is unavailable or when the higher stability offered by many concrete armor units is needed to resist high wave loads.

(f) Piles. Reinforced or prestressed concrete piles are used for piers and wharfs and to support the foundations of other coastal structures such as concrete seawalls placed on soil with low bearing capacity. Concrete piles exceed 36 m (118 ft) in length, and typically the piles have round, square, octagonal, or hollow cross sections (Moffatt and Nichol 1983).

(g) Floating Structures. Concrete pontoons are used for floating pontoon bridges, floating breakwaters in short-wave environments, wharfs, boat slips, and floating dry docks. In these applications, the individual units often are linked together to form the structure.

(h) Other Applications. Concrete is used extensively in construction of conventional land-based facilities that may be part of a coastal project. This may include roadways, bridges, foundations, drainage ponds, pipelines, ocean outfalls, and discharge structures. Concrete is also used to encase wooden or steel structural components to provide protection against biological and corrosive agents in seawater.

(2) Asphalt. Bituminous concrete (referred to as *"asphalt"* because asphalt is a primary ingredient) can be used in coastal construction as a binder or filler to stabilize rubble mounds or soils, as a sealer to reduce or prevent water flow, or as a wearing surface that can be repaired easily. Asphalt is also used as a preservative treatment or coating to protect wood and metal. Typical project elements that may use asphalt include the following:

(a) **Dikes**. Although asphalt is not widely used in the United States for coastal protection structures, the Dutch have made good use of asphalt to protect the slopes of earthen dikes.

(b) **Jetties and Breakwaters**. In the United States asphalt is used only as a binder or filler for rubble-mound structures, or as part of the crest roadway.

(c) **Revetments**. Asphalt can be used to bind revetment riprap together to form a stronger, impermeable armor layer. When wave action is slight, an asphalt layer alone is adequate to protect the revetment slope.

(d) **Roadways and Slope Protection**. Bituminous concrete is used extensively for road construction and surfaces supporting vehicular traffic, such as surfaces on wharfs and quays. Asphalt may be suitable for protecting eroding mild slopes against erosion or for lining drainage ponds and ditches.

b. Physical and mechanical properties of concrete. Portland cement concrete exists in a semi-liquid state while being mixed, transported, and placed into forms. The concrete then undergoes irreversible hardening into a durable form having excellent compressive strength properties and resistance to the harsh coastal environment. The materials used to manufacture concrete are reasonably inexpensive and exist in relative abundance throughout the world. Concrete has two main ingredients: aggregates, which comprise between 60 and 80 percent of the concrete volume; and paste, which makes up most of the remaining volume. Coarse aggregates (e.g., gravel) have diameters greater than 6 mm, whereas fine aggregates (e.g., sand) have diameters usually much less than 6 mm. The relative proportions of fine and coarse aggregates help determine concrete properties. Cement paste is portland cement and water mixed in proportions that relate directly to strength. Entrained air or special additives may occupy up to 8 percent of the volume of a concrete mixture. Aggregates should be hard, nonporous materials; and the water used in mixing should be reasonably clean and nearly free of silts or harmful chemicals, such as sulfates and alkalies. Seawater can be used if no freshwater source is available and no steel reinforcement is used in the concrete. However, concrete made with seawater has less strength than equivalent concrete made with fresh water. Several important concrete properties are listed below. Generally, the design engineer will not specify concrete mixture proportions, additives, etc.; but instead will request certain properties and minimum strengths, and the contractor will provide an appropriate concrete. Field tests and tests on sample cylinders are used to verify concrete compliance with specifications.

(1) Strength. Concrete strength is based on its capability to withstand compressive stresses. Concrete has only minor resistance to tensile stress (ranging between 7 and 10 percent of the compressive strength), and any structural member subjected to bending moments must contain steel reinforcement to resist tensile stresses. Usually the steel reinforcement is designed with the assumption that the concrete will not carry any of the applied tensile load. What little tensile strength concrete has is useful in reducing cracks that form due to shrinkage. For a particular type of portland cement, concrete strength is largely determined by the ratio of water to cement (by weight) used in mixing. Generally, concrete strength increases as water content decreases. Variations in strength for a given water-to-cement mixture are caused by aggregate properties such as maximum size, grading, shape, and strength; by entrained air content; and by types of concrete additives (called admixtures). Compressive strength is best determined by testing sample cylinders of the proposed concrete mixture, and most experienced concrete suppliers can provide accurate test results for their standard concrete mixtures. Five types of

Portland cement are available for use in coastal projects. They have the following general characteristics, as specified by ASTM Standard C-150 (ASTM C-150 1994). Other, more exotic types of concrete are available for specialized purposes.

(a) **Type I**. Cement used for ordinary structural concrete for foundations, roads, and foundations not subject to freezing/thawing conditions or marine exposure. Type IA concrete specifies air entrainment for freezing conditions.

(b) **Type II**. Mild sulphate-resisting cement that can be used in nonfreezing marine environments. Not as durable as Type V cement in seawater. Air entrainment in Type IIA concrete helps it tolerate freezing conditions.

(c) **Type III**. This cement provides high strength earlier in the curing process. After 7 days, Type III concrete reaches the same strength as Type I after 28 days. Type III should <u>NOT</u> be used for marine construction.

(d) **Type IV**. Provides low heat of hydration for use in structures such as dams or where heat buildup is undesirable.

(e) **Type V**. This cement has the greatest resistance to sulfates and should be used in all marine environments. Air entrainment is essential in freezing environments.

Typical compressive strengths of the above five types of concrete are shown on Table VI-4-6. Use of the tabulated values should be limited to preliminary design calculations, because actual strengths will vary greatly with materials, proportions, and curing conditions.

Table VI-4-6 Typical Compressive Strengths of Concrete (from CRC (1976))					
	Compressive Strength in MPa (lb/in.^{2})				
ASTM Type	7 days	28 days	3 months	1 year	5 years
I	20.7 (3000)[1]	29.6 (4300)	35.2 (5100)	37.9 (5500)	39.3 (5700)
II	17.9 (2600)	29.0 (4200)	35.8 (5200)	40.7 (5900)	44.1 (6400)
III	26.2 (3800)	32.4 (4700)	35.2 (5100)	37.2 (5400)	37.9 (5500)
IV	10.3 (1500)	24.1 (3500)	35.8 (5200)	41.4 (6000)	44.8 (6500)
V	17.2 (2500)	28.3 (4100)	36.5 (5300)	42.1 (6100)	46.2 (6700)
[1] Note that values in parentheses are of lb/in.^{2}					

As mentioned, water content in concrete mixtures is an important factor in concrete compressive strength. Table VI-4-7 presents the American Concrete Institute's (ACI 1986) suggested maximum permissible water-to-cement ratios for concrete when strength data from field experience or trial mixes are unavailable.

Table VI-4-7
Concrete Compressive Strength for Different Water-Cement Ratios (from Mehta (1991))

Compressive Strength (at 28 days)		Water-Cement Ratio (by weight)	
(MPa)	(lb/in.^{2})	Non-Air-Entrained Concrete	Air-Entrained Concrete
41.4	6,000	0.41	------
34.5	5,000	0.48	0.40
27.6	4,000	0.57	0.48
20.7	3,000	0.68	0.59
13.8	2,000	0.82	0.74

Concrete modulus of elasticity E_c, used in calculating compressive stresses due to bending, can be estimated by the following empirical formula (ACI 1986)

$$E_c = 33 \left(w_c\right)^{3/2} \left(f_c\right)^{1/2} \qquad \text{English Units} \tag{VI-4-5}$$

where

w_c = specific weight of concrete in lb/ft^3 (must be in the range 90-150 lb/ft^3)

f_c = compressive strength of concrete in lb/in.^{2}

E_c = modulus of elasticity in lb/in.^{2}

A metric equivalent of this nonhomogeneous equation is

$$E_{cm} = 1392 \left(w_{cm}\right)^{3/2} \left(f_{cm}\right)^{1/2} \qquad \text{SI Units} \tag{VI-4-6}$$

where

w_{cm} = specific weight of concrete (must be in the range 14-24 kN/m^3)

f_{cm} = compressive strength of concrete in kPa

E_{cm} = modulus of elasticity in kPa

The modulus of elasticity for non-prestressed steel reinforcement is generally equal to 200,000 MPa (29,000,000 lb/sq in.). In addition to portland cement, there are compounds known as *blended hydraulic cements* that are covered by the Standard Specification for Blended Hydraulic Cements (ASTM C-595, 1997). Blended cements in commercial production in the United States are Type IS, which contains 30 to 65 percent rapidly cooled, finely pulverized, blast-furnace slag, and Type IP, which contains 15 to 30 percent fine pozzolan (Mehta 1991). Blended cements have lower heat of hydration (resistance to thermal cracking), lower rate of strength development, and better chemical resistance than ordinary portland cement. However,

Mehta (1991) noted that these characteristics can be obtained by using ground granulated blast-furnace slag or pozzolan as a mineral admixture into portland cement mixtures. Another attraction of blended cements is lower cost.

(2) Durability. Durability is the capability of concrete to withstand the deteriorating effects of the environment without loss of functionality. The primary factors causing deterioration of concrete are weathering, chemical action, and wear.

(a) Damage by weathering is caused mainly by freeze/thaw cycles and by restrained expansion and contraction due to wetting and drying and temperature changes. Weathering resistance is better for air-entrained concrete because the air pockets relieve pressures developed by expanding water. High-density concrete with low permeability also has better weathering resistance.

(b) Chemical reactions between alkalies in cement and mineral constituents of concrete aggregates can cause large-scale random cracking, excessive expansion, and formation of large cracks. Concrete is also affected by acids, sulfates, chlorides, salt brine at high temperatures, and hot distilled water. Steel reinforcement will rust if cracks in the covering concrete allow water and oxygen to reach the steel. Steel corrosion is particularly problematic if exposed to salt water, often causing spalling of the concrete and exposure of the reinforcement.

(c) Wearing away of concrete is caused primarily by flow cavitation, abrasion by particles in flowing water, traffic, wind blasting, and floating ice impacts (Department of the Interior 1975). Low pressure areas can develop on concrete exposed to high-velocity flows, leading to cavitation erosion of the concrete surface. Even the highest strength concretes can succumb to cavitation, and the only solution is to avoid abrupt transitions adjacent to rapid flows.

Portions of concrete structures in proximity to active sand transport are susceptible to wear by abrasion. Wind-blown sand can also erode concrete, but this process occurs slowly. Impacts by floating ice, vessels, debris, or even waves can chip concrete surfaces, possibly weakening the structure or exposing steel reinforcement. Wear resistance against abrasion and impacts increases proportionally with compressive strength. Also, wear resistance increases with curing age up to 28 days, and special precautions may be needed to protect concrete while it is curing. Some special situations might require protecting the concrete surface with a layer of a less erosive material or a material more capable of absorbing impacts. Wear resistance is not appreciably affected by hydrated lime or inert powdered admixtures up to 20 percent of the concrete volume (La Londe and Janes 1961). Mehta (1991) provides a thorough overview of factors causing deterioration of concrete in the marine environment.

(3) Consistency. Water content in wet concrete is a key factor in how well the concrete flows when being poured into formwork. Other factors include aggregate angularity, size, and texture. Increasing water content produces concrete that flows easier and is less likely to leave voids in the concrete. However, the ease of handling is offset by reduced compressive strength and the potential for aggregate segregation during placement. Concrete consistency is judged by its "*slump.*" Wet concrete is placed in a special container which is then upturned and removed, leaving a free-standing mass of concrete. Slump is the vertical distance between the original

height of the container and the final resting height of the concrete pile. Large slump values corre-spond to wetter concrete mixtures and larger aggregates. The American Concrete Association recommends the minimum and maximum slump values shown in Table VI-4-8.

Table VI-4-8
Recommended Concrete Slump for Various Types of Construction (from Mehta (1991))

| | Slump | | | |
| | Maximum | | Minimum | |
Types of Construction	(cm)	(in.)	(cm)	(in.)
Reinforced foundation walls and footings	7.6	3	2.5	1
Plain footings, caissons, and substructure walls	7.6	3	2.5	1
Beams and reinforced walls	10.2	4	2.5	1
Building columns	10.2	4	2.5	1
Pavements and slabs	7.6	3	2.5	1
Mass concrete	5.1	2	2.5	1

(4) Workability. Concrete "workability" is a qualitative term used to describe a mixture's capability to be handled, transported, placed, and properly finished without any harmful segregation of the aggregates. Concrete plasticity and uniformity have much influence on the functionality and appearance of the finished structural component. The type of structure and concrete placement requirements determine to some extent what workability is needed. Heavily reinforced structures require a mixture that will totally encase the reinforcement when worked using conventional techniques (e.g., vibrated). Workability is influenced by properties of the aggregate (grading, shape, proportions), amount of cement, entrained air, admixtures, and consistency (Moffatt and Nichol 1983). Practical field experience is paramount in judging concrete workability.

(5) Watertightness. As concrete cures, small voids are created by water evaporation and by shrinkage of the cement paste. Additional cavities are present in air-entrained concretes. These tiny voids may be sufficiently interlinked to allow water to pass through under capillary action or hydrostatic pressure. Concrete can be made to be virtually impervious by exercising care in mixing and placement. Aggregates must be nonporous and surrounded by impervious cement paste, and the mixture must have a low water-to-cement ratio and have no purposely entrained air. During placement, the concrete should be worked thoroughly to eliminate any pockets of entrapped air. Care should be taken to avoid contaminating the concrete with foreign matter such as dirt clods, dry leaves, or discarded trash. Concrete becomes more impervious if it cures at a slow rate. Normal concrete should be kept moist for at least 7 days, and high-early-strength concrete should be wetted for at least 3 days to assure watertightness (La Londe and Janes 1961). Admixtures are available to enhance structure watertightness.

(6) Specific weight. The specific weight of typical concrete mixtures varies between 22-25 kN/m^3 (140-160 lb/ft^3). Reinforced concrete has a nominal specific weight of 23.6 kN/m^3 (150 lb/ft^3), stone concrete is 22 kN/m^3 (140 lb/ft^3), and cinder concrete is about 15.7 kN/m^3 (100 lb/ft^3). Specific weights of concretes made with lightweight aggregates depend on the weight and proportion of lightweight aggregate. Table VI-4-9 (Department of the Interior 1975)

shows average specific weights of normal fresh concrete for given water content and aggregate size and specific gravity. English units are provided in the top table, and metric equivalents (direct conversion of English unit values) are provided on the lower table.

				Table VI-4-9					
			Average Unit Weight of Fresh Concrete (from Department of the Interior (1975))						
		Average Values			Aggregate Specific Gravity Unit Weight				
Maximum Aggregate Size	Air Content (%)	Water Content	Cement Content	2.55	2.60	2.65	2.70	2.75
Unit Weight in lb/ft^3								
0.75 (in.)	6.0	283 (lb/yd^3)	566 (lb/yd^3)	137	139	141	143	145
1.5 (in.)	4.5	245 (lb/yd^3)	490 (lb/yd^3)	141	143	146	148	150
3.0 (in.)	3.5	204 (lb/yd^3)	408 (lb/yd^3)	144	147	149	152	154
6.0 (in.)	3.0	164 (lb/yd^3)	282 (lb/yd^3)	147	149	152	154	157
Unit Weight in kN/m^3								
20 (mm)	6.0	1.65 (kN/m^3)	3.29 (kN/m^3)	21.5	21.8	22.2	22.5	22.8
40 (mm)	4.5	1.43 (kN/m^3)	2.85 (kN/m^3)	22.2	22.5	22.9	23.3	23.6
75 (mm)	3.5	1.19 (kN/m^3)	2.37 (kN/m^3)	22.6	23.1	23.4	23.9	24.2
150 (mm)	3.0	0.95 (kN/m^3)	1.64 (kN/m^3)	23.1	23.4	23.9	24.2	24.7

(7) Volume change. Concrete shrinks as it cures and hardens, and it also expands and contracts with temperature and moisture content. Generally, expansion is not too serious a problem because it induces compressive stresses. However, overall expansion must be considered for structural components like constrained slender beams and slabs, which could buckle unless some allowance is made for excessive expansion. Contraction of concrete is a more serious problem because of the material's low tolerance of tensile stresses. Contraction cracks form in the surface, allowing water to penetrate the concrete. If water between the cracks freezes, the cracks are enlarged and damage occurs. Steel reinforcement helps distribute the concrete contraction more uniformly, resulting in smaller cracks. Most shrinkage occurs as new concrete cures because of the large amount of water in the paste. Shrinkage increases with increases in initial water content and entrained air and with compressibility of the aggregates. A 1-percent increase in water quantity increases shrinkage by 2 percent (La Londe and Janes 1961). Shrinkage that occurs in average concrete while curing to complete dryness is about equivalent to the shrinkage that would occur due to a temperature drop of 56EC (100EF).

c. Physical and mechanical properties of asphalt. Combining asphalt cement with various types of aggregates in different proportions can produce a wide range of bituminous concretes exhibiting different characteristics. This versatility makes Aasphalt@ a useful construction material for coastal projects. Usually, the aggregates used in bituminous concrete are durable, so the physical properties of the mixture stem largely from the properties of the asphalt cement. Asphalt is not affected by most chemicals, with the notable exception of other petroleum-based products that can act as solvents. The flexibility of asphalt allows bituminous concrete to conform to uneven surfaces and to adjust to differential movements. Asphalt mixtures can be made to be porous or impervious when placed. Impervious mixtures are often used to line drainage ditches or to waterproof structures. Asphalt mixtures have both plastic and elastic properties that are mainly a function of temperature. Asphalt mixes must be designed to achieve project objectives at an economical cost. Some of the factors considered when blending asphalt and aggregates include the following:

(1) There must be sufficient quantities of asphalt cement to ensure mixture durability under design load conditions.

(2) Proper type and size distribution of aggregates is needed to ensure a bituminous concrete that can tolerate loads without excessive deformation.

(3) Sufficient voids in the mixture are necessary to allow for a slight amount of additional compaction without loss of stability and without loss of impermeability.

(4) Good workability of the initial heated mixture allows easy placement of the asphalt compound without segregation of the aggregates.

Moffatt and Nichol (1983) provide additional descriptions of bituminous concrete and asphaltic compounds. Specific design guidance for various types of asphalt mixtures can be found in publications from the Asphalt Institute or from local asphalt contractors.

d. Concrete construction practices. An important and essential reference for the design and construction of concrete structures is the American Concrete Institute's *Manual of Concrete Practice* (ACI 1986), which is revised annually. Standard ACI 318 in the ACI manual provides building code requirements, along with a detailed commentary on code provisions.

(1) Transport and placement. Concrete can be mixed onsite or batched offsite and transported to the site by a number of different means including revolving drum trucks, barges, rail cars, conveyor belts, and pipelines. In all cases the objective is deliver the mix to the site without significantly altering the concrete's water-cement ratio, slump, air content, and distribution of aggregates.

(a) Concrete placement should strive to achieve the same objectives as stated above for transportation, and handling equipment should be chosen accordingly. Transport and placing capacity must allow the concrete to be kept plastic and free of cold joints while it is being placed in forms. Horizontal layers should not exceed 0.6 m (2 ft) in depth; and for monolithic structures, successive layers should be placed while the underlayer can still be vibrated to join the layers together, thus avoiding cold joints. When possible, concrete should be placed directly into the forms with minimum lateral movement, as the lateral movement tends to segregate the aggregates. Placement on sloping surfaces should begin at the toe and proceed upslope.

(b) Placement techniques should avoid high-velocity discharge or long drops, both of which contribute to aggregate segregation. Requests for addition of water to assist concrete flow down the chute should not be routinely granted without an assessment of how the concrete strength will be affected. Water should not be added to concrete that retains good workability and can be properly consolidated in place. Concrete with slump in excess of specifications should be rejected by the supervising engineer. Contractors should provide handling equipment sufficient to place concrete of specified consistency.

(c) Consolidation by screeding and vibrating removes air bubbles that are entrapped during placement. Vibrating also provides a more uniform distribution of solids and water in the concrete. (However, it is possible to over-vibrate concrete, which could result in a less homogeneous mixture.)

(d) Floor surfaces are more durable if steel troweling machines are used to finish the concrete as it sets. Troweling produces a low-maintenance, dense surface layer free of surface voids. Good vibrating adjacent to forms helps assure reasonably smooth surface finishes for nonhorizontal surfaces.

(2) Curing and formwork removal.

(a) Proper curing is essential for concrete to reach its design strength. Rapid loss of moisture must be prevented because water is needed for cement hydration, and the temperature should be controlled to assure the concrete attains its mature strength. In above-freezing conditions, water can be ponded on horizontal surfaces, and structural members such as columns can be covered with wet burlap or kept under constant "misting" with water. Where ambient temperatures are below freezing, fresh concrete must be protected from freezing with insulating blankets and cured with steam or electrically heated forms or infrared lamps (Mehta 1991). An alternate method is to raise the temperature of the mixture by heating the water and aggregates. Seawater should not be used to cure reinforced concrete; however, high-strength concrete can be exposed to seawater after 3 days of curing because by this time it is considered impermeable.

(b) Formwork should not be removed until the concrete has cured sufficiently to support the dead load and any live load that may be imposed during subsequent construction. To prevent damage to the surface, formwork should have been properly oiled or treated, and the concrete should be hard before forms are removed. Form removal may expose warm concrete to chilly winds that may cause cracks to form as contraction occurs.

(3) Reinforcement cover thickness. *The Manual of Concrete Practice* (ACI 1986) specifies minimum cover thicknesses for conventional and prestressed concrete structural elements. For coastal and offshore structures, steel reinforcing bars should have a minimum concrete cover of 50 mm (2 in.) for portions submerged or exposed to the atmosphere. The cover should be increased to 65 mm (2.5 in.) for portions of the structure in the splash zone or exposed to salt spray. Submerged prestressed members should have 75 mm (3 in.) of minimum cover, and 90 mm (3.5 in.) of cover in the splash zone. Stirrups may have 13 mm (0.5 in.) less cover than the minimums for reinforcement.

(4) Joints and sealants. Most concrete structural elements contain construction joints to compensate for volume changes, to allow for construction sequence, or to serve some other design purpose.

(a) **Contraction Joints** are control joints used to control the amount of cracking that occurs during contraction of the concrete. They are most commonly used to subdivide large, thin members like slabs into smaller units. The intent is for cracking to occur at the joint which can later be sealed with a flexible sealant.

(b) **Expansion Joints** are placed between concrete members to allow for expansion, thus avoiding crushing, buckling, or warping of slender members. Expansion joints also serve to isolate adjacent members so loads are not transferred between structural components, allowing differential movement between members. The joint is typically a clear space between member cross sections with keyways or dowels sometimes used to prevent lateral displacements of components. The joint can be sealed with a flexible sealant.

(c) **Construction Joints** are predetermined discontinuities in concrete to facilitate construction sequence. They can resemble either expansion or contraction joints. When continuous structural integrity is required, reinforcement is carried through the joint and efforts are made to bond adjacent components at the joint. Sealing of joints, particularly contraction joints, may be necessary to prevent water from entering the concrete and causing deterioration. In other cases, joint sealing is necessary to help retain backfill soil, contain a liquid, or prevent ice from forming in the joint. A variety of sealants are available for use in joints and cracks. These include sand grout, epoxies, oil-based mastics, bituminous compounds, metallic materials, thermoplastics, and others. Moffatt and Nichol (1983) give an overview of common sealants and their properties.

(5) Repairs. Post-construction repairs may be needed to seal curing cracks that are excessively wide or to repair damage incurred during form removal. Cracks can be filled with an appropriate epoxy, and surface voids left by entrapped air can be filled using concrete grout or mortar. Repair of deteriorating concrete structures is covered in Part VI-8, "Monitoring, Maintenance, and Repair of Coastal Projects."

e. Concrete for armor units. A unique application of concrete in coastal structures is artificial armor units placed as protection on rubble-mound structures in lieu of stone. Concrete armor units are cast in a variety of sizes and shapes (see Part VI-2-3, "Main Types of Armor Units." Because concrete armor units are usually unreinforced, they become vulnerable to tension breakage above a certain size, depending on type of unit (slender versus bulky) and other parameters. Steel reinforcement has been used in the past, but the cost of reinforcement is high. Large, slender, unreinforced concrete armor units have low reserve strength in tension beyond simply supporting their own self-weight, and any movement of placed units could cause breakage. Consequently, engineers strive to design the armor unit layer for no movement of the units after placement. High-strength concrete helps lower the risk of breakage, and methods are now available to access stress probabilities in terms of concrete strength (see Part VI-5-3-c, "Structural Integrity of Concrete Armor Units." Special precautions should be followed to ensure uniform, high-quality concrete is used in casting concrete armor units. The concrete should be properly vibrated to remove all voids which could substantially weaken the armor unit. Units must be cured properly before placing them on the armor layer. In particular, it is important to avoid formation of thermal cracks due to rapid curing. Any means to reduce high temperature gradients in the curing concrete will help reduce crack formation. Special equipment will be needed to handle, transport, and place the armor units.

    f. Environmental effects on concrete and asphalt.

    (1) Pollutants. Some pollutants may contain chemicals (sulfates and acids) in sufficient quantities to damage concrete. The chemicals must be in solution form to do harm. Naturally occurring sulfates may be present in soil or dissolved in groundwater adjacent to concrete structures. In general, concrete is not significantly affected by most pollutants found in the coastal zone. Bituminous asphalt is resistant to most chemicals with the exception of petroleum solvents, which can cause deterioration of the asphalt.

    (2) Water penetration. Water itself is not harmful to concrete, but it may carry sulfates or acids in solution that can have a detrimental effect. Salt water that penetrates to steel reinforcement causes the steel to corrode, weakening the structure. Also, the products of corrosion can expand and cause spalling. Periodic wetting and drying may cause cracks to form. Asphalt is usually considered to be impermeable and resistant to water penetration.

    (3) Waves and currents. Concrete structures designed under correct loading assumptions will not be affected by waves and currents. High flow velocities at abrupt transitions may cause flow cavitation, which can lead to deterioration of the concrete surface. Asphalt is not affected by waves and currents unless the waves are large enough to damage the asphalt layer directly or carry floating bodies that impact the asphalt.

    (4) Ice and temperature changes. Ice can damage concrete structures in two ways. Water that freezes in cracks will enlarge the cracks and eventually damage the concrete through spalling or outright fracturing. Impacts by floating ice or stresses induced by ice riding up on a structure can also damage concrete. Temperature changes cause expansion and contraction, but these effects are countered by proper design. Temperature control is a critical aspect of the curing process in order to obtain specified strength. Exposure of asphalt to temperatures above 163EC causes solvents to dissipate, resulting in deterioration. Ice does not cause problems with asphalt other than damage caused by floating ice. However, cold temperatures cause asphalt to become brittle.

    (5) Marine organisms. Concrete is one of the toughest materials used in coastal construction and has no food value for marine organisms. Neither marine organisms nor larger land animals have any effect on good concrete made with strong aggregates. Barnacles and marine plants can attach to concrete surfaces, but they have little effect outside of causing additional drag resistance to flows. The softness of bituminous asphalt makes it susceptible to damage from crustaceous organisms; but because asphalt is petroleum-based, it has no attraction for other animals.

    (6) Abrasion. Hard particles carried by wind and water can wear down concrete surfaces over time, but this process is so slow that it usually is inconsequential over the life of the structure. Bituminous asphalt resists wearing by small waterborne particles quite well; however, impacts by larger particles can cause damage.

(7) Seismic activity. Accelerations from earthquakes can be detrimental to concrete structures through direct inducement of stresses in the structure or through differential settlement caused by foundation damage. The *Manual of Concrete Practice* (ACI 1986) describes special provisions for the design of earthquake- resistant structures. The plasticity of asphalt allows it to flex and deform with earthquake motions rather than resisting through a rigid structure. This flexibility helps reduce damage to the asphalt project elements.

(8) Other effects. Concrete is not affected by sunlight, and it has good resistance to fire or extremely high temperatures. Human activity has little effect on the performance of concrete structures, and concrete is difficult to damage through vandalism. However, coastal structures provide ample canvas for graffiti artists, which may cause noticeable visual pollution. Fire is a real hazard for bituminous asphalt because it is petroleum-based. Usually there is not enough solvent in the asphalt binder to sustain fire; but in the presence of other combustible materials, asphalt will burn.

VI-4-5. Steel and Other Metals.

a. Use of metal in coastal construction. Many components of coastal projects are well-suited for fabrication using common metals. Construction requirements such as strength, availability, ease of construction, durability, and adaptability can often be met with metals such as steel, aluminum, copper, or various metal alloys. Consideration should be given to the economic benefits of yard fabrication of structural components as opposed to site fabrication.

(1) Steel.

(a) Steel has been used in marine construction since the late 1800s. Perhaps the most common use of steel is as concrete reinforcement in such structures as caissons, seawalls and bulkheads, paved working surfaces and roadways, and pretensioned piles. Steel reinforcement should be placed to provide adequate concrete coverage to protect it from the corrosive effects of water. (see Part VI-4-4-d-(3), "Reinforcement Cover Thickness").

(b) Another important marine application of steel is pilings. Pipe piles and H-piles are used to support foundations or as supports for coastal structures or fendering systems. Steel H-piles can be driven into hard strata or through soils containing obstructions such as rocks. Site-welding steel H-piles end to end allows deep penetration through soft soils down to bedrock. Pilings are sometimes encased in concrete to prevent corrosion. Steel sheet piles are used extensively in port and harbor facilities to construct seawalls and wharfs.

(c) Conventional steel framing is used for building construction in the marine environment in the same manner as inland; however, more attention is given to preventing corrosion. Steel is also used for fabricating specialty components such as fendering and mooring system components, structural framework, supports for navigation aids, chains, flow control gates, and storm surge barriers. Steel wire is used to construct wire cages for gabions and chain-link fencing. Steel bolts, plates, and fasteners are used to connect structure components of similar or different materials. Steel rods are used as bracing and as vertical retaining wall tiebacks. Steel wire rope is used to lash batter piles and for other purposes.

(d) In special applications, high-strength steel may be specified; but the cost will be greater, and there may be restrictions regarding onsite modification with cutting torches, which could weaken the steel. Drilling holes will also be more difficult. Stainless steel is not used in great quantities, but it is an important material for components that are openly exposed to salt water and must remain free of corrosion. Cast iron is used to fabricate special shapes such as mooring bollards.

(2) Aluminum alloys. Many aluminum alloys are resistant to corrosion, which makes them ideally suited for low-maintenance applications in exposed coastal regions. These applications include door and window frames in buildings, building roofing and siding, tread plates, decking, catwalks, railings, support framework, and architectural trim such as gutters and downpipes, facia, etc. Aluminum fasteners can be used in corrosive environments, but consideration must be given to aluminum's reduced strength capacity and fatigue characteristics when compared to steel. Aluminum alloys are also used as electrical conductors and in constructing specialty components, where light weight is an important criterion.

(3) Other metals and metal alloys. Other metals and alloys are used to a much lesser extent in coastal construction. Copper is used in electrical wiring and buses and for piping and sheathing. Because of its relatively noncorrosive nature, brass (copper and zinc) is used for hardware fittings and fasteners and often as survey monument marker plates. More exotic metals and alloys, such as monel (nickel-copper) and titanium are found only in project components that require the unique capabilities offered by these materials.

b. Physical and mechanical properties of metals. Much of the versatility of metals stems from their strength, durability, "workability," and competitive cost. Metals have a crystalline structure that results in a very homogeneous material with consistent strength properties throughout. During manufacture, metal can be formed into a variety of shapes and sizes, and a wide assortment of metal "stock" is available for design. Depending on the metal, its manufacture, and its shape, metal properties vary from rigid to flexible, ductile to brittle, soft to hard, and weak to strong. Undoubtedly, metals offer numerous options as design materials.

(1) Steel. Steel is the most commonly used metal, and it is available in a wide assortment of grades and sizes. Steel is an ideal material for construction because is can be easily joined, has high tensile strength, good ductility, and good toughness. Physical and mechanical properties of steel are well-documented in material handbooks, standards from the ASTM, and the *Manual of Steel Construction* (AISC 1980). Table VI-4-10, is an abbreviated list of steel specifications used in the United States along with typical applications for each grade of steel. Each ASTM specification sets out manufacturing guidelines such as steel ingredients and quantities, manufacturing processes, quality control, etc. Other countries may have similar steel specifications or have adopted those of the ASTM.

Table VI-4-10
Specifications and Applications for Steel Suitable for Marine Service (from Moffatt and Nichol
(1983))

ASTM Designation	Title of Standard	Application
A36-93a	Standard specification for structural steel.	Bridges, bulkheads, general structures.
A252-93	Standard specification for welded and seamless steel pipe piles.	Structures, forms for cast-in-place concrete piles.
A328-93	Standard specification for steel sheetpiling.	Sheetpiling, dock walls, and cofferdams.
A573-93	Standard specification for structural carbon steel plates of improved toughness.	Steel plates and sheetpiling.
A690-93	Standard specification for high-strength, low-alloy steel H-piles and sheetpiling for use in marine environments.	Dock walls, seawalls, bulkheads; providing 2 to 3 times greater resistance to seawater splash zone than ordinary carbon steel.
A709-93a	Standard specification for structural steel for bridges.	Carbon and high-strength, low-alloy steel plates and sheets.

(2) Aluminum alloys. Pure aluminum is soft and ductile and does not have sufficient strength for most commercial applications. Other metal elements are added to aluminum to create alloys having a variety of physical properties. Many alloys of aluminum have high corrosion resistance to marine atmosphere as well as good strength-to-weight ratios. Aluminum alloys are identified by numbers that are grouped together according to alloy components and manufacturing process.

(a) **1000 Series**. This series contains alloys that contain at least 99 percent aluminum. These alloys have high thermal and electrical conductivity, excellent corrosion resistance and workability, but they have the lowest structural strength. These alloys can only be hardened by cold working. Aluminum alloy 1350 is used for electrical wiring.

(b) **2000 Series**. This group contains alloys in which copper is the major alloying element. These alloys have less corrosion resistance than most other aluminum alloys.

(c) **3000 Series**. Alloys in which manganese is the major alloying element are in this series. These alloys generally cannot be heat-treated, but they can be hardened by cold working. Roofing and siding are usually 3004 aluminum.

(d) **4000 Series**. This group contains alloys in which silicon is the major alloying element. These alloys have a lower melting point, and thus they are used in welding and brazing wire to weld other aluminum alloys.

(e) **5000 Series**. This series contains alloys in which magnesium is the major alloying element. These alloys have moderate to high strength, good corrosion resistance in marine environments, and good welding characteristics.

(f) **6000 Series**. Aluminum alloys containing silicon and magnesium in about equal proportion comprise this group. These alloys are heat-treatable, have medium strength, and have good corrosion resistance. Windows, door frames, and lampposts are usually 6063 aluminum. Tread plate is usually heat-treated 6061 aluminum.

(g) **7000 Series**. This group contains alloys in which zinc is the major alloying element. These are heat-treatable alloys that have very high strength.

Aluminum alloys 5083, 5086, 5052, and 6061 are commonly used in marine environments. Although the 5000 series has best corrosion resistance, alloys from the 1000, 3000, and 6000 series are also used in coastal applications. Aluminum alloys can be used in the splash zone, but they are not recommended for continuous immersion. In addition to the numerical series specification, aluminum alloy designations also have a letter indicating the method of tempering.

(3) Other metals and metal alloys. Copper has high electrical and thermal conductivity, excellent corrosion resistance under normal atmospheric conditions, and it has good workability. Copper can be alloyed with other metals to improve strength, and to provide better corrosion and creep resistance. Well- known copper alloys include brass and monel. Best corrosion resistance in seawater comes from copper alloys that form thin corrosion films that protect the metal from further corrosion, even in flowing water. Metals and metal alloys other than those mentioned above are not commonly used in coastal engineering projects, but may appear in special applications.

(4) Galvanic reactions. When two dissimilar metals in electrical contact are placed in salt water, an electric potential is established and a process occurs that is referred to as "galvanic corrosion." The more galvanically active metal (anode) will corrode at a faster rate than it would by itself. The more noble metal of the pair (cathode) is protected from corrosion by the galvanic coupling. Table VI-4-11 lists common metals and alloys in a "*galvanic series*" for flowing seawater at ambient temperature. In the table, the most active metal is at the top and the entries are in decreasing order to the least active metal at the bottom. If two dissimilar metals must come in contract, several steps can be taken to reduce galvanic corrosion: (a) choose metals close together in Table VI-4-11; (b) electrically insulate the two metals at the point of contact; (c) coat the anode metal (higher in the galvanic series); and (d) place a more active third metal in electrical contact with the other two metals to provide cathodic protection via a sacrificial anode. Zinc is commonly used for sacrificial anodes. Periodically inspecting and replacing sacrificial anodes will extend the working life of components that have potential for galvanic corrosion.

c. Design values for structural metals. Industry adherence to uniform quality standards in the manufacture of metals and metal alloys has led to reliable design values. Table VI-4-12 presents recommended allowable tensile strengths and yield strengths for some grades of steel, aluminum, and other metals. Values for metals not listed in the table can usually be found in the ASTM Standards or in design manuals prepared by industry associations. Most designs utilize metal fabricated into standard cross sections, such as I-beams, H-beams, box sections, angles, pipe sections, etc. These "standard sections" have allowable design loads associated with particular loading conditions. For example, unbraced slender columns will fail in buckling before allowable compressive strength is exceeded. Design handbooks are available covering most of the standard structural cross sections. For structural steel, the *Manual of Steel Construction*, produced by the American Institute of Steel Construction (AISC (1980 or more recent edition)) is widely used by engineers in the United States.

Table VI-4-11

Galvanic Series in Flowing Seawater (2.4 TO 4.0 m/s) at Ambient Temperature (from Moffatt and Nichol (1983))

Magnesium

Zinc

Aluminum alloys

Calcium

Carbon steel

Cast iron

Austenitic nickel cast iron

Copper - nickel alloys

Ferritin and mortensitic stainless steel (passive)

Nickel copper alloys, 400, K-500

Austenitic stainless steels (passive)

Alloy 20

Ni - Cr - Mo alloy C

Titanium

Graphite

Platinum

Table VI-4-12
Tensile Stress Limits for Selected Metals and Alloys

Name or ASTM Code	Grade	Min. Yield Stress MPa (ksi)	Tensile Strength MPa (ksi)
Steel			
A36		250 (36)	400-550 (58-80)
A252	1	205 (30)	345 (50)
	2	240 (35)	414 (60)
	3	310 (45)	455 (66)
A328		270 (39)	485 (70)
A573	58	220 (32)	400-490 (58-71)
	65	240 (35)	450-530 (65-77)
	70	290 (42)	485-620 (70-90)
A690		345 (50)	485 (70)
A709	36	250 (36)	400-550 (58-80)
	50	345 (50)	450 (65)
	50W	345 (50)	485 (70)
	70W	485 (70)	620-760 (90-110)
	100-100W	620-690 (90-100)	690-895 (100-130)
Aluminum Alloys			
B209	3004-0	55 (8)	145-193 (21-28)
	5052-0	65.5 (9.5)	172-214 (25-31)
	5083-0	124 (18)	276-352 (40-51)
	5086-0	97 (14)	241-304 (35-44)
	6061-0	83 (12)	138 (20)
B241	6063-0	(18)	(19)
Copper Alloys			
Copper, B152, B124, B133	Annealed	69 (10)	221 (32)
	Cold-drawn	276 (40)	310 (45)
Yellow Brass, B36, B134, B135	Annealed	124 (18)	331 (48)
	Cold-drawn	379 (55)	483 (70)
Naval Brass, B21	Annealed	152 (22)	386 (56)
	Cold-drawn	276 (40)	448 (65)
Alum. Bronze, B169, B124, B150	Annealed	172 (25)	483 (70)
	Hard	448 (65)	724 (105)
Nickel Alloys			
Cast Nickel	As cast	172 (25)	393 (57)
A Nickel, B160, B161, B162	Annealed	138 (20)	483 (70)
	Cold-drawn	483 (70)	655 (95)
K Monel	Annealed	310 (45)	690 (100)
	Spring	965 (140)	1034 (150)

d. Metal protective treatments. Corrosion is the primary cause of steel deterioration in the coastal zone. Other metals suffer corrosion to a lesser extent. If steel corrosion is allowed to continue, structural steel will eventually weaken to a point where allowable stresses will be exceeded and the structure may fail. Exposed steel must be covered with a protective coating; and as the Navy is well aware, painting (and regular repainting) will help fight steel corrosion by seawater and salt air. In extreme cases, a thick coating of tar may provide adequate protection. Other protection methods include encasing the steel in concrete, such as in pier pilings; protecting the steel with a plastic coating, such as used for chain-link fencing; and applying a protective layer such as galvanizing or chrome plating. Abrasion by sand particles or vessel contact will strip away protective metal coatings with time, necessitating field maintenance where feasible or replacement in situations where the protective coating is factory-applied and cannot be mended in the field. Aluminum alloys can be "anodized" to provide greater corrosion resistance and some degree of galvanic protection when placed in contact with other metals. Copper, brass, and some other metals and metal alloys provide their own corrosion protection by forming a thin layer of corroded material. For example, brass will tarnish to a dull finish without losing any structural functionality.

e. Metal fasteners and connections. Metal structural components are usually fitted together in the factory or at the project site using a variety of methods including machine screws, bolts and bolted connector plates, electric arc welding, and brazing.

(1) Rivets and Bolts.

(a) Advances in welding and high-strength bolts have lessened the use of riveting in steel structure connections during the construction phase. However, riveting is an important option during the manufacturing of structural components. The ASTM Standards (ASTM A-502 (1994)) provide guidelines for the use of steel structural rivets in fabrication. Riveting is more common for joining aluminum structural components, particularly for heat-treated alloys, which could be weakened by welding. Aluminum riveting requires less skill than welding and it is a relatively fast method for making connections. Rivets are particularly useful for connecting dissimilar metals.

(b) Bolts are typically used in construction when welding is impractical or when the connections may need to be disassembled for maintenance or replacement. Less skill is needed to make bolted connections than for welding. Common practice is to oversize bolts exposed to corrosive marine environments to compensate for excessive corrosion. Bolts are made of carbon structural steel or high strength steel. Resistance of high- strength bolts to atmospheric corrosion is about twice that of carbon structural steel, which is an advantage in coastal construction. Allowable stresses for standard and high-strength bolts are given in the *Manual of Steel Construction* (AISC 1980) or other appropriate design manuals.

(2) Welding. Factory and onsite welding of metals is usually accomplished using either gas welding or arc welding. Oxyacetylene is suitable for welding carbon and alloy steel, cast iron, copper, nickel, aluminum, and zinc alloys. Hydrogen, methane, or propane gases are used to weld metal with lower melting temperatures such as aluminum, magnesium, zinc, and lead. Welding flux is needed for oxyacetylene welding of stainless steel, cast iron, and most nonferrous metals. Arc welding provides strong connections for structural carbon steel; and for

critical connections, welding quality can be checked using x-ray techniques. Underwater arc welding is feasible, but not recommended except in emergency situations or for temporary repairs. Underwater welds in mild steel develop about 80 percent of the strength of comparable dry welds, but they are only half as ductile. Underwater welding surfaces must be free of marine growth, rust, mill scale, and paint to assure high-quality welds.

f. Environmental effects on metal. Metals and metal alloys are very durable and stand up reasonably well to the coastal environment. Most metals are unaffected by ultraviolet radiation; however, sunlight may contribute to stress corrosion cracking in stainless steel. Toughness of carbon steel and some steel alloys decreases as temperature decreases; and in regions of extreme temperature change, design allowances for expansion and contraction must be made. Although metal is fire-resistant, yield strength is reduced as temperature rises, which may result in structure collapse.

(1) Abrasion. Metals can suffer abrasion by sand particles carried by water or wind. Although loss of metal through abrasion will be minor, a more critical problem is stripping of protective coatings or corrosion films down to bare metal, which then leads to rapid corrosion.

(2) Corrosion. Metal corrosion due to freshwater and saltwater exposure is a primary concern in coastal designs. Corrosiveness of water is dependent on water acidity, electrical conductivity, and most importantly, oxygen content. Fresh water polluted with acidic compounds may be more corrosive to carbon steel than seawater. Salt ions in seawater cause localized destruction of the protective oxide films (e.g., rust), thus reducing the corrosion resistance of the metal. Corrosion rates for carbon steel exposed to the air at the shoreline are 10 times greater than rates at locations 500 m inland from the shoreline. Pilings located in the splash zone can achieve two to three times the corrosion resistance of carbon steel if they are fabricated of high-copper-bearing, high-strength, low-alloy steel conforming to ASTM Standards (ASTM A-690 (1994)).

(3) Marine fouling. Fouling of immersed metal by marine plants and animals increases corrosion rates in some metals, such as carbon steel, stainless steel, and aluminum. In addition, marine growth may increase flow resistance, which induces greater loads on the structure. Copper and copper-nickel alloys have the best resistance to biofouling, and brass and bronze have good resistance. Biofouling can be decreased by application of antifouling paints. Metal placed in contact with soil can suffer corrosion due to either anaerobic or aerobic bacteria in the soil.

(4) Seismic effects. Properly designed metal structures can withstand seismic accelerations with little damage. Structural steel is well-suited for seismic designs because of its high tensile strength, good ductility, and consistent yield stress properties when loaded in tension, shear, and compression.

VI-4-6. Wood.

a. Use of wood in coastal construction. Wood can be used in coastal projects such as seawalls, revetments, bulkheads, piers, wharfs, sand fences, and floating platforms. Wood is also used for temporary constructions such as formwork, bracing, blocking, etc.

(1) Untreated lumber. Stock untreated lumber is typically used only during project construction or when the wooden components are expected to survive for only a few months. Typical short-term applications include concrete formwork, temporary bracing, machinery supports, and dunnage. Untreated lumber can be used as a permanent part of a project provided it is protected by a covering (e.g., interior framework) or painted and maintained for the life of the project. However, untreated wood will rapidly decay if it comes into direct contact with soil or seawater.

(2) Treated lumber. Stock treated lumber is used in situations where the wood comes into direct contact with the ground or water. Contact with soil leads to rot, fungus, or insect attack. A typical wood pressure treatment consists of chromated copper arsenate, but local project conditions and intended application may dictate a different treatment to prolong project life. Lumber submerged or periodically immersed in seawater should be pressure-treated with coal-tar creosote or a similar protective treatment.

(3) Piles and poles. Wood piles are frequently used in coastal construction in applications such as pile dolphins (clusters of wood piles lashed together), guide piles for floating structures, piles for channel markers, pile-supported seawalls and bulkheads, building foundations, piers, wharfs, fendering systems, trestles, jetties, and groins. Practically all wood piles are pressure-treated with coal-tar creosote to resist insects, marine borers, limnoria, rot, and fungus. Untreated wood piles are used only as temporary supports or to carry electric power and communication lines to the project site.

(4) Beams and stringers. Treated lumber beams and stringers are used to build load-bearing structures such as groins, bulkheads, jetties, pier decks, wharfs, bracing, and other structures related to shipborne commerce. Untreated beams are used only as support members within protected areas (e.g., buildings), when protected by paint, or as temporary support members.

(5) Plywood and laminated wood.

(a) Practically all plywood used in coastal projects is designed to withstand the effects of humidity and water immersion. Wet-use plywood can be used as building flooring and sheathing, wood-frame gusset plates, concrete forms, and sign boards. Special plywood treatment is needed in extreme conditions like saltwater immersion.

(b) Glued laminated wood provides stronger load-bearing structural members because of better quality control and ability to size members to a specific need. Laminated structural members may be suitable as columns, beams, and trusses, particularly in larger sizes and lengths that are difficult to find in stock lumber. Laminating glue must be waterproof and the wood must be treated with preservatives except in protected or sheltered applications.

b. Physical and mechanical properties of wood. Trees are classified either as hardwoods or softwoods. Hardwoods are typically broad-leafed trees which shed their leaves in the fall; whereas most softwoods are evergreens, which have needles or scale-like leaves. The physical and mechanical properties of wood vary significantly between different tree species. Solid wood substance has a specific gravity of about 1.5 regardless of species. However, part of the volume of dry wood consists of air-filled cavities, giving variation to the density of wood construction materials. In general, wood strength increases with density. The *Timber Construction Manual* (AITC 1985) provides tables listing unit weight and specific gravity for commercial lumber species at different moisture contents.

(1) Directional strength characteristics. Mechanical properties of wood are specified according to three principal axes: (a) longitudinal axis (parallel to the grain), (b) tangential axis (perpendicular to the grain and tangential to the growth rings), and (c) radial axis (perpendicular to the grain and growth rings). For design purposes, stresses are usually determined parallel to and perpendicular to the grain because there is little difference between tangential and radial stress properties.

(2) Loading configurations. Because wood strength differs along each of its three principal axes, structural design of load-bearing members must consider grain orientation when determining appropriate member dimensions necessary to resist the applied loads. Wood has the greatest strength when loaded so as to produce tension or compression parallel to the grain. The following stress conditions are likely to occur in wood structures:

(a) Tension and compression parallel to grain. Wood structures are commonly designed so the primary force loadings produce tension or compression parallel to the wood grain in structural members. Columns, piles, struts, and beams typically have the grain parallel to the longest dimension. The strength capacity will be reduced if the load is applied at an angle to the grain or if knots are present in the wood.

(b) Tension and compression perpendicular to grain. Wood has the least strength in resisting loads that induce tension stresses perpendicular to the wood grain. This type of loading should be avoided for all load-bearing members. Compression forces perpendicular to the wood grain tend to compress the wood at the surface, but generally this can be tolerated provided the displacement of adjoining members does not cause difficulties.

(c) Shear parallel to grain. Wood has the capacity to resist shearing forces parallel to the wood grain, but this capacity is greatly reduced by shakes, checks, and splits that may occur in wood members as they dry. Consequently, allowable shear stress design limits are reduced to compensate for the likelihood of imperfections in the wood.

(d) Shear perpendicular to grain. Solid wood components can withstand substantial shear stresses when loads are applied perpendicular to the grain. In most loading configurations, safe stress limits will be exceeded by compression perpendicular to the grain well before shear perpendicular to the grain becomes critical.

(3) Temperature and moisture effects. Wood does expand and contract with changes in temperature, but in most cases this expansion is not sufficient to create significant stresses. Wood shrinks as it dries and swells as moisture content increases. The amount of shrinkage or swelling is different in the radial, tangential, and longitudinal directions relative to the wood grain, with the most shrinkage occurring in the tangential direction. Generally, dimensional changes due to moisture content are greater for hardwoods than for softwoods. Allowances for swelling should be made when using kiln-dried lumber for project components that will usually

be saturated with water. Average shrinkage in the three principal axes for various wood species when kiln-dried is given in the *Timber Construction Manual* (AITC 1985).

c. Design values for structural lumber. Structural lumber is classified according to wood species, size, and intended use. Dimension lumber refers to rectangular-shaped pieces of smaller dimensions typically used as framing materials, joists, planks, etc. Dimension lumber is usually graded for strength in bending edgewise or flatwise, but it is also used in applications requiring tensile or compressive strength. Beams and stringers are larger rectangular cross-section pieces used primarily in construction that is more robust than simple house framing. Nominal cross-section dimensions are greater than 150 mm, and the members are graded for strength in bending in the widest dimension. Posts and timbers have square or nearly square cross sections with dimensions greater than 150 by 150 mm. Grading is based on the intended use as columns and posts with compression as the primary loading condition. Structural lumber design parameters are based on the natural strength of the wood species along with reductions to account for factors such as knots and their location, grain slope, location of checks and splits, and seasoning. Most lumber is visually graded according to specifications set forth by the American Society for Testing and Materials (ASTM D-2555 (1994) and ASTM D-245 (1994)). The *National Design Specification for Wood Construction* (National Forest Products Association 1991) is the principal reference for determining lumber engineering parameters for use in design. The *Timber Construction Manual* (AITC 1985) provides comprehensive guidance for designing wooden structures and structural components. Allowable working stresses for timber in coastal projects should be those for wood that is continuously wet or damp. Table VI-4-13 lists some general engineering characteristics for the more common softwoods and hardwoods found in the United States.

Table VI-4-13
General Characteristics of Common Wood (from Moffat and Nichol (1983))

Softwoods

General Characteristics	Douglas Fir	Redwood	Cedar	Spruce	Southern Pine
Shrinkage in volume from green to ovendry (pct)	10.9	11.5	11.2	10.4	12.4
Modulus of rupture (MPa)	43.7 (green)	51.4 (green)	41.6 (green)	31.4 (green)	44.1 (green)
Modulus of elasticity (GPa)	7.798 (green)	7.591 (green)	5.199 (green)	5.971 (green)	7.860 (green)

Hardwoods

General Characteristics	Oak	Maple	Ash	Birch	Greenheart
Shrinkage in volume from green to ovendry (pct)	12.7 to 17.7	12.0 to 14.5	11.7	15.0 to 16.8	3
Modulus of rupture (MPa)	49.5 to 73.8 (green) 54.2 to 77.3 (dry)	40.1 to 62.5 (green) 105.2 (dry)	41.4 to 68.9 (green) 108.0 (dry)	59.2 (green) 133.8 (dry)	123.4 (green) 206.8 (dry)
Modulus of elasticity (GPa)	6.047 to 11.63 (green) 9.108 to 14.36 (dry)	6.502 to 10.16 (green) 12.14 to 12.95 (dry)	7.102 to 10.20 (green) 9.618 to 12.26 (dry)	10.27 to 10.64 (green) 16.52 (dry)	20.00 (green)

d. Wood preservatives and treatment. Wood that has been correctly treated with preservative can increase the service life of the wood member by a factor of four to five over untreated wood. Wood treatment is practical and highly recommended for coastal projects, particularly in regions populated by marine borers and other natural enemies of wood. The most effective injected wood preservative for timber exposed to seawater or in direct contact with the ground appears to be creosote oil with a high phenolic content. Piles subject to marine-borer attack need a maximum creosote penetration and retention, and coal-tar solutions are recommended. If the borer infestation is severe, it may also be necessary to treat the pile with a waterborne salt preservative. Thorough descriptions of wood preservative treatments and standards are given in the Wood Preservers' Book of Standards (American Wood Preservers' Association 1984) and in Moffatt and Nichol (1983). Untreated timber piles should not be used unless the piles are protected from exposure to marine-borer attack. In some applications, untreated piles can be encased in a protective armor such as gunite. Field boring and cutting of treated lumber or piles should be avoided if possible. When unavoidable, cut surfaces should receive a careful field treatment of similar preservative to prevent (or at least retard) the onset of dry rot.

e. Wood fasteners and connectors. Except in rare occasions, wood structures in the coastal zone are held together by either metal fasteners and connectors or by adhesives.

(1) The most common metal connectors are nails, spikes, and bolts which are often used in conjunction with metal plates and brackets. Metal is also used for spike grids and split ring connectors to increase shear capacity at joints, and for miscellaneous components like bearing plates and straps. Chain, wire rope, and metal rods are sometimes used to secure and brace wood construction or to anchor wood portions to other parts of the structure. Metal connectors and fasteners are subject to rapid corrosion by water and damp air. Some metals will experience galvanic corrosion if placed in seawater. Protecting metal fasteners used in wood construction requires the same precautionary steps as detailed in Part VI-4-5, "Steel and Other Metals." Abrasion by sand particles or floating debris and chafing by objects such as mooring lines can quickly strip away protective coatings applied to corrodible metals. The process of selecting a protection system must consider these factors and recognize that any metal protection has a limited life in the marine environment. Periodic inspection and maintenance of metal fasteners and connectors is recommended, particularly for critical connections.

(2) Field use of adhesives in coastal wood construction is limited primarily to framework and sheathing for house-like structures in which the adhesive is protected from the harsh environment. Field application of adhesives to exposed wood construction should be limited to secondary joints where failure would not be catastrophic. Wooden components, such as plywood and laminated beams, use adhesives which are applied during manufacture in controlled factory settings. This assures well-mated surfaces, uniform adhesive application, and proper curing of the glued surfaces. Factory-fabricated wood components should use waterproof glue if the piece is expected to be immersed in water. Additional information on application and use of wood adhesives is given in Moffatt and Nichol (1983) and in several standards from the American Society for Testing and Materials (ASTM D-2559 (1994)).

f. Environmental effects on wood. Wood reacts to a number of environmental conditions encountered in the coastal zone.

(1) Water will penetrate wood and cause swelling with a corresponding strength reduction, and extended immersion will soften the wood fibers. Periodic wetting and drying causes uneven drying and may lead to development of cracks or provide conditions favorable to fungi that cause dry rot. Strong acids will hydrolyze wood and severely impact strength, but exposure to strong acids is rare and limited to accidental spills in places such as cargo handling areas.

(2) Water pollution may help preserve wood structures by reducing the oxygen that supports wood-attacking marine biota.

(3) Marine organisms are the principal cause of wood destruction for immersed timbers and piles, and the concentration of damaging biota will vary with location. Application of proper preservatives is essential for longer service life in these conditions. Marine plants also grow on immersed wood, but flora growth does not seem to harm the strength characteristics of the wood. However, slippery marine growth on wood decking may be hazardous to pedestrians.

(4) Dry wood can catch fire, but larger structural wood members will retain a substantial portion of their strength for a period of time as fire slowly chars the wood inward from the surface.

(5) Abrasion of wood by wind and sand can eventually lead to a reduction in strength due to a decrease in cross section; however, this process takes a long time to occur. Rubbing of wood surfaces by a harder material, such as a steel vessel, will also wear down the wood. Although such wear eventually damages the wood, the relative softness of wood compared to other materials is a beneficial design feature in some applications.

(6) The resilience of wood allows wood structures and structural members to absorb impact energy and rebound better than more rigid structures. This flexibility helps wood structures withstand wave and vessel impact loads and seismic accelerations.

(7) Many wood structures are constructed to serve human activities such as harbor facilities, wharfs, piers, etc. Eventually wood will begin to show wear from human and vehicular traffic, accidental impacts, vandalism, and other causes. Periodic monitoring is needed to assure that normal wear from human activities does not weaken a structure beyond a safe level. Immediate inspection is needed after any accident that may have caused structural damage.

VI-4-7. Geotextiles and Plastics.

a. Use of plastics in coastal construction. The term "plastic" is a generic label for a large number of synthetic materials composed of chainlike molecules called polymers. Plastics can be easily molded into shapes during manufacture. The most common use for plastic in coastal construction is in the form of geotechnical fabrics, which are the main focus of this section.

(1) Geotextile fabrics. Plastic filaments or fibers can be woven or needlepunched into strong fabrics called "geotextiles" that are often used as filter cloth beneath hard armor systems. Other names for these types of fabrics include filter fabrics, construction fabrics, plastic filter cloth, engineering fabrics, and geotechnical fabrics. The most frequent use of geotextiles in coastal construction is as a filter between fine granular sands or soils and overlying gravel or small stone that forms the first underlayer of a coastal structure such as a revetment. The purpose of the geotextile is to retain the soil while permitting flow of water through the fabric. Figure VI-4-5 illustrates typical usage of a geotextile fabric in coastal construction. Geotextile filters have several general advantages over conventional gravel filters (Barrett 1966):

(a) Filtering characteristics are uniform and factory controlled.

(b) Geotextile filter fabrics can withstand tensile stresses.

(c) Geotextile placement is more easily controlled, and underwater placement is likely to be more successful than comparable gravel filters.

(d) Inspection and quality control are quick and accurate.

(e) Local availability of filter materials is not a cost consideration, and often substantial savings can be realized.

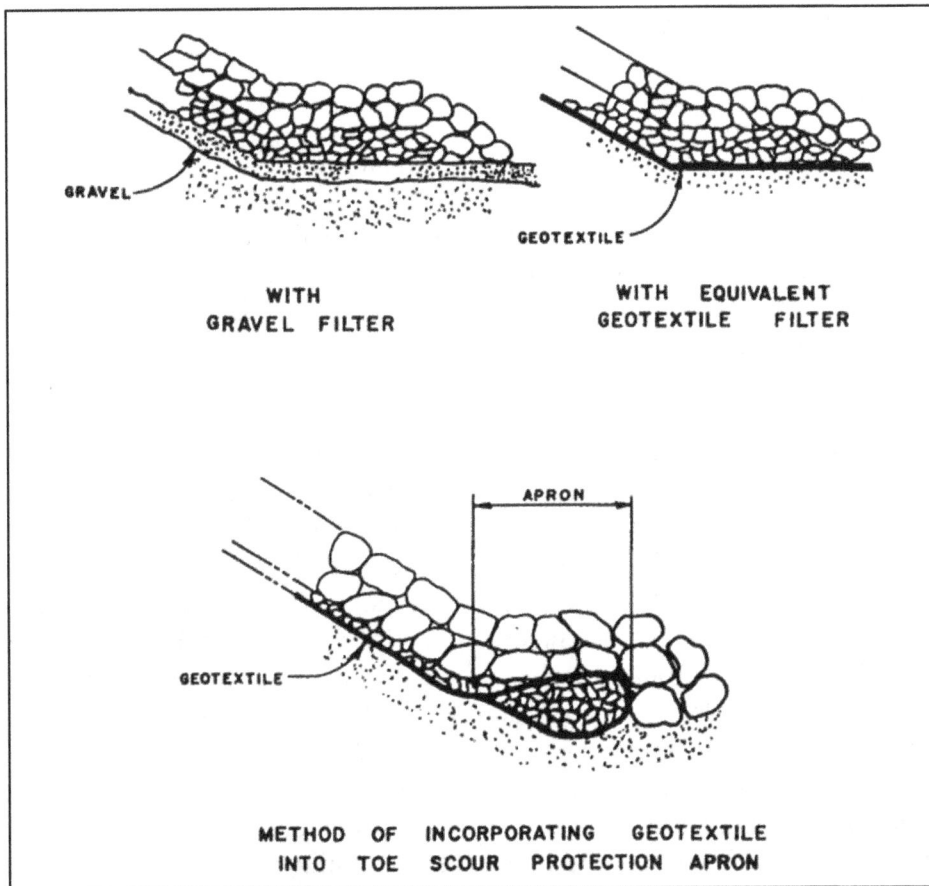

Figure VI-4-5. Typical use of geotextile fabric in coastal revetment (from Moffatt and Nichol (1983))

(2) Some potential disadvantages of geotextile filter fabrics are the following: (1) it is difficult to repair damaged fabric that underlays several layers of stone, (2) if improperly designed, some fabrics can be relatively impervious to rapid hydraulic transients, which could lead to uplift pressures over the fabric surface, and (3) the fabric is susceptible to undermining at the structure toe if not properly anchored. Coastal application of woven geotextile fabric began in the mid-1950s. In the United States geotextiles were first used as a filter for an ocean-front concrete block revetment in 1958. Dutch coastal engineers first used geotextiles in 1956, and they continued development of geotextiles as work began on the massive Delta Works Scheme (John 1987). During the 1960s, geotextiles became well-established as replacements for granular filters due in part to extensive use in the Delta Works Scheme. It is estimated that over 10 million cu m of geotextile were used in the Dutch flood protection project. Initially, use of geotextiles was not cost-effective, and applications were limited to sites that lacked local sources of good granular fill material. Presently, the use of geosynthetics has become more widespread. Although the most common geotextile application is to serve as a filter, these fabrics can also serve the following functions:

(a) Separate different soil layers.

(b) Reinforce soil banks against lateral movement.

(c) Control erosion of banks.

(d) Provide drainage.

(e) Cap and/or contain contaminated dredged material.

Generally, geotextile filter fabrics should allow water to flow through while retaining the soil. Other coastal applications, such as bank reinforcement, rely on high fabric tensile strength. Examples of geotextile use in coastal construction can be found in Barrett (1966), Dunham and Barrett (1974), Koerner and Welsh (1980), and John (1987).

(3) Other forms of plastic. High-strength plastic fabric such as nylon has been used successfully as flexible forms for concrete. Two layers of cloth are injection-filled with concrete or grout to form a mattress-like structure for slope protection. Similarly, grout-filled plastic tubes have been used in various shore protection schemes. Examples of the above applications are given by Koerner and Welsh (1980) and by Moffatt and Nichol (1983). Impervious plastic sheets made of polyethylene, vinyl, or rubber compounds are used as liners and covers to control water seepage or to contain pollutants. During construction, plastic sheets can be used to shield unprotected metal components from corrosive effects of salt air and rainwater or to keep lumber (or other construction materials) dry. Highly porous plastic mesh can be used as dune fencing to trap windblown sediment. High-strength plastics can be molded into almost any shape for specialized applications. For example, high- density polyethylene is often used for mooring fenders and guards because it has low surface friction, good toughness, and it resists abrasion and impact damage. Plastic is also used to line and protect pilings from corrosion and from attack by marine borers. Special fiber-reinforced polyester pipe can replace steel pipe in some applications, but at a greater cost. Standard polyvinyl chloride (PVC) pipe up to 25 cm in diameter can be used for applications such as electrical conduit, water supply, and drainage. Finally, epoxy resins can be used to affix bolts in concrete or stone, or the resin can be mixed with sand to produce a chemical grout for patching concrete.

b. Physical and mechanical properties of plastics.

(1) General characteristics. Most plastics are synthetic, carbon-based products of a chemical reaction that alters the characteristics of the original component materials. The resulting plastics are high polymers composed of monomer atoms joined together in molecular aggregations. During some stage of production, plastics are soft and can be formed into shapes by application of heat, pressure, or both.

(a) *Thermoplastics* can be repeatedly softened (up to melting) and hardened again by heating and cooling without changing the plastic properties. At low temperatures the material becomes brittle. Included in the thermoplastics group are polyethylene, polyvinyl chloride, acrylics, nylon, and polystyrene.

(b) *Thermosets* are plastics that go through a soft, pliable stage only once before irreversibly transforming to a permanently hardened material. Heating thermosets breaks down the plastic. Polyesters, epoxies, silicones, phenol-formaldehydes, and melamine-formaldehydes are examples of thermosetting plastics.

Physical characteristics of some plastics can be modified by combining various additives such as plasticizers and stabilizers with the plastic during manufacture. Plasticizers make plastics that are rigid or brittle at normal temperatures softer and more pliable. Stabilizers help to decrease the deteriorating effects of weather and ultraviolet light on plastics. Different plastics can be combined during manufacture to form *copolymers* that exhibit some of the beneficial properties of both components. Most structural plastic components are copolymers. Plastics can also be reinforced with high-strength fibers to attain greater yield strength. Plastic is a good construction material because it can be shaped into practically any form. In addition, there are several other characteristics that make plastic a good construction material. These are its corrosion, wear, and impact resistance; its light weight; its flexibility and energy absorption capability; and its electrical and thermal insulating qualities. Some properties vary greatly over the range of commonly available plastics.

(2) Geotextiles. Most geotextiles are made from one of the four main polymer families: polyester, polyamide, polypropylene, or polyethylene. Polyethylene has one of the simplest molecular structures, and its main attractions include low cost and chemical resistance. Polyamides (e.g., nylon) are roughly three times more expensive than polyethylene, and they exhibit moderate strength and chemical resistance characteristics. Polypropylenes are low cost and currently comprise the most widely used group of geotextiles (John 1987). Polyesters have the best tensile strength characteristics, the least long-term creep and high inherent ultraviolet light resistance, but these attributes come at a high cost. The relative differences between the four polymer families in terms of important physical properties are shown Table VI-4-14. Within each main group there are many subgroups that can have significantly different characteristics than those attributed to the group as a whole. In particular, strength properties vary with manufacturing method.

Table VI-4-14
Comparative Properties of Geotextile Materials (from John (1987))

Characteristic		Polyester	Polyamide	Polypropylene	Polyethylene
Comparative Properties					
Tensile strength		High	Medium	Low	Low
Elastic modulus		High	Medium	Low	Low
Strain at failure		Medium	Medium	High	High
Creep		Low	Medium	High	High
Unit weight		High	Medium	Low	Low
Cost		High	Medium	Low	Low
Resistance Properties					
U-V Light:	Stabilized	High	Medium	High	High
	Unstabilized	High	Medium	Medium	Low
Alkalis		Low	High	High	High
Fungus, vermin, insects		Medium	Medium	Medium	High
Petroleum products		Medium	Medium	Low	Low
Detergents		High	High	High	High

Engineering properties and overall suitability of geotextiles for specific applications depend as much on the fabric manufacture as the properties of the polymer. Fabrics are either woven, nonwoven, or a combination of the two. Several weaving methods are used in the manufacture of geotextiles, and each method achieves different results in the fabric. Fabrics made of monofilament yarns have relatively regular and uniform pore sizes and are more reliable for critical filtration applications where the higher cost is justified. Nonwoven geotextiles are made of discrete fibers bonded together by some method that often allows for a somewhat thicker porous fabric. Porosity may be achieved by punching holes in the fabric with needles to attain a more uniform filtering capability. Both woven and nonwoven fabrics have been used in coastal applications, but woven monofilament geotextiles are overwhelmingly preferred for coastal structures.

    c. Design requirements for geotextile fabrics.

    (1) General design requirements. Use of a geotextile as a filter cloth requires that the fabric be permeable to water without allowing passage of retained soil particles or clogging. Flow of water through the geotextile must be at a rate that prevents excessive head loss or buildup of hydrostatic pressure. An effective filter requires a geotextile suited to the retained soil grain size and slope, groundwater, wave and water level loading, and particulars of the overlying stone layers. Selection of a geotextile may be difficult because of the wide range of fabrics available from a number of manufacturers; however, the specification should be based on properties such as transmissivity, porosity, etc. It may help to examine past performance of particular geotextiles in similar projects. Some combination of the factors listed below may influence the selection of a suitable geotextile fabric (Moffatt and Nichol 1983).

    (a) *Tensile strength.* Fabric tensile strength is needed to resist tearing when subjected to dynamic loads from waves, currents, and constant movement of structure underlayers. For rubble structures, strong fabrics allow placement of larger stones directly on the geotextile, thus reducing the overall structure thickness. However, if large voids occur in the overlying structure layers, soil pressure and/or hydrostatic pressure may rupture the fabric. Fabrics that have sufficient Aburst strength@ will continue to retain the soil, thus reducing rehabilitation costs.

    (b) *Elongation at failure.* Excessive elongation will distort and enlarge the pores, changing the filtering characteristics and perhaps resulting in soil loss.

    (c) *Puncture resistance.* Geotextiles need good puncture resistance to survive placement of materials over the fabric during construction. The fabric also needs to resist puncturing due to movement of armor stone and underlayer stone as the structure settles or as it responds dynamically to wave action.

    (d) *Abrasion resistance.* Constant wave action on a coastal structure causes movement of materials adjacent to the geotextile, and the fabric must withstand this abrasion over the life of the structure. Special care must be taken during construction to avoid fabric abrasion as materials are placed on the geotextile.

    (e) *Durability.* Geotextiles must perform consistently over the life of the structure. Durability depends on the chemical composition and construction of the fabric, physical properties of the finished fabric, exposure to deteriorating environmental conditions, and physical abuse experienced during service.

(f) *Site-specific factors*. Some coastal applications may subject geotextile fabrics to freeze/thaw conditions or to high or low temperatures. It may be necessary to test the geotextile for survivability under these conditions. Also, fabric selection should account for any anticipated exposure to chemicals, acids, alkalis, or fuels.

(g) *Construction factors*. Placement of geotextiles in severe wave environments may be difficult, and fabric may be damaged or severely abraded during placement attempts. Excessive movement of underlayer materials by waves may severely damage the fabric before more stable armor layers can be placed. Construction methods may need to be modified to minimize adverse wave exposure.

(2) Recommended minimum geotextile physical properties. Moffatt and Nichol (1983) presented recommended minimum values for various geotextile engineering parameters under three different loading conditions for coastal projects. These recommendations are reproduced in Table VI-4-15. "Severe dynamic loading" refers to continued abrasive movement of materials adjacent to the fabric due to wave action. "Dynamic and static loading" results from more restrictive placement procedures that limit abrasion. "Stringent placement and drainage" refers to applications where placement and service life are nearly free of any abrasive movement of adjacent materials. Design values for specific candidate geotextiles should be determined according to test procedures given in the referenced ASTM standards in Table VI-4-15 (also summarized in Moffatt and Nichol (1983)).

(3) Design properties of commercial geotextile fabrics. There are numerous commercial manufacturers of geotextiles, and each manufacturer produces a variety of fabrics having differing engineering properties. The best listing of currently available geotechnical fabrics and their associated design properties is the annual Specifier's Guide of the *Geotechnical Fabrics Report* published by the Industrial Fabrics Association International. The annual *Specifier's Guide* details property specifications of over 500 geosynthetic products from more than 50 international producers. Some textbooks and manuals also list commercial products. For example, Koerner (1986) provides tables from the *1985 Geotechnical Fabrics Report*, and Ingold and Miller (1988) provide addresses, distributors, product lines, and design parameters for European-based producers of geotechnical fabric.

(4) Geotextile filtering and clogging criteria. Geotextile filters in coastal structures may be exposed to rapid flow fluctuations including turbulent flows, high hydrodynamic pressure differentials, and sudden or periodic runup and rundown. The selected geotextile must be able to retain the soil, yet have openings large enough to permit rapid drainage without clogging. Calhoun (1972) conducted extensive tests to develop engineering criteria for geotextile fabrics, and these criteria have been verified through numerous field applications. The capability of a geotextile to retain soil while allowing water to pass is termed the "*piping resistance*." Calhoun developed a procedure for determining the piping resistance based on the size of the retained soil and the *equivalent opening size* (EOS) of the geotextile. He also developed clogging criteria based on the fabric "*percent of open area*" (POA). Values of EOS and POA are determined using the procedures described in Table VI-4-16.

Table VI-4-15
Minimum Geotextile Fabric Physical Property Requirements (from Moffatt & Nichol (1983))

Property		Test Method	Category		
			A	B	C
			Severe Dynamic Loading	Dynamic & Static Loadings	Stringent Placement & Drainage
Tensile strength[4]	SPD[1]	ASTM D-4632	1.56 kN		0.89 kN
	BPD[2]			0.89 kN	
	WPD[3]		0.98 kN		0.44 kN
Elongation at failure		ASTM D-4632	<36 percent	36 percent	36 percent
Seam strength[4]		ASTM D-4884	0.87 kN	0.80 kN	0.36 kN
Puncture resistance		ASTM D-4883	0.53 kN	0.53 kN	0.29 kN
Burst strength		ASTM D-3786	3450 kPa	3790 kPa	1650 kPa
Abrasion resistance[4]	SPD	ASTM D-4886	0.44 kN		0.27 kN
	BPD			0.29 kN	
	WPD		0.29 kN		0.15 kN
Optional Requirements					
Freeze-thaw resistance		ASTM D-5034	90 percent of required strength		
High temperature survivability		ASTM D-5034	80 percent of required strength		
Low temperature survivability		ASTM D-5034	85 percent of required strength		
Effects of acids		ASTM D-5034	90 percent of required strength		
Effects of alkalies		ASTM D-5034	90 percent of required strength		
Effects of JP-4 fuel		ASTM D-5034	85 percent of required strength		

[1] SPD = Stronger principal direction.
[2] BPD = Both principal directions.
[3] WPD = Weaker principal direction.
[4] In accordance with the specifications for the tests for these properties, these forces are applied over a width of 25.4 mm (1 in.).

Table VI-4-16
Determination of EOS and POA for Geotextiles (from Moffatt and Nichol (1983))

Equivalent Opening Size (EOS)

Based on the Calhoun (1972) method, five unaged samples shall be tested. Obtain about 150 gm of each of the following fractions of a sand composed of sound, rounded-to-subrounded particles:

U.S. Standard Sieve Numbers

Sample 1. Passing #10 and Retained on #20
Sample 2. Passing #20 and Retained on #30
Sample 3. Passing #30 and Retained on #40
Sample 4. Passing #40 and Retained on #50
Sample 5. Passing #50 and Retained on #70
Sample 6. Passing #70 and Retained on #100
Sample 7. Passing #100 and Retained on #120

The cloth shall be affixed to a standard sieve having openings larger than the coarsest sand used, in such a manner that no sand can pass between the cloth and the sieve wall. The sand shall be oven-dried. Shaking of the sample will continue for 20 min. Determine by sieving (using successively coarser fractions) that fraction of sand of which 5 percent or less by weight passes the cloth. The equivalent opening size of the cloth sample is the Aretained on@ U.S. Standard Sieve number of this fraction.

Percent of Open Area (POA)

Each of five unaged samples should be placed separately in a 50- x 50-mm (2- x 2-in.) glass slide holder and the image projected with a slide projector on a screen. Select a block of 25 openings near the center of the image and measure to the nearest 25.4 microns (0.001 in.) the length and width of each of the 25 openings and the widths of two fibers adjacent to each opening. The percent open area is determined by dividing the sum of the open areas of the 25 openings by the sum of the total area of the 25 openings and their adjacent fibers.

For retention of coarse-grained soils containing 50 percent or less by weight of particles passing U.S. No. 200 sieve (0.074 mm diameter), the piping resistance *(PR)* for woven geotextile fabric is given by

$$PR = \frac{D_{85} \, of \ protected \ soil}{EOS} \qquad \text{(VI-4-7)}$$

where D_{85} is the effective grain size (in mm) for which 85 percent of the soil (by weight) has smaller grain size. (Note *EOS* is expressed in millimeters.) Ideally, the value of the piping resistance should be unity or slightly greater to promote drainage and prevent clogging. As values of *PR* increase, flow resistance through the fabric also increases. Adequate clogging resistance is provided by geotextiles having an effective POA equal to or greater than 4 percent. If a percentage of the geotextile's surface is covered by flat smooth materials (e.g., patio-type blocks) without an intervening gravel layer, the necessary fabric POA must be increased proportionately. For example if one third of the fabric is to be covered by flat blocks, then the necessary geotextile POA must be increased by a factor of 3 to 12 percent to give an effective POA of 4 percent. Geotextiles adjacent to finer soils in which more than 50 percent of the grains (by weight) pass through the U.S. No. 200 sieve should have an EOS no larger than a U.S. No. 70 sieve (0.210 mm). Geotextiles with an EOS smaller than a U.S. No. 100 sieve (0.149 mm) should not be used as filter in coastal projects.

d. Geotextile installation considerations. Practical experience with geotextile filters in coastal projects has provided general guidelines for geotextile installation and maintenance. However, unique site conditions may dictate alternate techniques.

(1) Geotextile placement. Successful use of geotextiles in coastal projects depends critically on initial placement of the fabric. The sequence of geotextile placement is determined somewhat by the specific project and application, but in general the following guidelines should be followed.

(a) Geotextiles should be laid loosely, free of wrinkles, creases, and folds. This allows the fabric to conform to irregularities in the soil when heavier materials are placed on the fabric. Placing the geotextile in a stretched condition under tensile stress should be avoided.

(b) Fabric placement on slopes subjected to wave action should begin at the slope toe and proceed upslope with the upslope panel overlapping the downslope panel. For slopes subjected to along-structure currents, upstream panels should overlap downstream panels.

(c) When the slope continues beyond the protective armor layers, the filter should be keyed into a trench at the upper portion of the structure. Similar termination of the filter can be used at the structure toe as illustrated in Figure VI-4-5.

(d) Horizontal underwater fabric placement should start at the shoreward end and proceed seaward. For scour protection the placement should start adjacent to the protected structure and proceed to the outer limit of the protection.

(e) Any overlying gravel layers must have sufficient permeability so as not to reduce flow through the geotextile.

(f) Steel securing pins (when needed) should be 5 mm (3/16 in.) in diameter and have a head capable of retaining a steel washer having a 3.8-cm outside diameter. Pin length should be a minimum of 45 cm (1.5 ft) for medium to high density soil, and longer for looser soils. Pins should be placed at the overlap mid-point. Pin spacing along the overlap should be a maximum of 0.6 m (2 ft) for slopes steeper than 1-on-3, 1.0 m (3 ft) for slopes between 1-on-3 and 1-on-4, and 1.5 m (5 ft) for slopes flatter than 1-on-4. Additional pins should be used as necessary to prevent geotextile slippage.

(g) Placement of overlying stones should begin at the toe and proceed upslope. Some projects may require stone placement in conjunction with geotextile placement to hold the fabric against wave or current action.

(h) Care must be exercised in placing the overlying stone layers to avoid puncturing the geotextile. Tables VI-4-17 and VI-4-18 provide construction drop height limitations for quarrystone revetments, block revetments, and subaqueous applications. Loading conditions are the same as described for Table VI-4-15. No stones over 440 N (100 lb) should be rolled downslope over the fabric.

Table VI-4-17 Construction Limitations: Quarrystone Revetment[1] (from Moffatt and Nichol (1983))			
	Category		
Parameter	A	B	C
Steepest slope	1V on 2H	1V on 2.5H	1V on 3H
Min. gravel thickness above filter	None	None	20 cm
Stone Adjacent to Geotextile			
Max. stone weight[3]	1.1 kN	0.78 kN	Gravel
Max. drop height	1 m	1 m	1.5 m
Max. stone weight	1.8 kN	1.3 kN	
Riprap weight range[4]	0.89 - 3.3 kN	0.22 - 2.2 kN 0.61 m	NA[2]
Max. drop height	0.61 m		
Max. stone weight	1.8 - 8.9 kN	1.3 - 8.9 kN	
Max. drop height	placed	placed	NA
Subsequent Stone Layer			
Max. stone weight	NA	NA	0.67 kN
Max. drop height			1.2 m
Max. stone weight	NA	NA	1.3 kN
Max. drop height	NA	NA	1 m
Max. stone weight	44 kN	44 kN	4.4 kN
Max. drop height	3 m	2.5	placed
Max. stone weight	> 44 kN	> 44 kN	> 4.4 kN
Max. drop height	placed	placed	placed

NOTE:
a. Stronger principal direction (SPD) and seams of the geotextile should be perpendicular to the shoreline.
b. There is no limit to the number of underlayers between the armor and the geotextile.

[1] This table may also be used for sand core breakwaters (a jetty, groin, or breakwater in which the core material consists of sand rather than stone).
[2] Not applicable.
[3] Weight of quarrystone armor units of nearly uniform size.
[4] Weight limits of riprap, quarrystone well-graded within wide size limits.

Table VI-4-18
Construction Limitations: Block Revetments and Subaqueous Applications
(from Moffatt and Nichol (1983))

Block Revetment[1]	Category		
	A	B	C
Precast Cellular Block[2]			
Steepest slope			
Individual blocks	1V on 2H	1V on 3H	NA[3]
Cabled blocks[4]	1V on 1.5H	1V on 2H	NA
Max. block weight	>3.1 kPa	3.1 kPa	NA
Interlocking Concrete Block[2]			
Steepest slope	NA	1V on 2H	1V on 2.5H
Min. gravel thickness above filter	NA	15.2 cm	15.2 cm
Max. block weight	NA	>3.1 kPa	3.1 kPa
Subaqueous Applications[5]			
Steepest slope	1V on 15H	1V on 15H	1V on 15H
Stone adjacent to geotextile:			
Max. stone weight	8.9 kN	8.9 kN	3.3 kN
Min. drop through water	1.5 m	1.5 m	1.5 m
Max. stone weight	>13.3 kN	>13.3 kN	>3.3 kN
Max. drop height	placed	placed	placed
Subsequent stone layer (s)			
Max. stone weight	No limit	No limit	No limit
Max. drop height	NCP[6]	NCP	NCP

[1] Stronger principal direction (SPD) and seams of the geotextile should be perpendicular to the shoreline.
[2] With flat base.
[3] Not applicable.
[4] Precast cellular blocks cabled together in a horizontal plane.
[5] No limit to the number of underlayers between the armor and the geotextile.
[6] As in normal construction practice: the geotextile does not require special limitations in these layers.

(2) Geotextile seams and joins. Geotextiles can be obtained in fairly long lengths, but width is limited by practical considerations related to manufacture and transportation. Wider panels reduce the number of fabric overlaps (which is the most probable cause of error during placement). Overlaps that are not subjected to tensile loading should be at least 45 cm (1.5 ft) and staggered in above water applications where placement can be well-controlled. Underwater geotextile overlaps should be at least 1 m (3 ft). Geotextile panels can be joined before or during placement by either sewing or cementing the panels together at the seams. Generally, sewing is preferred for onsite joins. The most appropriate guidance on field joining of specific geotextile fabrics should be available from the manufacturer. More general guidance is given in various geotextile textbooks (e.g., Ingold and Miller (1988), John (1987)).

(3) Geotextile repairs. Construction damage to geotextile filters is easily repaired by trimming out the damaged section and replacing it with a section of fabric that provides a minimum of 0.6 m (2 ft) overlap in all directions. The edges of the replacement fabric should be placed under the undamaged geotextile. If damage occurs to geotextile panels in which the fabric tensile strength is needed to reinforce soil slopes, the entire fabric panel should be replaced. Repairing damaged fabric underlying a rubble-mound structure is more difficult because the overlying stone layers must first be removed to expose the damaged filter cloth.

    e. Environmental effects on geotextiles and plastics.

(1) Chemical and biological effects. Plastics are generally considered not biodegradable, and they are relatively unaffected by chemicals found in normal concentrations in the coastal zone. However, some chemicals, such as alkalis and fuel products, can rapidly destroy some plastic compounds. Although plastics are impervious to biological attack, marine growth on plastic structure components may induce additional drag forces or hinder smooth operation of moving parts. Bacterial activity in the interstices of geotextiles can clog the fabric and increase its piping resistance.

(2) Ultraviolet radiation. Unless stabilizers have been added during manufacture, plastics will deteriorate when exposed to ultraviolet (UV) radiation. For most coastal structure filtering applications, geotextiles are exposed to sunlight for only a short period during construction before placement of overlayers, and the effects of UV radiation are minor. In some cases, it may be prudent to sequence construction to minimize exposure of geotextile fabric to sunlight. Geotextiles can be exposed to UV radiation if the armor layer is relatively thin, allowing sunlight to penetrate through voids in the armor layer. Similarly, precast armoring blocks may have holes that allow light penetration. Storm damage to structure armor layers can expose geotextile filters to sunlight for extended periods before repairs can be initiated. In the above situations, UV radiation will ultimately destroy the geotextile unless the fabric has been stabilized. The relative ultraviolet radiation resistance of untreated polymer types is shown in Table VI-4-14.

(3) Fire. Plastics will burn or disintegrate if exposed to fire or high temperatures, often releasing very poisonous gases. Some plastics will burn easily, some slowly, and others with great difficulty. Flame-retardant chemicals can be combined into the molecular structure of the plastic materials. Temperatures above the polymer's melting point will alter the filtering characteristics of geotextile fabrics.

(4) Other factors. Abrasion by overlying material (or debris in the case of exposed fabric) can tear fibers in geotextiles, weakening the fabric. Impact loading by waves, vessels, or other objects may puncture geotextile fabric, and ice formation may induce tensile stresses exceeding the material yield strength. Excessive ground motion accelerations due to seismic events may cause differential shifting of the armor layer or soil slope, resulting in tension failure of the geotextile filter. Finally, exposed geotextile or high- strength fabrics may be damaged by vandalism.

VI-4-8. <u>References</u>.

EM 1110-2-1204
Environmental Engineering for Coastal Shore Protection

EM 1110-2-1601
Hydraulic Design of Flood Control Channels

EM 1110-2-1612
Ice Engineering

EM 1110-2-1906
Laboratory Soils Testing

EM 1110-2-2301
Test Quarries and Test Fills

EM 1110-2-2302
Construction with Large Stone

EM 1110-2-5025
Dredging & Dredged Material Disposal

ACI 1986
American Concrete Institute. 1986 (revised annually). *Manual of Concrete Practice*, Part 3, Detroit, MI.

AISC 1980
American Institute of Steel Construction, Inc. 1980. *Manual of Steel Construction*, 8th ed., Chicago, IL.

AITC 1985
American Institute of Timber Construction. 1985. *Timber Construction Manual*, John Wiley and Sons, New York.

American Wood Preservers' Association 1984
American Wood Preservers' Association. 1984. *Book of Standards*, Stevensville, Maryland.

ASTM C-150
American Society for Testing and Materials. 1994. "Standard Specification for Portland Cement," ASTM C-150, *Annual Book of ASTM Standards*, Philadelphia, PA, Vol. 04.01, pp 125-129.

ASTM C-595
American Society for Testing and Materials. 1997. "Standard Specification for Blended Hydraulic Cement," ASTM C-595, *Annual Book of ASTM Standards*, Philadelphia, PA, Vol. 04.01, pp 301-306.

ASTM D-245
American Society for Testing and Materials. 1994. "Practice for Establishing Structural Grades and Related Allowable Properties for Visually Graded Lumber," ASTM D-245, *Annual Book of ASTM Standards*, Philadelphia, PA, Vol. 04.10, pp 88-105.

ASTM D-502
American Society for Testing and Materials. 1994. "Specification for Steel Structural Rivets," ASTM D-503, *Annual Book of ASTM Standards*, Philadelphia, PA, Vol. 15.08, pp 113-115.

ASTM D-690
American Society for Testing and Materials. 1994. "Specification for High-Strength Low-Alloy Steel H-Piles and Sheet Piling for Use in Marine Environments," ASTM D-690, *Annual Book of ASTM Standards*, Philadelphia, PA, Vol. 01.04, pp 349-350.

ASTM D-2555
American Society for Testing and Materials. 1994. "Standard Test Methods for Establishing Clear Wood Strength Values," ASTM D-2550, *Annual Book of ASTM Standards*, Philadelphia, PA, Vol. 04.10, pp 371-384.

ASTM D-2559
American Society for Testing and Materials. 1994. "Specification for Adhesives for Structural Laminated Wood Products for Use Under Exterior (Wet Use) Exposure Conditions," ASTM D-2559, *Annual Book of ASTM Standards*, Philadelphia, PA, Vol. 15.06, pp 154-158.

ASTM D-4254
American Society for Testing and Materials. 1994. "Test Method for Minimum Index Density and Unit Weight of Soils and Calculation of Relative Density," ASTM D-4254, *Annual Book of ASTM Standards*, Philadelphia, PA, Vol. 04.08, pp 543-550.

Barrett 1966
Barrett, R. J. 1966. "Use of Plastic Filters in Coastal Structures," *Proceedings of the 10th International Conference on Coastal Engineering*, American Society of Civil Engineers, Vol. 2, pp 1048-1067.

Calhoun 1972
Calhoun, C. C., Jr. 1972. "Development of Design Criteria and Acceptance Specifications for Plastic Filter Cloth," Technical Report S-72-7, U.S. Army Engineer Waterways Experiment Station, Vicksburg, MS.

CIRIA/CUR 1991
Construction Industry Research and Information Association (CIRIA) and Centre for Civil Engineering Research and Codes (CUR). 1991. "Manual on the Use of Rock in Coastal and Shoreline Engineering," CIRIA Special Publication 83/CUR Report 154, CIRIA, London and CUR, The Netherlands.

CRC 1976
Chemical Rubber Company. 1976. *Handbook of Tables for Applied Engineering Science*, 2nd ed.,
CRC Press, Cleveland, OH.

Department of the Interior 1975
Bureau of Reclamation. 1975. *Concrete Manual, A Manual for the Control of Concrete
Construction*, 8th ed., U.S. Government Printing Office, Washington, DC.

Dunham and Barrett 1974
Dunham, J. W, and Barrett, R. J. 1974. "Woven Plastic Cloth Filters for Stone Seawalls,"
Journal of the Waterways, Harbors, and Coastal Engineering Division, American Society of
Civil Engineers, Vol. 100, No. WW1, pp 13-22.

Eckert and Callender 1987
Eckert, J., and Callender, G. 1987. "Geotechnical Engineering in the Coastal Zone," Instruction
Report CERC-87-1, U.S. Army Engineer Waterways Experiment Station, Coastal Engineering
Research Center, Vicksburg, MS.

Ingold and Miller 1988
Ingold, T. S., and Miller, K. S. 1988. *Geotextiles Handbook*, Thomas Telford, Limited, London.

John 1987
John, N. W. 1987. *Geotextiles*, Blackie and Sons, Ltd., London.

Koerner 1986
Koerner, R. M. 1986. *Designing with Geosynthetics*, Prentice-Hall, Englewood Cliffs, NJ.

Koerner and Welsh 1980
Koerner, R. M., and Welsh, J. P. 1980. *Construction and Geotechnical Engineering Using
Synthetic Fabrics*, John Wiley and Sons, New York.

La Londe and Janes 1961
La Londe, W. S., Jr., and Janes, M. F. 1961. *Concrete Engineering Handbook*, McGraw-Hill
Book Company, New York.

Latham 1991
Latham, J. P. 1991. "In-Service Durability Evaluation of Armourstone," *Durability of Stone for
Rubble Mound Breakwaters*, O. T. Magoon and W. F. Baird, eds., American Society of Civil
Engineers, pp 6-18.

Lienhart 1991
Lienhart, D. A. 1991. "Laboratory Testing of Stone for Rubble Mound Breakwaters: An
Evaluation," *Durability of Stone for Rubble Mound Breakwaters*, O. T. Magoon and W. F. Baird,
eds., American Society of Civil Engineers, pp 19-33.

Lutton 1991
Lutton, R. J. 1991. "U.S. Experience with Armor-Stone Quality and Performance," *Durability of Stone for Rubble Mound Breakwaters*, O. T. Magoon and W. F. Baird, eds., American Society of Civil Engineers, pp 40-55.

Magoon and Baird 1991
Magoon, O. T., and Baird, W. F. 1991. "Durability of Armor Stone for Rubble Mound Coastal Structures," *Durability of Stone for Rubble Mound Breakwaters*, O. T. Magoon and W. F. Baird, eds., American Society of Civil Engineers, pp 3-4.

Mehta 1991
Mehta, P. K. 1991. *Concrete in the Marine Environment*, Elsevier Science Publishers, Ltd., London, England.

Moffatt and Nichol 1983
Moffatt and Nichol, Engineers. 1983. "Construction Materials for Coastal Structures," Special Report SR-10, U.S. Army Engineer Waterways Experiment Station, Coastal Engineering Research Center, Vicksburg, MS.

National Forest Products Association 1991
National Forest Products Association. 1991. *National Design Specification for Wood Construction: Structural Lumber, Glued Laminated Timber, Timber Piles, Connections; Recommended Practice*, Washington, DC.

Thomas and Hall 1992
Thomas, R. S., and Hall, B. 1992. *Seawall Design*, Butterworth-Heinemann, Ltd., Oxford.

Woodhouse 1978
Woodhouse, W. W., Jr. 1978. "Dune Building and Stabilization with Vegetation," Special Report SR-3, U.S. Army Engineer Waterways Experiment Station, Coastal Engineering Research Center, Vicksburg, MS.

VI-4-9. <u>Acknowledgments</u>.

Author: Dr. Steven A. Hughes, Coastal and hydraulics Laboratory (CHL), U.S. Army Engineer Research and development Center, Vicksburg, MS.

Reviewers: Dr. Hans F. Burcharth, Department of Civil Engineering, Aalborg University, Aalborg, Denmark; Douglas A. Gaffney, private consultant; Han Ligteringen, Delft University of Technology, The Netherlands; John H. Lockhart, Headquarters, U.S. Army Corps of Engineers, Washington, DC (retired); and Charlie Johnson, U.S. Army Engineer District, Chicago, Chicago, IL (retired).

VI-4-10. <u>Symbols</u>.

γ_d Unit dry weight of an in situ material (force/length^3)

γ_s Specific weight of stone (force/length^3)

A Cross-sectional area (length^2)

D_{85} Effective grain size for which 85 percent of the soil (by weight) has a smaller grain size (length)

D_r Relative density of noncohesive sands (Equation VI-4-1)

D_s Diameter of an equivalent-volume sphere (Equation VI-4-4)

e Void ratio of a cohesionless soil

E_c Concrete modulus of elasticity (force/length^2)

f_c Compressive strength of concrete (force/length)

Δh Head difference over the flow length (length)

K Empirical coefficient of permeability (length/time)

L Length of flow path (length)

PR Piping resistance (Equation VI-4-7)

Q Discharge through a uniform soil (Equation IV-4-3) (length^3/time)

R_c Relative soil compaction (Equation VI-4-2)

w_c Specific weight of concrete, (force/length^3)

W_s Stone weight (force)

CHAPTER 5

Fundamentals of Design

TABLE OF CONTENTS

List of Figures

List of Tables

CHAPTER VI-5

Fundamentals of Design

VI-5-1. Introduction.

    a. Overview.

    (1) Planning and design procedures for coastal projects are described in Part V-1, "Planning and Design Process." The engineering design steps related to a specific type of coastal structure can be schematized as follows:

    (a) Specification of functional requirements and structure service lifetime.

    (b) Establishment of the statistics of local short-term and long-term sea states as well as estimation of possible geomorphological changes.

    (c) Selection of design levels for the hydraulic responses: wave runup, overtopping, wave transmission, wave reflection (e.g., 20 percent probability of overtopping discharge exceeding 10^{-5} m^3/s A m during 1 hr in a 50-year period).

    (d) Consideration of construction equipment and procedures, and of availability and durability of materials (e.g., only land based equipment operational and available at reasonable costs, rock of sufficient size easily available).

    (e) Selection of alternative structure geometries to be further investigated (e.g., composite caisson structures, rubble structures with and without crown walls).

    (f) Identification of all possible failure modes for the selected structures (e.g., armor layer displacement).

    (g) Selection of design damage levels for the identified failure modes (e.g., 50 percent probability of displacement of 5 percent of the armor units within 50 years).

    (h) Conceptual design of the structural parts based on the chosen design levels for failure mode damage and hydraulic responses (e.g., determination of armor layer block size and crest height for a breakwater).

    (i) Evaluation of costs of the alternative structures and selection of preferred design(s) for more detailed analysis and optimization.

    (j) Detailed design including economical optimization and evaluation of the overall safety of the structure. This stage will involve scale model tests and/or advanced computational analyses for non-standard and major structures.

    (2) Items *c* and *g* are closely related to item *a*, and the failure modes mentioned in item *f* are dealt with in Part VI-2-4, "Failure Modes of Typical Structure Types."

EM 1110-2-1100 (Part VI)
Change 3 (28 Sep 11)

(3) The previous steps are a brief summary of the flow chart given as Figure V-1-2 in Part V-1-1. They are the steps most related to actual design of project structure elements. In all steps, the outlined design procedure should preferably involve a probabilistic approach which allows implementation of safety based on reliability assessments. The principles are explained in Part VI-6, "Reliability Based Design of Coastal Structures." The present Part VI-5 discusses the basic tools available for conceptual design related to wave-structure interactions (item *h* in the design process).

(4) Wave-structure interaction can be separated into hydraulic responses (such as wave runup, wave overtopping, wave transmission and wave reflection), and loads and response of structural parts. Each interaction is described by a formula, which in most cases is semiempirical in nature with the form based on physical considerations but the empirical constants determined by fitting to experimental data.

(5) The uncertainty and bias of the formula are given when known. Tables of available partial safety factors and the related design equations which show how the partial safety factors are implemented are given in Part VI-6, "Reliability Based Design of Coastal Structures."

b. Wave/structure interaction.

(1) Hydraulic response.

(a) Design conditions for coastal structures include acceptable levels of hydraulic responses in terms of wave runup, overtopping, wave transmission, and wave reflection. These topics are covered in Part VI-5-2, "Structure Hydraulic Response."

(b) The wave runup level is one of the most important factors affecting the design of coastal structures because it determines the design crest level of the structure in cases where no (or only marginal) overtopping is acceptable. Examples include dikes, revetments, and breakwaters with pedestrian traffic.

(c) Wave overtopping occurs when the structure crest height is smaller than the runup level. Overtopping discharge is a very important design parameter because it determines the crest level and the design of the upper part of the structure. Design levels of overtopping discharges frequently vary, from heavy overtopping of detached breakwaters and outer breakwaters without access roads, to very limited overtopping in cases where roads, storage areas, and moorings are close to the front of the structure.

(d) At impermeable structures, wave transmission takes place when the impact of overtopping water generates new waves at the rear side of the structure. With submerged structures, the incident waves will more or less pass over the structure while retaining much of the incident wave characteristics. Permeable structures like single stone size rubble mounds and slotted screens allow wave transmission as a result of wave penetration. Design levels of transmitted waves depend on the use of the protected area. Related to port engineering is the question of acceptable wave disturbance in harbor basins, which in turn is related to the movements of moored vessels. Where groins are included as part of a coastal protection scheme, it is desirable to ensure wave transmission (sediment transport) across the groins.

(e) Wave reflection from the boundary structures like quay walls and breakwaters determines to a large extent the wave disturbance in harbor basins. Also, maneuvering conditions at harbor entrances are highly affected by wave reflection from the breakwaters. Reflection causing steep waves and cross waves can be very dangerous to smaller vessels. Moreover, breakwaters and jetties can cause reflection of waves onto neighboring beaches and thereby increase wave impacts on beach processes.

(2) Wave loadings and related structural response.

(a) An important part of the design procedure for structures in general is the determination of the loads and the related stresses, deformations, and stability conditions of the structural members.

(b) In the case of rubble-mound structures exposed to waves, such procedures cannot be followed because the wave loading on single stones or blocks cannot be determined by theory, by normal scale model tests, or by prototype recordings. Instead a black box approach is used in which experiments are used to establish relationships between certain wave characteristics and the structural response, usually expressed in terms of armor movements. The related stresses, e.g., in concrete armor blocks, are known only for a few types of blocks for which special investigations have been performed. Rubble-mound structures are covered in Part VI-5-3, "Rubble-Mound Structure Loading and Response."

For vertical-front monolithic structures like breakwater caissons and seawalls it is possible either from theory or experiments to estimate the wave loadings and subsequently determine stresses, deformations, and stability. Vertical-front structures are covered in Part VI-5-4, "Vertical-Front Structure Loading and Response."

VI-5-2. Structure Hydraulic Response.

a. Wave runup and rundown on structures.

(1) Introduction.

(a) Wind-generated waves have wave periods which trigger wave breaking on almost all sloping structures. The wave breaking causes runup, R_u, and rundown, R_d, defined as the maximum and minimum water-surface elevation measured vertically from the still-water level (SWL), see Figure VI-5-1a.

(b) R_u and R_d depend on the height and steepness of the incident wave and its interaction with the preceding reflected wave, as well as the slope angle, the surface roughness, and the permeability and porosity of the slope. Maximum values of flow velocities and values of R_u and R_d for a given sea state and slope angle are reached on smooth impermeable slopes.

(c) Figure VI-5-1a illustrates the variation of the flow velocity vectors along an impermeable slope over the course of a wave cycle. Figure VI-5-1b illustrates this variation for a permeable slope. Both the magnitude and direction of the velocity vectors are important for stability of the armor units. Generally, the most critical flow field occurs in a zone around and just below still-water level (swl) where down-rush normally produces the largest destabilizing forces. Exceptions are slopes flatter than approximately 1:3.5 in which cases up-rush is more vulnerable. The velocity vectors shown in Figure VI-5-1b explain why reshaping breakwaters attain S-profiles.

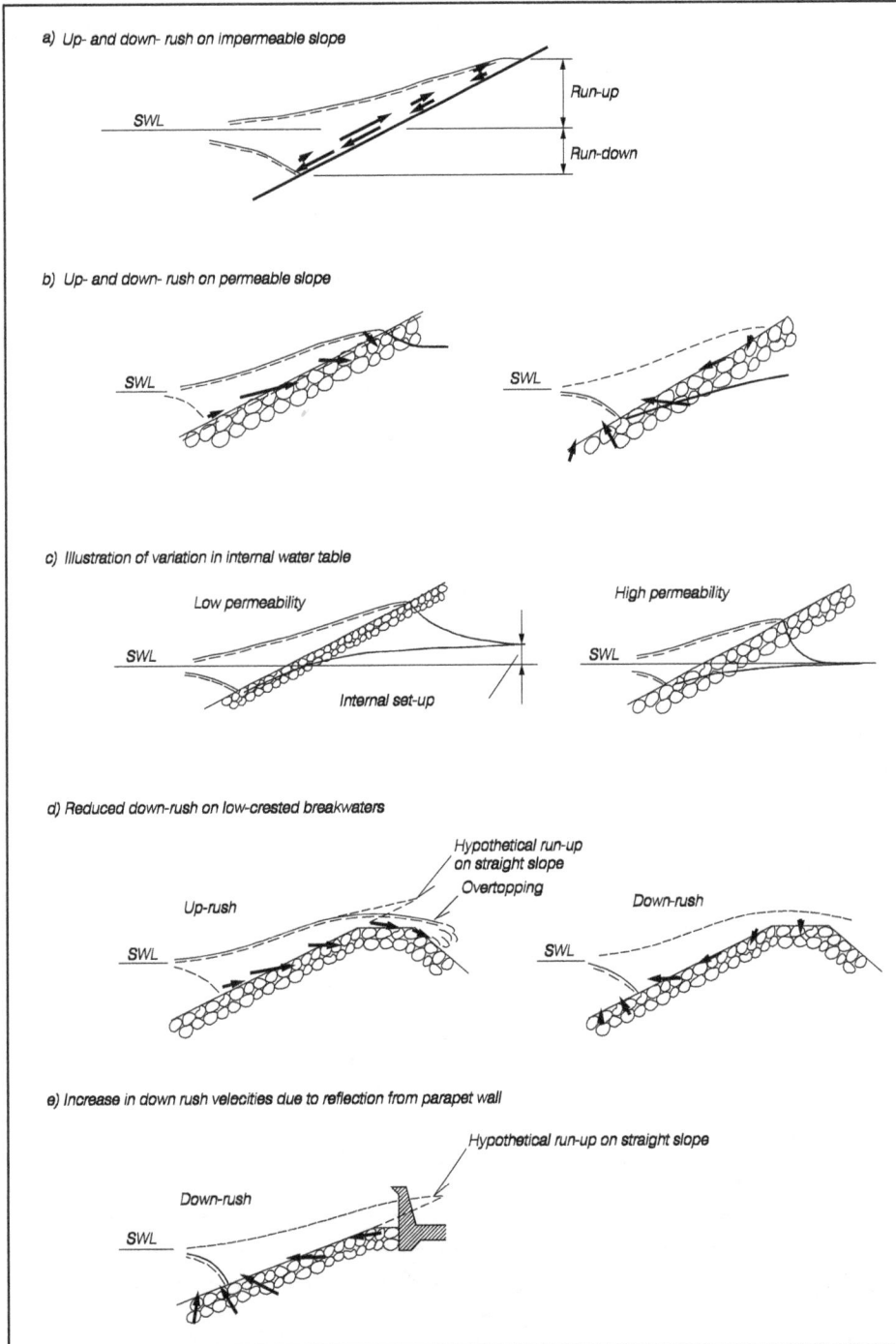

Figure VI-5-1. Illustration of runup and rundown (Burcharth 1993)

(d) Increase in permeability of the slope reduces the flow velocities along the slope surface because a larger proportion of the flow takes place inside the structure. The wave action will cause a rise of the internal water level (phreatic line) indicated in Figure VI-5-1c, leading to an increase in the mean pore pressures. The internal setup is due to a greater inflow surface area during wave runup than the outflow surface area during rundown. The mean flow path for inflow is also shorter than that for outflow. The rise of the phreatic line will continue until the outflow balances the inflow. The lower the permeability of the structure, the higher the setup as indicated on Figure VI-5-1c.

(e) Barends (1988) suggested practical formulae for calculation of the penetration length and the maximum average setup which occurs after several cycles. Two cases are considered: a conventional breakwater structure with open (permeable) rear side, and a structure with a closed (impermeable) rear side. The latter case causes the largest setup.

(f) An example of a numerical calculation of the internal flow patterns in a breakwater exposed to regular waves is shown in Figure VI-5-2. The strong outflow in the zone just below SWL when maximum rundown occurs is clearly seen.

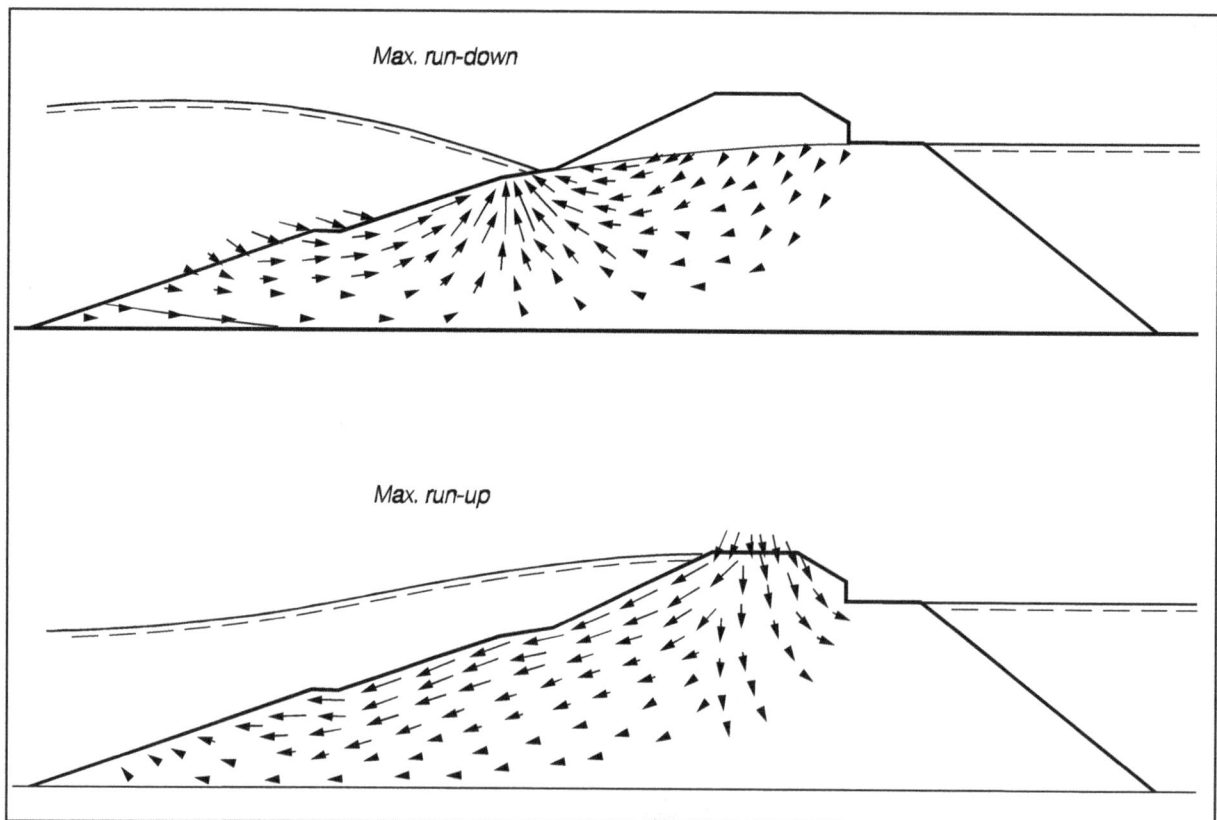

Figure VI-5-2. Typical velocity field for the porous flow in a breakwater. Numerical calculation (Barends et al. 1983)

(g) Increasing structure porosity also reduces the overflow velocities because a larger portion of the incoming water volume can be stored in the pores which then act as reservoirs.

The destabilizing forces on armor units are thereby reduced. This positive reservoir effect is reduced in the case of a large internal setup of the water table.

(h) Breakwaters with crest levels lower than the runup level are called low-crested breakwaters. Although the runup velocities are almost unchanged compared to nonovertopped slopes, the rundown velocities are reduced due to the overtopping of some part of the incoming wave as seen in Figure VI-5-1d. Greater overtopping reduces rundown, and thus, lessens the destabilizing flow forces on the armor units. Parapet walls which cut off the hypothetical runup wedge (shown in Figure VI-5-1e) will increase the down-rush velocities and thereby increase the destabilizing flow forces on the armor units.

(2) Surf similarity parameter (Iribarren number).

(a) Wave runup and rundown on a structure depend on the type of wave breaking. Breaker types can be identified by the so-called surf-similarity parameter, ξ (Battjes 1974b). The parameter ξ is also referred to as the breaker parameter or Iribarren number. The surf-similarity parameter was originally defined for regular waves as

$$\xi_o = \frac{\tan \alpha}{\sqrt{s_o}} \qquad\qquad\qquad \text{(VI-5-1)}$$

where

α = slope angle

s_o = deepwater wave steepness ($= H_o / L_o$)

H_o = deepwater wave height

L_o = deepwater wavelength ($= gT^2/2\pi$)

T = wave period

g = acceleration due to gravity

(b) The wave height H_b at the breaking point is sometimes substituted for H_o in which case the parameter is denoted by ξ_b. Breaker types and related ranges of ξ_o-values are given for impermeable slopes in Table VI-5-1. The boundaries of transition from one type of breaker to another are approximate.

Table VI-5-1
Types of Wave Breaking on Impermeable Slopes and Related ξ_o-Values

SPILLING	$\xi_o < 0.5$
PLUNGING	$0.5 < \xi_o < 3$
COLLAPSING	$\xi_o \approx 3 \text{ à } 3.5$
SURGING	$\xi_o > 3.5$

(c) For irregular waves the surf--similarity parameter is defined as

$$\xi_{om} = \frac{\tan \alpha}{\sqrt{s_{om}}} \quad \text{or} \quad \xi_{op} = \frac{\tan \alpha}{\sqrt{s_{op}}} \tag{VI-5-2}$$

where

$$s_{om} = \frac{H_s}{L_{om}} = \frac{2\pi}{g} \frac{H_s}{T_m^2}$$

$$s_{op} = \frac{H_s}{L_{op}} = \frac{2\pi}{g} \frac{H_s}{T_p^2}$$

and

H_s = significant wave height of incident waves at the toe of the structure

T_m = mean wave period

T_p = wave period corresponding to the peak of the wave spectrum

Note that s_{om} and s_{op} are fictitious wave steepnesses because they are ratios between a statistical wave height at the structure and representative deepwater wavelengths.

(d) The relative runup R_u/H is a function of ξ, the wave angle of incidence, and the slope geometry (profile, surface roughness, porosity). Differences in runup characteristics make it convenient to distinguish between impermeable and permeable slopes. Impermeable slopes belong to dikes, revetments, and breakwaters with either impermeable surfaces (e.g., asphalt, concrete) or rough surfaces (e.g., rubble stones, concrete ribs) on fine core materials. Permeable slopes belong typically to rubble-mound structures with secondary armor layers, filter layers, and quarryrun core.

(3) Wave runup and rundown on impermeable slopes. Runup on impermeable slopes can be formulated in a general expression for irregular waves having the form (Battjes 1974)

$$\frac{R_{ui\%}}{H_s} = (A\xi + C)\gamma_r\gamma_b\gamma_h\gamma_\beta \qquad (VI-5-3)$$

where

$R_{ui\%}$ = runup level exceeded by i percent of the incident waves

ξ = surf-similarity parameter, ξ_{om} or ξ_{op}

A, C = coefficients dependent on ξ and i but related to the reference case of a smooth, straight impermeable slope, long-crested head-on waves and Rayleigh-distributed wave heights

γ_r = reduction factor for influence of surface roughness ($\gamma_r = 1$ for smooth slopes)

γ_b = reduction factor for influence of a berm ($\gamma_b = 1$ for non-bermed profiles)

γ_h = reduction factor for influence of shallow-water conditions where the wave height distribution deviates from the Rayleigh distribution ($\gamma_h = 1$ for Rayleigh distributed waves)

γ_β = factor for influence of angle of incidence β of the waves ($\gamma_\beta = 1$ for head-on long-crested waves, i.e., $\beta = 0°$). The influence of directional spreading in short-crested waves is included in γ_β as well

(a) Smooth slope, irregular long-crested head-on waves. Van Oorschot and d'Angremond (1968) tested slopes of 1:4 and 1:6 for $\xi_{op} < 1.2$. Ahrens (1981a) investigated slopes between 1:1 and 1:4 for $\xi_{op} > 1.2$. Figure VI-5-3 shows the range of test results and the fit of Equation VI-5-3 for $R_{u2\ percent}$. Considerable scatter is observed, most probably due to the fact that the runs for ξ_{op}

> 1.2 contained only 100-200 waves. The coefficient of variation, $\sigma_{Ru} / \overline{R_u}$, seems to be approximately 0.15.

- The significant runup level $R_{us} = R_{u33\%}$ depicted in Figure VI-5-4 does not contain data for $\xi_{op} < 1.2$. The coefficient of variation appears to be approximately 0.1.

- The coefficients A and C together with estimates of the coefficient of variation for R_u are given in Table VI-5-2. It should be noted that data given in Allsop et al. (1985) showed runup levels considerably smaller than given here.

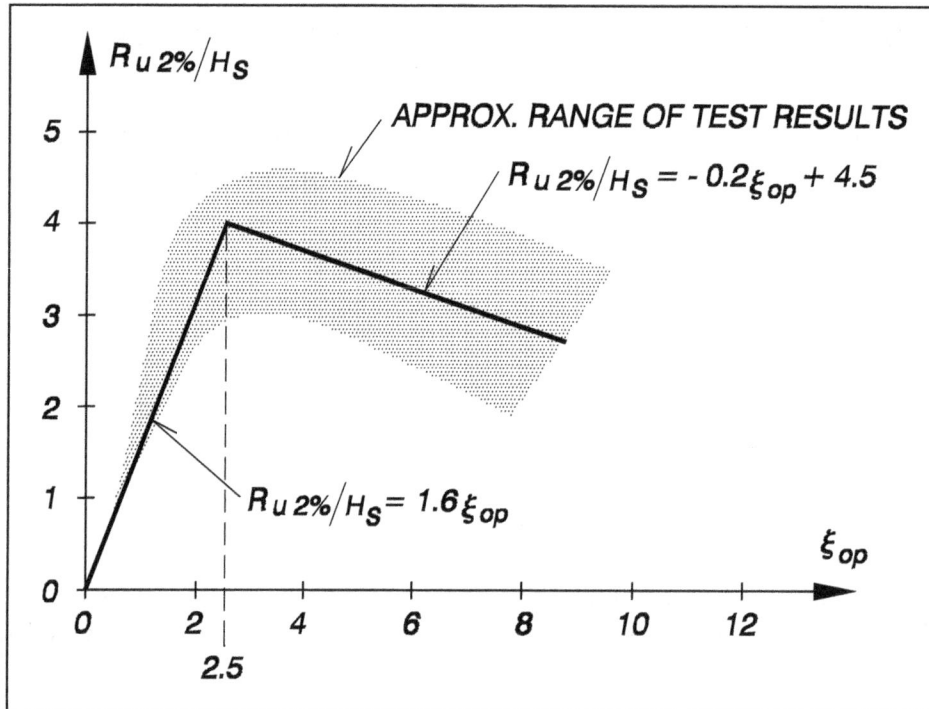

Figure VI-5-3. $R_{u2\%}$ for head-on waves on smooth slopes. Data by Ahrens (1981a) and Van Oorschot and d'Angremond (1968)

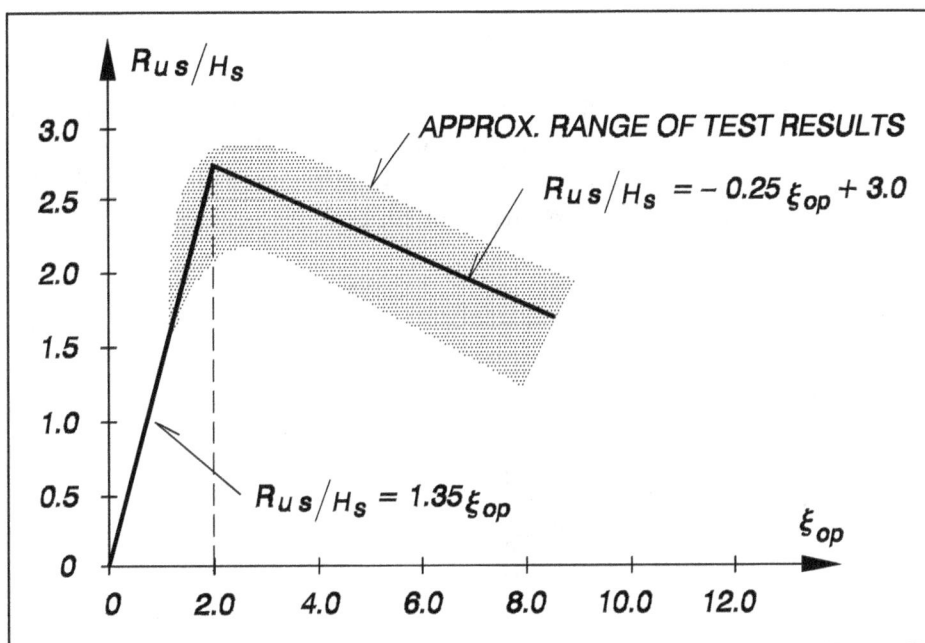

Figure VI-5-4. R_{us} for head-on waves on smooth slopes. Data by Ahrens
(1981a)

Table VI-5-2
Coefficients in Equation VI-5-3 for Runup of Long-Crested Irregular Waves on Smooth
Impermeable Slopes

ξ	R_u	ξ-Limits	A	C	σ_{Ru} / R_u
ξ_{op}	$R_{u2\,percent}$	$\xi_p \leq 2.5$	1.6	0	
		$2.5 < \xi_p < 9$	-0.2	4.5	≈ 0.15
	R_{us}	$\xi_p \leq 2.0$	1.35	0	
		$2.0 < \xi_p < 9$	-0.15	3.0	≈ 0.10

- Generally less experimental data are available for rundown. Rundown corresponding to $R_{d2\,percent}$ from long-crested irregular waves on a smooth impermeable slope can be estimated from

$$\frac{R_{d2\%}}{H_s} = \begin{cases} 0.33\xi_{op} & \text{for} \quad 0 < \xi_{op} \leq 4 \\ 1.5 & \text{for} \quad \xi_{op} > 4 \end{cases} \tag{VI-5-4}$$

- In the Dutch publication by Rijkswaterstaat Slope Revetments of Placed Blocks, 1990, the following expression was given for rundown on a smooth revetment of placed concrete block

$$\frac{R_{d2\%}}{H_s} = 0.5\xi_{op} - 0.2 \quad \text{(VI-5-5)}$$

• Another set of runup data for long-crested head-on waves on smooth slopes was presented by de Waal and van der Meer (1992). The data cover small scale tests for slopes 1:3, 1:4, 1:5, 1:6 and large scale tests for slopes 1:3, 1:6, 1:8. The surf-similarity parameter range for the small scale tests is $0.6 < \xi_{op} < 3.4$, and for the large scale tests $0.6 < \xi_{op} < 2.5$. The data are shown in Figure VI-5-5 and were used by de Waal and van der Meer (1992) and van der Meer and Janssen (1995) as the reference data for the evaluation of the γ-factors defined by Equation VI-5-3.

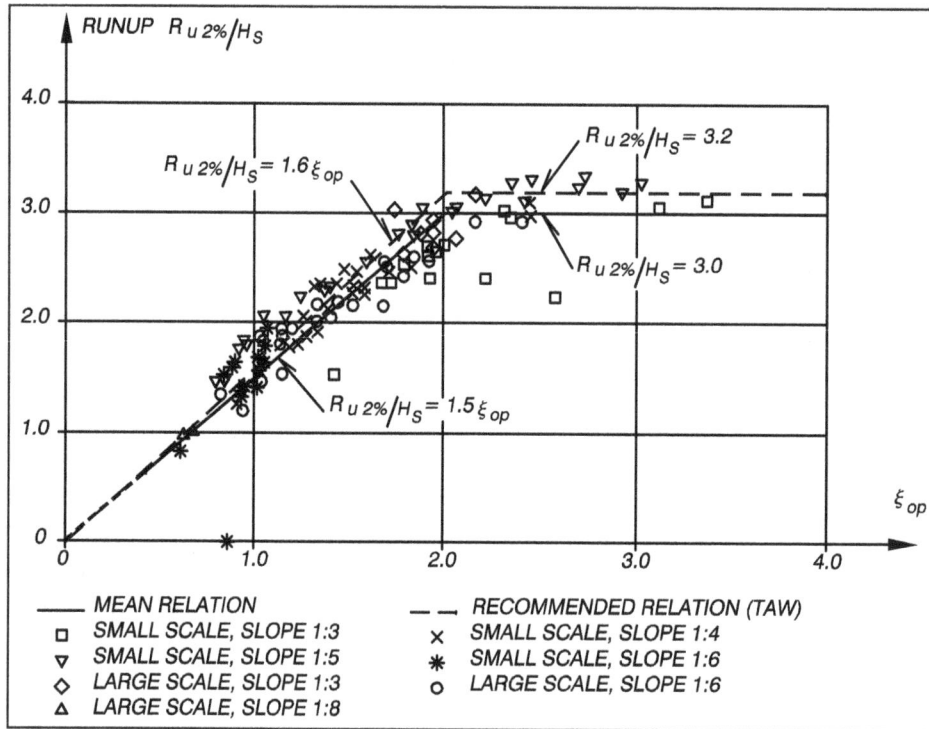

Figure VI-5-5. $R_{u2\%}$ for long-crested head-on waves on smooth slopes.
From de Waal and van der Meer (1992)

• The mean relationship, taken as the reference case for Equation VI-5-3, is shown with the solid line and is represented by the expression

$$\frac{R_{u2\%}}{H_s} = \begin{cases} 1.5\xi_{op} & \text{for} \quad 0.5 < \xi_{op} \leq 2 \\ 3.0 & \text{for} \quad 2 < \xi_{op} < 3-4 \end{cases} \qquad \text{(VI-5-6)}$$

• The dotted line includes a small safety factor, and this relationship is recommended for design by the Technical Advisory Committee on Water Defence in Holland.

• Based on a somewhat reduced data set compared to Figure VI-5-5, the uncertainty on Equation VI-5-6 is described by de Waal and van der Meer (1992) by assuming the factor 1.5 as a stochastic variable with a normal distribution and a coefficient of variation of 0.085.

– Influence of surface roughness on runup. The original values for γ_r given in Dutch publications and in the old *Shore Protection Manual* have been updated based on experiments including large-scale tests with random waves. These factors are given in Table VI-5-3. The new γ_r values taken from de Waal and van der Meer (1992) are valid for $1 < \xi_{op} < 3\text{-}4$. For larger ξ_{op}-values the γ_r factors will slowly increase to 1.

Table VI-5-3
Surface Roughness Reduction Factor γ_r in Equation VI-5-3, Valid for $1 < \xi_{op} < 3\text{-}4$

Type of Slope Surface	γ_r
Smooth, concrete, asphalt	1.0
Smooth block revetment	1.0
Grass (3 cm length)	0.90 - 1.0
1 layer of rock, diameter D, (H_s/D = 1.5 - 3.0)	0.55 - 0.6
2 or more layers of rock, (H_s/D = 1.5 - 6.0)	0.50 - 0.55
Roughness elements on smooth surface (length parallel to waterline = P, width = b, height = h)	
Quadratic blocks, $P = b$	
h/b b/H_s area coverage	
0.88 0.12 - 0.19 1/9	0.70 - 0.75
0.88 0.12 - 0.24 1/25	0.75 - 0.85
0.44 0.12 - 0.24 1/25	0.85 - 0.95
0.88 0.12 - 0.18 1/25 (above SWL)	0.85 - 0.95
0.18 0.55 - 1.10 1/4	0.75 - 0.85
Ribs	
1.00 0.12 - 0.19 1/7.5	0.60 - 0.70

   - Influence of a berm on runup. A test program at Delft Hydraulics was designed to clarify the influence of a horizontal or almost horizontal berm on wave runup. Figure VI-5-6 shows the range of tested profiles and sea states.

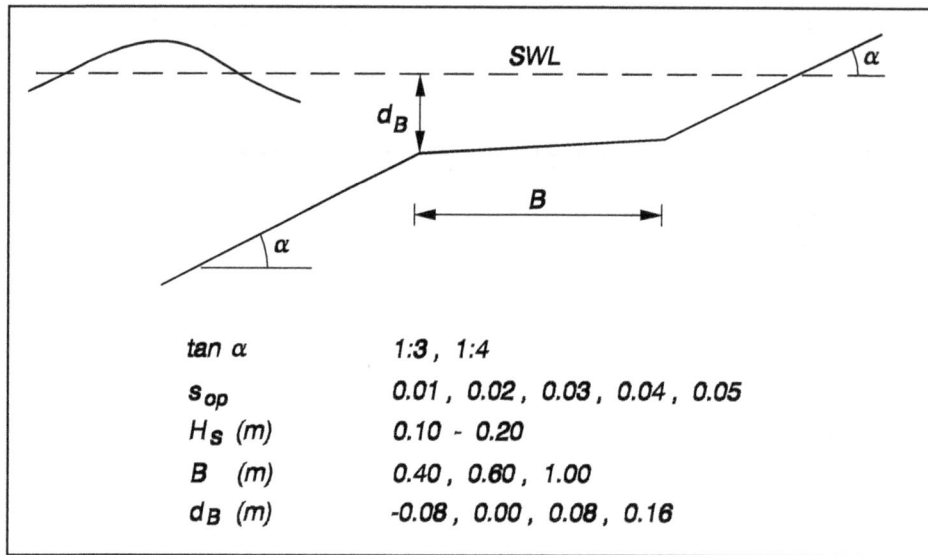

Figure VI-5-6. Parameters in berm test program at Delft Hydraulics

• According to de Waal and van der Meer (1992) the effect of a berm can be taken into account by the following formulation of the reference Equation VI-5-6

$$\frac{R_{u2\%}}{H_s} = \begin{cases} 1.5\xi_{op}\gamma_r\gamma_b\gamma_h\gamma_\beta = 1.5\xi_{eq}\gamma_r\gamma_h\gamma_\beta & \text{for} \quad 0.5 < \xi_{eq} \le 2 \\ 3.0\gamma_r\gamma_h\gamma_\beta & \text{for} \quad \xi_{eq} > 2 \end{cases} \tag{VI-5-7}$$

where ξ_{eq} is the breaking wave surf similarity parameter based on an equivalent slope (see Figure VI-5-7). The berm influence factor γ_b is defined as

$$\gamma_b = \frac{\xi_{eq}}{\xi_{op}} = 1 - r_B(1 - r_{dB}), \quad 0.6 \le \gamma_b \le 1.0 \tag{VI-5-8}$$

where

$$r_B = 1 - \frac{\tan\alpha_{eq}}{\tan\alpha}$$

$$r_{dB} = 0.5\left(\frac{d_B}{H_s}\right)^2, \quad 0 \le r_{dB} \le 1 \tag{VI-5-9}$$

and the *equivalent* slope angle α_{eq} and the *average* slope angle α are defined in Figure VI-5-7.

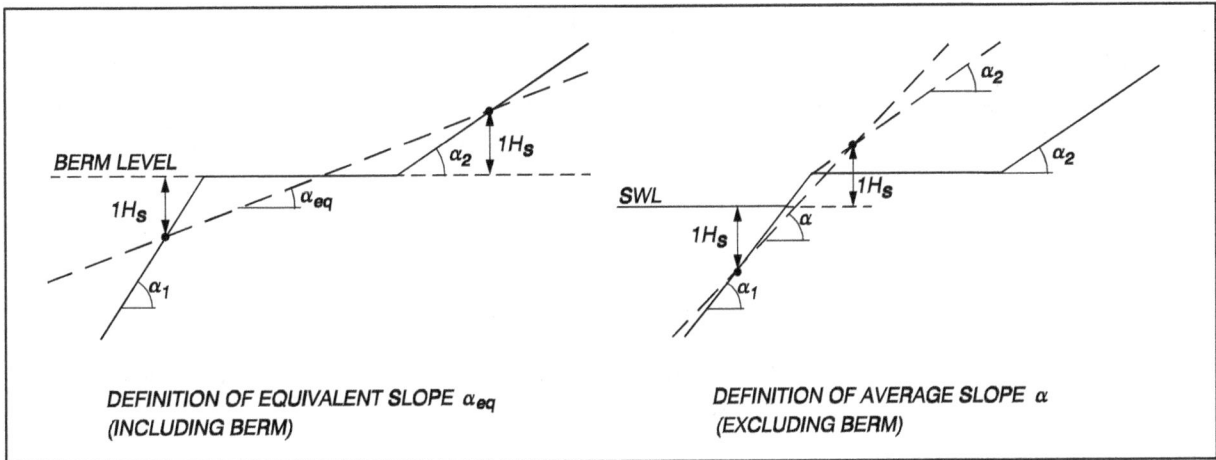

Figure VI-5-7. Definition of α_{eq} and α in Equation VI-5-9

- The influence of the berm can be neglected when the berm horizontal surface is positioned more than $H_s\sqrt{2}$ below SWL. If the berm horizontal surface lies higher than $d_B = H_s\sqrt{2}$ above SWL, then the runup can be set to $R_{u2\%} = d_B$ if $B/H_s \geq 2$. The berm is most effective when lying at SWL, i.e., $d_B = 0$. An optimum berm width B, which corresponds to $\gamma_b = 0.6$, can be determined from the formulae given by Equations VI-5-8 and VI-5-9.

- The use of ξ_{eq} in Equation VI-5-7 is evaluated in Figure VI-5-8 on the basis of the test program given in Figure VI-5-6, which implies $\gamma_r = \gamma_h = \gamma_\beta = 1$.

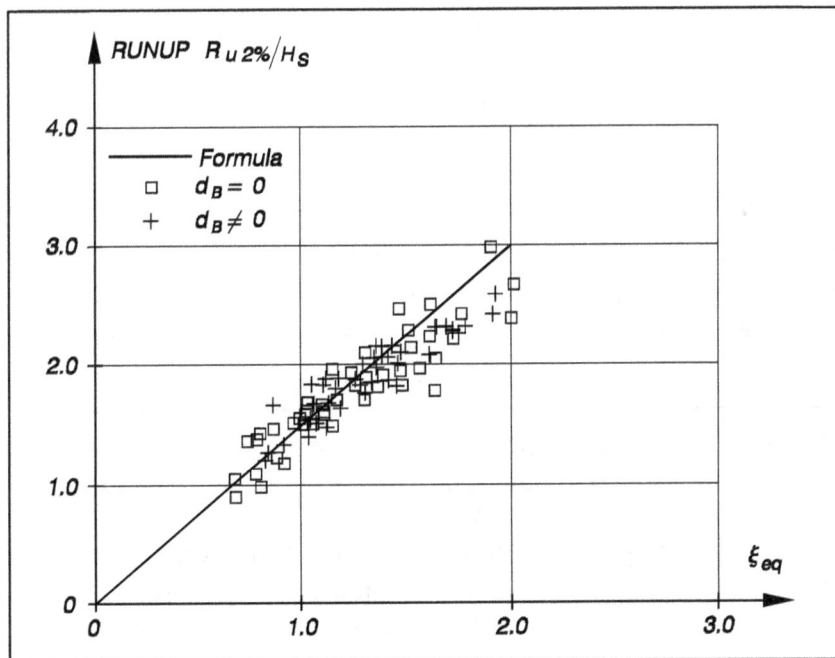

Figure VI-5-8. Evaluation of the use of ξ_{eq} to account for the influence of a berm

— Influence of shallow water on runup. Wave heights in Equation VI-5-7 are characterized by H_s which provides a unique definition for deep water conditions where wave heights are Rayleigh distributed. In shallow water where some waves break before they reach the structure, the wave heights will no longer be Rayleigh distributed. According to de Waal and van der Meer (1992), the influence factor can be estimated as

$$\gamma_h = \frac{H_{2\%}}{1.4 H_s} \tag{VI-5-10}$$

where the representative wave heights are specified for the water depth at the toe of the structure ($H_{2\%}/H_s = 1.4$ for Rayleigh distributed wave heights).

— Influence of angle of wave attack on runup. Both the angle of incidence and the directional spreading of the waves influence the runup. A test program for runup on smooth slopes at Delft Hydraulics, as specified in Figure VI-5-9, revealed the variations in the influence factor γ_β as given by Equation VI-5-11 and depicted in Figure VI-5-10.

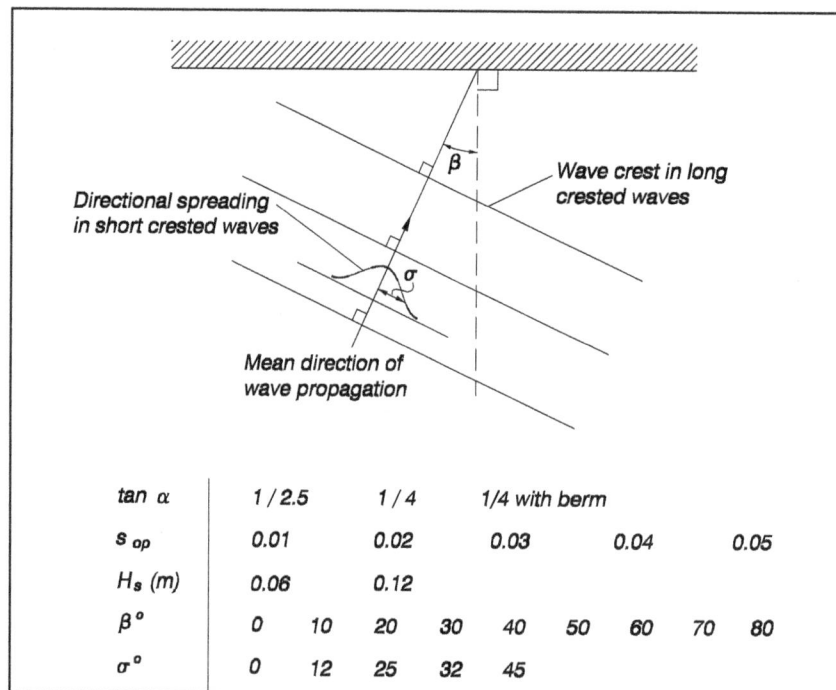

tan α	1 / 2.5		1 / 4		1/4 with berm			
s_{op}	0.01		0.02		0.03	0.04		0.05
H_s (m)	0.06		0.12					
$\beta°$	0	10	20	30	40	50	60	70	80
$\sigma°$	0	12	25	32	45			

Figure VI-5-9. Test program for wave runup on smooth slopes
conducted at Delft Hydraulics, de Waal and van der Meer (1992)

• Note that γ_β-values larger than 1 were obtained for long-crested waves in the range $10° \le \beta \le 30°$, and that values very close to 1 were obtained for short-crested waves for β up to $50°$.

• Based on the results, the following formulas for mean values of γ_β were given

Long-crested waves $= 1.0$ for $0^\circ \leq \beta \leq 10^\circ$
(mainly swell) $\gamma_\beta = \cos(\beta - 10^\circ)$ for $10^\circ < \beta \leq 63^\circ$
 $= 0.6$ for $\beta > 63^\circ$ (VI-5-11)

Short-crested waves $\gamma_\beta = 1 - 0.0022\,\beta$

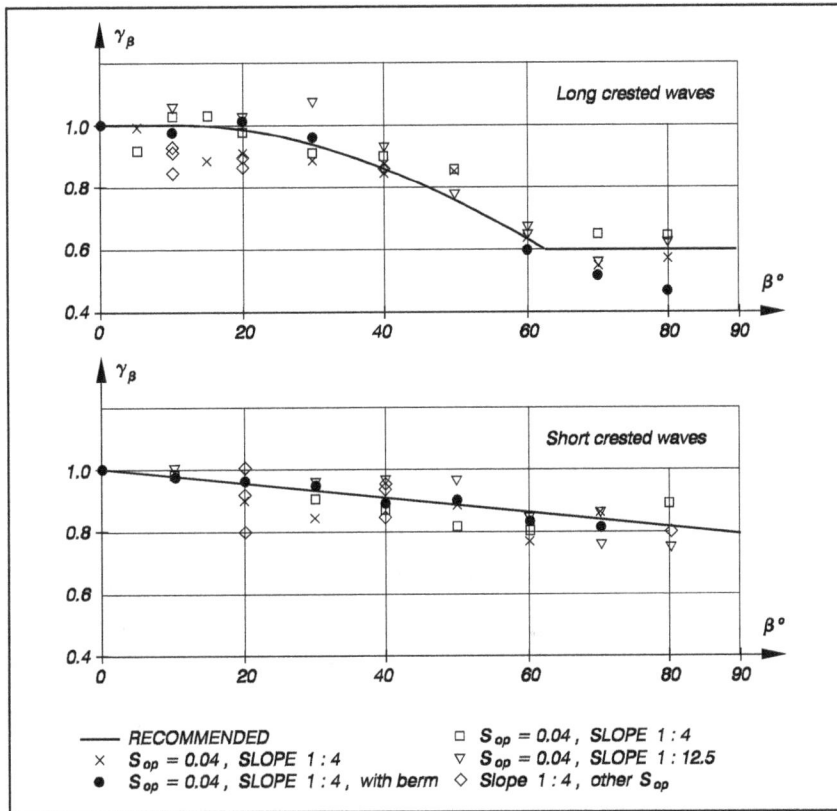

Figure VI-5-10. Influence of angle of incidence β and directional
spreading on runup on smooth slopes conducted at Delft
Hydraulics; de Waal and van der Meer (1992)

(b) Rock armored slopes, irregular long-crested head-on waves. Runup on rock armored impermeable and permeable slopes was studied by Delft Hydraulics in the test program given in Table VI-5-4.

Table VI-5-4
Test Program(van der Meer 1988)

Slope Angle cot α	Grading D_{85}/D_{15}	Spectral Shape	Core Permeability	Relative Mass Density	Number of Tests	Range $H_s/\Delta D_{n50}$	Range s_{om}
2	2.25	PM	none	1.63	1.9e+31	0.8-1.6	0.005-0.016
3	2.25	PM	none	1.63		1.2-2.3	0.006-0.024
4	2.25	PM	none	1.63		1.2-3.3	0.005-0.059
6	2.25	PM	none	1.63		1.2-4.4	0.004-0.063
3*	1.25	PM	none	1.62		1.4-2.9	0.006-0.038
4	1.25	PM	none	1.62		1.2-3.4	0.005-0.059
3	2.25	narrow	none	1.63		1.0-2.8	0.004-0.054
3	2.25	wide	none	1.63		1.0-2.4	0.004-0.043
3^1	1.25	PM	permeable	1.62		1.6-3.2	0.008-0.060
2	1.25	PM	permeable	1.62		1.5-2.8	0.007-0.056
1.5	1.25	PM	permeable	1.62		1.5-2.6	0.008-0.050
2	1.25	PM	homogeneous	1.62		1.8-3.2	0.008-0.059
2	1.25	PM	permeable	0.95		1.7-2.7	0.016-0.037
2	1.25	PM	permeable	2.05		1.6-2.5	0.014-0.032
2^2	1.25	PM	permeable	1.62		1.6-2.5	0.014-0.031
2^3	1.25	PM	permeable	1.62		1.4-5.9	0.010-0.046

PM Pierson Moskowitz spectrum [2] Foreshore 1:30
[1] Some tests repeated in Delta Flume [3] Low-crested structure with foreshore 1:30

- The core permeability in Table VI-5-4 refers to the structures shown in details *a, c* and *d* of Figure VI-5-11, taken from van der Meer (1988). The figure provides definition of a notational permeability parameter P which is used in various formulae by van der Meer to take into account the effect of permeability on response to wave action. The value $P = 0.4$ in Figure VI-5-11, detail *b*, is not identified by tests, but instead is an estimated value.

- The runup results from the test program described in Table VI-5-4 are presented in Figure VI-5-12.

- Note that $\xi_{om} = \tan \alpha / (2\pi H_s /gT_{om}^2)^{1/2}$, where T_{om} is the mean wave period, is used instead of ξ_{op}. By using T_{om} instead of T_{op} variations in the width of the wave spectrum are taken into account. The ratio $T_{om}/T_{op} = \xi_{om}/\xi_{op} = 0.79 - 0.87$ for Joint North Sea Wave Program (JONSWAP) spectra and 0.71 - 0.82 for Pierson-Moskowitz spectra.

• The central fit to the data for impermeable rock slopes was given by Delft Hydraulics (1989) as

$$\frac{R_{ui\%}}{H_s} = \begin{cases} A\xi_{om} & \text{for} \quad 1.0 < \xi_{om} \le 1.5 \\ B\left(\xi_{om}\right)^C & \text{for} \quad \xi_{om} > 1.5 \end{cases} \tag{VI-5-12}$$

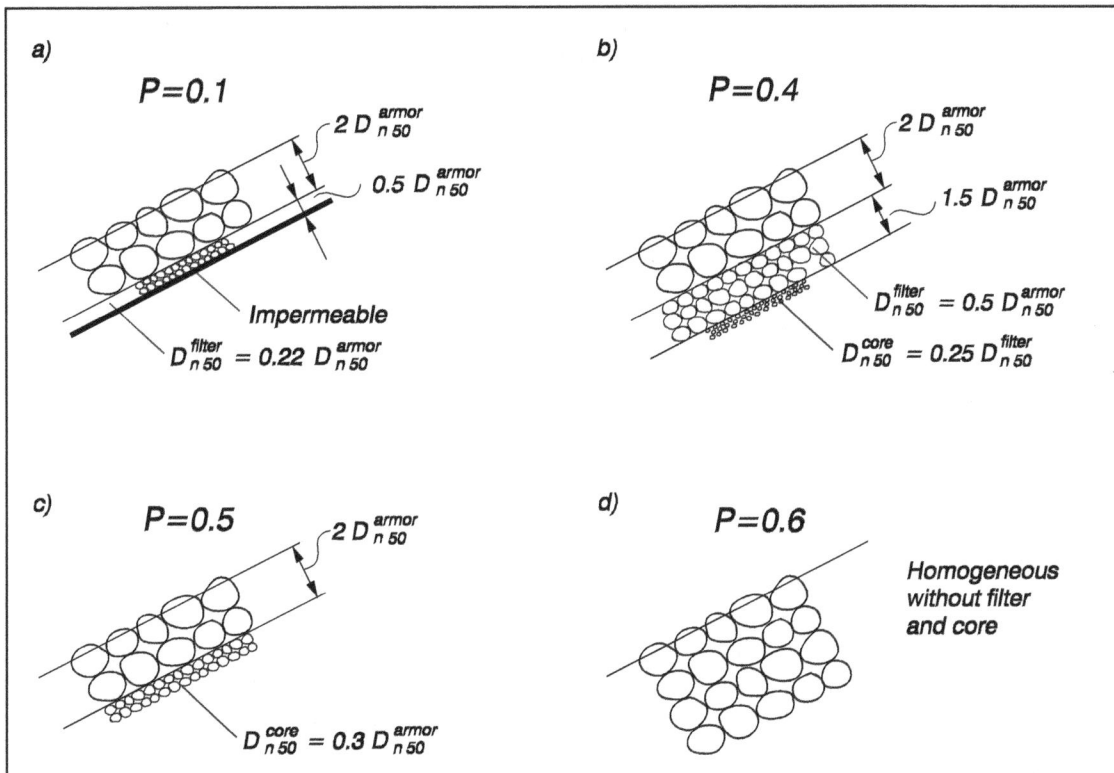

Figure VI-5-11. Notational permeability coefficients (van der Meer 1988)

• The coefficients A, B and C are given in Table VI-5-5. For impermeable slopes the coefficient of variation for A, B and C is 7 percent. Data presented by Ahrens and Heinbaugh (1988a) for maximum runup on impermeable riprap slopes are in agreement with the data represented by Equation VI-5-12.

• Equation VI-5-12 is valid for relatively deep water in front of a structure where the wave height distribution is close to the Rayleigh distribution. Wave breaking on a foreshore results in a truncation in the runup distribution which mainly results in lower runup heights for small exceedence probability levels. However, sometimes higher runup may occur according to observations in the Delft Hydraulics tests and recent tests conducted at Texas A&M University.

(4) Wave runup and rundown on permeable slopes. With respect to runup, permeable structures are defined as structures with core material of such permeability that wave induced porous flow and fluctuations of the internal phreatic line do vary with the frequencies of the waves. The storage capacity of the structure pores results in maximum runup that is smaller than for an equivalent structure with an impermeable core.

(a) Rock armored slopes, irregular long-crested head-on waves. Rock armored permeable slopes with notational permeability $P = 0.5$, as shown in detail c of Figure VI-5-11, were tested in irregular head-on waves by Delft Hydraulics in the program specified in Table VI-5-4. The results are shown in Figure VI-5-12, and the corresponding equation for the central fit to the data is given by

$$R_{ui\%}/H_s \quad \begin{aligned} &= A\,\xi_{om} & &\text{for } 1.0 < \xi_{om} \leq 1.5 \\ &= B\,(\xi_{om})^C & &\text{for } 1.5 < \xi_{om} \leq (D/B)^{1/C} \\ &= D & &\text{for } (D/B)^{1/C} \leq \xi_{om} < 7.5 \end{aligned} \qquad \text{(VI-5-13)}$$

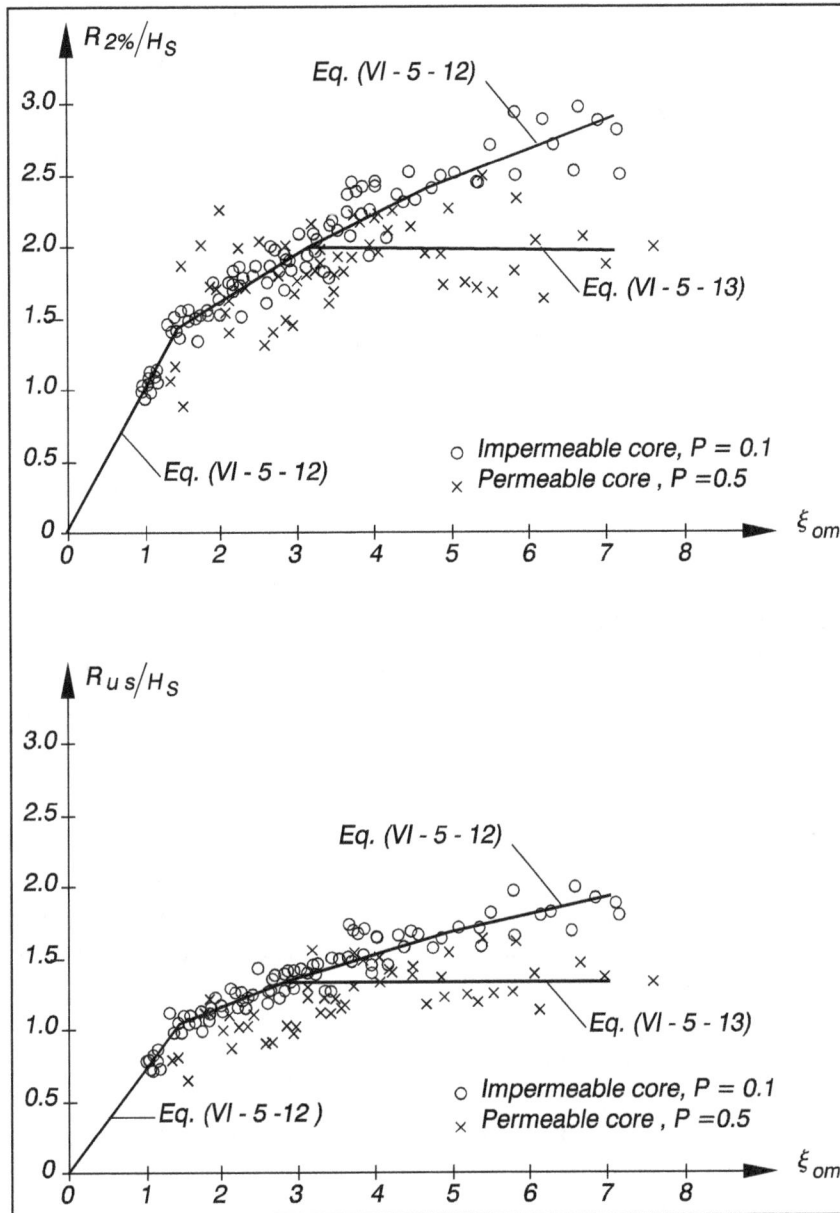

Figure VI-5-12. 2 percent and significant runup of irregular head-on waves on impermeable and permeable rock slopes. Delft Hydraulics (1989)

Table VI-5-5
Coefficients in Equations VI-5-12 and VI-5-13 for Runup of Irregular Head-On Waves on
Impermeable and Permeable Rock Armored Slopes

Percent [1]	A	B	C	D [2]
0.1	1.12	1.34	0.55	2.58
2.0	0.96	1.17	0.46	1.97
5	0.86	1.05	0.44	1.68
10	0.77	0.94	0.42	1.45
(significant)	0.72	0.88	0.41	1.35
50 (mean)	0.47	0.60	0.34	0.82

[1] Exceedence level related to number of waves
[2] Only relevant for permeable slopes

• The coefficients A, B, C and D are listed in Table VI-5-5. For permeable structures the coefficient of variation for A, B, C and D is 12 percent. Tests with homogeneous rock structures with notational permeability $P = 0.6$, as shown in detail d of Figure VI-5-11, showed results almost similar to the test results corresponding to $P = 0.5$ as shown in Figure VI-5-12.

• Equation VI-5-13 is valid for relatively deepwater conditions with wave height distributions close to a Rayleigh distribution. Wave breaking due to depth limitations in front of the structure cause truncation of the runup distribution and thereby lower runup heights for small exceedence probability levels. However, higher runup might also occur according to observations in the Delft Hydraulics tests, van der Meer and Stam (1992). The influence on runup for the shallow-water conditions included in the test program given in Table VI-5-4 were investigated for the rock armored permeable slope. However, no systematic deviations from Equation VI-5-13 were observed.

(b) Statistical distribution of runup. The runup of waves with approximately Rayleigh distributed wave heights on rock armored permeable slopes with $tan\ \alpha \ni 2$ were characterized by van der Meer and Stam (1992) with a best-fit two-parameter Weibull distribution as follows:

$$Prob\left(R_u > R_{up\%}\right) = \exp\left[-\left(\frac{R_{up\%}}{B}\right)^C\right] \ or \qquad \text{(VI-5-14)}$$

$$R_{up\%} = B\left(-\ln p\right)^{1/C} \qquad \text{(VI-5-15)}$$

where

$R_{up\%}$ = Runup level exceeded by p % of the runup

$$B = H_s\left[0.4\left(s_{om}\right)^{-1/4}\left(\cot\alpha\right)^{-0.2}\right] \qquad \text{(VI-5-16)}$$

$$C = \begin{cases} 3.0\left(\xi_{om}\right)^{-3/4} & \text{for} \quad \xi_{om} \le \xi_{omc} \text{ (plunging waves)} \\ 0.52P^{-0.3}\left(\xi_{om}\right)^{P}\sqrt{\cot\alpha} & \text{for} \quad \xi_{om} > \xi_{omc} \text{ (surging waves)} \end{cases}$$ (VI-5-17)

$$\xi_{omc} = \left(5.77P^{0.3}\sqrt{\tan\alpha}\right)^{[1/(P+0.75)]}$$ (VI-5-18)

$$s_{om} = \frac{2\pi H_s}{gT_{om}^2}$$

P = notational permeability, see Figure VI-5-11.

- It follows from Equation VI-5-15 that the scale parameter B is equal to $Ru_{37\%}$ (ln p = -1 for p = 0.37). If the shape parameter C is equal to 2, then Equation VI-5-14 becomes a Rayleigh distribution. The uncertainty on B corresponds to a coefficient of variation of 6 percent for $P < 0.4$ and 9 percent for $P \ge 0.4$.

- Rundown on rock slopes in the Delft Hydraulics test program listed in Table VI-5-4 gave the following relationship which includes the effect of structure permeability P (see Figure VI-5-11).

$$\frac{R_{d2\%}}{H_s} = 2.1\sqrt{\tan\alpha} - 1.2P^{0.15} + 1.5e^{-(60s_{om})}$$ (VI-5-19)

b. Wave overtopping of structures. Wave overtopping occurs when the highest runup levels exceed the crest freeboard, R_c as defined in Figure VI-5-13. The amount of allowable overtopping depends on the function of the particular structure. Certain functions put restrictions on the allowable overtopping discharge. For example access roads and installations placed on the crest of breakwaters and seawalls, berths for vessels as well as reclaimed areas containing roadways, storage areas, and buildings located just behind the breakwater are overtopping design considerations. Design criteria for overtopping should include two levels: Overtopping during normal service conditions and overtopping during extreme design conditions where some damage to permanent installations and structures might be allowed. Very heavy overtopping might be allowed where a breakwater has no other function than protection of harbor entrances and outer basins from waves. However, significant overtopping can create wave disturbances which could lead to damage of moored vessels. Fortunately, waves generated by overtopping usually have much shorter periods than the incident wave train.

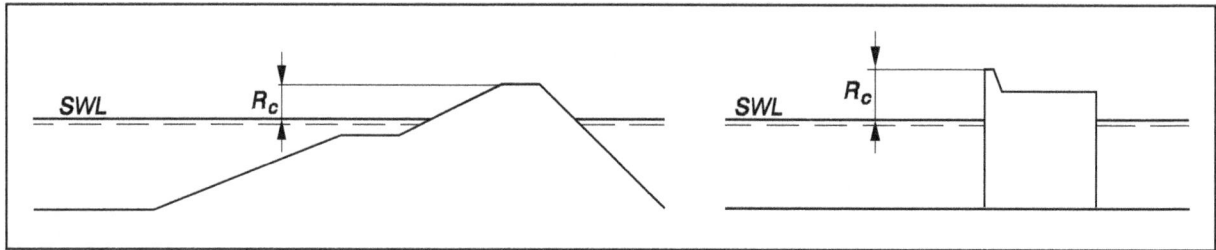

Figure VI-5-13. Definition of crest freeboard, R_c

(1) Admissible average overtopping discharge.

(a) The overtopping discharge from wind-generated waves is very unevenly distributed in time and space because the amount varies considerably from wave to wave. The major part of the overtopping discharge during a storm is due to a small fraction of the waves. In fact the local overtopping discharge (in m³/s per meter structure) from a single wave can be more than 100 times the average overtopping discharge during the storm peak. Nevertheless, most information on overtopping is given as the time averaged overtopping discharge, q, expressed in m³/s per meter of structure length. However, some limited information exists on the probability distribution of the volume of overtopping water per wave.

(b) Field studies of tolerable overtopping limits of dikes and revetments have been performed by Tsuruta and Goda (1968), Goda (1970), and Fukuda, Uno, and Irie (1974). Some critical values for overtopping of a breakwater were discussed by Jensen (1984), and Dutch Guidelines on river dikes indicated allowable overtopping rates for inner slopes. Delft Hydraulics tested admissible overtopping rates for grass dikes (Smith, Seijffert, and van der Meer 1994). De Gerloni et al. (1991), and Franco, de Gerloni, and van der Meer (1994) studied the effect of falling water jets on a person, simulating the conditions on breakwater crests. Endoh and Takahashi (1994) performed full-scale tests as well as numerical modeling of overtopping rates which endanger people.

(c) The information from these various studies is condensed in Table VI-5-6, which presents critical values of the average overtopping discharge, q. The values given in this table must be regarded only as rough guidelines because, even for the same value of q, the intensity of water hitting a specific location is very much dependent on the geometry of the structure and the distance from the front of the structure. The maximum intensities might locally be up to two orders of magnitude larger than the value of q. Moreover, what is regarded as acceptable conditions is to a large extent a matter of local traditions and individual opinions.

Table VI-5-6
Critical Values of Average Overtopping Discharges

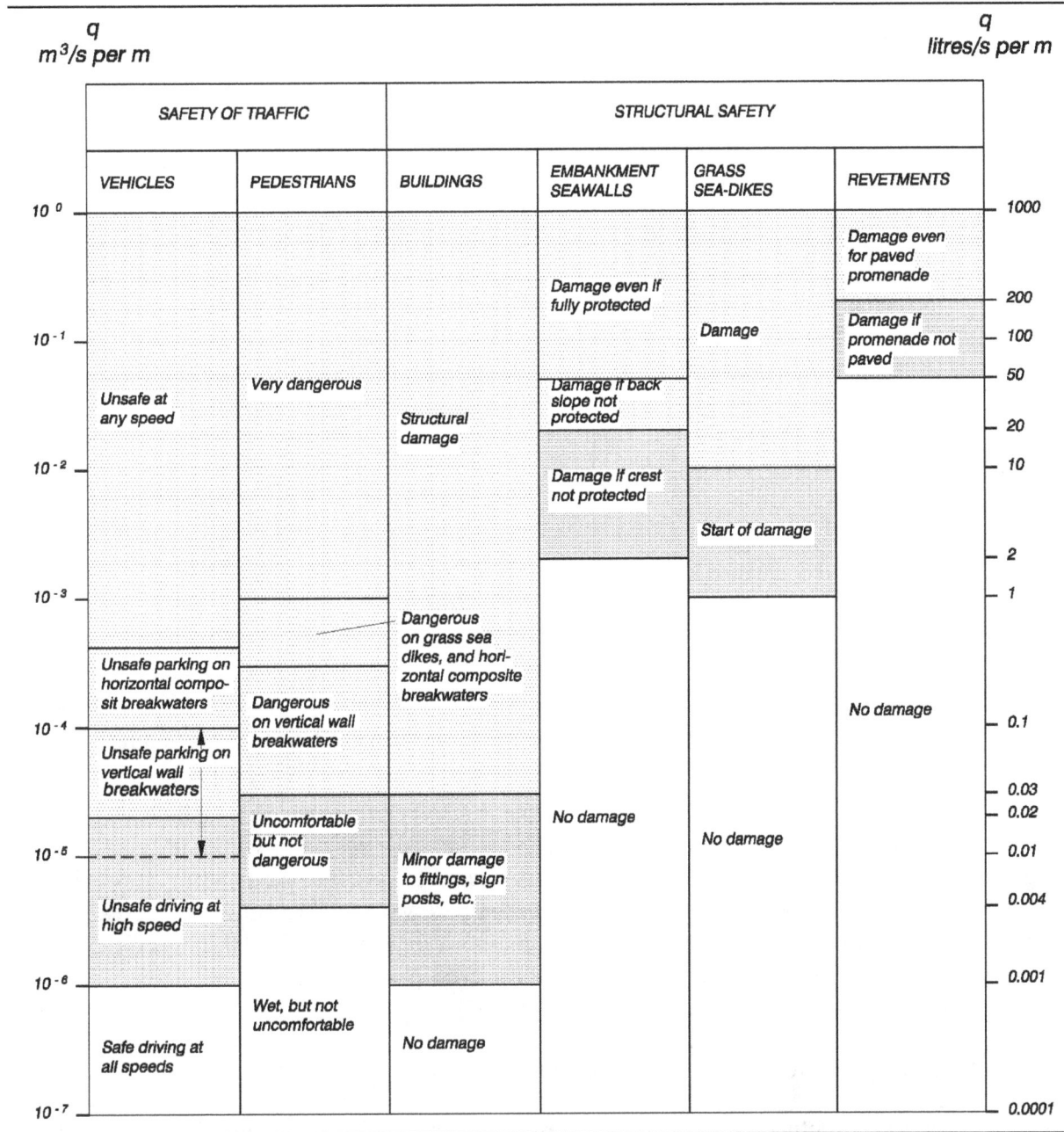

q m^3/s per m	SAFETY OF TRAFFIC		STRUCTURAL SAFETY				q litres/s per m
	VEHICLES	PEDESTRIANS	BUILDINGS	EMBANKMENT SEAWALLS	GRASS SEA-DIKES	REVETMENTS	
10^0						Damage even for paved promenade	1000
				Damage even if fully protected			200
10^{-1}					Damage	Damage if promenade not paved	100
		Very dangerous		Damage if back slope not protected			50
	Unsafe at any speed		Structural damage				20
10^{-2}				Damage if crest not protected			10
					Start of damage		
							2
10^{-3}			Dangerous on grass sea dikes, and horizontal composite breakwaters				1
	Unsafe parking on horizontal composit breakwaters						
		Dangerous on vertical wall breakwaters				No damage	
10^{-4}	Unsafe parking on vertical wall breakwaters						0.1
							0.03
		Uncomfortable but not dangerous		No damage			0.02
10^{-5}	Unsafe driving at high speed		Minor damage to fittings, sign posts, etc.		No damage		0.01
							0.004
10^{-6}							0.001
	Safe driving at all speeds	Wet, but not uncomfortable	No damage				
10^{-7}							0.0001

(d) The wind can carry spray long distances whereas solid (green) water is practically unaffected by the wind. It is important to consider spray because it can cause damage to goods placed on storage areas and can cause icing of vessel superstructures in cold regions.

(e) Overtopping occurs only if the runup level exceeds the freeboard, R_c, of the structure. Figure VI-5-14 shows the notation used to describe profile geometry for several structure types.

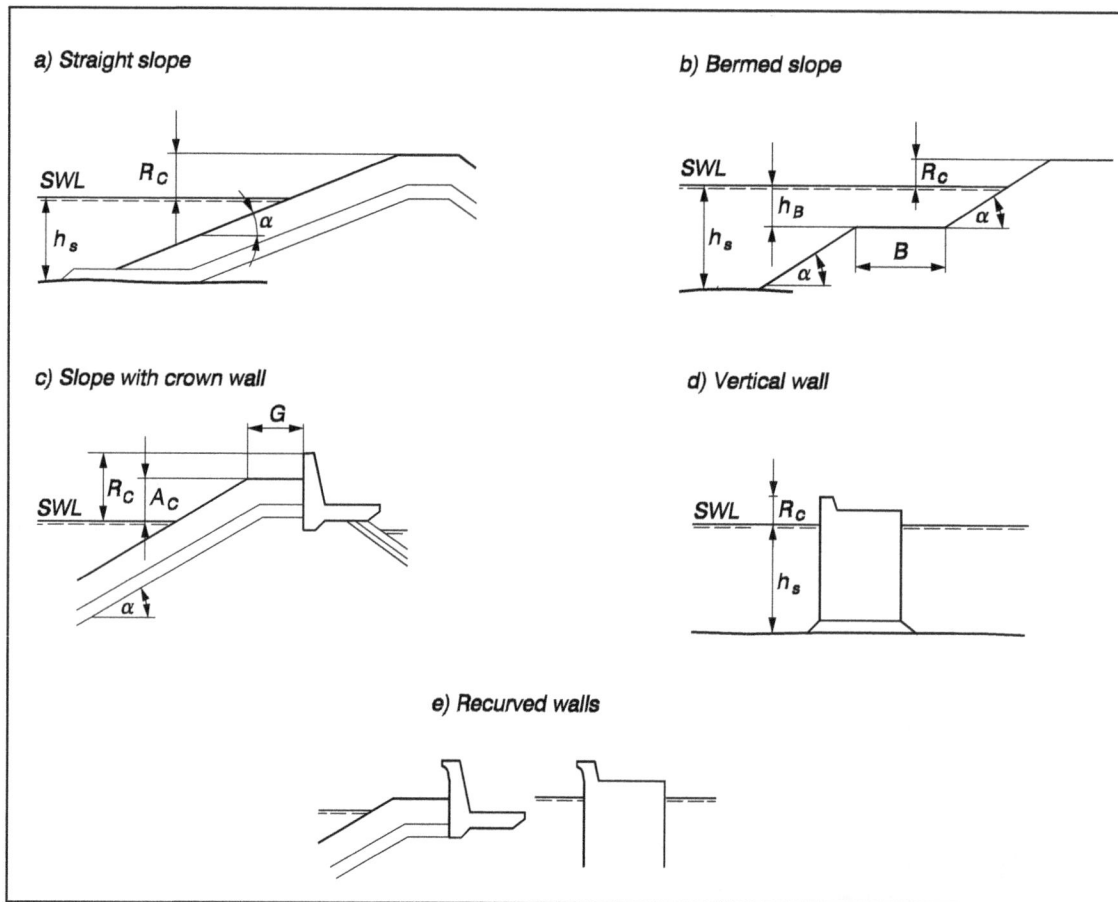

Figure VI-5-14. Structure profile geometrical parameters related to overtopping

(f) The relative freeboard, R_c/H_s, is a simple, but very important, dimensionless parameter for the prediction of overtopping. However, the wave period or wave steepness is also a significant parameter as are geometric parameters related to structure permeability, porosity and surface roughness. Under certain conditions a recurved wave wall as shown in Figure VI-5-14 e is effective in reducing overtopping. For small values of R_c/H_s (< 0.3) when the overtopping is excessive, the detailed geometry of the crest part of the structure becomes less important because the waves just travel over the structure.

(2) Average overtopping discharge formulas.

(a) Sloping structures. Formulae for overtopping are empirical because they are fitted to hydraulic model test results for specific breakwater geometries. In general the average overtopping discharge per unit length of structure, q, is a function of the standard parameters:

$$q = \text{function}\left(H_s, T_{op}, \sigma, \beta, R_c, h_s, g, \text{structure geometry}\right)$$

where

H_s = significant wave height

T_{op} = wave period associated with the spectral peak in deep water (alternately T_{om})

σ = spreading of short-crested waves

β = angle of incidence for the waves

R_c = freeboard

h_s = water depth in front of structure

g = gravitational acceleration

Two types of mathematical formulatons (models) for dimensionless overtopping dominate the literature, i.e.,

$$Q = a_e^{-(bR)} \qquad\qquad\qquad\qquad\qquad\qquad\qquad\qquad\text{(VI-5-20)}$$

and

$$Q = aR^{-b} \qquad\qquad\qquad\qquad\qquad\qquad\qquad\qquad\text{(VI-5-21)}$$

where Q is a dimensionless average discharge per meter and R is a dimensionless freeboard. Table VI-5-7 gives an overview of the models used in recent overtopping formulae along with the associated definitions for dimensionless discharge and freeboard.

(b) The fitted coefficients a and b in Equations VI-5-20 and VI-5-21 are specific to the front geometry of the structure and must be given in tables. So far no general model for the influence of front geometry exists except for rubble-mound slopes with a seawall (Pedersen 1996), in which case the front geometry (described by the front berm width B, berm crest height A_c, and slope angle α), as well as R_c, enters into R.

(c) Some formulae take into account the reduction in overtopping due to slope surface roughness, berm, shallow water, angle of wave incidence and shortcrestedness, and specific front geometries by dividing R by the respective reduction coefficients: γ_r (Table VI-5-3), γ_b (Equation VI-5-8), γ_h (Equation VI-5-10), γ_β (Equations VI-5-11, VI-5-26, VI-5-29), and γ_s (Table VI-5-13).

(d) Goda (1985) presented diagrams for wave overtopping of vertical revetments and block-mound seawalls on bottom slopes of 1:10 and 1:30. The diagrams are based on model tests with irregular long-crested head-on waves and express average discharge per meter width as a function of wave height, wave steepness, freeboard, and water depth.

- Sloping structures. Tables VI-5-8 to VI-5-12 pertain to sloping-front structures.

- Figure VI-5-15 shows the data basis for Equations VI-5-24 and VI-5-25 which includes the data of Owen (1980, 1982) for straight slopes, data of Führböter, Sparboom, and

Witte (1989) and various data sets of Delft Hydraulics. It is seen that Equation VI-5-24 contains some bias for small values of q.

- Vertical front structures.

  – Figure VI-5-16 shows the data used to establish Equation VI-5-28. Appropriate values of γ_β from Table VI-5-13 were used in plotting Figure VI-5-16; however γ_s was taken as unity (plain impermeable wall).

Table VI-5-7
Models for Average Overtopping Discharge Formulae

Authors	Structures	Overtopping model	Dimensionless discharge Q	Dimensionless freeboard R
Owen (1980,1982)	Impermeable smooth, rough, straight and bermed slopes	$Q = a\,exp(-b\,R)$	$\dfrac{q}{g\,H_s\,T_{om}}$	$\dfrac{R_c}{H_s}\left(\dfrac{s_{om}}{2\,\pi}\right)^{0.5}\dfrac{1}{\gamma}$
Bradbury and Allsop (1988)	Rock armored impermeable slopes with crown walls	$Q = a\,R^{-b}$	$\dfrac{q}{g\,H_s\,T_{om}}$	$\left(\dfrac{R_c}{H_s}\right)^2\left(\dfrac{s_{om}}{2\,\pi}\right)^{0.5}$
Aminti and Franco (1988)	Rock, cube, and Tetrapod double layer armor on rather impermeable slopes with crown walls, (single sea state)	$Q = a\,R^{-b}$	$\dfrac{q}{g\,H_s\,T_{om}}$	$\left(\dfrac{R_c}{H_s}\right)^2\left(\dfrac{s_{om}}{2\,\pi}\right)^{0.5}$
Ahrens and Heimbaugh (1988b)	7 different seawall/revetment designs	$Q = a\,exp(-b\,R)$	$\dfrac{q}{\sqrt{g\,H_s^3}}$	$\dfrac{R_c}{(H_s^2\,L_{op})^{1/3}}$
Pedersen and Burcharth (1992)	Rock armored rather impermeable slopes with crown walls	$Q = a\,R$	$\dfrac{q\,T_{om}}{L_{om}^2}$	$\dfrac{H_s}{R_c}$
van der Meer and Janssen (1995)	Impermeable smooth, rough straight and bermed slopes	$Q = a\,exp(-b\,R)$	$\dfrac{q}{\sqrt{g\,H_s^3}}\sqrt{\dfrac{s_{op}}{\tan\alpha}}$ for $\xi_{op} < 2$ $\dfrac{q}{\sqrt{g\,H_s^3}}$ for $\xi_{op} > 2$	$\dfrac{R_c}{H_s}\dfrac{\sqrt{s_{op}}}{\tan\alpha}\dfrac{1}{\gamma}$ for $\xi_{op} < 2$ $\dfrac{R_c}{H_s}\dfrac{1}{\gamma}$ for $\xi_{op} > 2$
Franco et al. (1994)	Vertical wall breakwater with and without perforated front	$Q = a\,exp(-b\,R)$	$\dfrac{q}{\sqrt{g\,H_s^3}}$	$\dfrac{R_c}{H_s}\dfrac{1}{\gamma}$
Pedersen (1996)	Rock armored permeable slopes with crown walls	$Q = R$	$\dfrac{q\,T_{om}}{L_{om}^2}$	$3.2\cdot 10^{-5}\dfrac{H_s^5\,\tan\alpha}{R_c^3\,A_c\cdot B}$

Table VI-5-8
Overtopping Formula by Owen (1980, 1982)

Straight and bermed impermeable slopes, Figures VI-5-14 a and b.
Irregular, head–on waves.

$$\frac{q}{g\,H_s\,T_{om}} = \mathbf{a}\,exp\left(-\mathbf{b}\,\frac{R_c}{H_s}\,\sqrt{\frac{s_{om}}{2\,\pi}}\,\frac{1}{\gamma_r}\right) \qquad\qquad (VI-5-22)$$

Coefficients in Eq VI-5-22
Straight smooth slopes.
Non-depth limited waves.

Slope	a	b
1 : 1	0.008	20
1 : 1.5	0.010	20
1 : 2	0.013	22
1 : 3	0.016	32
1 : 4	0.019	47

Surface roughness reduction
factor γ_r.
Updated γ_r-values are given
in Table VI-5-3.

Smooth impermeable (including smooth concrete and asphalt)	1.0
One layer of stone rubble on impermeable base	0.8
Gravel, gabion mattresses	0.7
Rock riprap with thickness greater than $2\,D_{n50}$	0.5 - 0.6

Coefficients in Eq VI-5-22
Bermed smooth slopes.
Non-depth limited waves.

Slope	h_B (m)	B (m)	$\mathbf{a}\cdot 10^4$	b
1 : 1	−4.0	10	64	20
1 : 2			91	22
1 : 4			145	41
1 : 1	−2.0	5	34	17
1 : 2			98	24
1 : 4			159	47
1 : 1	−2.0	10	48	19
1 : 2			68	24
1 : 4			86	46
1 : 1	−2.0	20	8.8	15
1 : 2			20	25
1 : 4			85	50
1 : 1	−2.0	40	3.8	23
1 : 2			5.0	26
1 : 4			47	51
1 : 1	−1.0	5	155	33
1 : 2			190	37
1 : 4			500	70
1 : 1	−1.0	10	93	39
1 : 2			340	53
1 : 4			300	80
1 : 1	−1.0	20	75	46
1 : 2			34	50
1 : 4			39	62
1 : 1	−1.0	40	12	49
1 : 2			24	56
1 : 4			1.5	63
1 : 1	0	10	97	42
1 : 2			290	57
1 : 4			300	80

Table VI-5-9
Overtopping Formula by Bradbury and Allsop (1988)

Straight rock armored slope with berm in front of crown wall, Figure VI-5-14c. Slope 1:2. Impermeable membrane at various depths as shown in the following figure. Berm width G is three or six stone diameters. Irregular, head–on waves.

$$\frac{q}{g\,H_s\,T_{om}} = \mathbf{a} \left[\left(\frac{R_c}{H_s}\right)^2 \sqrt{\frac{s_{om}}{2\,\pi}} \right]^{-\mathbf{b}} \qquad\qquad (VI-5-23)$$

Coefficients in Eq VI-5-23. Non–depth limited waves.

Note: "a" coefficients are shown multiplied by 10^9. For example, a value of 6.7 in the table represents $6.7(10)^{-9}$.

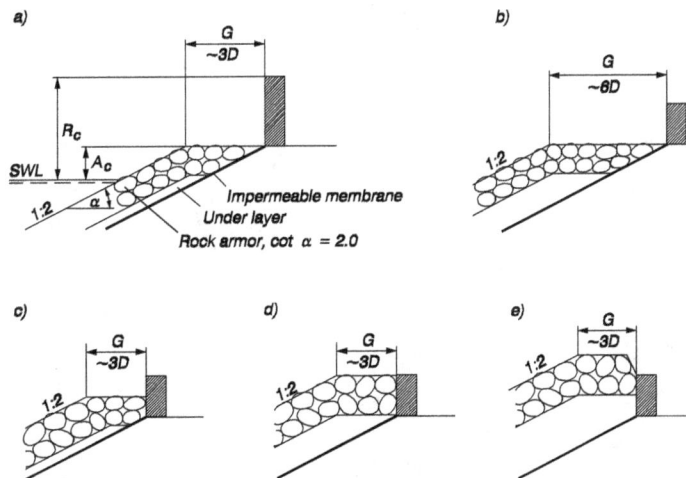

Section	G/H_s	G/R_c	A/R_c	$a \cdot 10^9$	b
a	0.79 - 1.7	0.75	0.28	6.7	3.5
		0.58	0.21	3.6	4.4
		1.07	0.39	5.3	3.5
		0.88	0.32	1.8	3.6
b	1.6 - 3.3	2.14	0.39	1.0	2.8
c	0.79 - 1.7	1.07	0.71	1.6	3.2
d	0.79 - 1.7	1.07	1.00	0.37	2.9
e	0.79 - 1.7	0.83	1.00	1.30	3.8

Table VI-5-10
Coefficients by Aminti and Franco (1988) for Overtopping Formula by Bradbury and Allsop
in Table VI-5-9

Straight slope with berm in front of crown wall, Figure VI-5-14c. Rock, cube, and tetrapod armor on rather impermeable core. Only one sea state tested (JONSWAP spectrum). Non–depth limited waves. Irregular, head–on waves.

Note: "a" coefficients are shown multiplied by 10^8. For example, a value of 17 in the table represents $17(10)^{-8}$.

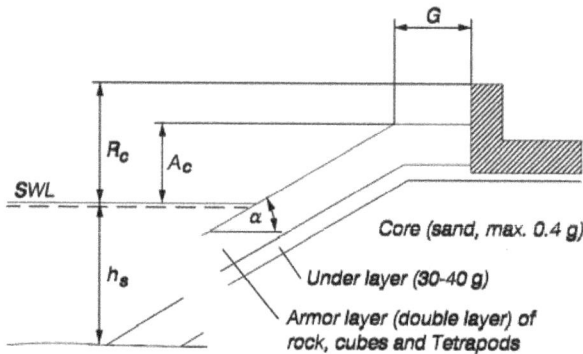

Tested ranges:
$H_s = 0.136\ m$
$T_{om} = 1.33\ s$
$S_{om} = 0.05$
$h_s/H_s = 2.9$
$\cot \alpha$ 1.33, 2.0
R_c/H_s 0.6 - 2.0
A_c/H_s 0.6, 0.75, 1.05
G/H_s 1.1, 1.85, 2.6 corresponding to width of 3, 5 and 7 stone diameters.

ARMOR	cot	G/H_s	$a \cdot 10^8$	b
ROCK	2.00	1.10	17	2.41
		1.85	19	2.30
		2.60	2.3	2.68
	1.33	1.10	5.0	3.10
		1.85	6.8	2.65
		2.60	3.1	2.69
CUBES	2.00	1.10	8.3	2.64
		1.85	15	2.43
		2.60	84	2.38
	1.33	1.10	62	2.20
		1.85	17	2.42
		2.60	1.9	2.82
TETRAPODS	2.00	1.10	1.9	3.08
		1.85	1.3	3.80
		2.60	1.1	2.86
	1.33	1.10	5.6	2.81
		1.85	1.7	3.02
		2.60	0.92	2.98

Table VI-5-11
Overtopping Formula by van der Meer and Janssen (1995)

Straight and bermed impermeable slopes including influence of surface roughness, shallow foreshore, oblique, and short-crested waves, Figures VI-5-14a and VI-5-14b.

$\underline{\xi_{op} < 2}$

$$\frac{q}{\sqrt{g\,H_s^3}}\sqrt{\frac{s_{op}}{\tan\alpha}} = 0.06\,exp\left(-5.2\,\frac{R_c}{H_s}\,\frac{\sqrt{s_{op}}}{\tan\alpha}\,\frac{1}{\gamma_r\,\gamma_b\,\gamma_h\,\gamma_\beta}\right) \qquad (VI-5-24)$$

application range: $0.3 < \dfrac{R_c}{H_s}\,\dfrac{\sqrt{s_{op}}}{\tan\alpha}\,\dfrac{1}{\gamma_r\,\gamma_b\,\gamma_h\,\gamma_\beta} < 2$

Uncertainty: Standard deviation of factor 5.2 is $\sigma = 0.55$ (See Figure VI-5-15).

$\underline{\xi_{op} > 2}$

$$\frac{q}{\sqrt{g\,H_s^3}} = 0.2\,exp\left(-2.6\,\frac{R_c}{H_s}\,\frac{1}{\gamma_r\,\gamma_b\,\gamma_h\,\gamma_\beta}\right) \qquad (VI-5-25)$$

Uncertainty: Standard deviation of factor 2.6 is $\sigma = 0.35$ (See Figure VI-5-15).

The reduction factors references are

$\quad \gamma_r \quad$ Table VI-5-3

$\quad \gamma_b \quad$ Eq VI-5-8

$\quad \gamma_h \quad$ Eq VI-5-10

$\quad$ Short-crested waves

$\quad \gamma_\beta = 1 - 0.0033\,\beta$

$\quad$ Long-crested waves (swell)

$$\gamma_\beta = \begin{cases} 1.0 & \text{for } 0° \leq \beta \leq 10° \\ \cos^2(\beta - 10°) & \text{for } 10° < \beta \leq 50° \\ 0.6 & \text{for } \beta > 50° \end{cases} \qquad (VI-5-26)$$

The minimum value of any combination of the γ-factors is 0.5.

Table VI-5-12
Overtopping Formula by Pedersen and Burcharth (1992), Pedersen (1996)

Rock armored permeable slopes with a berm in front of a crown wall, Figure VI-5-14c. Irregular, head-on waves.

$$\frac{q\,T_{om}}{L_{om}^2} = 3.2 \cdot 10^{-5} \left(\frac{H_s}{R_c}\right)^3 \frac{H_s^2}{A_c\,B\,\cot\alpha} \qquad (VI-5-27)$$

Notational permeability $P = 0.4$.

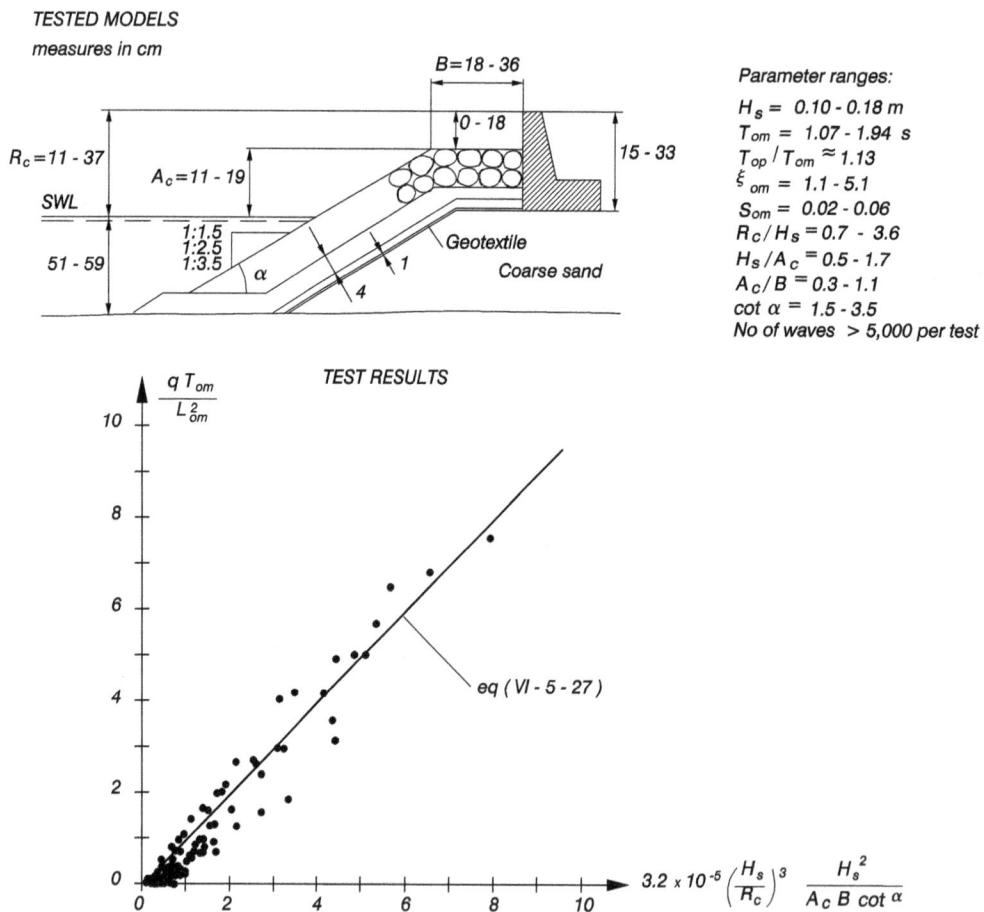

Some conservative bias of Eq VI-5-27 for small values of q is observed.

Table VI-5-13
Overtopping Formula by Franco and Franco (1999)

Impermeable and permeable vertical walls. Non-breaking, oblique, long- and short-crested waves.

$$\frac{q}{\sqrt{gH_s^3}} = 0.082 \exp\left(-3.0\frac{R_c}{H_s}\frac{1}{\gamma_\beta\gamma_s}\right) \qquad (VI-5-28)$$

Uncertainty: Standard deviation of factor 3.0 = 0.26 (see Figure VI-5-16).

Tested range:
$H_s = 12.5 - 14.0$ cm
$s_{op} = 0.04$ (wave steepness)
$\beta = 0° - 60°$ (angle of incidence)
$\sigma = $ app. 22° and app. 28° (directional spreading)
$R_c/H_s = 1.2$ and 1.6
$h_s/H_s = $ app. 4.4
$h_b/h_s = 0.21$

Tested cross sections:

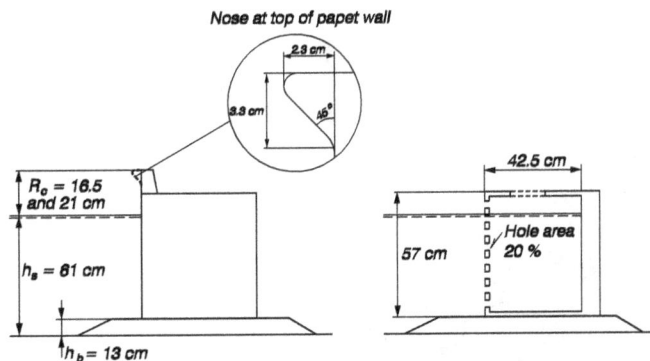

$$\gamma_\beta = \begin{cases} \left.\begin{array}{ll} \cos\beta & \text{for } 0° \leq \beta \leq 37° \\ 0.79 & \text{for } \beta > 37° \end{array}\right\} \text{Long-crested waves} & (VI-5-29) \\ \\ \left.\begin{array}{ll} 0.83 & \text{for } 0° \leq \beta \leq 20° \\ 0.83\cos(20° - \beta) & \text{for } \beta > 20° \end{array}\right\} \begin{array}{l}\text{Short-crested waves} \\ \sigma = 22° \text{ and } 28° \end{array} \end{cases}$$

Front geometry	γ_s
Plain impermeable wall	1.00
Plain impermeable wall with recurved nose	0.78
Perforated front (20% hole area) and deck	0.72 – 0.79
Perforated front (20% hole area) and open deck	0.58

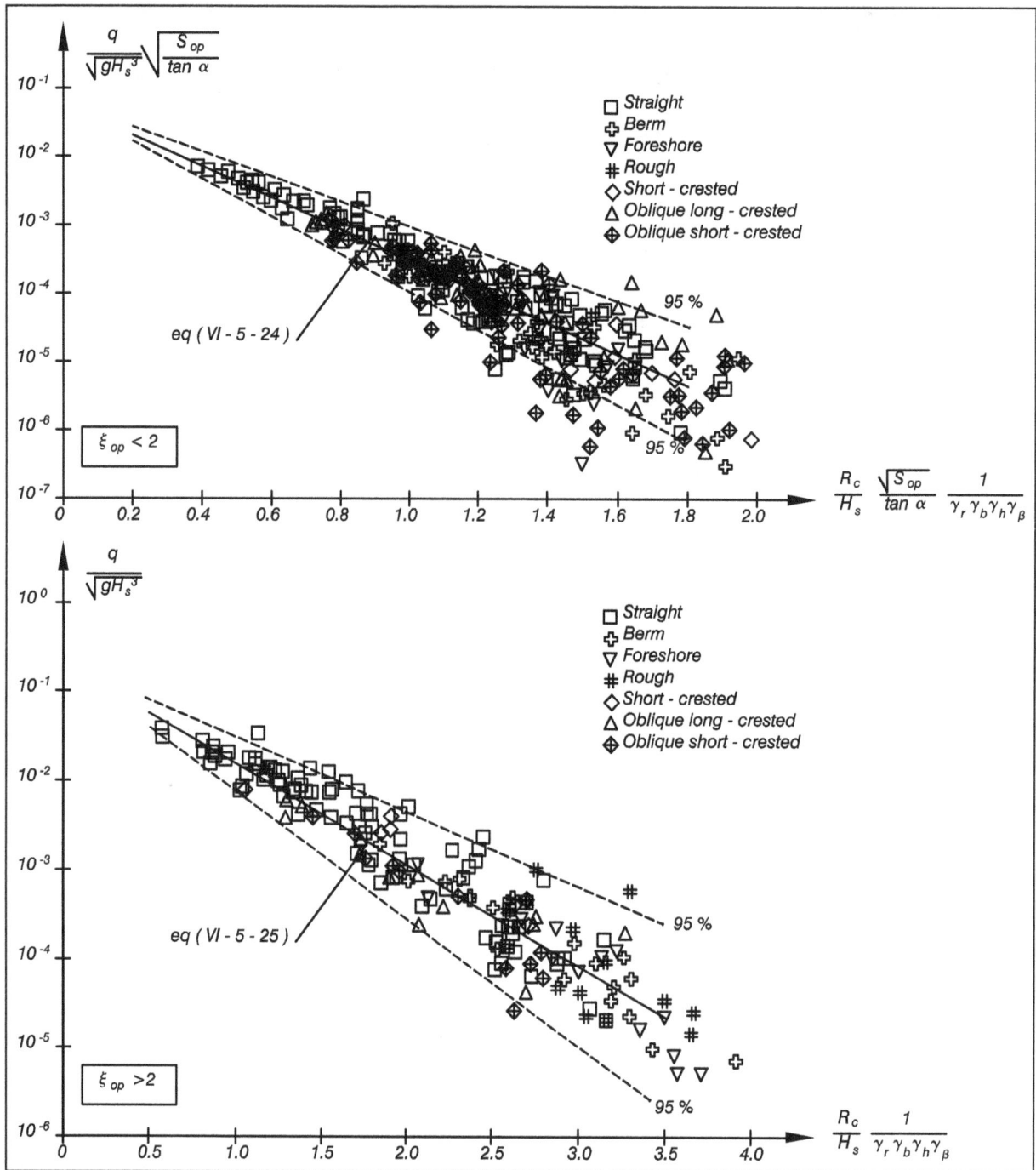

Figure VI-5-15. Wave overtopping data as basis for Equations VI-5-24 and VI-5-25. Fitted mean and 95 percent confidence bands (van dere Meer and Janssen 1995)

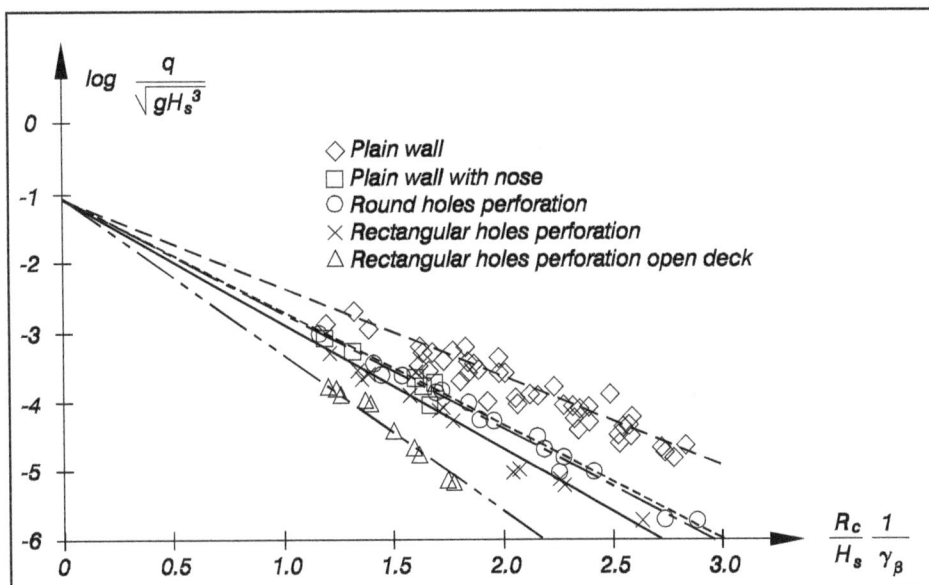

Figure VI-5-16. Vertical wall wave overtopping data plotted with $\gamma_s = 1.0$
(Franco and Franco 1999)

–   Figure VI-5-17 shows the same vertical wall overtopping data plotted with appropriate values of γ_β and γ_s from Table VI-5-13. The solid line is Equation VI-5-28.

(3) Overtopping volumes of individual waves. The average overtopping discharge q provides no information about the discharge intensity of the individual overtopping waves. However, such information is important because most damaging impacts on persons, vehicles, and structures are caused by overtopping of large single waves. The overtopping volume per wave has been recorded in model tests and it was found that the probability distribution function for overtopping volume per wave per unit width (V m^3/m) follows a Weibull distribution as given in Equation VI-5-30 (Franco, de Gerloni, and van der Meer 1994; van der Meer and Jansson 1995).

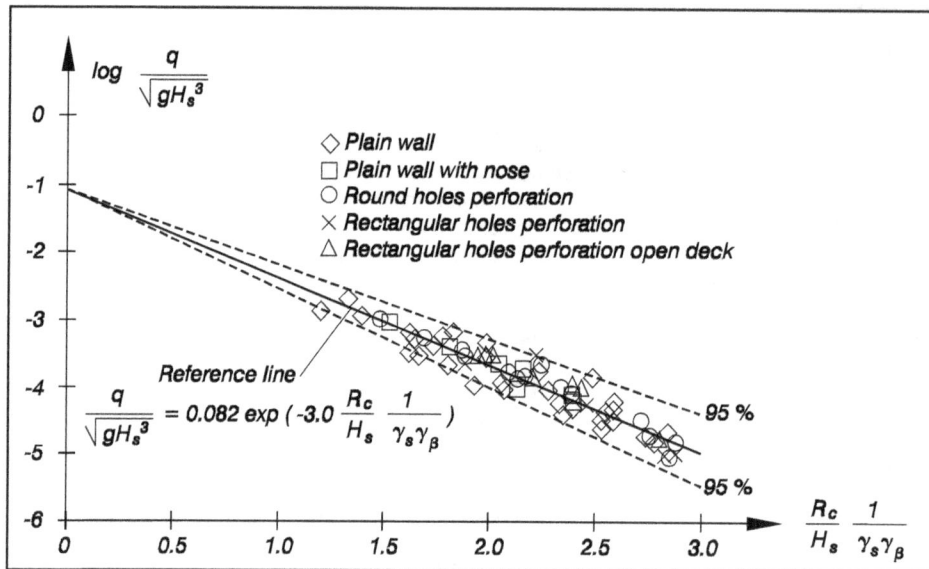

Figure VI-5-17. Vertical wall wave overtopping data with fitted mean and
95 percent confidence bands (Franco and Franco 1999)

$$prob(v > V) = \exp\left[-\left(\frac{V}{B}\right)^{3/4}\right] \text{ or} \qquad \text{(VI-5-30)}$$

$$V = B\left(-\ln\left[prob(v > V)\right]\right)^{4/3} \qquad \text{(VI-5-31)}$$

with

$$B = 0.84 \frac{T_m q}{P_{ow}} \qquad \text{(VI-5-32)}$$

where

$prob(v > V) =$ probability of individual wave overtopping volume per unit width, v, being larger than the specified overtopping volume per unit width, V

$T_m =$ average wave period (in units of seconds)

$q =$ average overtopping discharge per unit width (in units of m^3/s per m)

$P_{ow} =$ probability of overtopping per incoming wave ($= N_{ow} / N_w$)

$N_{ow} =$ number of overtopping waves

$N_w =$ number of incoming waves

If the runup levels follow a Rayleigh distribution, the probability of overtopping per incoming wave can be estimated as

$$P_{ow} = \exp\left[-\left(\frac{R_c}{cH_s}\right)^2\right]$$

(VI-5-33)

where

For sloping structure, irregular waves:
$c = 0.81\,\xi_{eq}\,\gamma_r\,\gamma_h\,\gamma_\beta$ with a maximum of $c = 1.62\,\gamma_r\,\gamma_h\,\gamma_\beta$
For vertical wall structure, irregular, impermeable,
long-crested, nonbreaking, head-on waves:
$c = 0.91$

(VI-5-34)

and

R_c = structure crest height relative to swl

H_s = significant wave height

A first estimate of the maximum overtopping volume per unit width produced by one wave out of the total number of overtopping waves can be calculated using the expression

$$V_{max} = B\left(\ln N_{ow}\right)^{4/3}$$

(VI-5-35)

c. Wave reflection.

(1) Introduction.

(a) Coastal structures reflect some proportion of the incident wave energy. If reflection is significant, the interaction of incident and reflected waves can create an extremely confused sea with very steep waves that often are breaking. This is a difficult problem for many harbor entrance areas where steep waves can cause considerable maneuvering problems for smaller vessels. Strong reflection also increases the sea bed erosion potential in front of protective structures. Waves reflected from some coastal structures may contribute to erosion of adjacent beaches.

(b) Non-overtopped impermeable smooth vertical walls reflect almost all the incident wave energy, whereas permeable, mild slope, rubble-mound structures absorb a significant portion of the energy. Structures that absorb wave energy are well suited for use in harbor basins.

(c) In general incident wave energy can be partly dissipated by wave breaking, surface roughness and porous flow; partly transmitted into harbor basins due to wave overtopping and penetration; and partly reflected back to the sea, i.e.

$$E_i = E_d + E_t + E_r \qquad\qquad\qquad\qquad\qquad\text{(VI-5-36)}$$

where E_i, E_d, E_t, and E_r are incident, dissipated, transmitted, and reflected energy, respectively.

(d) Reflection can be quantified by the bulk reflection coefficient

$$C_r = \frac{H_{sr}}{H_s} = \left(\frac{E_r}{E_i}\right)^{1/2} \qquad\qquad\qquad\text{(VI-5-37)}$$

where H_s and H_{sr} are the significant wave heights of incident and reflected waves, respectively, at that position; and E_i and E_r are the related wave energies.

(2) Reflection from non-overtopped sloping structures.

(a) Very long waves such as infragravity and tidal waves are almost fully reflected by any type of impervious structure. Wind-generated waves generally break on slopes (see Table VI-5-1) with the type of wave breaking given as a function of the surf-similarity parameter ξ, defined by Equation VI-5-2. Wave energy dissipation by wave breaking is much greater than dissipation due to surface roughness and porous flow for conventional coastal structures. Therefore, it is relevant to relate the bulk reflection coefficient, C_r, to ξ, (Battjes 1974b; Seelig 1983).

(b) The bulk reflection coefficient for straight non-overtopped impermeable smooth slopes and conventional rubble-mound breakwaters can be estimated from Equation VI-5-38 (Seelig 1983) given in Table VI-5-14. Figure VI-5-18 shows the fitting of the model test results by Allsop and Hettiarachichi (1988). Some scatter in the fitting can be seen.

Table VI-5-14
Wave Reflection Coefficients for Non-Overtopped Sloping Structures
Based on Seelig (1983) Equation

Head-on waves

$$C_r = \frac{a\,\xi^2}{(b+\xi^2)} \qquad \xi = \frac{\tan\alpha}{\sqrt{\frac{2\pi H}{gT^2}}} \qquad\qquad (VI-5-38)$$

For irregular wave H is replaced by H_s, T is replaced either by T_p (ξ_{op}) or T_m (ξ_{om}).

Fitted coefficients in Eq VI-5-38

Author	Structure	a	b
Seelig (1983) $2.5 \leq \xi \leq 6$	Impermeable, smooth, straight slopes, regular waves	1.0	5.5
Allsop and Hettiarachchi (1988) range of ξ or ξ_{op} shown in Figure VI-5-18	Dolosse, regular waves (ξ) Slope 1:1.5, 1:2, 1:3	0.56	10.0
	Cobs, regular waves (ξ) Slope 1:1.5, 1:2, 1:3	0.50	6.54
	Tetrapods and Stabit, irregular waves (ξ_{om}) Slope 1:1.33, 1:1.5, 1:2	0.48	9.62
	Shed and Diode, irregular waves (ξ_{om}) Slope 1:1.33, 1:1.5, 1:2	0.49	7.94
Allsop (1990) $3 \leq \xi_{om} \leq 6$	Smooth and impermeable	0.96	4.8
	1 layer rock and stone underlayer on impermeable slope (P=1)	0.64	7.22
	2 layer rock and stone underlayer on impermeable slope (P=1)	0.64	8.85
Benoit and Teisson (1994) $2.7 \leq \xi_{op} \leq 7$	2 layer rock armor $H_s : 0.03 - 0.09m$, $T_p = 1.3s$, $d = 0.4m$ Slope 1:1.33, 1:1.5, 1:2	0.6	6.6
Davidson, et al (1994) $8 \leq \xi_{op} \leq 50$	Field measurement on rock slope 1:1.1 Water depth h in meters		
	$h > 3.25$	0.65	25
	$2.5 \leq h \leq 3.25$	0.60	35
	$h < 2.5$	0.64	80

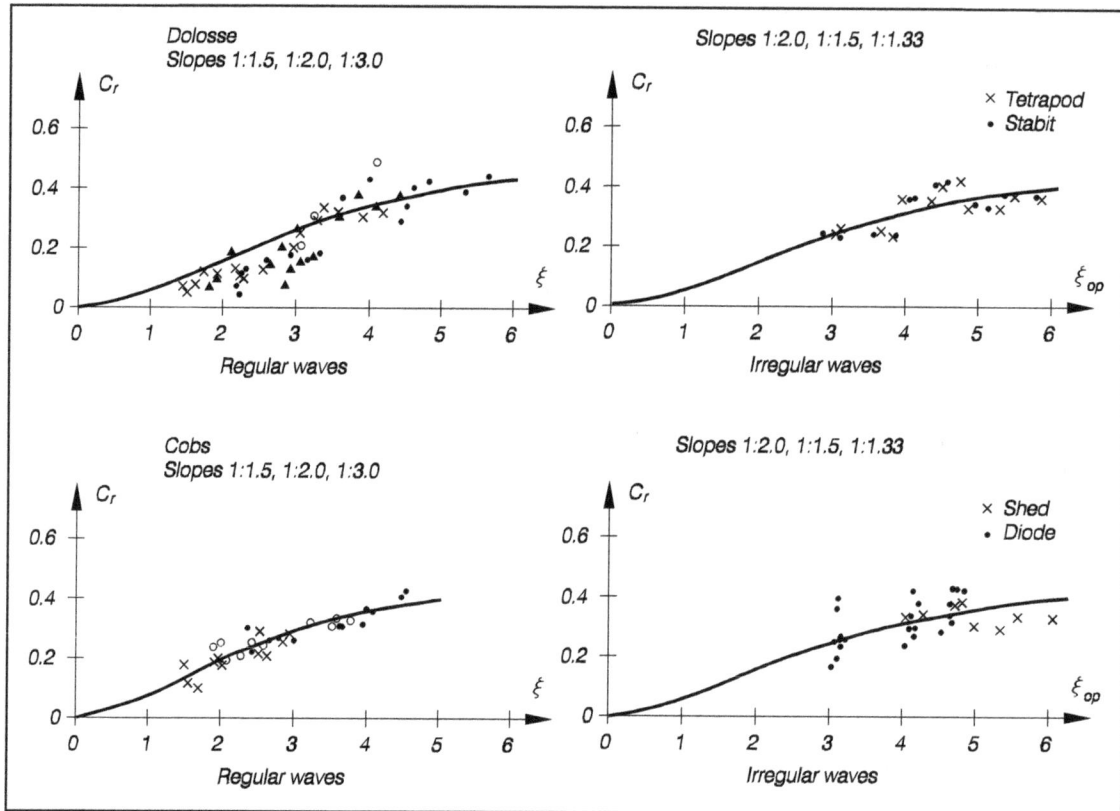

Figure VI-5-18. Reflection coefficients for concrete armor unit slopes. Head-on waves
(Allsop and Hettiarachchi 1988)

(c) An alternative formula to Equation VI-5-38 was given by Postma (1989), who analyzed van der Meer's (1988) reflection data (see Table VI-5-4) for non-overtopped rock slopes. Postma introduced the notational permeability P (shown on Figure VI-5-11), the slope angle α and the wave steepness s_{op} in the formula

$$C_r = 0.071(P)^{-0.082}(\cot\alpha)^{-0.62}(s_{op})^{-0.46}$$
(VI-5-39)

(d) The uncertainty of Equation VI-5-39 corresponds to a variational coefficient of 0.036.

(e) The effect of a berm in a slope is generally a reduction in C_r. Figure VI-5-19 shows C_r values for a rubble-mound structure with berms of varying width at SWL (Allsop 1990).

(3) Reflection from vertical walls.

(a) Bulk reflection coefficients for plain vertical breakwaters on seabed, for plain vertical breakwaters on rubble foundation, for horizontal composite breakwaters, for sloping top caissons, for single perforated screens, and for perforated caissons are given in Figures VI-5-20, VI-5-21, VI-5-22, VI-5-23, VI-5-24, and VI-5-25, respectively. They were obtained from scaled model tests with irregular, head-on waves. The effect of oblique waves and wave shortcrestedness on plain and perforated vertical wall caissons is shown in Figure VI-5-26.

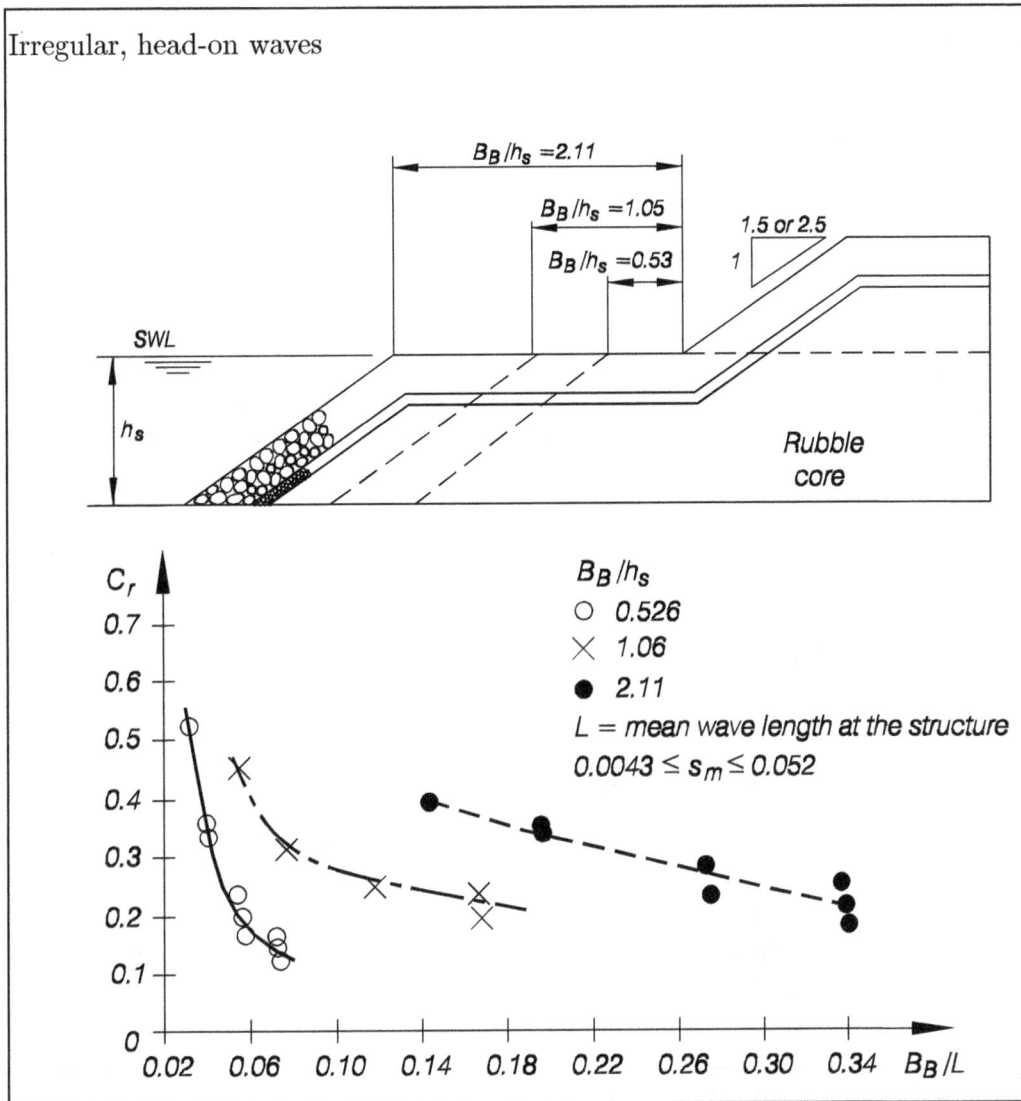

Figure VI-5-19. Wave reflection coefficients for rock armored slope with berm at
SWL (Allsop 1990)

(b) The influence of wave shortcrestedness and oblique wave approach on reflection
from plain impermeable and perforated vertical caissions is illustrated by Figure VI -5-26.

(4) Kinematics of reflected irregular waves.

(a) Close to highly reflective coastal structures incident and reflected waves interact with
some degree of "phase locking." This result is a partially standing wave field characterized by
nodes and antinodes. For the extreme case of perfectly reflected regular waves, a standing wave
field occurs with stationary nodes and antinodes. Reflecting irregular waves create a less
noticeable spatial variation of partially standing nodes and antinodes that decrease in magnitude
with distance from the structure.

Irregular, head-on waves

$$C_r = \begin{cases} 0.79 + 0.11 \frac{R_c}{H_s} & \frac{R_c}{H_s} \leq 1.0 \\ 0.90 & \frac{R_c}{H_s} > 1.0 \end{cases} \qquad (\text{VI}-5-40)$$

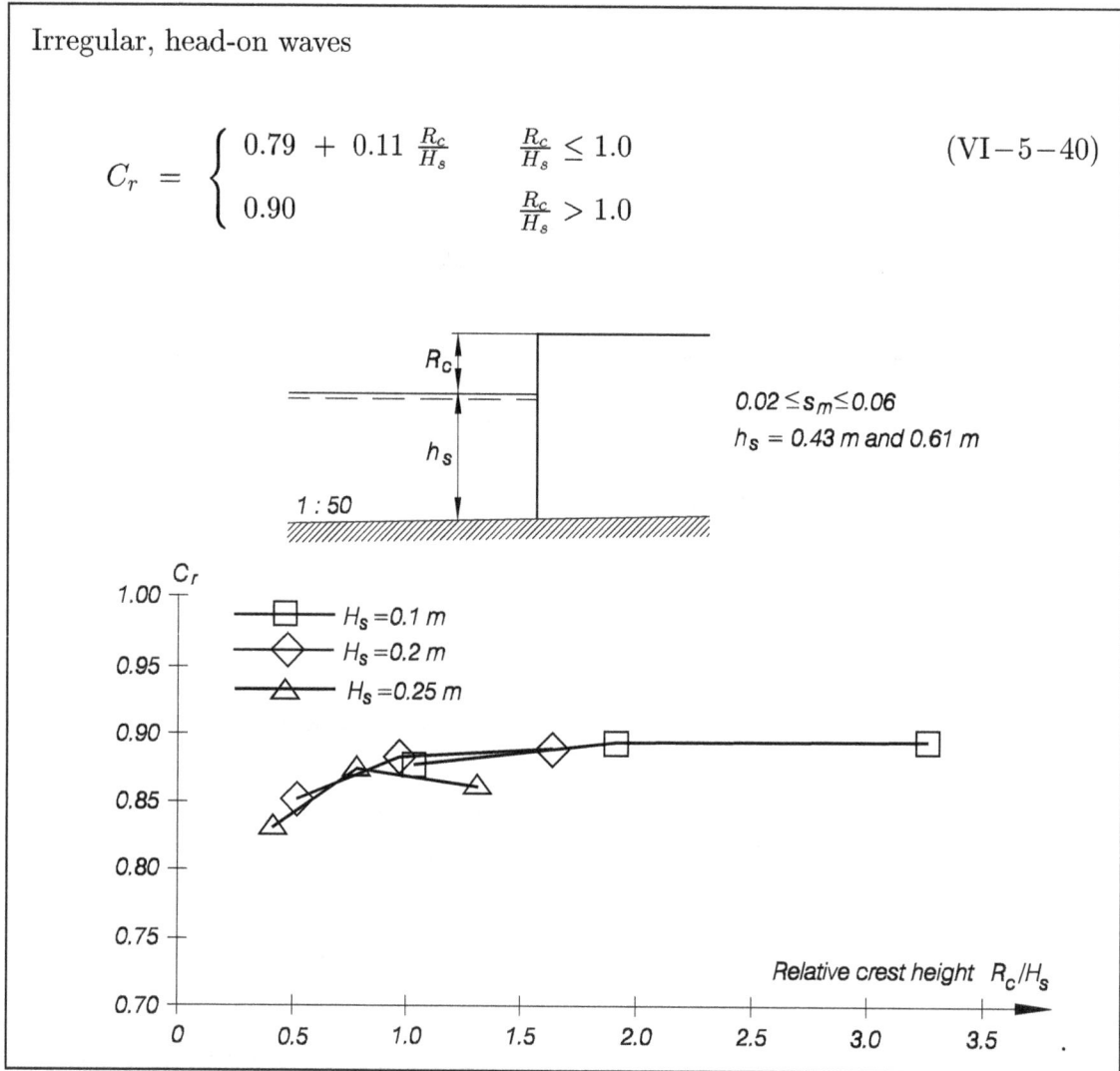

$0.02 \leq s_m \leq 0.06$

$h_s = 0.43$ m and 0.61 m

Figure VI-5-20. Wave reflection coefficients for plain vertical breakwater on 1:50 seabed (Allsop, McBride, and Columbo 1994)

(b) Assuming that the sea surface is comprised of a large number of linear wave trains that can be superimposed, the sea surface elevation adjacent to a reflective structure can be written as

$$\eta = \sum_{i=1}^{\infty} a_i \sqrt{1 + C_{ri}^2 + 2C_{ri}\cos(2k_i x + \theta_i)} \cos(\sigma_i t - \varepsilon_i) \qquad (\text{VI-5-42})$$

and the horizontal component of the wave orbital velocity is given as

$$u = \sum_{i=1}^{\infty} a_i \left(\frac{gk_i}{\sigma_i}\right) \frac{\cosh[k_i(h+z)]}{\cosh(k_i h)} \sqrt{1 + C_{ri}^2 - 2C_{ri}\cos(2k_i x + \theta_i)} \cos(\sigma_i t - \varepsilon_i) \qquad (\text{VI-5-43})$$

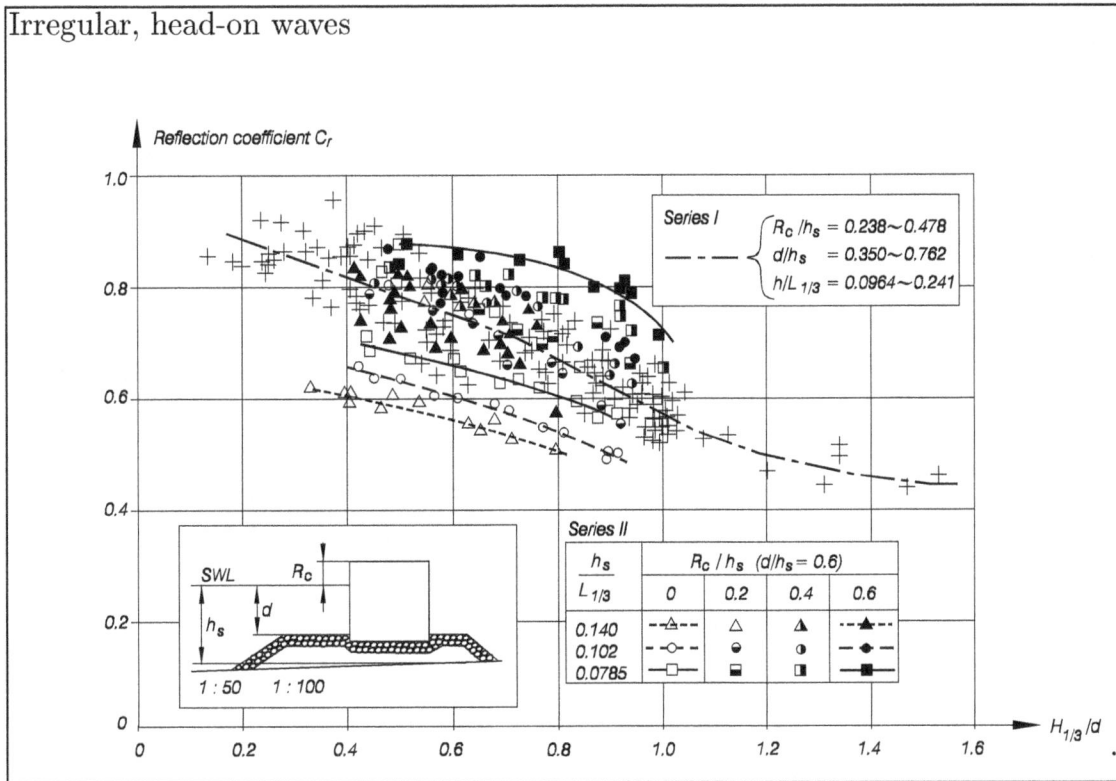

Figure VI-5-21. Wave reflection coefficients for plain vertical breakwater on rubble-mound foundation (Tanimoto, Takahashi, and Kimura 1987)

where

a_i = amplitude of ith incident wave component

k_i = wave number of ith incident wave component

σ_i = angular wave frequency of the ith incident wave component

g = gravitational acceleration

h = water depth

x = horizontal coordinate with positive toward the structure and x=0 located at the structure toe

z = vertical coordinate with z=0 at swl and z=-h at bottom

C_{ri} = reflection coefficient of ith incident wave component

θ_i = reflection phase angle of ith incident wave component

ε_i = random wave phase angle of ith incident wave component

Irregular, head-on waves

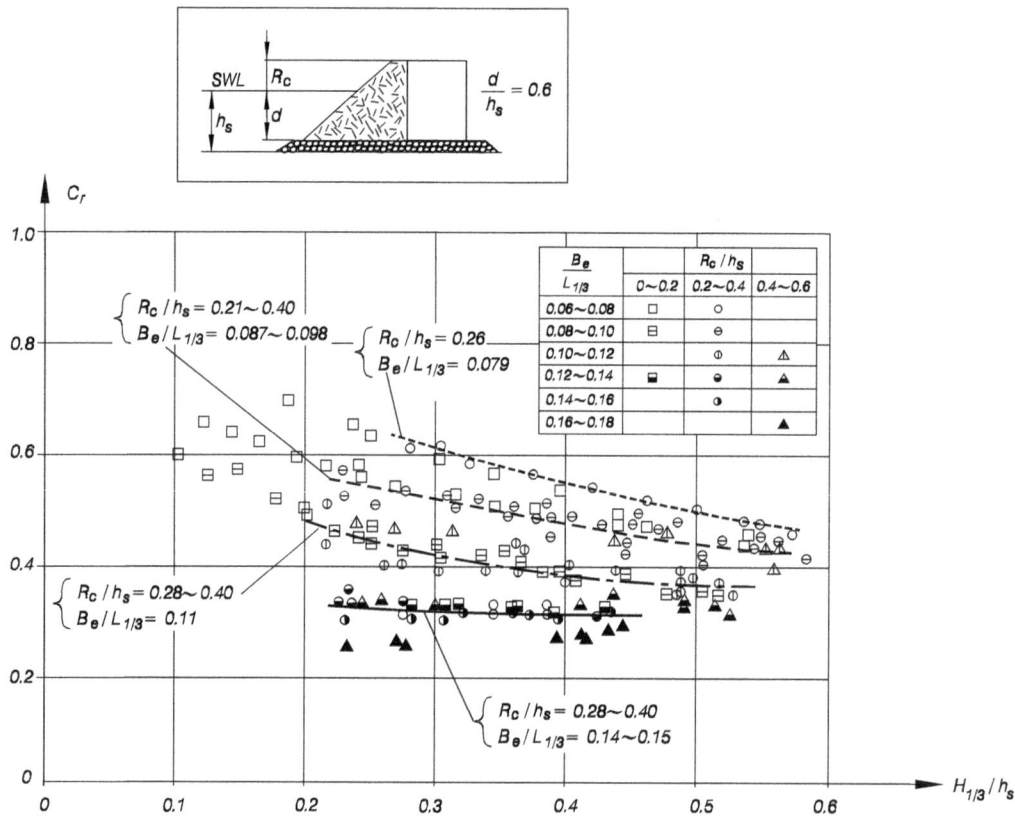

Figure VI-5-22. Wave reflection coefficients for horizontal composite breakwaters with tetrapod slope 1:1.5 (Tanimoto, Takahashi, and Kimura 1987)

$$B_e = b_0 - \left(\frac{\cot \alpha}{h_s + R_c}\right)\left(\int_{-d}^{R_c} \frac{cosh^2 2\pi(h_s + z)/L}{cosh^2 2\pi h_s/L} z\, dz + 0.5R_c^2\right) \qquad (VI-5-41)$$

where b_0 covering width at the still water level

       α slope angle of the structure measured from horizontal

       z upward distance away from the still water level

(c) These two equations strictly apply to the case of two-dimensional, nonbreaking, irregular waves propagating over a flat bottom and approaching normal to reflective structures. Similar expressions can be written for the case of oblique reflection of irregular, long-crested waves.

(d) The corresponding equation for estimating the root-mean-squared sea surface elevations is (Goda and Suzuki 1976)

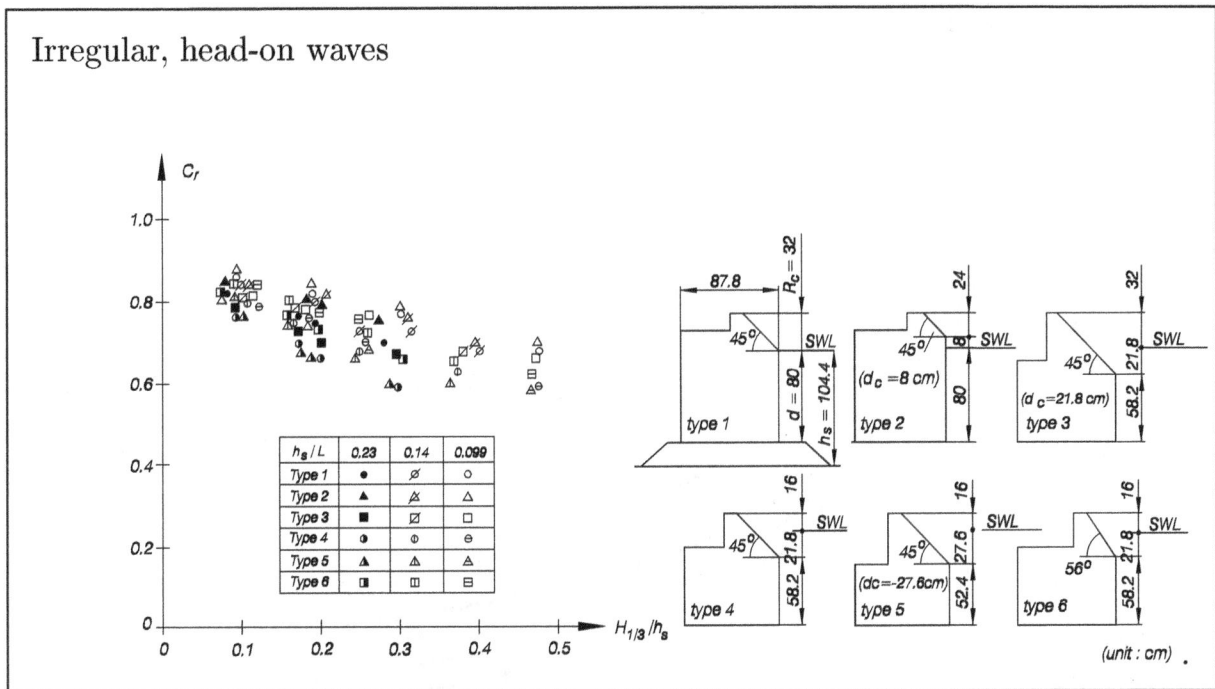

Figure VI-5-23. Wave reflection coefficients for sloping top breakwaters (Takahashi 1996)

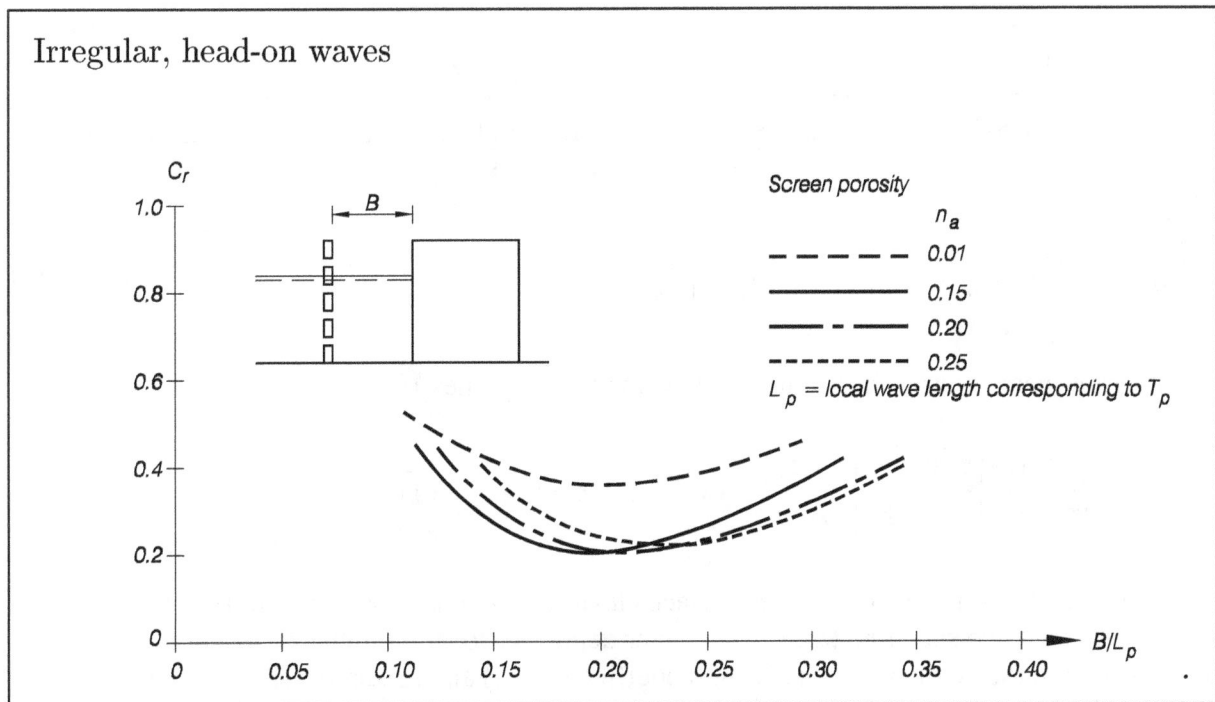

Figure VI-5-24. Wave reflection coefficients for perforated caissions (Allsop and Hettiarachchi 1988)

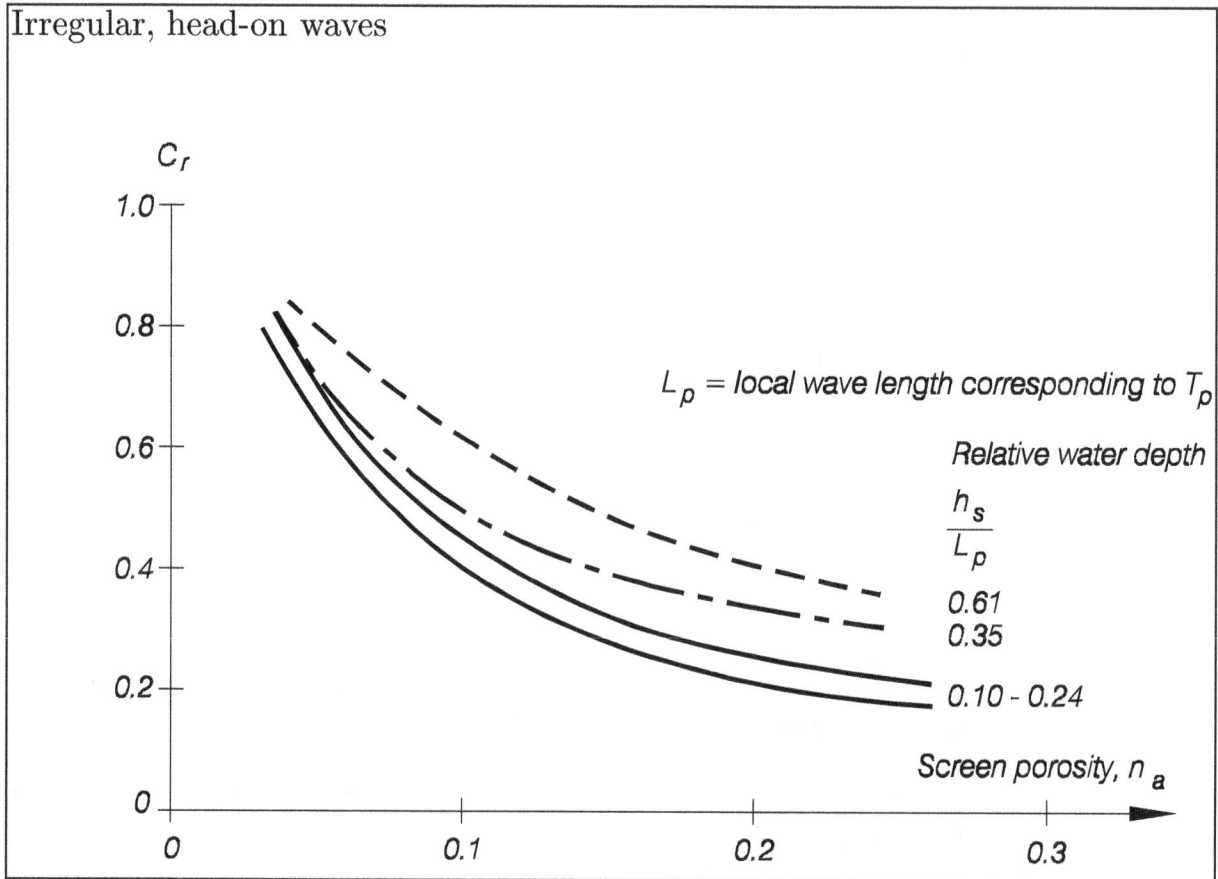

Figure VI-5-25. Wave reflection coefficients for single perforated screen (Allsop and Hettiarachchi 1988)

$$\eta_{rms}^2 = \sum_{i=1}^{\infty}\left[1 + C_{ri}^2 + 2C_{ri}\cos\left(2k_i x + \theta_i\right)\right]\frac{a_i^2}{2} \tag{VI-5-44}$$

and the root-mean-squared horizontal wave velocity is (Hughes 1992)

$$u_{rms}^2 = \sum_{i=1}^{\infty}\left(\frac{gk_i}{\sigma_i}\right)\frac{\cosh^2\left[k_i(h+z)\right]}{\cosh^2\left(k_i h\right)}\left[1 + C_{ri}^2 - 2C_{ri}\cos\left(2k_i x + \theta_i\right)\right]\frac{a_i^2}{2} \tag{VI-5-45}$$

(e) The root-mean-squared sea surface elevations and horizontal velocities are functions of the incident wave spectrum (a_i, k_i, σ_i), water depth (h), location in the water column relative to the structure toe (x, z), and the reflection coefficient (C_{ri}) and reflection phase angle (θ_i) associated with each wave component in the incident spectrum.

(f) For impermeable vertical walls the reflection coefficient C_{ri} is equal to unity for all wave componets and the reflection phase shift is $\theta_i = 0$, 2π, 4π, ... However, for sloping structures reflection is less than perfect, and it is necessary to estimate the reflection coefficient and phase angle as functions of wave component frequency.

Test set-up

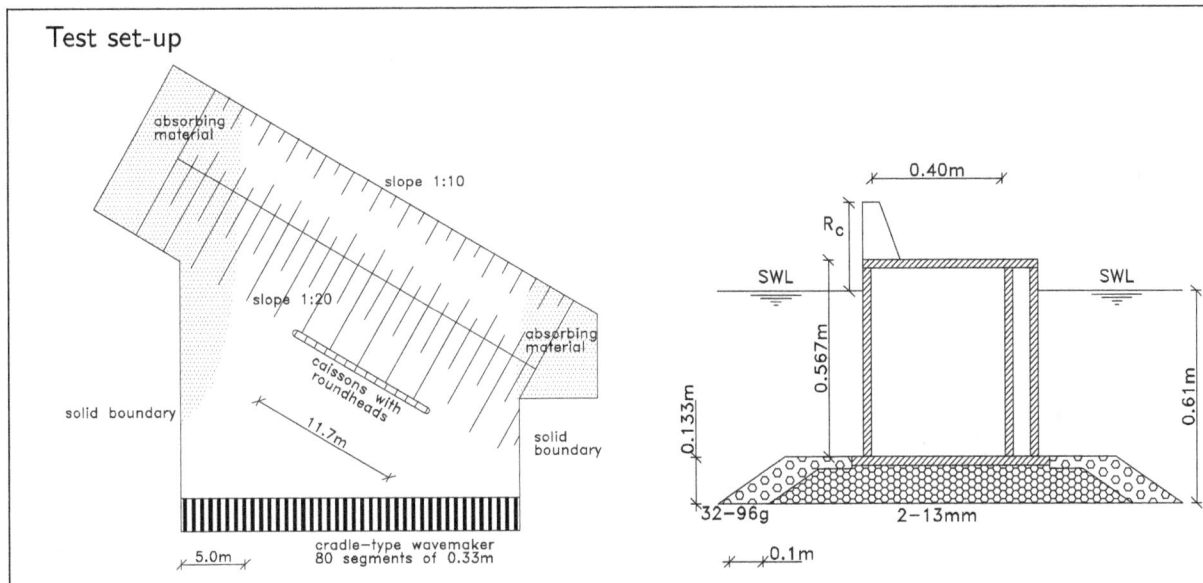

JONSWAP spectrum; $H_s = 14$ cm, $T_p = 1.5$ s
Gaussian spreading function; Spreading angle: 0^0(long-crested), 15^0, 30^0
Mean incident direction: 0^0 (head-on), 10^0, 20^0, 40^0, 60^0
Impermeable plain and perforated (porosity 25%, chambe width 0.4 m) vertical front

Test results

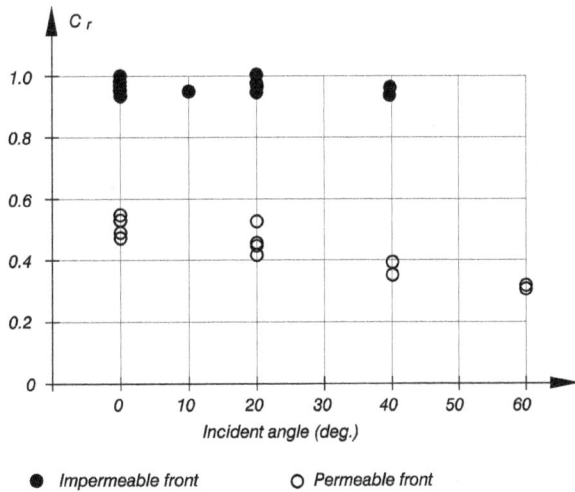

The data on the figure show that the reflection coefficients are almost independent
of the wave short-crestedness within the tested range. The reflection coefficients for
an impermeable plain vertical caisson are independent of wave obliquity, while it is
decreasing with wave incident angle for a perforated caission.

Figure VI-5-26. Wave reflection coefficients for impermeable and permeable vertical
breakwaters exposed to oblique, nonbreaking, short-crested waves (Helm-Petersen 1998)

(g) Empirical expressions for θ_i and C_{ri} for sloping impermeable and rubble-mound structures have been developed based on laboratory experiments (Hughes and Fowler 1995; Sutherland and O'Donoghue 1998a; Sutherland and O'Donoghue 1998b). The reflection phase for each incident wave component can be estimated from the following expression presented by Sutherland and O'Donoghue (1998a)

$$\theta_i = -8.84\pi\chi^{5/4} \tag{VI-5-46}$$

where

$$\chi = \frac{\sigma_i}{2\pi \tan \alpha}\sqrt{\frac{d_t}{g}} \tag{VI-5-47}$$

and

d_t = depth at the toe of the sloping structure

α = structure slope

The reflection coefficient for each incident wave component is estimated from recent results of Sutherland and O'Donoghue (1998b) by the empirical expressions

$$C_{ri} = \frac{\xi_\sigma^{2.58}}{7.64+\xi_\sigma^{2.58}} \text{ for smooth impermeable slopes} \tag{VI-5-48}$$

$$C_{ri} = \frac{0.82\xi_\sigma^2}{22.85+\xi_\sigma^2} \text{ for rubble-mound slopes} \tag{VI-5-49}$$

where

$$\xi_\sigma = \frac{\tan \alpha}{\sigma_i}\sqrt{\frac{2\pi g}{H_s}} \tag{VI-5-50}$$

and H_s is the significant wave height of the incident spectrum.

Figure VI-5-27 compares measured data to estimates of u_{rms} at middepth adjacent to a smooth, impermeable laboratory structure on a 1:2 slope. The estimates were made using the measured incident wave spectrum.

Sutherland and O'Donoghue (1997) showed that the two-dimensional expression for root-mean-square velocity can be extended to include the case of obliquely incident, long-crested waves.

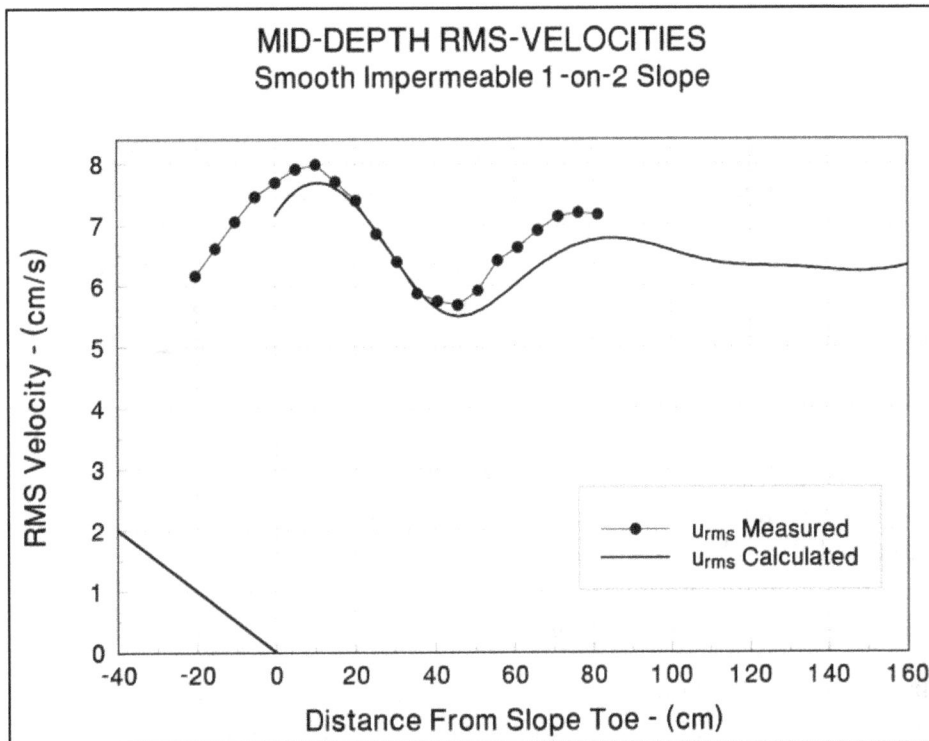

Figure VI-5-27. Measured versus estimated u_{rms} near smooth,
impermeable 1:2 slope (Hughes and Fowler 1995)

d. Wave transmission.

(1) Introduction.

(a) Wave action behind a structure can be caused by wave overtopping and also by wave penetration if the structure is permeable. Waves generated by the falling water from overtopping tend to have shorter periods than the incident waves. Generally the transmitted wave periods are about half that of the incident waves.

(b) Wave transmission can be characterized by a transmission coefficient, C_t, defined either as the ratio of transmitted to incident characteristic wave heights (e.g., H_{st} and H_s) or as the square root of the ratio of transmitted to incident time-averaged wave energy (e.g., E_t and E_i) as given in Equation VI-5-51.

$$C_t = \frac{H_{st}}{H_s} = \left(\frac{E_t}{E_i} \right)^{1/2}$$

(VI-5-51)

(c) Specific transmission coefficients for wave overtopping (C_{to}) and wave penetration (C_{tp}) could be defined as follows

$$C_{to} = \frac{H_{st}^{overtop}}{H_s}$$

(VI-5-52)

$$C_{tp} = \frac{H_{st}^{penetr.}}{H_s} \qquad (VI\text{-}5\text{-}53)$$

(d) However, in practice it is difficult to distinguish between $H_{st}^{overtop}$ and $H_{st}^{penetr.}$, and consequently, usual practice is to calculate C_t as defined by Equation VI-5-51.

(e) Values of C_t given in the literature are almost all from laboratory experiments, many of which were conducted at rather small scales. Some scale effects might have influenced the results, especially for the proportion of C_t related to wave penetration.

(2) Wave transmission through and over sloping structures.

(a) The total coefficient of wave transmission, C_t, for rock armored low-crested and submerged breakwaters, and reef breakwaters under irregular head-on waves are given in Figure VI-5-28 and Table VI-5-15.

(b) Figure VI-5-29 shows an example of the use of Equation VI-5-54.

(c) Breakwaters with complex types of concrete armor units, such as tetrapods or CORE-LOCS7 hereafter referred to as Core-Locs, generally have a more permeable crest than rock armored breakwaters, and this results in larger transmission coefficients.

(d) Detached breakwaters for coastal protection are placed in very shallow water and are often built entirely of armor blocks without underlayer and core. Such breakwaters are very permeable and C_{tp} can reach 0.8 in the case of complex armor units and small wave steepnesses.

(3) Wave transmission for vertical structures. Wave transmission for vertical breakwaters is mainly the result of wave overtopping. Therefore the ratio of the breakwater crest height (R_c) to the incident wave height (H_s) is the most important parameter. Wave transmission coefficients for plain vertical breakwaters, horizontal composite breakwaters, sloping top breakwaters and perforated walls are given in Table VI-5-16, Table VI-5-17, Figure VI-5-30, Figure VI-5-31, and Figure VI-5-32, respectively.

Irregular, head-on waves

Wave transmission coefficient by Allsop (1983)

Wave transmission coefficient by Powell and Allsop (1985)

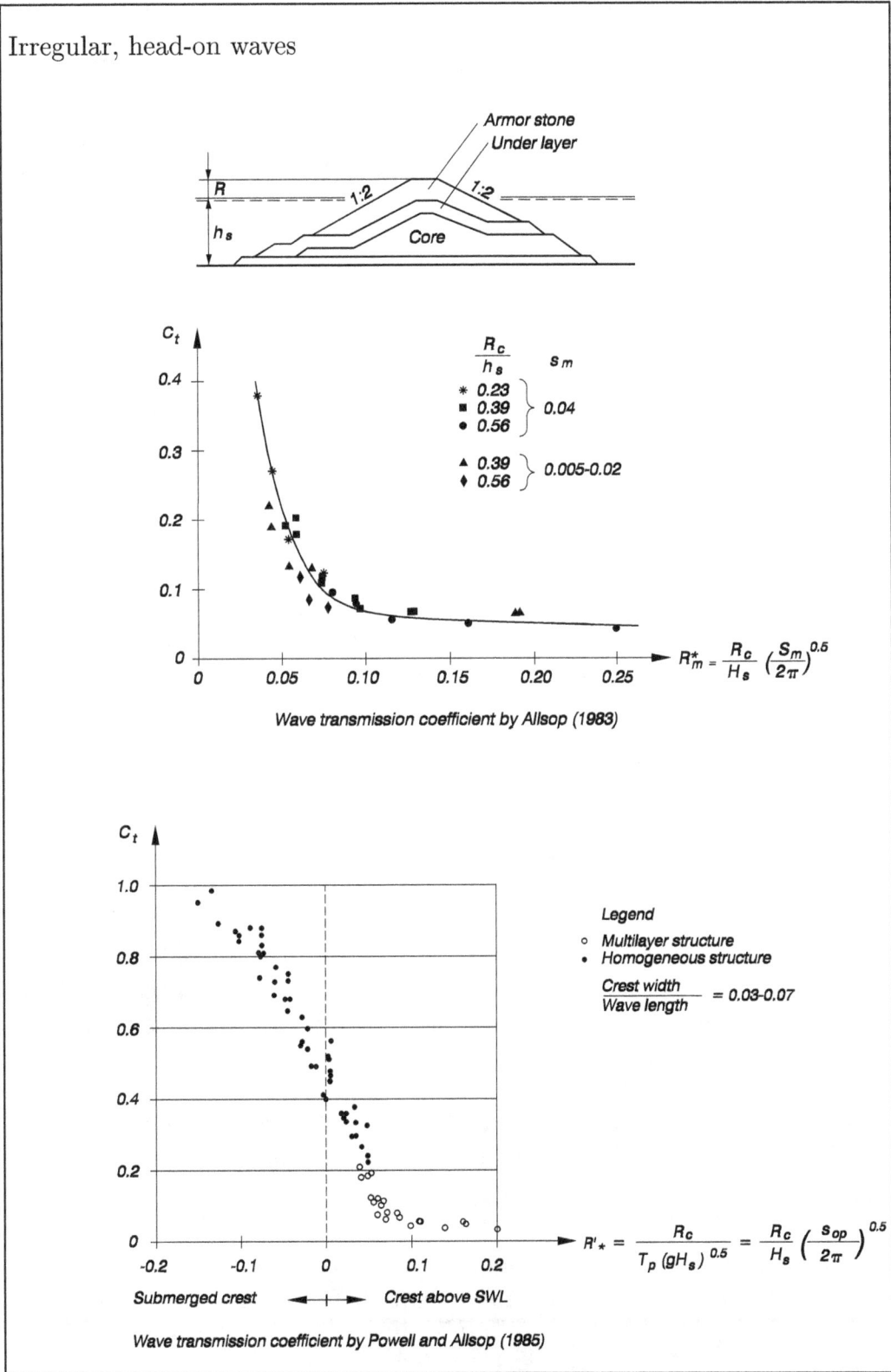

Figure VI-5-28. Wave transmission diagram by Allsop (1983) and Powell and Allsop (1985)

Table VI-5-15
Wave Transmission Formula by van der Meer and d'Angremond (1991) for Rock Armored
Low-crested, Submerged, and Reef Breakwaters

Conventional rock armored low-crested breakwaters with their crests above and below still water level, and reef breakwaters, i.e., a homogenous rock structure without filter layers and core. Irregular, head–on waves.

$$C_t = \left(0.031\,\frac{H_s}{D_{n50}} - 0.24\right)\frac{R_c}{D_{n50}} + b \qquad\qquad (VI-5-54)$$

maximum $C_t = 0.75$, minimum $C_t = 0.075$ conventional structure
maximum $C_t = 0.60$, minimum $C_t = 0.15$ reef type structure

where

$$b = \begin{cases} -5.42\,s_{op} + 0.0323\,\frac{H_s}{D_{n50}} - 0.0017\left(\frac{B}{D_{n50}}\right)^{1.84} + 0.51 & \text{conventional structure} \\ -2.6\,s_{op} - 0.05\,\frac{H_s}{D_{n50}} + 0.85 & \text{reef type structure} \end{cases}$$

H_s –   significant wave height of incident waves

D_{n50} –   median of nominal diameter of rocks for design conditions

R_c –   freeboard, negative for submerged breakwater

B –   width of the crest

s_{op} –   deepwater wave steepness corresponding to peak period.

The maximum and minimum limits of C_t must be interpreted as the valid ranges of Equation VI-5-54 and not as general limits since it is obvious that C_t can exceed these limits when the crest is deeply submerged. For example when the crest is deeply submerged ($R_c/D_{n50} < -6$) then C_t becomes close to unity.

The formula is based on model test results of Seelig (1980), Powell and Allsop (1985), Daemrich and Kahle (1985), Ahrens (1987), van der Meer (1988) and Daemen (1991).

Tested ranges : $1 < \frac{H_s}{D_{n50}} < 6$ $0.01 < s_{op} < 0.05$ $-2 < R_c/D_{n50} < 6$

Comparison between model test results and results calculated by Eq VI-5-54 gives a standard deviation of $\sigma(C_t) = 0.05$ corresponding to the 90% confidence limits given by $C_t \pm 0.08$.

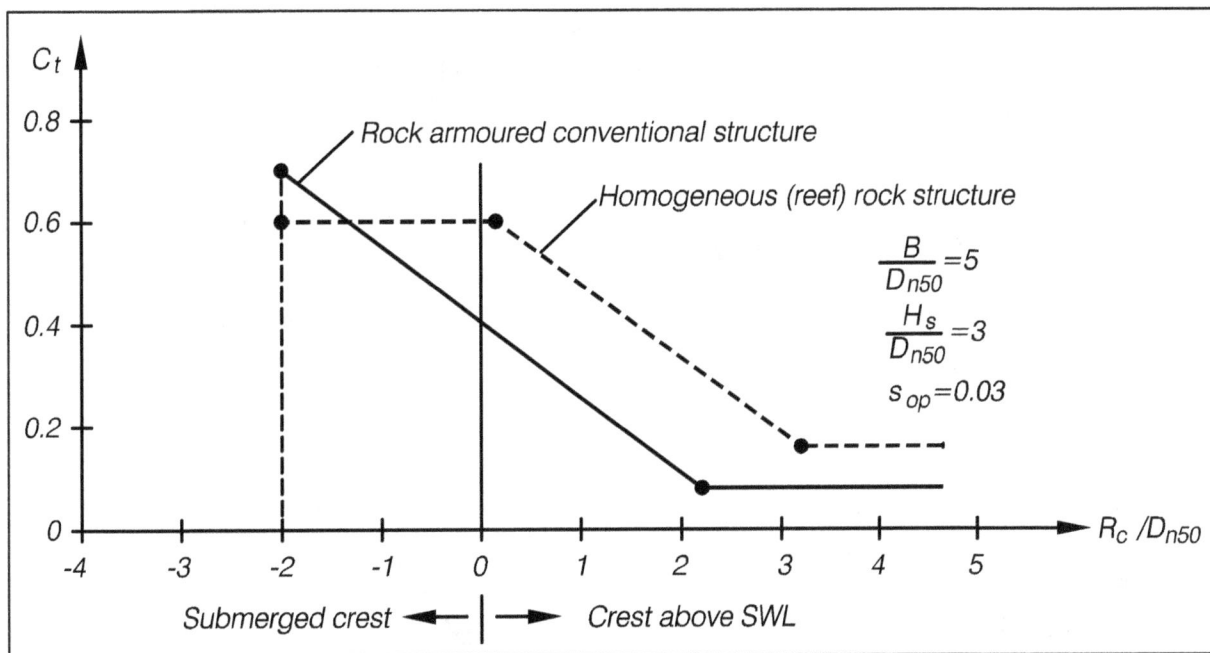

Figure VI-5-29. Example of total wave transmission coefficients, C_t, for conventional and reef type low-crested and submerged breakwaters, calculated from the van der Meer and d'Angremond (1991) formula given by Equation VI-5-54

Table VI-5-16
Wave Transmission Formula by Goda (1969)

Regular, head-on waves

$$C_t = \begin{cases} \left(\left(0.25 \left(1 - \sin\left(\frac{\pi}{2\alpha}\right) \left(\frac{R_c}{H} + \beta\right)\right) \right)^2 + 0.01 \left(1 - \frac{h_c}{h_s}\right)^2 \right)^{0.5} & \beta - \alpha < \frac{R_c}{H} < \alpha - \beta \\ 0.1 \left(1 - \frac{h_c}{h_s}\right) & \frac{R_c}{H} \geq \alpha - \beta \end{cases} \qquad (VI-5-55)$$

$\alpha = 2.2$

β is given by the figure

h_c is the vertical distance from water level to the bottom of caissons.

Formula is based on regular wave tests, but can be applied to irregular waves by using H_s for H

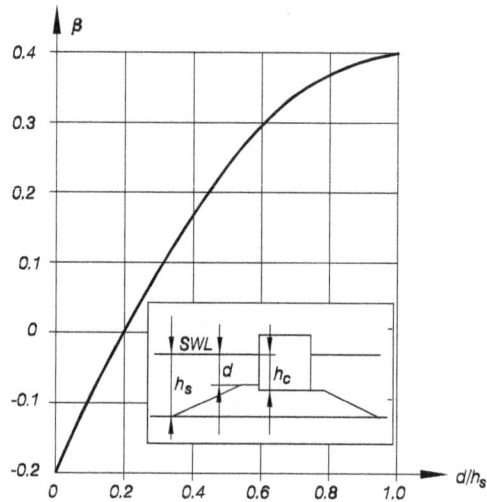

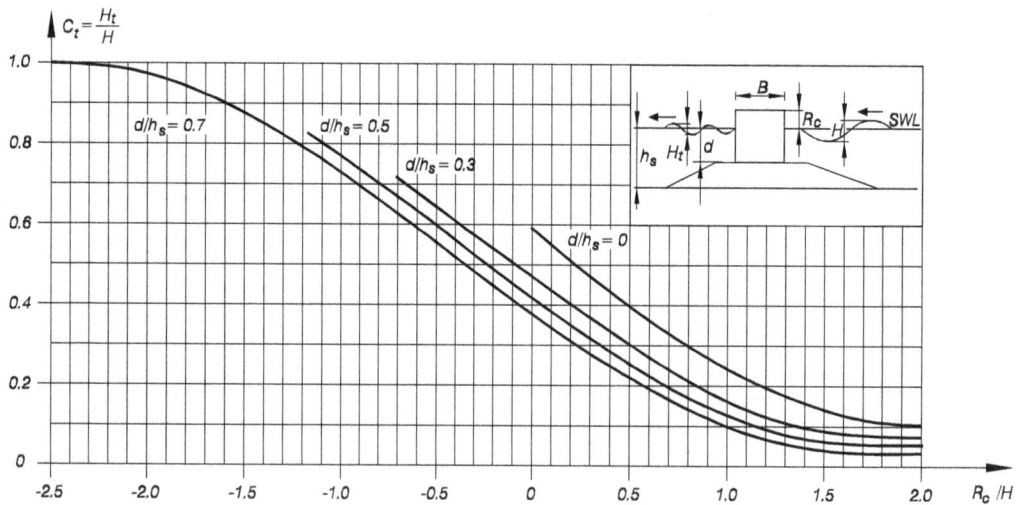

Table VI-5-17
Wave Transmission Formula by Takahashi (1996)

Irregular, head-on and oblique long-crested waves

$$C_t = \left[0.25\left[\left(1 - sin\frac{\pi}{4.4}\right)\left(\frac{R_c}{H_{1/3}} + \beta + \beta_s\right)\right]^2 + 0.01\left(1 - \frac{h_c}{h_s}\right)^2\right]^{0.5}$$

$$\text{valid for } \beta + \beta_s - 2.2 < \frac{R_c}{H_{1/3}} < 2.2 - \beta - \beta_s$$

$$(VI-5-56)$$

$$C_t = 0.1\left(1 - \frac{h_c}{h_s}\right) \qquad \text{valid for } \frac{R_c}{H_{1/3}} \geq 2.2 - \beta - \beta_s$$

where

$$\beta_s = -0.3\left[(R_c - 2\,d_c)/\left(H_{1/3}\,tan\theta\right)\right]^{0.5}$$

β is given by the figure

d_c is the elevation of the lower edge of the sloping face relative to still-water level, i.e., positive if over SWL and negative if under SWL.

θ is angle of wave incidence with 0° being normal incidence

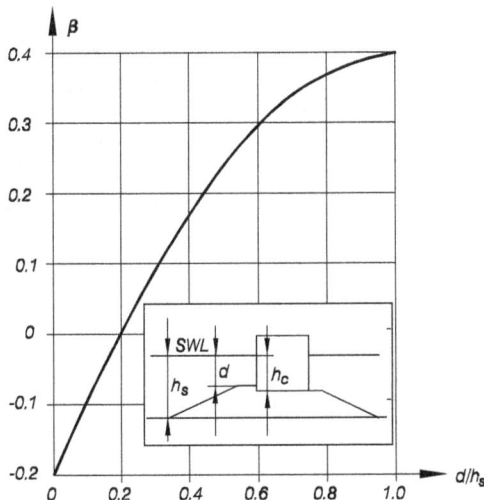

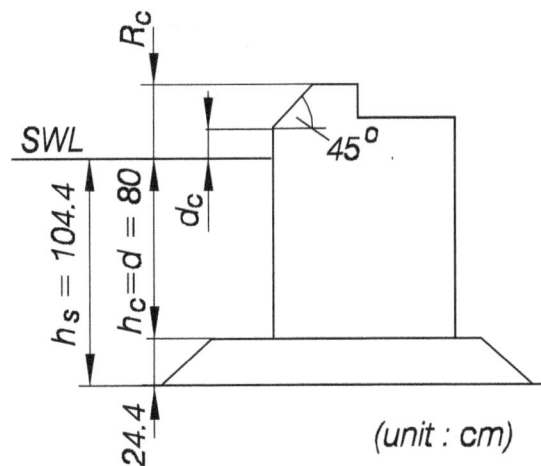

Irregular, head-on waves

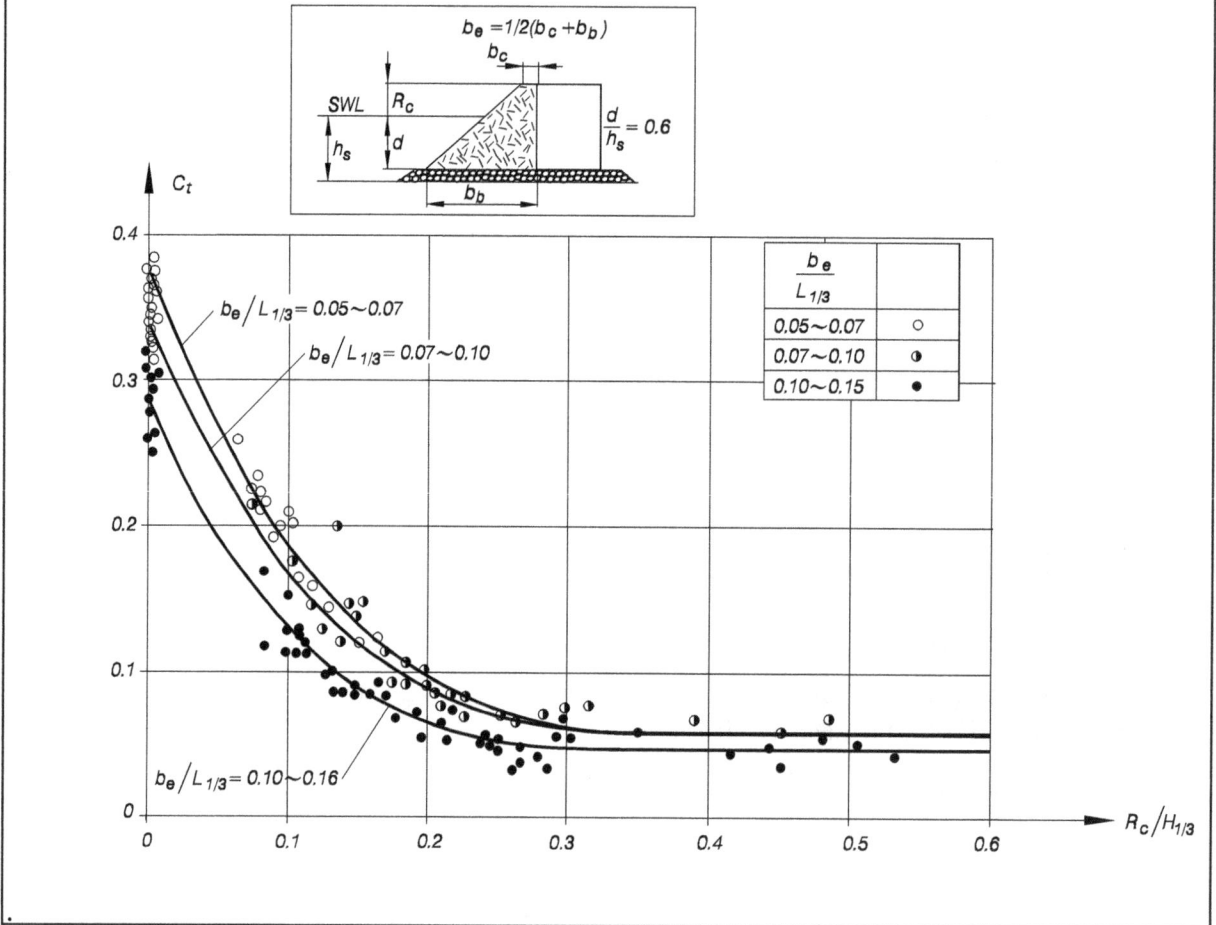

Figure VI-5-30. Wave transmission by overtopping of horizontal composite breakwaters armored with tetrapods (Tanimoto, Takanashi, and Kimura 1987)

Irregular, head-on waves

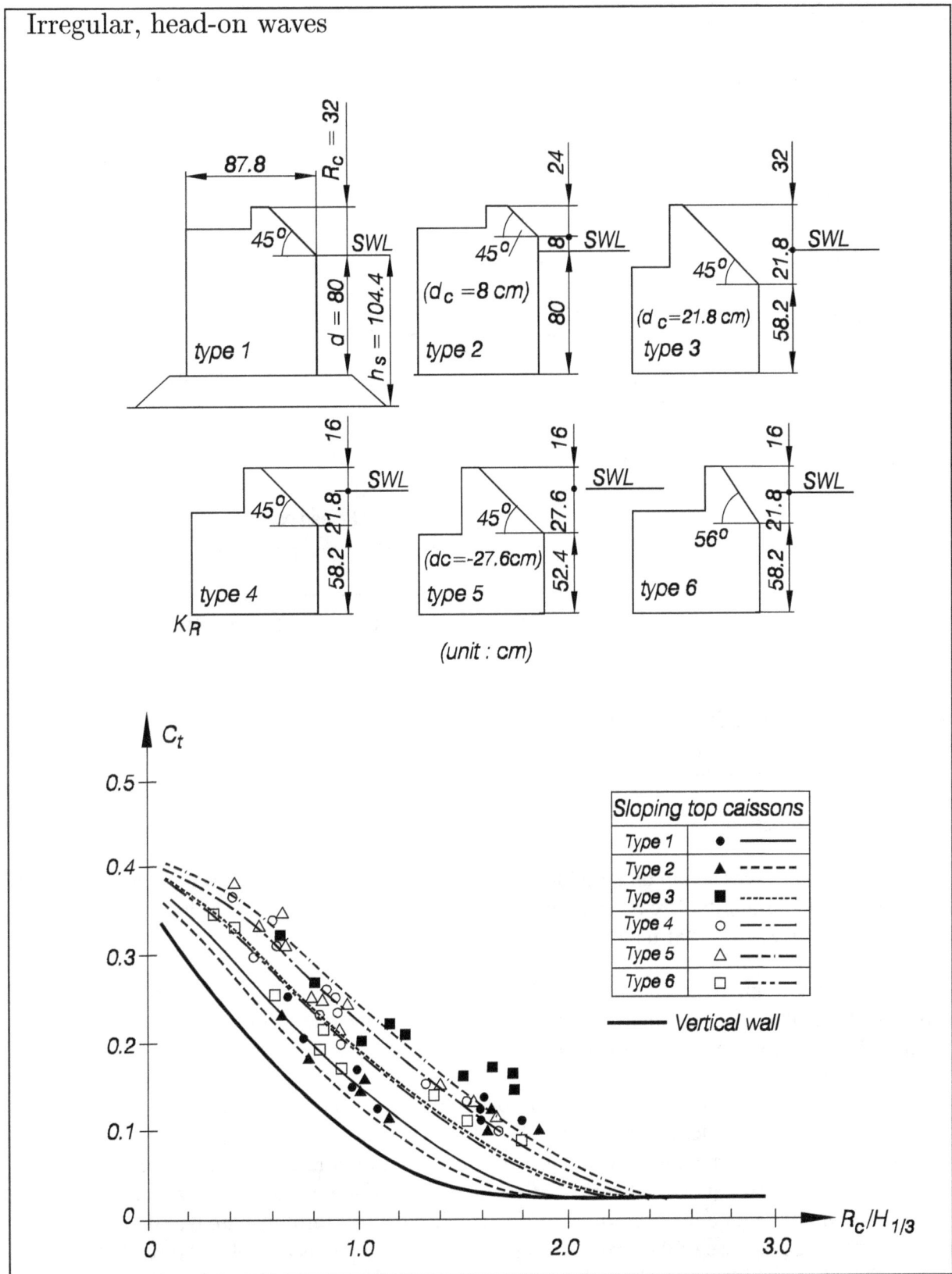

Figure VI-5-31. Wave transmission by overtopping of sloping top structures (Takahashi and Hosoyamada 1994)

Irregular, head-on waves

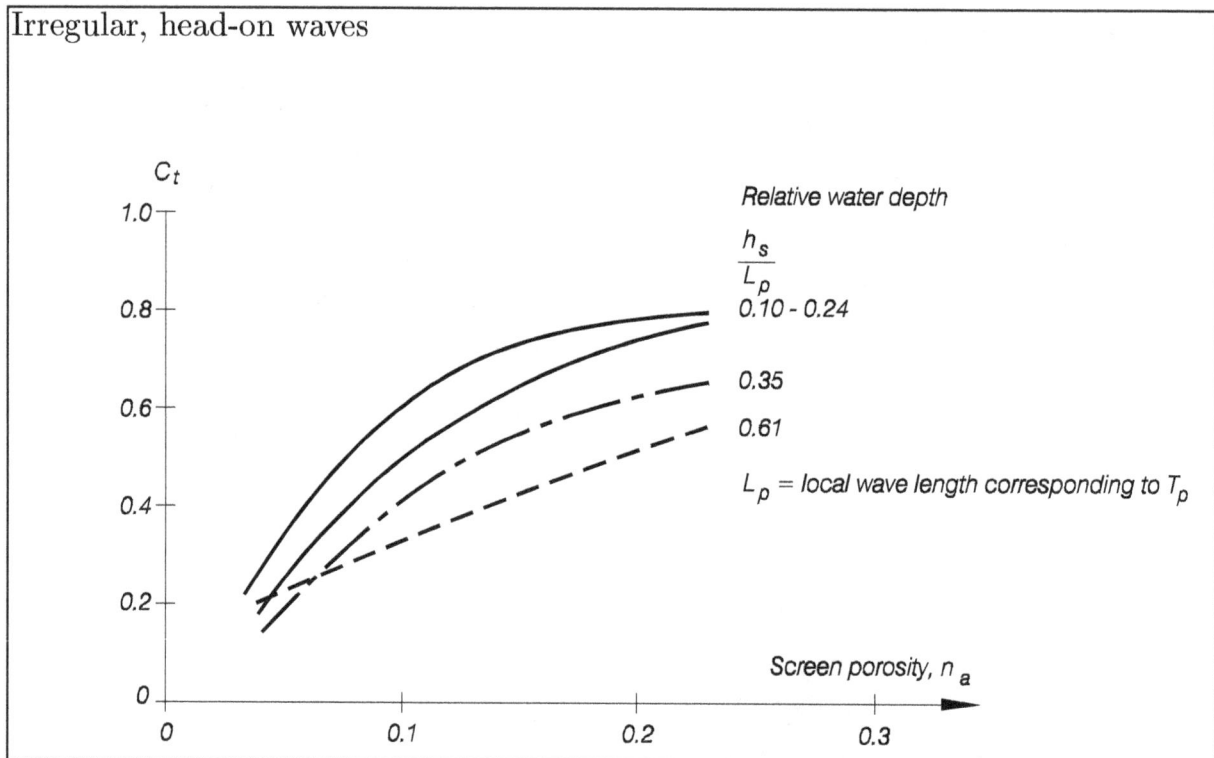

Figure VI-5-32. Wave transmission through perforated single wall (Allsop and Hettiarachchi 1988)

VI-5-3. Rubble-Mound Structure Loading and Response.

 a. Armor layer stability.

 (1) Introduction.

 (a) Wave forces acting on a rubble-mound slope can cause armor unit movement. This is called hydraulic instability. Breakage of armor units is another type of instability which is discussed in Part VI-5-3c, "Structural integrity of concrete armor units."

 (b) Armor unit movements can be rocking, displacement of units out of the armor layer, sliding of a blanket of armor units, and settlement due to compaction of the armor layer. Figure VI-5-33 shows the most typical armor layer failure modes.

 (c) The complicated flow of waves impacting armor layers makes it impossible to calculate the flow forces acting on armor units. Moreover, the complex shape of units together with their random placement makes calculation of the reaction forces between adjacent armor units impossible. Consequently, deterministic calculations of the instantaneous armor unit stability conditions cannot be performed, which is why stability formulae are based on hydraulic model tests. The response of the armor units in terms of movements are related directly to parameters of the incident waves, while treating the actual forces as a "black box" transfer function. However, some qualitative considerations of the involved forces can be used to explore the structure of stability formulae.

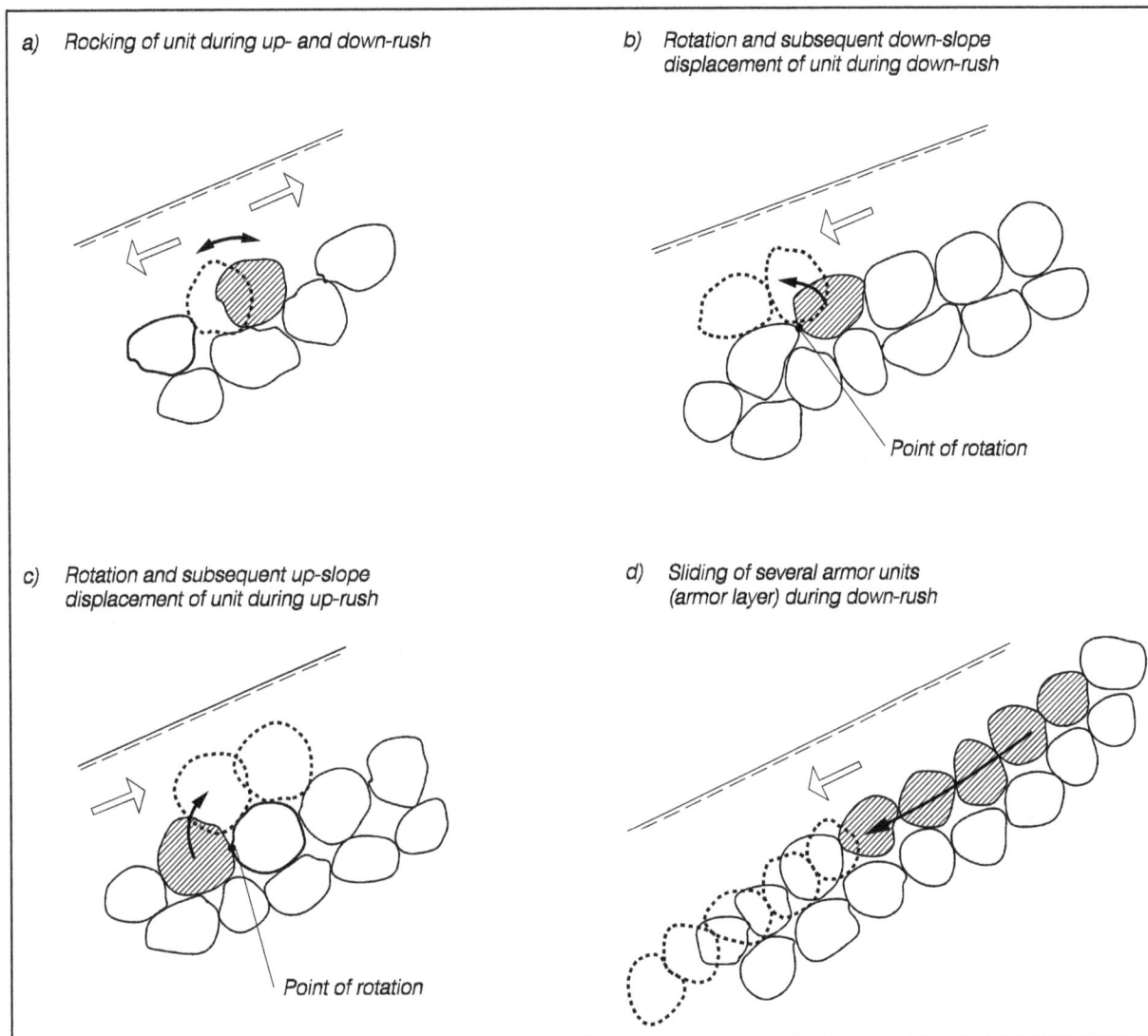

Figure VI-5-33. Typical armor layer failure modes (Burcharth 1993)

(2) Stability parameters and structure of stability formulae.

(a) The wave-generated flow forces on armor units might be expressed by a Morison equation containing a drag force F_D, a lift force F_L and an inertia force F_I. The stabilizing force is the gravitational force F_G. Assuming that at the stage of instability drag and lift force dominates the inertia force, a qualitative stability ratio can be formulated as the drag force plus the lift force divided by the gravity force

$$\frac{F_D + F_L}{F_G} \approx \frac{\rho_w D_n^2 v^2}{g(\rho_s - \rho_w)D_n^3} = \frac{v^2}{g\Delta D_n} \qquad \text{(VI-5-57)}$$

where D_n = (armor unit volume)$^{1/3}$ is the equivalent cube length, ρ_s and ρ_w are the mass densities of armor units and water, respectively, and v is a characteristic flow velocity. By inserting v.

$(gH)^{1/2}$ for a breaking wave height of H in Equation VI-5-57 the following stability parameter, N_s, is obtained.

$$N_s = \frac{H}{\Delta D_n} \qquad \text{(VI-5-58)}$$

where $\Delta = (\rho_s/\rho_w - 1)$. Non-exceedence of instability, or a certain degree of damage, can then be expressed in the general form

$$N_s = \frac{H}{\Delta D_n} \leq K_1^a K_2^b K_c^3 ... \qquad \text{(VI-5-59)}$$

where the factors depend on all the other parameters, except H, Δ and D_n, influencing the stability. Table VI-5-18 gives an overview of the sea state and structural parameters influencing armor layer stability. Also given are the combined parameters including wave height-period parameters commonly used in stability formulae. Stability formulae do not contain explicitly all the parameters shown in Table VI-5-18. This together with the stochastic nature of wave load and armor response introduces uncertainty in any stability formula. This uncertainty is in most cases included in Equation VI-5-59 in the form of a Gaussian distributed stochastic variable with a specified mean value and standard deviation.

(b) Simple geometrical considerations of the balance of the forces acting on an armor stone have been used to explore the right-hand side of Equation VI-5-59. Examples are:

$$\frac{H}{\Delta D_n} = K \cos\alpha \qquad \text{Svee (1962)}$$

$$\frac{H}{\Delta D_n} = (K \cot\alpha)^{1/3} \qquad \text{Hudson (1958, 1959)}$$

$$\frac{H}{\Delta D_n} = K(\tan\varphi\cos\alpha - \sin\alpha) \qquad \text{Iribarren (1938), Iribarren and Nogales (1954)}$$

where φ is the angle of repose of the armor. The coefficient K includes some level of damage as well as all other influencing parameters not explicitly included in the formulae.

(c) For armor units of complex shape and interlocking capability it is more difficult to make simple realistic force balance models. Qualitatively the difference between interlocking and noninterlocking armor is illustrated in the graphs of Figure VI-5-34, which show the influence of slope angle on the stabilizing effects of gravitational force, interlocking and surface friction. The interlocking effect is significant only for steeper slopes. Price (1979) performed dolos armor pullout tests in the dry that indicated maximum resistance occurs at slope of $\cot\alpha = 2$. As a further demonstration Burcharth and Thompson (1983) showed that dolos armor placed on a horizontal bed and exposed to oscillatory flow is not more stable than rock armor of similar weight.

Table VI-5-18
Parameters Influencing Hydraulic Stability of Armor Layers

Sea state parameters

- Characteristic wave heights: H_s, $H_{1/3}$, H_{mo}, $H_{1/10}$, etc.
- Characteristic wave length: L_m, L_{om}, L_p, etc.
- Characteristic wave steepness: s_m, s_{om}, s_p, etc.
- Wave assymmetricity
- Shape of wave spectrum: JONSWAP, P-M, TMA etc. and double peak spectra.
- Wave grouping
- Water depth, h
- Wave incident angle, β
- Number of waves, N_z
- Mass density of water, ρ_w

Structural parameters

- Seaward profile of the structure, including armor layer slope angle α, freeboard, etc.
- Mass density of armor units, ρ_s
- Grading of rock armor, d_{n50}, d_{n15}, d_{n85}
- Mass M and shape of armor units
- Packing density, placement pattern and layer thickness of main armor
- Porosity and permeability of underlayers, filter layer(s) and core

Combined parameters

$$\Delta = \frac{\rho_s}{\rho_w} - 1$$

$$N_s = \frac{H_s}{\Delta D_n} \qquad \text{Shore Protection Manual (1984)}$$

$$N_s^* = N_s\, s_p^{-1/3} \qquad \text{Ahrens (1987)}$$

$$H_0\, T_0 = N_s\, T_m \sqrt{\frac{g}{D_n}} \qquad \text{van der Meer (1988)}$$

$$\xi_m = \frac{\tan \alpha}{\sqrt{s_{om}}} \qquad \text{Battjes (1974b)}$$

where s_p the local wave steepness

T_m mean wave period

$$s_{om} = \frac{2\pi H_s}{g T_m^2}$$

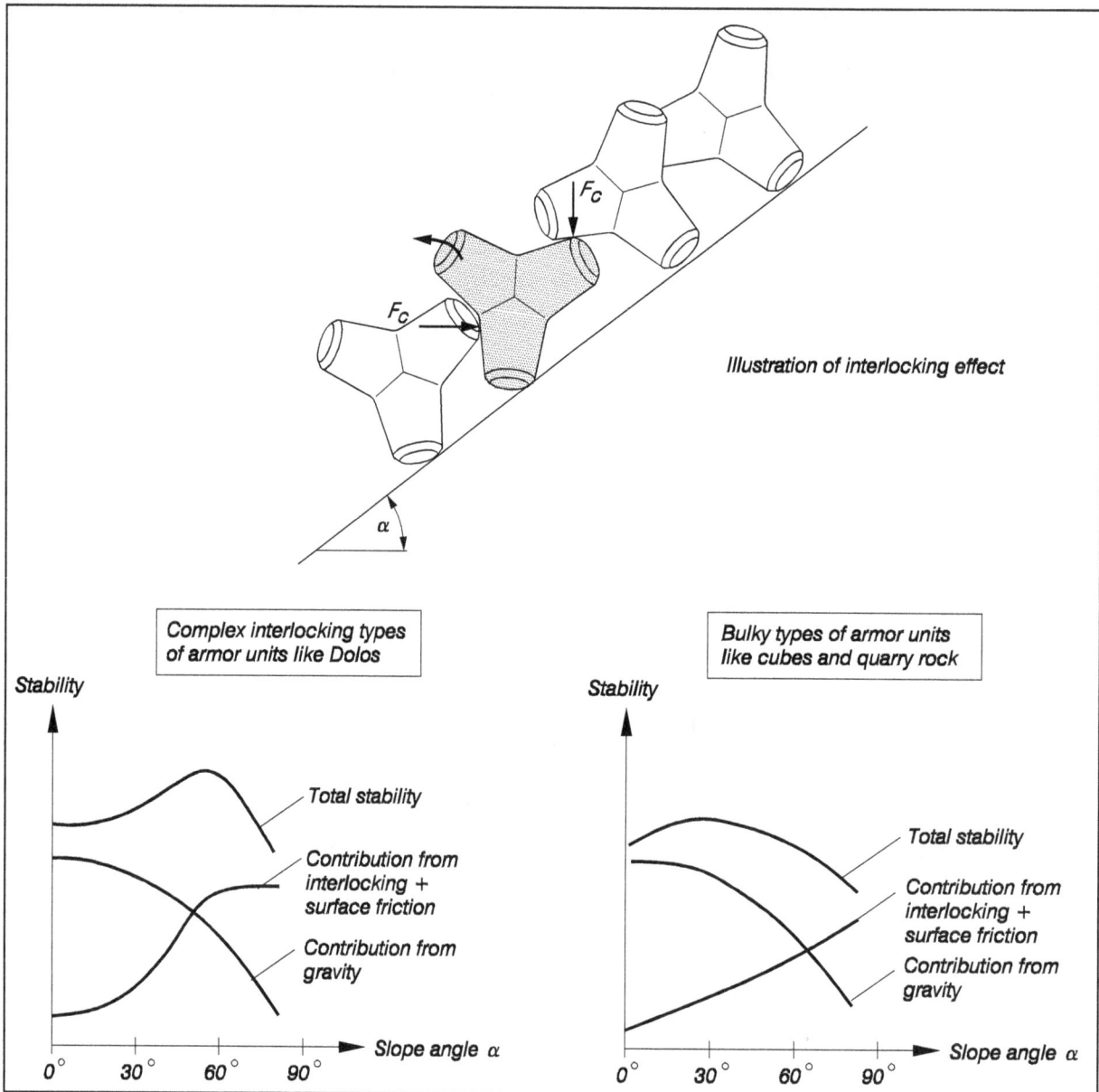

Figure VI-5-34. Illustration of influence of slope angle on the stabilizing effects of gravitational force, interlocking and surface friction (Burcharth 1993)

(3) Definition of armor layer damage.

(a) Damage to armor layers is characterized either by counting the number of displaced units or by measurement of the eroded surface profile of the armor slope. In both cases the damage is related to a specific sea state of specified duration. The counting method is based on some classification of the armor movements, for example:

- No movement.

- Single armor units rocking.

• Single armor units displaced from their original position by a certain minimum distance, for example D_n or h_a, where h_a is the length (height) of the unit

(b) Displacements can be in terms of units being removed out of the layer or units sliding along the slope to fill in a gap. In case of steep slopes, displacements could also be sliding of the armor layer due to compaction or loss of support.

(c) Damage in terms of displaced units is generally given as the relative displacement, D, defined as the proportion of displaced units relative to the total number of units, or preferably, to the number of units within a specific zone around swl. The reason for limiting damage to a specific zone is that otherwise it would be difficult to compare various structures because the damage would be related to different totals for each structure. Because practically all armor unit movements take place within the levels $\forall H_s$ around swl, the number of units within this zone is sometimes used as the reference number. However, because this number changes with H_s it is recommended specifying a H_s-value corresponding to a certain damage level (as proposed by Burcharth and Liu 1992) or to use the number of units within the levels swl $\forall n D_n$, where n is chosen such that almost all movements take place within these levels. For example for dolosse $n = 6$ is used.

(d) Damage D can be related to any definition of movements including rocking. The relative number of moving units can also be related to the total number of units within a vertical strip of width D_n stretching from the bottom to the top of the armor layer. For this strip displacement definition, van der Meer (1988) used the term N_{od} for units displaced out of the armor layer and N_{or} for rocking units. The disadvantage of N_{od} and N_{or} is the dependence of the slope (strip) length.

(e) Damage characterization based on the eroded cross-section area A_e around swl was used by Iribarren (1938) and Hudson (1958) (Table VI-5-19). Hudson defined D as the percent erosion of original volume. Iribarren defined the limit of severe damage to occur when erosion depth in the main armor layer reached D_n.

(f) Broderick (1983) defined a dimensionless damage parameter for riprap and rock armor given as

$$S = \frac{A_e}{D_{n50}^2} \qquad \text{(VI-5-60)}$$

which is independent of the length of the slope and takes into account vertical settlements but not settlements and sliding parallel to the slope. S can be interpreted as the number of squares with side length D_{n50} which fit into the eroded area, or as the number of cubes with side length D_{n50} eroded within a strip width D_{n50} of the armor layer. The damage parameter S is less suitable in the case of complex types of armor like dolosse and tetrapods due to the difficulty in defining surface profile. An overview of the damage parameters is given in Table VI-5-19.

Table VI-5-19
Definition of Damage Parameters D, N_{od} and S

---

1) Relative displacement within an area

$$D = \frac{\text{number of displaced units}}{\text{total number of units within reference area}}$$

Displacement has to be defined, e.g., as position shifted more than distance D_n, or displacements out of the armor layer.

The reference area has to be defined, e.g., as the complete armor area, or as the area between two levels, e.g., $SWL \pm H_s$, where H_s corresponds to a certain damage, or $SWL \pm nD_n$, where $\pm nD_n$ indicates the boundaries of armor displacements.

2) Number of displaced units within a strip with width D_n (van der Meer 1988)

$$N_{od} = \frac{\text{number of units displaced out of armor layer}}{\text{width of tested section } / D_n}$$

3) Relative number of displaced units within total height of armor layer (van der Meer 1988)

$\frac{N_{od}}{N_a}$, where N_a is the total number of units within a strip of horizontal width D_n

$\frac{N_{od}}{N_a} = D$ if in D the total height of the armor layer is considered, and no sliding $> D_n$ of units parallel to the slope surface takes place

4) Percent erosion of original volume (Hudson 1958)

$$D = \frac{\text{average eroded area from profile}}{\text{area of average original profile}} \times 100\%$$

5) Relative eroded area (Broderick 1983)

$S = A_e / D_{n50}^2$

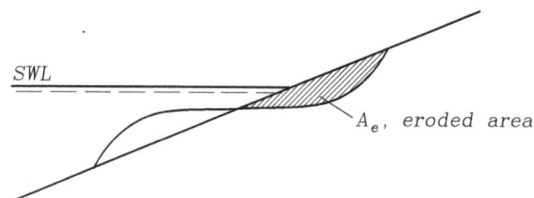

SWL

A_e, eroded area

If settlements are disregarded the following relationship between N_{od} and S is valid:

$$N_{od} = G(1-p)S \qquad \text{(VI-5-61)}$$

where p is the porosity of the armor layer and G is a factor dependent on the armor layer gradation. The range of p is approximately 0.4 - 0.6 with the lowest values corresponding to rock and the highest to dolosse. $G = 1$ for uni-size concrete armor and 1.2 - 1.6 for stone armor. It is seen that N_{od} is roughly equal to $S/2$. Unfortunately Equation VI-5-61 is not generally applicable because experience shows that the relationship depends on the armor slope angle. Table VI-5-20 shows examples of relationships between N_{od} and S as determined from model tests.

(g) A conventional damage level classification and the related values of the damage parameters D, N_{od} and S are given in Table VI-5-21.

(4) Armor layer damage progression.

(a) During the projected service life of a rubble-mound structure, damage to the armor layer may occur if design wave conditions are exceeded or the structure is exposed to repeated storms near the design conditions. Often it is not possible to mobilize and repair armor layer damage before the structure is impacted by additional severe storm waves that could worsen damage and possibly result in structure failure. A method for assessing armor layer damage progression due to multiple storms of differing wave conditions was developed by Melby and Kobayashi (1998a, 1998b) and Melby (1999). The method is based on seven long-duration physical model tests simulating various combinations of successive storms. The 1:2 sloping structure was protected with uniform armor stone (five tests) or wide-graded riprap (two tests). Irregular breaking wave conditions generally exceeding the design wave condition were used with the highest wave conditions causing moderate overtopping of the structure. Two water depths were used in the testing. The average damage as a function of time was given by Melby (1999) in terms of time domain wave parameters as

$$\overline{S}(t) = \overline{S}(t_n) + 0.025 \frac{(N_s)_n^5}{(T_m)_n^{1/4}} \left(t^{1/4} - t_n^{1/4} \right) \qquad \text{for } t_n \le t \le t_{n+1} \qquad \text{(VI-5-62)}$$

or in terms of frequency domain wave parameters

$$\overline{S}(t) = \overline{S}(t_n) + 0.0202 \frac{(N_{mo})_n^5}{(T_p)_n^{1/4}} \left(t^{1/4} - t_n^{1/4} \right) \qquad \text{for } t_n \le t \le t_{n+1} \qquad \text{(VI-5-63)}$$

with

$$\overline{S} = \frac{A_e}{D_{n50}^2} \qquad N_s = \frac{H_s}{\Delta D_{n50}} \qquad N_{mo} = \frac{H_{mo}}{\Delta D_{n50}} \qquad \Delta = \frac{\rho_a}{\rho_w} - 1 \qquad \text{(VI-5-64)}$$

Table VI-5-20
Examples of Experimentally Determined Relationships Between N_{od} and S

van der Meer (1988)
$$\begin{cases} \text{Cubes, slope 1:1.5} & N_{od} = (S - 0.4)/1.8 \\[2mm] \text{Tetrapod, slope 1:1.5} & N_{od} = (S - 1)/2 \\[2mm] \text{Accropode, slope 1:1.33} & N_{od} = (S - 1)/2 \end{cases}$$

Holtzhausen and Zwamborn (1991) Accropodes

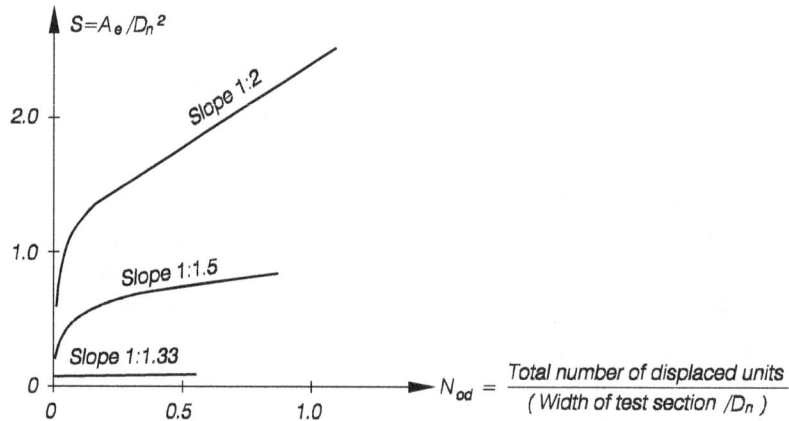

where t_n is the time at start of storm n, and t is time at end of storm n. (Time has the same units as wave period.) The wave parameters are local incident wave conditions not too far seaward of the structure toe, and the subscript n refers to those wave parameters associated with storm n. The standard deviation of average damage was given by the expression

$$\sigma_s = 0.5\overline{S}^{0.65} \tag{VI-5-65}$$

(b) For a specified sequence of storms of given duration Equation VI-5-62 or VI-5-63 is solved with the damage result from the previous storm being the initial damage for the next storm. Reasonable sequences of wave parameters and storm durations must be estimated using probabilistic methods based on long-term wave measurements or hindcasts.

Table VI-5-21
Damage Classification and Related Values of the Damage Parameters D, N_{od} and S

No damage	No unit displacement. Note that S might not be equal to zero due to settlement
Initial damage	Few units are displaced. This damage level corresponds to the *no-damage* level used in *Shore Protection Manual* 1977 and 1984 in relation to the Hudson formula stability coefficient, where the *no-damage* level is defined as 0-5% displaced units within the zone extending from the middle of the crest height down the seaward face to a depth below SWL equal to a H_s-value which causes the damage 0-5%.
Intermediate damage ranging from moderate to severe damage	Units are displaced but without causing exposure of the under or filter layer to direct wave attack
Failure	The underlayer or filter layer is exposed to direct wave attack

Damage level by D for two-layer armor

Unit	Slope	Initial damage	Intermediate damage	Failure	Reference
Rock[1]	1:2-1:3	0-5%	5-10%	≥20%	Jackson (1968)
Cube[2]	1:1.5-1:2		4%		Brorsen et al. (1974)
Dolosse[2]	1:1.5	0-2%		≥15%	Burcharth and Liu (1992)
Accropode[2,3]	1:1.33	0%	1-5%	≥10%	Burcharth et al. (1998)

[1] D is defined as percentage of eroded volume.
[2] D is defined as percentage of units moved more than D_n within the following level restricted areas: For rock see definition under initial damage, for cube $SWL \pm 6D_n$, for Dolosse $SWL \pm 6D_n$, for Accropodes between levels $SWL + 5D_n$ and $-9D_n$.
[3] One-layer armor cover layer.

Damage level by N_{od} for two-layer armor (van der Meer 1988)

Unit	Slope	Initial damage	Intermediate damage	Failure
Cube	1:1.5	0		2
Tetrapods	1:1.5	0		1.5
Accropode	1:1.33	0		0.5

Damage level by S for two-layer armor (van der Meer 1988)

Unit	Slope	Initial damage	Intermediate damage	Failure
Rock	1 : 1.5	2	3–5	8
Rock	1 : 2	2	4–6	8
Rock	1 : 3	2	6–9	12
Rock	1 : 4 − 1 : 6	3	8–12	17

Melby and Kobayashi also noted that average damage was related to the armor layer eroded depth, d_e, cover depth, d_c, and the upslope eroded length, l_e as defined in Figure VI-5-35.

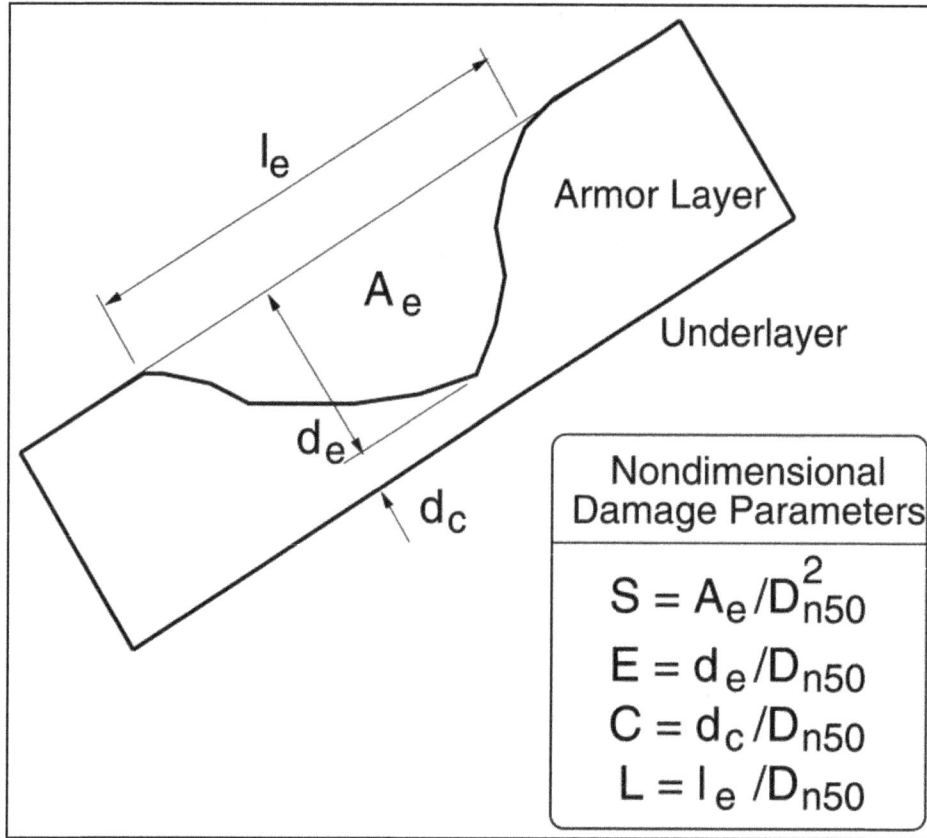

Figure VI-5-35. Damage parameters for structure armor layer (after Melby and Kobayashi 1998b)

In terms of the nondimensional parameters presented in Figure VI-5-35, these relationships were given as

$$\overline{E} = 0.46\overline{S}^{0.5} \qquad \sigma_E = 0.26 - 0.00007\left(\overline{S} - 7.8\right)^4$$

$$\overline{C} = C_o - 0.1\overline{S} \qquad \sigma_C = \sigma_{C_o} + 0.098 - 0.002\left(\overline{S} - 7\right)^2 \qquad \text{(VI-5-66)}$$

$$\overline{L} = 4.4\overline{S}^{0.5}$$

where σ_e and σ_c are the standard deviations of the average nondimensional eroded depth and cover depth, respectively; and C_o is the zero-damage cover layer thickness.

(c) The nondimensional eroded depth in Equation VI-5-66 could be used to estimate average damage in rock armor from an observed eroded depth after a severe storm. This estimate could then be used in Equation VI-5-62 or VI-5-63 to predict damage progression from subsequent storms.

(d) Although the previous damage progression relationships are based on a small number of laboratory experiments, they were formulated to be conservative in the estimates. The more difficult problem is to develop good realizations of storm sequences.

(5) Practical formulae for hydraulic stability of armor layers.

(a) Formulae for hydraulic stability of armor layers are almost exclusively based on small scale model tests. Large scale model tests for verification of small scale model test results have been performed in few cases. Adjustment of formulae due to prototype experience seems not to be reported in the literature.

(b) Generally small scale hydraulic tests of armor layer stability are assumed to be conservative if any bias is present. Nevertheless, armor stability formulae should be applied only for conceptual design, and the uncertainty of the formulae should be considered. When the formulae do not cover the actual range of structure geometries and sea states, preliminary designs should be model tested before actual construction. Major structures should always be tested in a physical model.

(c) Some of the factors by which armor stability formulae can be classified are as follows:

- Type of armor unit.

- Deep or shallow-water wave conditions.

- Armor layers crest level relative to wave runup and swl.

- Structures with and without superstructure.

(d) Type of armor unit distinguishes between rock armor, for which shape and grading must be defined, and uni-size concrete armor units.

(e) Deepwater conditions correspond to Rayleigh distributed wave height at the structure, i.e., depth-limited wave breaking does not take place. Shallow-water conditions correspond to non-Rayleigh distributed wave heights at the structure, i.e., depth limitations cause wave breaking in front of, or in the worst case, directly upon the structure.

(f) Overtopping affects the armor stability. When the crest is lower than the runup level, wave energy can pass over the structure. Thus, the size of the front slope armor can be reduced while the size of the crest and rear slope armor must be increased compared to non-overtopped structures. With respect to armor stability it is common to distinguish between

- Non-overtopped or marginally overtopped structures.

- Low-crested structures, i.e., overtopped structures but with crest level above swl.

- Submerged structures, i.e., the crest level is below swl.

(g) This section presents armor layer stability formulae for use in designing coastal structures. These stability formulae can be used in the context of reliability based design using the partial safety factors given in the tables of Part VI-6-6, "Partial Safety Factor System for Implementing Reliability in Design." Guidance for designing structure cross sections is given in Part VI-5-3*e*, "Design of Structure Cross Section," and complete design examples for specific structure types are given in Part VI-7, "Example Problems."

- Structure trunk stability. Stability formulae for front slope armor on structure trunks are presented in the following tables outlined as follows:

Armor Unit	Non-Overtopped	Overtopped	Submerged
Rock	Tables VI-5-22/23	Tables VI-5-24/26	Tables VI-5-25/26
Concrete cubes	Table VI-5-29		
Tetrapods	Table VI-5-30		
Dolosse	Table VI-5-31		
ACCROPODES [7]	Tables VI-5-32/33		
CORE-LOC [7]	Table VI-5-34		
Tribars	Table VI-5-36		

- Rear side armor stability. Information on rear side armor stability is given in Table VI-5-28. A formula for stability of reef breakwater is presented in Table VI-5-27. A formula for stability of armor in front of a vertical wall is presented in Table VI-5-35. Rubble-mound structure head stability is given in Tables VI-5-37/38. Parapet walls are placed on top of rubble-mound structures to reduce overtopping by deflecting the uprushing waves back into the sea. This generally reduces the front slope armor stability. A low wall behind a wide front armor berm will hardly affect the armor stability (see Figure VI-5-36a). On the other hand a high wall with a relatively deep foundation situated behind a narrow front armor berm will significantly reduce the armor stability (see Figure VI-5-36b).

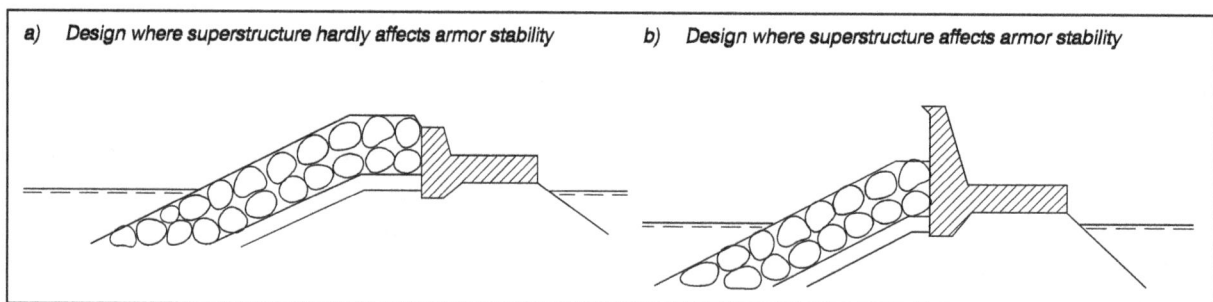

a) Design where superstructure hardly affects armor stability   b) Design where superstructure affects armor stability

Figure VI-5-36. Illustration of superstructure designs causing insignificant and significant reduction in front slope armor stability

- Front slope armor stability. No generally applicable formulae are available for reduction in front slope armor stability caused by parapet walls.

(h) All of the various armor stability criteria represented by the equations and empirical coefficients in Tables VI-5-22 to VI-5-36 were developed in laboratory physical models, most often at reduced scale. Although field experience has added validation to some of these stability formulae, designers should be aware of the following limitations when applying laboratory stability results to prototype conditions.

- Some of the earlier results were obtained using monochromatic waves, whereas most of the more recent model tests used irregular waves. Numerous studies have suggested that the monochromatic wave height leading to armor instability roughly corresponds to the significant wave height of irregular waves; however, not all studies have found this correspondence. For preliminary design for nonbreaking wave conditions always use a stability formula based on irregular wave testing if possible. For breaking wave conditions monochromatic wave stability results will be conservative.

- It is generally thought that the higher waves associated with wave groups are responsible for armor layer damage. Typically irregular wave stability model tests use wave trains with assumed random phasing of the spectral components. Over the course of the testing wave groups of differing characteristics impact the structure, and the assumption is that these wave groups are representative of nature. However, it is possible that nonrandom phasing occurs in nature, particularly in shallow water (Andrews and Borgman 1981). Therefore, use of regular wave stability results will be appropriate in some cases.

- Hand-built armor layers on laboratory structures could be tighter than are armor layers typically constructed in the prototype. This leads to unconservative stability results. In particular special placement of armor in the laboratory is unlikely to be reproduced as well on the job site, especially below the water surface where placement will be much more random. For this reason it may be advisable to use stability criteria for random placement as a basis for design.

- Armor stability formulae are intended for use in preliminary design phases and for estimating material quantities. When feasible, preliminary designs should be confirmed and optimized with hydraulic model tests.

Table VI-5-22
Rock, Two-Layer Armored Non-Overtopped Slopes (Hudson 1974)

Irregular, head-on waves

$$\frac{H}{\Delta D_{n50}} = (K_D \cot \alpha)^{1/3} \quad \text{or} \quad M_{50} = \frac{\rho_s H^3}{K_D \left(\frac{\rho_s}{\rho_w} - 1\right)^3 \cot \alpha} \qquad (VI-5-67)$$

where
H	Characteristic wave height (H_s or $H_{1/10}$)
D_{n50}	Equivalent cube length of median rock
M_{50}	Medium mass of rocks, $M_{50} = \rho_s D_{n50}^3$
ρ_s	Mass density of rocks
ρ_w	Mass density of water
Δ	$(\rho_s/\rho_w) - 1$
α	Slope angle
K_D	Stability coefficient

K_D-values by SPM 1977, $H = H_s$, for slope angles $1.5 \le \cot \alpha \le 3.0$. (Based entirely on regular wave tests.)

Stone shape	Placement	Damage, D [4]			
		0-5%		5-10%	10-15%
		Breaking waves [1]	Nonbreaking waves [2]	Nonbreaking waves	Nonbreaking waves
Smooth, rounded	Random	2.1	2.4	3.0	3.6
Rough angular	Random	3.5	4.0	4.9	6.6
Rough angular	Special [3]	4.8	5.5		

K_D-values by SPM 1984, $H = H_{1/10}$.

Stone shape	Placement	Damage, D [4] = 0-5%	
		Breaking waves [1]	Nonbreaking waves [2]
Smooth rounded	Random	1.2	2.4
Rough angular	Random	2.0	4.0
Rough angular	Special [3]	5.8	7.0

[1] Breaking waves means depth-limited waves, i.e., wave breaking takes place in front of the armor slope. (Critical case for shallow-water structures.)

[2] No depth-limited wave breaking takes place in front of the armor slope.

[3] Special placement with long axis of stone placed perpendicular to the slope face.

[4] D is defined according to SPM 1984 as follows: The percent damage is based on the volume of armor units displaced from the breakwater zone of active armor unit removal for a specific wave height. This zone extends from the middle of the breakwater crest down the seaward face to a depth equivalent to the wave height causing *zero* damage below still-water level.

Shore Protection Manual (1977) versus *Shore Protection Manual* (1984): When considering that $H_{1/10} = 1.27 H_s$ for Rayleigh distributed wave heights (non-depth-limited waves) it is seen that the recommendations of *Shore Protection Manual* (1984) introduce a considerable safety factor compared to the practice based on *Shore Protection Manual* (1977).

Uncertainty of the formula: The coefficient of variation of Eq VI-5-67 is estimated to be 18% by van der Meer (1988). Melby and Mlaker (1997) reported a coefficient of variation for K_D of 25% for stone and 20% for Dolosse.

Table VI-5-23
Rock, Two-Layer Armored Non-Overtopped Slopes (van der Meer 1988)

Irregular, head-on waves

$$\frac{H_s}{\Delta D_{n50}} = 6.2 \cdot S^{0.2} \, P^{0.18} \, N_z^{-0.1} \, \xi_m^{-0.5} \qquad \text{Plunging waves}: \; \xi_m < \xi_{mc} \qquad (VI-5-68)$$

$$\frac{H_s}{\Delta D_{n50}} = 1.0 \cdot S^{0.2} \, P^{-0.13} \, N_z^{-0.1} (\cot \alpha)^{0.5} \, \xi_m^P \qquad \text{Surging waves}: \; \xi_m > \xi_{mc} \qquad (VI-5-69)$$

$$\xi_m = s_m^{-0.5} \tan \alpha \qquad\qquad \xi_{mc} = \left(6.2 \, P^{0.31} \, (\tan \alpha)^{0.5} \right)^{1/(P+0.5)}$$

where
H_s Significant wave height in front of breakwater

D_{n50} Equivalent cube length of median rock

ρ_s Mass density of rocks

ρ_w Mass density of water

Δ $(\rho_s/\rho_w) - 1$

S Relative eroded area (see Table VI-5-21 for nominal values)

P Notional permeability (see Figure VI-5-11)

N_z Number of waves

α Slope angle

s_m Wave steepness, $s_m = H_s/L_{om}$

L_{om} Deepwater wavelength corresponding to mean wave period

Validity:

1) Equations VI-5-68 and VI-5-69 are valid for non-depth-limited waves. For depth-limited waves H_s is replaced by $H_{2\%}/1.4$.

2) For $\cot \alpha \geq 4.0$ only Eq VI-5-68 should be used.

3) $N_z \leq 7,500$ after which number equilibrium damage is more or less reached.

4) $0.1 \leq P \leq 0.6$, $0.005 \leq s_m \leq 0.06$, $2.0 \; tonne/m^3 \leq \rho \leq 3.1 \; tonne/m^3$

5) For the 8 tests run with depth-limited waves, breaking conditions were limited to spilling breakers which are not as damaging as plunging breakers. Therefore, Eqs VI-5-68 and VI-5-69 may not be conservative in some breaking wave conditions.

Uncertainty of the formula: The coefficient of variation on the factor 6.2 in Eq VI-5-68 and on the factor 1.0 in Eq VI-5-69 are estimated to be 6.5% and 8%, respectively.

Test program: See Table VI-5-4.

Table VI-5-24
Rock, Two-Layer Armored Overtopped, but Not Submerged, Low-crested Slopes

Powell and Allsop (1985) analyzed data by Allsop (1983) and proposed the stability formula

$$\frac{N_{od}}{N_a} = a \exp\left[b\, s_p^{-1/3} H_s / (\Delta D_{n50}) \right] \quad \text{or} \quad \frac{H_s}{\Delta D_{n50}} = \frac{s_p^{1/3}}{b} \ln\left(\frac{1}{a} \frac{N_{od}}{N_a} \right) \qquad (\text{VI}-5-70)$$

where values of the empirical coefficients a and b are given in the table as functions of freeboard R_c and water depth h. N_{od} and N_a are the number of units displaced out of the armor layer and the total number of armor layer units, respectively.

Values of coefficients a and b in Eq VI-5-70.

R_c/h	$a \cdot 10^{-4}$	b	wave steepness H_s/L_p
0.29	0.07	1.66	<0.03
0.39	0.18	1.58	<0.03
0.57	0.09	1.92	<0.03
0.38	0.59	1.07	>0.03

van der Meer (1991) suggested that the van der Meer stability formulae for non-overtopped rock slope, Eqs VI-5-68 and VI-5-69, be used with $f_i D_{n50}$ substituted for D_{n50}. The reduction factor f_i is given as

$$f_i = \left(1.25 - 4.8\, \frac{R_c}{H_s} \sqrt{\frac{s_{op}}{2\pi}} \right)^{-1} \qquad (\text{VI}-5-71)$$

where R_c is the freeboard, $s_{op} = H_s/L_{op}$, and L_{op} is deepwater wavelength corresponding to the peak wave period. Limits of Eq VI-5-71 are given by

$$0 < \frac{R_c}{H_s} \sqrt{\frac{s_{op}}{2\pi}} < 0.052$$

Table VI-5-25
Rock, Submerged Breakwaters with Two-Layer Armor on Front, Crest and Rear Slope (van der Meer 1991)

Irregular, head-on waves

$$\frac{h'_c}{h} = (2.1 + 0.1\ S)\ \exp(-0.14\ N_s^*) \qquad\qquad (VI-5-72)$$

where h Water depth

h'_c Height of structure over seabed level ($h - h'_c$ is the water depth over the structure crest).

S Relative eroded area

N_s^* Spectral stability number, $N_s^* = \frac{H_s}{\Delta D_{n50}} s_p^{-1/3}$

Uncertainty of the formula: The uncertainty of Eq VI-5-72 can be expressed by considering the factor 2.1 as a Gaussian distributed stochastic variable with mean of 2.1 and standard deviation of 0.35, i.e., a coefficient of variation of 17%.

Data source: Givler and Sorensen (1986): regular head-on waves, slope 1:1.5

van der Meer (1991): irregular head-on waves, slope 1:2

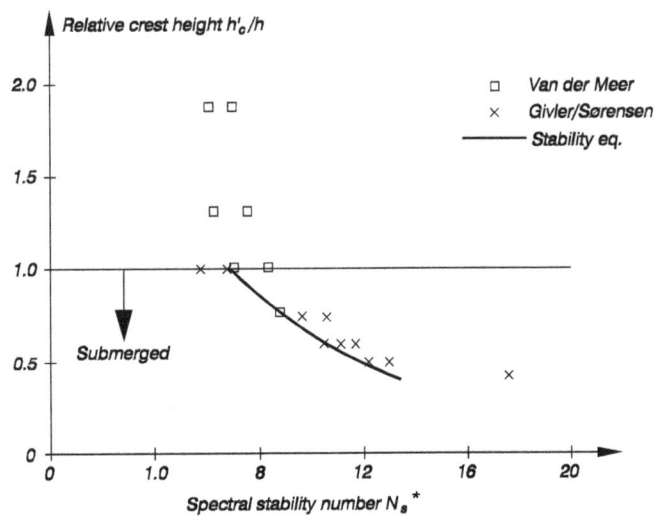

Table VI-5-26
Rock, Two-Layer Armored Low-Crested and Submerged Breakwaters (Vidal et al. 1992)

Tested trunk cross section

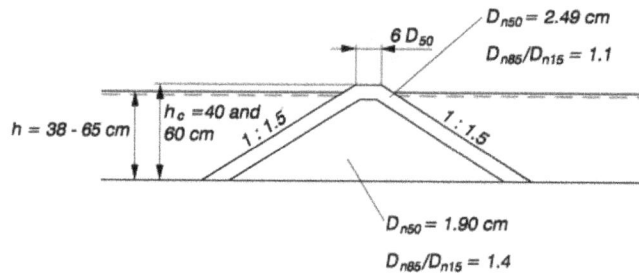

Tested ranges

Irregular, head-on waves

Spectral H_s=5-19 cm, T_p= 1.4 and 1.8 sec.

Free board : $-5\text{cm} \leq R_c = h_c - h \leq 6\text{cm}$

Dimensionless freeboard: $-2 \leq R_c/D_{n50} \leq 2.4$

Stability corresponding to initiation of damage, S=0.5-1.5

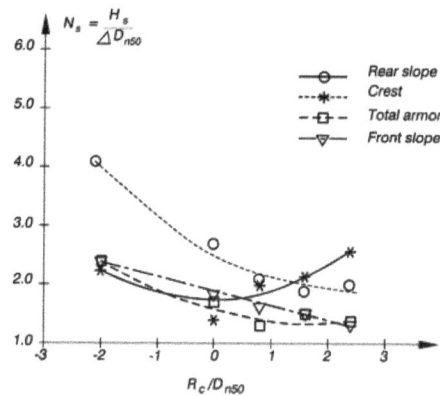

Stability corresponding to extraction of some rocks from lower layer , S=2.0-2.5.

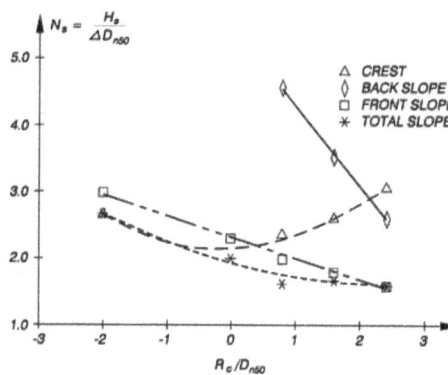

Table VI-5-27
Rock, Low-Crested Reef Breakwaters Built Using Only One Class of Stone

Irregular, head-on waves

van der Meer (1990)

Trunk cross section of reef breakwater

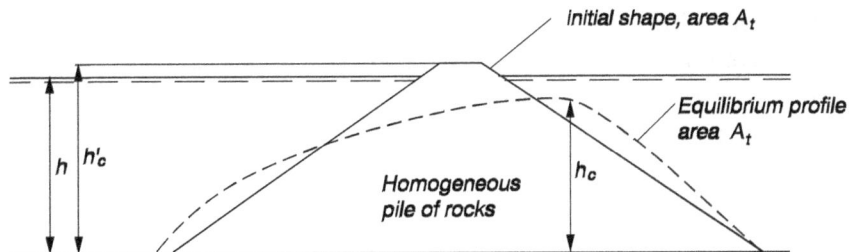

The equilibrium height of the structure

$$h_c = \sqrt{\frac{A_t}{\exp(aN_s^*)}} \quad \text{with a maximum of } h_c' \qquad (VI-5-73)$$

where A_t area of initial cross section of structure

h water depth at toe of structure

h_c' initial height of structure

$N_s^* = \frac{H_s}{\Delta D_{n50}} s_p^{-1/3}$

$a = -0.028 + 0.045 \frac{A_t}{(h_c')^2} + 0.034 \frac{h_c'}{h} - 6 \times 10^{-9} \frac{A_t^2}{D_{n50}^4}$

Data source: Ahrens (1987), van der Meer (1990)

Powell and Allsop (1985) analyzed data by Ahrens, Viggosson, and Zirkle (1982) and Ahrens (1984) and proposed the stability formula

$$\frac{N_{od}}{N_a} = a \exp\left[bH_s/(\Delta D_{n50})\right] \quad \text{or} \quad \frac{H_s}{\Delta D_{n50}} = \frac{1}{b}\ln\left(\frac{1}{a}\frac{N_{od}}{N_a}\right) \qquad (VI-5-74)$$

where values of the empirical coefficients a and b are given in the table as functions of freeboard R_c and water depth h. N_{od} and N_a are the number of displaced rocks and the total number of rocks in the mound, respectively.

Values of coefficients a and b in Eq (VI-5-74).

R_c/h	$a \cdot 10^4$	b
0.0	15	0.31
0.2	17	0.33
0.4	4.8	0.53

Valid for $0.0012 < H_s/L_p < 0.036$

Table VI-5-28
Rock, Rear Slope Stability of Two-Layer Armored Breakwaters Without Superstructures
(Jensen 1984)

Irregular, head-on waves

Jensen (1984) reported results from two case studies of conventional rock armored rubble-mound breakwaters with the main armor carried over the crests and the upper part of the rear slope. Crest width was approximately 3–4 stone diameters. Although Jensen points out that the results are very project dependent, these results could be useful for preliminary estimates. Wave steepness significantly influences the rear side damage.

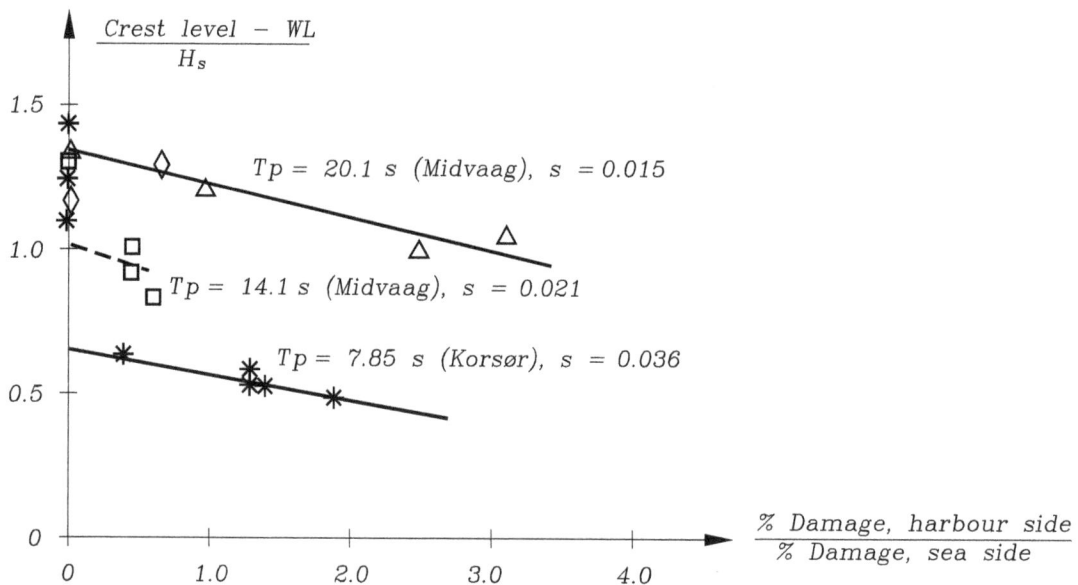

Table VI-5-29
Concrete Cubes, Two-Layer Armored Non-Overtopped Slopes

van der Meer (1988b)

$$N_s = \frac{H_s}{\Delta D_n} = \left(6.7\, N_{od}^{0.4} / N_z^{0.3} + 1.0\right) s_m^{-0.1} \qquad\qquad (VI-5-75)$$

where

H_s	Significant wave height in front of breakwater
ρ_s	Mass density of concrete
ρ_w	Mass density of water
Δ	$\left(\rho_s / \rho_w\right) - 1$
D_n	Cube length
N_{od}	Number of units displaced out of the armor layer within a strip width of one cube length D_n
N_z	Number of waves
s_{om}	Wave steepness, $s_{om} = H_s / L_{om}$

Valid for: Non-depth-limited wave conditions. Irregular head-on waves

Two layer cubes randomly placed on 1:1.5 slope

Surf similarity parameter range $3 < \xi_m < 6$

Uncertainty of the formula: corresponds to a coefficient of variation of approximately 0.10

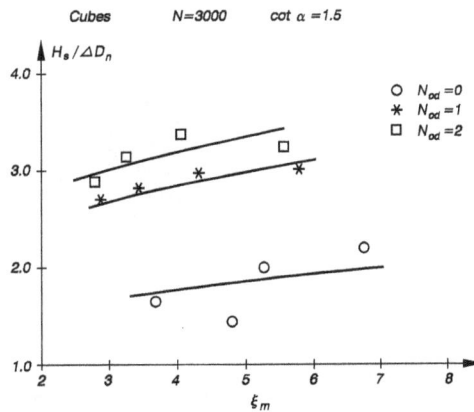

Brorsen, Burcharth, and Larsen (1974)

Brorsen, Burcharth, and Larsen gave the following average N_s and corresponding K_D-values for a two layer concrete cube armor, random placement, slope angles $1.5 \leq \cot\alpha \leq 2.0$ and non-depth-limited irregular waves

Damage level	$N_s = \dfrac{H_s}{\Delta D_n}$	K_D	
		slope $1:1.5$	slope $1:2$
Onset, $D = 0\%$	1.8 - 2.0	3.9 - 5.3	2.9 - 4.0
Moderate, $D = 4\%$	2.3 - 2.6	8.1 - 12	6.1 - 8.8

Table VI-5-30
Tetrapods, Two-Layer Armored Non-Overtopped Slopes

van der Meer (1988b) for non-depth-limited waves

$$N_s = \frac{H_s}{\Delta D_n} = \left(3.75\, N_{od}^{0.5} / N_z^{0.25} + 0.85\right) s_{om}^{-0.2} \qquad (VI-5-76)$$

where H_s Significant wave height in front of breakwater
ρ_s Mass density of concrete
ρ_w Mass density of water
Δ $\left(\rho_s / \rho_w\right) - 1$
D_n Equivalent cube length, i.e., length of cube with the same volume as Tetrapods
N_{od} Number of units displaced out of the armor layer within a strip width of one cube length D_n
N_z Number of waves
s_{om} Wave steepness, $s_{om} = H_s / L_{om}$

Valid for: Non-depth limited wave conditions, Irregular head-on waves
Two layer tetrapods on 1:1.5 slope

Surf similarity parameter range $3.5 < \xi_m < 6$

Uncertainty of the formula: corresponds to a coefficient of variation of approximately 0.10

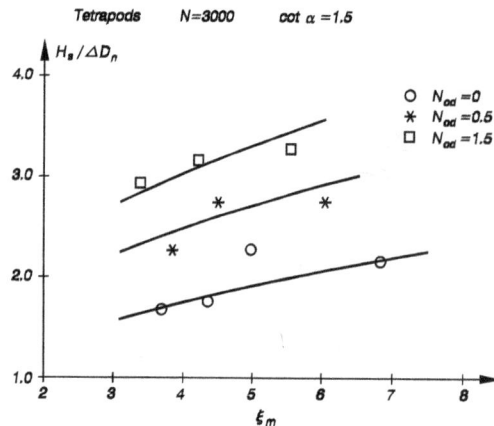

d'Angremond, van der Meer, and van Nes (1994) for depth-limited waves

$$N_s = \frac{H_{2\%}}{\Delta D_n} = 1.4 \left(3.75\, N_{od}^{0.5} / N_z^{0.25} + 0.85\right) s_{om}^{-0.2} \qquad (VI-5-77)$$

In deep water the ratio $H_{2\%}/H_s = 1.4$ for Rayleigh distributed waves. In shallow water this ratio decreases with decreasing relative water depth due to wave breaking.

Table VI-5-31
Dolos, Non-Overtopped Slopes (Burcharth and Liu 1992)

$$N_s = \frac{H_s}{\Delta D_n} = (47 - 72\,r)\,\varphi_{n=2}D^{1/3}N_z^{-0.1}$$

$$= (17 - 26\,r)\,\varphi_{n=2}^{2/3}N_{od}^{1/3}N_z^{-0.1} \qquad\qquad (VI-5-78)$$

where	H_s	Significant wave height in front of breakwater
	ρ_s	Mass density of concrete
	ρ_w	Mass density of water
	Δ	$(\rho_s/\rho_w) - 1$
	D_n	Equivalent cube length, i.e., length of cube with the same volume as dolosse
	r	Dolos waist ratio
	φ	Packing density
	D	Relative number of units within levels SWL $\pm$ 6.5 Dn displaced one dolos height h, or more (e.g., for 2% displacement insert $D = 0.02$)
	N_{od}	Number of displaced units within a strip width of one equivalent cube length D_n
	N_z	Number of waves. For $N_z \geq 3000$ use $N_z = 3000$

Valid for: Breaking and nonbreaking wave conditions
Irregular head-on waves, two layer randomly placed
dolosse with a 1:1.5 slope

$0.32 < r < 0.42$

$0.61 < \varphi < 1$

$1\% < D < 15\%$

$2.49 < \xi_0 < 11.7$

Slope angle: The effect of the slope angle on the hydraulic stability
is not included in Eq VI-5-78. Brorsen, Burcharth,
and Larsen (1974) found only a marginal influence for
slopes in the range 1:1 to 1:3.

Uncertainty of the formula: corresponds to a coefficient of variation of approximately 0.22.

(continued)

Table VI-5-31 (Concluded)

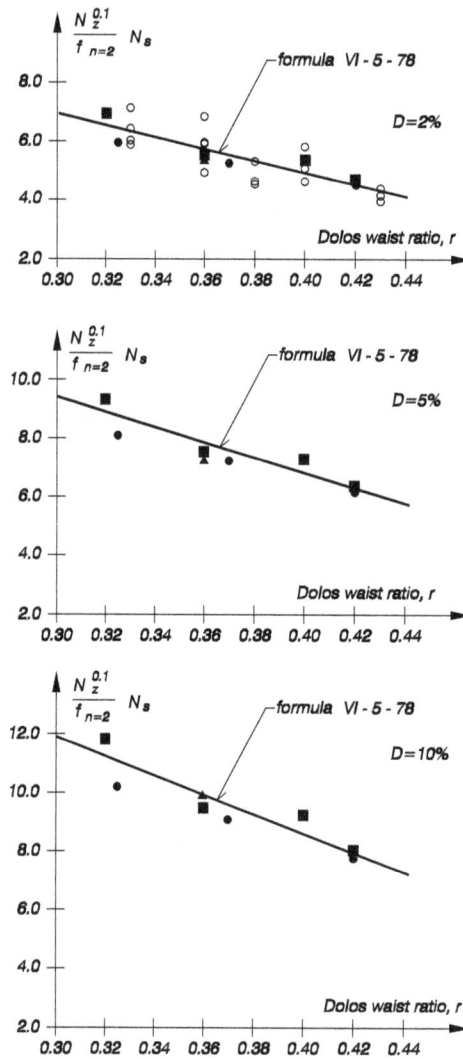

Legend:

Reference	$\phi_{n=2}$	Repeated No	Duration (min.)	ξ_{mo}
▲ Brorsen et al. (1974)	1 (App.)	2	60	2.49-5.37
■ Burcharth et al. (1986)	0.61-0.7	5 or 15	20	3.04-4.49
○ Holtzhausen et al. (1990)	1	3 or 8	60	2.91-7.6
● Burcharth et al. (1992)	0.74	20	5	3.23-11.7

Fit of hydraulic stability formula for a two-layer randomly placed dolos armor on a slope of 1:1.5. Damage levels, $D = 2\%$, 5% and 10% displaced units within levels SWL $\pm 6.5D_n$.

Table VI-5-32
ACCROPODE 7 (van der Meer 1988b)

$$N_s = \frac{H_s}{\Delta D_n} = \begin{cases} 3.7 & \text{no damage} \\ 4.1 & \text{failure} \end{cases} \qquad (VI-5-79)$$

where
H_s Significant wave height in front of breakwater

ρ_s Mass density of concrete

ρ_w Mass density of water

Δ $(\rho_s/\rho_w) - 1$

D_n Equivalent cube length, i.e., length of cube with the same volume as Accropode

Valid for: Irregular, head-on waves
Nonbreaking wave conditions
One layer of Accropodes on slope 1:1.33 placed in accordance with SOGREAH recommendations

No influence of number of waves were found except after start of failure.

Uncertainty of the formula: The standard deviation of the factors 3.7 and 4.1 is approximately 0.2.

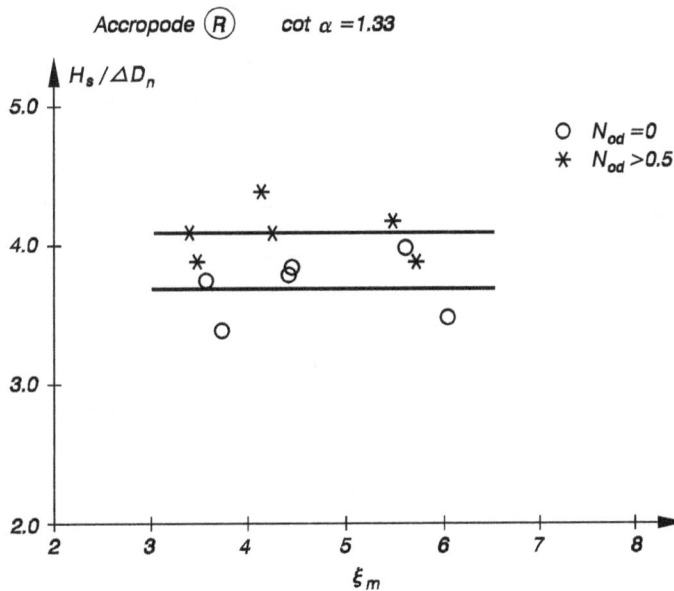

Accropode ® cot α =1.33

Table VI-5-33
ACCROPODE 7, Non-Overtopped or Marginally Overtopped Slopes (Burcharth et al. 1998)

$$N_s = \frac{H_s}{\Delta D_n} = A\left(D^{0.2} + 7.70\right) \quad \text{or} \quad D = 50\left(\frac{H_s}{\Delta D_n} - 3.54\right)^5 \qquad (VI-5-80)$$

where H_s Significant wave height in front of breakwater

ρ_s Mass density of concrete

ρ_w Mass density of water

Δ $(\rho_s/\rho_w) - 1$

D_n Equivalent cube length, i.e., length of cube with the same volume as Accropode

D Relative number of units displaced more than distance D_n

A Coefficient with mean value $\mu = 0.46$ and coefficient of variation
$\sigma/\mu = 0.02 + 0.05(1 - D)^6$, where σ is the standard deviation

Valid for: Irregular, head-on waves

Breaking and nonbreaking wave conditions

One layer of Accropodes on 1:1.33 slope placed in accordance with SOGREAH recommendations

Accropodes placed on filter layer and conventional quarry rock run

$3.5 < \xi_m < 4.5$ (minimum stability range, see figure)

No influence of number of waves were found except after start of failure.

Uncertainty of the formula: see figure and explanation of A.

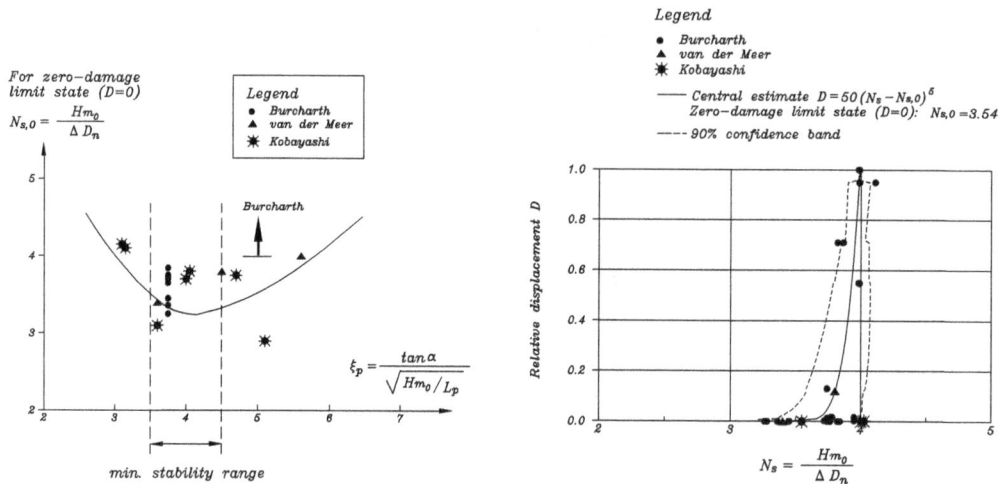

SOGREAH recommends for preliminary design the following K_D-values to be used in Eq VI-5-67.

$$K_D = \begin{cases} 15 & \text{Nonbreaking waves} \\ 12 & \text{Breaking waves} \end{cases}$$

Table VI-5-34
CORE-LOC 7, Non or Marginally Overtopped Slopes
(Melby and Turk 1994; Turk and Melby 1997)

Irregular, head-on waves

$$\frac{H}{\Delta D_{n50}} = (K_D \cot \alpha)^{1/3} \quad or \quad M_{50} = \frac{\rho_c H^3}{K_D (\frac{\rho_c}{\rho_w} - 1)^3 \cot \alpha} \qquad \text{(VI-5-81)}$$

where H Characteristic wave height (H_s)
         D_{n50} Equivalent length of cube having same mass as Core-Loc, $D_{50} = (M_{50}/\rho_c)^{1/3}$
         M_{50} Mass of Core-Loc armor unit, $M_{50} = \rho_c (D_{n50})^3$
         ρ_c Mass density of concrete
         ρ_w Mass density of water
         Δ (ρ_c /ρ_w) - 1
         α Slope angle
         K_D Stability coefficient

Trunk section stability. Melby and Turk (1994) found no reasonable ($K_D < 50$) irregular breaking or nonbreaking wave conditions that would destabilize the layer. For an armor layer exposed to regular depth-limited plunging to collapsing waves, $K_D = 16$ in Equation VI-5-81 is recommended for preliminary design of all trunk sections. The recommended value of K_D is conservative, and it represents a zero-damage condition with little to no armor unit rocking. Site specific physical model tests will usually yield higher values.

Head section stability. $K_D = 13$ is recommended for preliminary design of head sections exposed to both breaking and nonbreaking oblique and head-on waves.

Stability test parameters

Model parameters	M_{50} = 219 g; Depths: 36 and 61 cm; Height: 90 cm
Wave parameters	4.6 # H_{mo} # 36 cm; 1.5 # T_p # 4.7 sec
Structure slope, α	1V:1.33H and 1V:1.5H
Surf similarity parameter	2.13 # ξ_o # 15.9
Relative depth	0.012 # d/L_o # 0.175
Wave steepness	0.001 # H_{mo}/L_o # breaking

Placement. Core-Locs are intended to be randomly placed in a single-unit thick layer on steep or shallow slopes. They are well suited for use in repairing existing dolos structures because they interlock well with dolosse when properly sized (length of Core-Loc central flume is 92 percent of the dolosse fluke length).

Table VI-5-35
Tetrapods, Horizontally Composite Breakwaters (Hanzawa et al. 1996)

$$N_s = \frac{H_s}{\Delta D_n} = 2.32 \left(N_{od}/N_z^{0.5} \right)^{0.2} + 1.33 \qquad \text{(VI-5-82)}$$

where H_s Significant wave height in front of breakwater

ρ_s Mass density of concrete

ρ_w Mass density of water

Δ $(\rho_s/\rho_w) - 1$

D_n Equivalent cube length, i.e., length of cube with the same volume as Tetrapods

N_{od} Number of units displaced out of the armor layer within a strip width of one cube length D_n

N_z Number of waves

Test range: The formula was obtained by fitting of earlier model test results and five real project model tests

Irregular head-on waves
Water depth: 0.25 - 0.50 cm
Slope: 1:1.5
Foreshore : 1:15 - 1:100
Mass of Tetrapods: 90 - 700 g
H_s: 8 - 25.9 cm; T_s: 1.74 - 2.5 s; s_{om}: 0.013 - 0.04

Uncertainty of the formula: Not given. Tanimoto, Haranaka, and Yamazaki (1985) gave the standard deviation of N_{od} equal to $0.36 N_{od}^{0.5}$

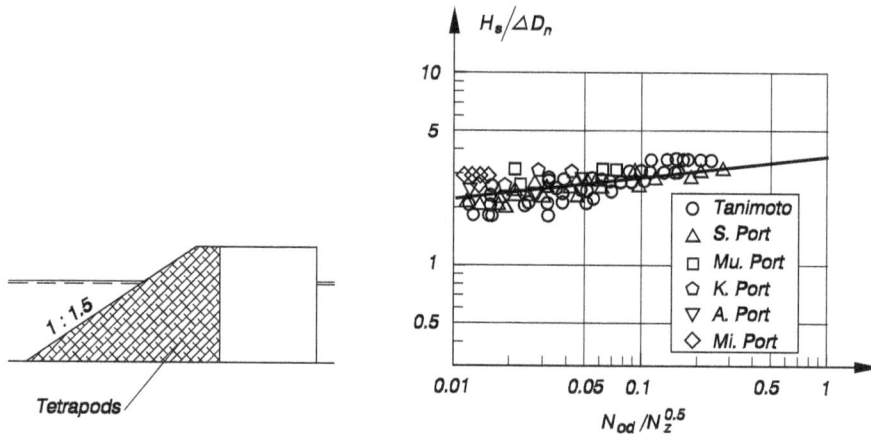

Table VI-5-36
Tribars, Non-Overtopped or Minor Overtopped Slopes, Random and Uniform Placement

Regular, head-on waves

$$\frac{H}{\Delta D_{n50}} = \left(K_D \cot \alpha\right)^{1/3} \quad \text{or} \quad M_{50} = \frac{\rho_s H^3}{K_D \left(\dfrac{\rho_s}{\rho_w} - 1\right)^3 \cot \alpha} \qquad \text{(VI-5-83)}$$

where
H	Characteristic wave height (H_s)
D_{n50}	Equivalent cube length of median rock
M_{50}	Median mass of stone armor unit, $M_{50} = \rho_s (D_{n50})^3$
ρ_s	Mass density of stone
ρ_w	Mass density of water
Δ	$(\rho_s/\rho_w) - 1$
α	Slope angle
K_D	Stability coefficient

Trunk section stability

K_D-values by *Shore Protection Manual* (1984), $H - H_{1/10}$, 0% to 5% damage

Placement	Layers	Breaking waves[1]	Nonbreaking waves[2]	Slope angle cot α
Random	2	9.0	10.0	1.5 – 3.0
Pattern-placed	1	12.0	15.0	(not given)
[1] Depth-limited breaking with waves breaking in front of and on the armor slope. [2] No depth-limited breaking occurs in front of the armor slope.				

• Design wave height considerations. In shallow water the most severe wave condition for design of any part of a rubble-mound structure is usually the combination of predicted water depth and extreme incident wave height and period that produces waves which would break directly on the structure. In some cases, particularly for steep foreshore wlopes, waves breaking offshore will strike directly on the structure. Goda (1985) recommended computing the design wave height a distance $5H_s$ from the structure toe to account for the travel distance of large breakers. A shallow-water coastal structure exposed to a variety of water depths, especially a shore-perpendicular structure such as a groin, should have wave conditions investigated for each range of water depths to determine the highest breaking wave that might impact any part of the structure. For example, a groin that normally experiences wave forces on its armor layer near the seaward end might become submerged during storm surges, and the worst breaking wave condition could occur on a more landward portion of the groin. The effect of oblique wave approach on armor layer stability has not yet been sufficiently quantified. Tests in the European Marine Science and Technology (MAST) program seemed to indicate relatively little reduction in damage for rock armored slopes subjected to oblique wave approach angles up to 60 deg compared to waves of normal incidence (Allsop 1995). The stability of any rubble-mound structure exposed to oblique wave attack should be confirmed with physical model tests.

(6) Structure head section stability.

(a) Under similar wave conditions the round head of a rubble-mound structure normally sustains more extensive and more frequent damage than the structure trunk. One reason is very high cone-overflow velocities, sometimes enhanced in certain areas by wave refraction. Another reason is the reduced support from neighboring units in the direction of wave overflow on the lee side of the cone as shown in Figure VI-5-37. This figure also illustrates the position of the most critical area for armor layer instability. The toe within the same area is also vulnerable to damage in shallow-water situations, and a toe failure will often trigger failure of the armor layer see Part VI-5-6*b*(2), "Scour at sloping structures."

(b) Table VI-5-37 presents stability criteria for stone and dolos rubble-mound structure heads subjected to breaking and nonbreaking waves without overtopping, and Table VI-5-38 gives stability criteria for tetrapod and tribar concrete armor units.

(c) The stability in the critical area of the roundhead might be improved by increasing the head diameter or adding a tail as shown in Figure VI-5-38. Besides obtaining better support from neighboring units, a reduction in wave heights by diffraction is also achieved before the waves reach the vulnerable rear side. Optimization of the slope angle and the layout geometry of cone roundheads can only be achieved by physical model tests because quantitative information on roundhead stability is limited.

(d) The armor layer at bends and corners is generally more exposed than in straight trunk sections. A convex bend or corner will often follow the seabed contours because construction in deeper water increases costs dramatically. Refraction might then cause an increase of the wave height as illustrated in Figure VI-5-39, which in turn increases wave runup and overtopping. Moreover, in sharper convex corners and bends the lateral support by neighbor blocks is reduced as in the case of roundheads. A concave bend or corner will often be exposed to larger waves than the neighboring trunk sections due to the concentration of wave energy by oblique reflection on the slope (Figure VI-5-39). Consequently, runup and overtopping will also be increased.

(7) Riprap armor stability.

(a) The previous armor stability formulations are intended for fairly uniform distributions of armor stone or for uniform size concrete armor units. Riprap armor is characterized by fairly wide gradations in rock size with a large size difference between the largest and smallest stones in the distribution. Use of graded riprap cover layers is generally more applicable to revetments than to breakwaters or jetties. A limitation on the use of graded riprap is that the design wave height should be less than about 1.5 m. At higher design wave heights uniform-size armor units are usually more economical.

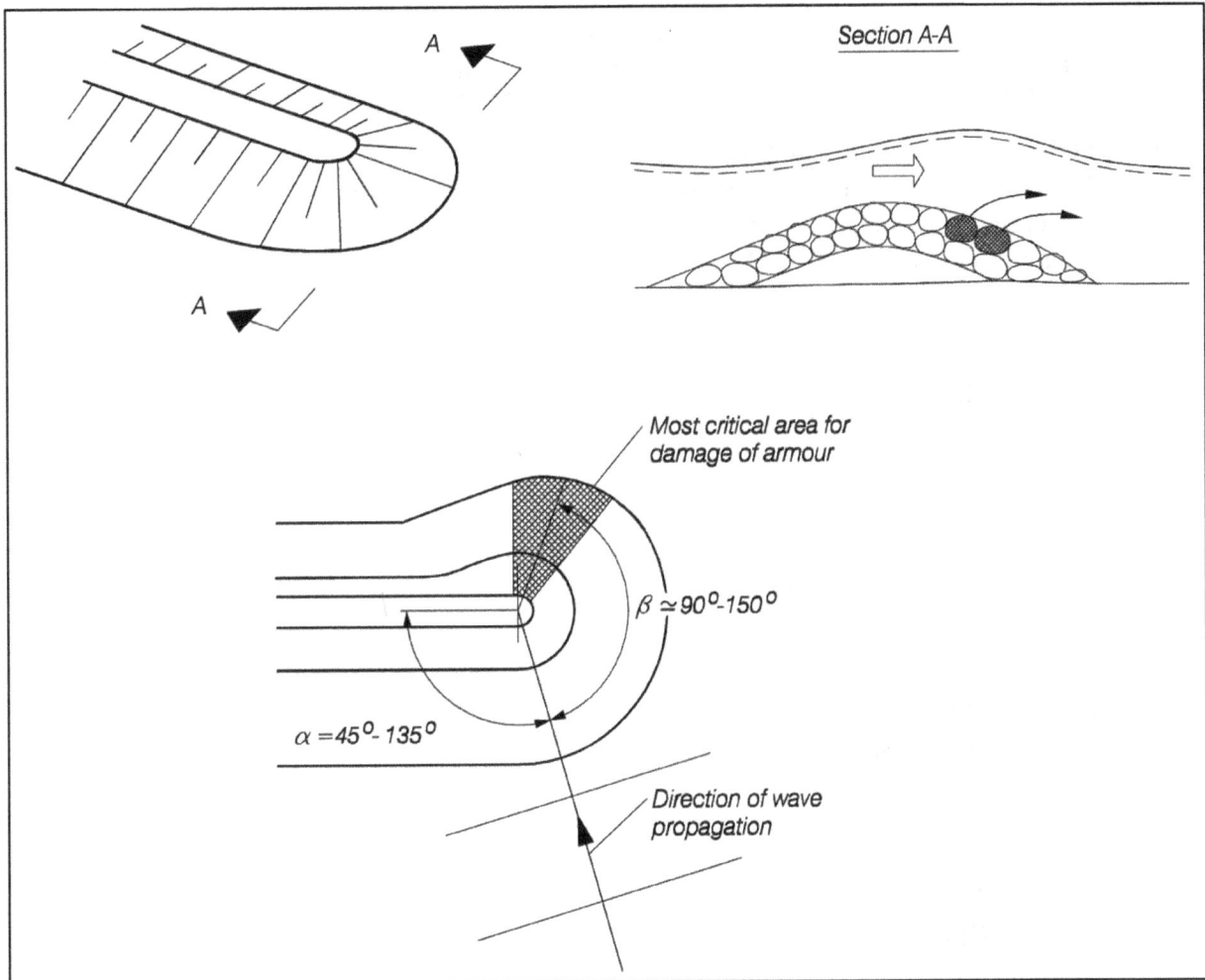

Figure VI-5-37. Illustration of critical areas for damage to armor layers in the round head
(Burcharth 1993)

Table VI-5-37
Rock and Dolos Breakwater Head Stability, No Overtopping (Carver and Heimbaugh 1989)

Rock and dolos armor, monochromatic waves

Mostly monochromatic waves with a few irregular wave cases

Breaking and nonbreaking waves

Incident wave angles: $0°$, $45°$, $90°$, $135°$ (note: $0°$ is wave crests perpendicular to trunk)

$$\frac{H}{\Delta D_{n50}} = A\xi^2 + B\xi + C_C \tag{VI-5-84}$$

where

$$\xi = \frac{\tan \alpha}{\left(H/L\right)^{1/2}}$$

and
H	Characteristic wave height
D_{n50}	Equivalent cube length of median rock
ρ_s	Mass density of stone
ρ_w	Mass density of water
Δ	(ρ_s/ρ_w) - 1
L	Local wavelength at structure toe
α	Structure armor slope
A, B, C_c	Empirical coefficients

Table of coefficients for use in Equation VI-5-84

Armor Type	A	B	C_c	Slope	Range of ξ
Stone	0.272	-1.749	4.179	1V to 1.5H	2.1 – 4.1
Stone	0.198	-1.234	3.289	1V to 2.0H	1.8 – 3.4
Dolos	0.406	-2.800	6.881	1V to 1.5H	2.2 – 4.4
Dolos	0.840	-4.466	8.244	1V to 2.0H	1.7 – 3.2

Notes: The curves giving the best fit to the data were lowered by two standard deviations to provide a conservative lower envelope to the stability results.

A limited number of tests using irregular waves produced corresponding results with T_p equivalent to the monochromatic period and H_{mo} equal to the monochromatic wave height.

Table VI-5-38
Tetrapod and Tribar Breakwater Head Section Stability, No Overtopping

Regular, head-on waves

$$\frac{H}{\Delta D_{n50}} = \left(K_D \cot\alpha\right)^{1/3} \qquad \text{or} \qquad M_{50} = \frac{\rho_s H^3}{K_D\left(\dfrac{\rho_s}{\rho_w} - 1\right)^3 \cot\alpha} \qquad \text{(VI-5-85)}$$

where H Characteristic wave height (H_s)

D_{n50} Equivalent cube length of median rock

M_{50} Median mass of stone armor unit, $M_{50} = \rho_s (D_{n50})^3$

ρ_s Mass density of stone

ρ_w Mass density of water

Δ (ρ_s / ρ_w) - 1

α Slope angle

K_D Stability coefficient

Head Section Stability.

K_D-values by *Shore Protection Manual* (1984), H = $H_{1/10}$, 0 percent to 5 percent damage

Armor Unit	Placement	Layers	Breaking Waves[1]	Nonbreaking Waves[2]	Slope Angle cot α
Tetrapod	Ramdom	2	5.0[3]	6.0	1.5
			4.5	5.5	2.0
			3.5	4.0	3.0
Tribar	Random	2	8.3	9.0	1.5
			7.8	8.5	2.0
			6.0	6.5	3.0
Tribar	Pattern	1	7.5	9.5	(not given)
[1] Depth-limited breaking with waves breaking in front of and on the armor slope. [2] No depth-limited breaking occurs in front of the armor slope. [3] K_D values shown in italics are unsupported by tests results and are provided only for preliminary design purposes.					

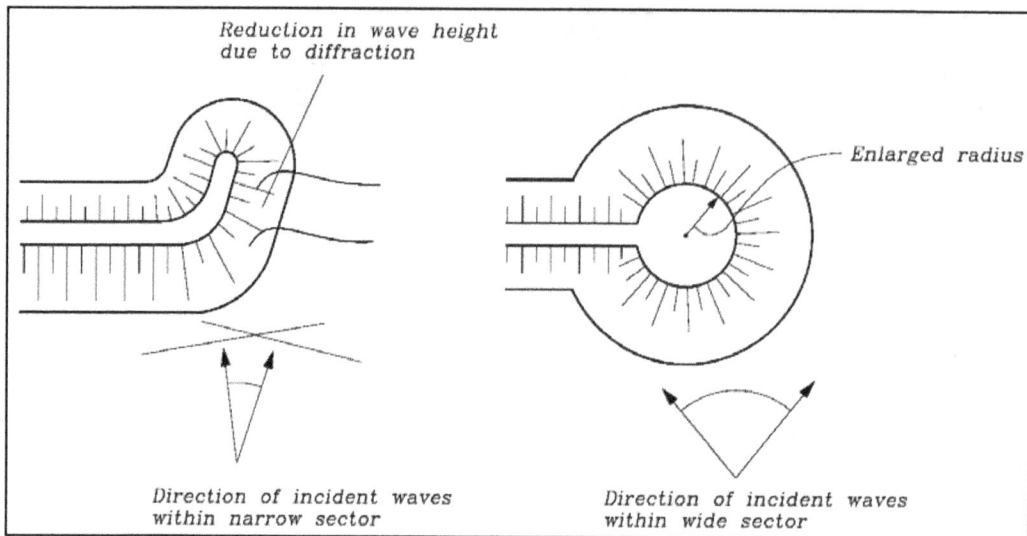

Figure VI-5-38. Illustration of improvement of round head stability by change of geometry (Burcharth 1993)

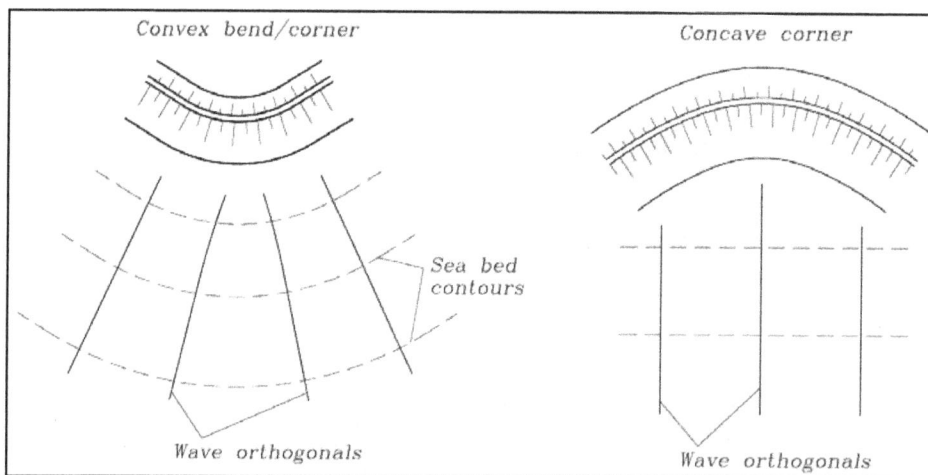

Figure VI-5-39. Convex and concave bends and corners

Generally, the maximum and minimum stone weights in riprap gradations should be limited to

$$W_{max} = 4.0W_{50} \qquad W_{min} = 0.125W_{50}$$

where W_{50} is the median stone weight. The median stone mass for a stable riprap distribution can be determined using the Hudson equation

$$M_{50} = \frac{\rho_r H_{10\%}^3}{K_{RR}\left(\frac{\rho_r}{\rho_w} - 1\right)^3 \cot\alpha} \qquad \text{(VI-5-86)}$$

where ρ_r is the mass density of the riprap, K_{RR} is the riprap stability coefficient, and the other variables are as defined for Equation VI-5-67 in Table VI-5-22. Recommended conservative stability coefficients (0 percent to 5 percent damage) are $K_{RR} = 2.2$ for breaking waves and $K_{RR} = 2.5$ for nonbreaking waves (Ahrens 1981b). Melby and Kobayashi (1998b) showed that deterioration of riprap and uniform armor with equivalent median stone weights was similar. Therefore, Equation VI-5-62 through VI-5-66 could be used to estimate damage progression for both narrow gradations and riprap. The van der Meer (1988) equation (see Table VI-5-23) can also be used to design riprap armor.

(b) An examination of riprap field performance at 14 different dams across the La Grande Hydroelectic complex in Quebec, Canada, generally confirmed the validity of Equation VI-5-86 (Belfadhel, Lefebvre, and Rohan 1996; also see discussion of this paper by van der Meer 1997). Design of riprap armor layer cross sections is covered in Part VI-5-3e, "Design of structure cross section." A complete design example for a riprap armored slope is included in Part VI-7, "Example Problems."

b. Granulated filters and geotextile filter stability. In coastal engineering, filter layers are defined as layers that protect the underlying base material or soil from erosion by waves and currents without excessive buildup of pore pressure in the underlying material. Filter functions can be achieved using either one or more layers of granulated material such as gravel or small stone of various grain sizes, geotextile fabric, or a combination of geotextile overlaid with granulated material. This section covers the function and design of granulated filters. Design criteria for geotextile filter cloth used in filter application are given in Part VI-4-7, "Geotextiles and Plastics." Design of rubble-mound structure underlayers is covered in Part VI-5-3e, "Design of structure cross section."

(1) Filter layer functions. Filter layers are designed to achieve one or more of the following objectives in coastal structures. They can prevent the migration of underlying sand or soil particles through the filter layer voids into the overlying rubble-mound structure layers. Leeching of base material could be caused by turbulent flow within the structure or by excessive pore pressures that can wash out fine particles. Without a filter layer, foundation or underlayer material would be lost and the stones in the structure layer over the filter would sink into the void resulting in differential settlement and decreased structure crest elevation.

Filter layers can aid in the distribution of structure weight. A bedding filter layer helps to distribute the structure's weight over the underlying base material to provide more uniform settlement. A levelled bedding layer also ensures a more uniform baseplate load on caisson structures.

Filter layers can also reduce the hydrodynamic loads on a structure's outer stone layers. A granular filter layer can help dissipate flow energy whereas a geotextile filter will not be as effective in this regard.

(a) Granulated filters are commonly used as a bedding layer on which a coastal structure rests, or in construction of revetments where the filter layer protects the underlying embankment. Filter layers are also needed in rubble-mound structures having cores composed of fine materials like sand or gravel. Stone blankets (used to prevent erosion due to waves and currents) also reduce leeching of the underlying sand or soil, but in this situation stability of the stone blanket material in waves and currents is an important design concern. Design of stone blankets is covered in Part VI-5-3f, "Blanket stability in current fields."

(b) It is advisable to place coastal structures on a bedding layer (along with adequate toe protection) to prevent or reduce undermining and settlement. When rubble structures are founded on cohesionless soil, especially sand, a filter blanket should be provided to prevent differential wave pressures, currents, and groundwater flow from creating an unstable foundation condition through removal of particles. Even when a filter blanket is not needed, bedding layers may be used to prevent erosion during construction, to distribute structure weight, or to retain and protect a geotextile filter cloth. Bedding layers are not necessary where depths are greater than about three times the maximum wave height, where the anticipated bottom current velocities are below the incipient motion level for the average-size bed material, or where the foundation is a hard, durable material such as bedrock.

(c) In some situations granular filters have several advantages over geotextile filters in coastal construction (Permanent International Association of Navigation Congresses (PIANC) 1992).

• The filter elements (stone, gravel, sand, etc.) are usually very durable.

• Granular filters provide a good contact interface between the filter and base material below and between the filter and overlying layers. This is important for sloping structures.

• Granular bedding layers can help smooth bottom irregularities and thus provide a more uniform construction base.

• The porosity of granular filters help damp wave energy.

• Self-weight of the filter layer contributes to its stability when exposed to waves and currents during construction whereas geotextiles may have to be weighted under similar conditions.

• The loose nature of the filter elements allows the filter to better withstand impacts when larger stones are placed on the filter layer during construction or the stones shift during settlement.

• Granular filter layers are relatively easy to repair, and in some instances may be self-healing.

• Filter materials are widely available and inexpensive.

(d) The major disadvantage of granular filters is the difficulty of assuring uniform construction underwater to obtain the required thickness of the filter layer.

(e) Placing larger armor stone or riprap directly on geotextile filter cloth is likely to puncture the fabric either during placement or later during armor settlement. Placing a granular filter layer over the geotextile fabric protects it from damage. In this application there is more flexibility in specifying the filter stone gradation because the geotextile is retaining the underlying soil.

(2) Granulated filter failure modes. Granular filter layers fail their intended function when:

(a) The base layer is eroded through the filter layer. Erosion can occur either by outgoing flow washing out particles perpendicular to the base/filter interface or by wave- and current-induced external flows parallel to the interface.

(b) The filter layer becomes internally unstable. Instability occurs in filters having a very wide gradation when the finer fraction of the filter grain-size distribution is flushed out of the layer between the coarser material. This could result in compaction of the filter layer, differential settlement of the overlayers, and gradual increase in layer permeability.

(c) The interface between adjacent granular layers becomes unstable, and lateral shearing motion occurs between layers constructed on a slope.

(d) The filter layer fails to protect the underlying geotextile fabric from punctures and loss of soil through the filter cloth.

(3) Granulated filter design criteria.

(a) Design criteria for granular filters were originally based on the geometry of voids between packed, uniform spheres. Allowances for grain-size distributions (and many successful field applications) led to the following established geometric filter design criteria. (Design guidance for exposed filter layers must also consider instability due to flow as discussed in Part VI-5-3f, "Blanket stability in current fields.")

- Retention criterion. To prevent loss of the foundation or core material by leeching through the filter layer, the grain-size diameter exceeded by 85 percent of the filter material should be less than approximately four or five times the grain-size diameter exceeded by the coarsest 15 percent of the foundation or underlying material, i.e.,

$$\frac{d_{15(\text{filter})}}{d_{85(\text{foundation})}} < (4 \text{ to } 5) \tag{VI-5-87}$$

The coarser particles of the foundation or base material are trapped in the voids of the filter layer, thus forming a barrier for the smaller sized fraction of the foundation material. The same criterion can be used to size successive layers in multilayer filters that might be needed when there is a large disparity between void sizes in the overlayer and particle sizes in the material under the filter. Filter layers overlying coarse material like quarry spall and subject to intense dynamic forces should be designed similar to a rubble-mound structure underlayer with

$$\frac{W_{50(\text{filter})}}{W_{50(\text{foundation})}} < \left(15 \text{ to } 20\right) \qquad \text{(VI-5-88)}$$

• Permeability criterion. Adequate permeability of the filter layer is needed to reduce the hydraulic gradient across the layer. The accepted permeability criterion is

$$\frac{d_{15(\text{filter})}}{d_{15(\text{foundation})}} > \left(4 \text{ to } 5\right) \qquad \text{(VI-5-89)}$$

• Internal stability criterion. If the filter material has a wide gradation, there may be loss of finer particles causing internal instability. Internal stability requires

$$\frac{d_{60(\text{filter})}}{d_{10(\text{filter})}} < 10 \qquad \text{(VI-5-90)}$$

• Layer thickness. Filter layers constructed of coarse gravel or larger material should have a minimum thickness at least two to three times the diameter of the larger stones in the filter distribution to be effective. Smaller gravel filter layer thickness should be at least 20 cm, and sand filter layers should be at least 10 cm thick (Pilarczyk 1990). These thickness guidelines assume controlled above-water construction. In underwater placement, bedding layer thickness should be at least two to three times the size of the larger quarrystones used in the layer, but never less than 30 cm thick to ensure that bottom irregularities are completely covered. Considerations such as shallow depths, exposure during construction, construction method, and strong hydrodynamic forces may dictate thicker filters, but no general rules can be stated. For deeper water the uncertainty related to construction often demands a minimum thickness of 50 cm.

• Bedding layer over geotextile fabric. In designs where a geotextile fabric is used to meet the retention criterion, a covering layer of quarry spalls or crushed rock (10-cm minimum and 20-cm maximum) should be placed to protect against puncturing by the overlying stones. Recommended minimum bedding layer thickness in this case is 60 cm, and filtering criteria should be met between the bedding layer and overlying stone layer.

(b) Examples of typical granular filters and bedding layers are illustrated in Lee (1972), who discussed and illustrated applications of granular and geotextile filters in coastal structures. Design of filters for block-type revetments with large holes in the cover layer can be found in the PIANC (1992) reference.

(c) The previous geometric granular filter criteria are widely accepted in practice, and they are recommended in cases when an appreciable pressure gradient is expected perpendicular to the soil/filter interface. However, these rules may be somewhat conservative in situations without significant pressure gradients and when flow is parallel to the filter layer.

(d) The need for reliable granular filter design guidance under steady flow and cyclic design conditions fostered research by Delft Hydraulics Laboratory in support of the Oosterschelde Storm Surge Barrier in The Netherlands. Stationary and cyclic flow both parallel to and perpendicular to the filter layer were investigated by de Graauw, van der Meulen, and van der Does de Bye (1984). They developed hydraulic filter criteria based on an expression for critical hydraulic gradient parallel to the filter/soil interface. This method assumes that erosion of base material is caused by shear stresses rather than groundwater pressure gradients; and where this is the case, the geometric filter requirements can be relaxed.

(e) The filter design guidance of de Graauw et al. was expressed in terms of the filter d_{15}, base material d_{50}, filter porosity, and critical shear velocity of the base material; and acceptable values for the critical gradient were given by graphs for each of the flow cases. Design of a hydraulic granular filter requires good understanding of the character of flow within the filter layer, e.g., steady flow in channels. In these cases the method of de Graauw et al. (1984) can be used. More recent research aimed at improving granular filter design criteria was reported by Bakker, Verheij, and deGroot (1994).

(4) Granulated filter construction aspects.

(a) Granular filter construction above water creates no special problems, and accurate placement is straightforward. However, constructing a filter beneath the water surface is somewhat more problematic. If small-size filter material with a wide gradation is dropped into place, there is a risk of particle segregation by size. This risk can be decreased by using more uniform material and minimizing the drop distance. Another problem is maintaining adequate layer thickness during underwater placement. This has led to the recommended layer thickness being greater than required by the geometric filter criteria. Finally, filter or bedding layers placed underwater are exposed to eroding waves and currents until the overlayers are placed. Depending on site-specific conditions, this factor may influence the construction sequence or the time of year chosen for construction.

(b) It is common practice to extend the bedding layer beneath rubble-mound structures at least 1.5 m beyond the toe of the cover stone to help reduce toe scour. Some low rubble-mound structures have no core, and instead are composed entirely of armor layer and underlayers. These structures should have a bedding layer that extends across the full width of the structure.

c. Structural integrity of concrete armor units.

(1) Introduction.

(a) Figure VI-5-40 shows examples of the wide variety of existing concrete armor units. These might be divided into the following categories related to the structural strength:

Massive or blocky   (e.g., cubes including Antifer type, parallelepiped block, grooved cube with hole)

Bulky   (e.g., seabee, Core-Loc[7], Accropode[7], Haro[7], dolos with large waist ratios)

Slender   (e.g., tetrapod, dolos with smaller waist ratios)

Multi-hole cubes   (e.g., shed, cob)

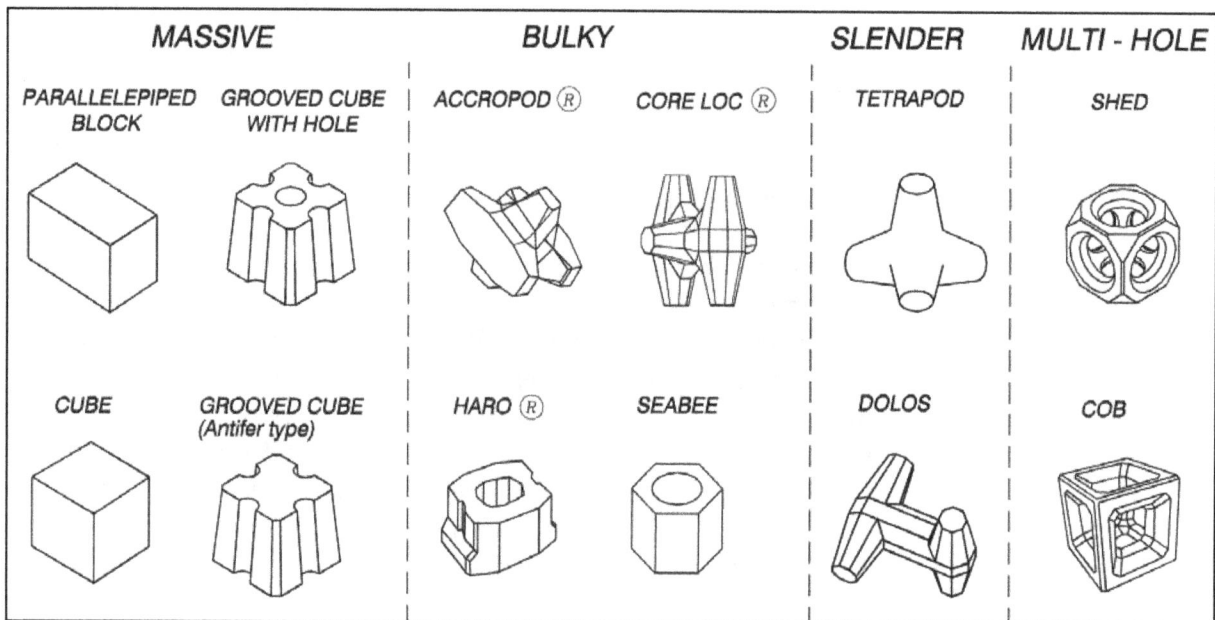

Figure VI-5-40. Examples of concrete armor units

(b) The units are generally made of conventional unreinforced concrete except the multi-hole cubes where fiber reinforcement is sometimes used.

(c) For slender units such as dolos with small waist ratios, various types of high-strength concrete and reinforcement (conventional rebars, prestressing, fibers, scrap iron, steel profiles) have been considered. However, reinforcement has only been used in few cases because it generally seems to be less cost-effective and because of the risk of rapid corrosion of the steel reinforcement.

(d) Hydraulic stability of armor layers is reduced if the armor units disintegrate causing reduction of the stabilizing gravitational force and possible interlocking effects. Moreover, broken armor unit pieces might be thrown around by wave action and thereby trigger additional breakage at an increased rate. In order to prevent this, it is necessary to ensure structural integrity of the armor units.

(e) Unreinforced concrete is a brittle material with a low tensile strength, f_T, on the order of 2-6 MPa and a compressive strength, f_C, which is one order of magnitude larger than f_T. Consequently, crack formation and breakage is nearly always caused by load induced tensile stresses, σ_T, that exceed f_T. The magnitude of f_T is therefore more important than f_C in armor unit concrete, and specifications should focus on achieving adequate values of f_T. It is important to note that f_T decreases with repeated load due to fatigue effects.

(f) The different categories of concrete armor units are not equally sensitive to breakage. Slender units are the most vulnerable because the limited cross-sectional areas give rise to relatively large tensile stresses. Some recent failures of breakwaters armored with tetrapods and dolosse were caused by breakage of the units into smaller pieces having less hydraulic stability than the intact armor units.

(g) Massive units will generally have the smallest tensile stresses due to the distribution of loads over large cross-sectional areas. However, breakage can take place if the units experience impacts due to less restrictive hydraulic stability criteria and if the concrete quality is poor with a low f_T. This latter point is related mainly to larger units where temperature differences during the hardening process can create tensile stresses which exceed the strength of the weak young concrete, thus resulting in microcracking of the material (thermal cracking). If massive units are made of good quality concrete and not damaged during handling, and if the armor layer is designed for marginal displacements, there will be no breakage problems. This statement also holds for the bulky units under the same precautions.

(h) The different types of loads on armor units and load origins are listed in Table VI-5-39.

(2) Structural design formulae for dolosse and tetrapods. Based on model tests with instrumented units, Burcharth (1993b), Burcharth and Liu (1995) and Burcharth et al. (1995b) presented a dimensional formula for estimation of the relative breakage of dolosse and tetrapods (fraction of total units) as presented in Table VI-5-40. Figures VI-5-41 and VI-5-42 compare the formulae to breakage data. Design diagrams for dolos were also presented in Burcharth and Liu (1992).

(a) Stress determination. Sturctural design methodologies for dolosse have also been proposed by Anglin et al. (1990) (see Table VI-5-41); Melby (1990, 1993); Zwamborn and Phelp (1990); and Melby and Turk (1992). The methods of Zwamborn and Phelp are based primarily on prototype failure tests, and therefore, are site specific.

- Melby (1990, 1993) provided a method to determine the design tensile stress for a dolos layer and discussed a computer program to compute this design stress. Figure VI-5-43 shows wave height in meters versus maximum flexural tensile stress in MPa for several dolos waist ratios and several Hudson stability coefficients. In this case, the wave height was used to determine a dolos weight using the Hudson stability equation. Figure VI-5-44 shows dolos weight in metric tons versus maximum flexural tensile stress in MPa for several dolos waist ratios. Both figures were generated using a tensile stress exceedance value of $E=2$ percent for the condition where the given stress level is exceeded in approximately 2 percent of the units on the slope. In addition, a structure slope of 1V:2H and a specific gravity of $\rho_a/\rho_w = 2.40$ were used to compute the stress level, although the effect of these parameters on the stress was negligible over typical ranges of these parameters. Further, Figure VI-5-44 was not affected by the choice of stability coefficient.

Table VI-5-39
Types and Origins of Loads on Armor Units (Burcharth 1993b)

TYPES OF LOADS		ORIGIN OF LOADS
Static		Weight of units
		Prestressing of units due to wedge effect and arching caused by movement under dynamic loads
Dynamic	Pulsating	Gradually varying wave forces
		Earthquake loads
	Impact	Collisions between units when rocking or rolling, collision with underlayers or other structural parts
		Missiles of broken units
		Collisions during handling, transport, and placing
		High-frequency wave slamming
Abrasion		Impacts of suspended sand, shingle, etc.
Thermal		Temperature differences during the hardening (setting) process after casting
		Freeze – thaw cycles
Chemical		Alkali-silica and sulphate reactions, etc.
		Corrosion of steel reinforcement

Table VI-5-40
Breakage Formula for Dolosse and Tetrapods (Burcharth 1993b, Burcharth and Liu 1995, Burcharth et al. 1995b, Burcharth et al. 2000)

$$B = C_0 \, M^{C_1} \, f_T^{C_2} \, H_s^{C_3} \qquad\qquad (VI\!-\!5\!-\!91)$$

where
	B	Relative breakage
	M	Armor unit mass in ton, $2.5 \le M \le 50$
	f_T	Concrete static tensile strength in MPa, $2 \le f_T \le 4$
	H_s	Significant wave height in meters
	C_0, C_1, C_2, C_3	Fitted parameters

The effects of static, pulsating, and impact stresses are included in the formula.

Fitted parameters for the breakage formula

	Waist ratio	Variational Coef. of C_0	C_0	C_1	C_2	C_3
Trunk of	0.325	0.188	0.00973	-0.749	-2.58	4.143
dolosse	0.37	0.200	0.00546	-0.782	-1.22	3.147
	0.42	0.176	0.01306	-0.507	-1.74	2.871
Round-head of dolosse	0.37	0.075	0.025	-0.65	-0.66	2.42
Trunk of tetrapods		0.25	0.00393	-0.79	-2.73	3.84

Summary of the hydraulic model tests

	Dolos trunk	Dolos round-head	Tetrapod trunk
Breakwater slope α	1.5	1.5	1.5
Foreshore	1:20	horizontal	1:50
Water depth at toe (cm)	23	50	30 and 50
Height of breakwater (cm)	60	70	55 and 75
Mass of units (kg)	0.187	0.187	0.290
Concrete density (tons/m^2)	2.3	2.3	2.3
Spectral peak period T_p (s)	1.5-3	1.5-2.5	1.3-2.8
Significant wave height H_s (cm)	5.7-17.7	5-13	8.8-27.3
Surf similarity parameter $\xi = (H_s/L_p)^{-0.5} \tan \alpha$	3-7.5	3.8-7.5	2.7-4.1

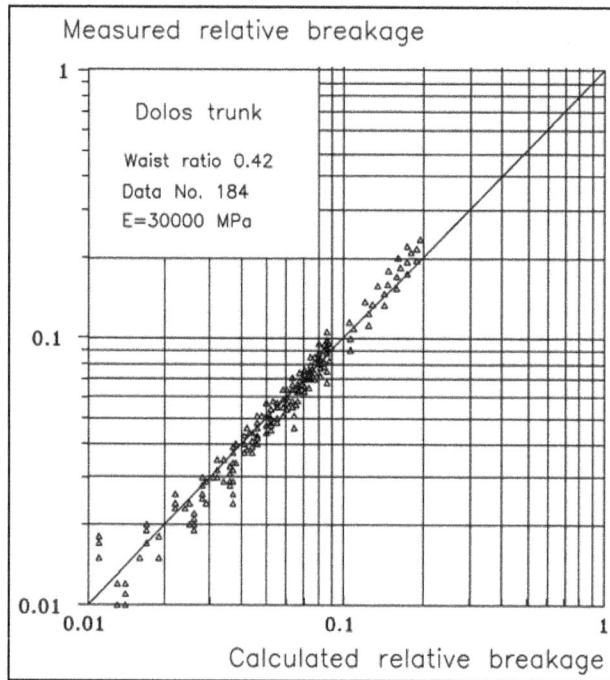

Figure VI-5-41. Breakage formula for dolosse
(Burcharth 1993b; Burcharth and Liu 1995;
Burcharth et al. 1995b)

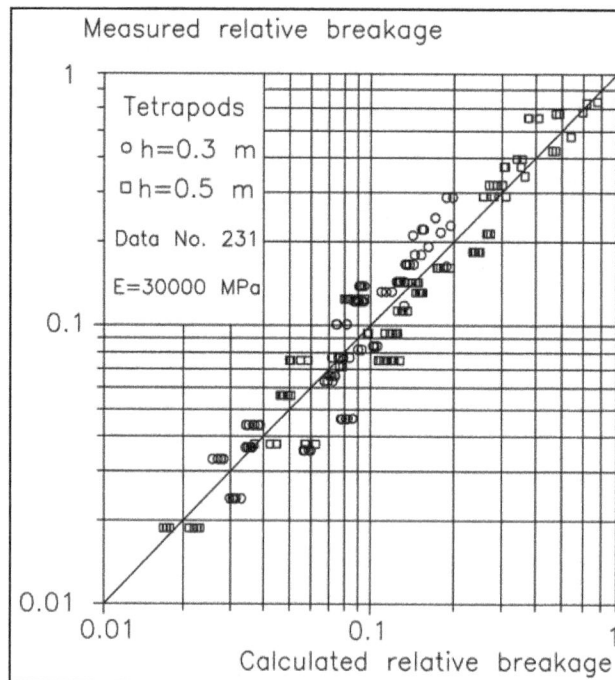

Figure VI-5-42. Breakage formula for tetrapods
(Burcharth 1993b, Burcharth and Liu 1995,
Burcharth et al. 1995b)

Table VI-5-41
Stress Prediction Formulae for Dolosse (Anglin et al. 1990)

Anglin et al. (1990) developed a dolos structural design methodology based on small scale measurements of strain in laboratory hydraulic models. Only the static stresses were considered. The criterion for allowable static tensile stress in a dolos at a vertical distance D_v down from the crest on a dry structure was proposed as

$$n(\sigma_s)_p < f_T \qquad\qquad\qquad \text{(VI-5-92)}$$

where

f_T = Prototype concrete static tensile strength (MPa)

$(\sigma_s)_p$ = Static principal stress in model dolos with probability of exceedance, p

n = Model scale factor

The static principal stress is estimated as

$$(\sigma_s)_p = 10^{\left(\log(\sigma_s)_{est} + 0.31\left[\Phi^{-1}(p)\right]\right)} \qquad\qquad\qquad \text{(VI-5-93)}$$

with

$$\log(\sigma_s)_{est} = -2.28 + 0.91\alpha + 0.30\left(\frac{D_v}{n} - 0.45\right) + 0.34l \qquad\qquad\qquad \text{(VI-5-94)}$$

and the model scale factor was given as

$$n = 9.43\left(\frac{W}{0.1549 w_a}\right)^{1/3} \qquad\qquad\qquad \text{(VI-5-95)}$$

and

α = Tangent of seaward armor slope

l = Layer (0 for top; 1 for bottom)

D_v = Vertical distance from crest to stressed dolos location

$\Phi^{-1}(p)$ = Tabulated inverse normal variate (see next page)

W = Prototype armor unit weight

w_a = Armor concrete specific weight

(Continued)

Table VI-5-41 (Concluded)

Values for the inverse normal variate in Eq VI-5-93 are given in the box to the right.

Probability of exceedance	$\Phi^{-1}(p)$
0.1	1.28
0.05	1.65
0.02	2.05
0.01	2.33

Equations VI-5-92 through VI-5-95 are limited to the range of values:

$0.4 \# \alpha \# 0.67$; 0.3 m $\# D_v /n \# 0.6$ m; $r = 0.32$ where r is the dolos waist ratio

Another model study examined the combined effects of static and quasistatic (wave-induced pulsating loads) under nonbreaking regular wave conditions, but did not include impact stresses. The criterion for allowable tensile stress in a dolos located a vertical distance, D_{swl}, from the swl was given as

$$n(\sigma_t)_p < f_T \qquad \text{(VI-5-96)}$$

where

$$(\sigma_t)_p = (\sigma_t)_{est} + 0.001\left[\Phi^{-1}(p)\right] \qquad \text{(VI-5-97)}$$

$$(\sigma_t)_{est} = 0.905(\sigma_s)_{est} + 0.639(\sigma_q)_{est} \qquad \text{(VI-5-98)}$$

$$\log(\sigma_q)_{est} = -2.36 + 0.15\alpha + 0.01\left(\frac{T}{\sqrt{n}}\right) + 0.29\left(\frac{D_{swl}}{n}\right) + 2.20\left(\frac{H}{n}\right) \qquad \text{(VI-5-99)}$$

and

$(\sigma_t)_p$ = Total static and pulsating principal stress in model armor unit with probability of occurrence,

$(\sigma_q)_p$ = Pulsating principal stress in model armor unit with probability of occurrence, p

$(\sigma_s)_p$ = Static principal stress with probability of occurrence, p, from Eq VI-5-94

H = Regular wave height

T = Regular wave period

D_{swl} = Vertical distance from swl to location of stressed dolos. (Positive above swl, negative below swl.)

n = Model scale factor from Eq VI-5-95

α = Tangent of seaward armor slope

$\Phi^{-1}(p)$ = Tabulated inverse normal variate from the preceding box

Equations VI-5-96 through VI-5-99 are limited in application to the range of values:
0.05 m $\# H/n \# 0.25$ m ; $0.4 \# \alpha \# 0.67$; 0.3 m $\# D_v /n \# 0.6$ m ;
1.25 s $\# T/(n)^{1/2} \# 2.5$ s ; -0.1 m $\# D_{swl} /n \# +0.1$ m

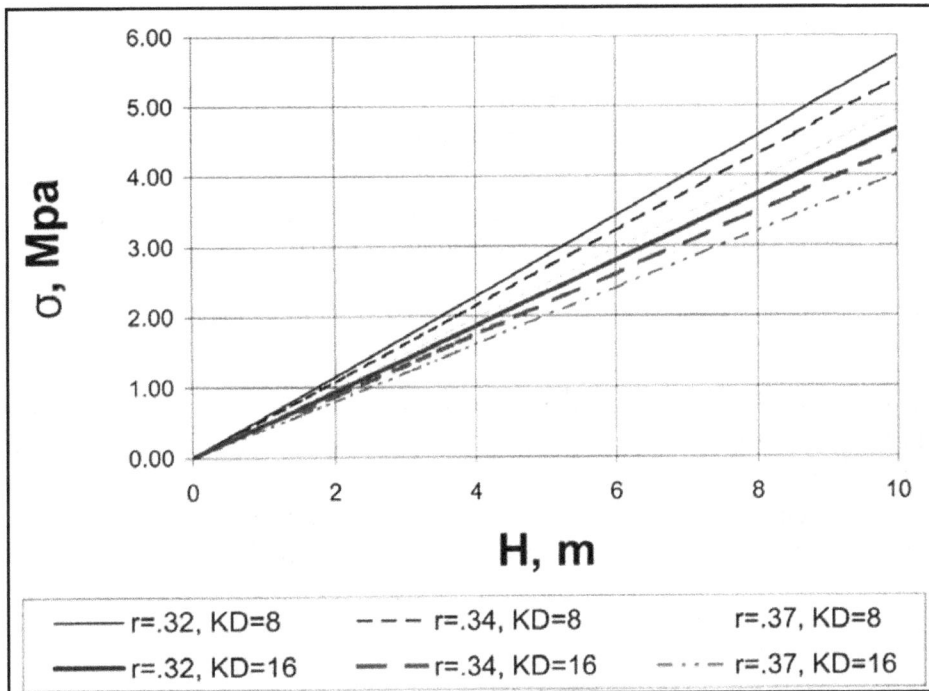

Figure VI-5-43. Wave height versus maximum flexural tensile stress for several dolos waist ratios

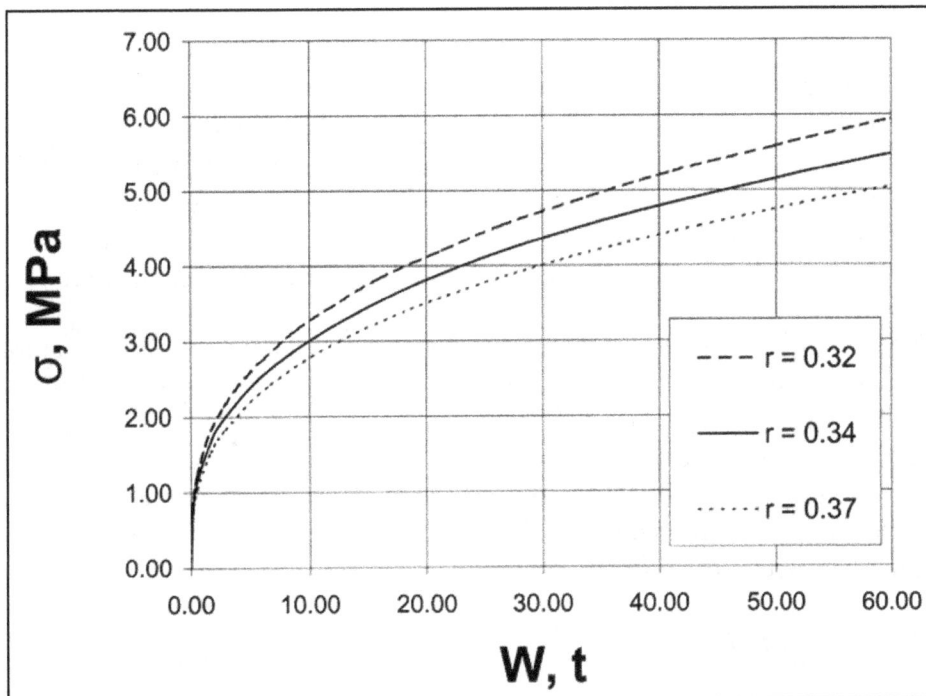

Figure VI-5-44. Dolos mass versus maximum flexural tensile stress for several dolos waist ratios

(b) Reinforced dolos design. Melby and Turk (1992) extended the method of Melby (1993) to include a level I reliability analysis and conventional reinforced concrete design methodology (American Concrete Institute (ACI) 1989). The following technique utilizes a probabilistic principal stress computed using any of the previous methods. These methods allow the designer to consider unreinforced concrete, conventional steel rebar reinforcement, or prestressing in a unified format. The basic design equation, following structural concrete design conventions, equates a factored strength with a factored load as

$$\gamma Q_n = \phi R_n \tag{VI-5-100}$$

where γ and φ are the load and strength factors, respectively, to account for uncertainty in nominal load Q_n and nominal strength R_n. Melby and Turk noted that the load factor ranges from 1.0 to 1.2 for typical values of exceedance probability for stress. American Concrete Institute (ACI) (1989) recommends $\varphi = 0.85$ for torsion. To facilitate reinforcement design, Melby and Turk assumed a circular cross section and decomposed Equation VI-5-100 into a flexure equation

$$\gamma S_M k_M \sigma_1 < \phi(0.7 M_{cr}) \tag{VI-5-101}$$

and a torsional equation

$$\gamma S_T k_T \sigma_1 < \phi(0.7 T_{cr} + T_s) \tag{VI-5-102}$$

where σ_1 is the principal stress, $S_M = 0.1053(rC)^3$ and $S_T = 0.2105(rC)^3$ are the section moduli for flexure and torsion, r is the dolos waist ratio, C is the dolos fluke length, and $k_M = k_T = 0.6$ are the moment and torque contribution factors, $M_{cr} = T_{cr} = 0.7 f_{ct}$ are the critical strengths of the concrete in moment and torque, f_{ct} is the concrete splitting tensile strength, and T_s is the strength contribution from the torsional steel reinforcement. The inequality in Equations VI-5-101 and VI-5-102 assures that the factored tensile strength will be greater than the factored tensile load.

• The technique for steel reinforcement design utilizes conventional structural design techniques. Torsional steel is specified first, and it is only required in the shank because the flukes are not likely to be twisted. Details are given in ACI 318-89 (ACI 1989). Assuming a circular section for the dolos shank, the amount of torsional steel is given as $T_s = R_h A_s f_y$, where R_h is the distance to the center of the section, A_s is the total area of steel intersecting the crack, and f_y is the yield strength of the steel. Substituting T_s into Equation VI-5-102 yields the equation for required torsional steel, i.e.,

$$A_s > \frac{\gamma(S_T k_T \sigma_1) - \phi(0.7 T_{cr})}{\phi f_y R_h} \tag{VI-5-103}$$

• The number of bars required is then given by $n = A_s / A_b$, where A_b is the cross-sectional area of hoop reinforcing bars, and the spacing is $s = 1.5 \pi R_h / n$, assuming the crack extends three-fourths of the distance around the circumference.

• For flexural reinforcement design, it is assumed that the concrete offers no resistance in tension. Nominal strength is reached when the crushing strain in the outer fiber of the concrete is balanced by the yield strain in the steel rebar. The balanced failure condition using the Whitney rectangular stress block is prescribed in ACI 318-89, Part 10 (ACI 1989). The solution requires an iterative approach because the neutral axis is a priori unknown. Assuming a rebar size, the neutral axis is located by solving the quadratic equation that results from balancing the compressive force moment from the Whitney stress block with the tensile force moment from the steel. Once the neutral axis is determined, the nominal moment from the steel can determined and substituted into Equation VI-5-101 to determine if the quantity of steel is adequate to balance the flexural design load. After determining the amount of flexural steel required, typical checks of compressive stress, shear, bond, minimum reinforcement, and temperature steel should be made as per ACI 318-89.

(c) Prestressed dolos design. Prestressing acts reduce principal stress. The principal stress reduction factor is given by

$$\xi = 0.5 \left(\left(k_M - \lambda \right) + \sqrt{\left(k_M - \lambda \right)^2 + 4k_T^2} \right)$$

(VI-5-104)

where λ is the ratio of applied precompressive stress to design principal stress. This equation was substituted into the moment-torque interaction relations to get design equations for torsion and flexure as follows:

$$\gamma \xi k_T \sigma_1 = 0.5\phi \sqrt{\frac{f_c}{1 + 4\left(\dfrac{k_M S_M}{k_T S_T} \right)^2}}$$

(VI-5-105)

$$\gamma \xi k_M \sigma_1 = 0.5\phi \sqrt{\frac{f_c}{1 + 0.25\left(\dfrac{k_T S_T}{k_M S_M} \right)^2}}$$

(VI-5-106)

where f_c is the concrete compressive strength. These equations are similar to Equations VI-5-101 and VI-5-102, but they are for prestressed concrete design. Details for determining prestressing steel requirements are given in ACI 318-89 (ACI 1989).

(3) Ultimate impact velocities end equivalent drop height.

(a) For evaluation of the placing technique during construction it is important to consider the ultimate impact velocities. The lowering speed of the crane at the moment of positioning of the units must be much slower than the values given in Table VI-5-42. The values of ultimate impact velocities given in Table VI-5-42 are rough estimates corresponding to solid body impact against a heavy rigid concrete base which causes breakage resulting in a mass loss of 20 percent or more. If the armor units are not dropped on a hard rigid surface but instead on soil or a rock underlayer, the ultimate impact velocities can be significantly higher than those given in Table VI-5-42.

Table VI-5-42
Approximate Values of Ultimate Rigid Body Impact Velocities for Concrete Armor Units
(Burcharth 1993b)

Armor Unit	Impact Velocity of the Unit's Center (m/s)	Equivalent Drop Height of the Unit's Center (m)
Cube < 5 tonne	5 - 6	1.2 - 1.8
20 tonne	4 - 5	0.8 - 1.2
50 tonne	3 - 4	0.4 - 0.8
Tetrapod	2	0.2
Dolos, waist ratio 0.42	2	0.2
Dolos, waist ratio 0.32	1 - 1.5	0.05 - 0.12

(b) When placing units underwater, a heavy swell might impose rather large horizontal velocities on a unit suspended from a crane. It is obvious from the values in Table VI-5-42 that free-fall dropping of concrete armor units by quick release from a crane should be avoided because even small drop heights can cause breakage. This is also true for underwater placement because the terminal free-fall velocity underwater exceeds the limiting values given in Table VI-5-42 except for very small massive types of units.

(4) Thermal stresses.

(a) As concrete cures, the heat of hydration increases the temperature. Because of the fairly low thermal conductivity of concrete and because of the poor insulation of conventional formwork (e.g., steel shutter), a higher temperature will be reached in the center part of the armor unit than on the concrete surface. The temperature difference will create differential thermal expansion, and internal thermal stresses will develop in the concrete. The temperature differences and resulting thermal stresses increase with the distance between the armor unit center and the surface of the unit. Tensile stresses can easily exceed the limited strength of the fresh young concrete thus causing formation of microcracks. Unfortunately, it is not possible to see thermal cracks because they will close at the surface due to the thermal contraction of the concrete as it cools.

(b) The curing process is very complicated and theoretically it can only be dealt with in an approximate manner, mainly because the description of creep and relaxation processes of the hardening concrete are not precise enough to avoid large uncertainties in the calculations. Calculations are performed by the use of special finite element computer programs for three-dimensional bodies. Necessary input is data on the concrete mix including the composition (type) of the cement, the concrete temperature when poured, the geometry of the units, the type of formwork (conductivity/insulation), the environmental climate (air temperature and wind velocities as function of time), and the cycling time for removal of the formwork. The output of the calculations is the development of stresses and related crack formation as function of time. Figure VI-5-45 shows an example of such a calculation for a 70-tonne cube.

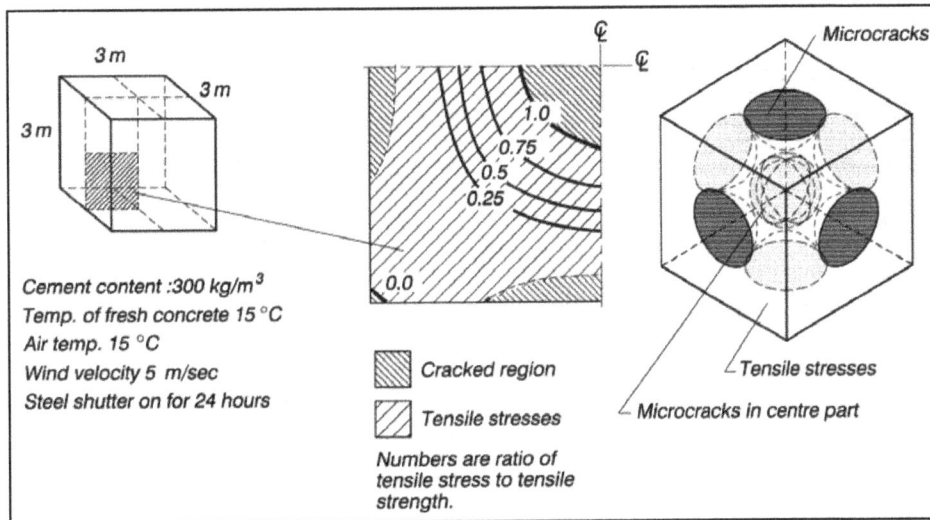

Figure VI-5-45. Example of calculation of thermal stresses and cracked
regions in a 70-tonne cube 100 hr after casting (Burcharth 1993b)

(c) The cube will have no visible sign of weakness, but it will be fragile and brittle
because the cracked regions at the surfaces and in the center will have almost zero tensile
strength and the noncracked regions will be in tension. This means that not only the strength, but
also the fatigue life and the resistance to deterioration, will be reduced.

(d) Thermal stress calculations are complicated and must be performed using numerical
models described in the concrete literature. However, a very important rule of thumb for
avoiding thermal cracks is that the temperature difference during curing should not exceed 20° C
between any two points within the concrete element. The temperature difference is easy to
monitor by placing/casting copper-constanting thermo-wire (e.g., 2 x 0.7 mm^2) in the concrete.
The wire insulation must be removed at the tips which are placed at positions in the center and
near the surface of the units where the temperatures are maximum and minimum, respectively.
Temperature readings can then be monitored by connecting a pocket instrument to the free wire
ends.

(e) There are several measures related to concrete technology for the prevention of
damaging thermal stresses, but they all involve some drawbacks as described by Table VI-5-43.

(f) Another way of dealing with the thermal stress problem is to keep the effective
dimensions of the armor units as small as possible. For cubes it can be done by making a hole as
was done in the hot-climate Bosaso Harbor project in Somalia. Figure VI-5-46 shows examples
of the temperature development in 30-tonne blocks with and without a hole. The reduced
temperature difference introduced by the hole is clearly seen by comparison of the two blocks
casted during winter time. In fact it was easier to keep the 20° C temperature difference limit in a
30-tonne unit with a hole than in a 7-tonne unit without a hole.

Table VI-5-43
Drawbacks Related to Crack-Reducing Procedures

Measure to Reduce Thermal Stresses	Drawback
Use of less cement	Reduced long-term durability due to higher porosity. Slower development of strength, longer cycle time for forms
Use of low-heat cement or retarder	Higher production costs due to slower development of strength, longer cycle time for forms, larger casing and stockpiling area needed
Cooling of water and aggregates	Higher production costs
Use of insulation during part of the curing period	Higher production costs

(5) Fatigue in concrete armor units.

(a) The strength of concrete decreases with the number of stress cycles. Each stress cycle larger than a certain stress range will cause partial fracture in some parts of the material matrix resulting in a decreased yield strength. Repeated loads cause an accumulative effect which might result in macro cracks, and ultimately, breakage of the structural element.

(b) The number of stress cycles caused by wave action will be in the order of 200 million during 50 years structural life in the North Atlantic area. About 10 million cycles will be caused by larger storm waves. In subtropical and tropical areas the number of storm wave cycles is generally one or two orders of magnitude less.

(c) Fatigue for conventional unreinforced concrete exposed to uniaxial and flexural stress conditions with zero mean stress is given in Table VI-5-44.

d. Toe stability and protection.

(1) Introduction.

(a) The function of a toe berm is to support the main armor layer and to prevent damage resulting from scour. Armor units displaced from the armor layer may come to rest on the toe berm, thus increasing toe berm stability. Toe berms are normally constructed of quarry-run, but concrete blocks can be used if quarry-run material is too small or unavailable.

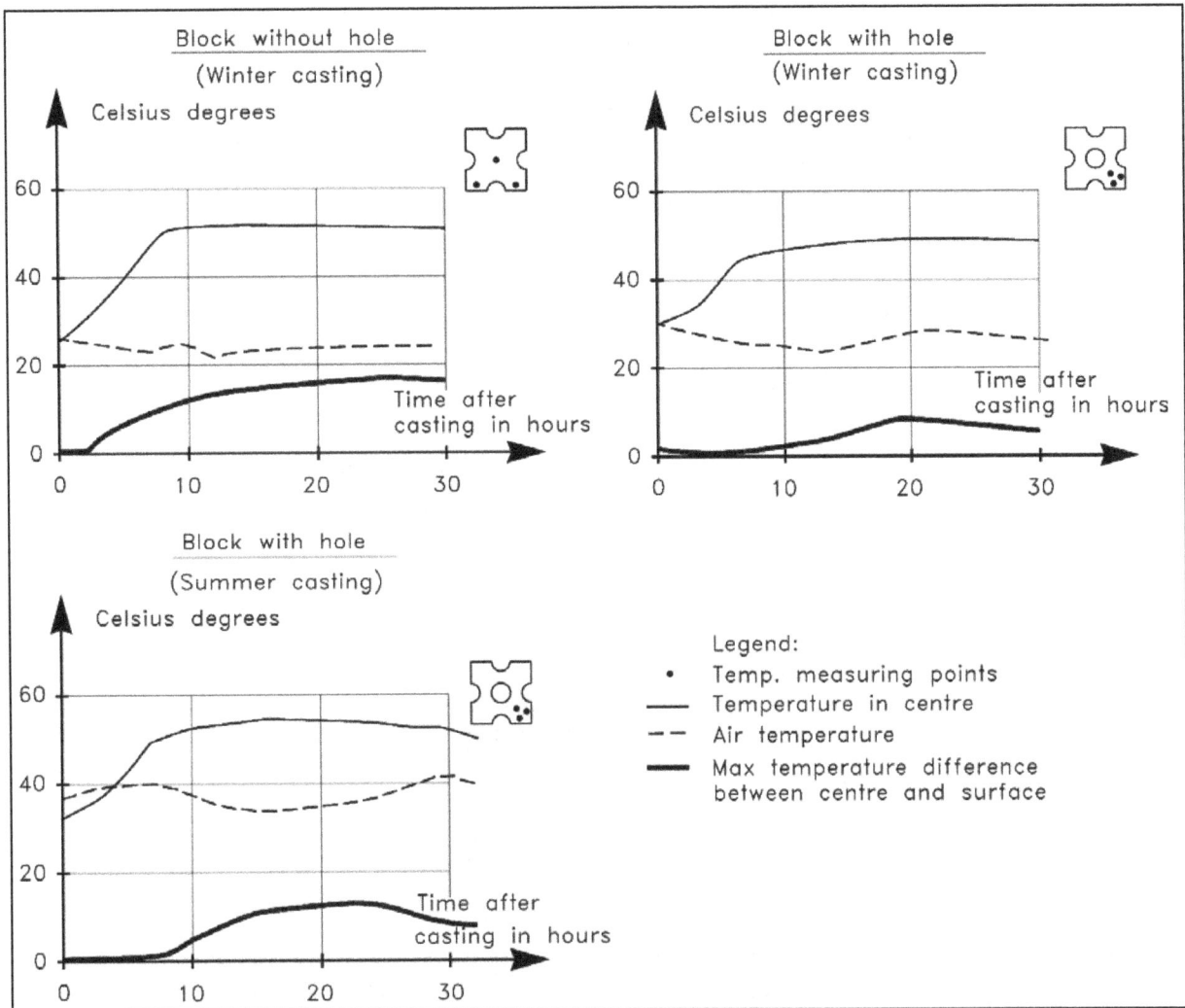

Figure VI-5-46. Examples of temperature development during curing in 30-tonne modified cubes with and without a hole (Burcharth et al. 1991)

(b) In very shallow water with depth-limited design wave heights, support of the armor layer at the toe is ensured by placing one or two extra rows of main armor units at the toe of the slope as illustrated in Figure VI-5-47a. This is a stable solution provided that scour does not undermine the toe causing the armor layer to slide as illustrated by Figure VI-5-48. In shallow water it is usually possible to use stones or blocks in the toe that are smaller than the main armor, as shown in Figure VI-5-47 b. In deep water, there is no need for the main armor to cover the slope at greater depths, and the toe berm can be constructed at a level above the seabed as illustrated by Figure VI-5-47c.

(c) Toe berm stability is affected by wave height, water depth at the top of the toe berm, width of the toe berm, and block density. However, wave steepness does not appear to be a critical toe berm stability parameter.

Table VI-5-44
Fatigue for Conventional Unreinforced Concrete Exposed to Uniaxial and Flexural Stress
Conditions With Zero Mean Stress (Burcharth 1984)

$$\frac{\Delta\,\sigma_N}{\Delta\,\sigma_{N=1}} = \frac{\text{Ultimate stress range for } N \text{ cycles}}{\text{Ultimate stress range for one cycle}}$$

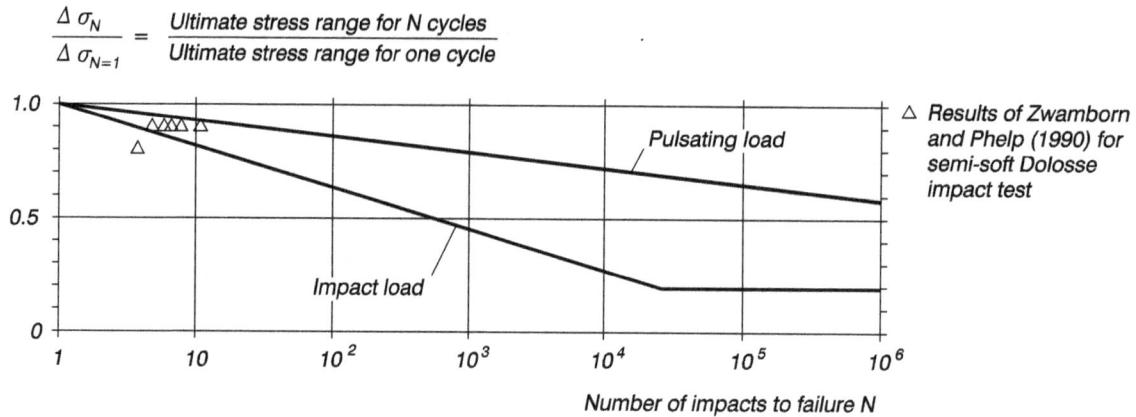

Number of impacts to failure N

It should be noted that the ultimate impact load strength for one stress cycle is on the order of 1.4 to 1.5 times the ultimate pulsating load strength in the case of uniaxial tension and compression, respectively. For flexural stresses a factor of approximately 1.4 should be applied. The ultimate pulsating load strength properties for one cycle can be taken to be equal to those found for static load conditions.

Data basis for fatigue curve

Author(s)	Loading condition	Specimens
Tepfers and Kutti (1979)	pulsating tension and compression	cubes of 150 mm length
Tait and Mills (1980)	pulsating tension	Dolosse of 300 mm height
Fagerlund and Larsson (1979)	impact compression	cylinders of 100 mm diameter
Zielinski, Reinhardt, and Körmeling (1981)	impact tension	cylinders of 74 mm diameter
Burcharth (1984)	impact flexural stress	Dolosse of 790 mm height

Zwamborn and Phelp (1990) performed drop tests with 9-ton Dolosse on a horizontal underlayer of quarry rock. This relatively soft base created a milder dynamic response than the solid rigid concrete base used by other authors. As seen from the figure Zwamborn and Phelp's data are in between the two curves which might be regarded as upper and lower limits for the fatigue effect.

(d) Model tests with irregular waves indicate that the most unstable location is at the shoulder between the slope and the horizontal section of the berm. The instability of a toe berm will trigger or accelerate the instability of the main armor. Lamberti (1995) showed that moderate toe berm damage has almost no influence on armor layer stability, whereas high damage of the toe berm severely reduces the armor layer stability. Therefore, in practice it is economical to design toe berms that allow for moderate damage.

a) VERY SHALLOW WATER

Main armor

Toe of main armor

b) SHALLOW WATER

Main armor

Toe armor stones
or blocks

c) DEEP WATER

Main armor

Toe berm armor stones
or blocks

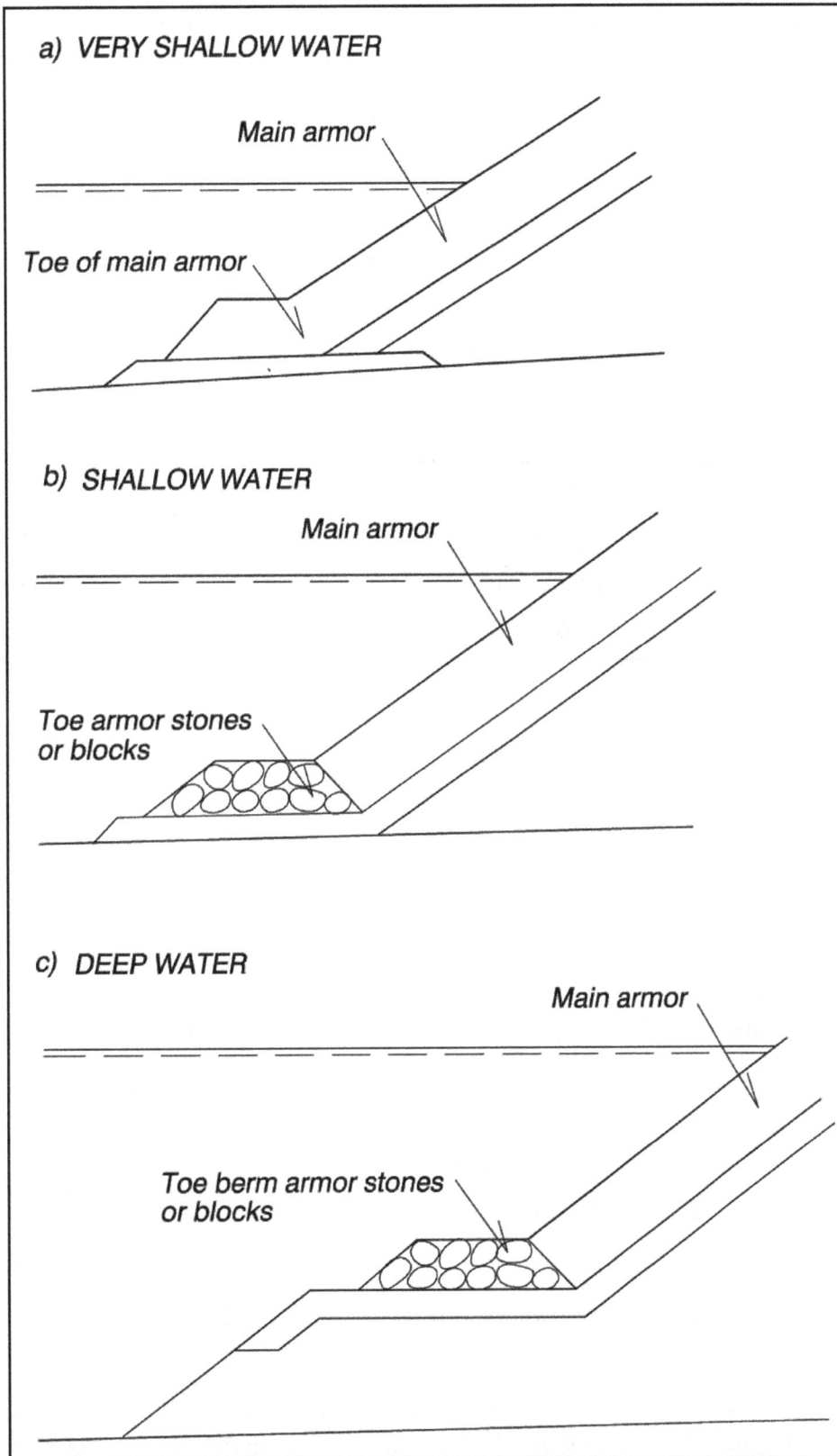

Figure VI-5-47. Typical toe and toe berm solutions in rubble-mound
breakwater design

(e) Rock seabeds often provide a poor foundation for the toe berm because of seaward sloping and/or rather smooth surfaces. Toe stability will be difficult to obtain, especially in shallow water with wave breaking at the structure (see Figure VI-5-48). Toe stones placed on hard bottoms can be supported by a trench or anchor bolts as sketched in Figure VI-5-49.

(f) Scour in front of the toe berm can also trigger a failure. The depth of toe protection required to prevent scour can be estimated from the scour depth prediction methods discussed in Part VI-5-6, "Scour and Scour Protection." Typical forms of scour toe protection are illustrated in Figure VI-5-50.

(2) Practical toe stability formulas for waves. Toe berm stability formulas are based exclusively on small scale physical model tests. These formulas are presented in the following tables.

Waves	Structure	Table
Regular, head-on and oblique	Sloping and vertical, trunk and head section	VI-5-45
Irregular, head-on	Trunk of sloping structure	VI-5-46 & VI-5-47
Irregular, head-on	Trunk of vertical structure	VI-5-48

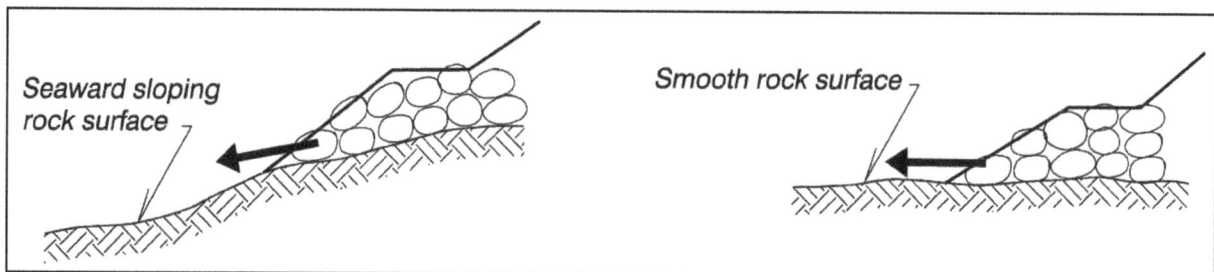

Figure VI-5-48. Example of potential instability of the stones placed on rock seabed

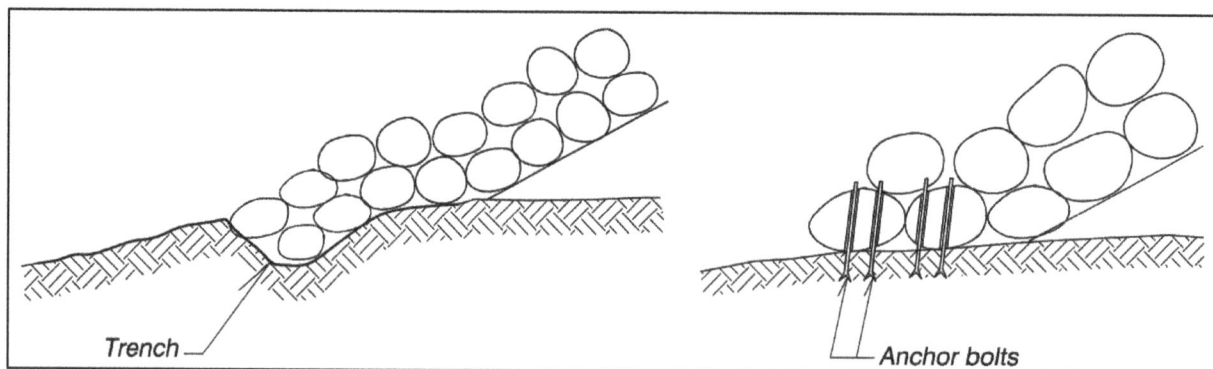

Figure VI-5-49. Support of the stones by a trench or anchor bolts

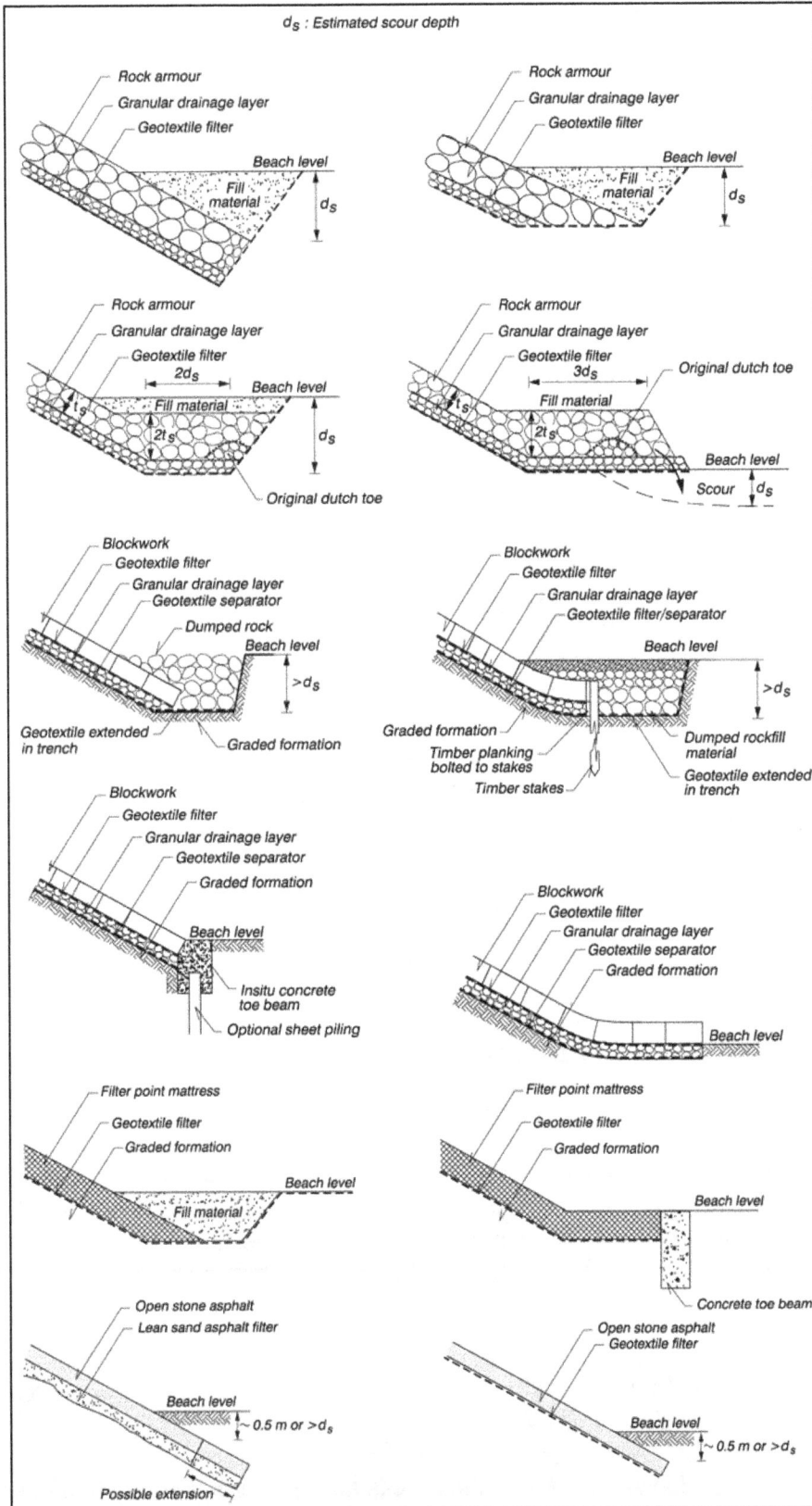

Figure VI-5-50. Typical seawall toe designs where scour is foreseen
(McConnell 1998)

Table VI-5-45
Stability of Toe Berm Tested in Regular Waves (Markle 1989)

Regular waves, head-on and oblique

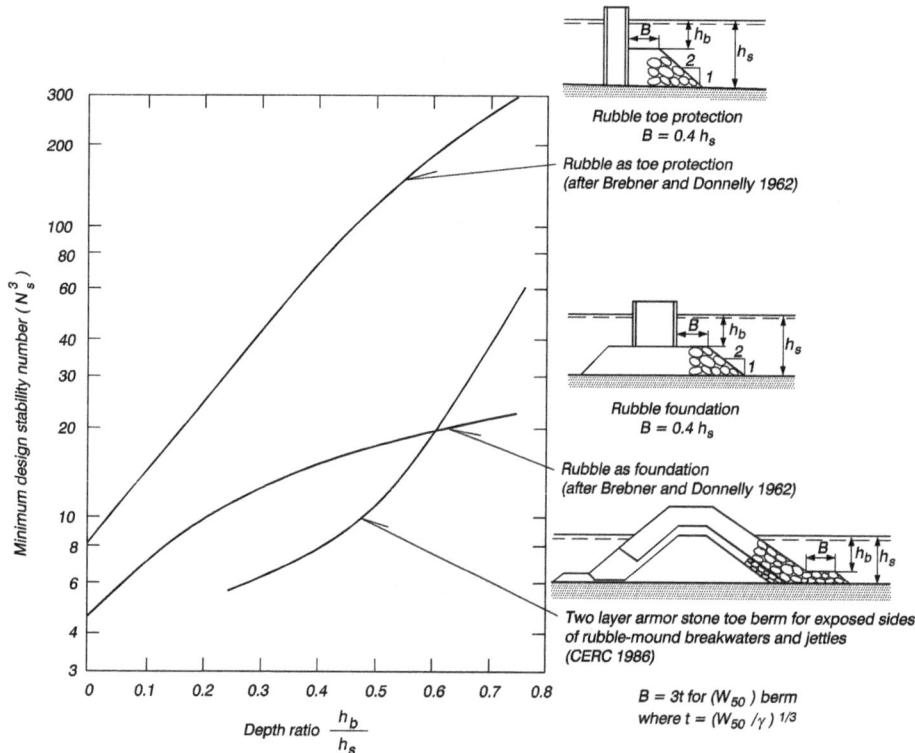

Rubble toe protection
B = 0.4 h_s

Rubble as toe protection
(after Brebner and Donnelly 1962)

Rubble foundation
B = 0.4 h_s

Rubble as foundation
(after Brebner and Donnelly 1962)

Two layer armor stone toe berm for exposed sides
of rubble-mound breakwaters and jetties
(CERC 1986)

$B = 3t$ for (W_{50}) berm
where $t = (W_{50}/\gamma)^{1/3}$

where N_s $N_s = H/(\Delta D_{n50})$
H Wave height in front of breakwater
Δ $(\rho_s/\rho_w) - 1$
ρ_s Mass density of stones
ρ_w Mass density of water
D_{n50} Equivalent cube length of median stone

Remarks: The curves in the figure are the lower boundary of N_s-values associated with acceptable toe berm stability (i.e., some stone movement occurs; but the amount of movement is minor and acceptable, which shows that the toe is not overdesigned)

(3) Foot protection blocks.

(a) Foot protection blocks have been applied to prevent foundation erosion at the toe of vertical structures as shown in Figure VI-5-51.

(b) According to Japanese practice the blocks are rectangular concrete blocks with holes (approximately 10 percent opening ratio) to reduce the antistabilizing pressure difference between the top and bottom of the blocks. Figure VI-5-52 shows a typical 25-tonne block.

Table VI-5-46
Stability of Toe Berm Formed by 2 Layers of Stone Having Density 2.68 tonnes/m³. Variable
Berm Width, and Sloping Structures (van der Meer, d=Angremond, and Gerding 1995)

$$N_s = \frac{H_s}{\Delta D_{n50}} = \left(0.24\,\frac{h_b}{D_{n50}} + 1.6\right) N_{od}^{0.15} \qquad\qquad (VI-5-107)$$

where H_s Significant wave height in front of breakwater

       Δ $(\rho_s/\rho_w) - 1$

       ρ_s Mass density of stones

       ρ_w Mass density of water

       D_{n50} Equivalent cube length of median stone

       h_b Water depth at top of toe berm

       N_{od} Number of units displaced out of the armor layer within a strip width of D_{n50}. For a standard toe size of about 3-5 stones wide and 2-3 stones high:

$$N_{od} = \begin{cases} 0.5 & \text{no damage} \\ 2 & \text{acceptable damage} \\ 4 & \text{severe damage} \end{cases}$$

For a wider toe berm, higher N_{od} values can be applied.

Tested cross sections

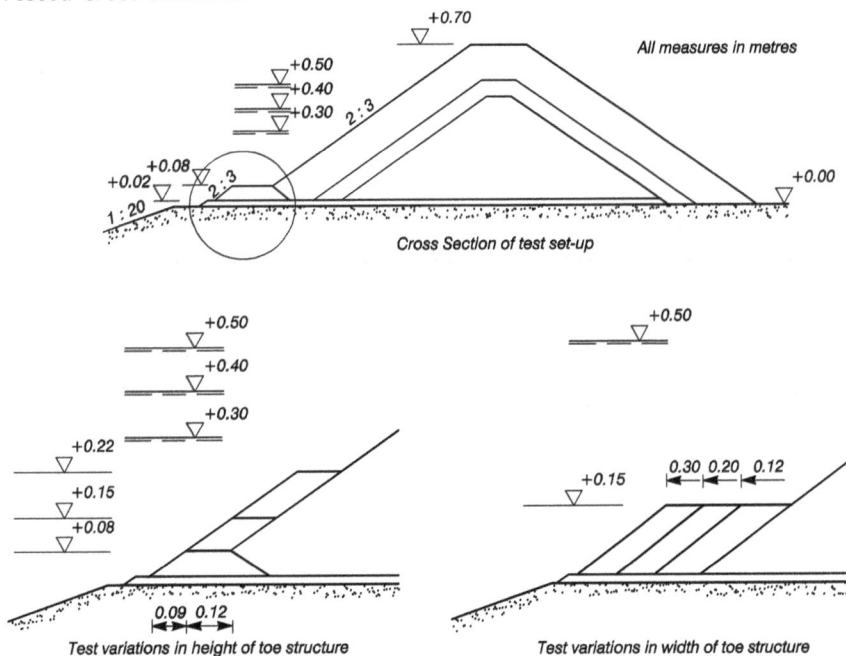

Cross Section of test set-up

Test variations in height of toe structure

Test variations in width of toe structure

(Continued on next page)

Table VI-5-46 (Concluded)

Valid for: Irregular head-on waves; nonbreaking, breaking and broken

Toe berm formed of two layers of stones with $\rho_s = 2.68 \ tonnes/m^3$ ($168 \ lb/ft^3$)

$0.4 < h_b/h_s < 0.9, \qquad 0.28 < H_s/h_s < 0.8, \qquad 3 < h_b/D_{n50} < 25$

where h_s is the water depth in front of the toe berm

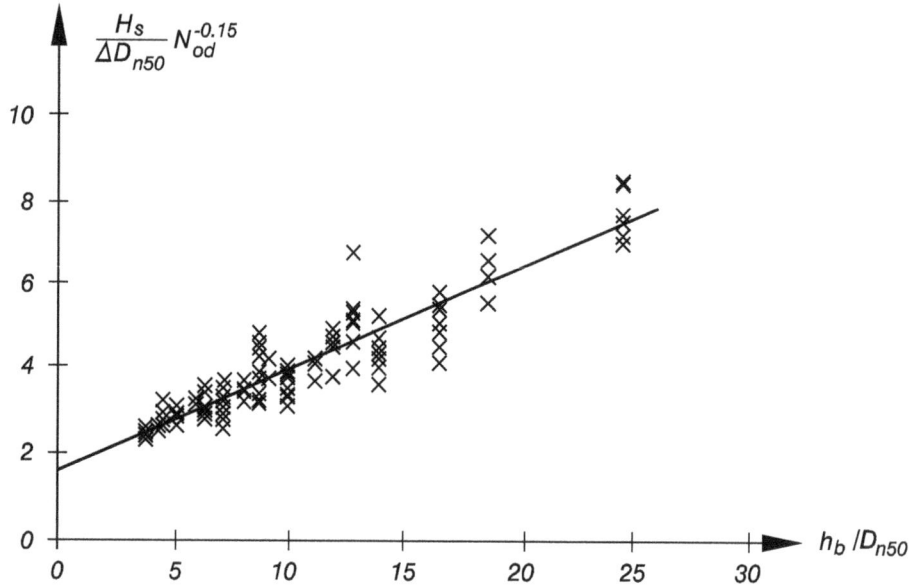

Uncertainty of the formula: corresponding to a coefficient of variation of approximately 0.10.

Table VI-5-47

Stability of Toe Berm Formed by Two Layers of Stones or Parallellepiped Concrete Blocks
(Burcharth et al. 1995a)

Equation VI-5-107 was modified so that it can be applied to the toe berm formed of stones having other densities or to parallellepiped concrete blocks.

$$N_s = \frac{H_s}{\Delta D_{n50}} = \left(0.4 \frac{h_b}{\Delta D_{n50}} + 1.6\right) N_{od}^{0.15} \quad \text{or} \quad \frac{H_s}{\Delta D_{n50}} = \frac{1.6}{N_{od}^{-0.15} - 0.4 \, h_b/H_s} \quad (VI-5-108)$$

Results of the stability tests with a toe berm made of 16.5-tonne parallellepiped concrete blocks are shown below. The negative influence of a high reflecting wave wall superstructure on the toe stability is demonstrated.

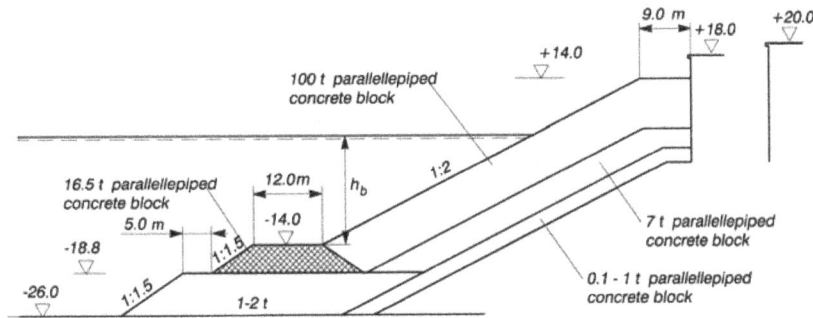

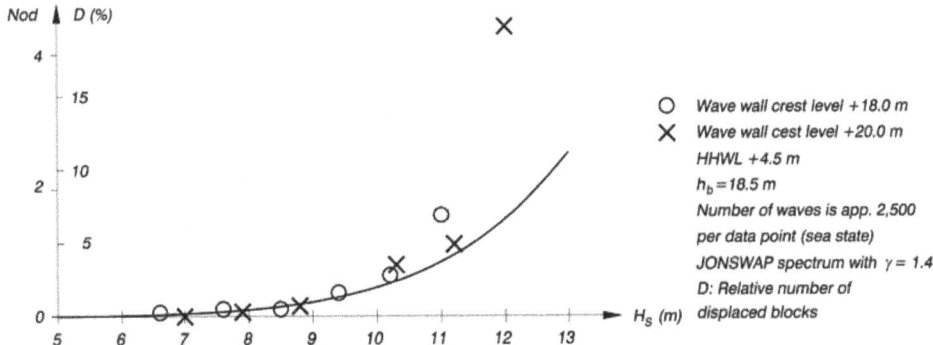

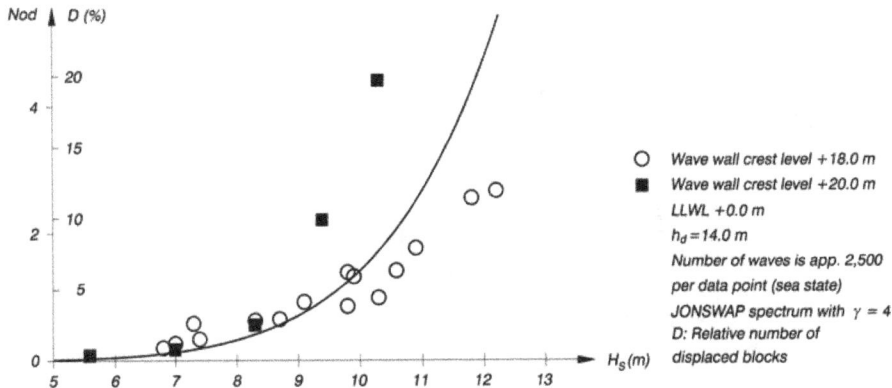

Table VI-5-48
Stability of Toe Berm Formed by Two Layers of Stones in Front of Vertical Impermeable Wall Structure

Madrigal and Valdés (1995) for two layers of quarrystones

$$N_s = \frac{H_s}{\Delta D_{n50}} = \left(5.8\frac{h'}{h_s} - 0.6\right) N_{od}^{0.19} \qquad (VI-5-109)$$

where
H_s	Significant wave height in front of breakwater
Δ	$(\rho_s/\rho_w) - 1$
ρ_s	Mass density of stones
ρ_w	Mass density of water
D_{n50}	Equivalent cube length of median stone
h'	Water depth at top of toe berm
h_s	Water depth in front of toe berm
N_{od}	Number of units displaced out of the armor layer within a strip width of D_{n50}

$$N_{od} = \begin{cases} 0.5 & \text{start of damage (1-3\% of units displaced)} \\ 2 & \text{acceptable damage (5-10\% of units displaced)} \\ 5 & \text{severe damage (20-30\% of units displaced)} \end{cases}$$

Tested cross sections

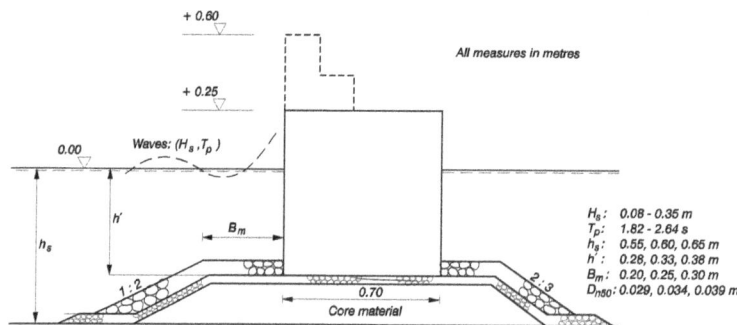

Valid for: Irregular head-on waves. Toe berm formed of two layers of stones with $\Delta = 1.65$
$0.5 < h_b/h_s < 0.8$; $7.5 < h_b/D_{n50} < 17.5$; $0.3 < B_m/h_s < 0.55$

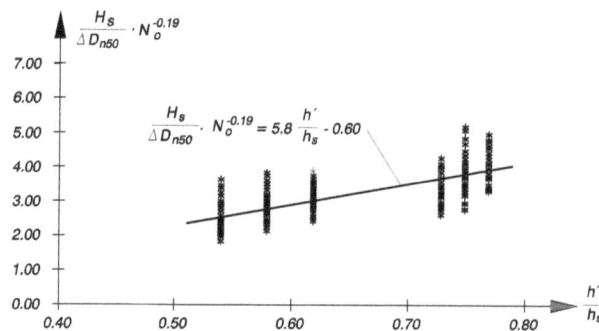

Uncertainty of the formula: Not given

(Continued)

Table VI-5-48 (Concluded)

Tanimoto, Yagyu, and Goda (1982), Takahashi, Tanimoto, and Shimosako (1990) for two layers of quarrystones

$$N_s = \frac{H_s}{\Delta D_{n50}} = \max\left\{ 1.8, \ 1.3\, \frac{1-\kappa}{\kappa^{1/3}} \frac{h'}{H_s} + 1.8\exp\left(-1.5\frac{(1-\kappa)^2}{\kappa^{1/3}} \frac{h'}{H_s}\right) \right\} \qquad (VI-5-110)$$

$$\kappa = \kappa_1\,\kappa_2$$
$$\kappa_1 = 2kh'/\sinh(2kh')$$
$$\kappa_2 = \max\left\{0.45\sin^2\theta\cos^2(kB\cos\theta), \ \cos^2\theta\sin^2(kB\cos\theta)\right\}$$

where
H_s Significant wave height in front of breakwater
Δ $(\rho_s/\rho_w) - 1$
ρ_s Mass density of stones
ρ_w Mass density of water
D_{n50} Equivalent cube length of median stone
h' Water depth on top of toe berm (excluding armor layer)
B Width of toe berm
k Wave number $k = 2\pi/L_p$
θ Wave incident angle ($\theta = 0^0$ for head-on)

Valid for: Irregular head-on and oblique waves
Toe berm formed by two layers of quarrystones
$\Delta = 1.65$

Uncertainty of the formula: Not given

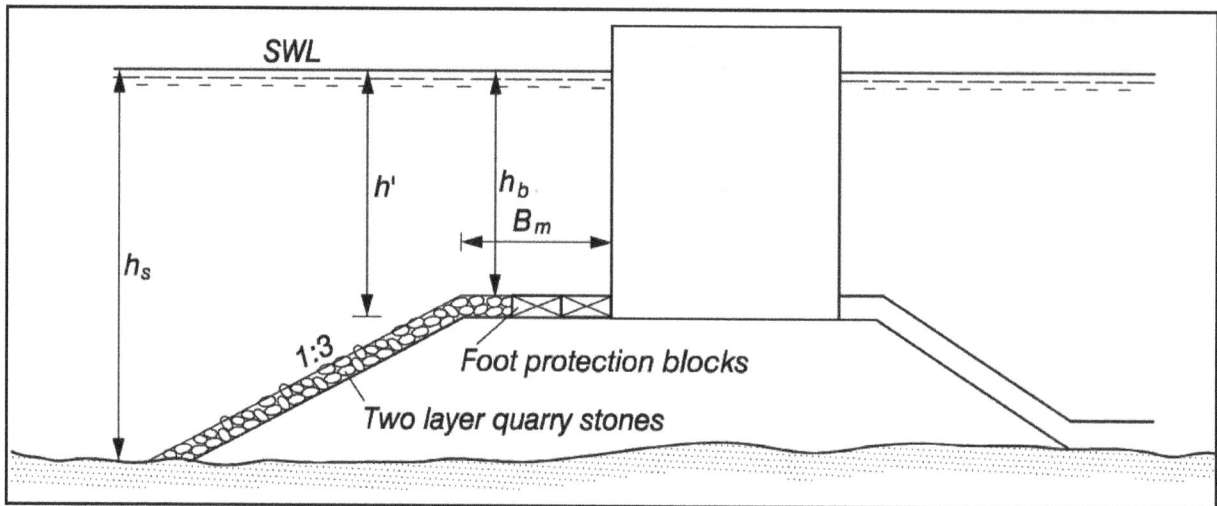

Figure VI-5-51. Illustration of foot protection blocks for vertical structures

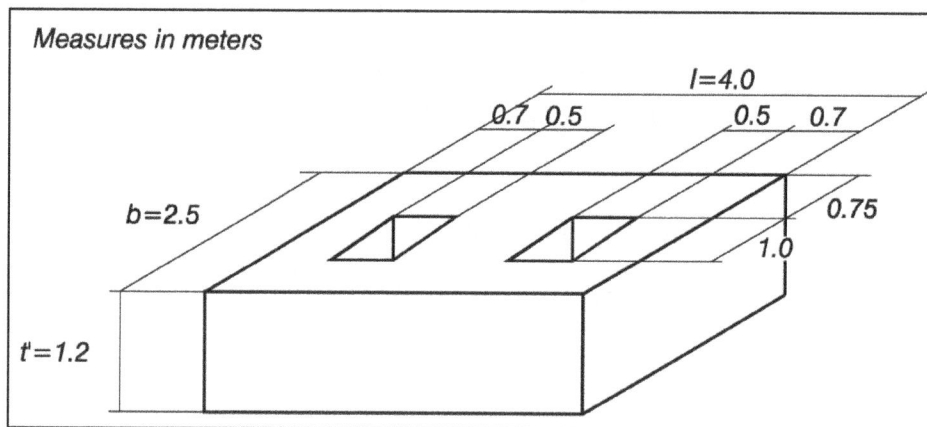

Figure VI-5-52. Example of Japanese foot protection block

Figure VI-5-53 shows a diagram taken from Takahashi (1996) for the determination of the necessary block thickness t' as functions of wave height H and the ratio of water depths h_b/h_s at the berm and in front of the structure as shown back on Figure VI-5-51.

(c) Stable foot protection blocks do reduce the pressure induced current in the mound, even when there are 10 percent openings in the blocks. Thus the risk of erosion of a sandy seabed underneath a thin rubble mound bedding layer is reduced too.

(4) Toe stability in combined waves and currents.

(a) Coastal structures, such as entrance jetties, are exposed to waves combined with currents running parallel to the structure trunk. In certain circumstances toe stability may be decreased due to the vectorial combination of current and maximum wave orbital velocity. For normal wave incidence the combined wave and current vector magnitude is not greatly increased. However, in the case of jetties where waves approach the jetty trunk at large oblique angles (relative to the normal), the combined velocity magnitude becomes large, and toe stability is jeopardized.

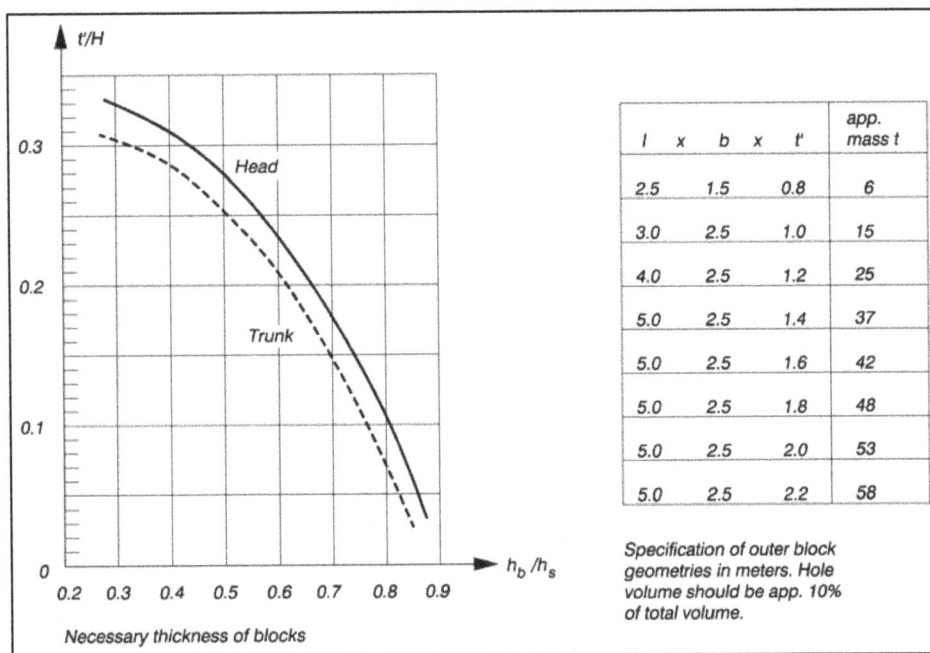

l	x	b	x	t'	app. mass t
2.5		1.5		0.8	6
3.0		2.5		1.0	15
4.0		2.5		1.2	25
5.0		2.5		1.4	37
5.0		2.5		1.6	42
5.0		2.5		1.8	48
5.0		2.5		2.0	53
5.0		2.5		2.2	58

Specification of outer block geometries in meters. Hole volume should be app. 10% of total volume.

Figure VI-5-53. Design of foot protection blocks according to Japanese practice

(b) Smith (1999) conducted 1:25-scale laboratory experiments to develop design guidance for jetty structures where oblique waves combine with opposing (ebb) currents. Smith found that small current magnitudes did not destabilize toes designed in accordance with guidance given by Markle (1989) and presented in Table VI-5-45. But damage did occur as currents were increased, and a pulsating effect was observed in the wave downrush as the wave orbital velocity combined with the ebb current.

(c) The test configuration had waves approaching at an angle of 70 deg from the normal to the structure trunk, and wave heights were adjusted until breaking occurred on the structure. This is fairly typical scenario for jettied entrance channels. Both regular and irregular wave conditions were used in the tests. Generally, less damage was recorded for equivalent irregular waves, but this was attributed to the relatively short duration of the wave runs during the experiments. The range of model parameters tested, and the prototype equivalents for the 1:25-scale model, are shown in the following tabulation. Generally, currents less than 15 cm/s in the model (0.75 m/s prototype) did not affect toe stability.

Parameter	Model Value	Prototype Equivalent
Depth	24 cm and 30 cm	6.1 m and 7.6 m
Wave Period	1.7 - 3.0 s	8.5 - 15.0 s
Ebb Current	0.0 - 46 cm/s	0.0 - 2.3 m/s
Wave Height	Breaking	Breaking

(d) Smith developed a procedure to modify Markle's toe stability criterion to account for currents flowing parallel to the structure. Strictly, the method is intended for situations where waves approach at a large angle from the normal (55-80 deg). Application to situations with wave approach more normal to the structure will yield conservative design guidance. The iterative procedure is outlined in Table VI-5-49.

    e. Design of structure cross-section.

    (1) Introduction.

    (a) A rubble-mound structure is normally composed of a bedding layer and a core of quarry-run stone covered by one or more layers of larger stone and an exterior layer or layers of large quarrystone or concrete armor units. Typical rubble-mound cross sections are shown in Figures VI-5-54 and VI-5-55. Figure VI-5-54 illustrates cross-section features typical of designs for breakwaters exposed to waves on one side (seaward) and intended to allow minimal wave transmission to the other (leeward) side. Breakwaters of this type are usually designed with crests elevated to allow overtopping only in very severe storms with long return periods. Figure VI-5-55 shows features common to designs where the breakwater may be exposed to substantial wave action from both sides, such as the outer portions of jetties, and where overtopping is allowed to be more frequent. Both figures show a more complex idealized cross section and a recommended cross section. The idealized cross section provides more complete use of the range of materials typically available from a quarry, but it is more difficult to construct. The recommended cross section takes into account some of the practical problems involved in constructing submerged portions of the structure.

    (b) Figures VI-5-54 and VI-5-55 include tables giving average layer rock size in terms of the stable primary armor unit weight, W, along with the gradation of stone used in each layer (right-hand column). To prevent smaller rocks in the underlayer from being pulled through an overlayer by wave action, the following criterion for filter design may be used to check the rock-size gradations given in Figures VI-5-54 and VI-5-55.

$$D_{15}\left(\text{cover}\right) \le 5D_{85}\left(\text{under}\right) \tag{VI-5-114}$$

where D_{85} *(under)* is the diameter exceeded by the coarsest 15 percent of the underlayer and D_{15} *(cover)* is the diameter exceeded by the coarsest 85 percent of the layer immediately above the underlayer.

    (c) Stone sizes are given by weight in Figures VI-5-54 and VI-5-55 because the armor in the cover layers is selected by weight at the quarry, but the smaller stone sizes are selected by dimension using a sieve or a grizzly. Thomsen, Wohlt, and Harrison (1972) found that the sieve size of stone corresponds approximately to

$$D_{sieve} \approx 1.15\left(\frac{W}{w_a}\right)^{1/3} \tag{VI-5-115}$$

Table VI-5-49
Stability Under Combined Waves and Currents (Smith 1999)

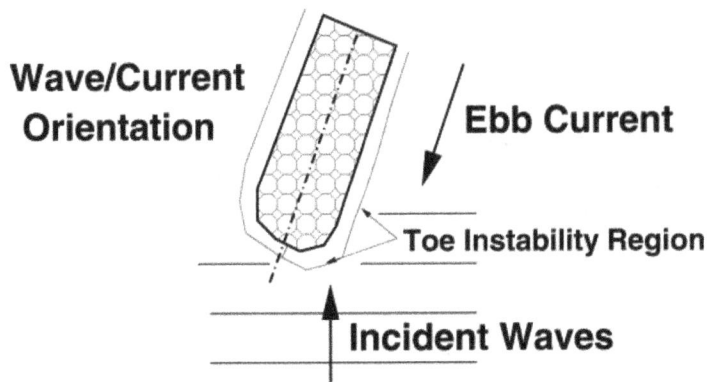

Wave/Current Orientation   **Ebb Current**

Toe Instability Region

Incident Waves

The current-modified stability number is caculated by the formula

$$\left(N_s\right)_c = a\left(\frac{U+u}{\sqrt{gh_s}}\right) \tag{VI-5-111}$$

where

$$u = \frac{g\,H\,T}{2\,L} \tag{VI-5-112}$$

$$a = 51.0\left(\frac{h_b}{h_s}\right) - 26.4 \tag{VI-5-113}$$

and

     u = m aximum wave orbital velocity in shallow water
     U = current magnitude
     g = gravity
     h_s = total water depth
     h_b = water depth over toe berm
     H = breaking wave height
     T = wave period
     L = local wavelength

Procedure: For a given wave condition, first calculate the stability number, N_s, using Markle's method from Table VI-5-45 for sloping rubble-mound structures. Then calculate a current-modified stability number from Equation VI-5-111. If $(N_s)_c > N_s$, the toe stone is unstable, and the procedure is repeated using a larger toe stone to calculate new values of N_s and h_b.

Uncertainty of the Formula: Unknown

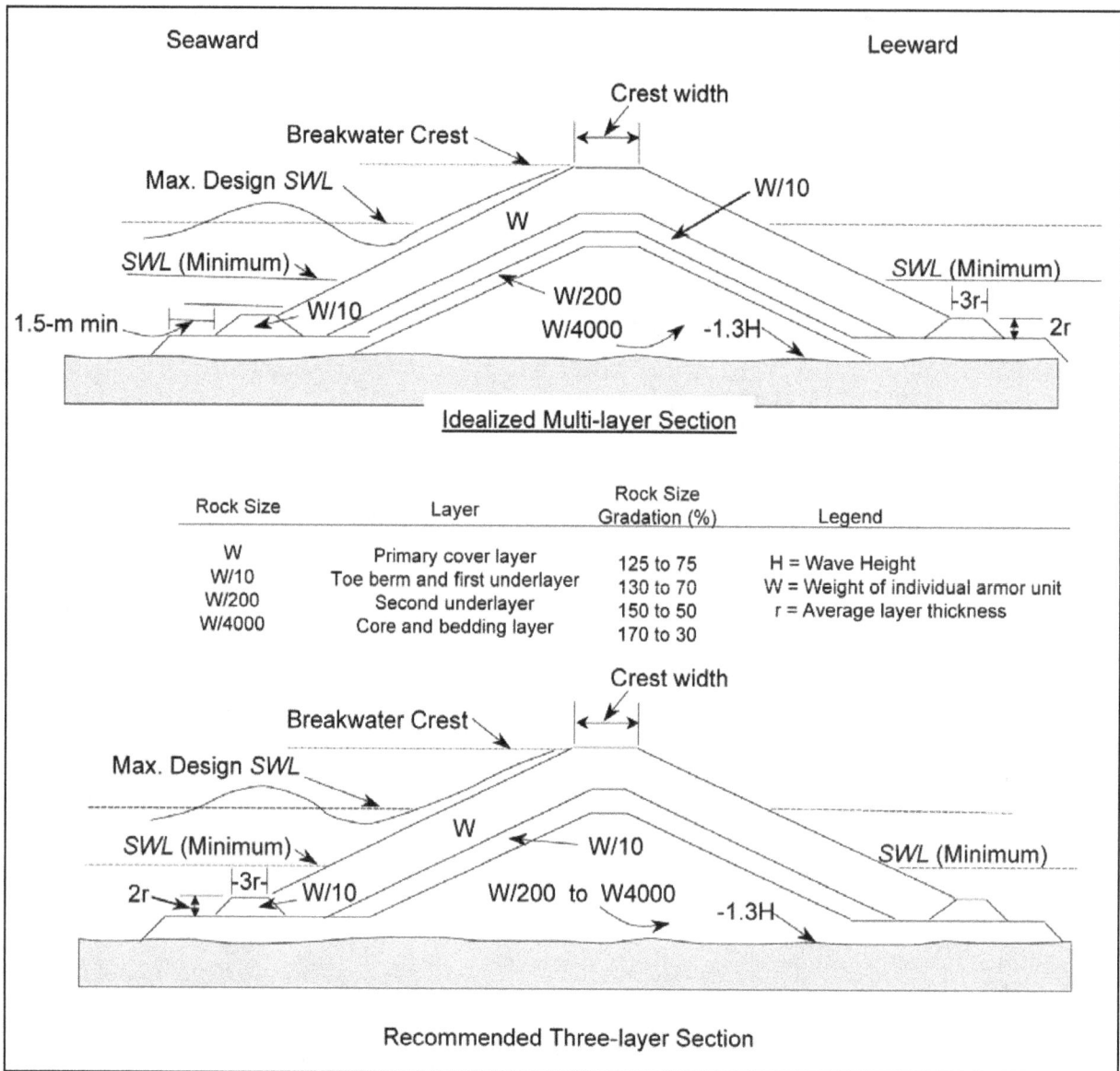

Figure VI-5-54. Rubble-mound section for seaward wave exposure with zero-to-moderate overtopping conditions

where W is the stone weight and w_a is the stone unit weight. Table VI-5-50 lists weights and approximate dimensions for a wide range of stone sizes having stone specific weight of 25.9 kN/m^3 (165 lb/ft^3). The dimensions listed for stone weighing several tons corresponds to the approximate size of the stone determined from visual inspection. Layer thickness should not be estimated as multiples of the dimensions given in Table VI-5-50 because that does not allow for stone intermeshing. Layer thickness is correctly estimated using Equation VI-5-117.

(d) Structure design is part of the overall project planning and design process as illustrated by the generic design diagrams given in Figures V-1-1 through V-1-3 in Part V-1-1-h. Figure VI-5-56 presents a logic diagram for preliminary design of rubble-mound structures.

Included in the diagram are three phases: structure geometry, evaluation of construction technique, and evaluation of design materials.

(e) As part of the design analysis indicated in the logic diagram of Figure VI-5-56, the following structure geometric features should be investigated:

- Crest elevation and width.

- Concrete cap for rubble-mound structures.

- Thickness of armor layer and underlayers.

- Bottom elevation of primary cover layer.

- Toe berm for cover layer stability.

- Structure head and leeside cover layer.

- Secondary cover layer.

- Underlayers.

- Bedding layers and filter blanket layer (see Part VI-5-3*b*, "Granulated and geotextile filter stability."

- Scour protection at toe see Part VI-5-6, "Scour and Scour Protection."

- Toe berm for foundation stability see Part VI-5-3*d*, "Toe stability and protection," and Part VI-5-5, "Foundation Loads."

(f) The following sections describe design aspects for the previously listed geometric features.

(2)   Crest elevation and width.

Table VI-5-50
Weight and Size Selection Dimensions of Quarrystone[1]

Weight mt (tons)	Dimension m (ft)	Weight kg (lb)	Dimension m (ft)	Weight kg (lb)	Dimension cm (in)	Weight kg (lb)	Dimension cm (in.)	Weight kg (lb)	Dimension cm (in.)
0.907 (1)	0.81 (2.64)	45.36 (100)	0.30 (0.97)	2.27 (5)	10.92 (4.30)	0.23 (0.5)	5.08 (2.00)	0.01 (0.025)	1.88 (0.74)
1.814 (2)	1.02 (3.33)	90.72 (200)	0.38 (1.23)	4.54 (10)	13.77 (5.42)	0.45 (1.0)	6.40 (2.52)	0.02 (0.050)	2.36 (0.93)
2.722 (3)	1.16 (3.81)	136.08 (300)	0.43 (1.40)	6.81 (15)	15.77 (6.21)	0.68 (1.5)	7.32 (2.88)	0.03 (0.75)	2.70 (1.06)
3.629 (4)	1.28 (4.19)	181.44 (400)	0.47 (1.54)	9.07 (20)	17.35 (6.83)	0.91 (2.0)	8.05 (3.17)	0.04 (0.100)	2.97 (1.17)
4.536 (5)	1.38 (4.52)	226.80 (500)	0.51 (1.66)	11.34 (25)	18.70 (7.36)	1.13 (2.5)	8.66 (3.41)	0.06 (0.125)	3.20 (1.26)
5.443 (6)	1.46 (4.80)	272.16 (600)	0.54 (1.77)	13.61 (30)	19.86 (7.82)	1.36 (3.0)	9.22 (3.63)	0.07 (0.150)	3.40 (1.34)
6.350 (7)	1.54 (5.05)	317.52 (700)	0.57 (1.86)	15.88 (35)	20.90 (8.23)	1.59 (3.5)	9.70 (3.82)	0.08 (0.175)	3.58 (1.41)
7.258 (8)	1.61 (5.28)	362.88 (800)	0.60 (1.95)	18.14 (40)	21.84 (8.60)	1.81 (4.0)	10.13 (3.99)	0.09 (0.200)	3.73 (1.47)
8.165 (9)	1.67 (5.49)	408.24 (900)	0.62 (2.02)	20.41 (45)	22.73 (8.95)	2.04 (4.5)	10.54 (4.15)	0.10 (0.225)	3.89 (1.53)
9.072 (10)	1.73 (5.69)	453.60 (1000)	0.64 (2.10)	22.68 (50)	23.55 (9.27)	2.27 (5.0)	10.92 (4.30)	0.11 (0.250)	4.04 (1.59)
9.979 (11)	1.79 (5.88)	498.96 (1100)	0.66 (2.16)	24.95 (55)	24.31 (9.57)				
10.866 (12)	1.84 (6.05)	544.32 (1200)	0.68 (2.23)	27.22 (60)	25.02 (9.85)				
11.793 (13)	1.89 (6.21)	589.68 (1300)	0.70 (2.27)	29.48 (65)	25.70 (10.12)				
12.700 (14)	1.94 (6.37)	635.04 (1400)	0.72 (2.35)	31.75 (70)	26.34 (10.37)				
13.608 (15)	1.98 (6.51)	680.40 (1500)	0.73 (2.40)	34.02 (75)	26.95 (10.61)				
14.515 (16)	2.03 (6.66)	725.76 (1600)	0.75 (2.45)	36.29 (80)	27.53 (10.84)				
15.422 (17)	2.07 (6.79)	771.12 (1700)	0.76 (2.50)	38.56 (85)	28.09 (11.06)				
16.330 (18)	2.11 (6.92)	816.48 (1800)	0.78 (2.55)	40.82 (90)	28.65 (11.28)				
17.237 (19)	2.15 (7.05)	861.84 (1900)	0.80 (2.60)	43.09 (95)	29.16 (11.48)				
18.144 (20)	2.19 (7.17)	907.20 (2000)	0.81 (2.64)	45.36 (100)	29.54 (11.63)				

[1] Dimensions correspond to size measured by sieve, grizzly, or visual inspection for stone of 25.9 kilonewtons per cubic meter unit weight. Do not use for determining structure crest width or layer thickness.

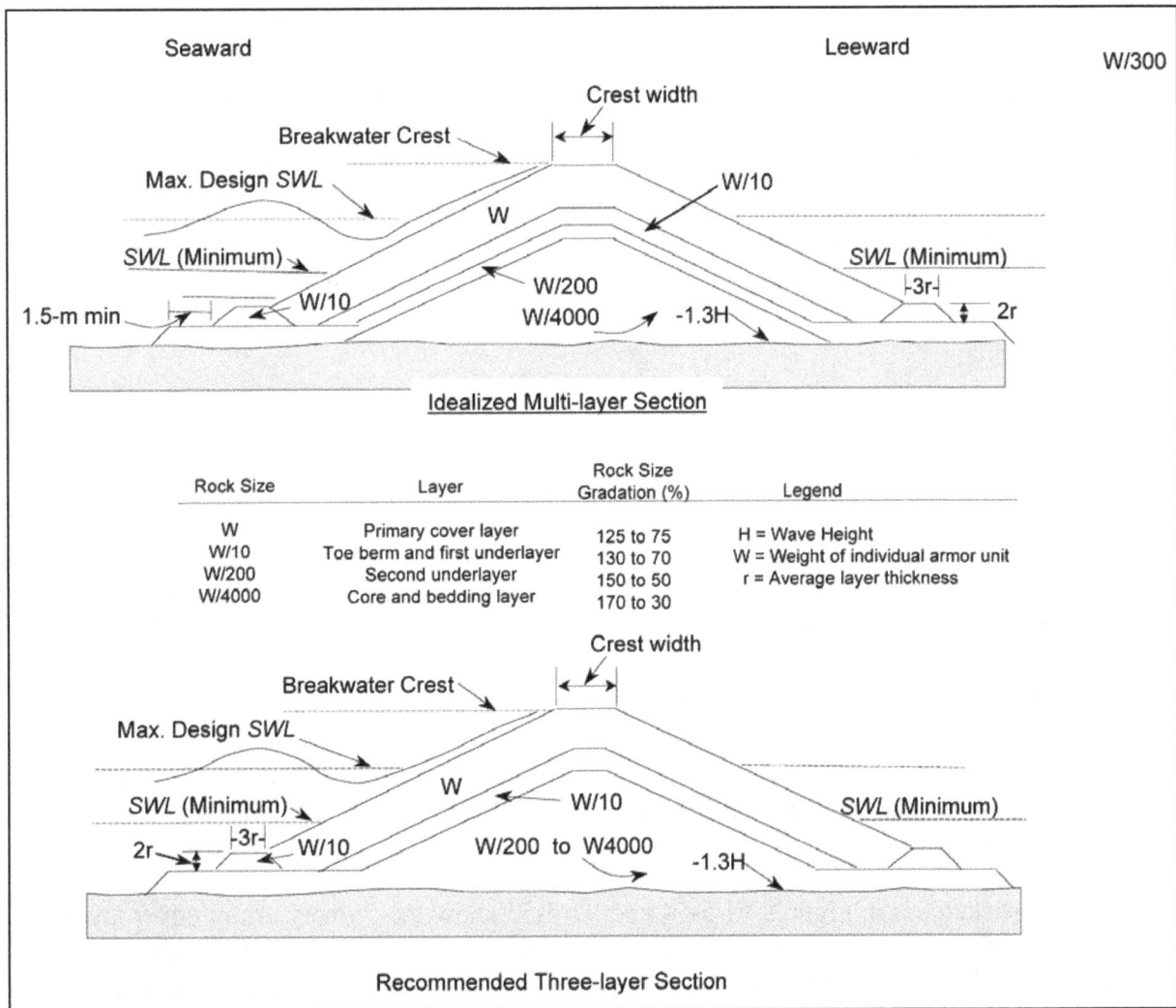

Figure VI-5-55. Rubble-mound section for wave exposure on both sides with moderate overtopping conditions

(a) Overtopping of a rubble-mound structure such as a breakwater or jetty usually can be tolerated if the waves generated by the overtopping do not cause damage behind the structure. Overtopping will occur if the crest elevation is lower than the wave runup, as estimated using the procedures in Part VI-5-2a, "Wave runup and rundown on structures." If the armor layer is chinked, or in other ways made smoother or less permeable, maximum runup will be increased.

(b) The selected crest elevation should be the lowest that provides the protection required. Excessive overtopping of a breakwater or jetty can cause choppiness of the water surface behind the structure and can be detrimental to harbor operations such as small craft mooring and most types of commercial cargo transfer. Overtopping of a rubble seawall or revetment can cause serious erosion behind the structure and flooding of the backshore area. Jetty overtopping is tolerable if it doesn't affect navigation in the channel. Signs warning pedestrians of overtopping dangers should be prominently posted on any publicly accessible structure designed for occasional wave overtopping.

(c) Crest width depends greatly on the degree of allowable overtopping; however, this dependency has not been quantified into general design guidance. The general rule of thumb for overtopping conditions is that minimum crest width should equal the combined widths of three armor units ($n = 3$) as determined by the formula

$$B = nk_\Delta \left(\frac{W}{w_a} \right)^{1/3}$$

(VI-5-116)

where

B = crest width

n = number of stones ($n = 3$ is recommended minimum)

k_Δ = layer coefficient from Table VI-5-51

W = primary armor unit weight

w_a = specific weight of armor unit material

Where there is no overtopping, crest width is not critical; but in either case the crest must be wide enough to accommodate any construction and maintenance equipment that might operate directly on the structure.

(d) The sketches in Figures VI-5-54 and VI-5-55 show the primary armor cover layer extending over the crest. Armor units designed according to the non-overtopping stability formulas in Part VI-5-3a, "Armor layer stability," are probably stable on the crest for minor overtopping. For low-crested structures where frequent, heavy overtopping is expected, use the appropriate stability formula given in the Part VI-5-3a tables for preliminary design. Physical model tests are strongly recommended to confirm the stability of the crest and backside armor under heavy overtopping conditions. Model testing is almost imperative to check the overtopping stability of concrete armor units placed on the crest which may be less stable than equivalent stone armor.

(3) Concrete cap for rubble-mound structures.

(a) Placed concrete may be added to the cover layer of rubble-mound jetties and breakwaters for purposes such as filling the interstices of stones in the cover layer crest and side slopes as far down as wave action permits, or as large monolithic blocks cast in place. Placed concrete may serve any of four purposes: to strengthen the crest, to deflect overtopping waves away from impacting directly on the leeside slope, to increase the crest height, and to provide roadway access along the crest for construction or maintenance purposes.

(b) Massive concrete caps have been used with cover layers of precast concrete armor units to replace armor units of questionable stability on an overtopped crest and to provide a rigid backup to the top rows of armor units on the slopes. To accomplish this dual purpose, the cap can be a slab with a solid or permeable parapet (Czerniak and Collins 1977; Jensen 1983) a slab over stone grouted to the bottom elevation of the armor layer, or a solid or permeable block (Lillevang 1977; Markle 1982). Massive concrete caps must be placed after a structure has settled or must be sufficiently flexible to undergo settlement without breaking up (Magoon et al. 1974).

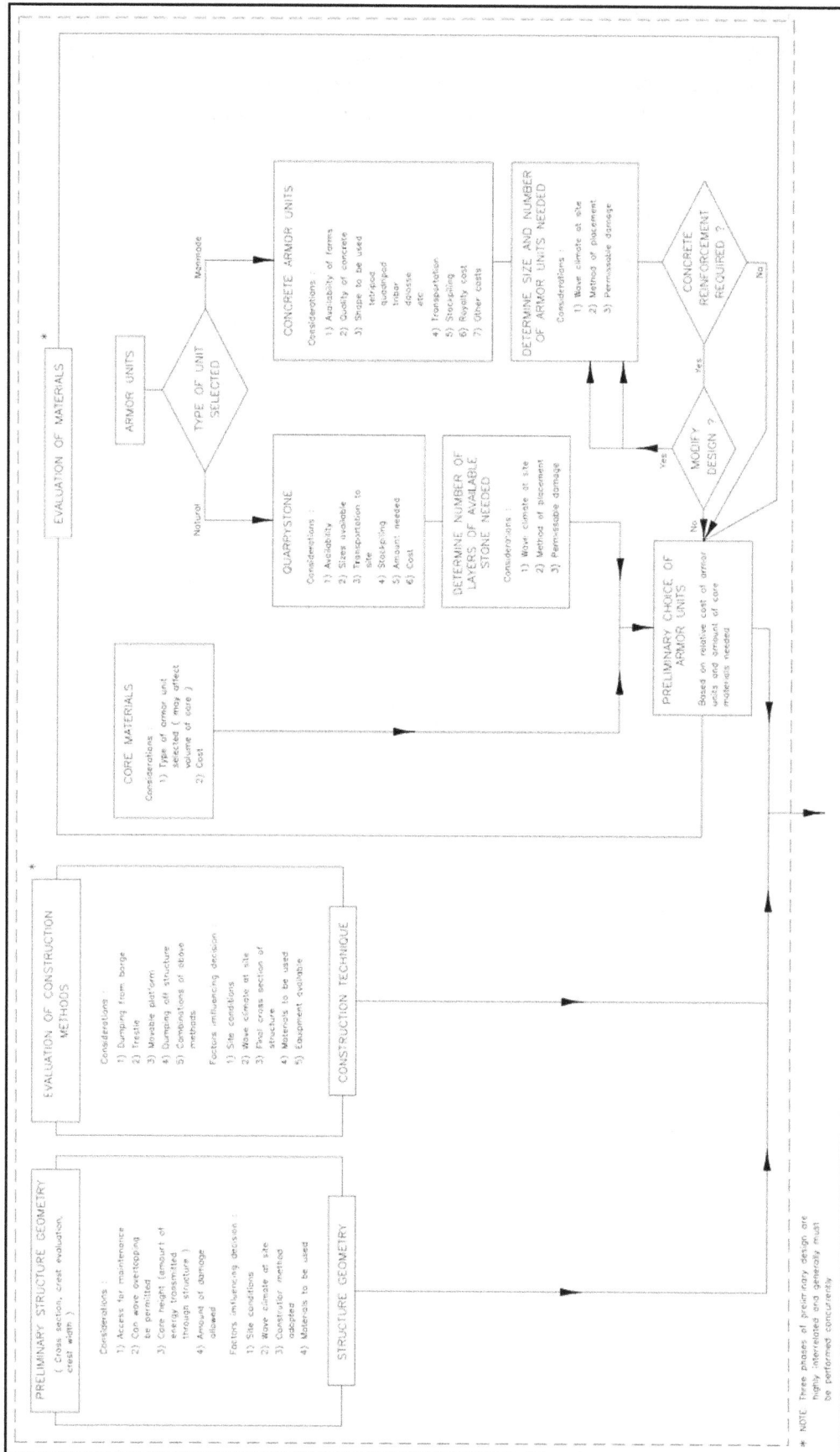

Figure VI-5-56. Logic diagram for preliminary design of rubble-mound structures

(c) Concrete caps with solid vertical or sloped walls reflect waves out through the upper rows of armor units, perhaps causing loss of those units. Solid slabs and blocks can trap air beneath them, creating uplift forces during heavy wave action that may crack or tip the cap (Magoon et al. 1974). A permeable cap decreases both of these problems. A parapet can be made permeable, and vertical vents can be placed through the slab or block itself (Mettam 1976). Lillevang (1977) designed a breakwater crest composed of a vented block cap placed on an unchinked, ungrouted extension of the seaward slope's underlayer, a permeable base reaching across the crest.

(d) Ribbed caps are a compromise between the solid block and a covering of concrete armor units. The ribs are large, long, rectangular members of reinforced concrete placed perpendicular to the axis of a structure in a manner resembling railroad ties. The ribs are connected by reinforced concrete braces, giving the cap the appearance of a railroad track running along the structure crest. This cap serves to brace the upper units on the slopes, yet is permeable in both the horizontal and vertical directions.

(e) Ribbed caps have been used on U.S. Army Corps of Engineers breakwaters at Maalea Harbor (Carver and Markle 1981), at Kahului (Markle 1982), on Maui, and Pohoiki Bay, all in the State of Hawaii.

(f) Waves overtopping a concrete cap can damage the leeside armor layer. The width of the cap and the shape of its lee side can be designed to deflect overtopping waves away from the structure's lee side (Czerniak and Collins 1977; Lillevang 1977; and Jensen 1983). Ribbed caps help dissipate waves.

(g) High parapet walls have been added to caps to deflect overtopping seaward and allow the lowering of the crest of the rubble mound itself. These walls present the same reflection problems described above and complicate the design of a stable cap (Mettam 1976; Jensen 1983). Hydraulic model tests by Carver and Davidson (1976, 1983) have investigated the stability of caps with high parapet walls proposed for Corps structures. Part VI-5-4d, "Stability of concrete caps and caissons against sliding and overturning," provides design guidance.

(h) To evaluate the need for a massive concrete cap to increase structural stability against overtopping, consideration should be given to the cost of including a cap versus the cost of increasing dimensions to prevent overtopping and for construction and maintenance purposes. A massive concrete cap is not necessary for the structural stability of a structure composed of concrete armor units when the difference in elevation between the crest and the limit of wave runup on the projected slope above the structure is less than 15 percent of the total wave runup. For this purpose, an all-rubble structure is preferable, and a concrete cap should be used only if substantial savings would result. Maintenance costs for an adequately designed rubble structure are likely to be lower than for any alternative composite-type structure. The cost of a concrete cap should also be compared to the cost of covering the crest with flexible, permeable concrete armor units, perhaps larger than those used on the slopes, or large quarrystone armor. Hydraulic model tests are recommended to determine the most stable and economical crest designs for major structures.

(i) Experience indicates that concrete placed in the voids on the structure slopes has little structural value. By reducing slope roughness and surface porosity, the concrete increases wave runup. The effective life of the concrete is short, because the bond between concrete and stone is quickly broken by structure settlement. Such filling increases maintenance costs. For a roadway, a concrete cap can usually be justified if frequent maintenance of armor slopes is anticipated. A smooth surface is required for wheeled vehicles; tracked equipment can be used on ribbed caps.

(4) Thickness of armor layer and underlayers.

(a) The thickness of the cover layer and underlayers is calculated using the formula

$$r = n k_\Delta \left(\frac{W}{w_a} \right)^{1/3}$$

(VI-5-117)

and the placing density (number of armor units per unit area) is estimated using the equation

$$\frac{N_a}{A} = n k_\Delta \left(1 - \frac{P}{100} \right) \left(\frac{w_a}{W} \right)^{2/3}$$

(VI-5-118)

where r is the average layer thickness, n is the number of quarrystone or concrete armor units in the thickness (typically $n = 2$), W is the weight of individual armor units, w_a is the specific weight of the armor unit material, and N_a is the required number of individual armor units for a given surface area, A. The layer coefficient (k_Δ) and cover layer average porosity (P) in percent were experimentally determined, and values are given in Table VI-5-51. Equations VI-5-117 and VI-5-118 can be used with either metric or English units.

(b) The specified placing or packing density must be strictly maintained during construction to assure proper interlocking, and therefore hydraulic stability, of the armor layer. During placement, packing density can be maintained by specifying a mean and allowable deviation for the centroidal distance (in three dimensions) between units, or it can be maintained by counting units in a specified area. For grid placement, each subsequent row of armor units is typically offset laterally from the previous lower row to avoid failure planes. To specify the placement grid, D_H is the distance between the centroids of two adjacent units on the same horizontal row and D_U is the distance between the centroids of units upslope in the plane of the structure slope. Values of D_H and D_U for specific armor sizes and packing density coefficients appropriate for Core-Loc and Accropod units can be obtained from the vendor or license holder. Within any matrix of armor units, every effort should be made to achieve maximum interlocking. The maximum centroidal distance D_{max} should not exceed 110 percent of the values specified. Greater spacing may jeopardize interlocking and the integrity of the armor layer.

Table VI-5-51
Layer Coefficient and Porosity for Various Armor Units

Armor Unit	n	Placement	Layer Coefficient k_Δ	Porosity P (percent)
Quarrystone (smooth)[1]	2	Random	1.02	38
Quarrystone (rough)[2]	2	Random	1.00	37
Quarrystone (rough)[2]	∃3	Random	1.00	40
Quarrystone (parallepiped)[3]	2	Special	--	27
Quarrystone[4]	Graded	Random	--	37
Cube (modified)[1]	2	Random	1.10	47
Tetrapod[1]	2	Random	1.04	50
Tribar[1]	2	Random	1.02	54
Tribar[1]	1	Uniform	1.13	47
dolos[5]	2	Random	0.94	56
Vol. < 5 m³ Core-Loc[6] 5 < Vol. < 12 m³ 12 < Vol. < 22m³	1	Random	1.51	60 63 64
Vol. < 5 m³ Accropod[7] 5 < Vol. < 12 m³ 12 < Vol. < 22m³	1	Random	1.51	57 59 62

[1] Hudson (1974)
[2] Carver and Davidson (1983)
[3] Layer thickness is twice the average long dimension of the parallelepiped stones. Porosity is estimated from tests on one layer of uniformly placed modified cubes (Hudson 1974).
[4] The minimum layer thickness should be twice the cubic dimension of the W_{50} riprap. Check to determine that the graded layer thickness is ∃1.25 the cubic dimension of the W_{max} riprap (see Equations VI-5-119 and VI-5-120).
[5] Carver and Davidson (1977)
[6] Turk and Melby (1997)
[7] Accropod informational brochure

(c) The thickness r of a layer of riprap is the greater of either 0.3 m, or one of the following, whichever of the three is greatest:

$$r = 2.0 \left(\frac{W_{50}}{w_a} \right)^{1/3}$$ (VI-5-119)

where W_{50} is the weight of the 50-percent size in the riprap gradation, or

$$r = 1.25 \left(\frac{W_{max}}{w_a} \right)^{1/3}$$ (VI-5-120)

where W_{max} is the heaviest stone in the gradation. The specified layer thickness should be increased by 50 percent for riprap placed underwater if conditions make placement to design dimensions difficult. The placing density of riprap is defined as the total weight of riprap placed (W_T) per unit area (A) of structure slope. Riprap placing density can be estimated as

$$\frac{W_T}{A} = r w_a \left(1 - \frac{P}{100} \right)$$ (VI-5-121)

(5) Bottom elevation of primary cover layer.

(a) When water depth is greater than 1.5 H (where H is the irregular wave height parameter used to determine a stable primary armor unit weight), the armor units in the cover layer should be extended downslope to an elevation below minimum SWL equal to the design wave height H as shown in Figure VI-5-54. For water depths less than 1.5 H extend the cover layer armor units to the toe as shown in Figure VI-5-55. Model tests to determine the bottom elevation of the primary cover layer and the type of armor placement should be conducted when feasible. Revetment cover layers located in shallow water should be extended seaward of the structure toe on sandy bottoms to serve as scour protection.

(b) Increased stability for special-placement parallelepiped stone (see higher K_D values in Table VI-5-22) can only be obtained if a toe mound is carefully placed to support the quarrystones with their long axes perpendicular to the structure slope. For dolosse it is recommended that the bottom rows of units in the primary cover layer be Aspecial placed@ on top of the secondary cover layer as shown in Figure VI-5-54, on top of the toe berm as shown in Figure VI-5-55, or on the bottom itself. This placement is highly dependent on wave conditions and water clarity. Site-specific model studies have placed the bottom layer of dolosse with vertical flukes away from the slope and the second row placed so that the units overlap the horizontal flukes of the bottom layer. This helps assure interlocking with the random-placed units farther up the slope (Bottin, Chatham, and Carver 1976), and provides better toe stability than random placement. The seaward dolosse in the bottom row should be placed with the bottom of the vertical flukes one-half the length of the units back from the design surface of the primary armor layer to produce the design layer thickness.

(c) Core-Loc units can be placed randomly along the toe, but experiments indicate a pattern placement along the toe is more stable and should be used when the breakwater is built in shallow, depth-limited conditions. For the bottom layer, individual Core-Loc units are set in a three-point stance in cannon fashion with the central fluke pointing seaward, up at a 45-deg angle like the cannon barrel. All toe units are placed side-by-side with minimal space between adjacent units. The second course of units is laid atop of the toe units such that they straddle each toe unit. Once the second row has been placed, all subsequent Core-Loc armor units are placed in a random matrix. While placing these units in a variety of random orientations, care must be taken to assure that all overlying units are interlocked with and constrain underlying units.

(6) Toe berm for cover layer stability.

(a) Structures exposed to breaking waves should have a quarrystone toe berm to protect the toe of the primary armor layer (see Figure VI-5-55). Design guidance for toe berm dimensions and stone size is given in Part VI-5-3d, "Toe stability and protection."

(b) The toe berm may be placed before or after the adjacent cover layer. For special-placement quarrystone or uniform-placement tribars, the toe berm serves as a base, and it must be placed first. When placed after the cover layer, the toe berm must be high enough to provide bracing up to at least half the height of the toe armor units. Usually, this requirement is exceeded by the design guidance recommended in Part VI-5-3d.

(7) Structure head and leeside cover layer.

(a) Armoring of the head of a breakwater or jetty should be the same on the leeside slope as on the seaside slope for a distance of about 15 to 45 m from the structure end. This distance depends on such factors as structure length and crest elevation at the seaward end. (See Tables VI-5-37 and VI-5-38 for sizing stable armor units for heads.)

(b) Design of leeside cover layers depends on the extent of wave overtopping, any waves or surges acting directly on the lee slope, structure porosity, and differential hydrostatic head resulting in uplift forces that may dislodge armor units on the back slope. If the crest elevation is established to prevent possible overtopping, the weight of armor units and the bottom elevation of the back slope cover layer should depend on the lesser wave action on the lee side (if any) and the porosity of the structure. Under minor overtopping the armor weight calculated for the seaward side primary cover layer should be used on the lee side down to at least the SWL or -0.5 H for preliminary designs. However, model testing may be needed to determine stable armor weights for overtopping wave impacts.

(c) For heavy overtopping of breaking waves in shallow water, the primary armor layer on the lee side should be extended to the bottom as shown in Figure VI-5-55. Where concrete caps are employed, stability of the leeside armor during overtopping should be verified with model tests. When both sides of a structure are exposed to similar wave action (groins and jetties), both slopes should have similar designs.

(8) Secondary cover layer.

(a) If the armor units in the primary and secondary cover layers are of the same material, the weight of armor units in the secondary cover layer, between -1.5 H and -2.0 H, should be greater than about one-half the weight of armor units in the primary cover layer. Below -2.0 H, the weight requirements can be reduced to about $W/15$ for the same slope condition (see Figure VI-5-54). If the primary cover layer is quarrystone, the weights for the secondary quarrystone layers should be ratioed from the weight of quarrystone that would be required for the primary cover layer. The use of a single size of concrete armor unit for all cover layers (i.e., upgrading the secondary cover layer to the same size as the primary cover layer) may prove to be economically advantageous when the structure is located in shallow water as shown in Figure VI-5-55 where the primary cover layer is extended down the entire slope.

(b) The secondary cover layer (shown in Figure VI-5-54 from elevation -1.5 H to the bottom) should be as thick as, or thicker than, the primary cover layer. As an example, cover layers of quarrystone of two-stone thickness ($n = 2$) will require a secondary cover layer thickness of $n = 2.5$ for the slope between elevations -H and -2.0 H, and a thickness of $n = 5$ for the slope below an elevation of -2.0 H. These layer thicknesses are based on the armor unit weight ratios given in Figure VI-5-54.

(c) The interfaces between the secondary cover layers and the primary cover layer are shown at the slope of 1-to-1.5 on Figure VI-5-54. Steeper slopes for the interfaces may contribute to the stability of the cover armor, but material characteristics and site wave conditions during construction may require using a flatter slope than shown in the figure.

(9) Underlayers.

(a) The first underlayer directly beneath the primary armor units (see Figures VI-5-54 and VI-5-55) should have a minimum thickness of two quarrystones ($n = 2$). The first underlayer stones should weigh about one-tenth of the weight of the overlying armor units ($W/10$) if the cover layer and first underlayer are both quarrystone, or the first underlayer is quarrystone and the cover layer is concrete armor units with a stability coefficient of $K_D \square 12$ (see Tables VI-5-29, VI-5-33, VI-5-34, VI-5-36). When the cover layer armor unit $K_D > 12$ (dolosse, Core-Loc, and uniformly-placed tribars) the first underlayer quarrystone weight should be about one-fifth the weight of the overlying unit ($W/5$). The larger size promotes increased interlocking between the first underlayer and the concrete armor units of the primary cover layer. Hydraulic model tests (Carver and Davidson 1977; Carver 1980) indicate for quarrystone armor units and dolosse on a breakwater trunk exposed to nonbreaking waves that the underlayer stone size could range from $W/5$ to $W/20$ with little effect on armor stability, wave runup or rundown. If the underlayer stone proposed for a given structure is available with a gradation in the range of $W/5$ to $W/20$, the structure should be model tested with that underlayer gradation to determine if this economical material will support a stable primary cover layer of planned armor units when exposed to the site design conditions.

(b) The second underlayer beneath the primary cover layer and upper secondary cover layer (above -2.0 H) should have a minimum equivalent thickness of two quarrystones and a weight about 1/20 the weight of the stones in the first underlayer. In terms of primary armor unit weight this is approximately 1/20 H $W/10 = W/200$ for quarrystone and some concrete armor units.

(c) The first underlayer beneath the lower secondary cover layer (below -2.0 H on Figure VI-5-54) should also have a minimum of two thicknesses of quarrystone. Stones in this layer should weigh about 1/20 of the immediately overlying armor unit weight. In terms of primary armor unit weight this is approximately 1/20 x $W/15 = W/300$ for units of the same material. The second underlayer for the secondary armor below -2.0 H can be as light as $W/6000$, or equal to the core material size.

(d) For the recommended cross section in Figure VI-5-54 when the primary armor is quarrystone and/or concrete units with $K_D \square 12$, the first underlayer and the cover layer below -2.0 H should have quarrystone weights between $W/10$ and $W/15$. If the primary armor is concrete armor units with $K_D > 12$, the first underlayer and cover armor below -2.0 H should be quarrystone with weights between $W/5$ and $W/10$.

(e) For graded riprap cover layers the minimum requirement for the underlayers (if one or more are required) is

$$D_{15}\left(\text{cover}\right) \le 5D_{85}\left(\text{under}\right)$$
(VI-5-122)

where D_{15} (cover) is the diameter exceeded by the coarsest 85 percent of the riprap or underlayer on top and D_{85} (under) is the diameter exceeded by the coarsest 15 percent of the underlayer or soil below (Ahrens 1981b). For a revetment where the riprap and the underlying soil satisfy the size criterion, no underlayer is necessary. Otherwise, one or more of the following is required.

(f) The size criterion for riprap is more restrictive than the general filter criterion given in Part VI-5-3b, "Granulated and geotextile filter stability." The riprap criterion requires larger stone in the lower layer to prevent the material from washing through the voids in the upper layer as cover layer stones shift during wave action. A more conservative underlayer than required by the minimum criterion may be constructed of stone with a 50-percent size of about $W_{50}/20$. This larger stone will produce a more permeable underlayer and should reduce runup and increase interlocking between the cover layer and underlayer. However, be sure to check the underlayer gradation against the underlying soil to assure the minimum criterion of Equation VI-5-122 is met.

(g) The underlayers should be at least three thicknesses of the W_{50} stone, but never less than 0.23 m (Ahrens 1981b). The thickness can be calculated using Equation VI-5-119 with a coefficient of 3 rather than 2. Because a revetment is placed directly on the soil or fill material of the bank it protects, a single underlayer also functions as a bedding layer or filter blanket.

f. Blanket stability in current fields. Stone blankets constructed of randomly-placed riprap or uniformly sized stone are commonly used to protect areas susceptible to erosion by fast-flowing currents. Blanket applications include lining the bottom and sloping sides of flow channels and armoring regions of tidal inlets where problematic scour has developed. Design of stable stone or riprap blankets is based on selecting stone sizes such that the shear stress required to dislodge the stones is greater than the expected shear stress at the bottom developed by the current.

   (1) Boundary layer shear stress.

   (a) Prandl established a universal velocity profile for flow parallel to the bed given by

$$\frac{u}{v_*} = \frac{1}{\kappa} \ln\left(\frac{y}{k_s}\right) + B \qquad\qquad\text{(VI-5-123)}$$

where

   κ = von Karman constant (= 0.4)

   y = elevation above the bed

   u = velocity at elevation y

   k_s = boundary roughness

   B = function of Reynolds number (= 8.5 for fully rough, turbulent flow)

   v_* = shear velocity (= $(\tau_o / \rho_w)^{1/2}$)

   τ_o = shear stress acting on the bed

   ρ_w = density of water

Equation VI-5-123 can be expressed in terms of the mean flow velocity, $\bar{u}$, by integrating over the depth, i.e.,

$$\frac{\bar{u}}{v_*} = \frac{1}{h}\int_0^h \frac{u}{v_*} dy = \frac{1}{\kappa} \ln\left(\frac{h}{k_s}\right) + B - \frac{1}{\kappa} \qquad\qquad\text{(VI-5-124)}$$

or

$$\frac{\bar{u}}{v_*} = 2.5 \ln\left(\frac{11h}{k_s}\right) \qquad\qquad\text{(VI-5-125)}$$

when fully rough turbulent flow is assumed, which is usually the case for flow over stone blankets. Equation VI-5-125 assumes uniform bed roughness and currents flowing over a distance sufficient to develop the logarithmic velocity profile over the entire water depth.

(b) Bed roughness k_s over a stone blanket is difficult to quantify, but it is usually taken to be proportional to a representative diameter d_a of the blanket material, i.e., $k_s = C_1 d_a$. Substituting for k_s and v_* in Equation VI-5-125 and rearranging yields an equation for shear stress given by

$$\tau_o = \frac{w_w}{g} \left[\frac{\bar{u}}{2.5 \ln\left(\dfrac{11h}{C_1 d_a}\right)} \right]^2 \qquad \text{(VI-5-126)}$$

where $w_w = \rho_w\, g$ is the specific weight of water.

(2) Incipient motion of stone blankets.

(a) Stone blankets are stable as long as the individual armor stones are able to resist the shear stresses developed by the currents. Incipient motion on a horizontal bed can be estimated from Shield's diagram (Figure III-6-7) for uniform flows. Fully rough turbulent flows occur at Reynolds numbers where Shields parameter is essentially constant, i.e.,

$$\Psi = \frac{\tau}{(\rho_a - \rho_w) g d_a} \approx 0.04 \qquad \text{(VI-5-127)}$$

where

τ = shear stress necessary to cause incipient motion

ρ_a = density of armor stone

Rearranging Equation VI-5-127 and adding a factor K_1 to account for blankets placed on sloping channel side walls gives

$$\tau = 0.04 K_1 \left(w_a - w_w \right) d_a \qquad \text{(VI-5-128)}$$

where w_a is the specific weight of armor stone ($= \rho_a\, g$), and

$$K_1 = \sqrt{1 - \frac{\sin^2 \theta}{\sin^2 \varphi}} \qquad \text{(VI-5-129)}$$

with

θ = channel sidewall slope

φ = angle of repose of blanket armor [. $40°$ for riprap]

(b) Equating Equations VI-5-126 and VI-5-128 gives an implicit equation for the stable blanket diameter d_a. However, by assuming the logarithmic velocity profile can be approximated by a power curve of the form

$$\ln\left(\frac{11h}{C_1 d_a}\right) \approx C_2 \left(\frac{h}{d_a}\right)^{\beta}$$

an explicit equation is found having the form

$$\frac{d_a}{h} = C_T \left[\left(\frac{w_w}{w_a - w_w}\right)^{\frac{1}{2}} \left(\frac{\bar{u}}{\sqrt{K_1 gh}}\right)\right]^{\frac{2}{(1-2\beta)}} \qquad \text{(VI-5-130)}$$

where all the constants of proportionality have been included in C_T. Equation VI-5-130 implies that blanket armor stability is directly proportional to water depth and flow Froude number, and inversely proportional to the immersed specific weight of the armor material. The unknown constants, C_T and β, have been empirically determined from laboratory and field data.

(3) Stone blanket stability design equation.

(a) Stable stone or riprap blankets in current fields should be designed using the following equation from Engineer Manual 1110-2-1601.

$$\frac{d_{30}}{h} = S_f C_s \left[\left(\frac{w_w}{w_a - w_w}\right)^{\frac{1}{2}} \left(\frac{\bar{u}}{\sqrt{K_1 gh}}\right)\right]^{\frac{5}{2}} \qquad \text{(VI-5-131)}$$

where

d_{30} = stone or riprap size of which 30 percent is finer by weight

S_f = safety factor (minimum = 1.1) to allow for debris impacts or other unknowns

C_s = stability coefficient for incipient motion
= 0.30 for angular stone
= 0.38 for rounded stone

(b) EM 1110-2-1601 presents additional coefficients for channel bends and other situations where riprap size must be increased due to flow accelerations. The methodology is also summarized in Maynord (1998). Equation VI-5-131 is based on many large-scale model tests and available field data, and the exponent and coefficients were selected as a conservative envelope to most of the scatter in the stability data. Riprap stone sizes as specified by Equation

VI-5-131 are most sensitive the mean flow velocity, so good velocity estimates are needed for economical blanket designs.

(c) Alternately, Equation VI-5-131 can be rearranged for mean flow velocity to give the expression

$$\bar{u} = \left(\frac{1}{s_f C_s}\right)^{\frac{2}{5}} \left(\frac{h}{d_{30}}\right)^{\frac{1}{10}} \left[g K_1 \left(\frac{w_a - w_w}{w_w}\right) d_{30} \right]^{\frac{1}{2}} \tag{VI-5-132}$$

(d) Equation VI-5-132, which is similar to the well-known Isbash equation, can be used to determine the maximum mean velocity that can be resisted by riprap having d_{30} of a given size. The main difference between Equation VI-5-132 and the Isbash equation is that the Isbash equation multiplies the term in square brackets by a constant whereas Equation VI-5-132 multiplies the square-bracketed term by a depth-dependent factor that arises from assuming a shape for the boundary layer. The Isbash equation is more conservative for most applications, but it is still used for fast flows in small water depths and in the vicinity of structures such as bridge abutments.

(e) By assuming the blanket stones are spheres having weight given by

$$W_{30} = \frac{\pi}{6} w_a d_{30}^3 \tag{VI-5-133}$$

where W_{30} is the stone weight for which 30 percent of stones are smaller by weight, Equation VI-5-131 can be expressed in terms of stone weight as

$$\frac{W_{30}}{w_a h^3} = \frac{\pi}{6} (S_f C_s)^3 \left[\left(\frac{w_w}{w_a - w_w}\right)^{\frac{1}{2}} \left(\frac{\bar{u}}{\sqrt{K_1 g h}}\right) \right]^{\frac{15}{2}} \tag{VI-5-134}$$

(4) Stone blanket gradation.

(a) All graded stone distributions (riprap) used for stone blankets should have distributions conforming to the weight relationships given below in terms of W_{30} or $W_{50\,min}$ (EM 1110-2-1601).

$$W_{50\,min} = 1.7\, W_{30} \tag{VI-5-135}$$

$$W_{100\,max} = 5 W_{50\,min} = 8.5 W_{30} \tag{VI-5-136}$$

$$W_{100\,min} = 2 W_{50\,min} = 3.4\, W_{30} \tag{VI-5-137}$$

$$W_{50\,max} = 1.5\, W_{50\,min} = 2.6\, W_{30} \tag{VI-5-138}$$

$$W_{15\,max} = 0.5\,W_{50\,max} = 0.75\,W_{50\,min} = 1.3\,W_{30} \qquad\qquad \text{(VI-5-139)}$$

$$W_{15\,min} = 0.31\,W_{50\,min} = 0.5\,W_{30} \qquad\qquad \text{(VI-5-140)}$$

(b) Recommended thickness of the blanket layer, r, depends on whether placement is submerged or in the dry as specified by the following formulas.

(c) For blankets placed above water, the layer thickness should be

$$r = 2.1\left(\frac{W_{50\,min}}{w_a}\right)^{\frac{1}{3}} = 2.5\left(\frac{W_{30}}{w_a}\right)^{\frac{1}{3}} \qquad\qquad \text{(VI-5-141)}$$

with a minimum blanket thickness of 0.3 m. Blankets placed below water should have layer thickness given by

$$r = 3.2\left(\frac{W_{50\,min}}{w_a}\right)^{\frac{1}{3}} = 3.8\left(\frac{W_{30}}{w_a}\right)^{\frac{1}{3}} \qquad\qquad \text{(VI-5-142)}$$

with a minimum blanket thickness of 0.5 m.

VI-5-4. <u>Vertical-Front Structure Loading and Response</u>.

    a. Wave forces on vertical walls.

    (1) Wave-generated pressures on structures are complicated functions of the wave conditions and geometry of the structure. For this reason laboratory model tests should be performed as part of the final design of important structures. For preliminary designs the formulae presented in this section can be used within the stated parameter limitations and with consideration of the uncertainties. Three different types of wave forces on vertical walls can be identified as shown in Figure VI-5-57.

    (a) Nonbreaking waves: Waves do not trap an air pocket against the wall (Figure VI-5-57a). The pressure at the wall has a gentle variation in time and is almost in phase with the wave elevation. Wave loads of this type are called pulsating or quasistatic loads because the period is much larger than the natural period of oscillation of the structures. (For conventional caisson breakwaters the period is approximately one order of magnitude larger.) Consequently, the wave load can be treated like a static load in stability calculations. Special considerations are required if the caisson is placed on fine soils where pore pressure may build up, resulting in significant weakening of the soil.

    (b) Breaking (plunging) waves with almost vertical fronts: Waves that break in a plunging mode develop an almost vertical front before they curl over (see Figure VI-5-57b). If this almost vertical front occurs just prior to the contact with the wall, then very high pressures are generated having extremely short durations. Only a negligible amount of air is entrapped, resulting in a very large single peaked force followed by very small force oscillations. The duration of the pressure peak is on the order of hundredths of a second.

(c) Breaking (plunging) waves with large air pockets: If a large amount of air is entrapped in a pocket, a double peaked force is produced followed by pronounced force oscillations as shown in Figure VI-5-57c. The first and largest peak is induced by the wave crest hitting the structure at point A, and it is similar to a hammer shock. The second peak is induced by the subsequent maximum compression of the air pocket at point B, and is it is referred to as compression shock, (Lundgren 1969). In the literature this wave loading is often called the ABagnold type.@ The force oscillations are due to the pulsation of the air pocket. The double peaks have typical spacing in the range of milliseconds to hundredths of a second. The period of the force oscillations is in the range 0.2-1.0 sec.

(2) Due to the extremely stochastic nature of wave impacts there are no reliable formulas for prediction of impulsive pressures caused by breaking waves. Determination of impact pressures in model tests is difficult because of scale effects related to the amount and size of air bubbles and size and shape of air pockets. Also the instrumentation, data sampling, and analyses need special attention to avoid bias by dynamic amplification and misinterpretation when scaling to prototype values. Another problem related to model tests is the sensitivity of the shock loads on the shape and kinematics of the breaking waves. This calls for a very realistic and statistically correct reproduction of natural waves in laboratory models.

(3) Impulsive loads from breaking waves can be very large, and the risk of extreme load values increases with the number of loads. Therefore, conditions resulting in frequent wave breaking at vertical structures should be avoided. Alternatives include placing a mound of armor units in front of the vertical wall structure to break the waves before they can break directly on the wall, or using a rubble-mound structure in place of the vertical wall structure.

(4) Frequent wave breaking at vertical structures will not take place for oblique waves with angle of incidence larger than 20 deg from normal incidence. Nor will it take place if the seabed in front of the structure has a mild slope of about 1:50 or less over a distance of at least several wavelengths, and the vertical wall has no sloping foundation at the toe of the wall.

(5) The use of a sloping-front face from about still-water level (swl) to the crest is very effective in reducing large impact pressures from breaking waves. In addition, the direction of the wave forces on the sloping part (right angle to the surface) helps reduce the horizontal force and the tilting moment. Structures with sloping tops might be difficult to optimize where large water level variations are present. Also, a sloping-front structure allows more overtopping than a vertical wall structure of equivalent crest height.

EXAMPLE PROBLEM VI-5-1

FIND:
Riprap distribution for a stable scour blanket over a nearly horizontal bottom

GIVEN:
The following information is known (English system units shown in parentheses)

Specific weight of riprap, $w_a = 25.9$ kN/m^3 (165 lb/ft^3)
Specific weight of water, $w_w = 10.05$ kN/m^3 (64 lb/ft^3)
Bottom slope, $\theta = 0$ deg i.e., $K_1 = 1.0$
Water depth, h = 6 m (19.7 ft)
Depth-averaged mean velocity, $\bar{u} = 2.5$ m/s (8.2 ft/s)
Stability coefficient, $C_s = 0.38$ i.e., rounded stone
Factor of safety, $S_f = 1.1$
Gravitational acceleration, $g = 9.81$ m/s^2 (32.2 ft/s^2)

SOLUTION:
From Equation VI-5-134

$$\frac{W_{30}}{w_a h^3} = \frac{\pi}{6}\left[(1.1)(0.38)\right]^3 \left[\left(\frac{10.05 kN/m^3}{[25.9-10.05]kN/m^3}\right)^{\frac{1}{2}}\left(\frac{2.5 m/s}{\sqrt{(1.0)(9.81 m/s^2)(6m)}}\right)\right]^{\frac{15}{2}} = 1.54(10)^{-6}$$

The W_{30} weight is found as

$$W_{30} = 1.54(10)^{-6} w_a h^3 = 1.54(10)^{-6}(25.9 kN/m^3)(6m)^3 = 0.0086 kN = \underline{8.6N}(1.9lb)$$

The rest of the riprap distribution is found using Equations VI-5-135 - VI-5-140, i.e.,

$W_{50max} = 2.6\,(8.6\,N) = 22.4\,N\ (5.0\,lb)$ $W_{50min} = 1.7\,(8.6\,N) = 14.6\,N\ (3.3\,lb)$
$W_{100max} = 8.5\,(8.6\,N) = 73.1\,N\ (16.4\,lb)$ $W_{100min} = 3.4\,(8.6\,N) = 29.2\,N\ (6.6\,lb)$
$W_{15max} = 1.3\,(8.6\,N) = 11.2\,N\ (2.5\,lb)$ $W_{15min} = 0.5\,(8.6\,N) = 4.3\,N\ (1.0\,lb)$

Blanket layer thickness for underwater placement is found using Equation VI-5-142

$$r = 3.8\left(\frac{0.0086 kN}{25.9 kN/m^3}\right)^{\frac{1}{3}} = 0.26m\,(0.86 ft)$$

The calculated value for blanket thickness is less than the minimum value, so use $\underline{r = 0.5\ m}$ (1.6 ft).

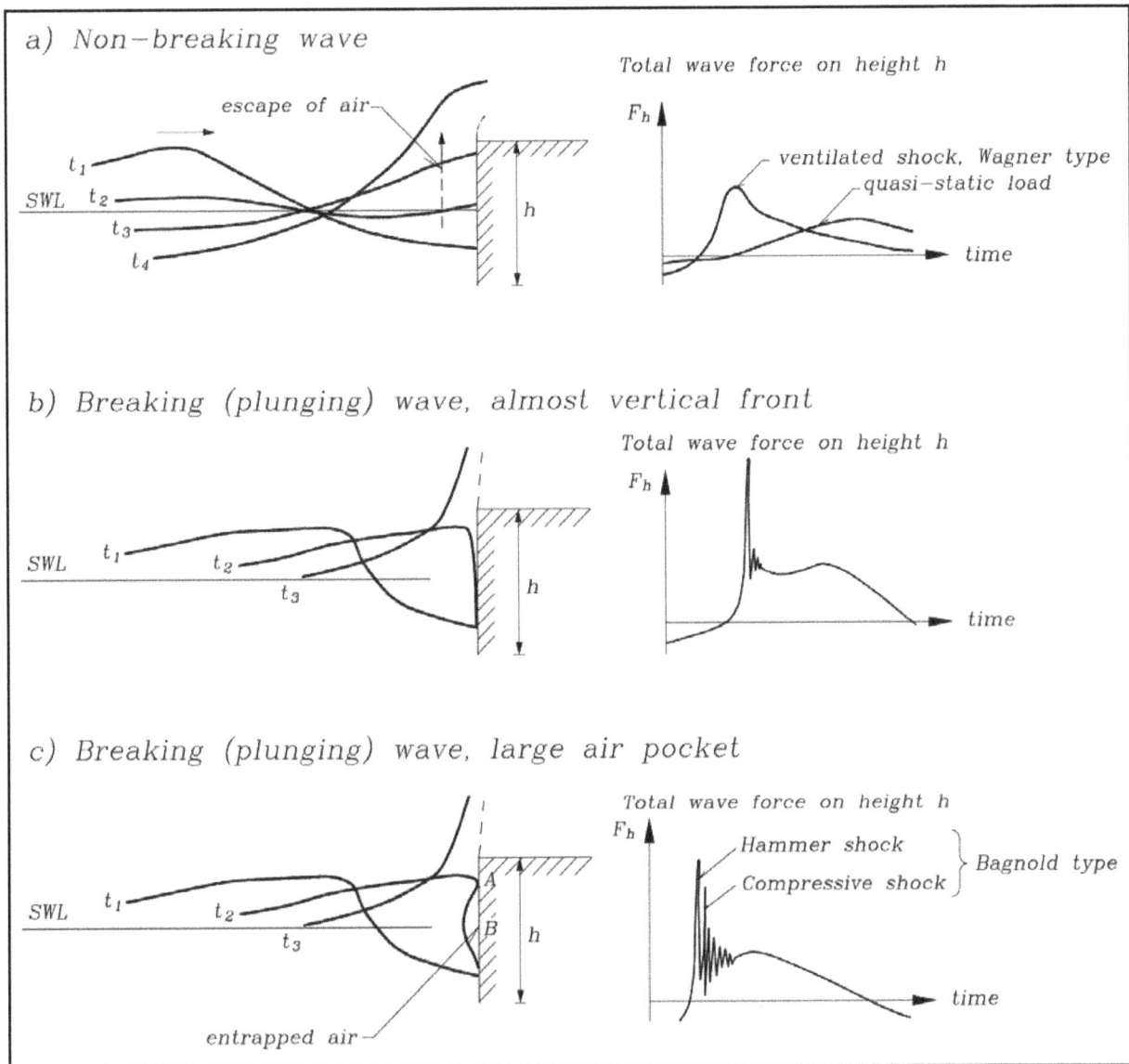

a) Non-breaking wave

Total wave force on height h

escape of air

F_h

ventilated shock, Wagner type
quasi-static load

time

b) Breaking (plunging) wave, almost vertical front

Total wave force on height h

F_h

time

c) Breaking (plunging) wave, large air pocket

Total wave force on height h

F_h

Hammer shock
Compressive shock
} Bagnold type

time

entrapped air

Figure VI-5-57. Illustration of vertical wall wave forces from nonbreaking and breaking waves

(6) It is important to investigate the effect of sloping rubble protection or any rubble foundation that extends in front of a vertical wall to make sure the slope does not trigger wave breaking, causing frequent impact loads on the wall.

(7) Figure VI-5-58 shows a system for identifying types of total horizontal wave loadings on the vertical-front structures as a function of structure geometry and wave characteristics (Kortenhaus and Oumeraci 1998). The system is based on two-dimensional model tests with irregular head-on waves. It should be noted that conditions for three-dimensional waves and oblique waves are different. Also note that the diagram does not cover situations where wave breaking takes place in a wider zone in front of the structure, i.e., typical shallow-water situations with depth-limited waves and seabeds flatter than 1:50.

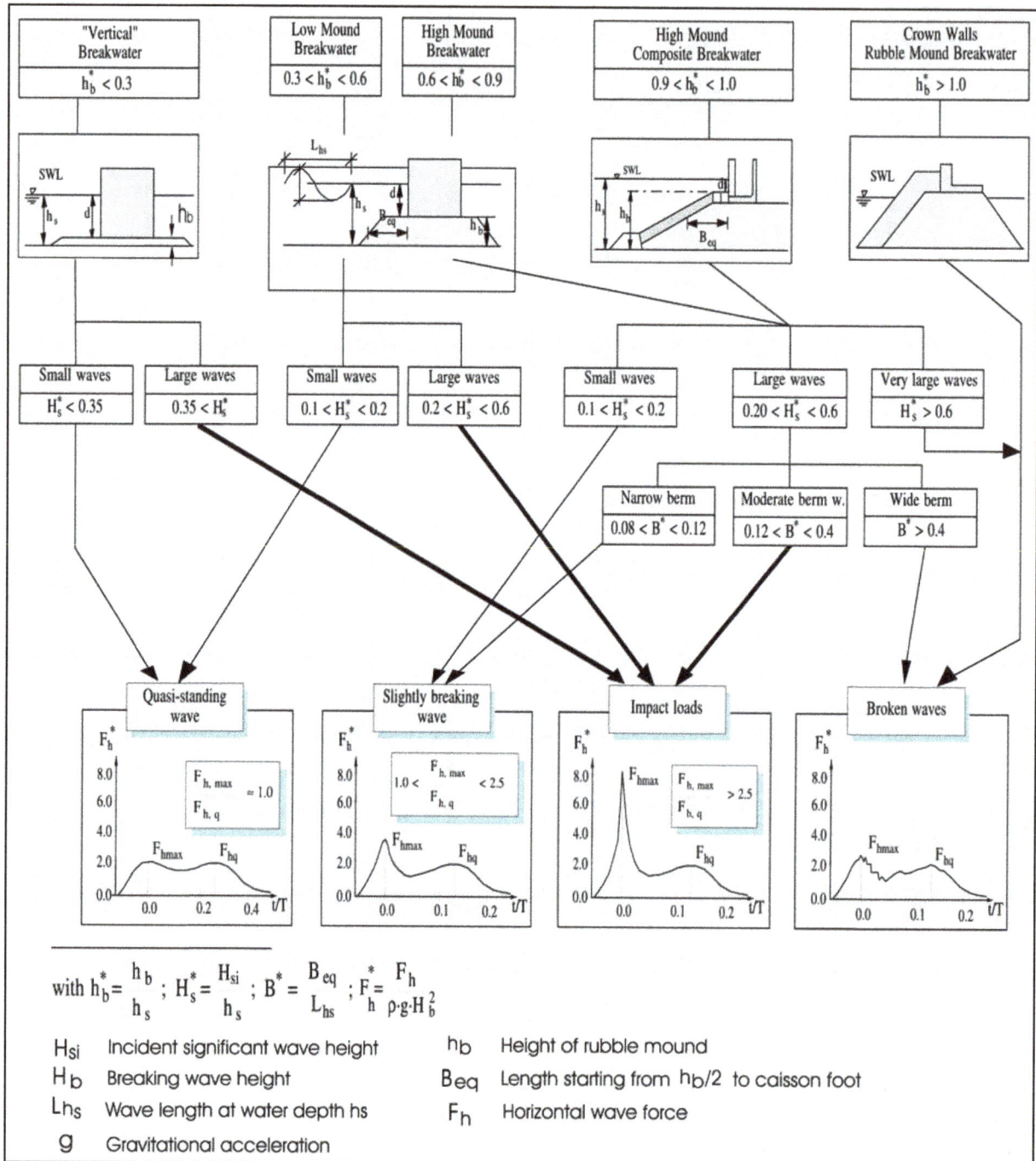

Figure VI-5-58. Identification of types of total horizontal wave loadings on vertical wall structure exposed to head-on long-crested irregular waves (Kortenhaus and Oumeraci 1998). Not valid if breaker zone is present in front of the structure

b. Wave-generated forces on vertical walls and caissons.

(1) Two-dimensional wave forces on vertical walls. Nonbreaking waves incident on smooth, impermeable vertical walls are completely reflected by the wall giving a reflection coefficient of 1.0. Where wales, tiebacks, or other structural elements increase the wall surface

roughness and retard the vertical water motion at the wall, the reflection coefficient will be slightly reduced. Vertical walls built on rubble bases will also have a reduced reflection coefficient.

(a) The total hydrodynamic pressure distribution on a vertical wall consists of two time-varying components: the hydrostatic pressure component due to the instantaneous water depth at the wall, and the dynamic pressure component due to the accelerations of the water particles. Over a wave cycle, the force found from integrating the pressure distribution on the wall varies between a minumum value when a wave trough is at the wall to a maximum values when a wave crest is at the wall as illustrated by Figure VI-5-59 for the case of nonovertopped walls or caissons.

(b) Notice in the right-hand sketch of Figure VI-5-59 the resulting total hydrodynamic load when the wave trough is at the vertical wall is less than the hydrostatic loading if waves were not present and the water was at rest. For bulkheads and seawalls this may be a critical design loading because saturated backfill soils could cause the wall to fail in the seaward direction (see Figures VI-2-63 and VI-2-71). Therefore, water level is a crucial design parameter for calculating forces and moments on vertical walls.

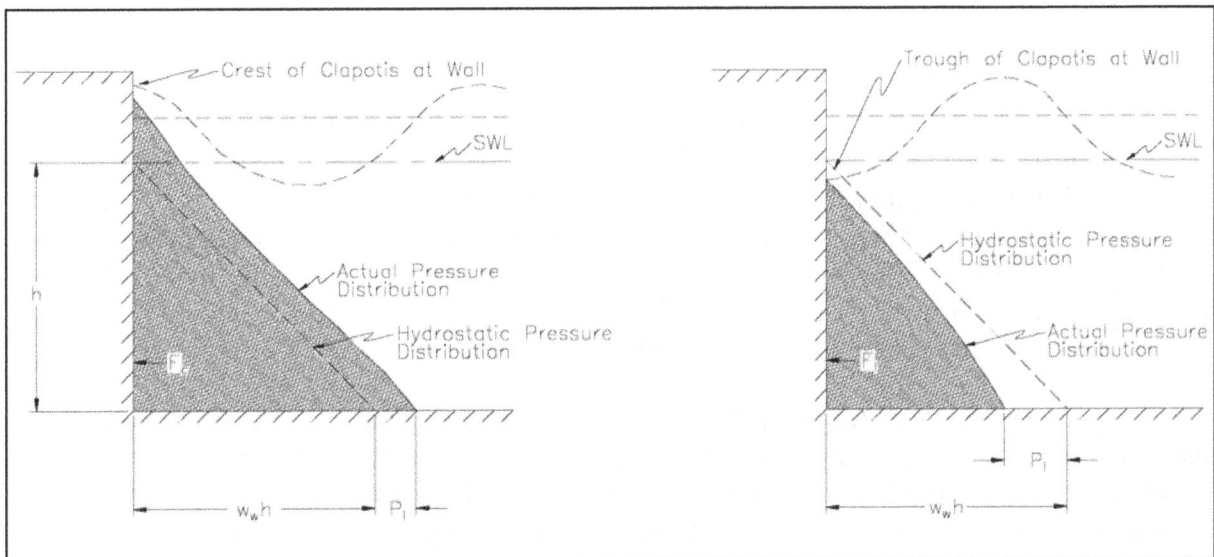

Figure VI-5-59. Pressure distributions for nonbreaking waves

(c) Wave overtopping of vertical walls provides a reduction in the total force and moment because the pressure distribution is truncated as shown schematically in Figure VI-5-60. Engineers should consider the ffect overtopping might have on land-based vertical structures by creating seaward pressure on the wall caused by saturated backfill or ponding water.

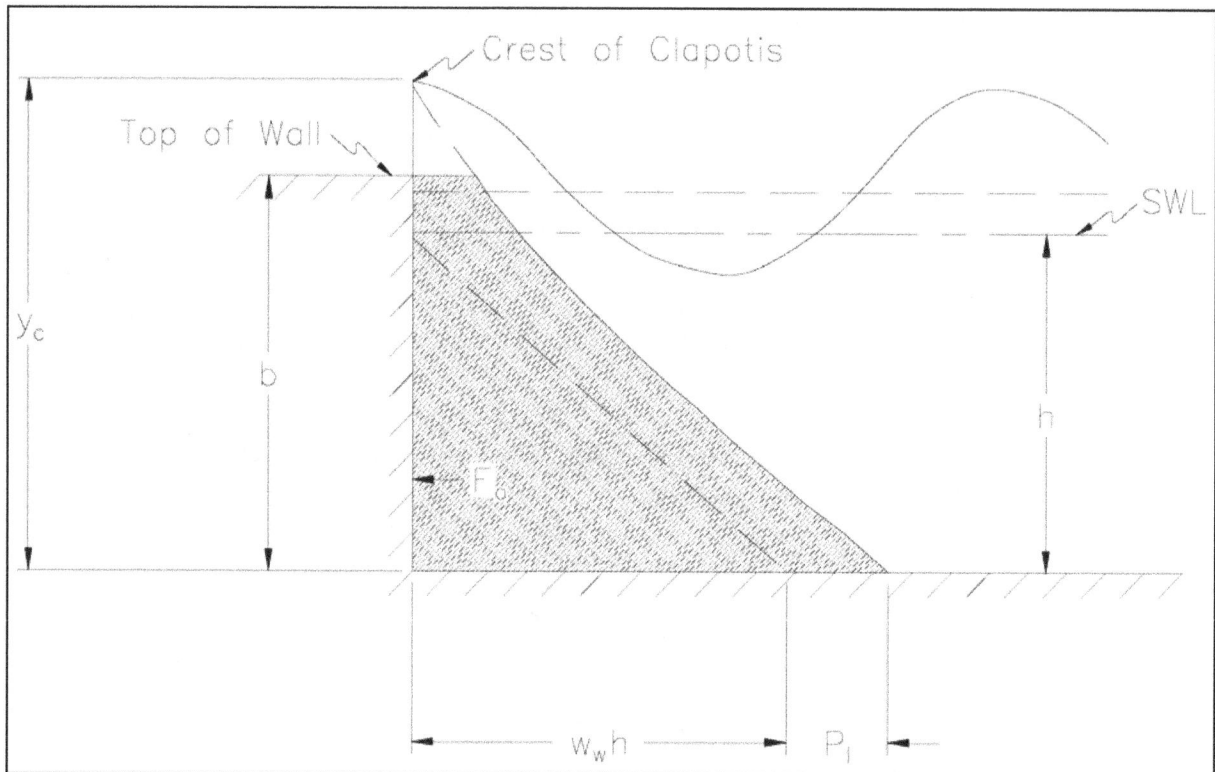

Figure VI-5-60. Pressure distributions on overtopped vertical wall

(d) This section provides formulae for estimating pressure distributions and corresponding forces and overturning moments on vertical walls due to nonbreaking and breaking waves. Most of the methodology is based on the method presented by Goda (1974) and extended by others to cover a variety of conditions. These formulae provide a unified design approach to estimating design loads on vertical walls and caissons.

(e) Important Note: All of the methods in this section calculate the pressure distribution and resulting forces and moments for only the wave portion of the hydrodynamic loading. The hydrostatic pressure distribution from the swl to the bottom is excluded (see Figure VI-5-59). For a caisson structure, the swl hydrostatic forces would exactly cancel; however, it will be necessary to include the effect of the swl hydrostatic pressure for vertical walls tied into the shoreline or an embankment.

(f) The formulae given in the following tables are exclusively based on small-scale model tests. They are presented as follows:

Formula	Waves	Structure	Table
Sainflou formula	Standing	Impermeable vertical wall	VI-5-52
Goda formula	2-D oblique	Impermeable vertical wall	VI-5-53
Goda formula, modified by Takahashi, Tanimoto, and Shimosako 1994a	Provoked breaking	Impermeable vertical wall	VI-5-54
Goda formula forces and moments	Provoked breaking	Impermeable vertical wall	VI-5-55
Goda formula modifed by Tanimoto and Kimura 1985	2-D head-on	Impermeable inclined wall	VI-5-56
Goda formula modified by Takahashi and Hosoyamada 1994	2-D head-on	Impermeable sloping top	VI-5-57
Goda formula modified by Takahashi, Tanimoto, and Shimosako 1990	2-D head-on	Horizontal composite structure	VI-5-58
Goda formula modifed by Takahashi, Tanimoto, and Shimosako 1994b	3-D head-on	Vertical slit wall	VI-5-59

Table VI-5-52
The Sainflou Formula for Head-on, Fully Reflected, Standing Regular Waves (Sainflou 1928)

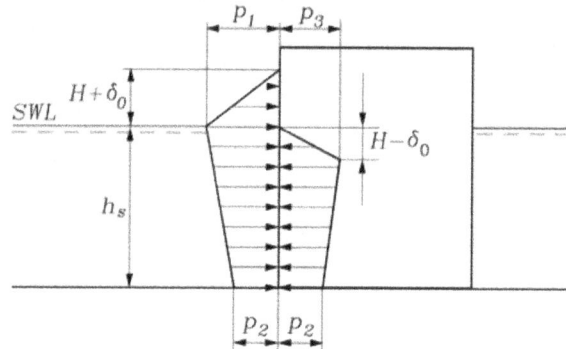

$$p_1 = (p_2 + \rho_w g h_s)\frac{H + \delta_o}{h_s + H + \delta_o} \qquad\qquad (VI-5-143)$$

$$p_2 = \frac{\rho_w g H}{\cosh(2\pi h_s/L)} \qquad\qquad (VI-5-144)$$

$$p_3 = \rho_w g(H - \delta_o) \qquad\qquad (VI-5-145)$$

$$\delta_o = \frac{\pi H^2}{L}\coth\frac{2\pi h_s}{L} \qquad\qquad (VI-5-146)$$

where H Wave height. In case of irregular waves, H should be taken as a characteristic wave height. In Japan $H_{1/3}$ is used, while in other countries $H_{1/10}$ might be used.

p_1 Wave pressure at the still water level, corresponding to wave crest

p_2 Wave pressure at the base of the vertical wall

p_3 Wave pressure at the still water level, corresponding to wave trough

δ_o Vertical shift in the wave crest and wave trough at the wall

ρ_w Water density

h_s Water depth at the foot of the structure

L Local wave length.

Remarks. The Sainflou formula for conditions under wave crest and wave trough were derived theoretically for the case of regular waves and a vertical wall. The formula cannot be applied in cases where wave breaking and/or overtopping takes place.

Table VI-5-53
Goda Formula for Irregular Waves (Goda 1974; Tanimoto et al. 1976) (Continued)

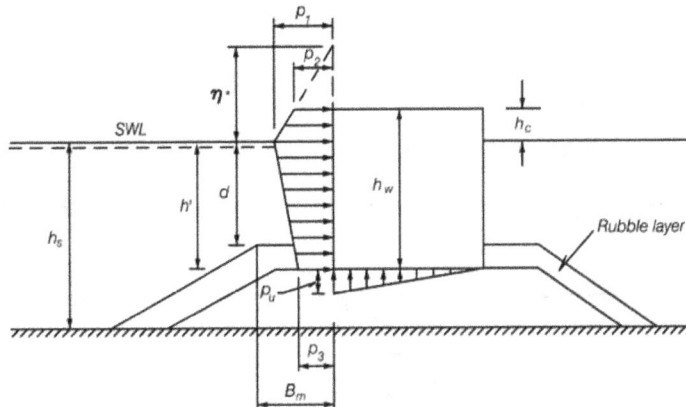

$$\eta^* = 0.75(1 + cos\beta)\, \lambda_1 H_{design} \qquad\qquad\qquad (VI-5-147)$$

$$p_1 = 0.5(1 + cos\beta)(\lambda_1 \alpha_1 + \lambda_2 \alpha_* cos^2\beta)\, \rho_w\, g\, H_{design} \qquad (VI-5-148)$$

$$p_2 = \begin{cases} \left(1 - \dfrac{h_c}{\eta^*}\right) p_1 & \text{for } \eta^* > h_c \\[2ex] 0 & \text{for } \eta^* \le h_c \end{cases} \qquad\qquad (VI-5-149)$$

$$p_3 = \alpha_3\, p_1 \qquad\qquad\qquad\qquad\qquad\qquad (VI-5-150)$$

$$p_u = 0.5(1 + cos\beta)\lambda_3 \alpha_1 \alpha_3 \rho_w g\, H_{design} \qquad\qquad (VI-5-151)$$

where

β Angle of incidence of waves (angle between wave crest and front of structure)

H_{design} Design wave height defined as the highest wave in the design sea state at a location just in front of the breakwater. If seaward of a surf zone Goda (1985) recommends for practical design a value of 1.8 H_s to be used corresponding to the 0.15% exceedence value for Rayleigh distributed wave heights. This corresponds to $H_{1/250}$ (mean of the heights of the waves included in 1/250 of the total number of waves, counted in descending order of height from the highest wave). Goda's recommendation includes a safety factor in terms of positive bias as discussed in Table VI-5-55. If within the surf zone, H_{design} is taken as the highest of the random breaking waves at a distance $5H_s$ seaward of the structure.

$$\alpha_* = \alpha_2$$

$$\alpha_1 = 0.6 + 0.5\left[\frac{4\pi h_s / L}{sinh\,(4\pi h_s / L)}\right]^2$$

$$\alpha_2 = \text{the smallest of } \frac{h_b - d}{3\,h_b}\left(\frac{H_{design}}{d}\right)^2 \text{ and } \frac{2d}{H_{design}}$$

$$\alpha_3 = 1 - \frac{h_w - h_c}{h_s}\left[1 - \frac{1}{cosh\,(2\pi h_s / L)}\right]$$

L Wavelength at water depth h_b corresponding to that of the significant wave $T_s \simeq 1.1 T_m$, where T_m is the average period.

h_b Water depth at a distance of $5H_s$ seaward of the breakwater front wall.

λ_1, λ_2 and λ_3 are modification factors depending on the structure type. For conventional vertical wall structures, $\lambda_1 = \lambda_2 = \lambda_3 = 1$. Values for other structure types are given in related tables.

(Continued)

Table VI-5-53. (Concluded)

Uncertainty and bias of formulae: See Table VI-5-55.

Tested ranges:

water depth (cm)	wave height (cm)	wave period (s)	mound height (cm)
35	7.1-31.2	2	0 & 15
45	6.7-41.6	1.7	0 & 25
45	7.6-32.8	1.3	0 & 25
45	9.2-22.9	1	0 & 25

The formulae have been calibrated with the cases of 21 slidings and 13 nonslidings of the upright sections of the prototype breakwaters in Japan.

Table VI-5-54
Goda Formula Modified to Include Impulsive Forces from Head-on Breaking Waves
(Takahashi, Tanimoto, and Shimosako 1994a)

The modification of Goda's formula concerns the formula for the pressure p_1 at the still water level (SWL). The coefficient α_* is modified as

$$\alpha_* = \text{largest of } \alpha_2 \text{ and } \alpha_I$$

$$\alpha_2 = \text{the smallest of } \frac{h_b - d}{3\,h_b}\left(\frac{H_{design}}{d}\right)^2 \text{ and } \frac{2d}{H_{design}}$$

$$\alpha_I = \alpha_{I0} \cdot \alpha_{I1}$$

$$\alpha_{I0} = \begin{cases} H_{design}/d & \text{for } H_{design}/d \le 2 \\ 2.0 & \text{for } H_{design}/d > 2 \end{cases}$$

$$\alpha_{I1} = \begin{cases} \dfrac{\cos \delta_2}{\cosh \delta_1} & \delta_2 \le 0 \\ \dfrac{1}{\cosh \delta_1 \cdot (\cosh \delta_2)^{\frac{1}{2}}} & \delta_2 > 0 \end{cases}$$

$$\delta_1 = \begin{cases} 20 \cdot \delta_{11} & \text{for } \delta_{11} \le 0 \\ 15 \cdot \delta_{11} & \text{for } \delta_{11} > 0 \end{cases}$$

$$\delta_{11} = 0.93\left(\frac{B_m}{L} - 0.12\right) + 0.36\left(\frac{h_s - d}{h_s} - 0.6\right)$$

$$\delta_2 = \begin{cases} 4.9 \cdot \delta_{22} & \text{for } \delta_{2\,2} \le 0 \\ 3 \cdot \delta_{22} & \text{for } \delta_{22} > 0 \end{cases}$$

$$\delta_{22} = -0.36\left(\frac{B_m}{L} - 0.12\right) + 0.93\left(\frac{h_s - d}{h_s} - 0.6\right)$$

where H_{design}, L, d, h_s, h_b, B_m are given in the figure and text of Table VI-5-53.

Range of tested parameters: Regular waves
bottom slope 0.01 $h_s = 42\,\text{cm}$ and $54\,\text{cm}$
$d = 7 - 39\,\text{cm}$ $B_m = 2.5 - 200\,\text{cm}$
$H = 17.2 - 37.8\,\text{cm}$ T - $1.8 - 3\,\text{sec.}$

Table VI-5-55
Resulting Wave Induced Forces and Moments, and Related Uncertainties and Bias When Calculated From Wave Load Equations by Goda and Takahashi

Per running meter of the breakwater the wave induced horizontal force, F_H, the uplift force, F_U, and the reduced weight of the vertical structure due to buoyancy, F_G, can be calculated from equations by Goda and Takahashi as follows:

$$F_H = U_{F_H} \left[\frac{1}{2}(p_1 + p_2)h_c + \frac{1}{2}(p_1 + p_3)h' \right] \qquad (VI-5-152)$$

$$F_U = U_{F_U} \cdot \frac{1}{2}p_u \cdot B \qquad (VI-5-153)$$

$$F_G = \rho_c \cdot g\, B \cdot h_w - \rho_w \cdot g\, B \cdot h' \qquad (VI-5-154)$$

where ρ_c Mass density of the structure

      ρ_w Mass density of the water

      U_{F_H} Stochastic variable signifying the bias and the uncertainty related to the horizontal force

      U_{F_U} Stochastic variable signifying the bias and the uncertainty related to the uplift force

      h' Submerged height of the wall from the toe to the still water level.

      B Vertical structure width

The corresponding moments at the heel of the caisson breakwater are:

$$M_H = U_{M_H} \left[\frac{1}{6}(2p_1 + p_3)h'^2 + \right.$$

$$\left. + \frac{1}{2}(p_1 + p_2)h' h_c + \frac{1}{6}(p_1 + 2p_2)h_c^2 \right] \qquad (VI-5-155)$$

$$M_U = U_{M_U} \cdot \frac{1}{3}p_u \cdot B^2 \qquad (VI-5-156)$$

$$M_G = \frac{1}{2}B^2 g\left(\rho_c h_w - \rho_w h'\right) \qquad (VI-5-157)$$

where U_{M_H} Stochastic variable signifying the bias and the uncertainty of the horizontal moment

      U_{M_U} Stochastic variable signifying the bias and the uncertainty of the uplift moment.

Uncertainty and bias of the Goda formulae in Table VI-5-53. Based on reanalysis by Juhl and van der Meer (1992), Bruining (1994), and van der Meer, Juhl, and van Driel (1994) of various model tests performed at Danish Hydraulic Institute and Delft Hydraulics. The mean values and standard deviations of the stochastic variables U are given as

Uncertainty and bias of horizontal wave induced force, uplift force, horizontal moment and uplift moment (vertical composite type)

Stochastic variable	Mean value	no model tests		model test performed	
		Stand. dev.		Stand. dev.	
X_i	μ_{X_i}	σ_{X_i}	$\frac{\sigma_{X_i}}{\mu_{X_i}}\%$	σ_{X_i}	$\frac{\sigma_{X_i}}{\mu_{X_i}}\%$
U_{F_H}	0.90	0.25	0.22	0.05	0.055
U_{F_U}	0.77	0.25	0.32	0.05	0.065
U_{M_H}	0.81	0.40	0.49	0.10	0.12
U_{M_U}	0.72	0.37	0.51	0.10	0.14

Table VI-5-56
Wave Loads on Impermeable Inclined Walls (Tanimoto and Kimura 1985)

Tanimoto and Kimura (1985) performed model tests and demonstrated that the Goda formula can be applied by projection of the Goda wave pressures calculated for a vertical wall with the same height (crest level) as illustrated in the figure.

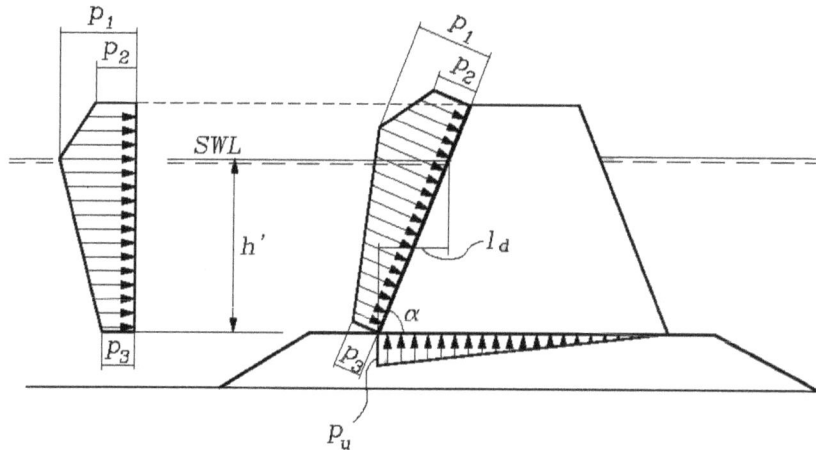

The wave induced uplift pressure on the base plate is reduced compared to the vertical face case. Consequently λ_3 for the calculation of p_u in the Goda formula is modified as

$$\lambda_3 = exp\left[-2.26(7.2\,\ell_d\,/L)^3\right] \qquad (VI-5-158)$$

where $\ell_d = h' \cot \alpha$ and L is the wavelength.

Eq VI-5-158 is valid for $\alpha \geq 70°$ and $\ell_d < 0.1L$.

Table VI-5-57
Wave Loads on Sloping Top Structures (Takahashi and Hosoyamada 1994)

Tested cross sections

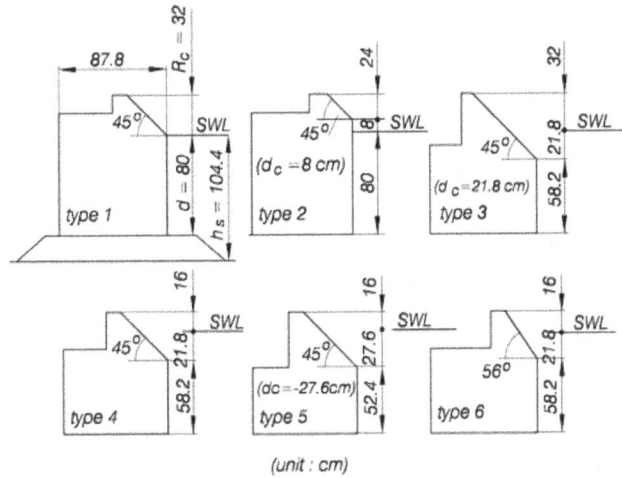

(unit : cm)

Pressure distribution

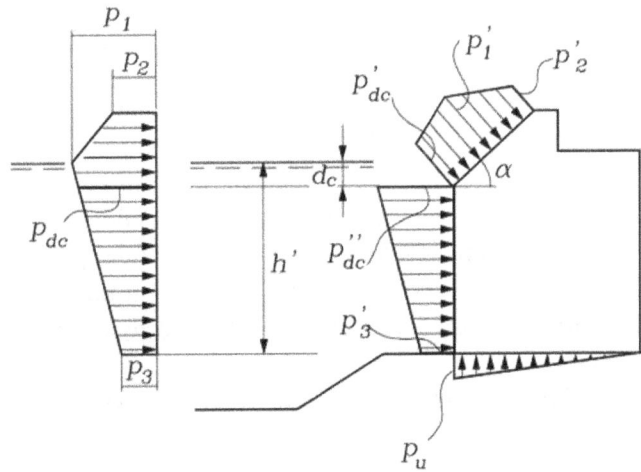

$$p_1' \;=\; \lambda_{SL}\, p_1 \sin\alpha \;,\; p_2' = \lambda_{SL}\, p_2 \sin\alpha \;,\; p_{d_c}' = \lambda_{SL}\, p_{d_c} \sin\alpha$$

$$(VI-5-159)$$

$$p_{d_c}'' \;=\; \lambda_V\, p_{d_c} \;,\; p_3' = \lambda_V\, p_3$$

where

$$\lambda_{SL} = \tfrac{1}{\sin^2\alpha}\, min[1.0 \;,\; max(\sin^2\alpha \;,\; 1 + 0.46\cos^2\alpha - 23\cos^2\alpha H/L)]$$

$$\lambda_V = min[1.0 \;,\; max(1.1. \;,\; 1.1 + 11 d_c/L) - 5.0 H/L]$$

$p_1,\; p_2,\; p_3,\; p_{d_c}$ & p_u are calculated from the Goda formula (Table VI-5-53)

Table VI-5-58
Wave Loads on Vertical Walls Protected by a Rubble-Mound Structure (Takahashi, Tanimoto, and Shimosako 1990)

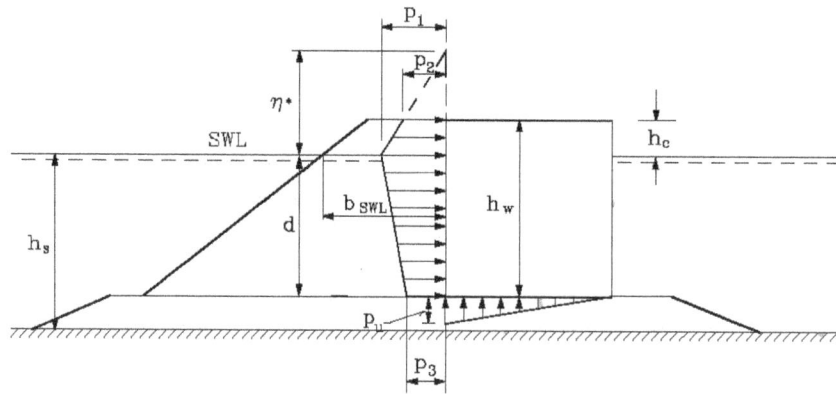

λ_1, λ_2 and λ_3 in the Goda formula are modified as:

$$\lambda_1 = \lambda_3 = \begin{cases} 1.0 & H_{design}/h_s < 0.3 \\ 1.2 - 0.67\,(H_{design}/h_s) & 0.3 \leq H_{design}/h_s \leq 0.6 \\ 0.8 & H_{design}/h_s > 0.6 \end{cases} \qquad (VI-5-160)$$

$$\lambda_2 = 0$$

Validity range: These values presume that the rubble consists of blocks of the complex types like Tetrapods and Dolosse. Also, the width of the block section at the top of the vertical wall should be no less than twice the height of a block. The front slope is approximately 1 : 1.5. The model tests cover the parameter intervals: $h_s/L_{1/3} = 0.07 - 0.11$ and $b_{SWL}/L_{1/3} = 0.046 - 0.068$.

Uncertainty and bias: From the test results the mean value, μ_{λ_1}, and the variational coefficient, $\frac{\sigma_{\lambda_1}}{\mu_{\lambda_1}}$, of λ_1 are estimated to be approximately

$$\mu_{\lambda_1} = \begin{cases} 0.90 & H_{design}/h_s < 0.4 \\ 0.90 - (H_{design}/h_s - 0.4) & 0.4 \leq H_{design}/h_s \leq 0.7 \\ 0.60 & 0.7 < H_{design}/h_s < 0.8 \end{cases} \qquad (VI-5-161)$$

and

$$\frac{\sigma_{\lambda_1}}{\mu_{\lambda_1}} = 5\% - 10\%$$

Table VI-5-59
Wave Pressures from Regular Head-on Waves on Caissons with Vertical Slit Front Face and
Open Wave Chamber (Tanimoto, Takahashi, and Kitatani 1981; Takahashi, Shimosako, and
Sakaki 1991) (Continued)

Cross sections

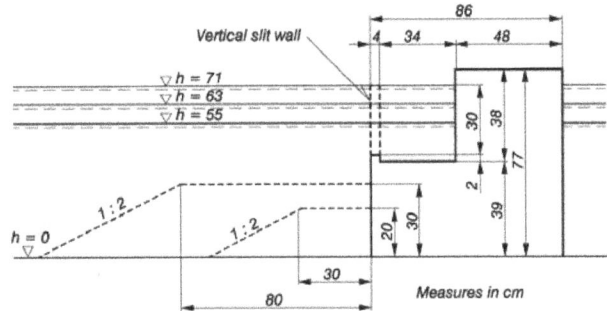

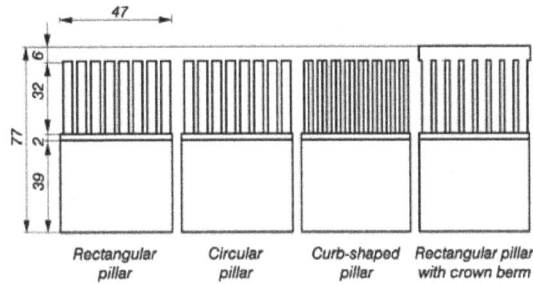

Tested wave range: Regular head-on waves
Incident wave height: 10-30 cm
Wave period: 1.5, 2.0, 2.5 s

Considered wave crest faces

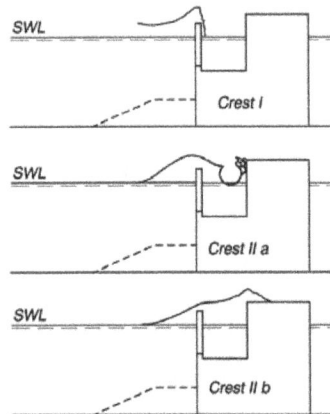

(Continued)

Table VI-5-59. (Concluded)

Pressure distribution

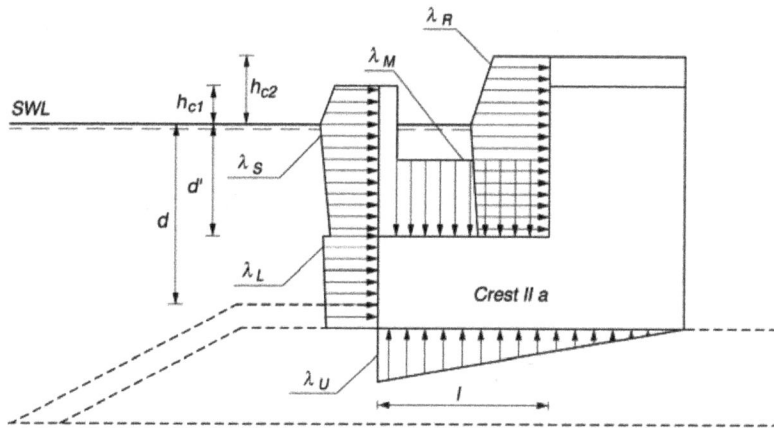

Pressure calculation: Use the Goda formulae with modified λ_1, λ_2 and λ_3 as given in the following table. For example, the wave pressure on the slit wall in the case of Crest-I is calculated by the Goda formulae with λ_1 and λ_2 replaced by λ_{S1} and λ_{S2}, respectively.

Modification factors for vertical slit wall caisson (From Takahashi et al. 1994b).

		Crest-I		Crest-IIa		Crest-IIb	
Slit wall	λ_{S1}	0.85		0.7		0.3	
	λ_{S2}	0.4	($\alpha^* \leq 0.75$)	0		0	
		$0.3/\alpha^*$	($\alpha^* > 0.75$)				
Impermeable part of front wall	λ_{L1}	1		0.75		0.65	
	λ_{L2}	0.4	($\alpha^* \leq 0.5$)	0		0	
		$0.2/\alpha^*$	$\alpha^* > 0.5$)				
Wave chamber rear wall	λ_{R1}	0		$20l/3L'$	$l/L' \leq 0.15$)	1.4	($H/h \leq 0.1$)
				1.0	($1/L' > 0.15$)	$1.6 - 2H/h$	($0.1 < H/h < 0.3$)
						1.0	($H/h \geq 0.3$)
	λ_{R2}	0		0.56	($\alpha^* \leq 25/28$)	0	
				$0.5/\alpha^*$	($\alpha^* > 25/28$)		
Wave Chamber bottom slab	λ_{M1}	0		$20l/3L'$	($l/L' \leq 0.15$)	1.4	($H/h \leq 0.1$)
				1.0	($l/L' > 0.15$)	$1.6 - 2H/h$	($0.1 < H/h < 0.3$)
						1.0	($H/h \geq 0.3$)
	λ_{M2}	0		0		0	
Uplift force	λ_{U3}	1		0.75		0.65	

In the calculation of α^* for the rear wall, α_1 should be replaced by α'_1 which is obtained with the parameters d', L' and B'_M instead of d, L and B_M respectively, where d' is the depth in the wave chamber, L' is the wavelength at water depth d, $B'_M = l - (d - d')$, and l is the width of the wave chamber including the thickness of the perforated vertical wall.

(g) Wave pressure distributions for breaking waves are estimated using Table VI-5-54, and the corresponding forces and moments are calculated from Table VI-5-55. Not included in this manual is the older breaking wave forces method of Minikin as detailed in the *Shore Protection Manual* (1984). As noted in the Shore Protection Manual the Minikin method can result in very high estimates of wave force, Aas much as 15 to 18 times those calculated for nonbreaking waves.@ These estimates are too conservative in most cases and could result in costly structures.

(h) On the other hand, there may be rare circumstances where waves could break in just the right manner to create very high impulsive loads of short duration, and these cases may not be covered by the range of experiment parameters used to develop the guidance given in Table VI-5-54. In addition, scaled laboratory models do not correctly reproduce the force loading where pockets of air are trapped between the wave and wall as shown in Figure VI-5-57. For these reasons, it may be advisable to design vertical-front structures serving critical functions according to Minikin's method given in *Shore Protection Manual* (1984).

(2) Vertical wave barriers.

(a) A vertical wave barrier is a vertical partition that does not extend all the way to the bottom as illustrated by the definition sketch in Figure VI-5-61. Wave barriers reduce the transmitted wave height while allowing circulation to pass beneath the barrier. A useful application for vertical wave barriers is small harbor protection.

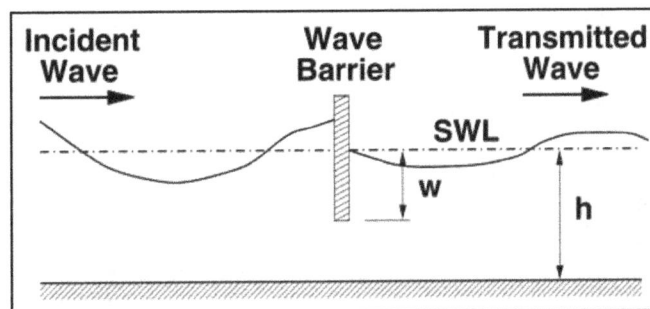

Figure VI-5-61. Wave barrier definition sketch

(b) Kriebel, Sollitt, and Gerken (1998) presented small- and large-scale laboratory measurements of forces on vertical wave barriers and found that existing methods for estimating wave forces on wave barriers overpredicted measured forces by about a factor of 2. They also presented an eigenvalue expansion method for calculating theoretical wave forces, and the predicted forces matched the experiment measurements within 10-20 percent. Both regular and irregular wave experiments were used in the analysis.

(c) Estimation of wave forces using the eigenvalue expansion method involves solving matrix equations for unknown coefficients under the physical constraints of no flow through the barrier and matching dynamic pressure in the gap beneath the barrier. However, this method must be programmed on a computer to obtain force estimates.

(d) An empirical equation for estimating forces on vertical wave barriers was developed for this manual based on the large-scale laboratory irregular wave measurements presented in Kriebel et al. (1998). Their experiments used solid vertical plates having penetration values of $w/h = 0.4$, 0.5, 0.6, and 0.7 placed in a 3-m water depth. Time series of total force on the plate were recorded, and significant force amplitudes per unit width of barrier were calculated from the zeroth-moment of the force spectra as

$$F_{mo} = \frac{1}{2}\left(4\sqrt{m_o}\right)\frac{1}{B}$$ (VI-5-162)

where m_o is the area beneath the measured force spectrum and B is the horizontal width of the barrier. The 1/2-factor arises because the force spectrum also includes forces directed seaward, which are approximately the same magnitude as the landward directed forces (Kriebel et al. 1998).

(e) The relative force measurements per unit width of barrier are shown in Figure VI-5-62. The significant force per unit width (F_{mo}) is nondimensionalized by the significant force per unit width (F_o) for a vertical wall extending over the entire depth, given by the equation

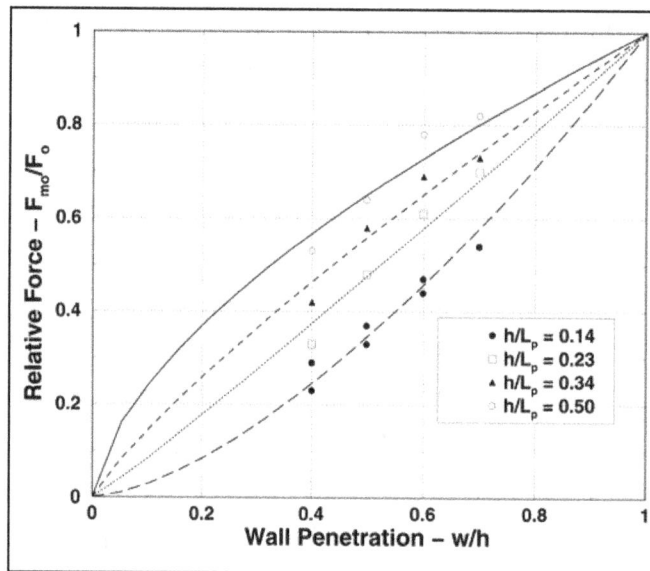

Figure VI-5-62. Best-fit to wave barrier force data

$$F_o = \rho g\, H_{mo}\, \frac{\sinh k_p h}{k_p \cosh k_p h}$$ (VI-5-163)

where

ρ = water density

g = gravity

H_{mo} = incident significant wave height

k_p = wave number associated with the spectral peak period, T_p

h = water depth at the barrier

(f) The lines in Figure VI-5-62 are best-fit curves of the form $F_{mo}/F_o = (w/h)^m$. The exponents (m) are plotted in Figure VI-5-63 as a function of relative depth, h/L_p, along with a best-fit power curve.

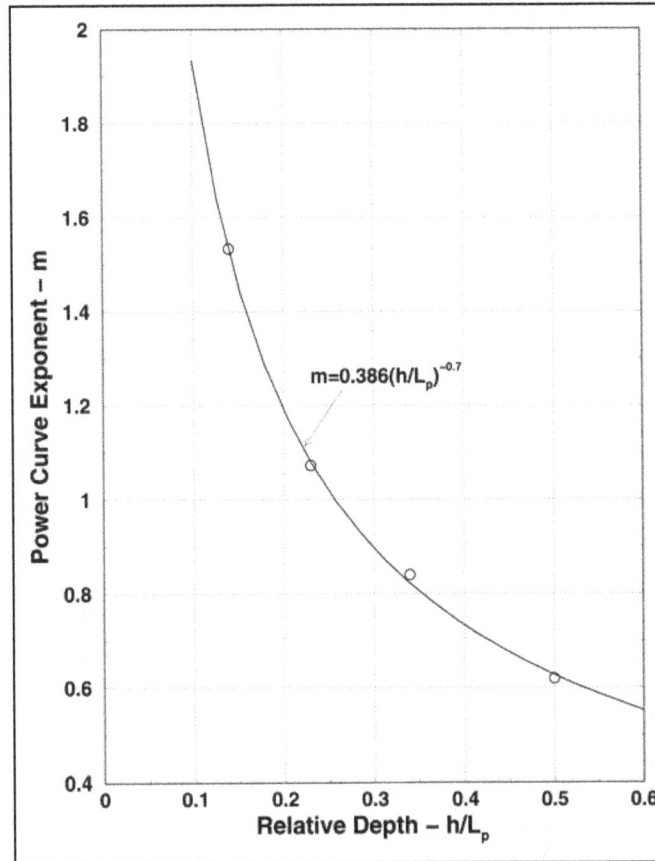

Figure VI-5-63. Power curve exponents

(g) The resulting empirical predictive equation is then given by

$$F_{mo} = F_o \left(w/h \right)^{0.386\left(h/L_p\right)^{-0.7}}$$ (VI-5-164)

where

F_{mo} = significant force per unit width of barrier

F_o = significant force per unit width of vertical wall (Equation VI-5-163) *w* = barrier penetration depth

h = water depth

L_p = local wavelength associated with the peak spectral period, T_p

(h) A comparison of the measured force values versus estimates based on the empirical Equation VI-5-164 is shown in Figure VI-5-64.

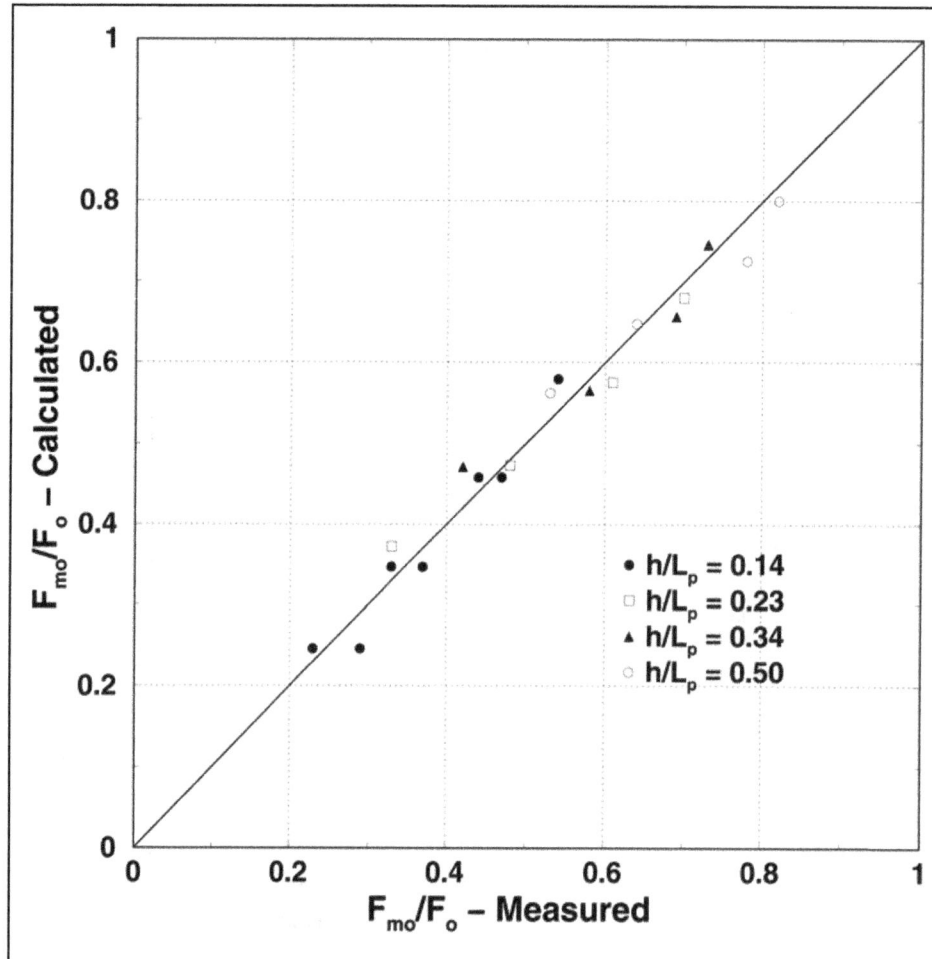

Figure VI-5-64. Comparison of Equation VI-5-139 to data used in empirical curve fits

(i) Use of Equation VI-5-164 should be limited to the range $0.4 < w/h < 0.7$ and $0.14 < h/L_p < 0.5$; however, estimates slightly outside the strict bound of the laboratory data are probably reasonable.

(j) The design force load on the vertical barrier should be the load corresponding to the design wave height, $H_{design} = 1.8\ H_s$ as recommended by Goda (1985). For Rayleigh distributed waves, $H_{design} = H_{1/250}$; and by linear superposition, we can assume that force amplitudes will also be Rayleigh distributed. Thus, the design force load is determined as

$$F_{design} = 1.8 F_{mo} \qquad\qquad (VI-5-165)$$

(3) Structure length and alignment effects on wave height.

(a) Diffraction at the head of a structure creates variations in wave heights along the structure. For a semi-infinite, fully reflecting structure exposed to nonbreaking long-crested regular waves, Ito, Tanimoto, and Yamamoto (1972) calculated the ratio of the wave height along the structure, H, to the incident wave height, H_I, as

$$\frac{H}{H_I} = \sqrt{(C+S+1)^2 + (C-S)^2} \qquad\qquad (VI-5-166)$$

where

$$C = \int_0^u \cos\left(\frac{\pi}{2}t^2\right)dt, \quad S = \int_0^u \sin\left(\frac{\pi}{2}t^2\right)dt, \quad u = 2\sqrt{\frac{2x}{L}}\sin\left(\frac{\alpha}{2}\right) \qquad (VI-5-167)$$

and x is the distance from the tip of the structure, L is the wavelength and α is the angle between the direction of wave propagation and the front alignment of the structure.

(b) Figure VI-5-65 shows an example of the wave height variation for regular head-on waves of period $T = 10$ s. Shown with the dotted line is the wave height variation calculated for nonbreaking long-crested irregular (random) waves (Bretschneider-Mitsuyasu spectrum, $T_{1/3} = 10$ s). The smoothing effect of random seas is clearly seen. At some locations the wave height exceeds twice the incident wave height expected for infinitely long vertical wall structures.

(c) For short-length breakwaters, the diffraction from both ends of the structure influences the wave height variation (see Goda 1985). Also note that experiments indicate that the theoretical assumption of complete reflection of waves from smooth vertical walls appears not fulfilled, because reflection coefficients on the order of 0.95 have been measured. (However, the methods for measuring reflection are less than perfect, as well.) Oblique waves create wave height variations different from those created by head-on waves. Concave and convex corners also affect the wave height variation along the structure (see Part VI-5-4e).

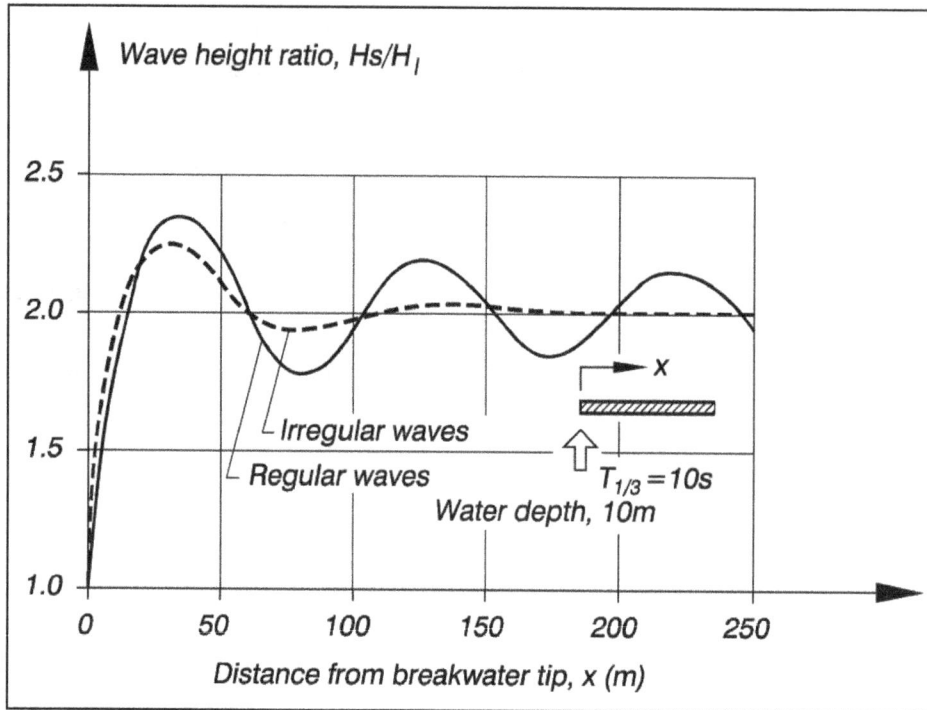

Figure VI-5-65. Variation of wave height along a semi-infinite, fully
reflecting breakwater exposed to head-on, long-crested waves
(from Goda 1985)

(4) Horizontal wave force reduction for nonbreaking waves.

(a) The effect of incident wave angle on the horizontal wave force exerted on a caisson is twofold. One effect is a force reduction, compared to head-on waves, due to the reduction of point pressure on the caisson, referred to as point-pressure force reduction. The second effect is a force reduction due to the fact that peak pressures do not occur simultaneously along the caisson, referred to as peak-delay force reduction. These two-force reduction effects will be present in short-crested waves because of spreading of the wave energy over a range of incident angles. Model test results Franco, van der Meer, and Franco (1996) with long-crested waves indicate that the point-pressure reduction can be estimated by the Goda formula.

(b) The peak-delay force reduction for oblique nonbreaking regular waves can be predicted by the Battjes formula (Battjes 1982)

$$r_F(L,\theta) = \frac{\text{max. force, wave incident angle } \theta}{\text{max. force, head-on wave}(\theta=0°)} = \frac{\sin\left(\frac{\pi L_s}{L}\sin\theta\right)}{\frac{\pi L_s}{L}\sin\theta} \tag{VI-5-168}$$

where L and L_s are the wavelength and the structure length, respectively, and θ is the wave incident angle. Equation VI-5-168 is depicted in Figure VI-5-66. (In the figure β is used instead of θ.)

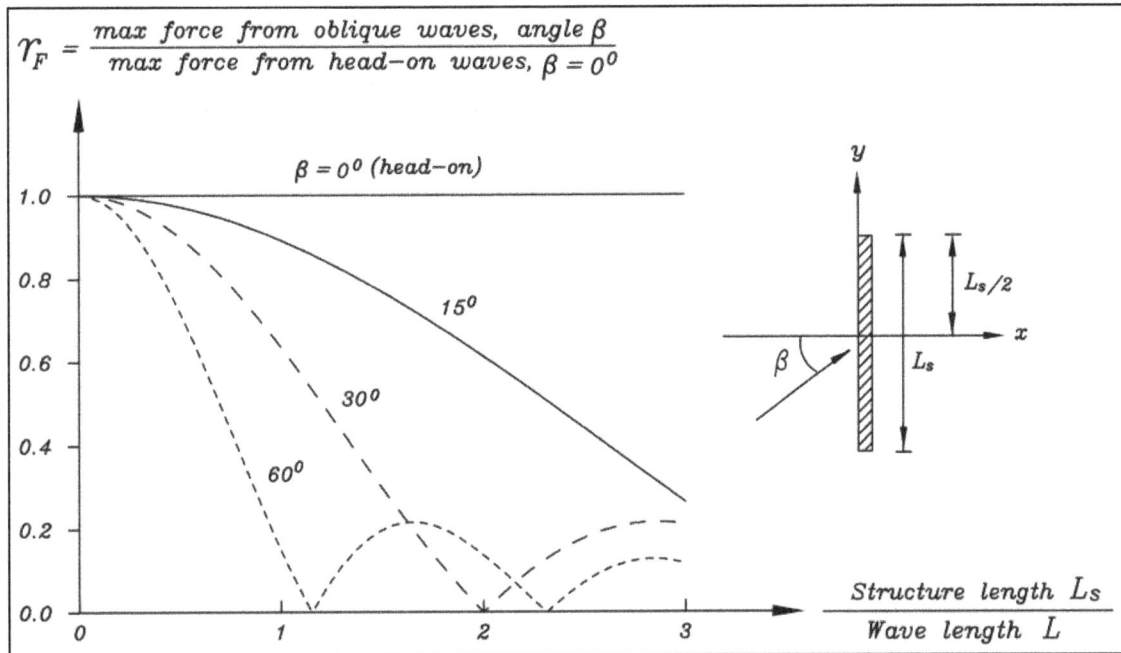

Figure VI-5-66. Peak-delay force reduction for oblique regular waves (Burcharth and Liu 1998)

(c) The peak-delay force reduction for oblique nonbreaking long-crested irregular waves can be estimated by the formula (Burcharth and Liu 1998)

$$r_F\left(L_p,\theta\right)=\frac{\text{characteristic wave force, wave incident angle }\theta}{\text{characteristic wave force, head-on wave }\left(\theta=0°\right)}=\left|\frac{\sin\left(\dfrac{\pi L_s}{L_p}\sin\theta\right)}{\dfrac{\pi L_s}{L_p}\sin\theta}\right| \qquad \text{(VI-5-169)}$$

where L_p is the wavelength corresponding to the peak frequency. For example, the characteristic wave force can be chosen as F_{max}, $F_{1/250}$, $F_{1\,percent}$, $F_{10\,percent}$, etc.

(d) In order to investigate the uncertainty and bias of Equation VI-5-169, a real-time calculation of the wave force on a caisson by nonbreaking long-crested irregular waves was performed by Burcharth and Liu (1998). The result is given in Figure VI-5-67.

(e) Figure VI-5-67 shows that Equation VI-5-169 gives a close estimate of the mean value of the peak-delay reduction. However, a large variation of the peak-delay force reduction factor corresponding to a low exceedence probability, e.g., $F_{1/250}$, was observed.

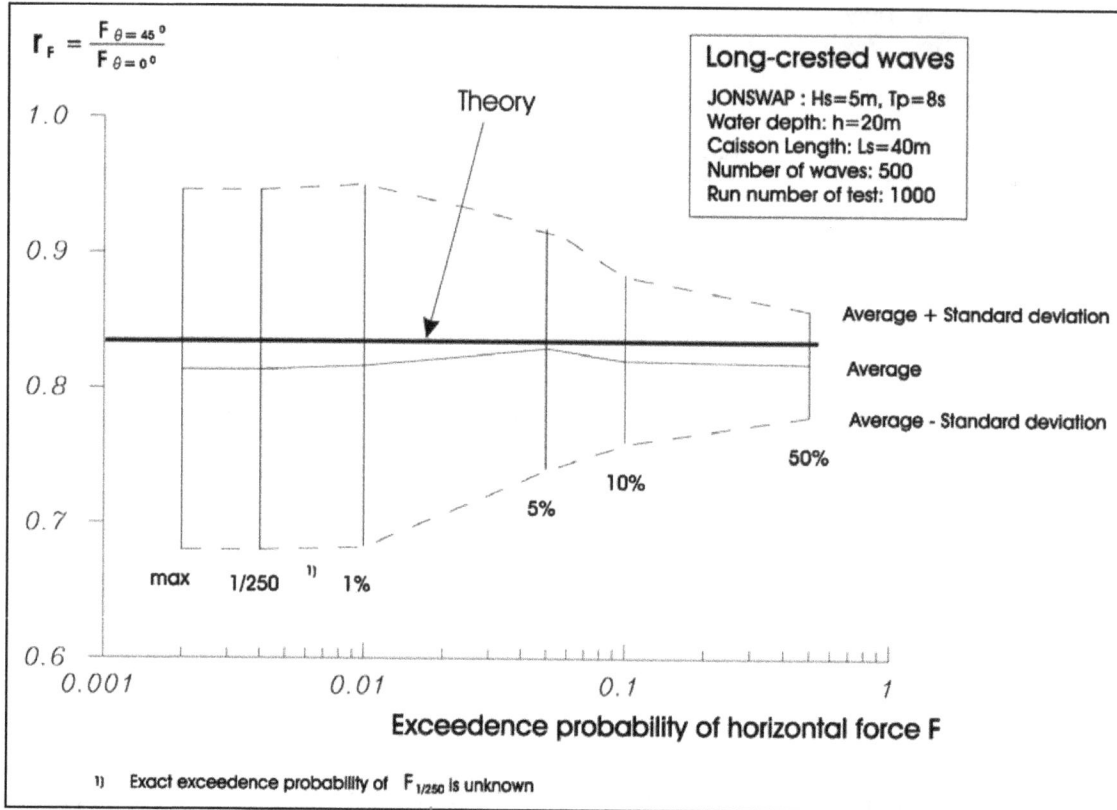

Figure VI-5-67. Numerical simulation of peak-delay reduction, long-crested waves. Example of uncertainty calculation for wave train with 500 waves (Burcharth and Liu 1998)

(f) The peak-delay force reduction for oblique nonbreaking short-crested waves can be estimated by the formula (Burcharth and Liu 1998)

$$r_F\left(\sigma,\theta_m\right)=\frac{\text{characteristic wave force, short-crested wave}}{\text{characteristic wave force, head-on long-crested wave}}\approx$$

$$\approx\left(\int_{-\pi}^{\pi}r_F\left(L_p,\theta\right)D\left(\sigma,\theta_m\right)d\theta\right)^{1/2}$$

(VI-5-170)

where $r_F\left(L_p,\theta\right)$ is given by Equation VI-5-169 and $D(\sigma,\theta_m)$ is the wave directional spreading function with the wave energy spreading angle σ and the mean wave incident direction θ_m. An example of Equation VI-5-170 is depicted in Figure VI-5-68.

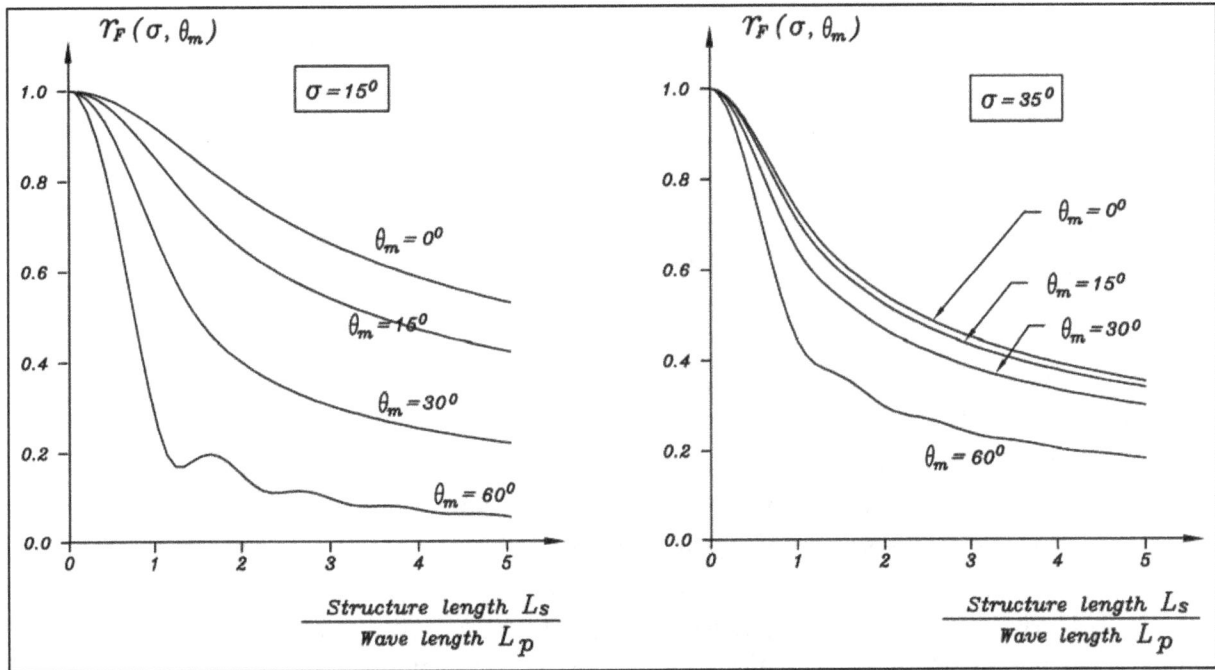

Figure VI-5-68. Example of peak-delay force reduction for short-crested waves (Burcharth and Liu 1998)

(5) Horizontal turning moment for nonbreaking waves. Oblique wave attack generates resultant wave forces acting eccentrically on the caisson front. The horizontal turning moment around the caisson center caused by oblique regular waves can be estimated by the formula (Burcharth 1998)

$$r_M = \frac{\text{max. moment, wave incident angle } \theta}{(\text{head-on max. force}) \times (\text{structure length})}$$

$$= \frac{1}{2} \left| \frac{\cos\left(\dfrac{\pi L_s}{L} \sin\theta\right)}{\dfrac{\pi L_s}{L} \sin\theta} - \frac{\sin\left(\dfrac{\pi L_s}{L} \sin\theta\right)}{\left(\dfrac{\pi L_s}{L} \sin\theta\right)} \right| \tag{VI-5-171}$$

Equation VI-5-171 is depicted in Figure VI-5-69. The maximum horizontally turning moment around caisson center under arbitrary wave incident angle is

$$M_{max} = 0.22 F_{\theta=0°} L_s \tag{VI-5-172}$$

where $F_{\theta=0°}$ is the maximum head-on wave force.

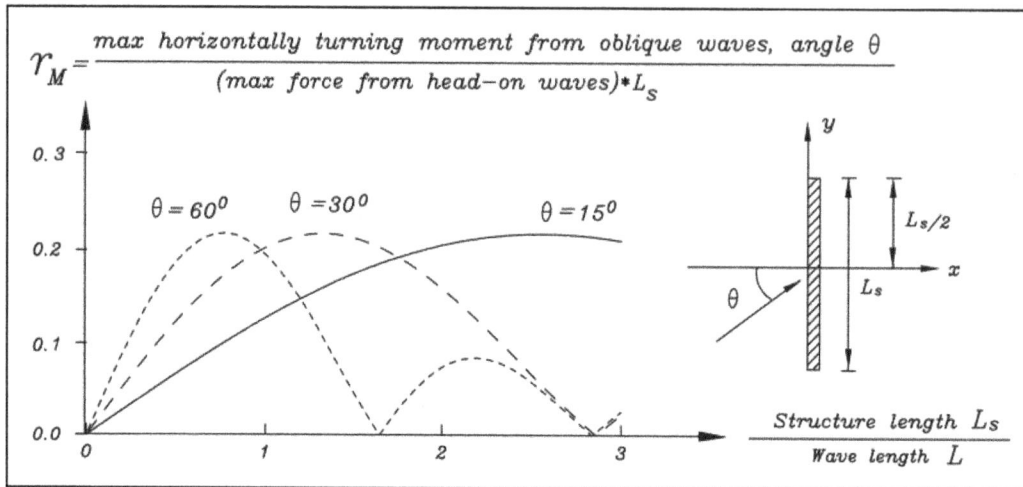

Figure VI-5-69. Nondimensional amplitude of horizontal turning moment around the center of the caisson exposed to oblique nonbreaking regular waves

(6) Horizontal wave force reduction for breaking waves. Short-crested waves break in a limited area and not simultaneously along the whole caisson, which results in an even larger force reduction in comparison with nonbreaking waves. Figure VI-5-70 shows an example of force reduction from model tests with short-crested, breaking, head-on waves, where the force reduction r_F is defined as

$$r_F = \frac{F_{1/250}, \text{short-crested wave, mean wave incident angle } \theta_m}{F_{1/250}, \text{long-crested head-on wave}} \qquad \text{(VI-5-173)}$$

(7) Broken wave forces.

(a) Shore structures may be located where they are only subjected to broken waves under the most severe storm and tide condition. Detailed studies relating broken wave forces to incident wave parameters and beach slope are lacking; thus simplifying assumptions are used to estimate design loads. Critical designs should be confirmed with physical model tests.

(b) Model tests have shown approximately 78 percent of the breaking wave height ($0.78\,H_b$) is above the still-water line when waves break on a sloping beach (Wiegel 1964). The broken wave is assumed to decay linearly from the breakpoint to the intersection of the swl with the beach slope, where the wave height is reduced to a height of $H_{swl} = 0.2\,H_b$ for beach slopes in the range $0.01 \# \tan\beta \# 0.1$ (Camfield 1991). The water mass in the broken wave is assumed to move shoreward with velocity equal to the breaking wave celery by linear theory, i.e., $C = (gh_b)^{1/2}$.

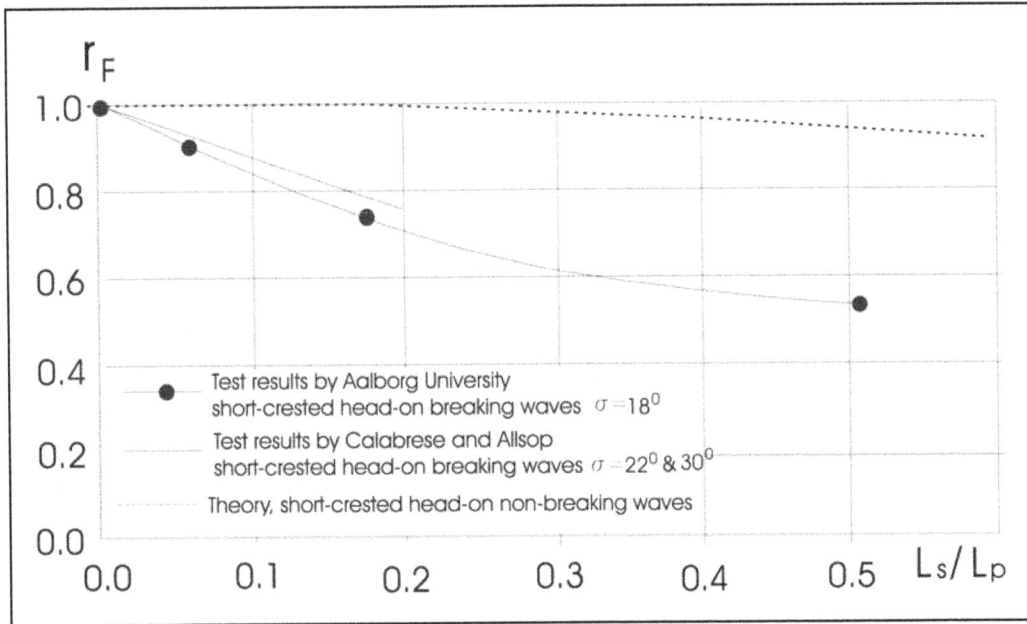

Figure VI-5-70. Example of force reduction from model tests with short-crested breaking waves (Burcharth 1998, Calabrese and Allsop 1997)

- Vertical wall seaward of the shoreline. Vertical walls situated seaward of the SWL/beach intersection are subjected to wave pressures composed of dynamic and hydrostatic pressures as illustrated in the sketch of Figure VI-5-71. The wave height at the wall, H_w, is determined by similar triangles to be

$$H_w = \left(0.2 + 0.58\frac{h_s}{h_b}\right)H_b \qquad\qquad \text{(VI-5-174)}$$

where h_s is the water depth at the wall, and h_b is the water depth at wave breaking.

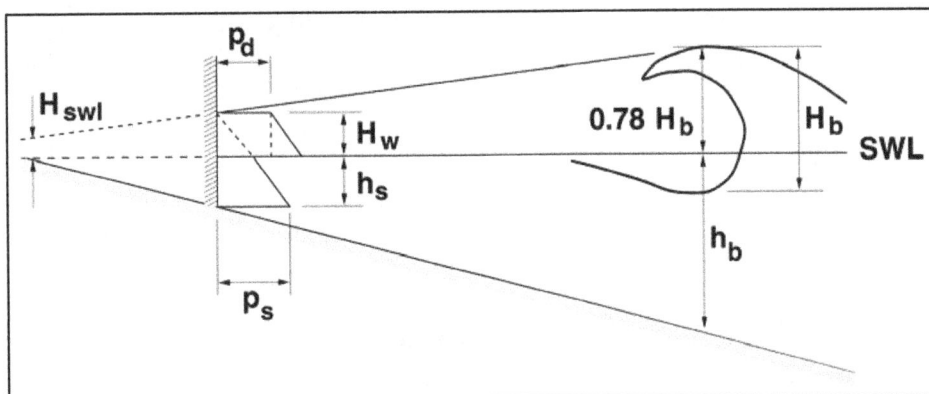

Figure VI-5-71. Broken wave forces on wall seaward of shoreline

- Above the swl, the dynamic component of the pressure is given as

$$p_d = \frac{1}{2}\rho C^2 = \frac{1}{2}\rho g h_b \qquad \text{(VI-5-175)}$$

and the corresponding force per unit horizontal length of the wall is

$$R_d = p_d H_w = \frac{\rho g h_b H_w}{2} \qquad \text{(VI-5-176)}$$

where ρ is the density of water. The overturning moment per unit horizontal length about the toe of the wall due to the dynamic pressure is given by

$$M_d = R_d\left(h_s + \frac{H_w}{2}\right) \qquad \text{(VI-5-177)}$$

- The hydrostatic pressure varies from zero at a height H_w above the SWL to a maximum at the base of the wall given by

$$P_s = \rho g\left(h_s + H_w\right) \qquad \text{(VI-5-178)}$$

- The hydrostatic force per unit horizontal width of the wall is calculated as

$$R_s = \frac{\rho g}{2}\left(h_s + H_w\right)^2 \qquad \text{(VI-5-179)}$$

and the corresponding hydrostatic overturning moment per unit width is

$$M_s = R_s\left(\frac{h_s + H_w}{3}\right) = \frac{\rho g}{6}\left(h_s + H_w\right)^3 \qquad \text{(VI-5-180)}$$

- The total force and moment per unit horizontal width of wall is the summation of dynamic and hydrostatic components, i.e.,

$$R_T = R_d + R_s \qquad \text{(VI-5-181)}$$

$$M_T = M_d + M_s \qquad \text{(VI-5-182)}$$

- Any backfilling with sand, soil or stone behind the wall will help resist the hydrodynamic forces and moments on the vertical wall.

- Vertical wall landward of the shoreline. Landward of the intersection of the SWL with the beach and in the absence of structures, the broken wave continues running up the beach slope until it reaches a maximum vertical runup height, R_a, that can be estimated using the procedures given in Part II-4-4, "Wave Runup on Beaches." If a vertical wall is located in the runup region, as shown in Figure VI-5-72, the surging runup will exert a force on the wall that is related to the height, H_w, of the surge at the wall.

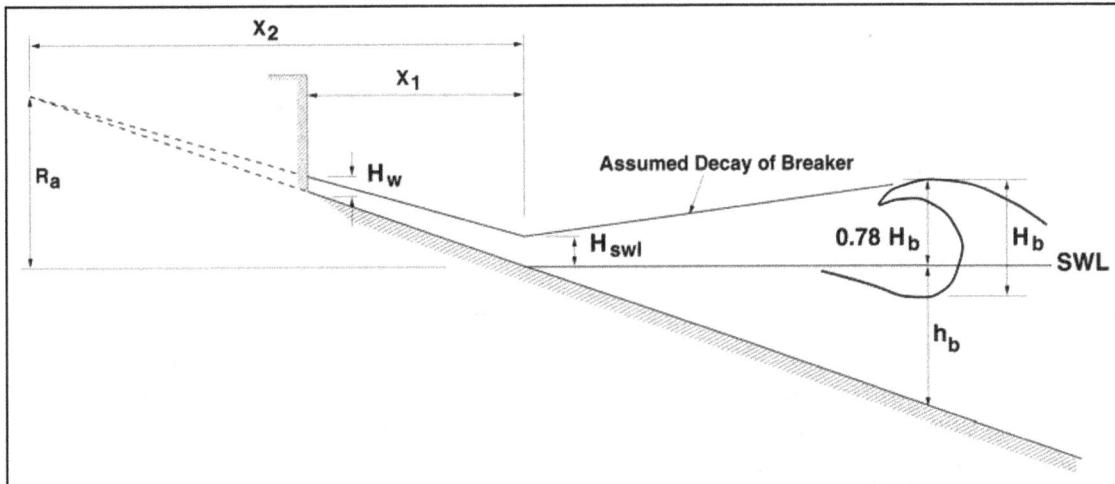

Figure VI-5-72. Broken wave forces on wall landwater of shoreline

• Camfield (1991) assumed a linear decrease in the runup surge over the distance X_2 shown in Figure VI-5-72 which yielded the following expression for surge height at the wall

$$H_w = H_{SWL}\left(1-\frac{X_1}{X_2}\right) = 0.2H_b\left(1-\frac{X_1\tan\beta}{R_a}\right) \qquad \text{(VI-5-183)}$$

where $H_{SWL} \cdot 0.2\ H_b$ and β is the beach slope angle. The force of the surge per unit horizontal width of the vertical wall was approximated by Camfield (1991) based on the work of Cross (1967) to be

$$F_{surge} \approx 4.5\rho g H_w^2 \qquad \text{(VI-5-184)}$$

or when combined with Equation VI-5-158

$$F_{surge} \approx 0.18\rho g H_b^2\left(1-\frac{X_1\tan\beta}{R_a}\right)^2 \qquad \text{(VI-5-185)}$$

• This approximate method is intended for use on plane slopes in the range $0.01 \# \tan\beta \# 0.1$. The methodology does not apply to steeper slopes or composite slopes. No estimates are given for the pressure distribution or the resulting overturning moment on the vertical wall.

c. Wave-generated forces on concrete caps.

(1) Wave loads on concrete caps occur only if the runup reaches the wall. The load is very dependent, not only on the characteristics of the waves, but also on the geometry (including the porosity) of the seaward face of the structure.

(2) The wave forces on a monolithic superstructure exposed to irregular waves are of a stochastic nature. The pressure distributions and the related resultant forces at a given instant are schematized in Figure VI-5-73. Not included in the figure is the distribution of the effective stresses on the base plate.

(3) The wave-generated pressure, p_w, acting perpendicular to the front of the wall is the pressure that would be recorded by pressure transducers mounted on the front face. The distribution of p_w is greatly affected by very large vertical velocities and accelerations which often occur. F_w is the instantaneous resultant of the wave generated pressures.

(4) The instantaneous uplift pressure, p_b, acting perpendicular to the base plate is equal to the pore pressure in the soil immediately under the plate. The resultant force is F_b. At the front corner (point f) the uplift pressure p_b^f, equals the pressure on the front wall. At the rear corner (point r) the uplift pressure, p_b^r, equals the hydrostatic pressure at point r. The actual distribution of p_b between p_b^f and p_b^r depends on the wave-generated boundary pressure field and on the permeability and homogeneity of the soil.

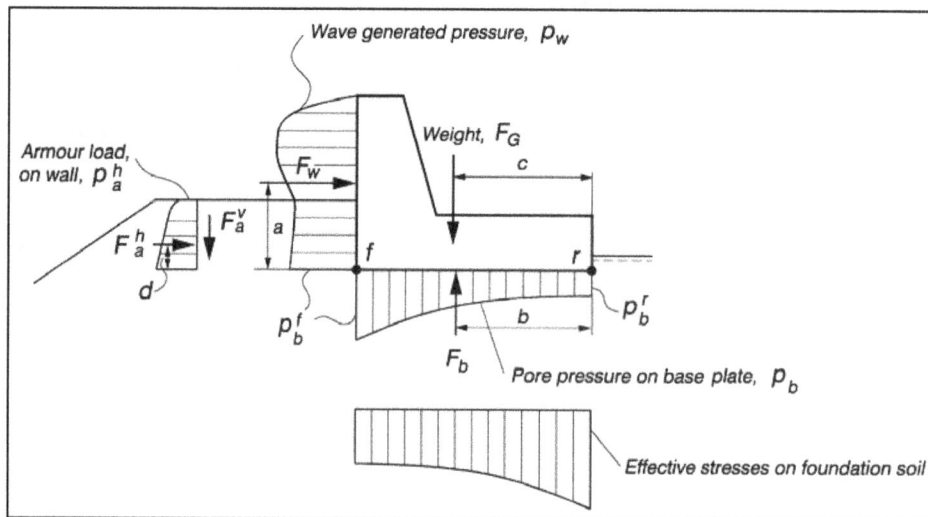

Figure VI-5-73. Illustration of forces on a superstructure

The distribution cannot be determined in normal wave flume scale tests because of strong scale effects related to porous flow. However, the corner pressures p_b^f and p_b^r can be measured or estimated, and in case of homogeneous and rather permeable soils and quasi-static conditions, a safe estimate on the most dangerous uplift can be found assuming a linear pressure distribution between a maximum value of p_b^f and a minimum value of p_b^r as shown in Figure VI-5-74a. If a blocking of the porous flow is introduced on the seaside of the base, the assumption of a linear distribution will be even safer as illustrated by Figure VI-5-74b. On the other hand a blockage under the rear end of the base plate might cause the linear assumption to be on the unsafe side as illustrated by Figure VI-5-74c. Note, that in case b and c the resultant of the base plate pressure is not vertical.

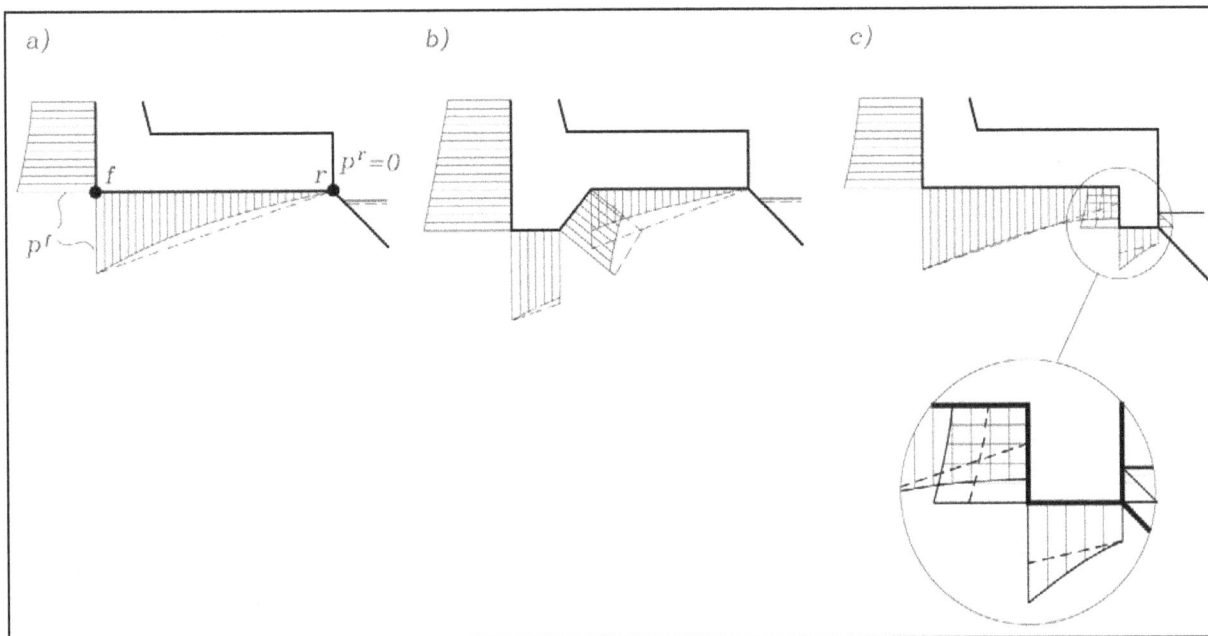

Figure VI-5-74. Illustration of comparison between base plate pore pressure distributions (under quasi-static porous flow conditions) and the approximated linear distribution

(5) Armor and filter stones resting against the front of the wave wall will introduce an armor load, p_a, on the front through the contact points. Both a normal soil mechanics load and a proportion of the dynamic wave loads on the armor contribute to p_a. The resultant force F_a is generally not perpendicular to the front wall due to friction between the soil and the wall, and must be split into the two orthogonal components F_a^h and F_a^v. In the case of high walls (low front berms) F_a is insignificant compared to the wave load, F_w.

(6) The load will in general be dynamic but is normally treated as quasi-static due to a rather smooth variation in time over a wave period. However, if wave breaking takes place directly on the wall face some short duration, but very large, slamming forces can occur, especially if the front face is almost vertical at the moment when the wave collides with the wall as shown in Figure VI-5-75. Such forces are also called impact or impulsive forces.

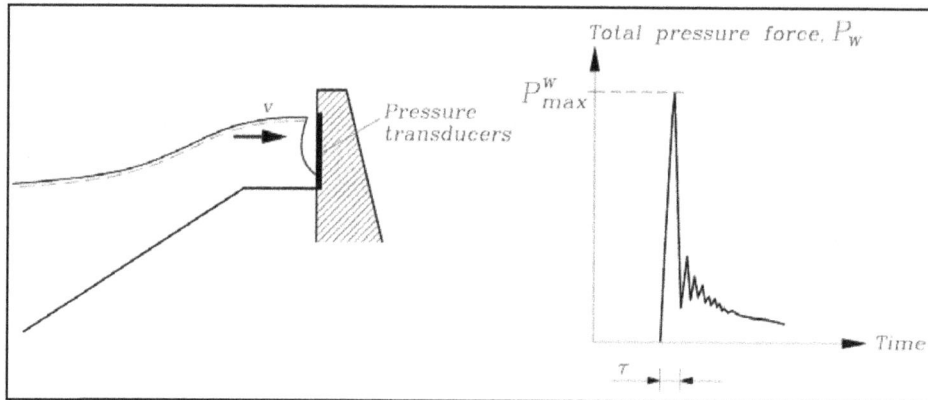

Figure VI-5-75. Impulsive pressure force caused by wave breaking on the wave wall

(7) Wave slamming on the wall can be avoided and the quasi-static wave loads reduced by increasing the crest level and/or the width of the front berm as shown by Figure VI-5-76. Wave slamming on the front of the wall will not occur in configurations *c* and *d*.

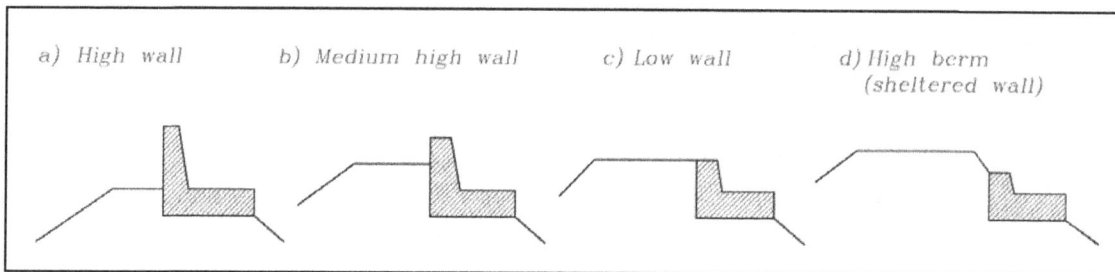

Figure VI-5-76. Typical crown wall configurations

(8) The wave loadings on a crown wall can be assessed only by physical model tests or by prototype recordings. However, no prototype results have been reported in the literature and most model test results are related to specific crown wall configurations.

(9) Table VI-5-60 shows an empirical formula for horizontal wave load given by Jensen (1984) and Bradbury et al. (1988). Table VI-5-61 shows empirical formulae for horizontal wave load, turning moment and uplift pressure presented by Pedersen (1996). The formulae are based on small scale model tests with head-on irregular waves. Predictions are compared to measurements in Figure VI-5-77.

Table VI-5-60
Horizontal Wave Force on Concrete Caps (Jensen 1984; Bradbury et al. 1988)

$$\frac{F_{h,0.1\%}}{\rho_w\, g\, h_w\, L_{op}} = \alpha + \beta \frac{H_s}{A_C} \qquad (VI-5-186)$$

where
$F_{h,0.1\%}$ Horizontal wave force per running meter of the wall corresponding to 0.1% exceedence probability
ρ_w Mass density of water
h_w Crown wall height
L_{op} Deepwater wavelength corresponding to peak wave period
H_s Significant wave height in front of breakwater
A_c Vertical distance between MWL and the crest of the armor berm
α, β Fitted coefficient, see table

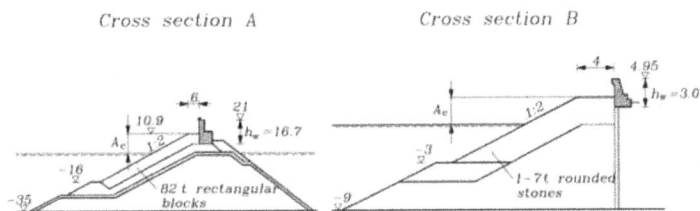

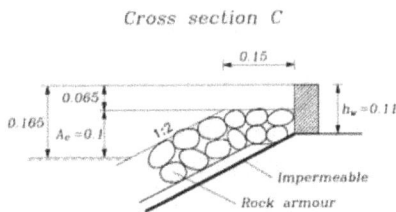

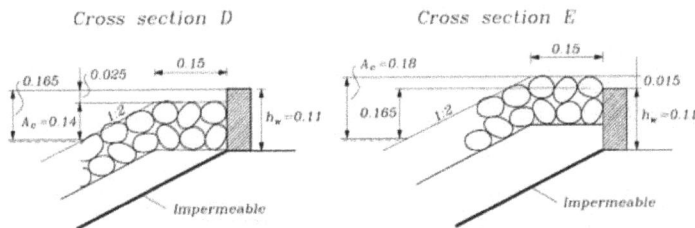

All measures in meters.

Cross section	Parameter ranges in tests			0.1% exceedence values of coefficients in Eq (VI-5-186)		Coefficient of variation	Reference
	A_c (m)	$s_{op} = \frac{H_s}{L_{op}}$	$\frac{H_s}{A_c}$	α	β		
A	5.6 - 10.6	0.016 - 0.036	0.76 - 2.5	−0.026	0.051	0.21	Jensen (1984)
B	1.5 - 3.0	0.05 - 0.011	0.82 - 2.4	−0.016	0.025	0.46	
C	0.10	0.023 - 0.07	0.9 - 2.1	−0.038	0.043	0.19	Bradbury, et al. (1988)
D	0.14	0.04 - 0.05	1.43	-0.025	0.028	—	
E	0.18	0.04 - 0.05	1.11	-0.088	0.011	—	

Table VI-5-61
Horizontal Wave Force, Uplift Wave Pressure and Turning Moment on Concrete Caps
(Pedersen 1996)

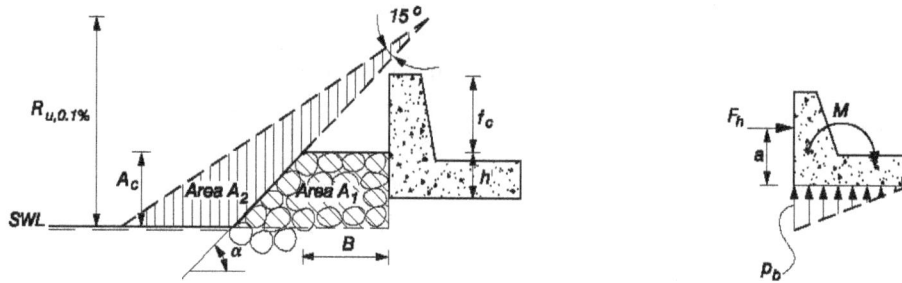

$$F_{h,0.1\%} = 0.21\sqrt{\frac{L_{om}}{B}}\left(1.6 p_m y_{eff} + A\frac{p_m}{2}h'\right) \qquad (VI-5-187)$$

$$M_{0.1\%} = a \times F_{h,0.1\%} = 0.55(h' + y_{eff})F_{h,0.1\%} \qquad (VI-5-188)$$

$$p_{b,0.1\%} = 1.00\ A\ p_m \qquad (VI-5-189)$$

where
$F_{h,0.1\%}$	Horizontal wave force per running meter of the wall corresponding to 0.1% exceedence probability
$M_{0.1\%}$	Wave generated turning moment per running meter of the wall corresponding to 0.1% exceedence probability
$p_{b,0.1\%}$	Wave uplift pressure corresponding to 0.1% exceedence probability
L_{om}	Deepwater wavelength corresponding to mean wave period
B	Berm width of armor layer in front of the wall
p_m	$p_m = \rho_w g (R_{u,0.1\%} - A_c)$
$R_{u,0.1\%}$	Wave runup corresponding to 0.1% exceedence probability

$$R_{u,0.1\%} = \begin{cases} 1.12 H_s \zeta_m & \zeta_m \le 1.5 \\ 1.34 H_s \zeta_m^{0.55} & \zeta_m > 1.5 \end{cases}$$

$$\zeta_m = \tan\alpha / \sqrt{H_s/L_{om}}$$

α	Slope angle of armor layer
A_c	Vertical distance between MWL and the crest of the armor berm
A	$A = \min\{A_2/A_1, 1\}$, where A_1 and A_2 are areas shown in the figure
y_{eff}	$y_{eff} = \min\{y/2, f_c\}$

$$y = \begin{cases} \frac{R_{u,0.1\%}-A_c}{\sin\alpha}\frac{\sin 15^0}{\cos(\alpha-15^0)} & y > 0 \\ 0 & y \le 0 \end{cases}$$

h'	Height of the wall protected by the armor layer
f_c	Height of the wall not protected by the armor layer

Uncertainty of the formulae

factor in the formulae	0.21	1.6	0.55	1.00
standard deviation σ	0.02	0.10	0.07	0.30

Tested range: See Table VI-5-12

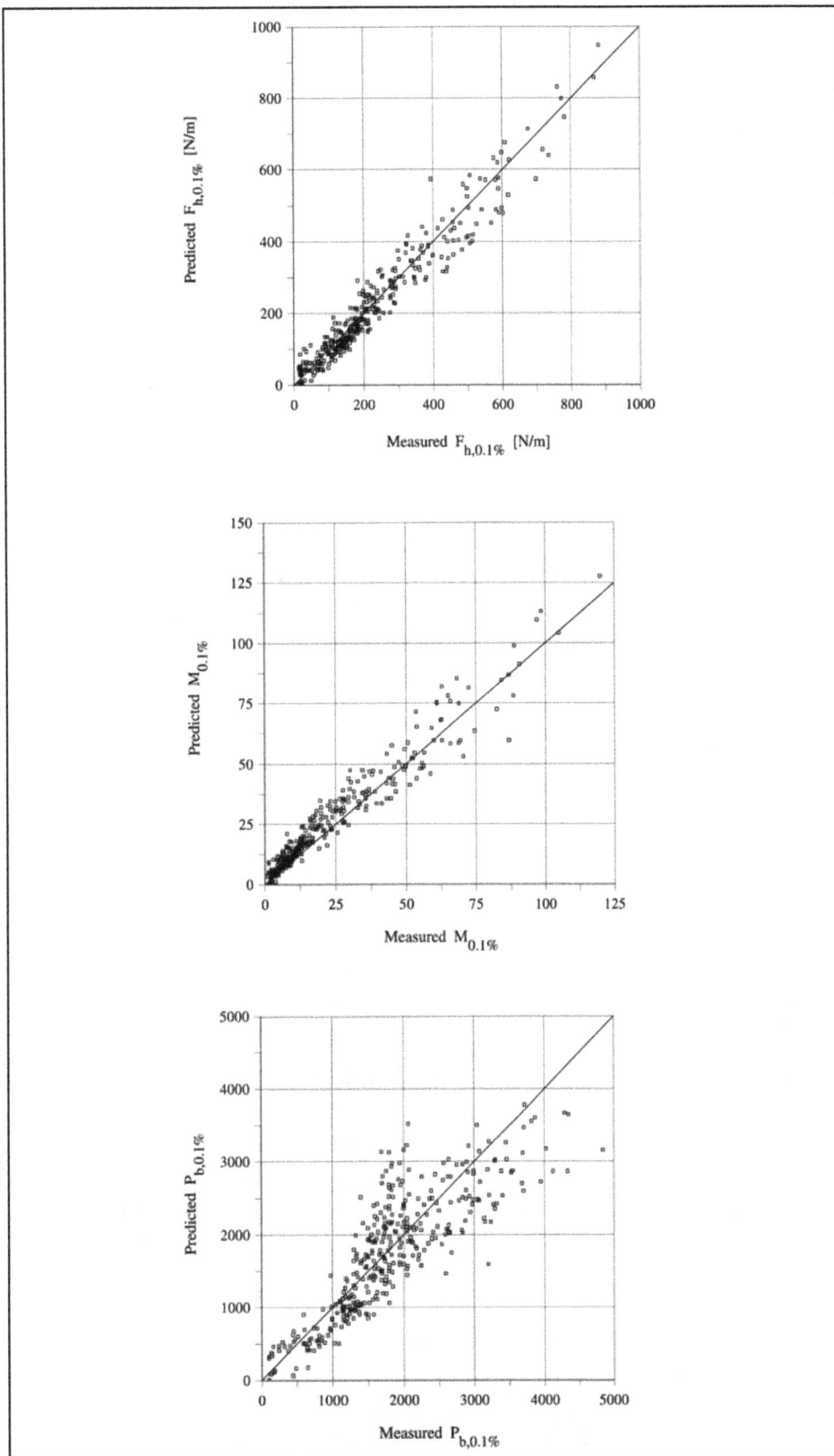

Figure VI-5-77. Comparison of predictions to measurements using the methods in Table VI-5-61 (from Pedersen 1996)

   d. Stability of concrete caps and caissons against sliding and overturning.

   (1) Stability against sliding between the caisson base and the rubble foundation requires

$$(F_G - F_U)\mu \geq F_H \qquad \text{(VI-5-190)}$$

where

   μ = friction coefficient for the base plate against the rubble stones

   F_G = buoyancy-reduced weight of the caisson

   F_U = wave induced uplift force

   F_H = wave induced horizontal force

   (2) Overturning can take place only when the heel pressure under the caisson is less than the bearing capacity of the foundation. If the caisson is placed on rubble stones and sand it is unlikely that overturning will occur. Instead there will be soil mechanics failure. Overturning is a realistic failure mode only if the caisson is placed on rock or on very strong clay, in which case breakage of the caisson is likely to occur.

   (3) Stability against overturning is maintained if

$$M_{FG} \geq M_{FU} + M_{FH} \qquad \text{(VI-5-191)}$$

where

   M_{FG} = stabilizing moment around the heel by buoyancy-reduced weight of the caisson

   M_{FU} = antistabilizing moment by wave induced uplift force

   M_{FH} = antistabilizing moment by wave induced horizontal force

   (4) The value of the friction coefficient μ has been investigated in models and in prototype studies. For a plane concrete slab resting on quarried rubble stones, Takayama (1992) found as an average a static friction coefficient of $\mu = 0.636$ and a coefficient of variation of 0.15. Table VI-5-62 taken from Stückrath (1996), presented experimental test results of friction coefficients conducted in Japan.

   (5) French tests (Cété-Laboratoire Régional Norde-Pas de Calais 1990) give a somewhat lower friction coefficient as shown in Table VI -5-63.

(6) Morihira, Kihara, and Horikawa[1] investigated the dynamic friction coefficient between caissons with different bottom patterns and rubble foundation with different levelling as shown in Table VI-5-64.

Table VI-5-62
Experimental Test Results of Friction Coefficient Conducted in Japan
(taken from Stückrath 1996)

No.	Stone type	Stone size (mm)	Condition of mound	μ	Average of μ
1	Crushed stone	30	Screeded surface	0.460-0.801	-
2	Rubble stone	120	Not screeded	0.564-0.679	0.624
3	Rubble stone	50	Surface smoothed with smaller stone	0.45-0.69	-
4	Rubble stone	30-80	Screeded	0.77-0.89	0.82
5	Cobble stone	30-50	Not screeded	0.69-0.75	0.70
6	Crushed stone	20-30	Not screeded	0.607-0.790	0.725
7	Crushed stone	10-50	Not screeded	0.486-0.591	0.540
8	Crushed stone	13-30	Not uniform	0.41-0.56	-

Table VI-5-63
Experimental Test Results of Friction Coefficient
(Cété-Laboratoire Régional Norde-Pas de Calais 1990)

Vertical Load (tonne)	Normal Stress (tonne/m^2)	Horizontal Force (tonne)		Friction Coefficient μ	
		Smooth	Corrugated	Smooth	Corrugated
Natural Sea Gravel 20-80 mm					
24.1	10.5	12.6	13.7	0.53	0.58
18.4	8	10.3	11.3	0.56	0.62
Crushed Gravel 0-80 mm					
24.1	10.5		10.4		0.43
18.4	8		8.6		0.47

---

[1] Personal Communication, 1998, M. Marihira, T. Kihara, and H. Horikawa. "On the Friction Coefficients Between Concrete Block Sea Walls and Rubble-Mound Foundations."

Table VI-5-64
Dynamic Friction Coefficient Between Caisson Bottom and Rubble-Mound (Morihira, Kihara, and Harikawa, personal communication 1998)

Levelling method for rubble mound

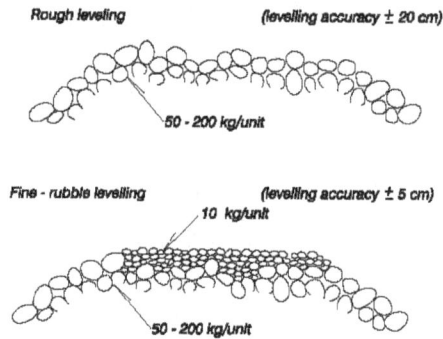

Bottom pattern of caissons

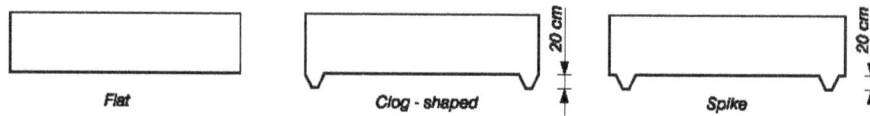

Table of Dynamic friction coefficients

Levelling	Bottom pattern	μ_{max}	$\overline{\mu_{max}}$	μ_{const}	μ			
					S=5 cm	S=10 cm	S=20 cm	S=30 cm
rough	flat	0.75	0.70	0.70	0.53	0.59	0.65	0.70
		0.73	0.70	0.70	0.70	0.70	0.70	0.70
	clog-shaped	1.19	1.13	1.16	0.76	0.91	0.98	1.08
		1.11	1.02	1.01	0.76	0.90	1.01	1.00
	spike	0.85	0.79	0.80	0.62	0.80	0.80	0.80
		0.97	0.81	0.84	0.70	0.70	0.83	0.95
	clog-shaped with	1.45	1.36	>1.4	1.11	1.30	1.41	
	foot protection	1.34	1.19	>1.3	0.94	1.09	1.28	
fine	flat	0.68	0.63	0.65	0.63	0.64	0.64	0.55
		0.70	0.60	0.60	0.59	0.60	0.60	0.60
	clog-shaped	1.18	1.08	1.08	0.95	1.03	1.08	1.08
		1.15	1.01	1.06	0.90	0.94	0.97	1.04
	spike	0.87	0.78	0.82	0.72	0.72	0.75	0.82
		1.04	0.87	0.82	0.78	0.95	1.01	0.85

μ_{max} dynamic friction coefficient corresponding to maximum tensile load
$\overline{\mu_{max}}$ dynamic friction coefficient corresponding to the average of the peak tensile loads
μ_{const} dynamic friction coefficient corresponding to constant tensile load
μ dynamic friction coefficient corresponding to caisson displacement S
S caisson displacement

e.   Waves at structure convex and concave corners. Many projects have coastal structures featuring concave or convex bends or sharp corners corresponding to structure realignment. Usually, the location and curvature of corners are determined by functional design factors, such as harbor layout or proposed channel alignment, or by site considerations, such as bathymetry. Regardless of the functional design motivation, structure bends and corners must meet or exceed the same design criteria as the rest of the structure. The orientation of bends and corners relative to the incident waves may cause changes in the local wave characteristics due to refraction, reflection, and focusing effects. Changes in wave heights could affect armor stability on the corner section, and local crest elevation may have to be heightened to prevent increased overtopping. Convex corners and bends are defined as having an outward bulge facing the waves, whereas concave corners and bends have a bulge away from the waves. Figure VI-5-78 illustrates convex and concave configurations for vertical-wall structures. Similar definitions are used for sloping-front structures.

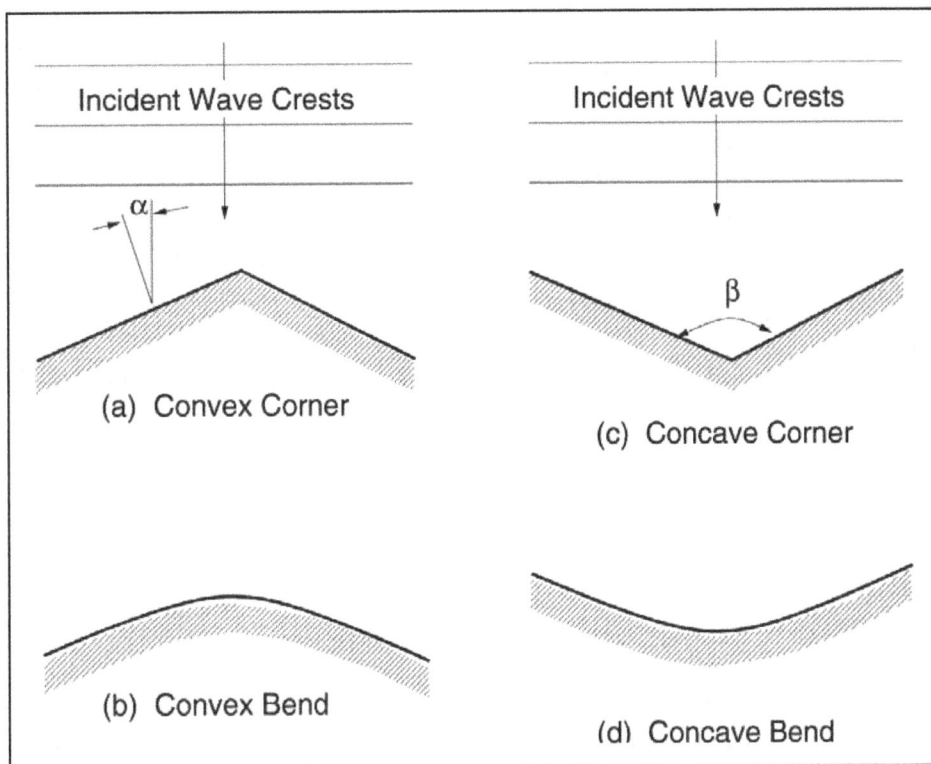

Figure VI-5-78. Convex and concave corners and bends at vertical walls

(1) Waves at convex corners.

(a) Vertical structures with convex corners. Waves approaching vertical walls with sharp convex corners such as depicted in Figure VI-5-78a will be almost perfectly reflected if the wall is impervious. This results in a diamond-like wave pattern of incident and reflected waves with the wave crests and troughs at the wall appearing to move along the wall. The maximum wave height at the wall depends on the incident wave height, H_i, angle of wave approach, α, and wave nonlinearity.

- Perroud (1957) performed laboratory tests of solitary waves obliquely reflected by a vertical wall. He observed "normal reflection" with the angle of reflection nearly the same as the incident wave angle for cases where the incident wave angle, α (defined in Figure VI-5-78), was less than about 45 deg. This is the same result given by linear wave theory for oblique reflection. The reflected wave height was just slightly less than the incident wave height for small incident angles, and it decreased as angle of incidence increased. This is contrary to linear wave theory. The maximum wave height at the wall was about twice the incident wave height up to $\alpha = 45$ deg, similar to linear wave theory for oblique reflection.

- For wave incident angles between about 45 deg and 70 deg Perroud observed a phenomenon referred to as "Mach reflection" in acoustics. Mach reflection of water waves is a nonlinear effect characterized by the presence of a reflected wave and a "Mach" wave with its crest propagating perpendicular to the vertical wall. The reflected wave height is significantly less than the incident wave height, and the angle of the reflected wave becomes less than the incident wave angle. The Mach reflection wave grows in length as it moves along the wall, and the maximum wave height, known as the "Mach stem" occurs at the wall.

- Figure VI-5-79 presents Perroud's (1957) averaged results for solitary waves obliquely reflected by a vertical wall. The upper plot shows the wave height at the wall in terms of the incident wave height for increasing angle of wave incidence. The ratio of reflected to incident wave height is shown in the lower plot. These plots are also given by Wiegel (1964) along with additional plots showing the decrease in reflected wave angle for Mach reflection and the increasing length of the Mach reflection wave with distance along the wall. (Note: In Wiegel (1964) the plots are given in terms of a differently defined angle of wave incidence i which is related to α via ($i = 90^\circ - \alpha$).)

- The speed of the Mach stem, C_M, was given as (Camfield 1990)

$$C_M = \frac{C}{\sin \alpha} \tag{VI-5-192}$$

where C is the incident wave celerity.

- For angles of incidence greater than 70 deg from normal, Perroud observed that the wave crest bends so it is perpendicular to the vertical wall, and no discernible reflected wave appears. The wave height at the wall decreases with continuing increase in angle of incidence as indicated in Figure VI-5-79a.

- Keep in mind that the experimental results were obtained for Mach reflection of solitary waves. This implies that the results represent the shallow-water limiting case. The Mach reflection effect will decrease for smaller amplitude waves in deeper water.

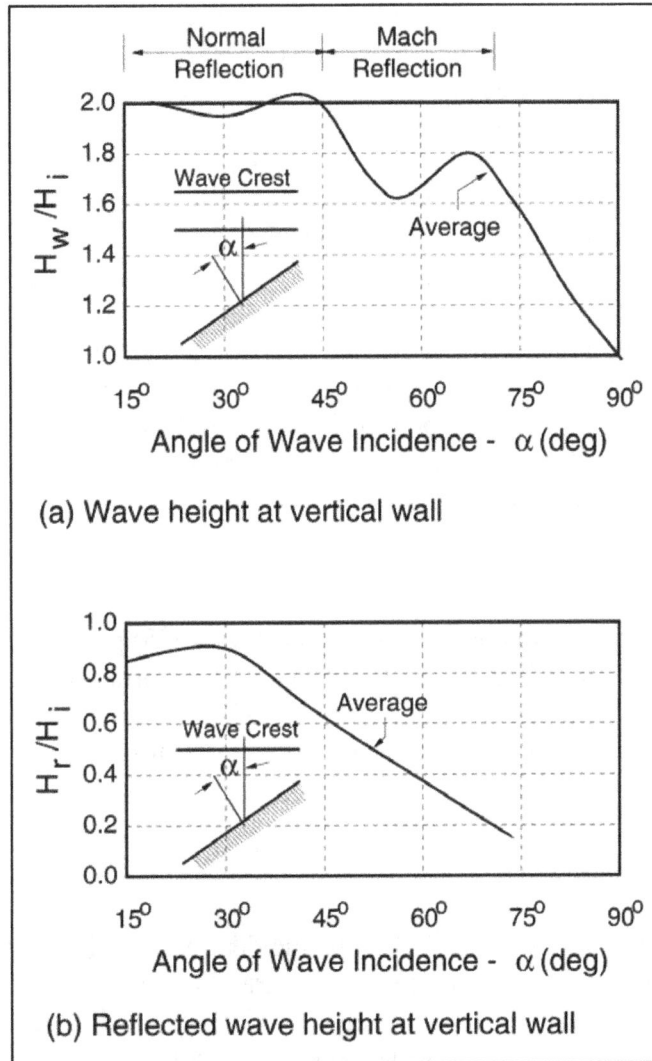

Figure VI-5-79. Mach reflection at a vertical wall
(after Wiegel 1964)

• Vertical walls with bends rather than sharp corners (Figure VI-5-78b) produce somewhat more complicated wave reflection patterns. Along the structure bend, the local angle of wave incidence varies, as does the reflected wave angle. Consequently, accurate estimates of maximum wave height along the vertical bend are best accomplished using laboratory tests or capable numerical wave models. Estimates from Figure VI-5-79 using the local angle of wave incidence should provide a reasonable approximation for mild bends. Vertical walls with very short radii bends are analogous to the seaward portion of large diameter vertical cylinders, and wave estimation techniques used in the offshore engineering field should be appropriate.

(b) Sloping structures with convex corners. The majority of coastal structures have impermeable or rubble-mound sloping fronts. Convex corners and bends for sloping-front structures are defined the same as illustrated in Figure VI-5-78 for vertical walls. Sharp corners are more likely on smooth, impermeable slopes whereas rubble-mound structures will have more rounded bends. Chen (1961) conducted experiments with solitary waves approaching smooth, impermeable slopes at oblique angles. For steep slopes the resulting wave behavior was similar to vertical walls with the onset of Mach reflection at larger angles of wave incidence. As the wall slope decreased, a large horizontal eddy formed over the slope. Further decreasing of the structure slope led to wave breaking along the slope. Generally, the onset of wave breaking depends on structure slope, incident wave angle, and the ratio of wave height to water depth (H/h). Chen's experiments used only one value of H/h so this relationship was not quantified. Rubble-mound structures with convex corners and bends may have armor stability problems for short-radius bends. In this case the bend is similar to the head section of a breakwater or jetty structure. Sakaiyama and Kajima (1997) conducted model tests of armor stability at convex bends in a structure protecting a manmade island. They found that armor stability increased as the bend radius increased. In many cases, armor stability at bends and corners is confirmed with physical model tests before construction begins. For short-radius bends an alternative is to use armor stability guidance developed for head sections. Increasing the bend radius will increase armor stability, but the tradeoff is greater quantities of construction materials.

(2) Waves at concave corners.

(a) Vertical structures with concave corners. Goda (1985) provided a simple formula for estimating the increased wave height at the apex of a concave corner of angle β formed by two impermeable vertical walls as illustrated by Figure VI-5-78c. A horizontal bottom is assumed. Provided the walls are sufficiently long, the wave height is estimated as

$$\frac{H_c}{H_i} = \frac{2\pi}{\beta}$$ (VI-5-193)

where H_c is the wave height in the corner, H_i is the incident wave height, and the angle β is expressed in radians. For β = π the corner becomes a straight wall, and $H_c/H_i = 2$. However, as β becomes small, H_c increases to unreasonable values, and steepness-limited wave breaking will occur. Therefore, estimates of maximum waves at concave corners using Equation VI-5-193 should never be greater than the steepness-limited wave at that location. Goda stated the formula is also applicable to random waves. The wave height varies greatly along the walls due to interference between incident and reflected waves. For certain combinations of wall angle β and incident wave angle, the wave height at some position along the wave may be greater than at the corner apex (Goda 1985). Goda also described a more involved procedure for estimating wave heights associated with directionally spread irregular waves. Perfectly reflecting vertical structures with concave bends (see Figure VI-5-78d) will have higher wave heights than straight walls with normal wave incidence. Wave height will depend on the radius of curvature, with greater heights expected for smaller radius bends. No simple formulas are available to estimate wave heights at concave bends; but a conservative estimate can be made by approximating the bend as a corner formed by two straight walls, and then applying Equation VI-5-193. Alternately, wave heights could be determined using an appropriate numerical model.

(b) Sloping structures with concave corners. There do not appear to be any simple, reliable engineering procedures for estimating wave height variations at sloping structures with concave corners or bends. For steep-sloped, impermeable structures, the previously described method for vertical walls will provide a conservative estimate. For milder slopes, the engineer should expect wave runup on the slope to be higher than would occur on straight structures because of the convergence of the incident wave crests. Generally, milder structure slopes, longer radii of curvature, and increased structure porosity will all contribute to a decrease in wave runup on the slope. Critical bends and corners should be tested in a physical model. If available, appropriate numerical models could also be used.

f. Uplift forces. The fluid induced force on a structure/object in the vertical (z-coordinate) direction is typically referred to as the "uplift" force (or "lift' force). The uplift force derives from various physical reasons depending on whether the structure is submerged or above water.

(1) Submerged or partially submerged structure.

(a) In the case of submerged or partially submerged structures in nonmoving fluids (i.e., a horizontal cylinderical object such as a timber cross-bracing in a pier or an outfall pipe), there is a buoyancy force which is equal to the volume of the fluid displaced by the structure/object times the specific weight of the fluid. This buoyancy force acts through the center of gravity of the displaced fluid volume in a vertically upward direction. The point through which the buoyant force acts is referred to as the center of buoyancy. The equation for this force component is given (Fox and McDonald 1985) as the integration over the volume of displaced fluid, i.e.,

$$F_B = \int_V (\rho_w g) \, dV \qquad\qquad \text{(VI-5-194)}$$

where

F_B = buoyancy force (positive upwards)

ρ_w = density of water

g = acceleration of gravity

V = volume of displaced fluid

(b) For example, the buoyancy force acting on a fully submerged 1-m-diameter sphere is

$$F_B = \gamma_w \left(\frac{\pi D^3}{6}\right) = (10.1 \, kN/m^3)(0.524 \, m^3) = 5.29 \, kN$$

where D is the sphere diameter and γ_w is the specific weight of salt water. The buoyancy force is directly countered by the gravitational force (weight) acting on the object. A net upward force occurs if the density of the submerged body is less than the water in which it is submerged.

(c) Additional vertically directed forces on the submerged or partially submerged solid body in the case of a moving fluid are due to the integration of the vertical component of pressure forces over the surface of the structure while neglecting elevation changes (Fox and McDonald 1985), i.e.,

$$F_L = -\int_S p_s (\overset{1}{dA} \cdot \overset{r}{n_z}) \tag{VI-5-195}$$

where

F_L = lift force (positive upwards)

p_s = pressure on solid body surface due to moving fluid (does not include hydrostatic pressure difference due to elevation changes over the surface

$\overset{1}{dA}$ = differential surface area element of solid body with direction outward normal to surface

$\overset{1}{n_z}$ = normal unit vectory in the positive z-direction (upwards)

(d) In the case of steady flow in the horizontal x-direction, an uplift force (often referred to as a lift force) develops when the flow field around the solid body has streamlines that are closer together above the body than below it (i.e., the "Bernoulli effect") creating a lower pressure above than below the solid body. This uplift force is analogous to the aerodynamic lift force that keeps an airplane aloft. Pipelines or outfalls lying on the seabed are examples of objects that could experience an uplift force due to the distortion of streamlines created by the protrusion of the pipeline/outfall in the flow field. Where the structure/object is only partially submerged and there is no flow over the top of the structure/object, the lift force will be acting vertically downward (i.e., negative lift force) due to the compression of streamlines (and hence lower pressure) under the structure/object.

(e) Uplift force computations on solid objects can be made via potential flow theory for simple geometry cases where there is low velocity flow (i.e., no flow separation). For the more typical design situation of turbulent flow over a solid body with flow separation, vortex shedding, and possibly a complex boundary imposed flow field, experimental laboratory measurements must be relied on to evaluate the uplift force. For steady flow situations, empirical uplift force coefficients (lift coefficients) are a function of the flow Reynold=s number, "roughness" of the solid body, and the boundary imposed flow field around the body.

(f) When the fluid is unsteady, (e.g., oscillatory wave motion) the time-varying uplift force is estimated in the same manner as for steady flow only the computation becomes even more intractable due to the unsteady nature of the flow. In oscillatory flow over a solid body, vortices are shed with frequency and phase shifting that is dependent on the Keulegan-Carpenter number. For this situation uplift force computations and determination of empirical uplift force coefficients for the solid bodies in the flow are based on experimental laboratory measurements, often combined with numerical calculations.

(g) Oscillatory flow empirical uplift force coefficients are a function of the Keulegan-Carpenter number of the flow, the Reynolds number, "roughness of the structure/object, and boundary imposed flow field" (e.g., Sarpkaya and Isaacson 1981). Where vortex shedding occurs at or near the natural frequency of the object in the flow, a large amplitude dynamic response, called vortex-induced vibration, may occur, causing much larger forces than predicted by the static approach previously discussed.

(h) Uplift forces induced by both steady and oscillatory currents need to be considered where the characteristic width of structure to wavelength ratio is small (e.g., $D/L < 0.2$ in the case of circular cylinders of diameter, D). The equation for calculation of lift force in this situation is simplified as given in the following equation (Fox and McDonald 1985, Rouse 1950; and Sarpkaya and Isaacson 1981):

$$F_L = C_L \, A_n \, \gamma_w \left(\frac{u^2}{2g} \right)$$
(VI-5-196)

where

C_L = empirical lift coefficient

A_n = projected area of solid body normal to the flow direction

γ_w = specific weight of water

g = gravitational acceleration

u = magnitude of flow velocity (lift will be perpendicular to flow direction)

(i) In the case of both steady and oscillatory currents, the velocity components of the currents must be added vectorially to provide the velocity to utilize in the previous equation.

(j) When the size of the solid structure/object is large enough to modify the incident wave field by wave diffraction and/or wave scattering, Equation VI-5-196 cannot be used to determine lift forces. For large structures, transverse and inline forces must be computed using diffraction theory (Wiegel 1964, Sarpkaya and Isaacson 1981). Typically, diffraction theory is implemented using numerical models that determine the pressure on the solid body surface and then integrate over the surface to determine the total force.

(2) Emergent structures.

(a) In the situation where the structure/object is above water (i.e., a horizontal structural member) and subjected to oscillatory wave action, intermittant approximately vertical directed impact forces occur when the level of the water reaches the structure/object. The uplift force on a structure/object in this scenerio cannot be theoretically derived due to the complex fluid structure interaction. Instead, engineers must rely on laboratory measurements or empirical impact force ("slamming") coefficients derived from laboratory testing. The uplift force for this situation is approximated as

$$F_U = C_U \, A_z \, \gamma_w \left(\frac{w^2}{2g} \right)$$
(VI-5-197)

where

    C_U = laboratory derived slamming coefficient

    A_z = projected area of solid body in the horizontal plane

    w = vertical component of flow velocity at level of object

    (b) A slamming coefficient approach to calculation of this type of uplift force is utilized primarily for slender members (for which the Morrison equation is utilized for the inline force computation). The wave theory utilized to calculate the vertical velocity at the level of the structure may depend on what level of approximation is desired and/or whether a monochromatic wave theory or irregular (linear) wave theory is utilized for the computation. A particular problem in evaluation of Equation VI-5-197 is estimating the velocity field at the structure. For even the most simple calculations an assumption that the structure does not influence the wave flow field must be made. Most uplift impact (slamming) force coefficients are derived from experimental laboratory measurements. Sarpkaya and Isaacson (1981) discussed experimental results for rigidly mounted horizontal circular cylinders subject to slamming forces, and they noted laboratory measured slamming force coefficients (C_U) ranging from 4.1 to 6.4.

    (c) Typical coastal structures on which uplift forces may need to be calculated that do not fit into any of the previous catagories are caisson or monolithic concrete type breakwaters. These structures have additional complications with regard to calculation of uplift forces because they are situated on permeable foundations of rock or sand making theoretical calculations for the uplift forces very difficult. In this situation, empirical or semiempirical formula (based on laboratory testing) are utilized to provide preliminary design calculations. Typically, design conditions will not be the same as tested in past laboratory tests; therefore, uplift forces may need to be determined by testing the design in a physical model.

    (d) Goda (1985) provided empirical formulae with which to make simple (uplift) dynamic component wave force calculations on the base of composite foundation vertical caisson (or monolithic concrete) breakwaters. The dynamic component of uplift force is assumed to be triangular over the base of the structure. The empirical formulae utilized are based on a limited number of laboratory tests and should only be utilized for preliminary calculations. Variables not in the empirical guidance but very important to the pressure distribution under the structure base are foundation permeability and structure width. High permeability and narrow structure widths could lead to uplift forces considerably in excess of Goda=s (1985) empirical guidance.

    (e) Uplift forces on docks and piers are also of concern to coastal engineers although limited information exists for the computation of forces on these types of structures. When the wave crest height exceeds the underside level of the pier or dock, the structure will be subjected to uplift forces in both transverse directions. The computation of uplift force in this situation is difficult due to the modification of the flow field by the structure and the nonlinear boundary conditions at the water surface that must be accommodated. Typically, laboratory experiments augmented by numerical modeling must be utilized to evaluate these types of uplift forces.

French (1969) measured (in a laboratory experiment) transverse (positive and negative uplift) forces due to a solitary wave moving perpendicular to a pier and found that negative uplift forces often exceeded the positive uplift forces for the situations addressed. Lee and Lai (1986) utilized a numerical model to calculate wave uplift forces on a pier; and they noted that under certain conditions of bottom slope and solitary wave height to water depth combinations, positive uplift pressures can be larger than those calculated utilizing hydrostatic pressure for the given depth of immersion. In the situation where a vertical wall abuts the platform and wave reflection takes place (e.g., a dock structure), the positive uplift appears to be significantly increased while the negative uplift is reduced compared to the pier (i.e., no wave reflection) case.

(f) Bea et al. (1999) examined wave forces on the decks of offshore platforms in the Gulf of Mexico. They summarized results from a performance study of platforms that had been subjected to hurricane wave loadings on their lower decks. Modification to guidelines of the American petroleum industry were discussed and validated. Bea et al. provides up-to-date references related to wave forces on decks of offshore platforms that may be useful for similar calculations for docks and piers.

VI-5-5. Foundation Loads.

a. Introduction.

(1) This section assumes the reader has a general knowledge about soil mechanics and foundation design because only limited basic information is given with emphasis on coastal structure foundations. The soil parameter values and empirical expressions given in this section are suitable for feasibility studies and preliminary design calculations prior to any direct soil parameter measurements being performed in the field or laboratory. The same applies for final design calculations in small projects where specific geotechnical investigations cannot be performed. In general, calculations for detailed design should be based on specific analysis of the local soil mechanics conditions. Moreover, the most relevant and accurate methods of analysis should be applied.

(2) The main objective of this section is to present two important geotechnical aspects related to the design and geotechnical stability of breakwaters, dikes and seawalls:

(a) Assurance of safety against failure in soils contained within structures, rubble-mound structures, and in foundation soils.

(b) Assurance of limited (acceptable) deformations in soils contained within structures, rubble-mound structures, and in the foundation soil during structure lifetime.

(3) Related to these two aspects are the geotechnical failure modes illustrated in Part VI-2-4:

(a) Slip surface and zone failures, causing displacement of the structure and/or the subsoil.

For rubble-mound structures and dikes see Figures VI-2-25, VI-2-41, and VI-2-51.
For monolithic structures see Figures VI-2-54, VI-2-55, VI-2-64, and VI-2-66.
For tied wall structures see Figures VI-2-69, VI-2-70, VI-2-71, and VI-2-72.

(b) Excess settlement due to consolidation of subsoil and rubble foundation, causing lowering of the crest of the structure as shown in Figures VI-2-42 and VI-2-53.

(4) Slip surface and zone failures are the result of insufficient soil bearing capacity caused by unforeseen external loadings and/or degradation of soil strength. Such failures generally lead to pronounced settlement and damage or collapse of the structure. Potential for such failure makes it important to implement proper safety factors in the design.

(5) Excess settlement due to consolidation is caused by misjudgment of subsoil characteristics and, in the case of larger rubble-mound structures, the core materials. If evenly distributed, the settlement lowers the crest level, which causes an increase in overtopping and might reduce structure functionality. Differential settlements can cause damage to the structure itself, for example breakage of concrete superstructures, cracking of long concrete caissons, or creating weaknesses in the armor layer.

(6) A significant difference between geotechnical stability of coastal structures and common land based structures is the presence of wave action on the structure and its foundation. Another difference is the wave- induced pore pressure variation which will be present in wave exposed porous structures and seabed soils. The wave load introduces stress variations in the soils that can lead to degradation in soil strength due to pore pressure build-up. The designer has to show that at any stage throughout the structure lifetime the soil stresses should not exceed the soil strength. This calls for prediction of short and long-term stress and strength development in the soils. Distinction is made between cases with gradually varying wave forces caused by nonbreaking waves and cases with short-duration impulsive wave forces due to waves breaking directly on the structure. The first case is referred to as cyclic loading; the second case is dynamic loading, which includes dynamic amplication.

(7) This section is organized into the following sections containing basic information about the soil and related hydromechanic processes:

Part/Chapter/Section Heading	Section Topic
VI-5-5-b. Soil and Rock Properties	Basic definitions and related typical parameter values. Deformation characteristics of soils are discussed as well.
VI-5-5-c. Strength Parameters	Soil parameter definitions and typical soil strength values.
VI-5-5-d. Hydraulic Gradients and Flow Forces in Soils.	Includes the Forchheimer equation and estimates on wave induced internal set-up and pore pressure gradients in breakwater cores.
VI-5-5-e. Cyclic loading of soils.	Discussion of drainage conditions, transmission of wave loads to the foundation soil, and degradation of soil strength and generation of residual pore pressure when exposed to wave induced cyclic loading.
VI-5-5-f. Dynamic Loading of Soils Under Monolithic Structures.	Evaluation of dynamic amplification of foundation forces and deformations caused by impulsive wave forces.
VI-5-5-g. Slip Surface and Zone Failures.	Stability of slopes, bearing capacity of quarry rock foundations and subsoils. Stability of soil retaining structures is not discussed.
VI-5-5-h. Settlement.	Short discussion of immediate and consolidation settlement.

b. Soil and rock properties.

(1) Grain sizes. Table VI-5-65 gives the fractional limits according to International Standards Organization (IS), and Comité Européen de Normalisation (CEN).

Table VI-5-65
Fractional Limits of Grain Sizes According to ISO/CEN

Main Group	Grain Size, mm	Sub-Groups	Grain Size, mm
Boulders	> 200		
Cobbles	60 – 200		
Gravel	2 – 60	Coarse	20 - 60
		Medium	6 - 20
		Fine	2 - 6
Sand	0.06 - 2.0	Coarse	0.6 - 2.0
		Medium	0.2 - 0.6
		Fine	0.06 - 0.2
Silt	0.002 - 0.06	Coarse	0.02 - 0.06
		Medium	0.006 - 0.02
		Fine	0.002 - 0.006
Clay	< 0.002		

(2) Bulk density. The bulk density is defined by the relation

$$\rho = m / V \tag{VI-5-198}$$

where m is total mass and V is total volume. Typical bulk densities are given in Table VI-5-66.

Table VI-5-66
Typical Bulk Density Values

| Soil Type | Bulk Density, ρ (tonne/m^3) | |
	Water-Saturated	Above Water Table
Peat	1.0 - 1.1	(often water-saturated)
Dy and gyttja	1.2 - 1.4	(often water-saturated)
Clay and silt	1.4 - 2.0	(often water-saturated)
Sand and gravel	2.0 - 2.3	1.6 - 2.0
Till	2.1 - 2.4	1.8 - 2.3
Rock fill	1.9 - 2.2	1.4 - 1.9

The unit weight is given by

$$\gamma = \rho\, g = \rho\,(9.81\, kN / m^3)$$

(3) Volume of voids. The volume of voids is either expressed in terms of

$$porosity\ n = V_p / V \qquad or \qquad void\ ratio\ e = V_p / V_s \tag{VI-5-199}$$

where V is the total volume and V_p and V_s are the volume of voids and solids, respectively.

(a) The porosity of coarse-grained soils is strongly dependent on the grain size distribution, the shape of the grains, and the compaction. Typical values of e and n for granular soils are given in Table VI-5-67.

Table VI-5-67
Typical Values of Void Ratio e and Porosity n for Granular Soils

| Material | Void Ratio | | Porosity | |
	e_{min}	e_{max}	n_{min}	n_{max}
Uniform spheres	0.35	0.92	0.26	0.48
Uniform sand	0.40	1.00	0.29	0.50
Sand	0.50	0.80	0.33	0.44
Silty sand	0.30	0.90	0.23	0.47
Uniform silt	0.40	1.1	0.29	0.52

(b) For cohesive soils the range of e (and n) is much larger than for granular soils. For clays e can range between 0.2 and 25.

(4) Relative density. The relative density is defined as

$$D_r = \frac{e_{max} - e}{e_{max} - e_{min}} 100\% \qquad (\text{VI-5-200})$$

where

e_{min} = void ratio of soil in most dense condition

e_{max} = void ratio of soil in loosest condition

e = in-place void ratio

Table VI-5-68 provides a density characterization of granular soils on the basis of D_r.

Table VI-5-68	
Density Characterization of Granular Soils	
Relative Density D_r (percent)	Descriptive Term
0 - 15	very loose
15 – 35	loose
35 - 65	medium
65 – 85	dense
85 – 100	very dense

(5) Plasticity index. The plasticity index I_p relates to cohesive soils and indicates the magnitude of water content range over which the soil remains plastic. The plasticity index is given by

$$I_p = w_l - w_p \qquad (\text{VI-5-201})$$

where w is the water content, i.e., the ratio of weight of water to the weight of solids in a soil element, and subscripts l and p refer to liquid and plastic limits, respectively.

(6) Total and effective stresses. The total stresses on a section through a soil element can be decomposed into a normal stress σ, and a shear stress τ as illustrated by Figure VI-5-80.

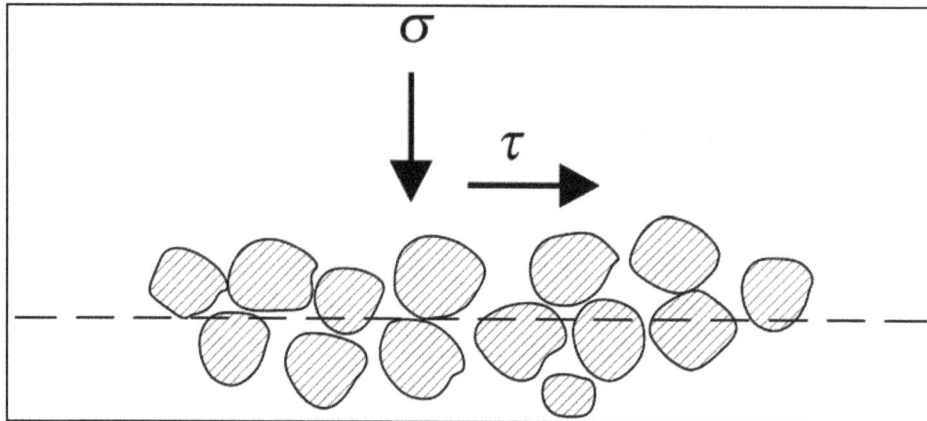

Figure VI-5-80. Total stresses in a soil element

(a) Because the soil is a three-phase medium consisting of solids and voids filled with water and/or gas it is seen that the total normal force is the sum of the contact forces between the grains and the pore pressure, u. In terms of stresses (force per unit area) we define

$$\sigma = \sigma' + u \qquad \text{(VI-5-202)}$$

where σ is total stress, σN is effective stress and u the pore pressure. Because of the small area of the contact points it can be assumed that u is acting over the whole unit area of the section.

(b) Water and gas cannot resist shear stress so the total shear stress, τ, is set equal to the effective shear stress, τN, i.e., the stress carried by the grains,

$$\tau = \tau' \qquad \text{(VI-5-203)}$$

(c) It follows from Equation VI-5-202 and Equation VI-5-203 that the ability of the soil to resist failure depends on the strength of the grain skeleton, which in turn depends on the effective stresses. This means that under constant normal stress, an increase in the pore pressure will lower the soil strength. For coastal structures changes in pore pressure are normally caused by changes in seawater level and by wave action.

(7) Geostatic stress. The geostatic stress is the stress caused by the weight of the soil when the ground surface is horizontal and the nature of the soil has only slight variation in the horizontal directions. For homogeneous soil the vertical geostatic stress is given by

$$\sigma_v = z\,\gamma, \quad \textit{based on total stress}$$
$$\sigma'_v = z\,\gamma', \quad \textit{based on effective stress} \qquad \text{(VI-5-204)}$$

where z is the depth, and γ and γN are the total and the submerged unit weights of the soil, respectively. In other words, σ_v and σN_v vary linearly with depth.

(8) Stresses within soil deposits. The coefficient of lateral stress, K, is the ratio of horizontal to vertical effective stress, i.e.,

$$K = \frac{\sigma_h - u}{\sigma_v - u} = \frac{\sigma'_h}{\sigma'_v} \qquad\qquad (VI-5-205)$$

K_o is the coefficient of lateral stress at rest. For sand deposits created by sedimentation values of K_o are typically in the range 0.4 - 0.5.

(9) Stresses due to externally applied surface loads. Although soil is an elastic plastic material, the theory of elasticity is often used to compute stresses from externally applied loads. (Examples are settlement calculations and verification of deformation amplification by dynamic loading.) Furthermore, most of the useful solutions from this theory assume that the soil is homogeneous and isotropic. Soil seldom, if ever, fulfills these assumptions. However, the engineer has little choice but to use the results from the elasticity theory together with engineering judgement. The assumption of elastic behavior is rather good if the applied stresses are low compared to stresses at failure. Diagrams for estimation of stresses induced by uniform loading on circular areas, rectangular areas and strip areas are given in most geotechnical textbooks, see for example Hansbo (1994) and Lambe and Whitman (1979).

(10) Overconsolidation ratio. A soil element that is at equilibrium under the maximum stress it has ever experienced is normally consolidated, whereas a soil at equilibrium under a stress less than the maximum stress to which it was once consolidated is termed overconsolidated. The ratio between the maximum past pressure and the actual pressure is the overconsolidation ratio (OCR). A value of OCR = 1 corresponds to normally consolidated clay where the soil tries to reduce volume (contract) when loaded further, whereas $OCR > 1$ corresponds to overconsolidated clay which tends to increase volume (dilate) under applied loads.

(11) Deformation moduli. Although soils generally exhibit plastic deformations during failure, the theory of elasticity is still widely used (for example relating soil response to dynamic loadings and stress distributions under static loads). Assuming soil behaves as an elastic material, the deformation characteristics can be expressed in terms of the moduli given in Table VI-5-69.

(a) Typical values of Poisson's ratio, v, for conditions after initial loading are given in Table VI-5-70. Exact determination of v is of less importance, because practical engineering solutions are generally not sensitive to v.

(b) The nonlinear deformation characteristics of soil makes it necessary to use secant values of the deformation moduli, as shown in Figure VI-5-81 which illustrates results from shear and compression tests. Uniaxial and confined compression tests exhibit a similar reaction. Secant values relate to stress levels being some fraction of the maximum (failure) stress. Distinction is made between initial loading where relative large deformations occur, and repeated (cyclic) loading where permanent deformations decrease and eventually disappear.

Table VI-5-69
Deformation of Moduli for Elastic Material

Young's modulus

$$E = \frac{\sigma_1}{\varepsilon_1}$$

Uniaxial loading of elastic cylinder

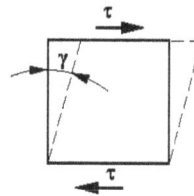

Poisson' ratio

$$\nu = -\frac{\varepsilon_2}{\varepsilon_1}$$

Shear modulus

$$G = \frac{\tau}{\gamma} = \frac{E}{2(1+\nu)}$$

Shear applied to elastic cube

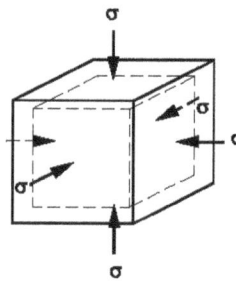

Bulk modulus

$$K = \frac{\sigma}{\varepsilon_{vol}} = \frac{2G(1+\nu)}{3(1-2\nu)}$$

ε_{vol} is volumetric strain, i.e. the relative change in volume V

Isotropic compression of elastic cube

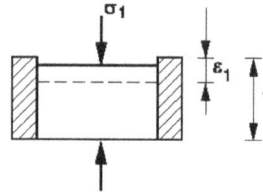

Constrained modulus

(Oedometer modulus)

$$M = \frac{\sigma_1}{\varepsilon_1} = \frac{2G(1-\nu)}{1-2\nu}$$

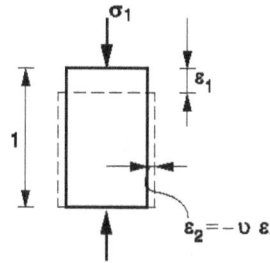

Confined compression

Table VI-5-70
Typical Values of Poisson's Ratio, *v*

Soil	v
Dry Sand	0.35
Partially saturated sand and clay	0.4
Saturated sand and clay	0.5

(c) Young modulus for sand varies with the void ratio, strength and shape of the grains, the stress history and the loading rate. Table VI-5-71 gives some example values of the secant Young's modulus corresponding to quasi-static loadings of 50 percent of the peak deviator stress and 101.3 kN/m^2 (1 atm) confining stress (Lambe and Whitman 1979).

(d) Young's modulus for clay varies with stress level, level of consolidation, and rate of strain. Table VI-5-72 provides typical values given by Richardson and Whitman (1964) corresponding to quasi-static loadings.

(e) It follows from Figure VI-5-81 that the deformation moduli depend on the strain level and the type of loading.

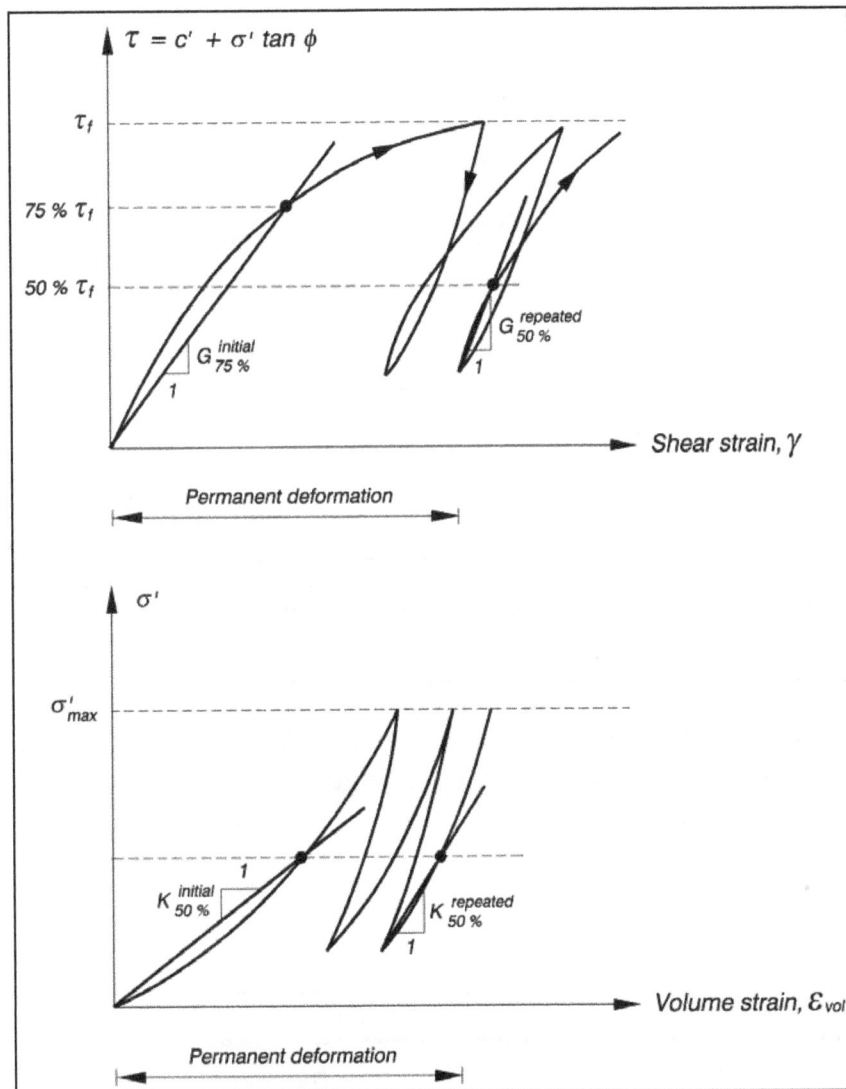

Figure VI-5-81. Illustration of shear modulus G and bulk modulus K for granular soils exposed to initial and repeated (cyclic) loadings

(f) Typical values of shear modulus G, bulk modulus K and oedometer modulus M for quartz sand is given in Table VI-5-73 corresponding to initial loading ($\sigma N \# 300$ kN/m^2) and subsequent unloading and reloading (mean $\sigma N = 100$ kN/m^2).

Table VI-5-71
Example Values of Secant Young's Modulus E in MN/m^2 for Sand

Material	Loading	Packing Density	
		Loose	Dense
Angular	Initial	15	35
	Repeated	120	200
Rounded	Initial	50	100
	Repeated	190	500

Table VI-5-72
Typical Values of Secant Young's Modulus, E, for Clay

Level of Consolidation	Strain Rate	E/σ	
		Safety Level 3[1]	Safety Level 1.5
Normal	1 percent / 1 min.	250	160
	1 percent / 500 min.	120	60
Over	1 percent / 1 min.	450	200
	1 percent / 500 min.	250	140

[1] Deviator stress equal to 33 percent of peak deviator stress.

Table VI-5-73
Typical Secant Values of Deformation-Moduli G, K and M for Quasi-Static Loaded Quartz Sand (Centre for Civil Engineering Research and Codes (CUR) 1995)

Parameter	Initial Loading	Repeated Loading
G (MN/m^2)	4 - 40	20 - 400
K (MN/m^2)	10 - 100	50 - 1000
M (MN/m^2)	15 - 150	80 - 500

Note: Higher values valid for dense sand, lower values valid for very loose sand.

(g) The shear modulus G is independent of drained or undrained conditions, and the value of G for clays is dependent on the type of clay (plasticity index), the type of loading, the stress level, and the OCR. Figure VI-5-82 shows the range of G over the static undrained shear strength, c_u, as a function of the shear strain for some saturated clays (not further characterized).

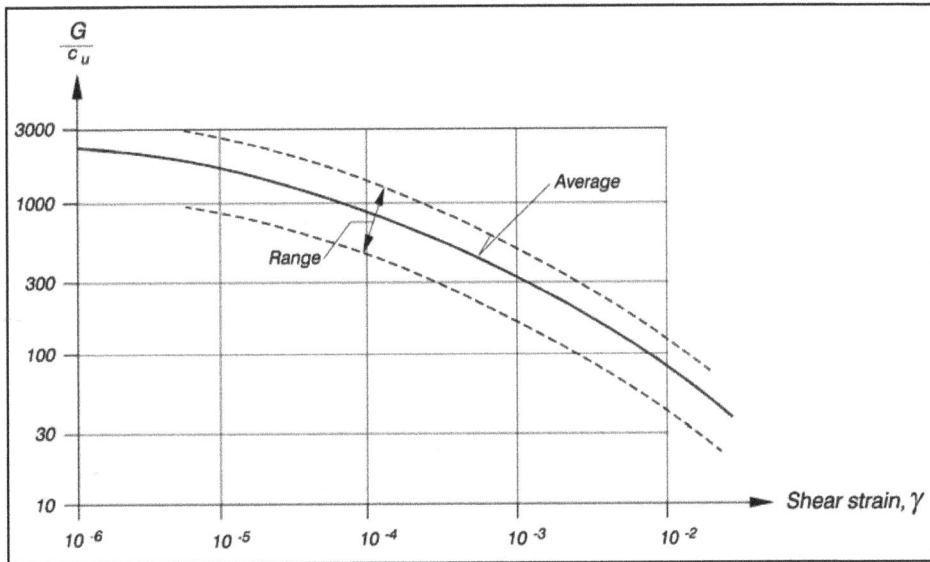

Figure VI-5-82. In-situ secant values of shear modulus G for quasi-static loaded saturated clays (after Seed and Idriss 1970)

(h) The significant influence of OCR and cyclic loading on G is shown in Figure VI-5-83 which presents results for Norwegian Drammen clay with plasticity index I_p of 27 percent and a clay content of 45-55 percent. These results were based on stress controlled DSS tests and resonant column tests. In Figure VI-5-83 the parameter σ_u^{DSS} is the undrained static DSS shear strength for two hours of loading to failure. The stress τ_{cy} is the shear stress amplitude in the symmetric cyclic loading. N is number of load cycles.

(i) The shear modulus G is an important parameter in soil response to dynamic loadings that might be caused by waves and earthquakes. In quasi-static loading tests, such as simple shear and triaxial tests, the lower limit for strain measurements is approximately 10^{-3}, whereas in bender element and resonant column tests strains down to 10^{-6} can be recorded. Thus in practice, the maximum value G_{max} which can be identified corresponds to a shear strain of approximately 10^{-6}. Formulae for G_{max} are given as follows:

- Sand (Hardin and Black 1968)

$$G_{max} = \begin{cases} \dfrac{6908\,(2.17 \text{-} e)^2}{1+e}\sqrt{p'} & round-grained \\[3mm] \dfrac{3230\,(2.97 \text{-} e)^2}{1+e}\sqrt{p'} & angular-grained \end{cases}$$

(VI-5-206)

- Gravel (Seed et al. 1986). They found G_{max} values approximately 2.5 times larger than for sand.

- Clay (Hardin and Drnevich 1972)

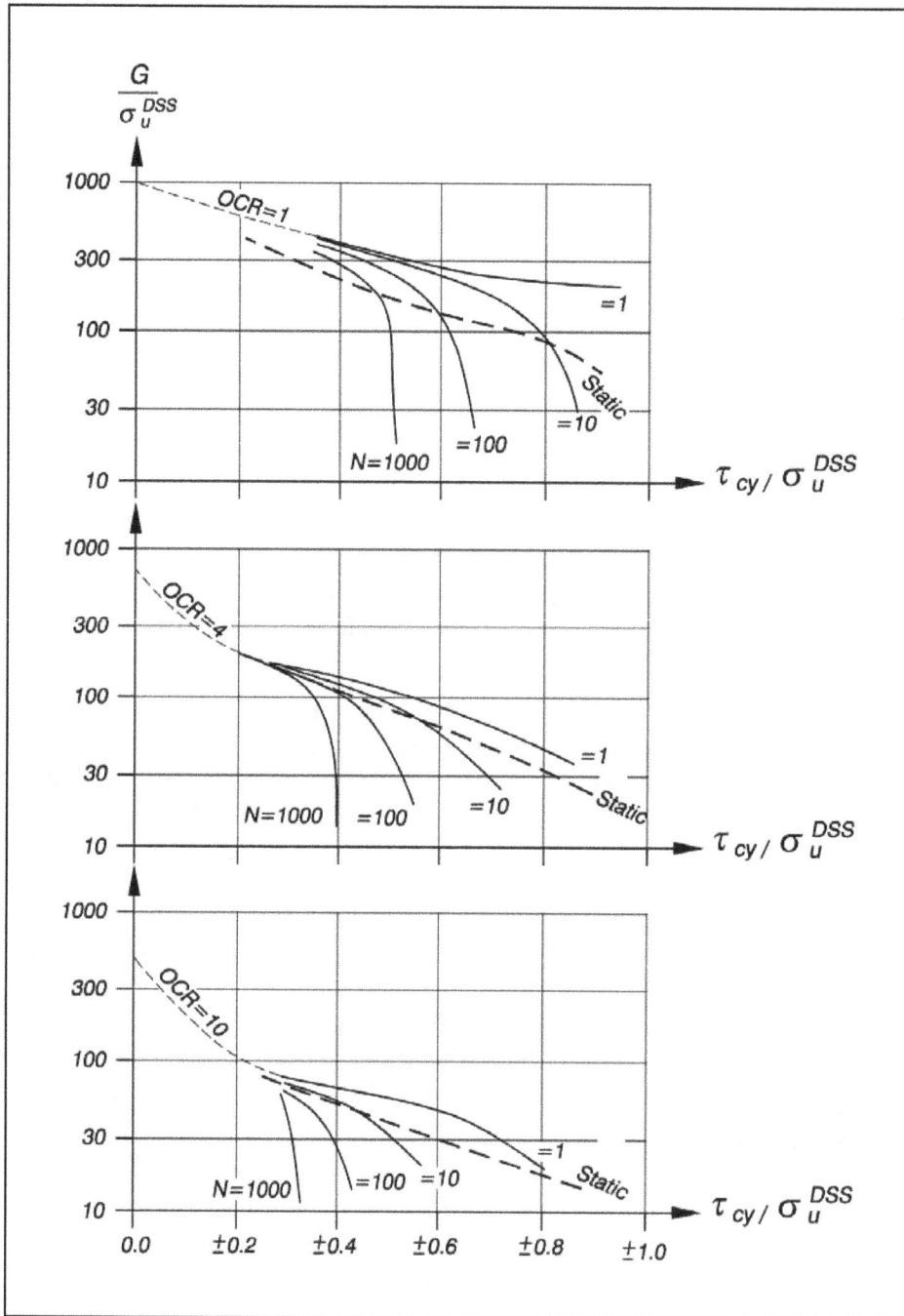

Figure VI-5-83. Static and secant cyclic shear modulus, G, for Drammen clay (Andersen, Kleven, and Heien 1988)

$$G_{max} = \frac{3230\,(2.97 - e)^2}{1+e}\,(OCR)^K\,\sqrt{p'} \qquad \text{(VI-5-207)}$$

where

e = void ratio

p' = mean effective stress, $1/3(\sigma_1 N + \sigma_2 N + \sigma_3 N)$ to be inserted in kN/m^2 to obtain G_{max} in kN/m^2

OCR = overconsolidation ratio

K = constant dependent on the plasticity index

Plasticity Index (percent)	0	20	40	60	80	∃ 100
K	0	0.18	0.30	0.41	0.48	0.50

Hardin (1978) proposed for both granular and cohesive soils that

$$G_{max} = \frac{625}{0.3 + 0.7\,e^2} (OCR)^K \sqrt{p_a\,p'} \tag{VI-5-208}$$

where p_a is atmospheric pressure ($101.3\ kN/m^2$). The ratio between G and G_{max} as function of the shear strain for sand and gravel is given in Figure VI-5-84.

(12) Damping ratio. The damping ratio D signifies the decrease in the displacement amplitude z_n of the oscillations and is defined by

$$D = \frac{\delta}{2\pi} = \frac{1}{2\pi} \ln\left(\frac{z_n}{z_n + 1}\right) \tag{VI-5-209}$$

where δ is the logarithmic decrement. Figure VI-5-85 shows damping ratios for sands and clays.

c. Strength parameters.

(1) Mohr-Coulomb failure criterion.

(a) The strength parameters of soil and rock fill constitute the basis for analysis of soil bearing capacity and wall pressures. Failure occurs when shear stresses reach an upper limit represented by the envelope to the Mohr failure circles, as shown in Figure VI-5-86.

(b) The Mohr envelope is generally curved for drained conditions. Figure VI-5-87 shows two commonly applied straight-line approximations to curved envelopes found from drained triaxial tests. Figure VI-5-87 demonstrates that the straight-line approximation is good only in the vicinity of the σN_f-value for which the tangent to the circle is constructed. The approximation in Figure VI-5-87a is given by the Mohr-Coulomb equation

$$\tau_f = c' + \sigma_{f'} \tan \varphi_{t'} \tag{VI-5-210}$$

where cN is the cohesion intercept, v_tN is the effective tangent angle of friction, and $\sigma_f N$ is the effective stress at failure as specified by Equation VI-5-204.

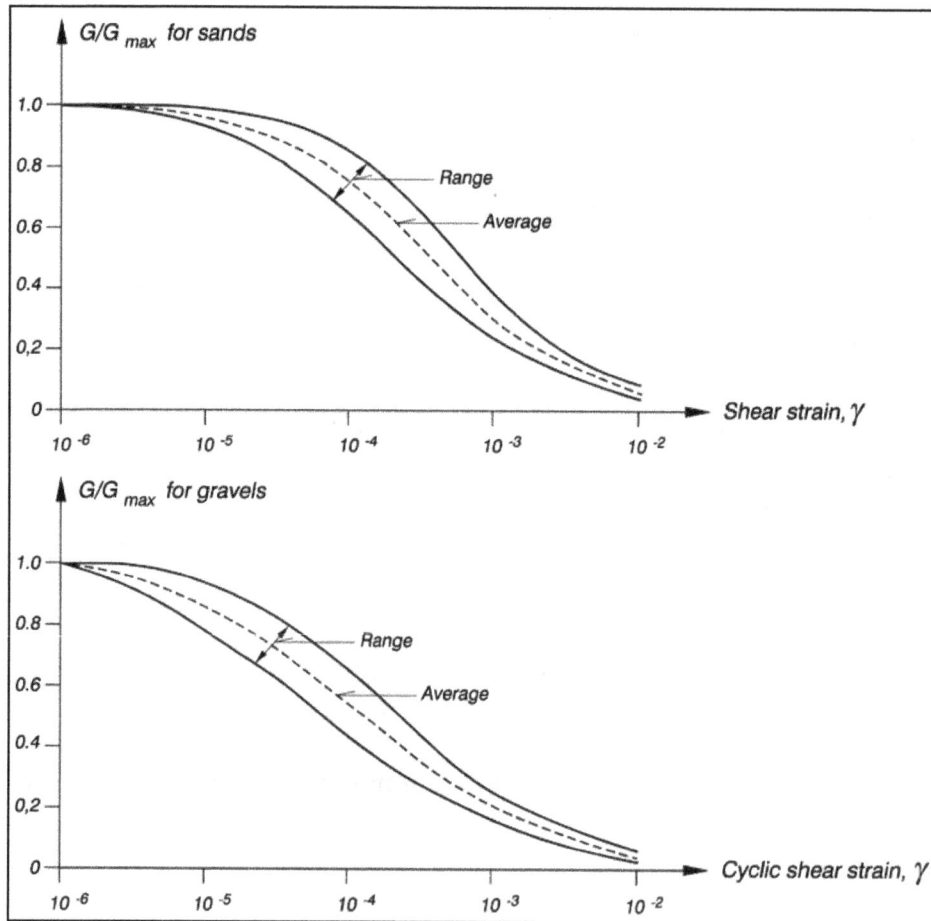

Figure VI-5-84. Values of G/G_{max} for sands and gravels
(after Seed et al. 1986)

(2) Noncohesive soils.

(a) The failure criterion approximation shown in Figure VI-5-87b corresponding to the equation

$$\tau_f = \sigma_{f'} \tan \varphi_{s'} \qquad \text{(VI-5-211)}$$

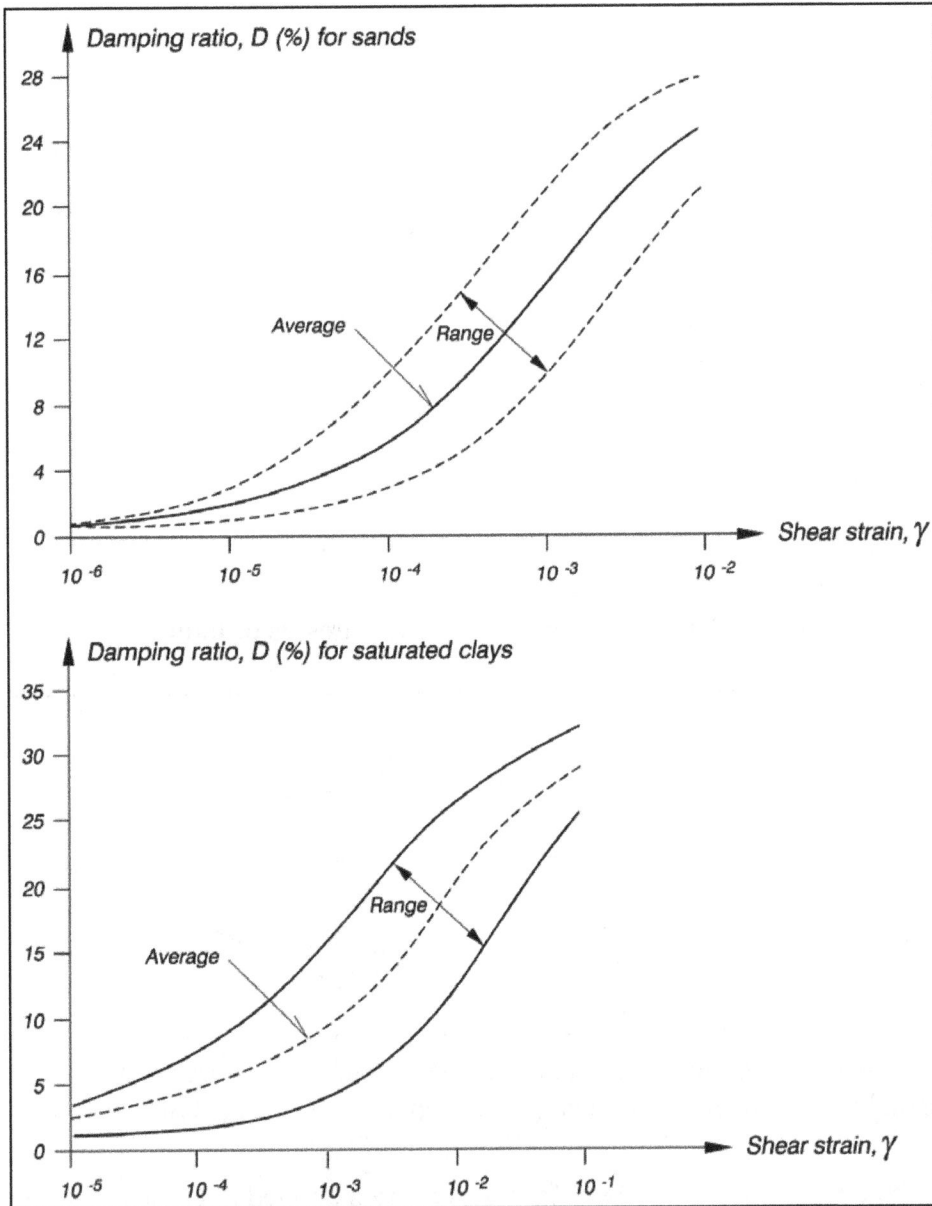

Figure VI-5-85. Damping ratios for sands and saturated clays (Seed 70)

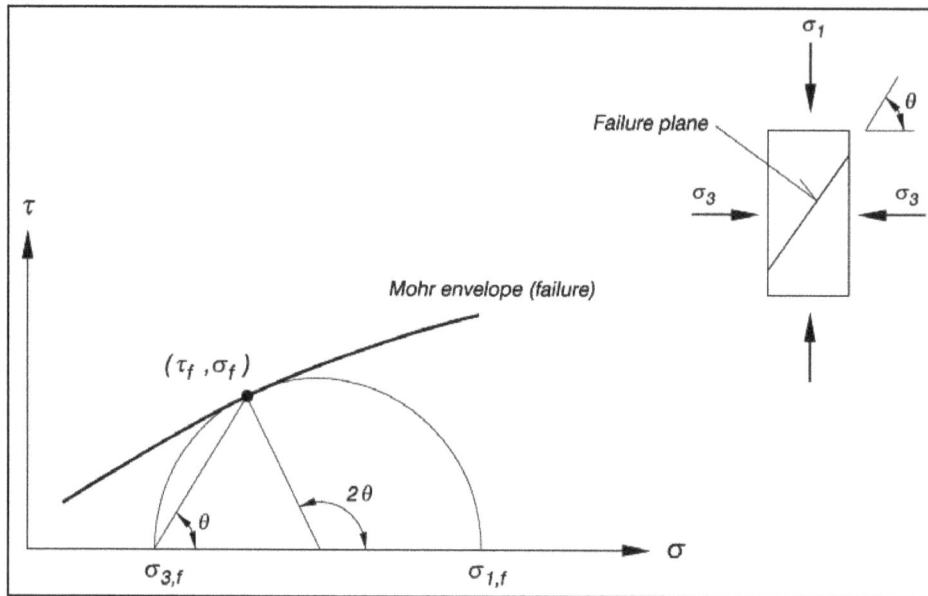

Figure VI-5-86. Mohr envelope for stresses of failure

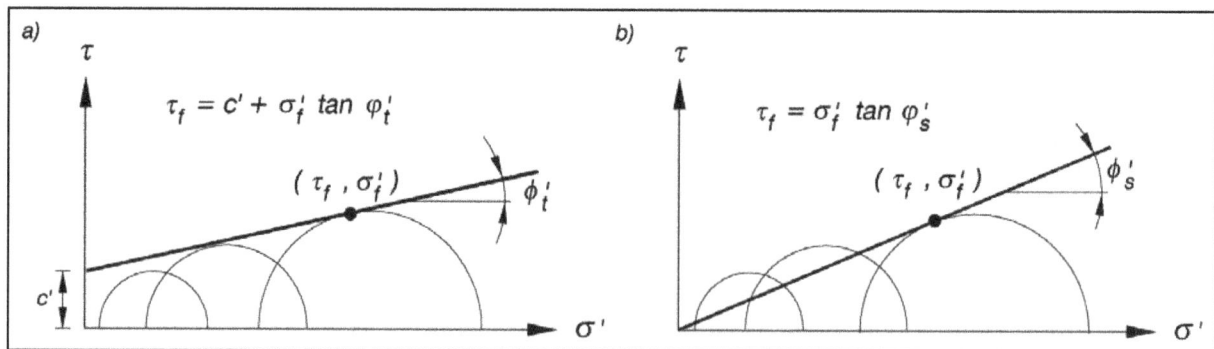

Figure VI-5-87. Illustration of straight-line approximations to curved Mohr envelopes corresponding to drained conditions: (a) Tangent formulation, (b) Secant formulation

where $\nu_s N$ is the effective secant angle of friction, has been applied to granular soils ever since the early studies by Coulomb. The equation is accurate only for relatively small values of $\sigma_f N$. However, for well graded quartz sand the limit for reasonable accuracy may be as high as 1,000 kN/m². In general the equation should be applied only to a limited stress range around the $\sigma_f N$ value corresponding to $\nu_s N$. Otherwise, for very high stress ranges the strength of a granular soil or rockfill can only be satisfactorily represented by Equation VI-5-210, or a curved Mohr envelope. Another way to represent the nonlinear strength relation is to treat $tan\ \nu N$ as a variable that depends on the confining pressure as indicated in Figure VI-5-87, which shows that νN is a function of the actual effective stress level.

(b) The angle of friction νN in granular materials depends on the grain-size distribution, size and shape of the grains, and on the porosity. Well graded materials exhibit higher friction than uniformly graded materials. Sharp edged angular grains give higher friction than rounded grains, and the friction angle will be higher in densely packed than it is in loose soils.

(c) Typical angles of friction for granular soils like quartz sand and quarried granite rock fill are given in Table VI-5-74 and Figure VI-5-88.

Table VI-5-74
Typical Values of Triaxial Test Friction Angle v_s for Quartz Sand

Relative Density	Friction Angle from Triaxal Tests v_s (degrees)
Very loose	-
Loose	29 - 35
Medium	33 - 38
Dense	37 - 43
Very dense	-

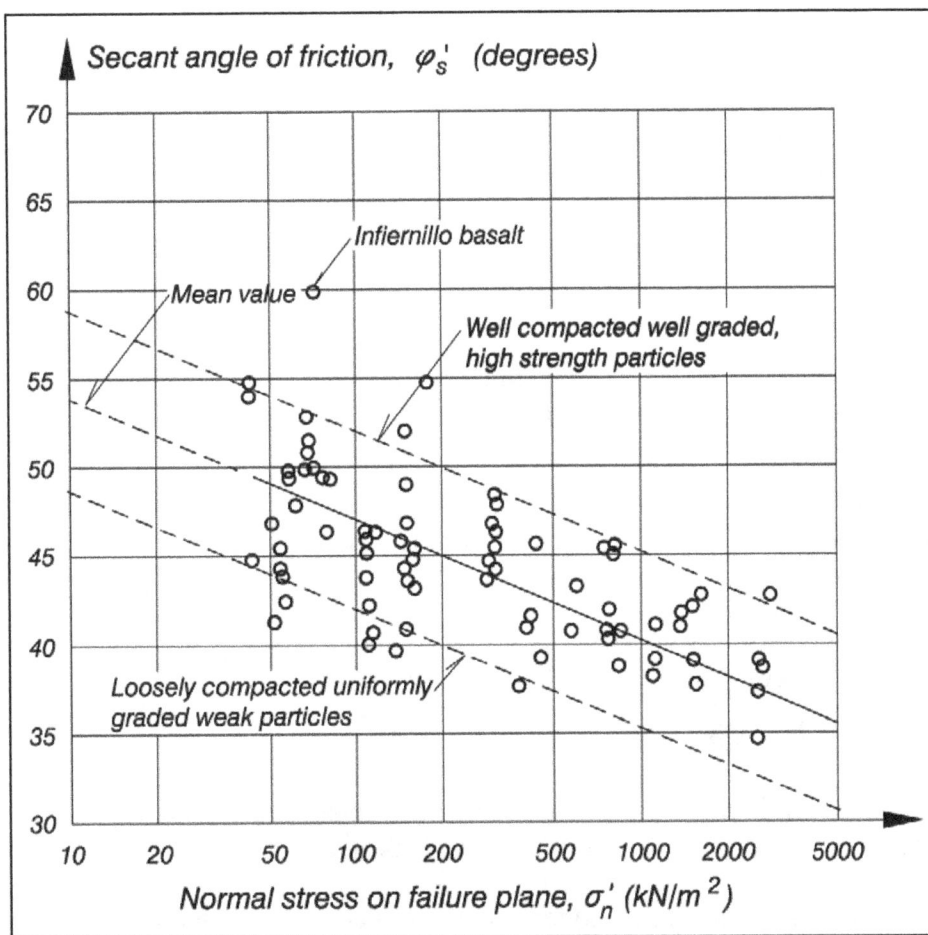

Figure VI-5-88. Angle of friction in rock fill of different grading and porosity with maximum diameter in the range 70-200 mm (after Leps 1970 and Kjaernsli, Valstad, and Høeg 1992)

(d) Steenfelt and Foged (1994) reported secant angles of friction $v_s N = 45° - 62.2°$ at normal stress on failure plane $\sigma_n N = 77 - 273$ kN/m^2 for Hyperite crushed stone of mass density

3.1 tonne/m^3, d_{50} = 15 - 16 mm and d_{max} = 64 mm. This compares well with the Infiernillo basalt data in Figure VI-5-88.

(3) Dilatancy.

(a) Shearing of frictional soils under drained conditions generally involves volume changes in terms of dilation or contraction. A crude visualization of dilatancy in plane strain is shown in Figure VI-5-89.

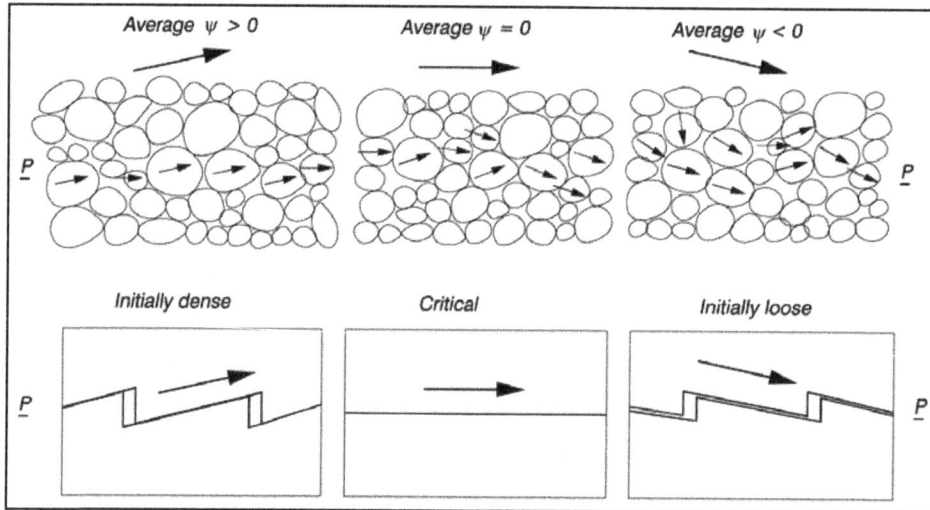

Figure VI-5-89. Crude visualization of dilatancy and angle of dilation ψ
(Bolton 1979)

(b) The volume changes associated with stress as it increases toward maximum strength (see v_sN in Equation VI-5-211) depend on the effective stress level and the initial density, which is given by porosity n or void ratio e. The volume changes are quantified by the angle of dilation, ψ, defined by

$$\sin \psi = -\frac{\dot{\varepsilon}_1 + \dot{\varepsilon}_3}{\dot{\varepsilon}_1 - \dot{\varepsilon}_3} = \frac{\dot{\varepsilon}_{vol}}{\dot{\varepsilon}_{vol} - 2\dot{\varepsilon}_1} \tag{VI-5-212}$$

where $\dot{\varepsilon}_1$, and $\dot{\varepsilon}_3$ are strain rates in principal stress directions 1 and 3, and $\dot{\varepsilon}_{vol}$ is the volume strain rate. The strain rates can be found from triaxial tests.

(c) The angle of friction corresponding to the critical (also called ultimate) condition where the soil strains without volume changes (see Figure VI-5-89) is denoted the critical angle of friction, vN_{crit}. The parameter vN_{crit} appears to be a material constant because it depends on the mineralogy, grading and shape of the grains for the soil in question, but seems independent on the relative density or porosity. Typical values of vN_{crit} are given in Table VI-5-75.

(d) An average value of vN_{crit} for sand is 32 deg. For quarried rockfill a somewhat higher value is found. Steenfelt (1992) stated that a simple bench test for vN_{crit}, offering an accuracy of about 1°, is the angle of repose of a loosely tipped heap of dry material subjected to excavation at the foot.

The contribution of dilation to the strength of the material is suggested as follows by Bolton (1986)

$$\varphi_{m'ax} - \varphi_{c'rit} = 0.8\,\psi_{max} = \begin{cases} 5° I_r & plane\ strain \\ 3° I_r & triaxial\ strain \end{cases} \tag{VI-5-213}$$

Table VI-5-75
Critical Value of Angle of Friction, νN_{crit} (Steenfelt 1992)

Material	d_{50} (mm)	d_{max} (mm)	νN_{crit} (deg)
Quartz sand,	0.17	-	27.5 - 32
dry and saturated	0.24	-	29 - 33.3
	0.52 - 0.55	-	33.5
	0.88	-	31.9
Rock fill, quarried granitic gneiss	-	9.5 - 80	39.1

where

$$I_R = D_r\,(A - \ln\,p') - 1 \tag{VI-5-214}$$

and

$\nu N_{max} = \nu N_s$ for triaxial strain, as given by Equation VI-5-211

D_r = relative density

pN = mean effective stress, $1/3(\sigma_1 N + \sigma_2 N + \sigma_3 N)$ in kN/m

A = material constant, 10 for quartz and feldspar, and 8 for limestone

Typical values of ψ_{max} for quartz granular materials are given in Table VI-5-76.

Table VI-5-76
Typical Values of ψ_{max} for Quartz Sand and Quarried Granitic Gneiss

Relative Density	Angle of Dilation, ψ_{max} (deg)
Loose	-2 to +3
Medium	+3 to +8
Dense	+8 to +13

(4) Cohesive soils.

(a) The shear strength of cohesive soils like clay and organic mineral soils is due to both friction (between coarser grains and between aggregates formed by clay particles) and cohesion within the material (sorption forces). The shear strength of clay normally refers to the static shear strength from undrained strain controlled tests with a monotonic load increase lasting 1-3 hours to failure. This so-called undrained shear strength, c_u and the related failure envelope are illustrated in Figure VI-5-90.

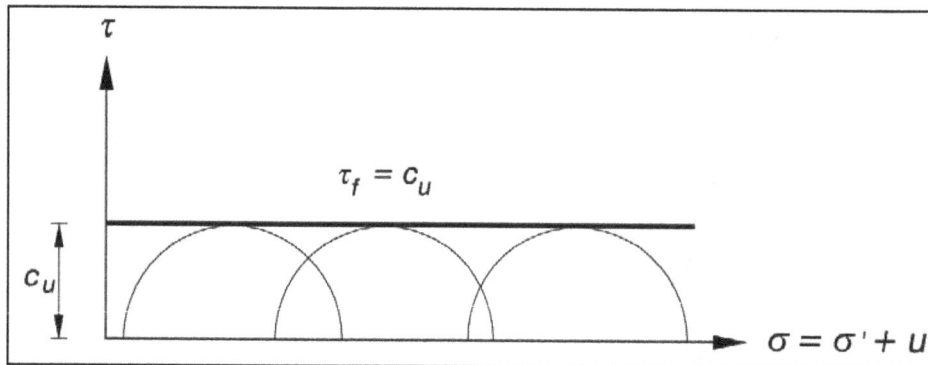

Figure VI-5-90. Failure criterion for a water-saturated clay in undrained condition defined from Mohr envelope

(b) For a specific clay with a given stress history, c_u depends solely on the initial effective stress conditions before the loading. Thus, the increase in σ in Figure VI-5-90 is equal to the increase in the pore pressure, u. In addition, the c_u-value and the deformation characteristics depend on the overconsolidation ratio, OCR, defined in Part VI-5-5b, as well as on the rate and number of loadings, as discussed in Part VI-5-5e on cyclic loading. Failure analysis related to cohesive soils in undrained conditions is performed on the basis of total stresses, σ, as opposed to analysis of noncohesive soils which is based on effective stresses, σN.

(c) The relative density of cohesive types of soils cannot be determined, and for this reason these soils are usually classified according to shear strength properties (see Table VI-5-77).

Table VI-5-77 Classification of Clay According to Undrained Shear Strength, c_u		
Descriptive Term	c_u (kN/m^2) (Hansbo 1994)	c_u (kN/m^2) (Tomlinson 1980)
Very soft	< 20	< 25
Soft	20 - 40	25 - 50
Firm	40 - 75	50 - 100
Stiff	75 - 150	100 - 200
Very stiff	> 150	> 200

(d) It should be noted that development of large shear stresses often involves soil deformations which might be damaging to the function of the structure. This is true especially for normally consolidated clay. For such cases the failure criterion must be defined as a strain level instead of the stress level, c_u.

(e) Cohesive soils are also classified according to their sensitivity to loss of strength when disturbed. The sensitivity, S_t, is defined as the ratio between the undrained shear strength of a specimen in undisturbed and in remoulded states. S_t is important for the estimation of shear strength reduction in case of disturbance due to activities such as piling and excavation. Fall-cone tests can be used to determine values of S_t. Soils are termed slightly sensitive when $S_t < 8$, moderately sensitive when $8 \# S_t \# 30$, and highly sensitive when $S_t > 30$. The last range includes quick clays for which $S_t \ni 50$.

d. Hydraulic gradient and flow forces in soils.

(1) Hydraulic gradient.

(a) If the seawater level and the groundwater level are horizontal and not moving, the pore water will be in static equilibrium corresponding to the hydrostatic pressure distribution and constant head, h. Any deviation from this stage causes a change in h, and generates a flow governed by the hydraulic gradient i, which is given by

$$i = \frac{\Delta h}{\Delta l} \tag{VI-5-215}$$

where Δh is the difference in hydraulic head over the distance Δl. The hydraulic head is defined as

$$h = z + \frac{u}{\gamma_w} \tag{VI-5-216}$$

where z is a vertical coordinate, u is the pore pressure, and $\gamma_w = \rho_w g$ is the unit weight of the water (ρ_w is the mass density of water and g is gravity).

(b) A flow force of $i\gamma_w$ will act on the grains in the direction of the hydraulic gradient, i. The effective unit weight, $\gamma_s N$, of a saturated soil can then be defined as

$$\gamma_{s'} = \gamma - \gamma_w \pm i\gamma_w \tag{VI-5-217}$$

where γ = unit weight of dry soil, the plus sign is used for vertical downward flow, and the minus sign is used for vertical upward flow. For an upward flow, if $i = (\gamma - \gamma_w)/\gamma_w$, then $\gamma_s N = 0$, corresponding to a total loss of soil bearing capacity, referred to as the limit stage of fluidization or liquefaction. The flow forces in the soil have to be included in the work or force balance equations for the failure limit states, either by including the flow force $i\gamma_w$ on all internal parts of the soil elements, or by including the pore pressures along the boundaries of the soil elements.

(c) The bulk flow velocity v introduced by i may be calculated by the one-dimensional extended Forchheimer equation

$$i = Av + B|v|v + C\frac{\delta v}{\delta t} \qquad\qquad \text{(VI-5-218)}$$

where the coefficients A, B and C depend on the soil and water characteristics, i.e., grain size and shape, gradation, porosity, viscosity and the Reynolds number. The last term in Equation VI-5-218 can be neglected because it has only minor influence for wave-induced flow in cores, subsoils and rubble foundations related to coastal structures.

(d) Figure VI-5-91 illustrates the variation of A and B in Equation VI-5-218. Table VI-5-78 presents expressions of A and B as well as related flow coefficients found from experiments as listed in Burcharth and Anderson (1995). Considerable scatter in the flow coefficients is observed.

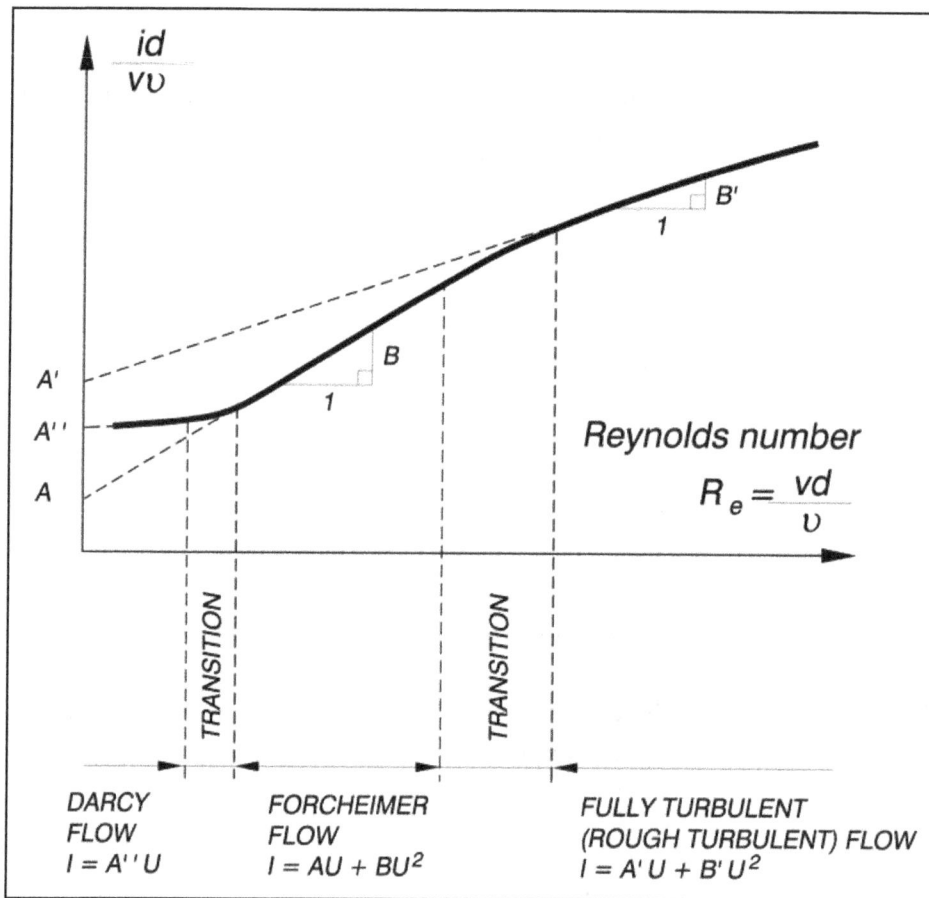

Figure VI-5-91. Representation of flow regimes for stationary porous flow based on a Forchheimer equation formulation (Burcharth and Anderson 1995)

(2) Permeability.

(a) For $R_e < 1$, Equation VI-5-219 in Table VI-5-78 is most often presented as the Darcy equation

$$v = k\,i \tag{VI-5-220}$$

where k is a dimensional quality referred to as the permeability coefficient. Comparing the first term in Equation VI-5-219 with Equation VI-5-220 gives

$$k = \frac{n^3}{\alpha\,(1-n)^2}\,\frac{g\,d^2}{\nu} \tag{VI-5-221}$$

(b) Equation VI-5-221 can be applied for fine materials like clay, silt, and fine sand ($d \leq 0.2$ mm) whereas for coarser material the nonlinear Equation VI-5-219 must be applied. It should be noted that α (and thereby k) depends on the Reynolds number and the soil gradation.

(c) Typical values of k are given in Table VI-5-79 for rather uniform sands. Order of magnitude values of k for stone materials are given in Table VI-5-80.

Table VI-5-78
One-Dimensional Porous Flow Equation

One-dimensional steady porous flow equation and related coefficients. Burcharth and Anderson (1995).

$$i = \alpha\frac{(1-n)^2}{n^3}\frac{\nu}{gd^2}v + \beta\,\frac{1-n}{n^3}\frac{1}{gd}v^2 \tag{VI-5-219}$$

where n Porosity
d Characteristic grain diameter, e.g. d_{50}
ν Kinematic viscosity ($1.3.10^{-6}$ m^2/s for water at 10^oC)
g Gravitational constant
v Bulk flow velocity

Typical values of α and β in Eqn. VI-5-219 for uniform sand and quarried rock materials

$$\frac{d_{85}}{d_{15}} = 1.3 - 1.9$$

Flow range	$Re = \frac{dv}{\nu}$	α	β
Darcy	< 1	300 - 400	0
Forchheimer	10 - 150	300	3.0 - 3.6
Fully turbulent	1,000 - 12,000	1,000 - 10,000	3.6 - 2.4[1]

[1] Smallest values of β correspond to largest R_e.

Table VI-5-79
Typical Values of Permeability, k, for Fine Materials

Material	Packing	k (m/s)
Coarse sand	loose	10^{-2}
	dense	10^{-3}
Medium sand	loose	10^{-3}
	dense	10^{-4}
Fine sand	loose	10^{-4}
	dense	10^{-5}
Silty sand	-	10^{-6}
Sandy clay	-	10^{-7}

Table VI-5-80
Typical Values of Permeability, k, for Stone Materials

Gradation Diameter Range (mm)	k (m/s)
100 - 300	0.3
10 - 80	0.1

(3) Wave-induced internal setup. Wave action on a pervious slope causes a fluctuating internal water table (phreatic surface) and a setup as indicated in the figure in Table VI-5-81. The reason for the setup is that inflow dominates outflow due to larger surface area and longer duration. The setup increases if the shore side of the structure is impermeable, e.g., a rubble revetment built in front of a clay cliff.

(a) The setup can be estimated by a method (Barends 1988) presented in Table VI-5-81. The method is based on a linearization of the Forchheimer equation, where the permeability k for sands can be estimated from Table VI-5-79. For quarry-run materials, where linearization is less suitable, Equation VI-5-219 should be used. Order of magnitude values are given in Table VI-5-80.

(b) Besides storage of water due to internal setup of the phreatic level, also some storage due to compressibility of the soil rock skeleton and water-air mix can occur. However, for conventional structures such elastic storage will be insignificant compared to the phreatic setup storage.

(4) Pore pressure gradients in sloping rubble-mound structures.

(a) The horizontal wave-induced pressure gradient in the core of a rubble-mound breakwater can be estimated by the method of Burcharth, Liu, and Troch (1999) as presented in Table VI-5-82. The method is mainly based on pore pressure recordings from a prototype and large and small scale model tests.

Table VI-5-81
Wave Induced Set-up in Sloping Rubble Mound Structures (Barends 1988)

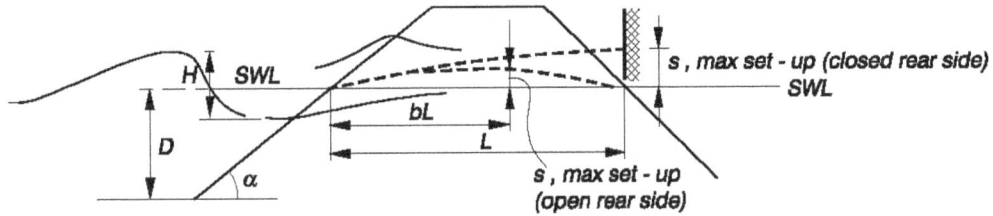

$$s/D = \sqrt{(1 + \xi\, F)} - 1 \qquad \text{for large waves, i.e. } H \leq D$$

$$s/H = \sqrt{(1 + \xi\, F)} - 1 \qquad \text{for small waves, i.e. } H \ll D$$

where ξ = $0.1cH^2/(n\lambda\, D \tan \alpha)$
  H = Height of incoming wave
  n = Porosity of structure
  c = Infiltration factor > 1. The magnitude is uncertain (Barends 1988)
       used $c \simeq 1.3$ to make calculations fit to conventional
       scale model test results
  λ = $0.5\sqrt{c\, k\, DT\, /n}$
  α = Slope angle
  k = Average permeability
  T = Wave period
  F = Function dependent on rear side conditions
       (open or closed) as given in diagram.
       The parameter b in the diagram defines position
       of maximum setup in the open case.

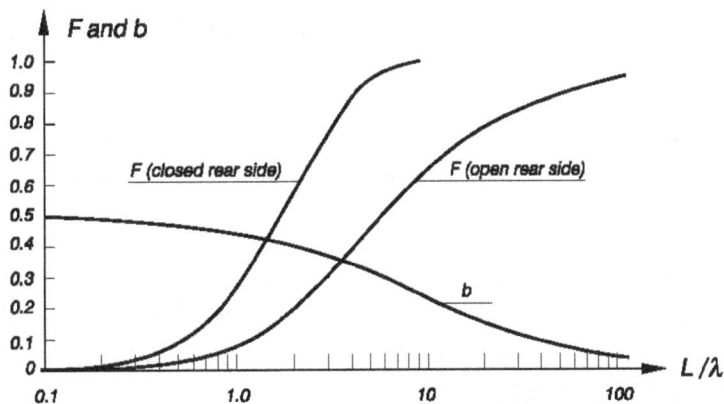

Table VI-5-82
Horizontal Wave Induced Pore Pressure Gradients in the Core of Rubble-Mound Breakwaters
(Burcharth, Liu, and Troch 1999)

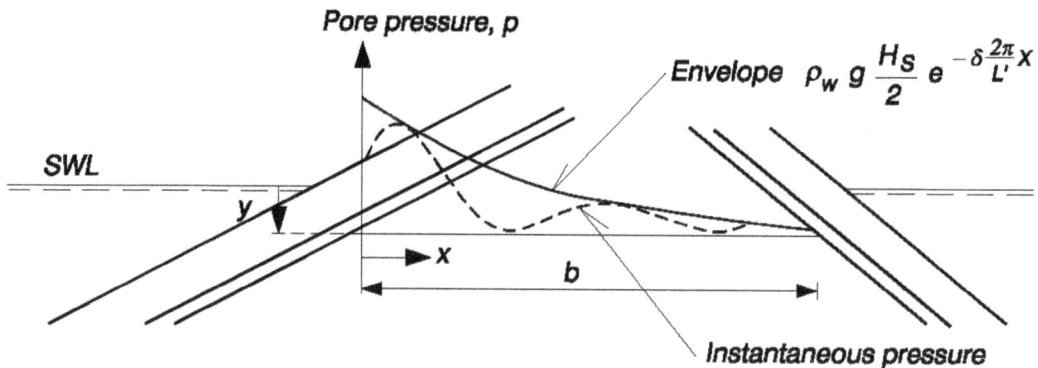

The horizontal pressure gradient

$$
\begin{aligned}
I_x &= \frac{1}{\rho_w\, g}\, \frac{d_p}{dx} \\
&= -\frac{\pi\, H_s}{L'}\, e^{-\delta \frac{2\pi}{L'}\, x}\, \left[\delta\, \cos\,(\frac{2\pi}{L'}\, x + \frac{2\pi}{T_p}\, t) + \sin\,(\frac{2\pi}{L'}\, x + \frac{2\pi}{t_p}\, t)\right]
\end{aligned}
\qquad (VI\text{-}5\text{-}222)
$$

where H_s Significant height of incoming waves
δ $0.014\, \frac{n^{0.5}\, L_p^{\,2}}{H_s\, b}$, damping coefficient
n Porosity
L_p Wavelength corresponding to peak of spectrum
b Width of core at the level of interest
L' $L_p/\sqrt{1.4}$, wave length in core valid for
$h/L_p < 0.5$ (where h is the water depth)
x Horizontal coordinate, $x = o$ at the seaward core boundary
T_p Period corresponding to peak of the wave spectrum
t Time
ρ_w Mass density of water

(b) Equation VI-5-222 is valid only for rather permeable core materials ($d_{50} \ni 50$ mm) and for normal breakwater cross sections with open rear side, i.e., no excess pressure. Additionally, Equation VI-5-222 holds for the region between swl and level $SWL + 2H_s$, i.e., $0 \# y \# 2H_s$. In each point within this region the larger pressure gradients will be of the same order of magnitude as the horizontal gradient.

e. Cyclic loading of soils. An essential part of the design of monolithic coastal structures is to ensure that the foundation soil or rubble base has sufficient capacity to carry both the static gravity loads and the wave-induced loads with an adequate safety margin and without excessive deformations. The bearing capacity under combined static and cyclic loads may be significantly smaller than under purely static loads. The strength of soils exposed to cyclic loading is influenced not only by the stress level and the stress variations but also by the soil drainage capability. Pore pressure build-up and related loss of strength might take place in rather impervious soils where the time scale of drainage or consolidation is larger than the time scale of the load cycles. The following sections discuss evaluation of drainage conditions under cyclic loading, approximation of wave- induced irregular loading in terms of equivalent cyclic loading, and estimation of strength and deformation of soils exposed to cyclic loading.

(1) Time scale of drainage and consolidation.

(a) In saturated soil, the immediate effect of a load-induced stress increment will be a similar increase in the total stress σ and the pore pressure u (see Equation VI-5-202), i.e., the loading will be carried solely by the pore water. The soil skeleton will not carry the extra load until it has rearranged itself. This can happen only if some pore water is squeezed out, due to the very small compressibility of the water compared to that of the skeleton. In permeable materials such as stone blankets this happens immediately, while in clay it can be a very slow process. The related decrease in volume is termed consolidation.

(b) The degree of consolidation is defined as

$$U = \frac{s_t}{s_\infty} \tag{VI-5-223}$$

where s_t is the settlement (decrease in layer thickness) at time t, and s_4 is the final settlement reached when the soil skeleton is fully carrying the load. For coastal structures the dominating live load is caused by wave loading that varies in time. The time scale of consolidation has to be compared to the time scale of the loading to estimate U and thereby the effective stress in the soil.

(c) For the one-dimensional case Terzaghi showed that U in terms of average degree of consolidation is a function of the dimensionless time factor (Terzaghi and Peck 1944)

$$T_c = \frac{k\,M}{\gamma_w\,H^2}\,t = \frac{C_V}{H^2}\,t \tag{VI-5-224}$$

where

C_V = coefficient of consolidation ($= kM/\gamma_w$)

k = permeability (see Table VI-5-79)

M = oedometer modulus

γ_w = unit weight of water

t = time

H = drainage distance, which is equal to layer thickness for one side drainage, and equal to half the layer thickness for double side drainage.

(d) Full consolidation (i.e., U=100 percent) is in principle never reached. Consolidation of U=99 percent corresponds to T_c-. 2, whereas U=95 percent corresponds to T_c . 1.2. The necessary time for almost 100 percent consolidation is approximated in practice as

$$t_{U(100\%)} = \frac{2\gamma_w H^2}{k M}$$
(VI-5-225)

(e) By comparing t_U with the rise time of the wave-induced load, t_{rise}, it is possible to classify the wave loading and to estimate whether drained, partially drained or undrained conditions will be present. This criterion is given in Table VI-5-83.

$\dfrac{t_{rise}}{t_{U(100\ percent)}}$	Type of Loading	Soil Condition
$\gg 1$	Quasi-stationary	Completely drained
-1	Nonstationary	Partially drained
$\ll 1$	Nonstationary	Undrained

Table VI-5-83
Classification of Loading and Soil Conditions

(f) Typical wave loadings from nonbreaking waves on coastal structures have periods in the range T . $2(t_{rise})$ = 3-20 sec. Using the $t_{U(100\ percent)}$ values in Table VI-5-84, if follows from Table VI-5-83 that sand subsoil under virgin loading should generally be regarded as undrained, except for coarse sand which in some cases might be regarded as partially drained. Under subsequent wave loadings fine sand should still be regarded as undrained, whereas medium sand typically might be regarded as partially drained, and coarse sand would be considered drained.

(g) Very short duration impulsive loadings from waves breaking on structures have load rise times on the order of t_{rise} = 0.01 - 0.05 s (see Figure VI-5-101); and in this case all soils, including quarry-rock rubble foundations, have to be regarded as undrained.

(2) Wave load transmission to monolithic structure foundations.

(a) Wave loads transmitted to the foundation soil/rubble by monolithic structures, such as caissons and superstructure parapet walls, depend on the period of the wave load as well as the mass of the structure and the deformation characteristics of the soil/rubble.

Example 5-2. Calculation of $t_{U(100\ percent)}$ for quartz sand.

The elastic plastic component of M for initial loading corresponding to mean normal effective stress $\sigma N\# 300$ kPa is found to be

$$M = \begin{cases} 15 & MPa & loose\ sand \\ 150 & MPa & dense\ sand \end{cases}$$

The elastic component of M found by unloading and reloading at $\sigma N = 100$ kPa is found to be

$$M = \begin{cases} 80 & MPa & loose\ sand \\ 500 & MPa & dense\ sand \end{cases}$$

The drainage distance H is given as 5 m. Using these typical M-values together with the k-values given in Table VI-5-79, Equation VI-5-225 gives the consolidation times presented in Table VI-5-84.

		Table VI-5-84	
		Example of Consolidation Times for Sand	
		$t_{U(100\ percent)}$ (s)	
Material	Packing	Initial Deformation	Elastic Deformation
Coarse sand	loose	3	0.6
	dense	3	1
Medium sand	loose	30	6
	dense	30	10
Fine sand	loose	300	60
	dense	300	100

(b) The natural period $T_{n,s}$ of typical monolithic structures would normally be in the range 0.2 - 2 sec. If the period of the loading, T, is close to $T_{n,s}$ then dynamic amplification occurs resulting in increased loading of the foundation. Design wave loading can be separated into pulsating loads from nonbreaking waves and impulsive loads from waves breaking on the structure (see Figure VI-5-57). The pulsating loads have periods corresponding to the wave period, i.e., normally in the range 5-20 sec, which is much larger than $T_{n,s}$. Consequently, such low frequency loading is assumed to be transmitted to the foundation with unchanged frequency.

(c) Figure VI-5-92 illustrates how the resultant foundation load force of a wave-loaded caisson changes size, direction, and position during the wave cycle. The variation of the force resultant can be given by fully correlated time series of a tilting moment and a horizontal force. Figure VI-5-92 also illustrates the wave- induced stress variations in two soil elements (shown as hatched boxes).

(d) The initial shear stress τ_i prior to the installation of the structure is assumed to act under drained conditions, and the soil is assumed fully consolidated under this stress. $\Delta\tau_s$ is the change in the average shear stress due to the submerged weight of the structure. Depending on the type of soil, $\Delta\tau_s$ will initially act under undrained conditions, but as the soil consolidates, this shear stress will also be applied under drained conditions. In the case of rubble-mound foundations the consolidation will be instantaneous. For sand foundations drainage will occur rapidly, as indicated by Table VI-5-84, and it is reasonable to assume that the soil will consolidate before the structure experiences design wave loading. In addition, it is unlikely that pore pressures will accumulate from one storm to the next. For clays, consolidation occurs much more slowly, varying from months for silty-sandy very stiff clays to many years for soft clays. The amount of settlement and the corresponding increase in effective stresses, is calculated by ordinary consolidation theory the same as for structures on dry land.

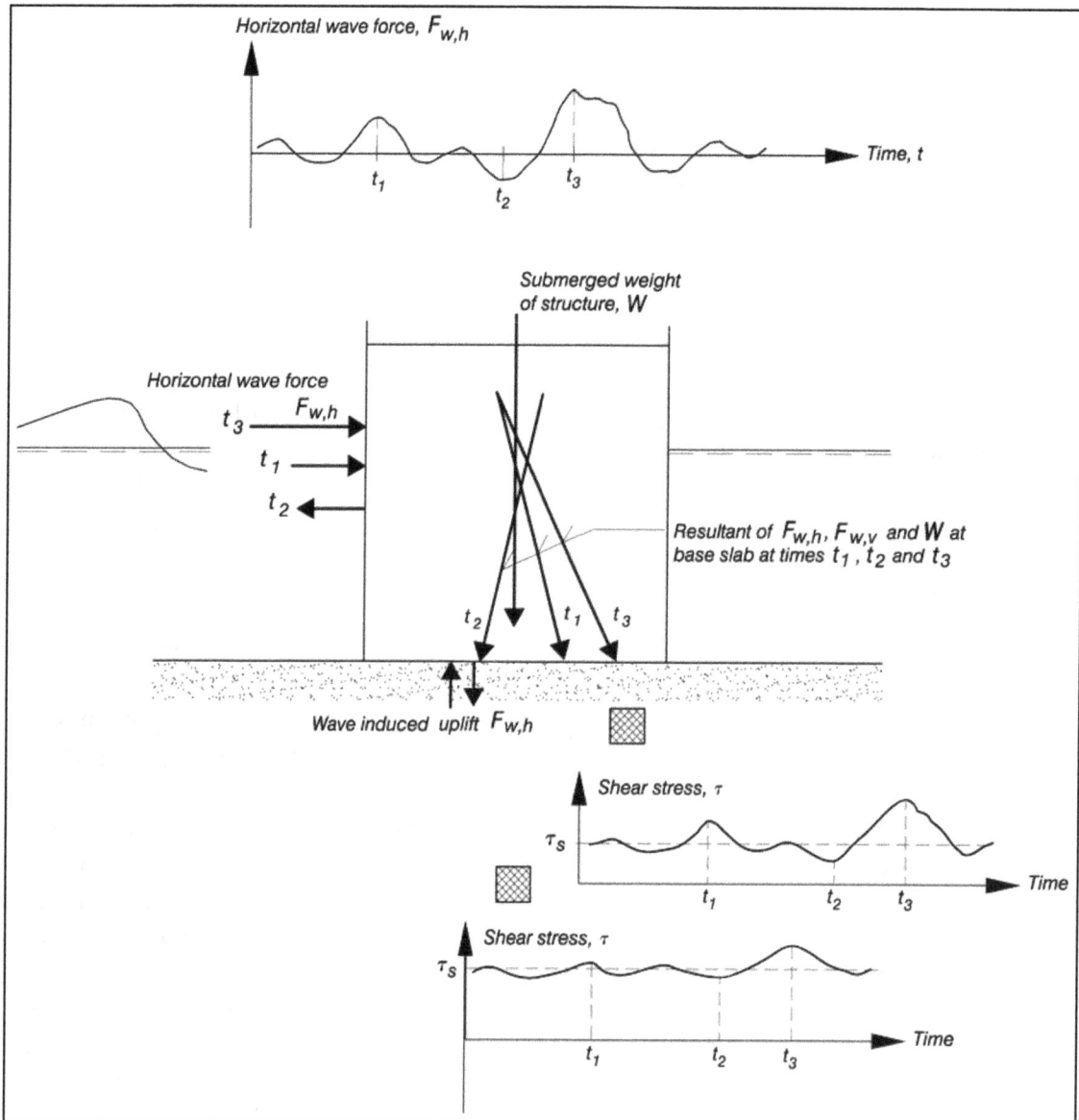

Figure VI-5-92. Illustration of wave induced forces on caisson foundation and related stress variations in the subsoil

(e) The effective static shear stress before wave loading is given by

$$\tau_s = \tau_i + \Delta\tau_s \tag{VI-5-226}$$

(f) The initial shear stress, τ_i, is determined by the submerged weight of the soil as $\tau_i = 0.5 (1 - K_o) p_o N$, where K_o is the coefficient of earth pressure at rest, and $p_o N$ is the vertical effective overburden pressure. $\Delta\tau_s$ can be estimated from Newmark's influence diagrams, assuming homogeneous, isotropic and elastic soil (e.g., see Hansbo 1994 and Lambe and Whitman 1979). This is usually a good approximation if the soil is not close to failure. A rough rule of thumb is a load spreading of 1 (horizontal) to 2 (vertical).

(g) The behavior of the soil when exposed to the cyclic loading can be studied in triaxial tests or direct simple shear (DSS) tests. The irregular wave loading F_W during the design storm might be approximated by equivalent cyclic wave loadings, causing cyclic shear stress variations with amplitude τ_{cy} as given in Figure VI-5-93. However, it is more correct if the real stress variations in the subsoil, as illustrated in Figure VI-5-92, are approximated by an equivalent cyclic variation. The stress τ_{cy} should be determined by finite element analysis.

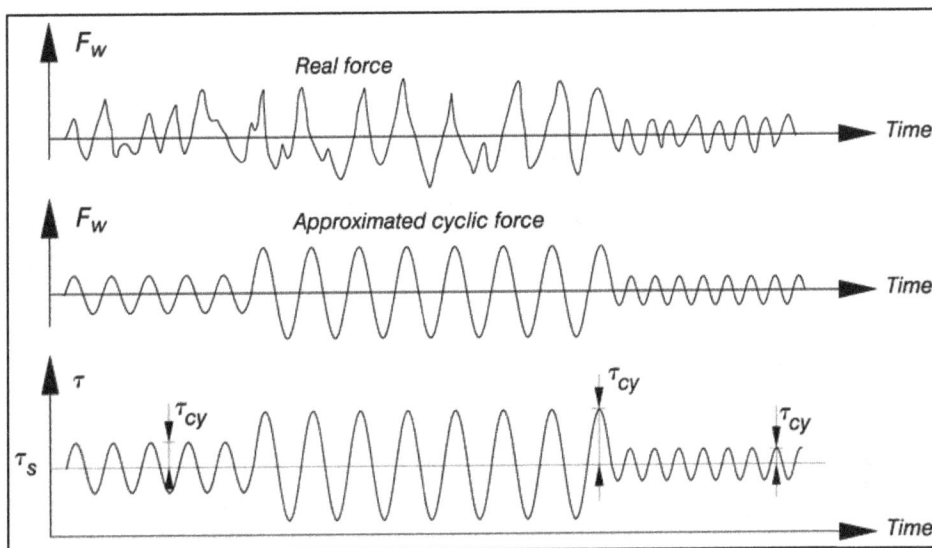

Figure VI-5-93. Illustration of approximate cyclic wave loading and
related cyclic shear stress variation in a subsoil element during a
storm sequence

(h) The criterion for determination of the equivalent cyclic stress in terms of τ_{cy} and number of cycles N_{eqv}, is that the approximation gives the same effect as the actual load history. Procedures to determine N_{eqv} were presented by Andersen (1981, 1983). For sands, N_{eqv} may be computed by accumulating the permanent pore pressure generated during the cyclic load history, taking into account that drainage is likely to occur during the design storm. Calculation of the pore pressure accumulation can be performed using pore pressure diagrams established from cyclic stress-controlled laboratory tests. The dissipation of the permanent pore pressure due to both drainage towards free boundaries and grain redistribution can be determined by finite element analysis or, for idealized situations, by closed-form solutions. In principle, the cyclic shear strength of clays could also be computed by accumulating the permanent pore pressure. However, measurements in clays are more difficult to acquire than in sands. In addition, short-term drainage will not take place in clays; consequently, it is preferable to use the shear strain as a measure of the cyclic strength for clays. Moreover, for situations where the cyclic shear moduli under undrained conditions are of primary interest, the shear strain will also be a more direct parameter than the pore pressure.

(i) The stress conditions in the soil beneath structures subjected to combinations of static and cyclic loads are very complex even though the irregular loadings are approximated by equivalent cyclic loadings. Advanced finite element numerical modeling is the obvious tool for calculation of stress and strain development provided the model is carefully verified against documented test cases. As an alternative, a practical approximate method is presented by Andersen (1991) and Andersen and Høeg (1991). This method is based on the stress path philosophy in which laboratory tests are performed to simulate the stress conditions in few typical soil elements along potential failure surfaces as illustrated in Figure VI-5-94. The elements follow various stress paths which might be approximated to triaxial or direct simple shear (DSS) types of loading corresponding to various conditions of average stresses, τ_s and cyclic shear stresses, τ_{cy}. Additionally, the number of cycles to failure, N_f, and the shear strains are determined in the tests.

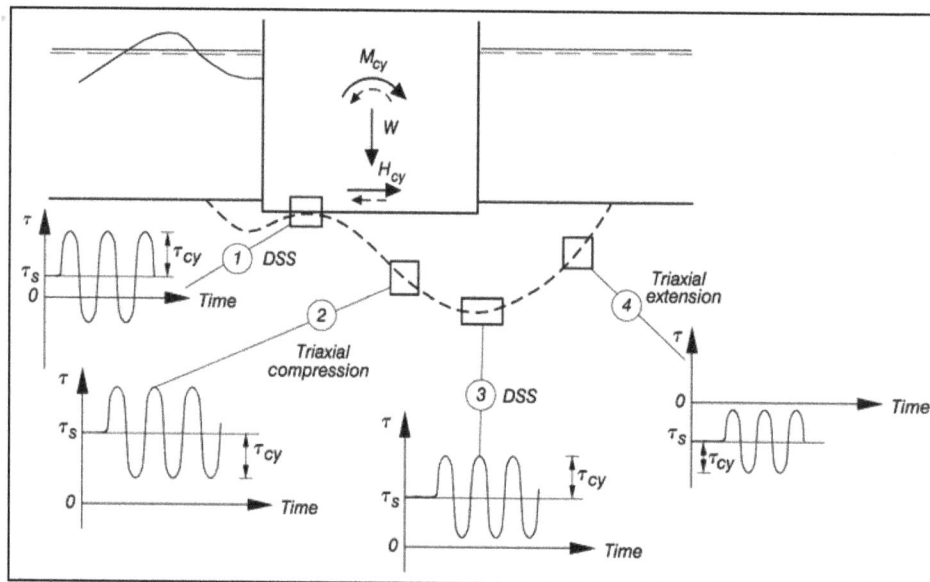

Figure VI-5-94. Simplified stress conditions for some elements along a potential failure surface (Andersen 1991)

(3) Noncohesive soil exposed to wave-induced cyclic loadings.

(a) For noncohesive soils, cyclic stress variations can either lead to strengthening of the soil or to soil weakening and eventual liquefaction due to pore pressure build-up. The outcome depends on soil permeability, average shear stress τ_s, wave-induced shear stress variations, and soil compaction. Pore pressure build-up does not happen in coarse materials like gravel and rubble foundation materials because of almost instant drainage. Consequently, only sand-sized noncohesive soils will be considered in the following discussion.

(b) Cyclic loading of soil specimens can be performed in undrained triaxial tests using a cell height-to-width ratio of one and lubricated cap and base, thus assuring uniform stress-strain conditions in the sample (Rowe and Barden 1964; Bishop and Green 1965; and Jacobsen 1967). From such tests the phenomena depicted in Figure VI-5-95 can be observed.

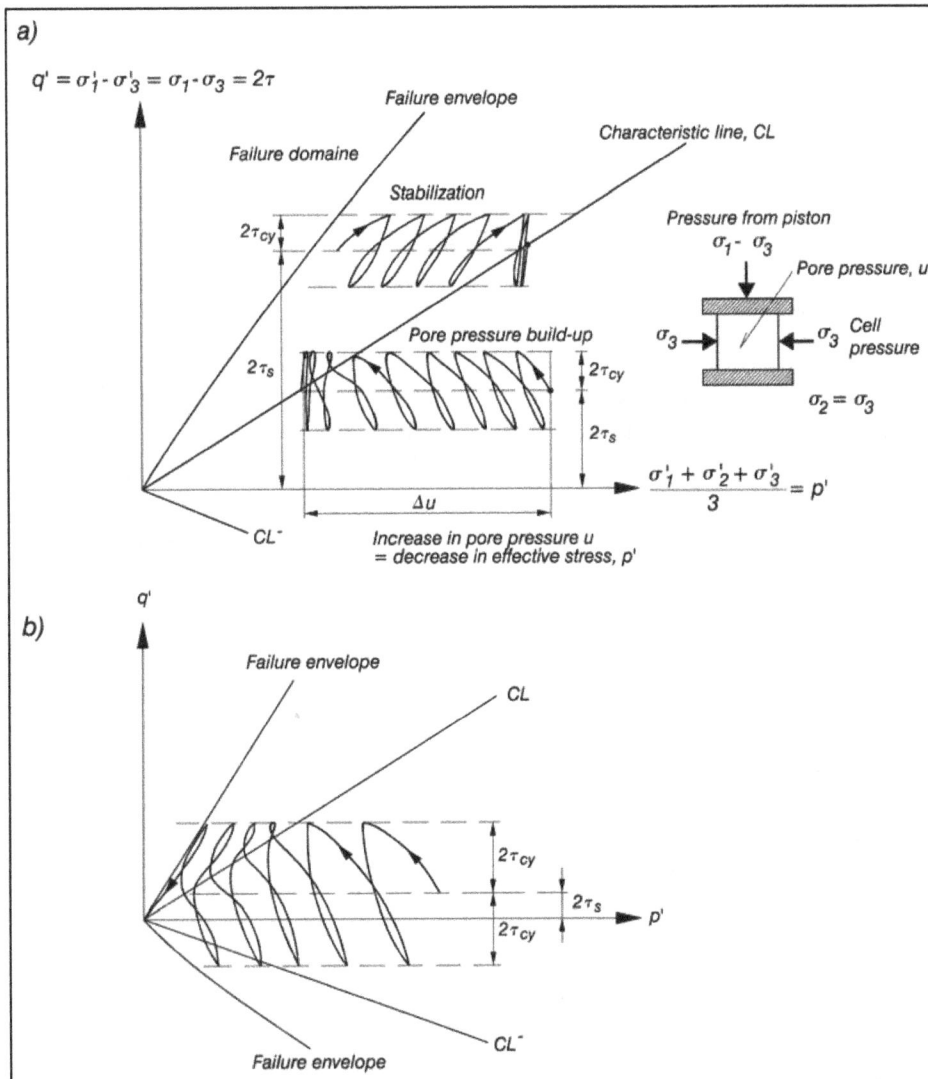

Figure VI-5-95. Illustration of (a) stabilization and pore pressure build-up, and (b) liquefaction undrained triaxial test on sand

(c) The shear stress τ is given by

$$\tau = \frac{\sqrt{3}}{2}\sqrt{J_2}$$

(VI-5-227)

where

$$J_2 = \frac{1}{6}\left[(\sigma_{1'} - \sigma_{2'})^2 + (\sigma_{2'} - \sigma_{3'})^2 + (\sigma_{1'} - \sigma_{3'})^2 \right] \tag{VI-5-228}$$

and $\sigma_1 \unicode{x2267} \sigma_2 \unicode{x2267} \sigma_3$ are the effective stresses in three orthogonal directions.

(d) The average effective stress level is given by

$$p' = \frac{\sigma_{1'} + \sigma_{2'} + \sigma_{3'}}{3} = \frac{\sigma_1 + \sigma_2 + \sigma_3}{3} - u \tag{VI-5-229}$$

where σ is total stress and u is the pore pressure, as in Equation VI-5-202. In undrained triaxial tests with cell pressure $\sigma_2 = \sigma_3$ the piston generated stress (deviator stress) is

$$q' = \sigma_{1'} - \sigma_{3'} = \sigma_1 - \sigma_3 = 2\tau \tag{VI-5-230}$$

(e) In the $q - p$ diagram of Figure VI-5-95 the characteristic line (CL) separates stress domains where deviator stress fluctuations cause dilation and contraction. The CL signifies a stable state where further cyclic loadings will not lead to hardening or softening of the soil. Figure VI-95a shows that if the average stress τ_s is situated above the CL, the cyclic test will generate negative pore pressures leading to stabilization (hardening) of the soil.

(f) If τ_s is situated below the CL, cyclic tests will generate positive pore pressures and decreasing effective stress (softening). With small τ_s and large stress fluctuations τ_{cy}, liquefaction will occur as shown in Figure VI-5-95b if the stress path touches the CL⁻ line.

The equations for the CL and CL⁻ lines are

$$\text{CL:} \quad q' = \frac{6 \sin \varphi_{c'rit}}{3 - \sin \varphi_{c'rit}} p' \tag{VI-5-231}$$

$$\text{CL}^-: \quad q' = \frac{-6 \sin \varphi_{c'rit}}{3 + \sin \varphi_{c'rit}} p' \tag{VI-5-232}$$

where φ_{crit} is the critical angle of friction, as given in Table VI-5-75. φ_{crit} is independent of the relative density or porosity and is very close to 30 deg for sand in the range $d_{50} = 0.14 - 0.4$ mm (Ibsen and Lade 1998). The number of cycles to failure can be determined from a series of triaxial or DSS laboratory tests conducted with various combinations of τ_s and τ_{cy}.

(g) The previous discussion of the effect of cyclic loading is related to undrained conditions in laboratory tests. The assumption of undrained conditions is either true or on the safe side with respect to soil strength properties. However, sands in nature may experience partial drainage during a storm. The amount of drainage depends upon the permeability of the sand and the drainage boundary conditions. The drainage can be significant and should be considered in design because experience from laboratory tests has shown that the soil structure and the resistance to further pore pressure generation may be significantly altered when the excess pore pressure due to cyclic loading dissipates (Bjerrum 1973; Andersen et al. 1976; Smits, Anderson, and Gudehus 1978). Cyclic loading with subsequent pore pressure dissipation is referred to as precycling.

(h) Moderate precycling in sands may lead to significant reduction in pore pressure generation under further cyclic loading, even in dense sands. Precycling may occur during the first part of the design storm. The beneficial effect of precycling might be taken into account in cyclic testing of sand in the laboratory by applying some precycling prior to the main cycling. As previously mentioned, the shear strength that the soil can mobilize to resist the maximum load (wave) depends on the effective stresses in the soil, and thus on the excess pore pressure that is generated during the storm. The shear strength also depends on whether the soil is contractive or dilative. If the soil is dilative and saturated, a negative pore pressure is generated when the soil is sheared under undrained conditions. This will give higher shear strength than achieved for drained conditions. However, for sands one should be careful about relying fully on higher shear strength caused by negative pore pressure due to uncertainty about the amount of drainage that might take place. The amount of drainage during a cycle and the residual pore pressure at the end of a storm might be estimated from calculations with finite element programs. Examples of design diagrams based on such calculations are pressented in de Groot et al. (1996). A method valid for the estimation of the changes in pN in sand as function of the number of cycles was given in Ibsen (1999).

(4) Cohesive soil exposed to wave-induced cyclic loadings.

(a) The shear strength, c_u, of clay normally refers to undrained strain controlled tests of approximately 1-3 hr duration to reach failure. Clays will be practically undrained during a storm, and possibly also over a seasonal period including several storms. Because c_u for a specific clay in undrained conditions depends solely on the initial effective stress conditions before the loading, there will be only insignificant changes in c_u as long as drainage of the clay has not taken place.

(b) The stress-strain behavior of a specific clay determined from samples is affected by the test method, OCR, τ_s, τ_{cy}, N and the stress rate (load frequency). During the cyclic loading the pressure build-up causes a reduction of the effective stresses as illustrated in Figure VI-5-96. Figures VI-5-96a and VI-5-96b show development of failure by cyclic loading. Figure VI-5-96c shows stabilization of effective stress after 25 cycles.

(c) After a certain number of cycles, the failure envelope will be reached and large shear strains developed. The cyclic shear strength can be defined as

$$\tau_{f,cy} = (\tau_s + \tau_{cy}) \tag{VI-5-233}$$

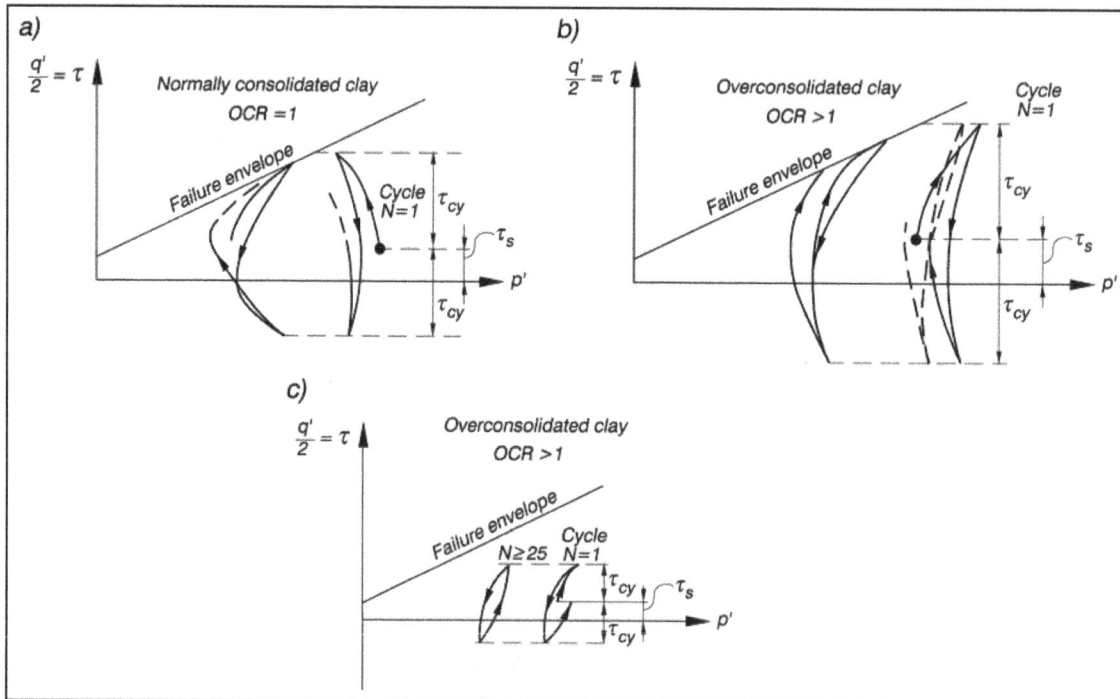

Figure VI-5-96. Illustration of effective stress paths for clay samples in undrained triaxial tests

(d) It is very difficult to determine accurately the change in pore pressure, and therefore, also the change in effective stresses in triaxial and DSS tests. Consequently, to determine the relationship between the shear strength c_u and τ_s, τ_{cy}, and number of cycles, N, it is better to examine the load increase to failure in normal static tests for samples already exposed to various ranges of cyclic loadings. From the load increase the actual c_u-value after a specific exposure in terms of τ_s, τ_{cy}, and N can then be estimated. Examples and information on such post-cyclic static shear strength are presented in Andersen (1988). For Norwegian Drammen clay, being a plastic clay with plasticity index $I_p = 27$ percent, it was found that cyclic loading causing large cyclic shear strains also caused significant reduction in the static shear strength. The reduction increases with the number of cycles. It was also found that the reduction is generally less than 25 percent as long as the cyclic shear strains are less than 3 percent and the number of cycles less than 1,000. This holds for OCR-values of 1, 4, and 10. Figure VI-5-97 shows an example of stress-strain behavior of Drammen clay. This example shows the importance of modeling the type of loading correctly when trying to determine the stress-strain behavior or the shear modulus in situ from laboratory tests.

(e) The number of cycles to failure, N_f, can be determined from a series of triaxial or DSS laboratory tests applying various combinations of τ_s and τ_{cy}. Due to the very large shear strain at failure, it is often appropriate to define failure as a lower strain level, the value of which must depend on the type and function of the structure. The test results can conveniently be plotted in diagrams as shown in Figure VI-5-98, where failure is taken when either the cyclic strain, γ_{cy}, or the average strain, γ_s, reaches 15 percent.

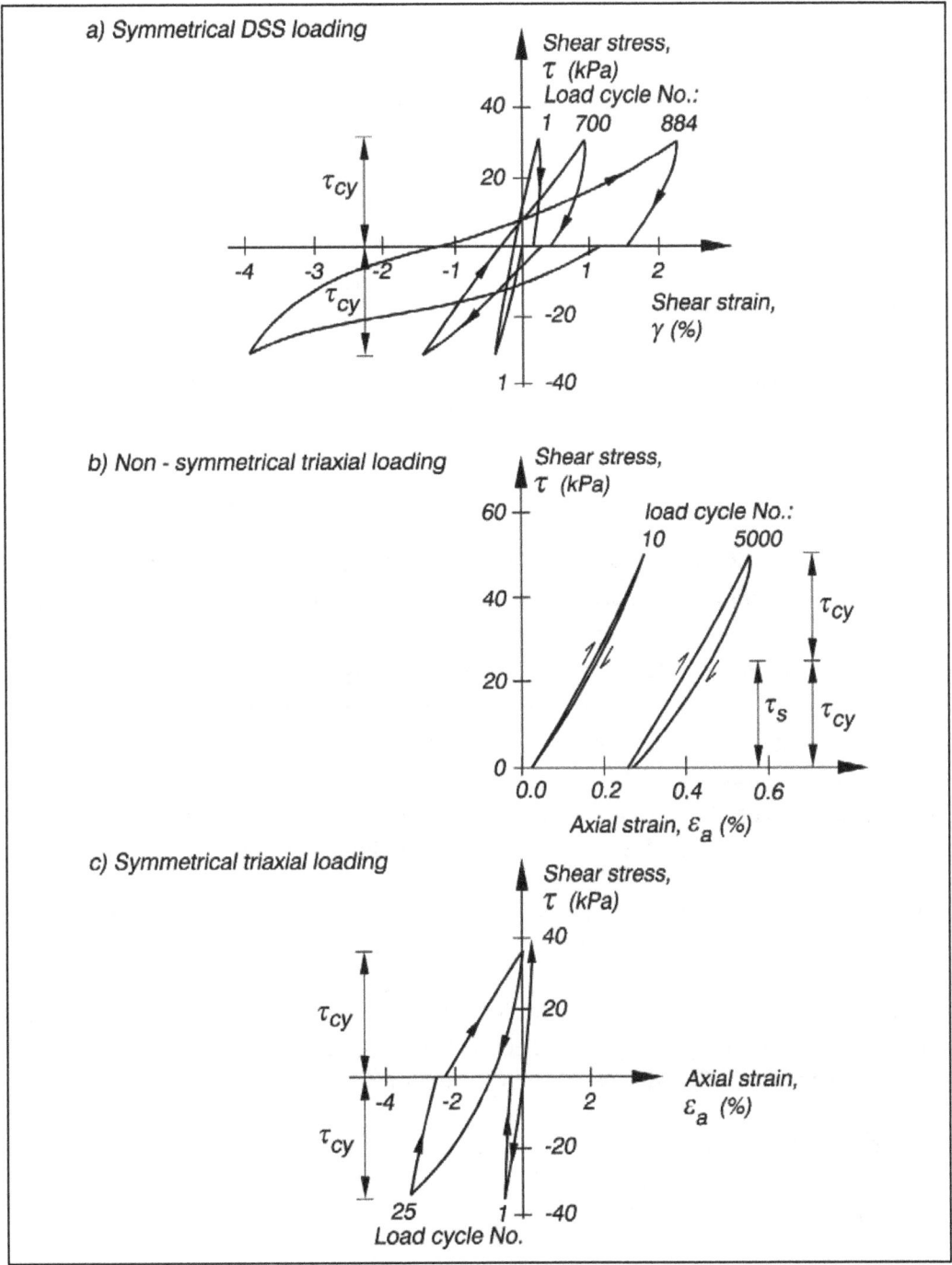

Figure VI-5-97. Stress strain behavior of Drammen clay (I_p = 27 percent) under various cyclic loading conditions corresponding to OCR = 4 (from Norwegian Geotechnical Institute 1992)

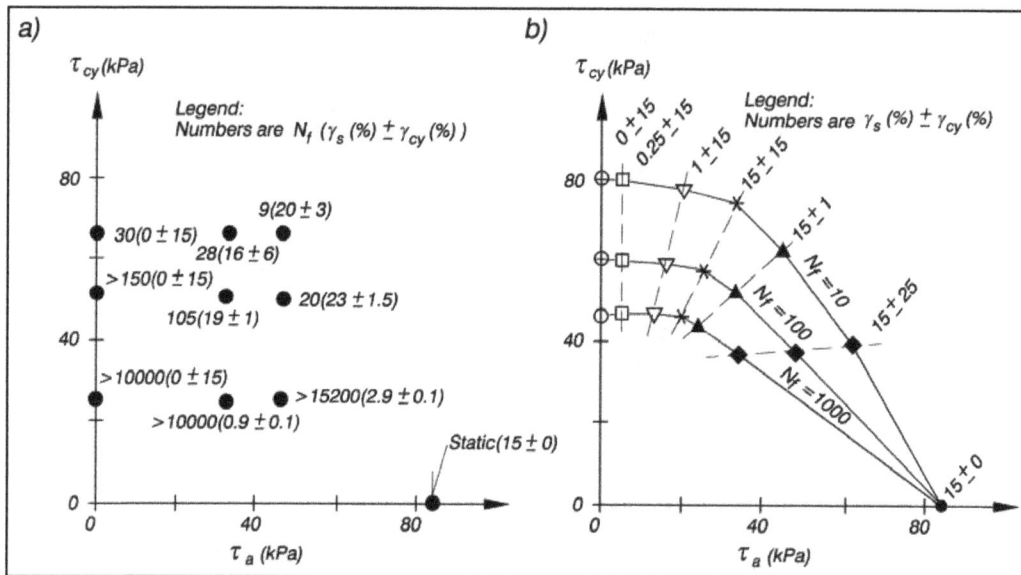

Figure VI-5-98. Result of cyclic tests on normally consolidated Drammen clay, with $OCR = 1$ and $I_p = 27$ percent (from Norwegian Geotechnical Institue 1992)

(f) In Figure VI-5-98 N_f is number of cycles to failure defined as either the cyclic strain γ_{cy} or the average strain γ_s reaching 15 percent. Figure VI-5-98a shows individual test results, and Figure VI-5-98b shows interpolated curves based on the individual tests. A diagram like Figure VI-5-98b can be transformed to normalized form using the vertical effective stress σ_{vc}N at the end of the cycling (consolidation), and the undrained static shear strength, σ_u, measured in strain-controlled tests. Figure VI-5-99 shows an example based on both triaxial and DSS tests.

(g) In Figure VI-5-99 σ_u^E, σ_u^C, and σ_u^{DSS} are undrained static shear strength in triaxial compression and extension tests and in DSS tests, respectively.

(h) By replotting the data from Figure VI-5-99 it is possible to show the relationship between the cyclic shear strength, $\tau_{f,cy}$, as defined by Equation VI-5-233, and N_f, σ_{vc}N and the undrained static shear strengths. An example is shown in Figure VI-5-100.

(i) A simple diagram for approximate correction of the static failure load to take into account the effect of cyclic loading in static calculations is presented in de Groot et al. (1996) for Drammen clay ($OCR = 1$, $= 4$ and $= 40$).

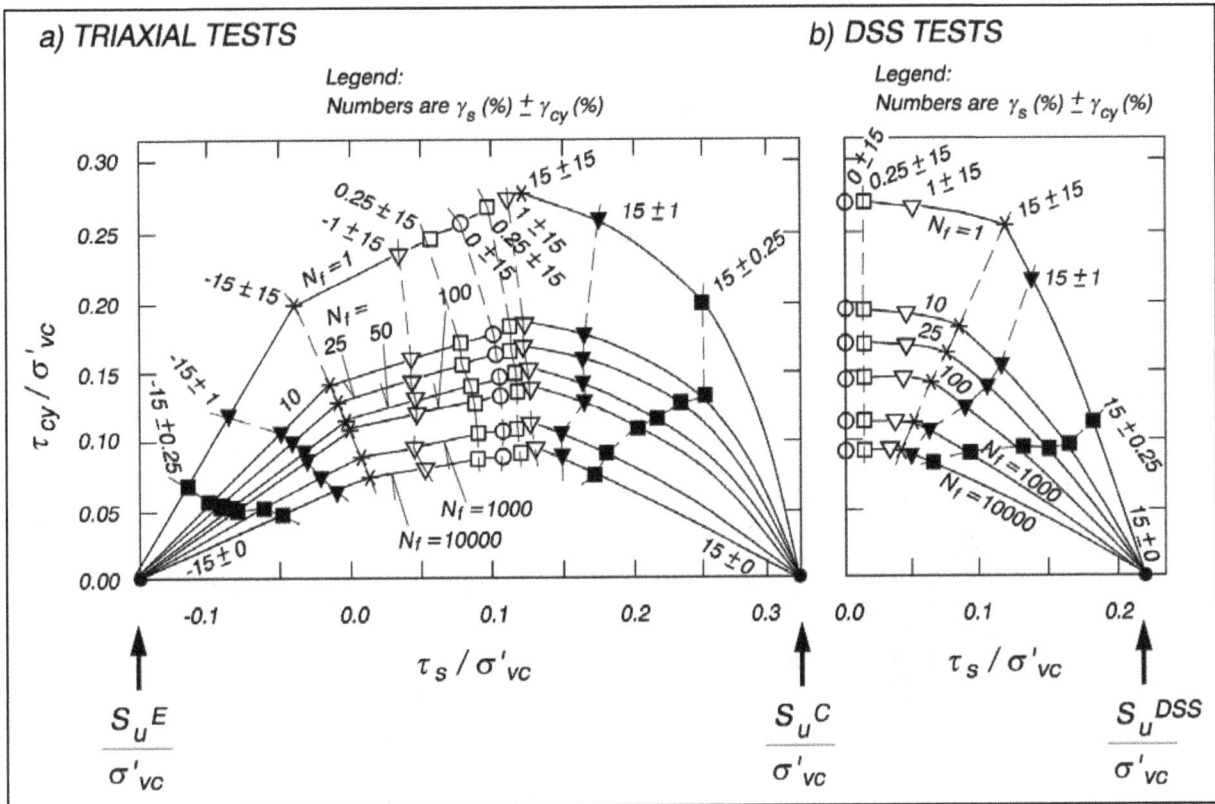

Figure VI-5-99. Example of normalized diagrams for cyclic loading of Drammen clay with *OCR* = 1, in triaxial tests (a), and DSS tests (b) (from Norwegien Geotechnical Institute 1992)

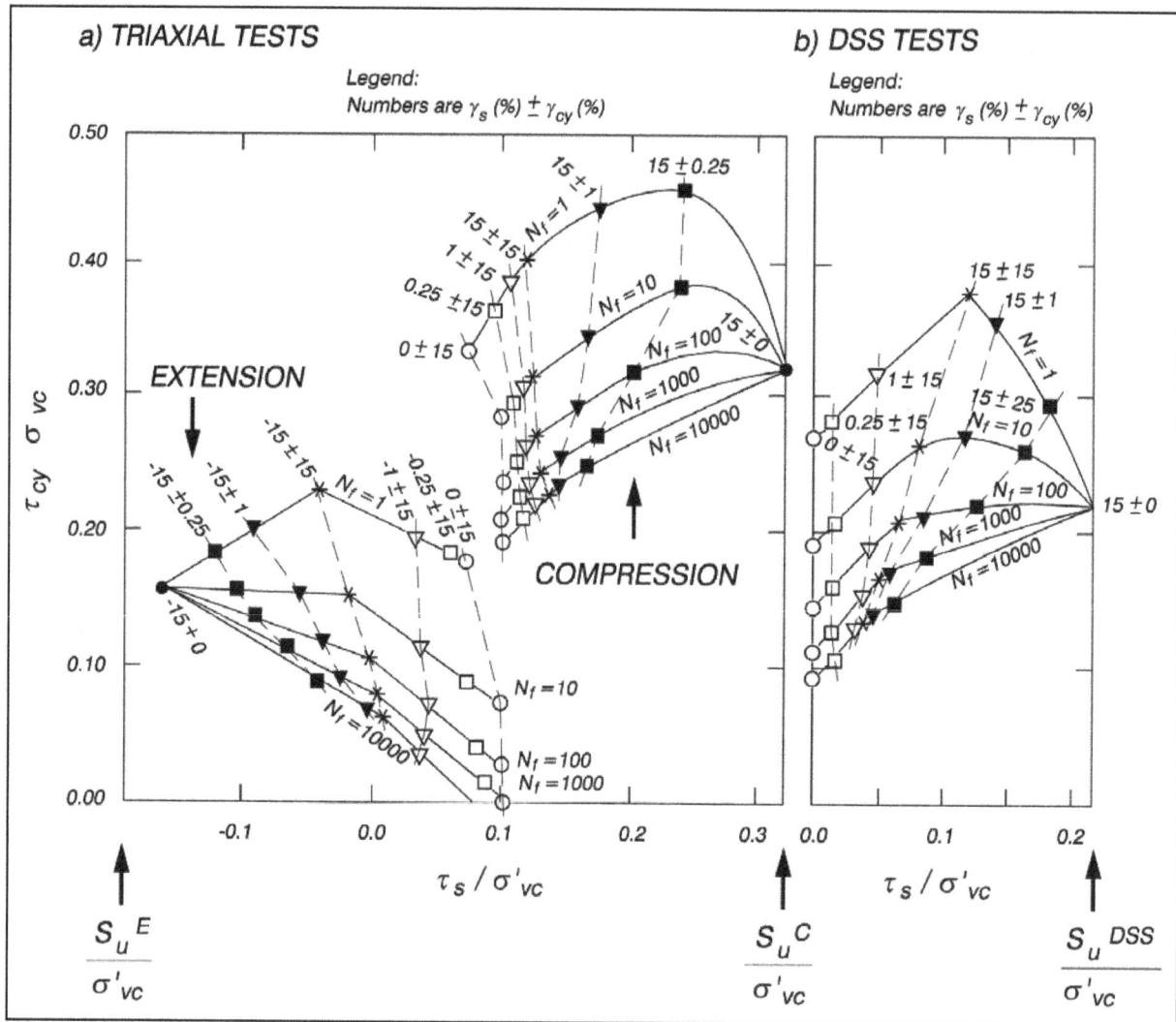

Figure VI-5-100. Cyclic shear strength of Drammen clay with $OCR = 1$
(from Norwegian Geotechnical Institute 1992)

    f. Dynamic loading of soils under monolithic structures.

    (1) Dynamic loading of soils and rubble rock foundations occurs when wave wall superstructures and vertical wall breakwaters are exposed to impulsive loads from waves breaking at the structures, as shown in Figure VI-5-56. The impulsive load magnitude can be very large, but the loads have very short durations with load periods in the range 0.1-1.0 sec for the peaked part of the loading. Because the natural period of some structures often are within (or close to) the same period range, dynamic amplification of the wave load and corresponding structure movements might occur.

    (2) When moderately loaded, the soil and rubble rock will react approximately as an elastic material; whereas under severe loading, permanent deformations will occur, corresponding to plastic behavior.

(3) Determination of impulsive wave forces caused by waves breaking directly on
vertical wall structures is extremely uncertain. The same can be said about the related loading on
the foundation. In addition, breaking wave loads can be very large; therefore, direct wave
breaking on the structure should be avoided. If necessary, the geometry or position of the
structure should be changed to avoid large impulsive wave forces. In cases where the wave load
is known, it is possible to obtain some estimates of the effect on the foundation as explained in
the following paragraphs.

(4) The actual time of the wave loading is an important factor in the dynamic
amplification. Model studies by Bagnold (1939) and Oumeraci (1991) showed that the load
history of forces from waves breaking on vertical walls can be approximated with a church-roof
like time-history as sketched in Figure VI-5-101.

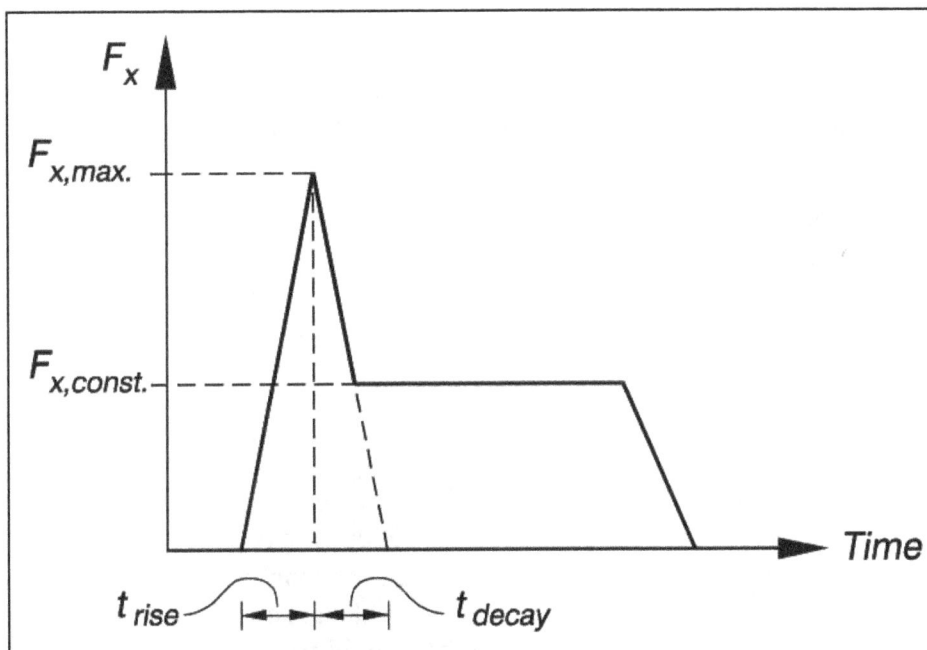

Figure VI-5-101. Approximation to horizontal wave load history for
waves breaking directly on vertical walls

(5) For the elastic case it is possible to get a crude estimate on the dynamic amplification
by modeling the soil-structure system as a rigid body resting on a linear elastic half-space,
idealized by a lumped mass system where the geodynamic response is represented by a
spring-dashpot model. A two-degrees-of-freedom system allowing only translatory motion, x, in
the horizontal direction and rotation, v, about the center of gravity, C_g, is commonly considered
(see Figure VI-5-102).

(6) The effect of any impulsive loading can be found by solving the equations of motion
for the complete translatory and rotational motion, provided the stiffness and damping
coefficients are known. However, for practical design purposes a simple static approach can be
accomplished by assuming an equivalent static load which will induce the same motions of the
structure as those found from a dynamic calculation. The following definitions of dynamic load

factors, Ω, show how the equivalent static force and motions are related to the dynamic force and motions.

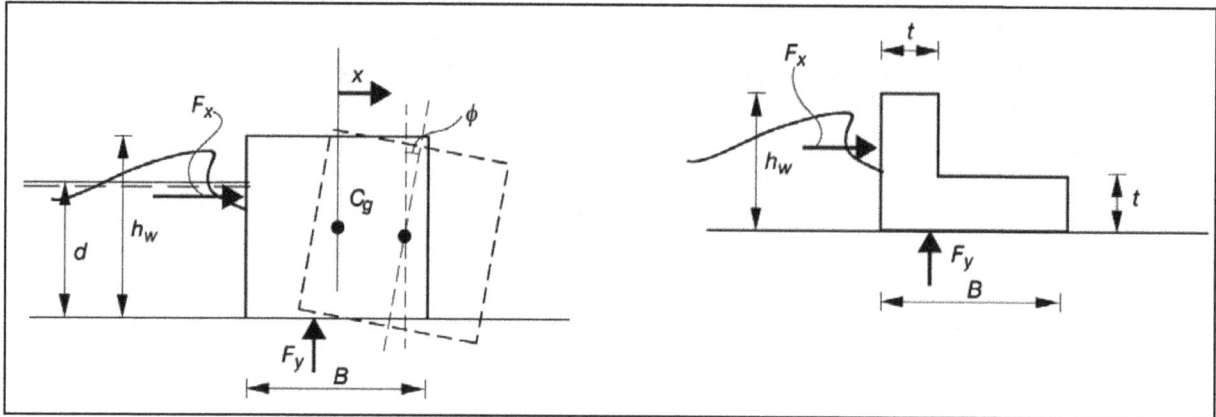

Figure VI-5-102. Definition of translatory and rotational motions and dimensions for caisson structure and parapet wave wall exposed to dynamic loading

$$\Omega_x = \frac{\max \, dynamic \, translation}{\max \, static \, translation} = \frac{\max\{x\}}{F_{x,\max} / kx} = \frac{F_{x,equiv}}{F_{x,\max}}$$

(VI-5-234)

$$\Omega_\varphi = \frac{\max \, dynamic \, rotation}{\max \, static \, rotation} = \frac{\max\{\varphi\}}{M_{\max} / k_\varphi} = \frac{M_{equiv}}{M_{\max}}$$

where $F_{x,\max}$ is defined in Figure VI-5-101, k_x and k_v are stiffness coefficients, and $M_{\max}$ is the maximum wave-load-induced moment around the center of gravity. The moment also includes wave-generated uplift forces, F_y. If Ω_x, Ω_v, k_x, k_v, and the maximum wave loading F_x and M_v are known, then the maximum motions and related equivalent static wave loadings can be determined. The vertical motion is of little interest for monolithic structures under predominantly horizontal wave loading.

(7) Pedersen (1996, 1997) presented diagrams of Ω_x and Ω_v for caissons and wave wall superstructures with square footings (i.e., $B \times B$ shown in Figure VI-5-102) exposed to the type of loading shown in Figure VI-5-101. The soil was modeled as a linear elastic half-space. Pedersen used results of Lysmer and Richardt (1966) and Hall (1967) to obtain expressions for optimized constant values of stiffness and damping coefficients. An example of Pedersen's diagrams for caisson structures is shown in Figure VI-5-103 for load history $t_{rise}/t_{decay} = 1$ under triangular loading. T_{nd} is the coupled, damped natural period of the caisson. Pedersen showed that the constant part of the wave loading following the peak has little influence on the response if $F_{x,const} \# 0.5 \, F_{x,max}$.

(8) Due to the many uncertainties and simplifying assumptions, diagrams such as shown in Figure VI-5-103 should be used only for judging the possibility of dynamic amplification. If dynamic amplification factors are found to be close to or greater than 1, then a detailed dynamic analysis should be performed or the structure design should be changed.

g. Slip surface and zone failures.

(1) Slip surface and zone failure calculations are based on limit state calculations related to assumed or approximate rupture figures. Two different solutions are applied:

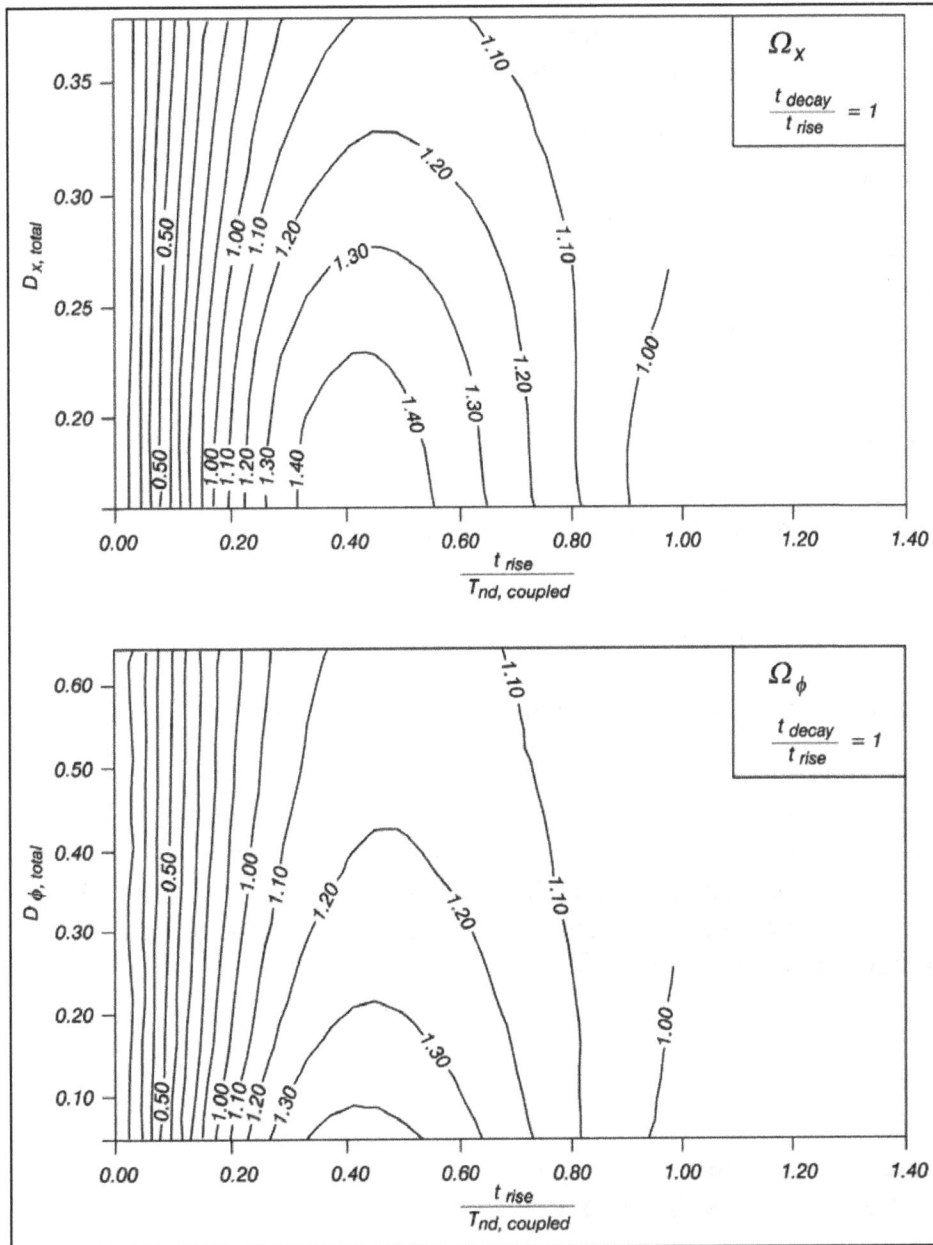

Figure VI-5-103. Amplification factors for translatory and rotational
motions for caisson structure with square footing and triangular load shape
(Pedersen 1997)

(a) Statically admissible solutions are defined by stress distributions that satisfy equilibrium for stresses and loads for all involved soil elements. In homogeneous soils with sufficiently simple boundary conditions, e.g., straight and uniformly loaded boundaries, these types of approximate solutions may represent a simple and efficient solution technique. Many standard formulas and calculation methods in soil mechanics for bearing capacity and earth pressure problems are derived from statically admissible solutions. However, even slight modifications of the boundary conditions, and especially the introduction of inhomogeneous soil properties, may make a realistic solution of this type extremely complicated. Consequently, statically admissible solutions do not represent a generally applicable solution method, even if a limited number of standard cases are known and are widely used.

(b) Kinematically admissible solutions are defined by displacement fields that satisfy the boundary conditions for displacements as well as the associated flow rule (normality condition) within the theory of plasticity. Satisfying the flow rule makes the use of work equations possible. The flow rule requires the angle of friction v and the angle of dilation ψ to be equal, although this is not true for frictional materials. To overcome this problem Hansen (1979) proposed to set $\psi = v = v_d$ where the modified angle of friction v_d is defined by

$$\tan \varphi_d = \frac{\sin \varphi \cos \psi}{1 - \sin \varphi \sin \psi} \tag{VI-5-235}$$

(c) When applying v_d it follows that both statically and kinematically admissible solutions will always be on the safe side. Otherwise statically admissible solutions will either be correct or on the safe side, whereas kinematically admissible solutions, according to the upper bound theorem, will either be correct or on the unsafe side.

(2) Experience indicates that solutions based on realistic rupture figures are in both cases generally close to the true situaton.

(3) For a given structure it is necessary to identify the most critical rupture figure, defined as the one which provides the lowest bearing capacity. For example, if work equations are used, then the rupture figure corresponding to the lowest ratio of work of stabilizing forces W_s to work of destabilizing forces W_d is the critical rupture figure. In any case in order to prevent failure and to have some safety the condition

$$\min\left(\frac{W_s}{W_d}\right) \geq 1 \tag{VI-5-236}$$

must be fulfilled. If not, the structure design has to be modified or the soil strength improved (by preloading, compaction, or installation of drains), or the soil must be replaced.

(4) For a number of standard cases the rather complicated equations related to statically and kinematically admissible solutions have been simplified to practical force equations, formulae, and diagrams (e.g., the determination of foundation bearing capacity and soil pressures on walls). The formulae and diagrams are based not only on the basis of theoretical solutions but also on model tests and field experience. This compensates for non-exact kinematically admissible solutions.

(a) Stability of slopes.

- Slope instability failure modes for coastal structures are schematized by the various slip failure surfaces shown in the figures in Part VI-2-4b. Slope instability is a conventional soil mechanics problem which is dealt with in almost every handbook on geotechniques and foundation engineering, e.g., Terzaghi and Peck (1944), Taylor (1958), Lambe and Whitman (1979), Anderson and Richards (1985), and Hansbo (1994). However, the conventional treatment of the subject does not pay attention to wave loadings which characterize the special conditions for coastal structures.

- Direct wave action on a permeable slope increases the antistabilizing forces because the runup presents an extra load and creates fluctuating pore pressures and related antistabilizing hydraulic gradients in the structure. In addition, both waves and tides create pore pressure gradients in porous seabeds.

- Slope instability rarely occurs in conventionally designed rubble-mound structures. Stability problems can occur if the structure is placed on weak soils or on soil with weak strata because the slip failure plane passes through weaker materials. Very large breakwaters with steep slopes might be suspectable to stability problems within the structure itself especially if exposed to earthquake loading. Another type of failure related to rubble-mound slopes is sliding of one layer over another layer which is caused by reduced shear strength at the interface between two layers of narrow graded materials of different particle size and shape, e.g., armor layer and filter layer. If geotextiles are used, the interface shear strength is significantly reduced.

- The two load categories pertinent to coastal structure slope stability are listed below:

Long-term stability	Permanent loads, i.e., weight of structure and soils, permanent surface loads, and average loads from groundwater.
Short-term stability	Permanent loads as well as variable loads from waves (direct wave loading and seepage forces), seismic activity and vehicles. Ice loads are usually not dangerous to slope stability.

- For each of the load cases it is important to apply the relevant soil strength parameters. This includes consideration of soil strength degradation related to variable loadings, as discussed in Part VI-5-5e of this chapter.

- Variable loads from waves and the related seepage forces should be considered for the two instantaneous load situations depicted in Figure VI-5-104. The pore pressures and the related hydraulic pressure gradient and seepage forces in a homogeneous, isotropic breakwater structure can be estimated from flow nets if the Darcy equation (Equation VI-5-220) is taken as valid, or calculated using advanced numerical models. In Figure VI-5-104 the seabed is assumed to be impermeable compared to the breakwater. This is usually a good approximation for rubble-mound structures built of quarry materials.

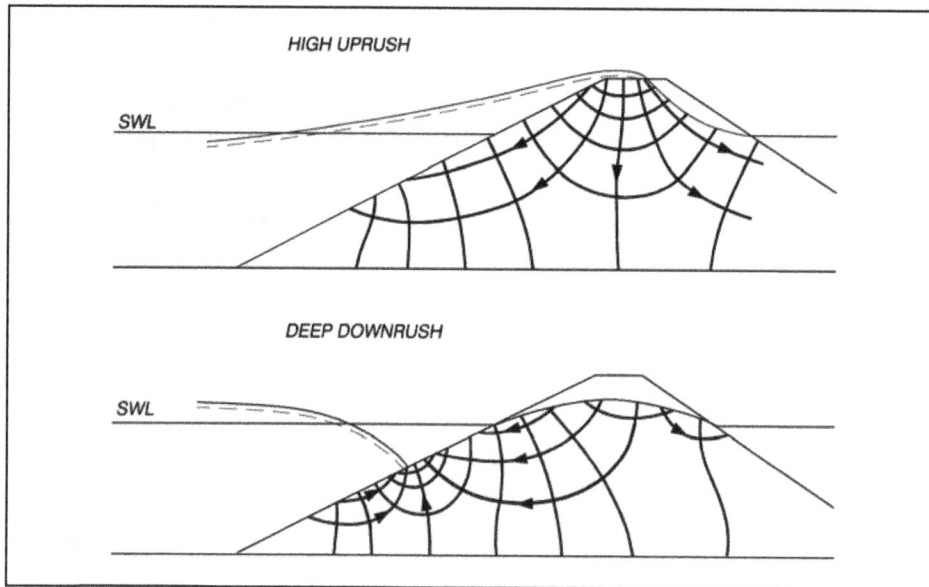

Figure VI-5-104. Illustration of flow nets in a homogeneous isotropic
breakwater for two instantaneous wave load situations

- The pore pressure variation in a homogeneous seabed due to water level changes caused by tides and waves can be estimated by the method of de Rouck (1991) as shown in Table VI-5-85. The pore pressure in deeper strata corresponds to the hydrostatic pressure at mean water level. However, some seepage forces are created due to the reduction in pressure at the seabed surface beneath a wave trough during low tide. Tidal variations only cause vertical seepage forces due to the long tidal wavelength. However, short waves also cause horizontal seepage forces that are generally smaller than the vertical seepage forces. Figure VI-5-105 illustrates the flow net related to wave action.

- Equation VI-5-238 in Table VI-5-85 assumes that the compressibility of seawater is negligible compared to that of the grain skeleton, which is almost always the case. The pore pressure variations in the seabed underneath a rubble-mound structure can be determined from Equation VI-5-238 by estimating u_0 along the seabed surface using flow nets similar to those illustrated in Figure VI-5-104.

- It follows from Equation VI-5-238 that the attenuation of u with depth z decreases with more permeable and stiffer soil and with longer wave periods. Pore pressure variations due to tides ($T = 12h\ 25\ min$) are only very slightly attenuated in sand, but there is a significant attenuation in clay. Pore pressure variations due to wind generated waves ($T < 20\ s$) are strongly attenuated, even in sand.

- Seismic loads are usually taken into account by adding the seismic related horizontal inertia forces to the forces acting on the soil along with additional hydrodynamic forces which might result from the displacement of the soil body. Possible seabed scour should be taken into account when defining the bottom topography.

Table VI-5-85
Wave and Tide Induced Pore Pressures in Permeable Seabeds (de Rouck 1991)

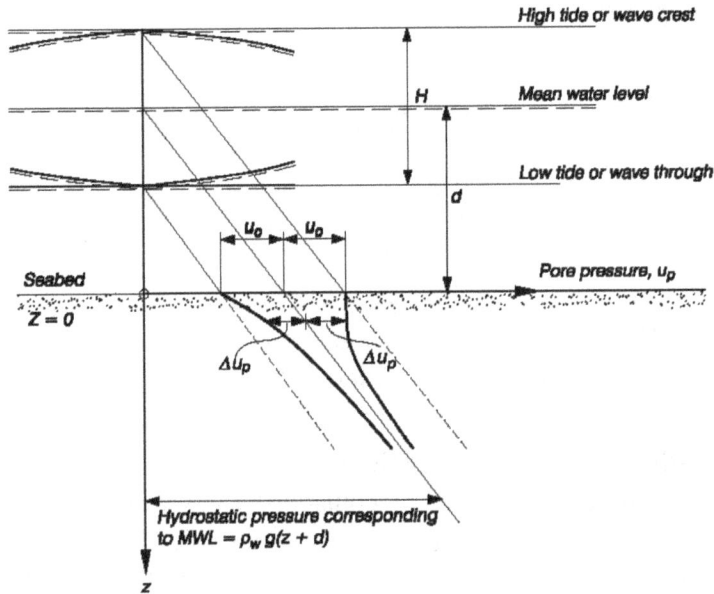

The pore pressure in depth z is given by

$$u_p = \rho_w g(z + d) + \Delta u_p \qquad\qquad (VI-5-237)$$

$$\Delta u_p = u_0 e^{-Az}\cos\left(\frac{2\pi X}{L} + \frac{2\pi t}{T} - Az\right) \qquad\qquad (VI-5-238)$$

where Δu_p Pore pressure deviation caused by wave or tide

        u_0 Bottom pore pressure amplitude
             $= \rho_w g\,\frac{H}{2}$ for tides
             $= \dfrac{\rho_w g H}{2\cosh\left(\frac{2\pi d}{L}\right)}$ for waves
             assuming linear wave theory

        ρ_w Mass density of water

        g Gravitational acceleration

        d Mean water depth

        x Horizontal coordinate

        L Wavelength

        t Time

        T Wave period

        $A = \left(\frac{\rho_w g \pi}{k E_{oed} T}\right)^{0.5}$

        k Darcy permeability coefficient

      E_{oed} Oedometric compression modulus of soil

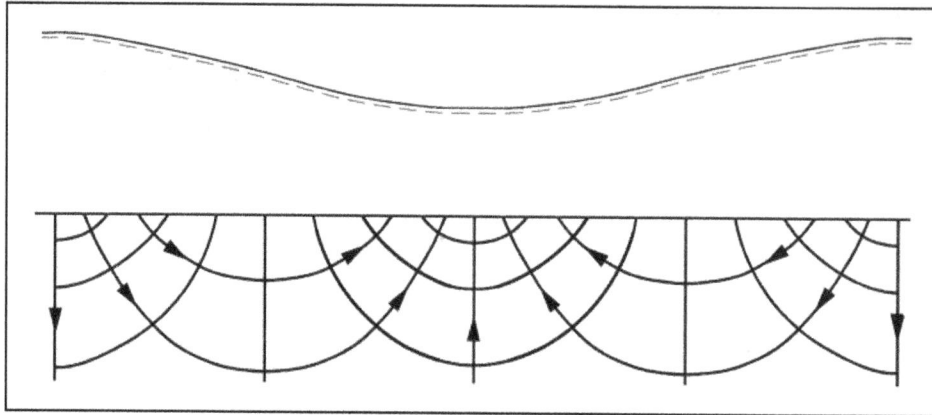

Figure VI-5-105. Illustration of instantaneous flow net in a homogeneous
isotropic seabed under wave action

• For the two-dimensional case, simple methods of estimating slope stability have been
developed. The stability can be investigated by considering the equilibrium of the soil body
confined by the failure surface as illustrated in Figure VI-5-106. The ratio between the
"stabilizing" and "driving" rotational moments, M_s and M_D, determined from all forces acting on
the free soil body, is a measure of the stability.

• In Figure VI-5-106, W is the total weight of the soil element including pore water, S
is the horizontal seismic inertia force, τ and σN are shear stress forces and effective normal stress
forces, respectively, u_s is the water pressure along the surface of the slope, and u_p is the pore
water pressure along the failure circle. The variables τ and σN usually vary along the failure
circle. The parameter u_s is determined by the mean water level and the wave action. At the time
of maximum runup a good approximation would be a hydrostatic pressure distribution, i.e.,
$u_s = \rho_w h$ where ρ_w is the water mass density and h is the local instantaneous water depth. The
variable u_p can be determined from flow nets sketched for the instantaneous wave action
situation, or from numerical models (Barends et al. 1983). Another, but in fact identical,
formulation of the force balance indicated in Figure VI-5-106 would be to subtract the effect of
hydrostatic water pressure corresponding to the mean water level from W, u_s and u_p.

• A safety factor F for the slope stability can be expressed as

$$F = \frac{M_s}{M_D} = \frac{moment\ of\ stabilizing\ forces}{moment\ of\ driving\ forces} \tag{VI-5-239}$$

or as

$$F = \frac{available\ shear\ strength}{shear\ strength\ required\ for\ stability} \tag{VI-5-240}$$

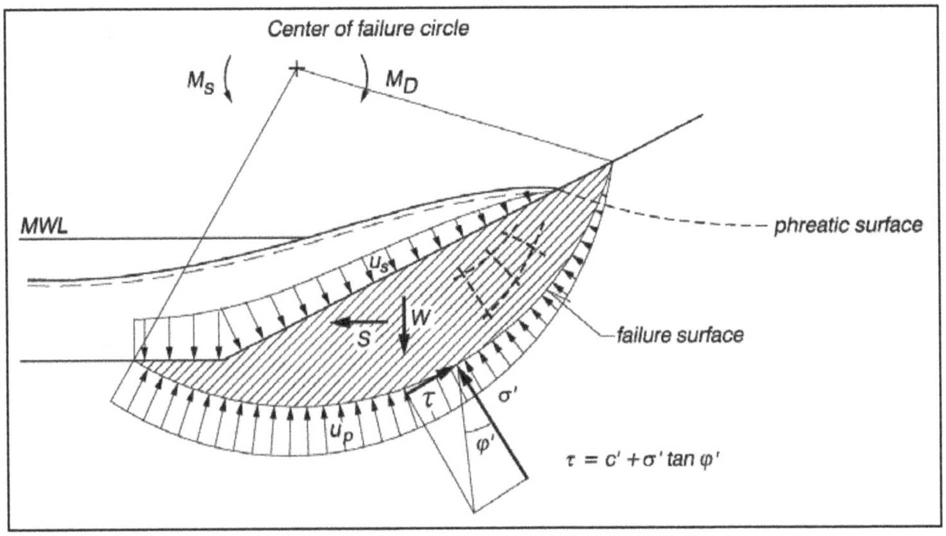

Figure VI-5-106. Illustration of forces to be considered in slope stability analysis

• If the failure surface is circular then the resultant force of the pore pressure u_p goes through the center of the circle and will not contribute to M_D. In this case it is common to define a safety factor as

$$F = \frac{\textit{moment of shear strength along failure circle}}{\textit{moment of weight of failure mass and surface loads}} \qquad \text{(VI-5-241)}$$

• The minimum value of F has to be identified by varying the position of the center of the failure circle and the radius. Also, F must be larger than unity to assure stability. The determination of the actual (minimum) safety factor for a given slope requires usually many trial failure surfaces calculations. It is important to notice that F is not a general safety factor because it depends on the applied definition. A specific value of F does not express a unique safety level.

• Various hydraulic load situations must be evaluated, such as a rapid run-down situation in which the phreatic surface in the slope material remains in a high position due to slow drainage (see Figure VI-5-104). This load situation, which occurs when rather impermeable materials are used, might be approximated and treated like rapid (instantaneous) drawdown known from earth dam design. Morgenstern and Price (1965) provide stability charts of F (Equation VI-5-239) as a function of slope angle, ratio of drawdown height over water depth, and soil strengths cN and vN.

• The critical circular failure surface and the related safety factor F can be determined directly following the method of Janbu (1954a, 1954b) for the case of homogeneous soil, stationary water table and undrained conditions, i.e., the soil strength is given by the undrained shear strength c_u. Hansbo (1994) presented diagrams for determination of F as function of slope geometry, water level, c_u, and surface load.

• A unique solution when determining slope stability for soils with an internal angle of friction, v, cannot be obtained because of four unknowns and only three equations of static equilibrium. If νN is constant along the failure surface, one solution to the problem is to substitute the circle with a logarithmic spiral, i.e.,

$$r = r_1 \exp(\omega \tan \_') \tag{VI-5-242}$$

in which the radius vector forms an angle νN with its normal at each point of the curved surface. The unknown frictional forces along the failure surface now pass through the center of the spiral as shown in Figure VI-5-107.

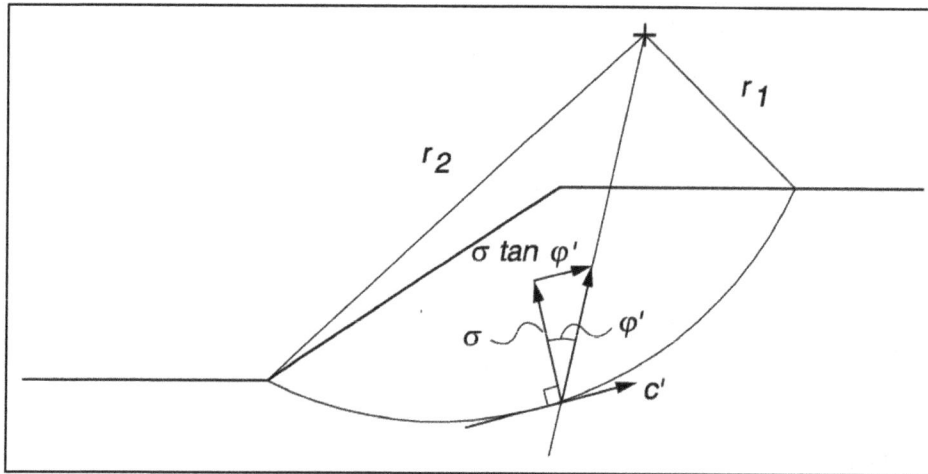

Figure VI-5-107. Illustration of logarithmic spiral

The stabilizing moment due to friction and cohesion, both taken as constants, is given by

$$M_s = \frac{1}{2} c' (r_1^2 - r_2^2) \cot \varphi' \tag{VI-5-243}$$

- The logarithmic spiral is not kinematically admissible as is the case for a circular (or straight line) failure plane. However, the deviation between the two curves is not significant in most cases.

- The simple methods illustrated in Figures VI-5-106 and VI-5-107 cannot be applied to inhomogeneous soils in which the soil strength parameters cN and νN vary along the failure surface. This situation arises when the slip surface goes through both the rubble-mound and seabed soil, or through layered parts of the rubble structure where the interfacial friction angles are different (smaller) from the friction angle of the rubble. Moreover, if weak strata are present, then the slip surface will not be circular or log-spiral shaped because the failure surface tends to go through the weak layers as illustrated in Figure VI-5-108.

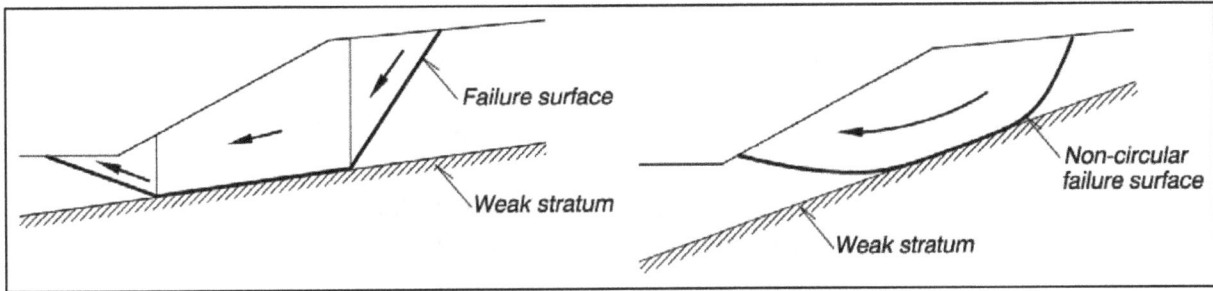

Figure VI-5-108. Illustration of failure surface in case of weak stratum

• For inhomogeneous conditions, slope stability is generally analyzed by the method of slices. The soil body is separated into fictitious vertical slices having widths that are determined such that cN and vN can be assumed constant within a slice. Slope stability is analyzed by considering all the forces acting on each slice, as shown by Figure VI-5-109. The failure surface that gives the lowest stability has to be identified by trial calculations. In Figure VI-5-109, W is the total weight of the slice including surface load, u_p is the total pore water pressure at the bottom of the slice, and the parameters P and T are the horizontal and vertical forces, respectively, on the sides of the slice.

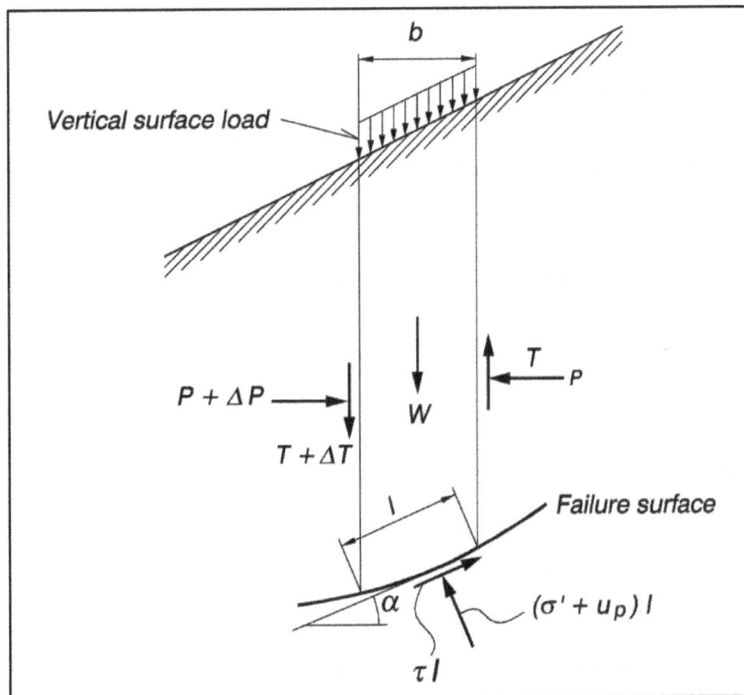

Figure VI-5-109. Illustration of forces on a soil slice in
the method of slices slope stability analysis

• Several approximate methods exist for determining F, as defined by Equation VI-5-241. The most commonly applied methods are the ordinary method of slices and the simplified method of slices by Bishop. Both methods are based on the assumption of circular-cylindrical failure surfaces. The reasonableness of this assumption should be considered in light of the comments about weak strata.

• The Ordinary Method of Slices, also known as the method of Fellinius (1936), assumes that the resultant of the forces P and T acting upon the sides of any slice have zero resultant force in the direction normal to the failure direction are for that slice. It is also assumed that the failure surface is circular-cylindrical. The related safety factor is given by

$$F = \frac{\sum_{i=1}^{n}\left[c_i l_i + \left(W_i \cos\alpha_i - u_{pi}l_i\right)\tan\varphi_{i'}\right]}{\sum_{i=1}^{n} W_i \sin\alpha_i} \tag{VI-5-244}$$

If cN and vN are taken as constants, Equation VI-5-244 simplifies to

$$F = \frac{c'L + \tan\varphi'\sum_{i=1}^{n}\left(W_i \cos\alpha_i - u_{pi}l_i\right)}{\sum_{i=1}^{n} W_i \sin\alpha_i} \tag{VI-5-245}$$

where L is the total length of the circular failure surface. The values of F calculated by Equations VI-5-244 or VI-5-245 fall below the lower bound of solutions that satisfy static analysis. Thus, the method is on the safe side. The method of slices was further developed by Janbu (1954a) and Bishop (1955).

• The Simplified Method of Slices by Bishop (1955) is valid for a circular-cylindrical failure surface, and it assumes that the forces acting on the sides of any slice have zero resultant in the vertical direction, i.e., ΔT in Figure VI-5-109 is zero. The related safety factor, defined by Equation VI-5-241, is

$$F = \frac{R\sum_{i=1}^{n}\left[c_1 b_i + \left(W_i - u_{pi}b_i\right)\tan\varphi_{i'}\right]\Big/\left[\left(1 + \tan\alpha_i \tan\varphi_{i'}/F\right)\cos\alpha_i\right]}{M_D + R\sum_{i=1}^{n} W_i \sin\alpha_i} \tag{VI-5-246}$$

where R is the radius of the failure surface circle and M_D is the driving moment of any load not included in Figure VI-5-109. Because F is implicitly given, an iteration procedure must be used; however, convergence of trials is very rapid.

• The Method of Slices by Janbu (1954a, 1973) is for more complicated situations where circular-cylindrical slip surfaces cannot be used, and a method for composite failure surfaces of arbitrary shape must be applied. The method is based on a combination of equations expressing moment and force equilibrium of each slice, and an iteration method for calculating F must be used.

• Most slope failures are three-dimensional. An approximate treatment of a three-dimensional slope failure is illustrated in Figure VI-5-110. The safety factors, F_1, F_2, and F_3, for three parallel cross-sections are computed. An estimate of the safety factor, F, for the whole body can then be estimated as the weighted safety factor using the total free body soil weights, W_1, W_2, and W_3, above the failure surface in each cross section as the weighting factors.

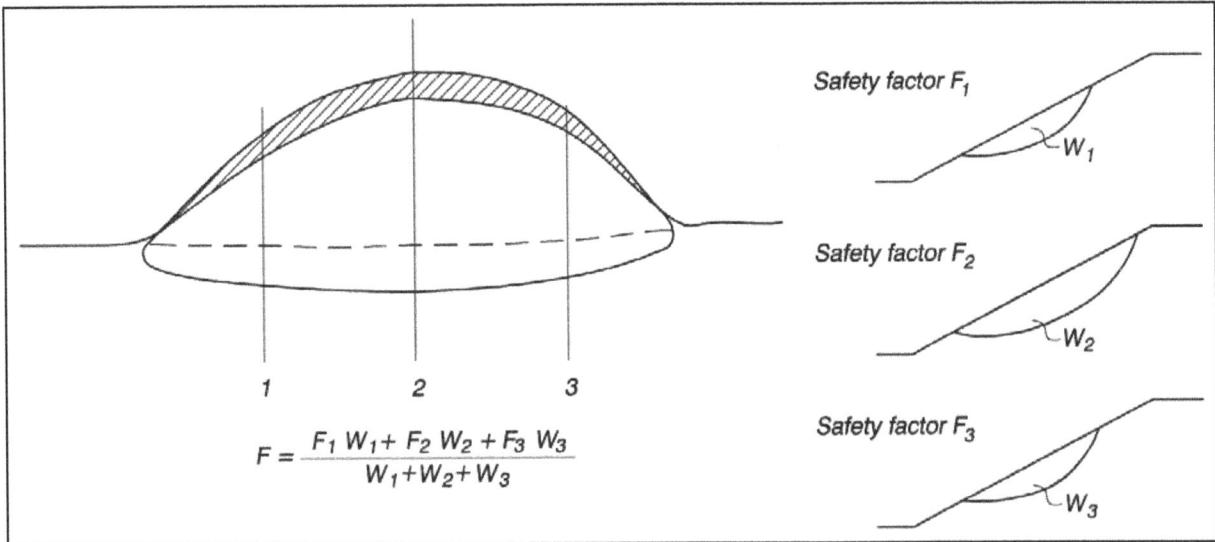

Figure VI-5-110. Illustration of safety factor F for three-dimensional slope failure

(b) Bearing capacity.

• The bearing capacity of a foundation is the load, transferred through the foundation - soil interface, that will initiate soil failure. Thus, bearing capacity is related to the ultimate limit state. The bearing capacity of the foundation of monolithic structures or structure elements like caissons and parapet concrete superstructures must be analyzed, and sufficient safety must be implemented in the design. Typical bearing capacity failure modes are shown in Part VI-2-4, "Failure Modes of Typical Structure Types."

• Rubble-mound breakwater structures placed on weak seabed soils might suffer from insufficient seabed bearing capacity. This can be investigated by the slip surface analysis explained in the previous section on slope stability.

• Bearing capacity calculations are based on zone failure analysis. In the case of homogeneous soil conditions the vertical bearing capacity of strip footings and individual rectangular footings can be estimated by formulae developed by Meyerhof (1951, 1963) and Brinch Hansen (1961, 1970), presented in Tables VI-5-86 and VI-5-87. The formulae, which represent a further development of Prandtl's and Terzaghi's theories for concentrically loaded horizontal footings, are valid for static loading and homogeneous soil conditions within the space of the zone failures.

Table VI-5-86
Bearing Capacity Formula for Rectangular Concentrically Statically Loaded Horizontal Footings (Meyerhof 1951, 1963)

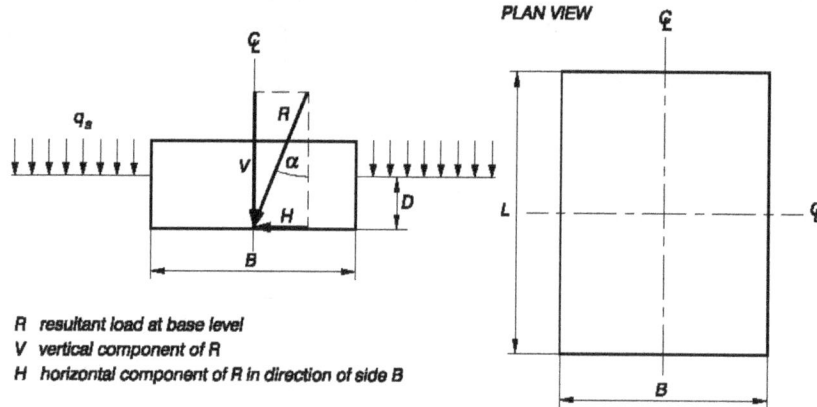

R resultant load at base level
V vertical component of R
H horizontal component of R in direction of side B

Ultimate vertical bearing capacity per unit area of the footing:

$$q_u = \frac{Q^{N_u}}{BL} = \frac{1}{2}\,\bar{\gamma}'\,B\,N_\gamma\,s_\gamma\,d_\gamma\,i_\gamma + q'\,N_q\,s_q\,d_q\,i_q + c\,N_c\,s_c\,d_c\,i_c \qquad (VI-5-247)$$

Q_u Ultimate (maximum) value for the vertical component of the load
B Width of footing
L Length of footing, always $\geq B$
D Minimum depth of footing below soil surface
$\bar{\gamma}'$ Average effective weight of soil from base level to depth B under base level
q' Effective overburden pressure at base level (contribution from surface load q_s and effective weight of soil above base level)
c Shear strength of soil. c_u for undrained conditions, and c' (effective) for drained conditions
φ' Effective friction angle of soil determined by plain strain tests. Friction angle φ'_{triax} determined by triaxial test should be replaced by $\varphi' = (1.1 - 0.1\,B/L)\varphi'_{triax}$

Bearing capacity factors:

$$N_\gamma = (N_q - 1)\tan(1.4\,\varphi')$$
$$N_q = \exp(\pi\tan\varphi')\tan^2(45^0 + \varphi'/2) = \exp(\pi\tan\varphi')(1 + \sin\varphi')/(1 - \sin\varphi')$$

$$N_c = \left\{ \begin{array}{l} (N_q - 1)\cot\varphi' \\ \pi + 2 \end{array} \right. \quad \text{for undrained conditions } (\varphi' = 0)$$

Shape coefficients:

$$s_\gamma = s_q = \left\{ \begin{array}{ll} 1.0 & \text{for } \varphi' = 0^0 \\ 1 + 0.1\,(B/L)\tan^2(45^0 + \varphi'/2) & \text{for } \varphi' \geq 10^0 \end{array} \right.$$

$$s_c = 1 + 0.2\,(B/L)\tan(45^0 + \varphi'/2)$$

Depth coefficients:

$$d_\gamma = d_q = \left\{ \begin{array}{ll} 1.0 & \text{for } \varphi' = 0 \\ 1 + 0.1\,(D/B)\tan(45^0 + \varphi'/2) & \text{for } \varphi' \geq 10^0 \end{array} \right.$$

$$d_c = 1.0 + 0.2\,(D/B)\tan(45^0 + \varphi'/2)$$

Inclination coefficients:

$$i_\gamma = \left\{ \begin{array}{ll} 1.0 & \text{for } \varphi' = 0 \\ (1 - \alpha/\varphi')^2 & \text{for } \varphi' \geq 10^0 \end{array} \right. \qquad \alpha = \arctan(H/V)$$

$$i_q = i_c = (1 - \alpha/90^0)^2$$

Table VI-5-87
Bearing Capacity Formula[1] for Rectangular Statically Loaded Horizontal Footing
(Brinch Hansen 1961, 1970)

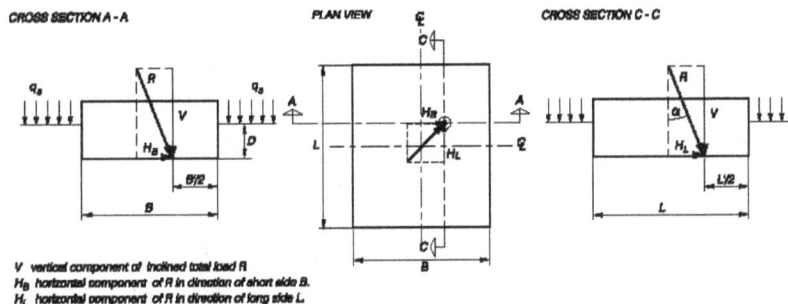

V   vertical component of inclined total load R
H_B horizontal component of R in direction of short side B.
H_L horizontal component of R in direction of long side L.

Ultimate vertical bearing capacity per unit area of effective footing:

$$q_u = \frac{Q_u}{B'L'} = \frac{1}{2} \bar{\gamma}' \, B' \, N_\gamma \, s_\gamma \, d_\gamma \, i_\gamma + q' \, N_q \, s_q \, d_q \, i_q + c \, N_c \, s_c \, d_c \, i_c \qquad (VI-5-248)$$

Q_u Ultimate (maximum) value for the vertical component of the load

$B' =$ $B - 2e_B$, effective width of footing, $B' \geq 0.4B$

$L' =$ $L - 2e_L$, effective length of footing, $L' \geq 0.4L$

D Minimum depth of footing below soil surface

$\bar{\gamma}'$ Average effective weight of soil from base level to depth B under base level

q' Effective overburden pressure at base level (contribution from surface load q_s and effective weight of soil above base level)

c Shear strength of soil. c_u for undrained conditions, and c' (effective) for drained conditions.

φ' Effective friction angle of soil determined by plain strain tests. Friction angle, φ'_{triax}, determined by triaxial tests should be replaced by $\varphi' = 1.1\varphi'_{triax}$.

Bearing capacity factors:

$N_\gamma = 1.5(N_q - 1) \tan \varphi'$

$N_q = \exp(\pi \tan \varphi')\tan^2(45^0 + \varphi'/2) = \exp(\pi \tan \varphi')(1 + \sin \varphi')/(1 - \sin \varphi')$

$N_c = \begin{cases} (N_q - 1)\cot \varphi' \\ \pi + 2 , \end{cases}$ for undrained conditions ($\varphi' = 0$)

Shape coefficients:

$s_\gamma = 1 - 0.4 \, B'/L'$, must always be ≥ 0.6

$s_q = 1 + \sin \varphi' \, B'/L'$

$s_c = 1 + 0.2 \, B'/L'$

Depth coefficients:

$d_\gamma = 1$

$d_q = 1 + 2 \tan \varphi'(1 - \sin \varphi')^2 \arctan(D/B')$

$d_c = 1 + 0.4 \arctan(D/B')$

Inclination coefficients:

$i_\gamma = (1 - \frac{0.7H_B}{V+B' \, L' \, c' \, \cot \varphi'})^5$
$i_q = (1 - \frac{0.5H_B}{V+B' \, L' \, c' \, \cot \varphi'})^5$ $\Big\}$ if the quantity inside the bracket becomes negative then the bearing capacity is negligible.

$i_c = \begin{cases} i_q - \frac{1-i_q}{N_q-1} \simeq i_c & \text{for } \varphi' \neq 0 \\ 0.5(1 + (1 - \frac{H_B}{B' \, L' \, c_u})^{0.5}) & \text{for } \varphi' = 0 \end{cases}$

[1] Failure can take place either along the long side or the short side of the footing. The formulae given above correspond to the first case. For the second case substitute L' for B', B' for L', and H_L for H_B.

• Brinch Hansen (1970) extended his formula to cover also the bearing capacity of statically loaded footings with inclined base in the vicinity of a slope. The formula which is termed the general bearing capacity formula is presented in Table VI-5-88 as an addition to the formula in Table VI-5-87.

• If foundation zone failures penetrate into more than one type of uniform soil then the formulae given in Tables VI-5-86, VI-5-87 and VI-5-88 cannot be applied, and the bearing capacity must be estimated by trial and error calculations in which the most critical rupture figure providing the lowest bearing capacity is identified.

• Eccentricity of the load, R, can, according to Meyerhof (1953), be taken into account by calculating the ultimate bearing capacity for a fictitious centrically loaded footing with width BN and length LN given by

$$B' = B - 2 e_B \quad and \quad L' = L - 2 e_L \qquad \text{(VI-5-249)}$$

where e_B and e_L are the eccentricity of R in the directions of the width and length of the footing, respectively, as shown in Figure VI-5-111. Values of BN must always be smaller than LN in the calculation of q_u when using Equation VI-5-247. Moreover, the eccentricities are limited to $BN ∃$ 0.4 B and $LN ∃$ 0.4 L corresponding to e smaller than 0.3 times the width of the footing. Otherwise a failure configuration underneath the unloaded part of the footing might develop. This situation is not covered by Equation VI-5-247. For the case of inclined loading, the method does not apply if horizontal sliding of the foundation occurs.

• For the case of nonhorizontal foundation base and ground surface, Brinch Hansen (1967, 1970) introduced a base inclination coefficient, b, and a ground inclination coefficient, g, in his bearing capacity formula to obtain a more general formula. In the context of coastal structures, sloping base and sloping ground surface are mostly relevant for cohesionless rubble materials as indicated by Figure VI-5-112, which shows a wave wall superstructure and a caisson on a high rubble-mound foundation. Also shown is the simplified geometry of the wave wall superstructure base and of the rear side of the mound foundation to be applied in the Brinch Hansen formula for cohesionless materials given in Table VI-5-88.

• Where the foundation inclined loading has a large horizontal component, the passive pressure P indicated in Figure VI-5-113 should be included in the force balance instead of using the depth coefficients in the calculation of the bearing capacity with Equations VI-5-248 and VI-5-250.

• Note that the bearing capacity formulae given in Tables VI-5-86, VI-5-87, and VI-5-88 are all approximations. Consequently, for final design more detailed bearing capacity calculations are recommended.

Table VI-5-88
General Bearing Capacity Formula for Rectangular Statically Loaded Inclined Footing on
Cohesionless Soil in Vicinity of Slope (Brinch Hansen 1961, 1967, 1970)

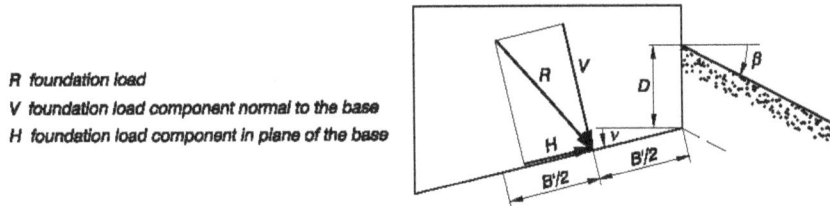

R foundation load
V foundation load component normal to the base
H foundation load component in plane of the base

Ultimate bearing capacity per unit area of the footing:

$$q_u = \frac{Q_u^N}{B' \, L'} = \frac{1}{2} \, \bar{\gamma}' \, B' \, N_\gamma \, s_\gamma \, d_\gamma \, i_{\gamma b} \, b_\gamma \, g_\gamma \; + q' \, N_q \, s_q \, d_q \, i_q \, b_q \, g_q \qquad (VI-5-250)$$

Q_u^N is the ultimate (maximum) value for the load component normal to the base.

The formula is identical to Eq VI-5-248 except for the missing c-term ($c_u = 0$) and for the
addition of the coefficients b and g, and a modified i_γ - coefficient.

Base inclination coefficients:

$$b_\gamma = \exp(-2.7 \, \nu \, \tan \varphi')$$

$$b_q = \exp(-2 \, \nu \, \tan \varphi')$$

Ground inclination coefficients:

$$g_\gamma = g_q = (1 - 0.5 \tan \beta)^5$$

Modified load inclination coefficients:

$$i_{\gamma b} = (1 - (0.7 - \nu^0/450^0) \, H/V)^5$$

Limitations:

The angles ν and β must be possitive but $\nu + \beta$ must not exceed 90^0. β must be
smaller than φ'.

- Publications of PIANC provide the limit state equations for rupture figures related to
the two-dmensional case of a statically loaded monolithic structure with horizontal base placed
on a rubble foundation overlaying a seabed of sand or clay.

- Following Equation VI-5-236, the limit state equations are defined as

$$g = W_s - W_d \geq 0 \qquad (VI-5-251)$$

• A related measure of safety can be defined as

$$F = \frac{W_s}{W_d}$$

(VI-5-252)

• For more accurate estimations of three-dimensional bearing capacity, it is necessary to use advanced finite element calculations.

• The given bearing capacity formulae for statically loaded foundations could be applied for dynamic loadings using a dynamic amplification factor on the load as discussed in Part VI-5-5*f*, Equation VI-5-234. Such simplified methods can be used in conceptual design, but detailed design of large structures should use more accurate methods if there is a risk of dynamic load amplification.

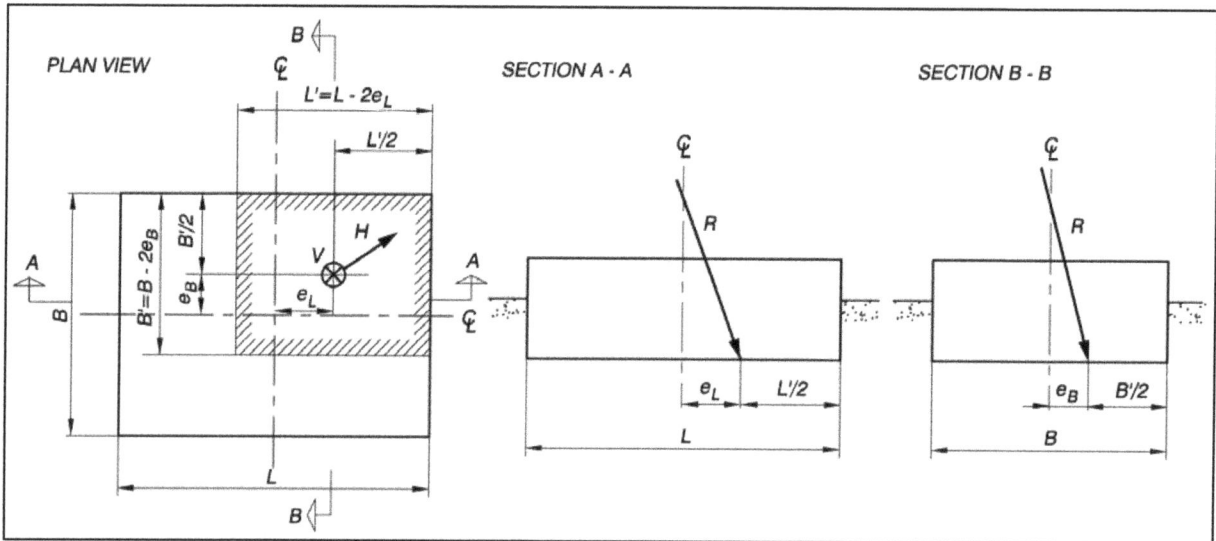

Figure VI-5-111. Illustration of fictitious footing to replace real footing under eccentric loading conditions

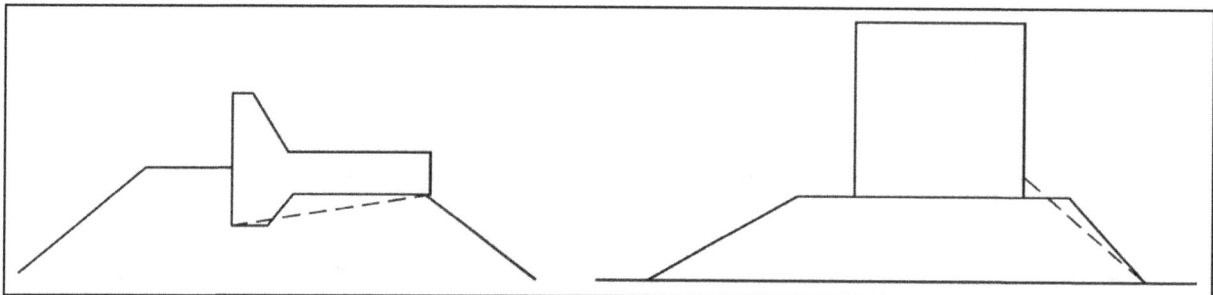

Figure VI-5-112. Simplified base and rear slope geometries to be applied in the general bearing capacity formula Table VI-5-86

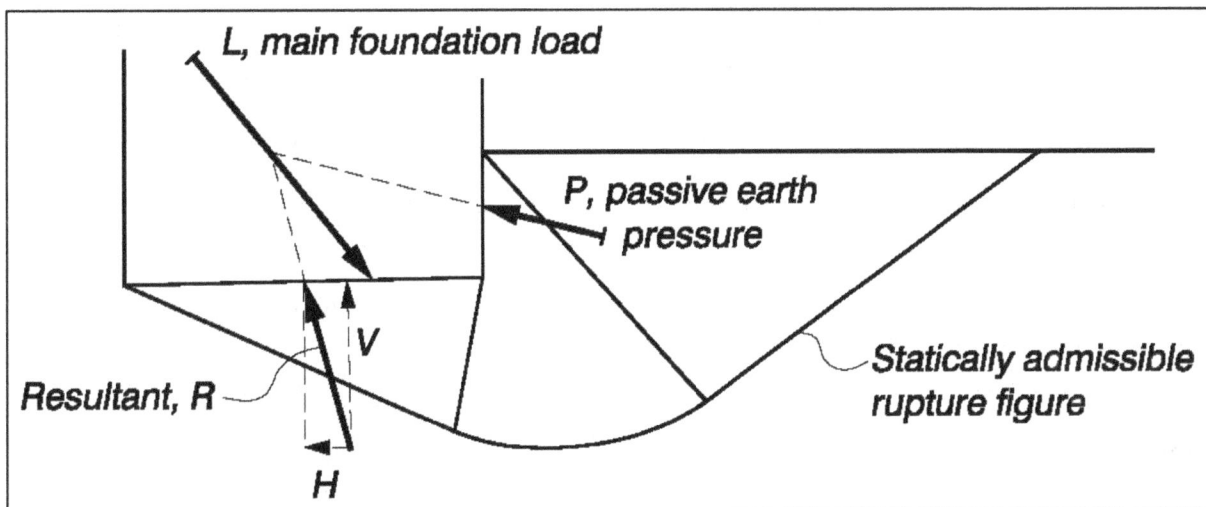

Figure VI-5-113. Illustration of passive earth pressure *P* to be included in the determination of the foundation load resultant *R* in place of the depth coefficients in Equations VI-5-248 and VI-5-250

h. Settlement.

(1) For coastal structures, settlement is related both to the seabed soils and to the structure mound materials. The mound materials are generally cohesionless permeable materials such as quarrystones, quarry-run, gravel, and coarse sand. The seabed soils are in most cases fine and less permeable materials such as sand, silt, and clay, quite often layered. Soft and muddy deposits exist in many places, especially in estuaries, deltas, and river outlets. Settlement is the direct result of volume reduction of the soil mass, and it is caused by escape of water from the voids between particles and compression of the particle skeleton.

(a) Vertical settlement of coastal structures is generally of concern where the foundation is on soft seabed materials, or at deepwater mound structures where the high mound can settle significantly. The latter case is also a concern for the foundation of caissons on high rubble mounds.

(b) Differential settlement is a problem where it might lead to damage of roads and installations placed on the structures. Damage to joints between caissons could also be due to differential settlements.

(2) Structure settlement increases vulnerability to wave overtopping by lowering the crest level of the structure. Thus, the expected total vertical settlement during the structure service lifetime has to be estimated, and the construction crest level increased accordingly.

(3) Poor seabed materials which cause large settlement and stability problems might necessitate soil improvement by methods such as preloading, compaction, installation of drains, or soil replacement. Also, it may be possible to select the type and design of structure that gives a minimum foundation load.

(4) The consequence of foundation loading on settlement depends to a great extent on the loading time relative to the consolidation time. The following three categories can be identified:

(a) Drained loading, when the consolidation time is much less than the loading time.

(b) Undrained loading, when the consolidation time is much greater than the loading time.

(c) Partially drained loading, when the consolidation time and the loading time are of the same order of magnitude.

(5) This description of the loading corresponds to the classification given in Table VI-5-83 in Part VI-5-5e(1) where consolidation time is discussed.

(6) Foundation loads related to coastal structures are given as follows:

(a) Loads from the weight of structure materials or structure elements placed during the construction phases. The expected loading time would be in the range from minutes to days to months.

(b) Weight of the completed structure including permanent external loads.

(c) Loads from wave action, traffic loads, and other live loads. The loading times would be in the range from seconds to hours. The wave loads will be cyclic.

(7) Generally the permeability of stone materials and coarse sand is so large that deformation problems related to the previously listed loadings can be handled as drained problems. On the other hand, the permeability of clay is so low that the conditions will always be undrained. For fine sand and silt with permeabilities between coarse sand and clay, it is not possible to make such general statements as each case must be examined. However, it is most likely that conditions during wave loadings will be undrained.

(8) Settlements are usually devided into immediate (instantaneous) settlement, primary consolidation settlement, and creep (also denoted secondary consolidation).

(a) Instantaneous settlement occurs rapidly almost in phase with the application of the load.

(b) Primary consolidation settlement is the deformation that occurs in saturated or partially saturated low permeability soils when the load carried by excess pore water pressure is gradually transferred to the soil skeleton with corresponding simultaneous excess pore water dissipation.

(c) Secondary consolidation settlement is a long-term creep phenomenon due to shear. It might continue for a long time after completion of primary consolidation.

(9) All three settlement components are relevant to low permeability materials, whereas only immediate and secondary consolidation settlements occur with high permeability materials with drained soil conditions.

(10) The starting point in calculation of settlement of the seabed soils is understanding the in situ stress distributions just after the loading is applied and estimating the relationship between stresses and soil deformations. The in situ stress distributions are generally calculated assuming elastic material and using methods such as the procedure given by Steinbrenner (1936) or by means of the influence diagrams by Newark (1942). The empirical 2:1 load spreading method might also be used. It should be noted that fill material used for rubble-mound structures is completely flexible whereas a caisson constitutes a stiff footing.

(11) Instantaneous settlement is estimated from the deformation moduli determined either by laboratory experiments with representative small soil specimens or by in situ tests such as plate loading tests, pressure meter tests, or other standard test procedures.

(12) Primary consolidation settlement is generally determined from consolidation theory by the use of the oedometer modulus and the permeability. During the construction phase, the load on the foundation is time-varying. Because the consolidation due to every load increment proceeds independently of the preceding load increment, the total settlement can be computed by superposition. Consolidation and the related settlement within the structure lifetime are caused almost entirely by the weight of the structure. Occasional loading from waves and other live loads can normally be disregarded in this context except where the wave-generated cyclic loadings cause significant volume changes of the soil (see Part VI-5-5e).

(13) Secondary settlement of seabed soils is difficult to estimate. It will usually be much smaller than the sum of the instantaneous and the primary consolidation settlements.

(14) Mound material such as quarrystones and quarry-run used for the construction of rubble-mound breakwaters is usually tipped from dumpers or barges. Most of the anticipated settlement takes place during the construction phase, especially if heavy vehicles such as dumpers pass over the already placed material. Settlement will then typically be in the order of 2 - 5 percent of the height of the mound. High quarrystone foundations for caisson breakwaters might need compaction to reduce the risk of unacceptable differential settlements.

VI-5-6. Scour and Scour Protection. Any coastal project built on erodible sand or soil may be susceptible to damage resulting from scour. This section describes scour problems that affect coastal projects, gives procedures for estimating maximum depth of scour for specific situations, and presents design guidance for scour protection. The available scour prediction methods presented here assume the erodible bed is composed of noncohesive sediment.

a.   Scour problems in coastal engineering. In the most general definition, scour is the erosive force of moving water. This broad definition of scour includes any erosion of sediment under any circumstances, such as beach profile change and inlet channel migration. A more specific definition of scour is used in reference to coastal engineering projects: Scour is the removal by hydrodynamic forces of granular bed material in the vicinity of coastal structures. This definition distinguishes scour from the more general erosion; and as might be expected, the presence of a coastal structure most definitely contributes to the cause of scour. Scour that occurs at coastal projects can lead to partial damage, or in some cases, complete failure of all or portions of the structure. Scour-induced damage happens at sloping-front structures when scour undermines the toe so it can no longer support the armor layer, which then slides downslope (see Figure VI-2-37). Scour impacts vertical-front caissons and other gravity-type structures if the structure is undermined to the point of tilting as illustrated by Figure VI-2-58. Monolithic gravity seawalls can also settle and tilt as a result of scour (see Figure VI-2-64). Scour at vertical sheetpile walls can result in seaward rotation of the sheetpile toe due to pressure of the retained soil as shown by Figure VI-2-69. Coastal structure damage or failure brought about by scour impacts coastal projects in several ways including: project functionality is decreased; costs will be incurred to repair or replace the structure, and scour related damage is often difficult and expensive to repair; upland property being protected by the structure may be lost or inundated; clients and cost-sharing partners will lose confidence in the project's capability to perform as required.

(1) Physical processes of scour.

(a) Scour will occur anywhere the hydrodynamic shear stresses on the bottom are high enough to initiate sediment transport. Clear water scour occurs when bottom shear stresses are high only in a localized portion of the bed; outside the local region sediment is not moving. This occurs mostly in uniform, steady flow situations. In live bed scour bottom shear stresses over the entire bed exceed the level for incipient motion with locally higher shear stresses where greater scour occurs. An equilibrium is reached when the volume of sediment being removed from the scour hole is exactly equal to sediment being deposited in its place. Understanding the physical processes involved in scour is difficult because the shear stresses responsible for scour are developed by waves, currents, or combined waves and currents, that usually are heavily influenced by the presence of a coastal structure. Because of the distinct influence coastal structures exert on the hydrodynamics, structural aspects such as geometry, location, and physical characteristics (roughness, permeability, etc.) impact the scour process. Therefore, modifying some physical characteristic of a structure may reduce scour potential.

(b) Typical structure and hydrodynamic conditions leading to scour include the following (acting singularly or in combination):

• Localized increases in peak orbital wave velocities due to combined incident and reflected waves

• Particular structure orientations or configurations that focus wave energy and increase wave velocity or initiate wave breaking

• Structure orientations that direct currents along the structure or cause a flow acceleration near the structure

• Flow constrictions that accelerate the fluid

• Breaking wave forces that are directed downward toward the bed or that generate high levels of turbulence capable of mobilizing sediment

- Wave pressure differentials and groundwater flow that produce a "quick" condition, allowing material to be carried off by currents

- Flow separation and creation of secondary flows such as vortices

- Transitions from hard bottom to erodible bed

(c) Even if the hydrodynamic aspects of scour were fully understood, there remains the difficulty of coupling the hydrodynamics with sediment transport. Consequently, most scour prediction techniques consist of rules of thumb and fairly simplistic empirical guidance developed from laboratory and field observations.

(d) Depending on the circumstances, scour can occur rapidly over short time spans (e.g., energetic storm events), or as a gradual loss of bed material over a lengthy time span (months to years). In the short-term case sediment is probably transported primarily as suspended load, whereas bedload transport is more likely during episodes of long-term scour. Scour may be cyclic with infilling of the scour hole occurring on a regular basis as the flow hydrodynamics undergo seasonal change.

(e) Most scour holes and trenches would eventually reach a stable configuration if the same hydrodynamic conditions persisted unchanged over a sufficient time span. Such an equilibrium is more likely to occur for scour induced primarily by current regimes than by wave action. It is difficult to determine if observed scour development at a particular coastal project represents an equilibrium condition. The scour might be the result of energetic flow conditions that subsided before the full scour potential was realized. Or it is possible the scour was initially greater, and infilling of the scour hole occurred prior to measurement. Finally, there is the possibility that the observed scour is simply the partial development of an ongoing long-term scour process.

(2) Common scour problems. Common coastal engineering situations where scour may occur are illustrated on Figure VI-5-114 and described as follows.

(a) Scour at coastal inlet structures.

- Kidney-shaped scour holes are sometimes present at the tip of one or both inlet jetty structures. These scour holes are usually permanent features of the inlet structure system, but there have been instances where seasonal infilling occurs due to longshore sediment transport. In some cases scour holes have been deep enough to result in partial collapse of the jetty head, while in other cases the scour holes have resulted in no structure damage. Hughes and Kamphuis (1996) observed in movable-bed model experiments that the primary hydrodynamic process responsible for kidney-shaped scour holes appears to be flood currents rounding the jetty head and entering the channel. Sediment mobilization, rate of scour, and extent of scour are increased by wave action, particularly waves that are diffracted around the jetty tip into the navigation channel. Waves breaking across the jetty head in the absence of currents will also cause scour of a lesser magnitude (Fredsøe and Sumer 1997).

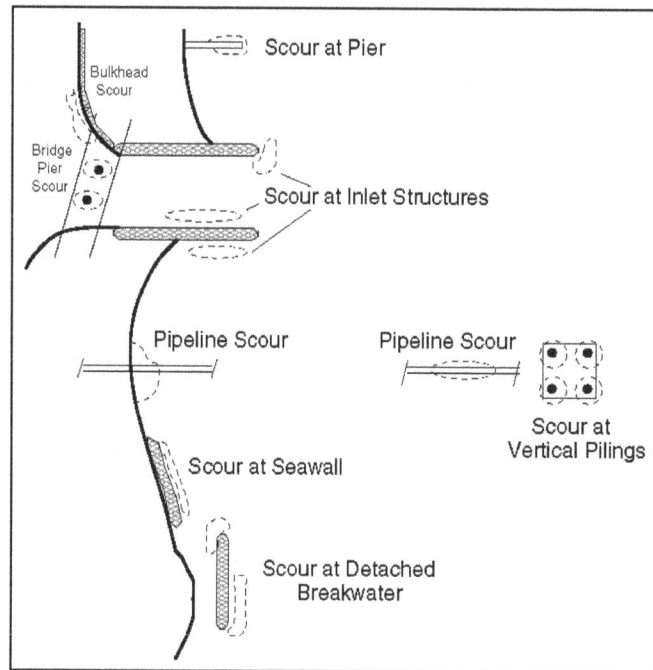

Figure VI-5-114. Coastal scour problems

• Substantial scour trenches are known to form along the channel-side toes of jetty structures. These trenches are caused either by migration of the navigation channel (by unknown causes) to a position adjacent to the jetty toe or by ebb-flow currents that are redirected by the jetty structure. Hughes and Kamphuis (1996) argued that ebb flows deflected by a jetty are analogous to plane jet flow exiting a nozzle with similar geometry. As the flow cross section decreases, the flow velocity increases proportionately to maintain the ebb flow discharge.

• Scour trenches can also form along the outside toe of the updrift jetty. These trenches might be formed by the seaward deflection of longshore currents that causes a local flow acceleration adjacent to the jetty toe, or the scour may stem from high peak orbital velocities resulting from the interaction of obliquely incident and reflected waves. A likely scenario is scour hole formation due to both hydrodynamic processes with the waves mobilizing sediment and the current transporting the material seaward. Scour trenches on the outside toe of a jetty may be seasonal at locations experiencing seasonal reversal of predominant wave direction.

• Scour holes occur regularly around bridge pilings and piers that span coastal inlets. Generally, this situation is similar to scour that plagues bridge piers on inland waterways. Additional factors complicating scour at inlet bridge piers are the unsteady and reversing nature of tidal flows, and the possible exposure to waves and storm surges.

(b) Scour at structures in deeper water.

• Scour can occur at the toes of vertical-faced breakwaters and caissons placed in deeper water. Wave-induced scour results from high peak orbital velocities developed by the interaction of incident and reflected waves. If a particular structure orientation results in increased currents along the structure toe, scour potential will be significantly enhanced.

Localized liquefaction due to wave pressure differentials and excess pore pressure within the sediment may cause sediment to be removed by reduced levels of bottom fluid shear stress.

• Characteristic scour patterns may occur around the vertical supporting legs (usually cylinders) of offshore platforms. Under slowly-varying boundary layer flow conditions, the platform leg interrupts the flow causing formation of a horseshoe vortex wrapped around the structure just above the bed. This secondary flow intensifies the bottom fluid shear stresses, and erodes sediment. The quasi-equilibrium scour hole closely resembles the shape of the horseshoe vortex. In the absence of currents, waves can cause scour in the shape of an inverted, truncated cone around the vertical cylinder provided the bottom orbital velocities are sufficiently high.

• Pipelines laid on the sea bottom are susceptible to scour action because the pipe cross section obstructs the fluid particle motion developed by waves and currents.

(c) Scour at structures in shallow water.

• Piers and pile-supported structures in shallow water react to currents and waves just as in deep water. However, the shallow depth means that orbital velocities from shorter period waves can cause scour. Therefore, vertical piles are vulnerable to scour caused by a wider range of wave periods than in deeper water.

• Scour can occur along the seaward toe of detached breakwaters due to wave reflection. The scour process will be enhanced in the presence of transporting currents moving along the breakwater. Scour holes may be formed at the ends of the breakwater by diffracted waves. In shallow water, breaking waves can create high turbulence levels at the structure toe.

• Vertical-front and sloping-front seawall and revetments located in the vicinity of the shoreline can be exposed to energetic breaking waves that produce downward-directed flows and high levels of turbulence which will scour the bed. Scour could also be produced by flows associated with wave downwash at less permeable sloping structures.

• Vertical bulkheads are usually not exposed to waves capable of producing scour; however, it is possible for scour to occur by local current accelerations.

• Scour around pipelines will occur by the same mechanisms as in deeper water with shorter period waves becoming more influential as water depth decreases. Buried pipelines traversing the surfzone can be at risk if beach profile erosion exposes the pipeline to pounding wave action and strong longshore currents.

• Depending on specific design details, coastal outfalls may develop scour patterns that jeopardize the structure.

(d) Other occurrences of scour.

• Any type of flow constriction caused by coastal projects has the potential to cause scour. For example, longshore currents passing through the gap between a jetty and a detached breakwater at Ventura Harbor, CA, accelerated and caused scour along the leeside toe of the detached breakwater (Hughes and Schwichtenberg 1998).

- Storm surge barriers, sills, and other structures founded on the sea floor can experience scour at the downstream edge of the structure. Small pad foundations can be undermined by waves and currents.

- Structure transition points and termination points may produce local flow accelerations or may focus wave energy in such a way that scour occurs.

- Scour may occur as a transient adjustment to new construction. For example, Lillycrop and Hughes (1993) documented scour that occurred during construction of the terminal groin at Oregon Inlet, North Carolina. Despite maintenance of a scour blanket in advance of construction, the project required 50 percent more stone because of the scour.

b. Prediction of scour. There have been many theoretical and laboratory studies conducted examining various aspects of scour related to coastal projects. Some studies focussed on discovering the physical mechanisms responsible for scour, whereas other studies were directed at developing engineering methods for predicting the location and maximum depth of scour. In the following sections usable engineering prediction methods are presented for estimating scour for specific coastal structure configurations and hydrodynamic conditions. To a large extent the predictive equations have been empirically derived from results of small-scale laboratory tests, and often the guidance is fairly primitive. In some situations the only predictive capability consists of established rules of thumb based on experience and field observation. A comprehensive discussion of scour mechanisms, theoretical developments, and experiment descriptions is well beyond the scope of this manual. However, there are several publications containing detailed overviews of scour knowledge for many situations of interest to coastal engineers (e.g., Hoffmans and Verheij 1997; Herbich 1991; and Sumer and Fredsøe 1998a). In the following sections, appropriate citations of the technical literature are provided for more in-depth study.

(1) Scour at vertical walls. Occurrence of scour in front of vertical walls can be conveniently divided into two cases: nonbreaking waves being reflected by a vertical wall, and breaking waves impacting on a vertical wall. In either case, waves can approach normal to the wall or at an oblique angle.

(a) Nonbreaking waves. Nonbreaking waves are more prevalent on vertical-front structures located in deeper water and at bulkhead structures located in harbor areas. Almost all the energy in incident waves reaching a vertical-front structure is reflected unless the structure is porous. Close to the structure, strong phase locking exists between incident and reflected waves, and this sets up a standing wave field with amplified horizontal particle velocities beneath the water surface nodes and minimal horizontal velocities beneath the antinodes. The bottom sediment responds to the fluid velocities by eroding sediment where bottom shear stresses are high and depositing where stresses are low.

- Normally incident nonbreaking waves. Researchers have identified two characteristic scour patterns associated with nonbreaking waves reflected by a vertical wall (de Best, Bijker, and Wichers 1971; Xie 1981; Irie and Nadaoka 1984; Xie 1985). Fine sand is transported primarily in suspension, and in this case scour occurs at the nodes of the sea surface elevation with deposition occurring at the antinodes. Coarse sediment is moved primarily as bed load so that scour occurs midway between the sea surface nodes and antinodes with deposition usually centered on the nodes of the standing wave pattern.

   –   Uniform, regular waves produce a repeating pattern of scour and deposition as a function of distance from the toe of the vertical wall as illustrated in the upper portion of Figure VI-5-115. For fine sand maximum scour nearest the wall occurs a distance $L/4$ from the wall where L is the wavelength of the incident wave. Irregular waves produce a similar scour pattern for fine sand as shown in the lower portion of Figure VI-5-115. However, phase-locking between incident and reflected irregular waves decreases with distance from the wall with the maximum scour depth for fine sand approximately located a distance $L_p/4$ from the vertical wall, where L_p is the wavelength associated with the peak spectral frequency using linear wave theory.

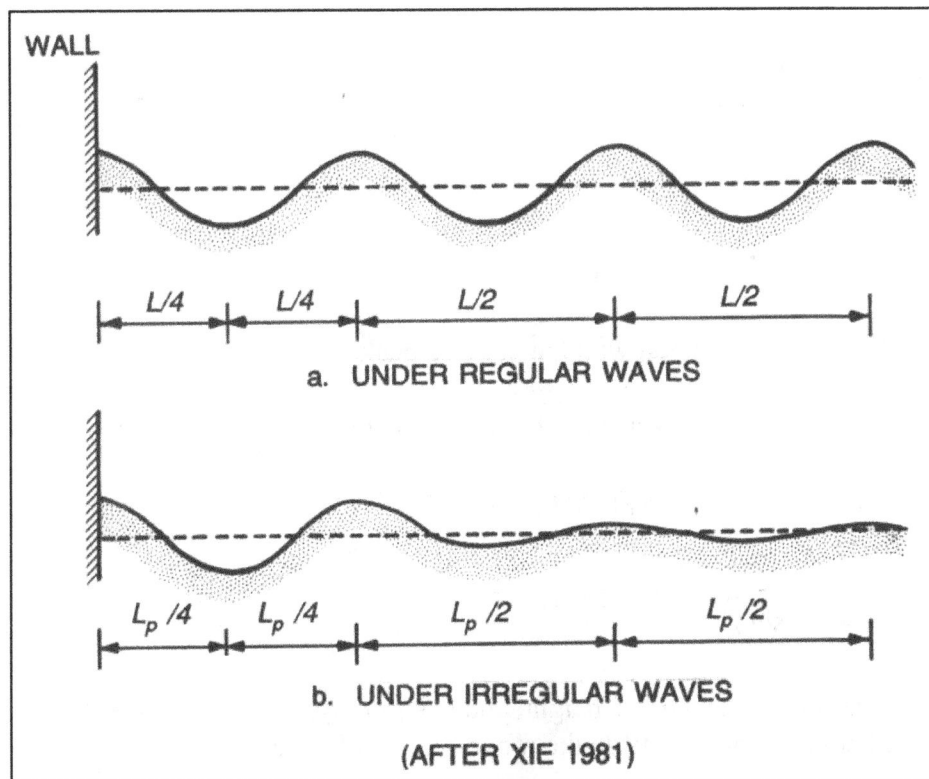

Figure VI-5-115. Regular and irregular wave-scoured profiles at a vertical-front structure

   –   Based on results from 12 movable-bed model tests, Xie (1981, 1985) proposed an empirically-based equation to estimate maximum scour for normally incident, nonbreaking, regular waves incident upon an impermeable vertical wall. The equation was given as:

$$\frac{S_m}{H} = \frac{0.4}{[\sinh(kh)]^{1.35}}$$

(VI-5-253)

where

S_m = maximum scour depth at node ($L/4$ from wall)

H = incident regular wave height

h = water depth

k = incident regular wave number ($k = 2\pi/L$)

L = incident regular wavelength

- A similar laboratory-based prediction empirical equation for the more appropriate case of normally incident, nonbreaking irregular waves was given by Hughes and Fowler (1991) as

$$\frac{S_m}{(u_{rms})_m T_p} = \frac{0.05}{[\sinh(k_p h)]^{0.35}}$$ (VI-5-254)

where

T_p = wave period of the spectral peak

k_p = wave number associated with the spectral peak by linear wave theory

$(u_{rms})_m$ = root-mean-square of horizontal bottom velocity

- The value of $(u_{rms})_m$ was given by Hughes (1992) as

$$\frac{(u_{rms})_m}{g\,k_p T_p H_{mo}} = \frac{\sqrt{2}}{4\pi \cosh(k_p h)}\left[0.54\cosh\left(\frac{1.5 - k_p h}{2.8}\right)\right]$$ (VI-5-255)

where H_{mo} is the zeroth-moment wave height, and g is gravity. (Equation VI-5-255 is empirically based and should not be applied outside the range $0.05 < k_p h < 3.0$.)

- Equation VI-5-255 is plotted on Figure VI-5-116 along with the movable-bed model experiment results. The dashed line is an equivalent to Equation VI-5-254. Scour predicted for irregular waves is significantly less than scour predicted for regular waves, and in many cases the predicted maximum scour does not represent a threat to the structure toe due to its location $L_p/4$ from the wall. Also, any effect related to sediment size is missing from these formulations (other than the stipulation of fine sand). Therefore, sediment scale effects may have influenced laboratory results causing less scour than might occur at full scale.

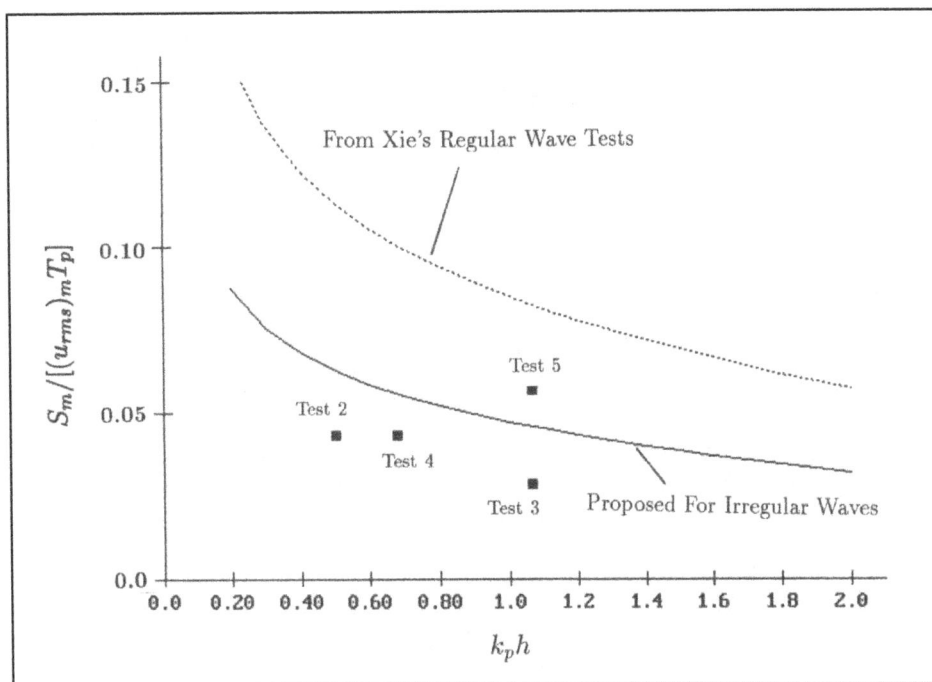

Figure VI-5-116. Scour prediction for nonbreaking waves at
vertical wall (Hughes and Fowler 1991)

– The relatively minor scour depths predicted for nonbreaking waves may be a direct result of scale effects or it may be related to the two-dimensionality of the laboratory experiments. In the wave flume an equilibrium profile is reached even though sediment is still constantly in motion. At an actual project site strong currents running parallel to a vertical-front structure could remove sediment put into motion by the standing wave pattern. If this occurs, scour will continue until a new live-bed equilibrium is reached. Sato, Tanaka, and Irie (1968) gave field examples of scour attributed to along-structure currents acting in conjunction with bed agitation by waves. Unfortunately, there are no scour prediction methods covering this possibility.

• Obliquely incident nonbreaking waves. Obliquely approaching incident nonbreaking waves will also be nearly completely reflected by a vertical wall. The resulting combined incident and reflected waves resemble a short-crested, diamond pattern that propagates in a direction parallel to the wall. (See Hsu (1991) for development of theories related to obliquely reflected long-crested waves.) Just as in the case of normal wave incidence, partial nodes and antinodes develop on lines parallel to the structure at distances that are a function of the wave properties and incident wave angle. However, obliquely reflected waves also generate a mass transport component parallel to the vertical structure which may contribute to enhanced scour along the structure. Silvester (1991) summarized laboratory results of scour at highly reflective (but not necessarily vertical-front) structures caused by obliquely incident long-crested regular and irregular waves. It was observed that obliquely incident waves tended to scour more than equivalent normally incident waves, and irregular waves scour at a slower rate and somewhat more uniformly than regular waves. No engineering methods are presently available to estimate scour caused by obliquely incident, nonbreaking irregular waves reflected by a vertical wall.

• Scour at the head of a vertical breakwater. Sumer and Fredsøe (1997) conducted small-scale movable-bed experiments to investigate scour around the circular head of a vertical breakwater aligned parallel to the wave crests. They discovered that scour around the breakwater head is due mainly to the lee-wake vortices, similar to wave-induced scour at vertical piles. Maximum scour depths from different sized breakwaters corresponded remarkably well with the associated Keulegan-Carpenter number, which is defined as

$$KC = \frac{U_m T}{B}$$

(VI-5-256)

where

U_m = maximum wave orbital velocity at the bed (in the absence of a structure)

T = regular wave period

B = diameter of the vertical breakwater circular head

- Sumer and Fredsøe presented the following empirical equation to predict maximum scour depth (S_m) as a function of the Keulegan-Carpenter number and diameter of the breakwater head:

$$\frac{S_m}{B} = 0.5 \, C_u \left[1 - e^{-0.175(KC-1)} \right]$$

(VI-5-257)

in which C_u is an uncertainty factor with a mean value of unity and a standard deviation of $\sigma_u = 0.6$. This empirical expression was developed for the data range $0 < KC < 10$. However, beyond $KC = 2.5$, data from only one breakwater diameter were used. Irregular waves will probably not scour as deeply, so the empirical equation could be considered conservative.

- Sumer and Fredsøe (1997) also investigated scour at the heads of squared-ended vertical breakwaters, perhaps representative of caissons. They found similar planform extent of scour, but depth of scour was greater by about a factor of 2. No empirical design equation was given for this situation, but it is possible to make estimates directly from the curve in their paper or from the simple equation

$$\frac{S_m}{B} = -0.09 + 0.123 \, KC$$

(VI-5-258)

which fits the data reasonably well. However, this expression is based on very limited laboratory data, and scour estimates should be considered tentative.

- The angle of obliquely incident waves on scour around the vertical breakwater head was also shown to be a factor in scour magnitude, and the addition of even small currents moving in the direction of wave propagation significantly increased depth of scour. No design guidance was suggested that included currents and wave angle. Sumer and Fredsøe analyzed scale effects in their laboratory experiments and concluded that scour holes at full scale will be slightly smaller than equivalent scaled-up model results. Design of scour protection for vertical breakwater heads is discussed in Part VI-5-6*c*, "Design of scour protection."

(b) Breaking waves. Scour caused by waves breaking on vertical-front structures has been a topic of numerous studies. (See Powell 1987; Kraus 1988; and Kraus and McDougal 1996 for overviews of the literature.) Scour caused by breaking waves is generally greater than for nonbreaking waves, and there is more likelihood of scour leading to structure damage. Spilling or plunging breaking waves can break directly on the vertical wall or just before reaching the wall. The physical mechanisms responsible for scour by breaking waves are not well understood, but it is generally thought that the breaking process creates strong downward directed flows that scour the bed at the base of the wall. For example, the re-entrant tongue of a plunging wave breaking before it reaches the structure generates a strong vortex motion that will mobilize sediment at the toe. A wave impacting directly on the vertical face will direct water down at the toe in the form of a jet. Sediment mobilization and transport is dominated by turbulent fluid motions rather than fluid shear stresses, and air entrained in the breaking wave also influences the erosion process (Oumeraci 1994). Figure VI-5-117 illustrates scour and profile change fronting a vertical seawall.

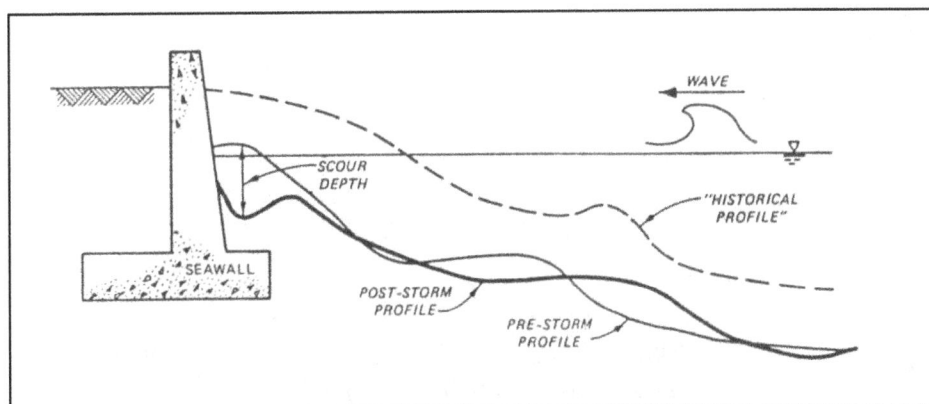

Figure VI-5-117. Scour due to breaking waves at a vertical seawall (Kraus 1988)

• Rules of thumb. There are several accepted rules-of-thumb pertaining to scour of noncohesive sediment at vertical walls. For the case of normally incident breaking waves with no currents:

– The maximum scour depth at a vertical wall (S_m) is approximately equal to the nonbreaking wave height (H_{max}) that can be supported by the water depth (h) at the structure, i.e.,

$$S_m = H_{max} \quad \text{or} \quad S_m \approx h \qquad (VI-5-259)$$

– Maximum scour occurs when the vertical wall is located around the plunge point of the breaking wave.

– Reducing the wall reflection reduces the amount of scour.

• Irregular breaking wave scour prediction. Predictive equations for estimating maximum scour at vertical walls due to normally incident regular breaking waves were proposed by Herbich and Ko (1968) and Song and Schiller (1973). Powell (1987) discussed shortcomings of these two methods and concluded the empirical equations were not useful for design purposes.

– Fowler (1992) also examined the Song and Schiller relationship using data from midscale movable-bed model tests using irregular waves, and reasonable correspondence was noted between measurements and predictions. Fowler then combined his irregular wave scour data with regular wave data from Barnett and Wang (1988) and from Chesnutt and Schiller (1971) as shown in Figure VI-5-118.

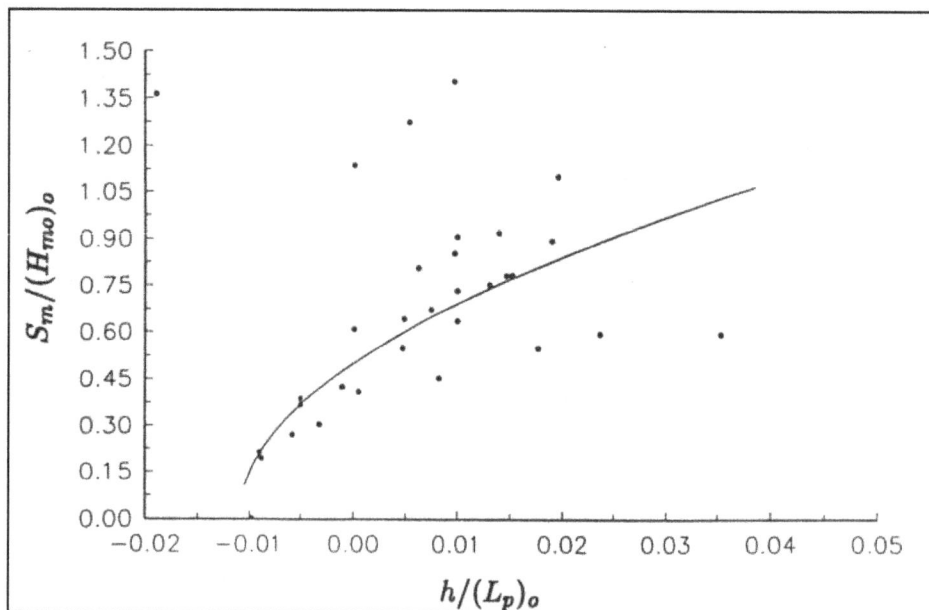

Figure VI-5-118. Relative scour depth as a function of relative depth at a vertical wall (Fowler 1992)

– The following empirical equation (solid line on Figure VI-5-118) was proposed for estimating maximum scour of noncohesive sediment due to normally incident breaking irregular waves with a mild approach slope.

$$\frac{S_m}{(H_{mo})_o} = \sqrt{22.72 \frac{h}{(L_p)_o} + 0.25}$$

(VI-5-260)

where

S_m = maximum scour depth

$(H_{mo})_o$ = zeroth-moment wave height in deep water

h = pre-scour water depth at the vertical wall

$(L_p)_o$ = deepwater wavelength associated with the peak spectral wave period, T_p, i.e., $(L_p)_o =$ $(g/2\pi) T_p^2$

    – Fowler noted that application of this empirical equation is limited by the data to values of relative depth and relative steepness within the ranges

$$0.011 < \frac{h}{(L_p)_o} < 0.045 \quad and \quad 0.015 < \frac{(H_{mo})_o}{(L_p)_o} < 0.040 \qquad \text{(VI-5-261)}$$

    – Fowler's predictive equation does not include any parameters relating to sediment properties, which are expected to have some influence in the scouring process. However, sediment transport induced by waves breaking against a vertical wall will not be very dependent on Shields parameter due to the turbulent nature of the entraining flow, and this would decrease the influence of sediment grain size. Also, the previous scour estimation method assumes no current flow along the vertical wall.

    – Scour of cobble (or shingle) beaches fronting vertical walls is discussed by Carpenter and Powell (1998). They provided dimensionless design graphs to predict maximum scour depth as a function of significant wave height, wave steepness, and local water depth. Their results were based on laboratory movable-bed model tests, which were correctly scaled due to the relatively large size of cobbles compared to sand.

    (2) Scour at sloping structures. Scour at the toe of sloping-front structures is thought to be a function of structure slope and porosity, incident wave conditions, water depth, and sediment grain-size. Despite considerable research into the processes responsible for wave-induced scour at sloping structures, there are no generally accepted techniques for estimating maximum scour depth or planform extent of scour (Powell 1987; Fowler 1993). However, progress is being made in development of numerical models to predict scour at sloping-front structures. Engineering use of such numerical models should consider model input requirements, representation of structure characteristics (particularly reflection parameters), and documented validation against field or laboratory experiments conducted at larger scales. Nonbreaking irregular waves impinging on a sloping structure will create a standing wave field similar to a vertical structure except the variation between the sea surface elevation nodes and antinodes is less pronounced, and the location of the node nearest the structure toe varies with wave condition and structure reflection properties (Hughes and Fowler 1995; O'Donoghue and Goldsworthy 1995; Losada, Silva, and Losada 1997). Erosion of fine sediment is expected to occur at the nodal location, but no empirical estimation method has been proposed.

    (a) Rules of thumb. In lieu of easily applied semi-empirical scour estimation tools, simple rules-of-thumb serve as engineering guidelines for scour at sloping-front structures.

    • Maximum scour at the toe of a sloping structure is expected to be somewhat less than scour calculated for a vertical wall at the same location and under the same wave condition. Therefore, a conservative scour estimate is provided by the vertical wall scour prediction equations, i.e., $S_m < H_{max}$.

• Depth of scour decreases with structure reflection coefficient. Therefore, structures with milder slopes and greater porosity will experience less wave-induced scour.

• Scour depths are significantly increased when along-structure currents act in conjunction with waves.

• Obliquely incident waves may cause greater scour than normally incident waves because the short-crested waves increase in size along the structure (Lin et al. 1986) due to the mach-stem effect. Also, oblique waves generate flows parallel to the structure.

(b) Scour at head of sloping breakwater. Fredsøe and Sumer (1997) conducted small-scale movable-bed model experiments to investigate mechanisms responsible for wave-induced scour around the conical heads of sloping-front breakwater structures. The experiments were similar in many respects to the companion study of scour at the ends of vertical breakwaters (Sumer and Fredsøe 1997). For most tests the rubble-mound breakwater head was approximated as an impermeable, smooth structure constructed of steel frames covered with sheet metal and having a slope of 1:1.5. The breakwater head was aligned parallel to the incident irregular waves. Observed scour was attributed to two different mechanisms; steady streaming of flow around the breakwater head, and waves breaking across the breakwater head and impinging on the leeside bed.

• Scour holes caused by steady streaming formed at the breakwater toe on the seaward curve of the breakwater head. An estimation of maximum scour depth (S_m) was developed as a function of the Keulegan-Carpenter number (KC) and given by Fredsøe and Sumer (1997) as

$$\frac{S_m}{B} = 0.04\, C_u \left[1 - e^{-4.0\,(KC-0.05)} \right] \tag{VI-5-262}$$

in which C_u is an uncertainty factor with a mean value of unity and a standard deviation of $\sigma_u = 0.2$. The Keulegan-Carpenter number is calculated as given by Equation VI-5-256 using the peak spectral wave period, T_p, as the period, T, and the breakwater head diameter at the bed as B.

• Fredsøe and Sumer suggested that U_m be calculated from linear wave theory as the bottom velocity found using a wave height of

$$H = \frac{1}{\sqrt{2}} H_s \tag{VI-5-263}$$

where H_s is the significant wave height. A similar expression for predicting deposition was also presented.

• The second scour mechanism is caused by waves breaking across the sloping front of the breakwater head. The geometry of the steep breakwater face causes lateral water motion that forms the tongue of the plunging breaker into a rounded re-entrant jet that impacts the bed at a steep angle and mobilizes sediment. This creates a scour hole at the breakwater toe on the leeside of the rounded head with the maximum depth located approximately at the intersection of breakwater head and trunk. Fredsøe and Sumer presented the following empirical equation for maximum scour depth (S_m) due to plunging breaking waves

$$\frac{S_m}{H_s} = 0.01 \, C_u \left(\frac{T_p \sqrt{g \, H_s}}{h} \right)^{3/2}$$ (VI-5-264)

where C_u is an uncertainty factor with a mean value of unity and a standard deviation of $\sigma_u = 0.34$, h is water depth, and the other parameters are as defined previously.

- As noted by Fredsøe and Sumer, these equations were developed for impermeable, smooth breakwater heads. The permeability and roughness of rubble-mound breakwaters will effectively decrease both scour mechanisms, thus scour estimates may be somewhat conservative. The previous empirical expressions for predicting maximum scour depths are based on a limited number of data points derived primarily from laboratory experiments, and the equations should be considered tentative until additional studies are conducted. Also, scour is caused by waves only; superimposed currents are expected to increase appreciably maximum scour depth. Design of scour protection for sloping-front breakwater heads is discussed in Part VI-5-6c, "Design of scour protection."

(3) Scour at piles. The majority of methods for estimating scour at vertical piles were developed for piles with circular cross section, which are widely used in coastal and offshore engineering applications. However, there are estimation techniques for piles with noncircular cross sections and for specialized structures such as noncircular bridge piers and large bottom-resting structures. Scour at small vertical piles (pile diameter, D, is less than one-tenth of the incident wavelength) is caused by three simultaneously acting mechanisms: formation of a horseshoe-shaped vortex wrapped around the front of the pile; vortex shedding in the lee of the pile; and local flow accelerations due to streamline convergence around the pile. The pile does not significantly affect the incident wave. Large diameter piles, in which the diameter is greater than one-tenth of the incident wavelength, do have an impact on the incident waves which are reflected by the pile and diffracted around the pile. The key parameters governing scour formation appear to be current magnitude, orbital wave velocity, and pile diameter. Less important parameters are sediment size and pile shape (if the pile has noncircular cross section). For detailed descriptions of the physical mechanisms responsible for scour at vertical piles see Niedoroda and Dalton (1986) or some of the following references. A general, and somewhat conservative, rule-of-thumb is: Maximum depth of scour at a vertical pile is equal to twice the pile diameter. This rule-of-thumb appears to be valid for most cases of combined waves and currents. Smaller maximum scour depths are predicted by the equations in the following sections. Estimation formulas for maximum scour depth have been proposed for the cases of currents only, waves only, and combined waves and currents. The flow problem and associated sediment transport are beyond a complete theoretical formulation, and even numerical modeling attempts have not been able to describe fully the scour process at vertical piles (see Sumer and Fredsøe 1998a for a summary of numerical modeling approaches).

(a) Scour at small diameter vertical piles. Vertical piles with diameter, *D*, less than one-tenth of the incident wavelength constitute the vast majority of pile applications in coastal engineering. Even cylindrical legs of some offshore oil platforms may fall into this category.

• Pile scour by currents. Many scour estimation formulas have been proposed for scour caused by unidirectional currents without the added influence of waves. A formulation widely used in the United States is the Colorado State University (CSU) equation developed for bridge piers (e.g., Richardson and Davis 1995) given by the expression

$$\frac{S_m}{h} = 2.0\, K_1 K_2 \left(\frac{b}{h}\right)^{0.65} F_r^{0.43} \tag{VI-5-265}$$

where

S_m = maximum scour depth below the average bottom elevation

h = water depth upstream of the pile

b = pile width

F_r = flow Froude number [$F_r = U/(gh)^{1/2}$]

U = mean current velocity magnitude

K_1 = pile shape factor

K_2 = pile orientation factor

– Equation VI-5-265 is a deterministic formula applicable for both clear water scour and live bed scour, and it represents a conservative envelope to the data used to establish the empirical coefficients. The shape factor, K_1, is selected from Figure VI-5-119, and the orientation factor, K_2, can be determined from the following equation given by Froehlich (1988).

$$K_2 = \left(\cos\theta + \frac{L}{b}\sin\theta\right)^{0.62} \tag{VI-5-266}$$

where L/b is defined in Figure VI-5-119 and θ is the angle of pile orientation. K_2 equals unity for cylindrical piles. Other modifying factors have been proposed to account for sediment gradation and bed forms, but these factors have not been well established. An additional factor is available for use when piles are clustered closely together. See Richardson and Davis (1995) and Hoffmans and Verheij (1997) for details.

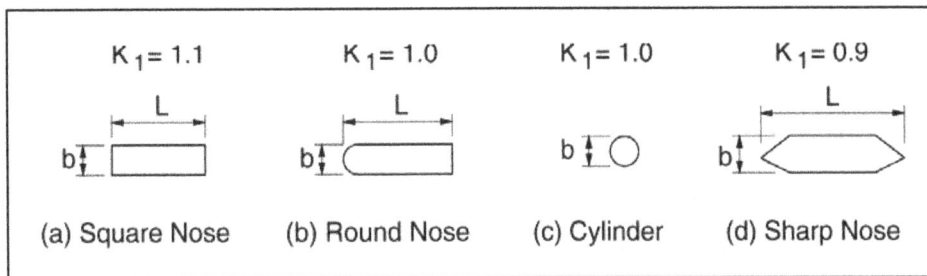

Figure VI-5-119. Correction factor, K_1, for pile/pier shape

    –   Johnson (1995) tested seven of the more commonly used scour prediction equations against field data and found that the CSU equation (Equation VI-5-265) produced the best results for $h/b > 1.5$. At lower values of h/b a different empirical formulation offered by Breusers, Nicollet, and Shen (1977) provided better results.

    –   Johnson (1992) developed a modified version of the CSU empirical equation for use in reliability analysis of failure risk due to scour at cylindrical piles. Her formula represents a best-fit to the data rather than a conservative envelope. An application example is included in her 1992 paper.

    •   Pile scour by waves. The physical processes associated with wave-only scour around vertical piles are reasonably well described qualitatively (See Sumer and Fredsøe (1998a) for a comprehensive review and listing of many references.)

    –   In an earlier paper Sumer, Christiansen, and Fredsoe (1992a) established an empirical equation to estimate scour at a vertical pile under live bed conditions. They used small- and large-scale wave flume experiments with regular waves, two different sediment grain sizes, and six different circular pile diameters ranging from 10 cm to 200 cm. Maximum scour depth (S_m) was found to depend only on pile diameter and Keulegan-Carpenter number (KC), as expressed by Equation VI-5-256 with pile diameter, D, as the denominator. The experimental data of Sumer, Christiansen, and Fredsoe (1992a) are shown plotted in Figure VI-5-120, and the solid line is the predictive equation given by

$$\frac{S_m}{D} = 1.3 \left[1 - e^{-0.03(KC-6)} \right]$$

(VI-5-267)

where D is cylindrical pile diameter. No live-bed scour occurs below values of $KC = 6$, which corresponds to onset of horseshoe vortex development. At values of $KC > 100$, $S_m/D \rightarrow 1.3$, representing the case of current-only scour.

    –   Independent confirmation of Equation VI-5-267 was presented by Kobayashi and Oda (1994) who conducted clear water scour experiments. They stated that maximum scour depth appeared to be independent of Shields parameter, grain size diameter, and whether scour is clear-water or live-bed.

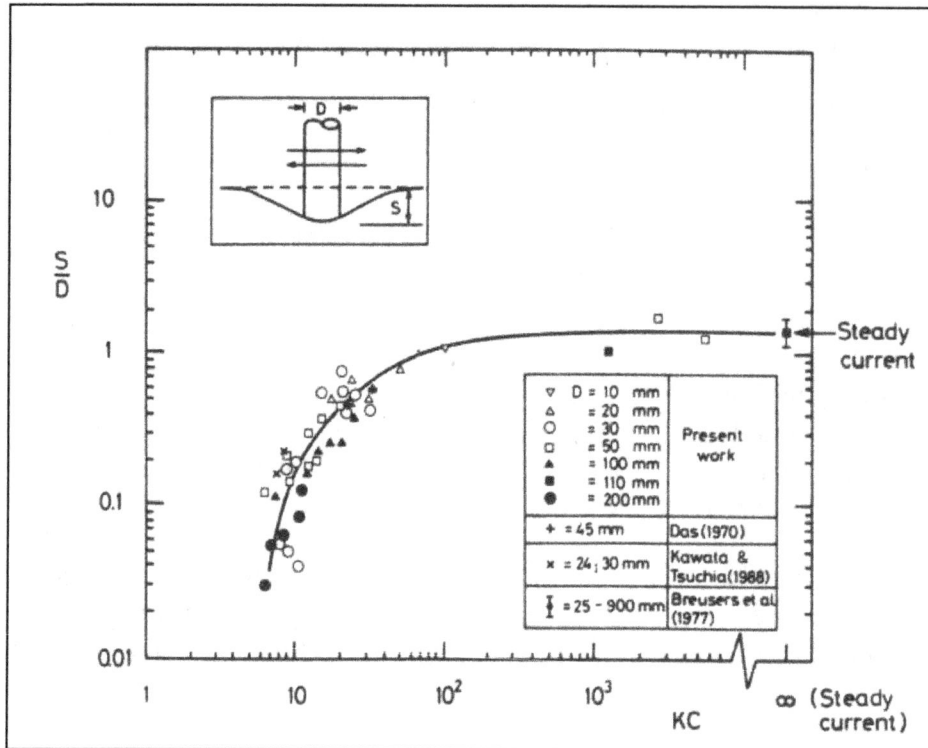

Figure VI-5-120. Wave-induced equilibrium scour depth at a vertical pile

    – In an extension to their 1992 study, Sumer, Christiansen, and Fredsoe (1993) conducted additional regular wave live-bed scour experiments using square piles oriented with the flat face 90 deg and 45 deg to the waves. The following empirical equations for maximum scour were obtained as best-fits to the observed results:

Square pile 90 deg to flow:

$$\frac{S_m}{D} = 2.0 \left[1 - e^{-0.015(KC-11)} \right] \quad \textit{for KC}\_11 \tag{VI-5-268}$$

Square pile 45 deg to flow:

$$\frac{S_m}{D} = 2.0 \left[1 - e^{-0.019(KC-3)} \right] \quad \textit{for KC}\_3 \tag{VI-5-269}$$

    – Scour for the square pile oriented at 45 deg begins at lower values of *KC*, but the maximum scour at large *KC* values approaches $S_m/D = 2$ regardless of orientation.

    – Studies on the time rate of scour development were reported by Sumer, Christiansen, and Fredsoe (1992b), Sumer et al. (1993), and Kobayashi and Oda (1994). Recent research on wave scour around a group of piles was summarized by Sumer and Fredsøe (1998a, 1998b).

• Pile scour by waves and currents. Kawata and Tsuchiya (1988) noted that local scour depths around a vertical pile were relatively minor compared to scour that occurs when even a small steady current is added to the waves. Eadie and Herbich (1986) conducted small-scale laboratory tests of scour on a cylindrical pile using co-directional currents and irregular waves. They reported the rate of scour was increased by adding wave action to the current, and the maximum scour depth was approximately 10 percent greater than what occurred with only steady currents. This latter conclusion contradicts Bijker and de Bruyn (1988) who found that nonbreaking waves added to steady currents slightly decreased ultimate scour depth whereas adding breaking waves caused increased scour to occur. Eadie and Herbich also noted that the inverted cone shape of the scour hole was similar with or without wave action, and the use of irregular versus regular waves appeared to influence only scour hole geometry and not maximum scour depth. They developed a predictive equation using results from approximately 50 laboratory experiments, but no wave parameters were included in the formulation. Finally, they pointed out that their conclusions may hinge on the fact that the steady current magnitude exceeded the maximum bottom wave orbital velocity, and different results may occur with weak steady currents and energetic waves.

• Earlier work by Wang and Herbich (1983) did provide predictive equations that included wave parameters along with current, pile diameter, sediment properties, and water depth. However, there were some unanswered questions about scaling the results to prototype scale. Consequently, until further research is published, maximum scour depth due to waves and currents should be estimated using the formulations for scour due to currents alone (Equation VI-5-265).

(b) Scour at large diameter vertical piles. Rance (1980) conducted laboratory experiments of local scour at different shaped vertical piles with diameters greater than one-tenth the incident wavelength. The piles were exposed to coincident waves and currents. Rance provided estimates of maximum scour depth as functions of pile equivalent diameter, D_e, for different orientations to the principal flow direction. (D_e is the diameter of a cylindrical pile having the same cross-sectional area as the angular pile.) These formulas are given in Figure VI-5-121.

(c) Maximum scour occurs at the corners of the square piles. Estimates of extent of scour are useful for design of scour blankets. Sumer and Fredsøe (1998a) provided additional information about flow around large piles.

(4) Scour at submerged pipelines. Waves and currents can scour material from beneath pipelines resting on the bottom, leading to partial or even complete burial of the pipeline. In most situations pipeline burial is usually considered a desirable end result. However, if the pipeline spans soil types having different degrees of erodibility, differential scour may result in sections of the pipeline being suspended between bottom hard points, and this could lead to pipeline failure. Onset of scour beneath a pipeline resting on, or slightly embedded in, the bottom occurs initially as piping when seepage beneath the pipeline increases and a mixture of sediment and water breaks through (Chiew 1990). Onset of scour is followed by a phase of rapid scour called tunnel erosion in which the bed shear stresses are increased four times above that of the undisturbed sand bed. Tunnel erosion is followed by lee-wake erosion in which the lee-wake of

the pipeline appears to control the final equilibrium depth and shape of the downstream scour. Equilibrium depth of scour beneath the pipeline is usually defined as the distance between the eroded bottom and the underside of the pipeline as illustrated on Figure VI-5-122. Overviews of pipeline scour knowledge and citations to the extensive literature are included in Sumer and Fredsøe (1992, 1998a) and Hoffmans and Verheij (1997). Only the established empirical equations for estimating scour depth are included in the following:

(a) Pipeline scour by currents. In steady currents the equilibrium scour depth beneath a pipeline is thought to be a function of pipe diameter, pipe roughness, pipe Reynolds number, and Shields parameter. For clear water scour, when mean flow velocity, U, is less than the critical velocity, U_c, maximum scour depth can be calculated using the following equation from Hoffmans and Verheij (1997)

$$\frac{S_m}{D} = \frac{\mu}{2}\left(\frac{U}{U_c}\right)$$

(VI-5-270)

where

$$\mu = \left(\frac{k_s}{12\,D}\right)\ln\left(\frac{6\,D}{k_s}\right)$$

(VI-5-271)

and

D = pipe diameter

h = water depth

U = depth averaged flow

U_c = critical depth-averaged flow velocity

k_s = effective bed roughness, $k_s = 3\ d_{90}$ (k_s must have the same units as D)

When $U/U_c > 1$, live-bed scour occurs, and in this case Sumer and Fredsøe (1992) stated that pipe Reynolds number only influences flow around smooth pipes and the influence of Shields parameter is minor. They recommended the simple equation for predicting maximum equilibrium scour depth. The 0.1-value represents the standard deviation of the data, so a conservative estimate of scour would be $S_m/D=0.7$.

$$\frac{S_m}{D} = 0.6 \pm 0.1$$

(VI-5-272)

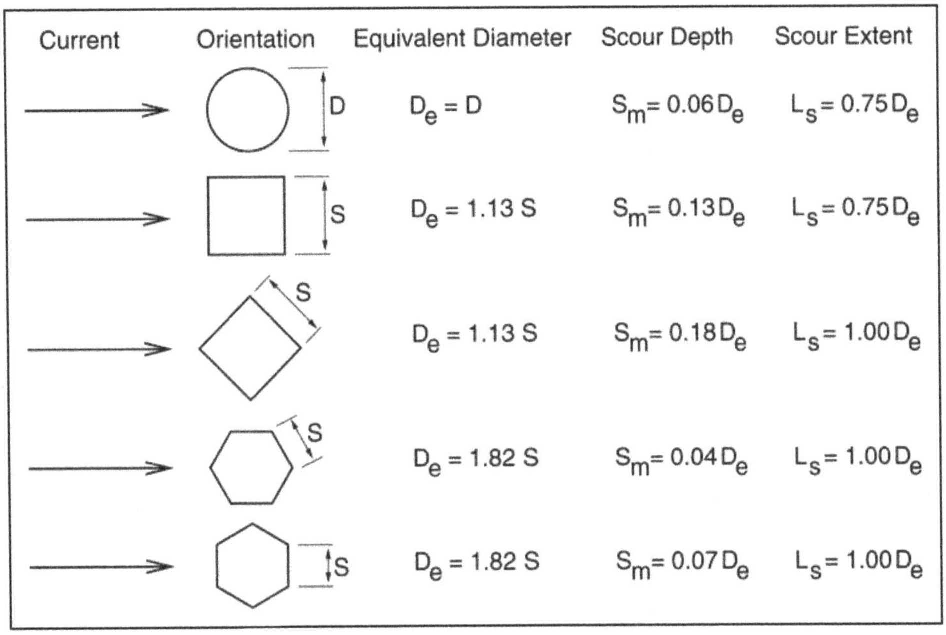

Figure VI-5-121. Wave and current scour aroundlarge vertical piles (Rance 1980)

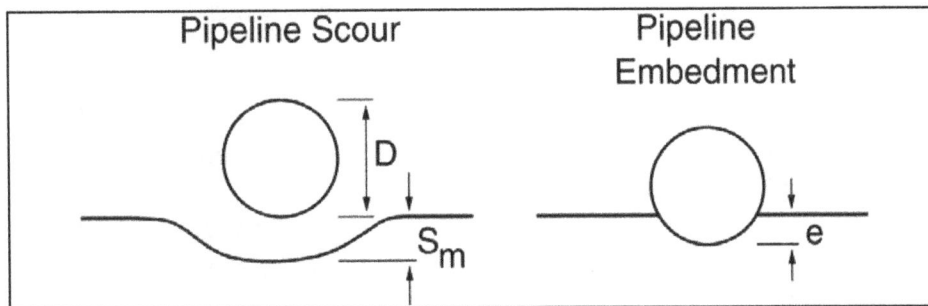

Figure VI-5-122. Pipeline scour and pipeline embedment

(b) Pipeline scour by waves. Oscillatory bottom velocities under waves create piping conditions beneath pipelines in the same manner as steady currents. Sumer and Fredsøe (1991) gave a criterion for onset of scour under waves based on a small number of laboratory experiments. This criterion is

$$\frac{e_{cr}}{D} = 0.1 \ln(KC) \qquad \text{(VI-5-273)}$$

where e_{cr} is the critical embedment (depth of pipeline burial beyond which no scour occurs), and KC is the Keulegan-Carpenter number, given by Equation VI-5-256 with D as the denominator. Scour is unlikely to occur for values of $e_{cr}/D > 0.5$ (half buried pipe). Sumer and Fredsøe (1990) studied scour beneath a bottom-resting pipeline under wave action. Their laboratory data, combined with that of an earlier researcher, indicated that live-bed scour was strongly related to Keulegan-Carpenter number and pipe diameter, while only weakly influenced by Shields parameter and pipe roughness. The data were well represented over a wide range of Keulegan-Carpenter number (2 < KC < 300) by the empirical expression

$$\frac{S_m}{D} = 0.1\sqrt{KC} \qquad\qquad\qquad\qquad \text{(VI-5-274)}$$

Klomb and Tonda (1995) presented a modified version of Equation VI-5-274 that included allowance for partial embedment, e, of the pipeline, i.e.,

$$\frac{S_m}{D} = 0.1\sqrt{KC}\left(1 - 1.4\,\frac{e}{D}\right) + \frac{e}{D} \qquad\qquad \text{(VI-5-275)}$$

with scour depth taken relative to the undisturbed bed. Equation VI-5-275 is valid for values of $e/D < 0.5$ (Hoffmans and Verheij 1997).

(c) Pipeline scour by waves and currents. Sumer and Fredsøe (1996) conducted laboratory tests of pipeline scour due to combined waves and currents covering a range of KC from 5 to about 50 with codirectional currents. The general trend, regardless of the value of KC, was for scour depth to initially decrease as current is increased from zero. At higher values of current, maximum scour depth approaches the value given by Equation VI-5-272 for currents alone. Sumer and Fredsøe (1996) provided empirical design equations based on the laboratory experiments; but for values of KC between 40 and 50 maximum scour depth is almost the same as the estimate for currents alone.

(d) Pipelines in the nearshore. Pipelines traversing the surfzone may be damaged if exposed to breaking waves and strong longshore currents. Little design guidance is available other than the fact that additional scour will occur once the pipeline is exposed. The burial depth for a pipeline through the nearshore should exceed in all places the expected bottom profile lowering that might occur over the life of the pipeline. This can be estimated using profile-change models or from long-term beach profile data.

(5) Other scour problems. Some coastal projects may include structural elements or hydrodynamic flow conditions that are typically associated with inland waterways or estuaries. Structures such as storm surge barriers, discharge control structures, or large pad footings may experience scour around their foundations due to currents or combined waves and currents.

(a) Hoffmans and Verheij (1997) provided a summary of techniques for estimating maximum scour for a number of situations that may be applicable to coastal projects:

- Scour downstream of sills and stone blankets due to currents.

- Scour downstream of hard bottoms due to horizontal submerged jets.

- Scour at control structures due to plunging jets.

- Scour at two- and three-dimensional culverts.

- Scour at abutments and spur dikes.

(b) See Hoffmans and Verheij (1997) for further details and associated technical literature.

c. Design of scour protection. Toe protection in the form of an apron is needed to prevent toe scour which may destabilize or otherwise decrease the functionality of a coastal structure. The apron must remain intact under wave and current forces, and it should be flexible enough to conform to an initially uneven sea floor. Scour apron width and required stone size for stability are related to wave and current intensity, bottom material, and structure characteristics such as slope, porosity, and roughness. Design guidance for scour protection is based largely on past successful field experience combined with results from small-scale laboratory tests. Special attention is needed where scour potential is enhanced such as at structure heads/ends, at transitions in structure composition, or at changes in structure alignment. This section provides general design guidance for scour aprons; however, this guidance should be considered preliminary. Projects requiring absolutely stable scour blankets should have proposed designs tested in a physical model. Hales (1980) surveyed scour protection practices in the United States and found that the minimum scour protection was typically an extension of the structure bedding layer and any filter layers. The following minimum rules-of-thumb resulted from this survey: minimum toe apron thickness - 0.6 m to 1.0 m (1.0 m to 1.5 m in northwest U.S.); minimum toe apron width - 1.5 m (3 m to 7.5 m in northwest U.S.); material - quarrystone to 0.3 m diameter, gabions, mats, etc. These rules-of-thumb are inadequate when the water depth at the toe is less than two times the maximum nonbreaking wave height at the structure or when the structure reflection coefficient is greater than 0.25 (structures with slopes greater than about 1:3). Under these more severe conditions use the scour protection methods summarized in the following sections for specific types of coastal structures.

(1) Scour protection for vertical walls.

(a) Vertical-front structures consist of large caisson-type gravity structures, gravity retaining walls, and cantilevered or anchored sheet-pile retaining walls. Toe protection design for larger vertical-front gravity structures subjected to waves is covered in Part VI-5-3d, "Toe stability and protection."

(b) For cantilevered or anchored retaining walls, Eckert (1983) proposed toe protection in the form of a scour apron constructed of quarrystone. The main purpose of the apron is to retain soil at the toe and/or to provide sufficient weight to prevent slip failure (see Figures VI-2-69 and VI-2-70). From geotechnical considerations the width (W) of the scour apron should be approximately

$$W = \frac{d_e}{\tan(45^o - \phi/2)} \approx 2.0\, d_e \qquad\qquad \text{(VI-5-276)}$$

where d_e is the depth of sheet-pile penetration below the seabed, and φ is the angle of internal friction of the soil (varies from about 26 deg to 36 deg). The width of the scour apron based on hydrodynamic criteria was given by Eckert as the greater of

$$W = 2.0\, H_i \qquad or \qquad W = 0.4\, d_s \qquad\qquad \text{(VI-5-277)}$$

where H_i is the incident wave height and d_s is the depth at the structure toe. Selected scour apron design width will be the greater of Equations VI-5-276 and VI-5-277.

(c) Eckert (1983) noted that gravity retaining walls do not require the apron to be as wide as needed for cantilevered walls. In this case, he recommended that scour apron width be about the same as the nonbreaking incident wave height.

(d) Determining the toe apron quarrystone size depends on the hydrodynamic conditions. They are as follows:

- Waves. If retaining walls are exposed to vigorous wave conditions, the toe quarrystone should be sized using the guidance given by Figure VI-5-45 (Part VI-5-3d "Toe stability and protection," and the apron thickness should be equal to either two quarrystone diameters or the minimum given in the prior rules-of-thumb, whichever is greater.

- Currents. If strong currents flow adjacent to the wall, toe quarrystone should be sized using the guidance provided in Part VI-5-3f, "Blanket stability in current fields."

- Waves and Currents. If both waves and strong currents impact the toe adjacent to a vertical retaining wall, estimate the size of the apron quarrystone for the waves alone and for the current alone. Then increase whichever is larger by a factor of 1.5 (Eckert 1983).

(e) In Sumer and Fredsøe's (1996) study of scour around the head of a vertical breakwater, laboratory tests were conducted to establish a relationship for the width of a scour apron that provides adequate protection against scour caused by wave-generated lee-wake vortices. Their empirical formula was given as

$$\frac{W}{B}=1.75\,(KC-1)^{1/2} \tag{VI-5-278}$$

where B is the diameter of the vertical breakwater circular head and KC is the Keulegan-Carpenter number given by Equation VI-5-256. Sumer and Fredsøe cautioned that this estimation of apron width may be inadequate in the presence of a current or for head shapes other than circular. Scour apron stone sizes are determined using the methods outlined in Part VI-5-3d, "Toe stability and protection."

(2) Scour protection for sloping structures.

(a) Scour protection for sloping structures exposed to waves is typically provided by the toe protection. Part VI-5-3d, "Toe stability and protection," presents guidance on the design of toe protection. Additional scour protection is sometimes needed at sloping-front structures to prevent scour by laterally-flowing currents. Strong tidally-driven currents adjacent to navigation jetties can scour deep trenches that may destabilize the jetty toe and result in slumping of the armor layer. Because prediction of the location and extent of potential scour is not well advanced, scour blankets are often not installed until after realization that scour has occurred. Depending on the scour hole configuration, it may be necessary to backfill the scour hole before placing a scour blanket, and the necessary extent of the protection is determined in part by the extent of the existing scour, by past experience, and by the judgment of the engineer. An understanding of the flow regime will help assure that the scour problem will not reoccur downstream of the scour protection blanket. Stone size for scour protection from currents is given in Part VI-5-3f, "Blanket stability in current fields." Bass and Fulford (1992) described the design and installation of scour protection along the south jetty of Ocean City Inlet in Maryland.

(b) Fredsøe and Sumer's (1997) laboratory study of wave-induced scour at the rounded heads of rubble-mound structures included design suggestions for scour protection. The width of the scour apron from the structure toe to outer edge was given by

$$\frac{W}{B} = A_1 (KC) \tag{VI-5-279}$$

where B is the breakwater head diameter at the bed and KC is given by Equation VI-5-256. Complete scour protection is provided with $A_1 = 1.5$ whereas a value of $A_1 = 1.1$ will result in relatively minor scour at the outer edge with a depth equal to about $0.01\ B$. Scour apron stone size are determined using the methods outlined in Part VI-5-3d, "Toe stability and protection."

(3) Scour protection for piles.

(a) Vertical piles and piers exposed only to currents can be protected against scour by placement of scour aprons constructed of stone or riprap, gabions, concrete mattresses, or grout-filled bags. Riprap aprons should be designed according to the relationships given in Part VI-5-3f, "Blanket stability in current fields." Options other than riprap or stone should be tested in physical models.

(b) Based on an earlier report by Bonasoundas (in German), Hoffmans and Verheij (1997) recommended that minimum width for the horizontal extent of the scour apron around circular piers be specified as a function of pile diameter, B. Upstream of the pile, and to both sides, apron width is $2.5\ B$. Downstream the apron elongates to a width of $4.0\ B$ as illustrated on Figure VI-5-123. Elongation in both directions is necessary for alternating tidal currents.

(c) An alternative recommendation was given by Carstens (1976) who found that scour apron width was a function of maximum scour depth (S_m) at the pile, i.e.,

$$\frac{W}{S_m} = \frac{F_s}{\tan \phi} \tag{VI-5-280}$$

where φ is the bed material angle of repose and F_s is a factor of safety.

(d) General recommendations for specifying apron width for different shaped piers and pilings, or for groups of piles, are lacking. In these cases laboratory model tests are needed to assure adequate scour protection. Past experience on other successful projects or case histories reported in the literature can also serve as design guidance (e.g., Edge et al. 1990; Anglin et al. 1996).

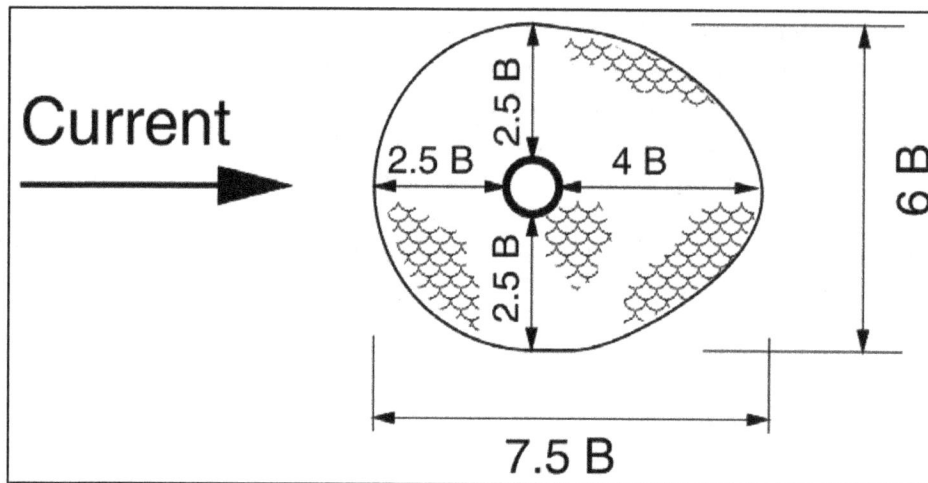

Figure VI-5-123. Scour apron for vertical pile in a current

(e) Similar protective measures can be deployed to prevent scour around piles by wave action. However, guidance is also lacking on how to design stable scour aprons in wave environments (Sumer and Fredsøe 1998a), and the best recourse is site-specific model tests. As a rule-of-thumb, the horizontal extent of the apron should be approximately twice the predicted scour depth.

(4) Scour protection for submerged pipelines.

(a) Submerged pipelines can be protected by either burying the pipeline in a trench or by covering the pipeline with a stone blanket or protective mattress. Protected pipelines are less susceptible to trawler damage and less likely to suffer damage caused by differential scour that leaves portions of the pipeline suspended between support points.

(b) Outside the active surfzone, burial depth is a function of local wave and current climate, sediment properties, and liquefaction potential. Usually the excavated material can be used as backfill provided it is sufficiently coarse to avoid buildup of excessive pore pressures which could lead to liquefaction and vertical displacement of the pipeline (Sumer and Fredsøe 1998a). Pipelines traversing the surfzone should be buried at an elevation lower than the anticipated erosion that would occur over the projected service life of the pipeline. Generally, stone blankets or mattresses are not considered effective protection in the surfzone because the elements must be designed to withstand the intense action of breaking waves.

(c) Pipelines resting on the bottom can be protected from being undermined by stabilizing the adjacent bed with a stone blanket having a horizontal width less than the extent of expected scour. Hjorth (1975) reported that covering at least the bottom half of the pipeline, as shown in the upper part of Figure VI-5-124, provides sufficient protection as evidenced by field experience. The alternative is to cover the pipeline completely with a stone blanket consisting of two or more filter layers as illustrated by the lower sketch of Figure VI-5-124. Stability of the uppermost stone layer requires that the shields parameter (Equation III-6-43) based on stone diameter must be less than the critical value for incipient motion. Stone blanket placement can be accomplished by dumping stone from the surface, provided the falling stones are not so large as to damage the pipeline on impact.

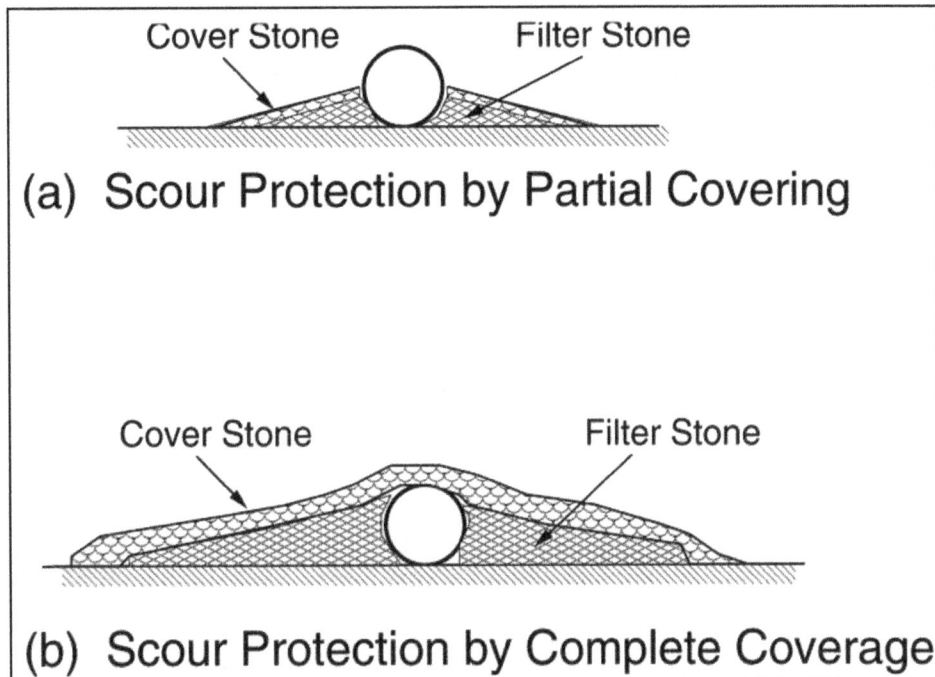

(a) Scour Protection by Partial Covering

(b) Scour Protection by Complete Coverage

Figure VI-5-124. Stone blanket scour protection for submerged pipelines

(d) Various types of scour mattresses have also been used effectively to protect pipelines. Mattresses may be economical when stone is not readily available; however, special mattress placing equipment is usually required. Hoffmans and Verheij (1997) illustrated several types of mattresses.

VI-5-7. <u>Wave Forces on Slender Cylindrical Piles</u>.

    a. Introduction.

    (1) Frequent use of pile-supported coastal and offshore structures makes the interaction of waves and piles of significant practical importance. The basic problem is to predict forces on a pile due to the wave-associated flow field. Because wave-induced flows are complex, even in the absence of structures, solution of the complex problem of wave forces on piles relies on empirical coefficients to augment theoretical formulations of the problem. This section is meant to be only an introduction to estimating forces and moments on slender cylindrical piles. For more detailed analysis see the literature related to ocean engineering and the design of offshore facilities.

    (2) Variables important in determining forces on circular piles subjected to wave action are shown in Figure VI-5-125. Variables describing nonbreaking, monochromatic waves are the wave height H, water depth d, and either wave period T, or wavelength L. Water particle velocities and accelerations in wave-induced flows directly cause the forces. For vertical piles the horizontal fluid velocity u and acceleration du/dt and their variation with distance below the

free surface are important. The pile diameter D and a dimension describing pile roughness elements k are important variables describing the pile. In this discussion the effect of the pile on the wave-induced flow is assumed negligible. Intuitively, this assumption implies that the pile diameter D must be small with respect to the wavelength L. Significant fluid properties include the fluid density ρ and the kinematic viscosity v. In dimensionless terms, the important variables can be expressed as follows:

$$\frac{H}{gT^2} = \text{dimensionless wave steepness}$$

$$\frac{d}{gT^2} = \text{dimensionless water depth}$$

$$\frac{D}{L} = \text{ratio of pile diameter to wavelength (assumed small)}$$

$$\frac{k}{D} = \text{relative pile roughness}$$

$$\frac{HD}{Tv} = \text{a form of the Reynolds number}$$

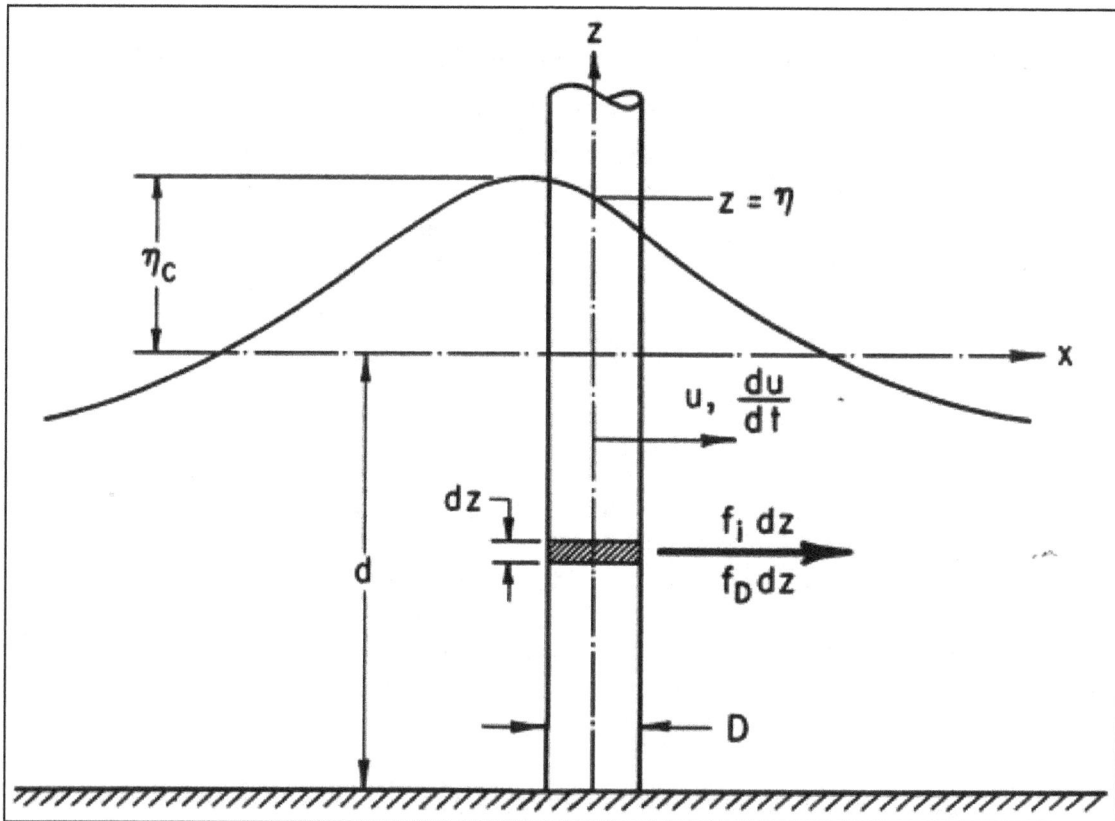

Figure VI-5-125. Definition sketch of wave forces on a vertical cylinder

(3) Given the orientation of a pile in the flow field, the total wave force acting on the pile can be expressed as a function of these dimensionless parameters. The variation of force over the length of the pile depends on the mechanism by which the water particle velocities and accelerations cause the forces. The following analysis relates the local forces acting on a section of pile element of length dz to the local fluid velocity and acceleration that would exist at the center of the pile if the pile were not present. Two dimensionless force coefficients, an inertia (or mass) coefficient C_M and a drag coefficient C_D, are used to establish the wave-force relationships. These coefficients are determined by experimental measurements of force, velocity, and acceleration or by measurement of force and water surface profiles, with accelerations and velocities inferred by assuming an appropriate wave theory.

(4) In the following section it is initially assumed that the force coefficients C_M and C_D are known to illustrate calculation of forces on vertical cylindrical piles subjected to monochromatic waves. Selection of C_M and C_D follows in Part VI-5-7c. Experimental data are available primarily for the interaction of nonbreaking waves and vertical cylindrical piles; and consequently, specific design guidance can be given for this common situation.

b. Vertical cylindrical piles and nonbreaking waves.

(1) Basic concepts. Morison et al. (1950) suggested that the horizontal force per unit length of a vertical cylindrical pile subjected to waves is analogous to the mechanism by which fluid forces on bodies occur in unidirectional flow, and this force can be expressed by the formulation

$$f = f_i + f_D = C_M \rho \frac{\pi D^2}{4} \frac{du}{dt} + C_D \frac{1}{2} \rho D u \,|u\,| \qquad \text{(VI-5-281)}$$

where

f_i = inertial force per unit length of pile

f_D = drag force per unit length of pile

ρ = mass density of fluid

D = pile diameter

u = horizontal water particle velocity at the axis of the pile (calculated as if the pile were absent) total

$\dfrac{du}{dt}$ = horizontal water particle acceleration at the axis of the pile (calculated as if the pile were absent)

C_D = drag hydrodynamic force coefficient

C_M = inertia or mass hydrodynamic force coefficient

(a) The inertia force term f_i is of the form obtained from an analysis of the force on a body in an accelerated flow of an ideal nonviscous fluid. The drag force term f_D is the drag force exerted on a cylinder in a steady flow of a real viscous fluid. The drag force f_D is proportional to u^2 and acts in the direction of the velocity u. To retain the correct direction sign, u^2 is written as $u\,|u\,|$. Although these remarks support the soundness of the formulation of the problem as given by Equation VI-5-281, it should be emphasized that expressing total force by the terms f_i and f_D is an assumption justified only if it leads to sufficiently accurate predictions of wave force as evidenced by ample measurements.

(b) Because the quantities u and du/dt in Equation VI-5-281 are defined as the values of these parameters at the axis of the pile, it is apparent that the influence of the pile on the flow field a short distance away from the pile has been neglected. Using linear wave theory MacCamy and Fuchs (1954) analyzed theoretically the problem of waves passing a circular cylinder. Their analysis assumed an ideal nonviscuous fluid and led to an inertia force having the form given for f_i under special conditions. Although their theoretical result is valid for all ratios of pile diameter to wavelength, D/L, the inertia force was found to be nearly proportional to the acceleration du/dt for small values of D/L (where L is wavelength calculated by linear theory). This theoretical result provides an indication of how small the pile should be for Equation VI-5-281 to apply, and the restriction is given as

$$\frac{D}{L} < 0.05 \tag{VI-5-282}$$

where L is calculated by linear wave theory. This restriction will seldom be violated for slender pile force calculations; however, the restriction may be important when applying Equation VI-5-281 to larger structures such as cylindrical caissons.

(c) To apply Equation VI-5-281 it is necessary to choose an appropriate wave theory for estimating u and du/dt from values of wave height H, wave period T, and water depth d; and for that particular wave condition appropriate values of C_D and C_M must be selected.

(2) Calculation of forces and moments. For structural design of a single vertical pile, it is often unnecessary to know in detail the distribution of forces over the height of the pile. Instead, the designer needs to know the total maximum force and the total maximum moment about the mud line ($z = -d$) acting on the pile. The total time-varying force and the time-varying moment acting about the mud line is found by integrating Equation VI-5-281 between the bottom and the free surface, i.e.,

$$F = \int_{-d}^{\eta} f_i\, dz + \int_{-d}^{\eta} f_D\, dz = F_i + F_D \tag{VI-5-283}$$

$$M = \int_{-d}^{\eta} (z+d)\, f_i\, dz + \int_{-d}^{\eta} (z+d)\, f_D\, dz = M_i + M_D \tag{VI-5-284}$$

In general form these quantities may be written

$$F_i = C_M \rho g \frac{\pi D^2}{4} H K_i \qquad \text{(VI-5-285)}$$

$$F_D = C_D \frac{1}{2} \rho g D H^2 K_D \qquad \text{(VI-5-286)}$$

$$M_i = C_M \rho g \frac{\pi D^2}{4} H K_i d S_i = F_i d S_i \qquad \text{(VI-5-287)}$$

$$M_D = C_D \frac{1}{2} \rho g D H^2 K_D d S_D = F_D d S_D \qquad \text{(VI-5-288)}$$

in which C_D and C_M have been assumed constant, and where K_i, K_D, S_i, and S_D are dimensionless parameters that depend on the specific wave theory used in the integrations. In the following sections values of the inertia coefficient C_M and drag coefficient C_D are assumed to be known constants. (Part VI-5-7c covers estimation of C_M and C_D.)

(a) Linear wave theory. The force on a slender cylindrical pile can be estimated using linear wave theory, but the result is limited to situations where linear wave theory provides a reasonable approximation of the wave kinematics. This implies small amplitude waves and greater depths. Also recall that any wave force on the pile above the swl will not be included in the estimate. Nevertheless, it is instructive to examine Equation VI-5-281 when linear wave theory is applied.

- With the pile center line located at $x = 0$, as shown in Figure VI-5-125, the equations from Part II-1, "Water wave mechanics" for surface elevation (Equation II-1-19), horizontal component of local fluid velocity (Equation II-1-22), and horizontal component of local fluid acceleration (Equation II-1-24) are respectively

$$\eta = \frac{H}{2} \cos\left(\frac{2\pi t}{T}\right) \qquad \text{(VI-5-289)}$$

$$u = \frac{H}{2} \frac{gT}{L} \frac{\cosh[2\pi(z+d)/L]}{\cosh[2\pi d/L]} \cos\left(\frac{2\pi t}{T}\right) \qquad \text{(VI-5-290)}$$

$$\frac{du}{dt} \approx \frac{\partial u}{\partial t} \approx \frac{g\pi H}{L} \frac{\cosh[2\pi(z+d)/L]}{\cosh[2\pi d/L]} \sin\left(-\frac{2pit}{T}\right) \qquad \text{(VI-5-291)}$$

- Introducing Equations VI-5-290 and VI-5-291 for u and du/dt into Equation VI-5-281 gives the following expressions for the inertia force and drag force.

$$f_i = C_M \rho g \frac{\pi D^2}{4} H \left[\frac{\pi}{L} \frac{\cosh[2\pi(z+d)/L]}{\cosh[2\pi d/L]} \right] \sin\left(-\frac{2pit}{T}\right) \qquad \text{(VI-5-292)}$$

$$f_D = C_D \frac{1}{2} \rho g D H^2 \left[\frac{gT^2}{4L^2} \left(\frac{\cosh\left[2\pi(z+d)/L\right]}{\cosh\left[2pid/L\right]} \right)^2 \right] \cos\left(\frac{2\pi t}{T}\right) \left| \cos\left(\frac{2\pi t}{T}\right) \right| \qquad \text{(VI-5-293)}$$

- Equations VI-5-292 and VI-5-293 show that the two force components vary with elevation z on the pile and with time t. The inertia force f_i is maximum for sin $(-2\pi t/T) = 1$, which corresponds to $t = -T/4$ for linear wave theory. Thus, the maximum inertia force on the pile occurs $T/4$ seconds before the passage of the wave crest that occurs at $t = 0$ (see Equation VI-5-289). The maximum value of the drag force component f_D coincides with passage of the wave crest at $t = 0$.

- The magnitude of the maximum inertia force per unit length of pile varies with depth the same as the horizontal acceleration component (Equation VI-5-291). The maximum value occurs at the swl ($z = 0$) and decreases with depth. The same trend is true for the maximum drag force per unit length of pile except the decrease with depth is more rapid because the depth attenuation factor (cosh $[2\pi(z+d)/L]/\cosh[2\pi d/L]$) is squared in Equation VI-5-293.

- The total time-varying force and the time-varying moment acting about the mudline is found for linear wave theory by integrating Equations VI-5-283 and VI-5-284 between the bottom and the swl ($z = 0$) using the expressions for f_i and f_D given by Equations VI-5-292 and VI-5-293, respectively. The integration results in total force and moment components given by Equations VI-5-285 through VI-5-288 with values of the dimensionless parameters K_i, K_D, S_i, and S_D given by

$$K_i = \frac{1}{2} \tanh\left(\frac{2\pi d}{L}\right) \sin\left(-\frac{2\pi t}{T}\right) \qquad \text{(VI-5-294)}$$

$$\begin{aligned} K_D &= \frac{1}{8} \left(1 + \frac{4\pi d/L}{\sinh\left[4\pi d/L\right]} \right) \cos\left(\frac{2\pi t}{T}\right) \left| \cos\left(\frac{2\pi t}{T}\right) \right| \\ &= \frac{1}{4} n \cos\left(\frac{2\pi t}{T}\right) \left| \cos\left(\frac{2\pi t}{T}\right) \right| \end{aligned} \qquad \text{(VI-5-295)}$$

$$S_i = 1 + \frac{1 - \cosh\left[2\pi d/L\right]}{(2\pi d/L)\sinh\left[2\pi d/L\right]} \qquad \text{(VI-5-296)}$$

$$S_D = \frac{1}{2} + \frac{1}{2n} \left(\frac{1}{2} + \frac{1 - \cosh\left[4\pi d/L\right]}{(4\pi d/L)\sinh\left[4\pi d/L\right]} \right) \qquad \text{(VI-5-297)}$$

where

$$n = \frac{C_g}{C} = \frac{1}{2} \left(1 + \frac{4\pi d/L}{\sinh\left[4\pi d/L\right]} \right) \qquad \text{(VI-5-298)}$$

- The maximum values for total inertia force and moment are found by taking $t = -T/4$ in Equations VI-5-294 and VI-5-296, respectively. Likewise, the maximum values for total drag force and moment are found by taking $t = 0$ in Equations VI-5-295 and VI-5-297, respectively. A conservative design approach would be to sum the individual maximum inertia and drag components that occur during a wave cycle to get total maximum force and moments. However, the individual maximums do not occur simultaneously, so the real maximum total force and moment wil be somewhat less. The correct method is to calculate the time-varying sum of inertia and drag components, and then use the maximum sum that occurs over the wave cycle. The time at which the maximum occurs may vary depending on the selected values for C_M and C_D.

- Although linear wave theory provides a nice closed-form solution for forces and moments on slender cylindrical piles, in practice the hydrodynamics associated with the steeper design wave conditions will not be well predicted by linear wave theory. Even more critical is the fact that linear theory provides no estimate of the force caused by that portion of the wave above the swl, an area where the horizontal velocities and accelerations are the greatest. An ad hoc adjustment is to assume a linear force distribution having a maximum value of force estimated at the still-water line and a value of zero at the crest location of the linear wave ($H/2$ above the swl). Most likely, the design wave will be nonlinear with steep wave crests and with much of the wave height above the swl, and it would be well advised to use an appropriate nonlinear wave theory in the force and moment calculation.

(b) Nonlinear wave theory.

- Design conditions for vertical cylindrical piles in coastal waters will most likely consist of nonlinear waves characterized by steep crests and shallow troughs. For accurate force and moment estimates, an appropriate nonlinear wave theory should be used to calculate values of u and du/dt corresponding to the design wave height, wave period, and water depth.

- The variation of f_i and f_D with time at any vertical location on the pile can be estimated using values of u and du/dt from tables such as Stoke's fifth-order wave theory (Skjelbriea et al. 1960) or stream-function theory (Dean 1974). Computer programs based on higher order monochromatic wave theories may be available to ease the task associated with using tabulated wave kinematics.

- The separate total maximum inertia force and moment and total drag force and moment on a vertical cylindrical pile subjected to nonlinear waves can be estimated using Equations VI-5-285 through VI-5-288. Values for K_i, K_D, S_i, and S_D in Equations VI-5-285 - VI-5-288 are given by K_{im}, K_{Dm}, S_{im}, and S_{Dm}, respectively, in the nomograms shown in Figures VI-5-126 through VI-5-129. (Note: In the nomograms the subscript m is used to denote maximum.) These nomograms were constructed using stream-function theory (Dean 1974), and they provide the maximum total force and total moment for the inertia and drag components considered separately rather than the combined total force and moment. The curves in Figures VI-5-126 to VI-5-129 represent wave height as a fraction of the breaking wave height. For example, curves labeled $1/2\ H_b$ represent $H/H_b = 1/2$. Breaking wave height is obtained from Figure VI-5-130 for values of d/gT^2 using the curve labeled Breaking Limit.

- For linear waves, the maximum inertia force occurs at t = -T/4 and the maximum drag force occurs at $t = 0$. However, for nonlinear waves the times corresponding to maximum inertia and drag forces are phase dependent and not separated by a constant quarter wavelength as in linear wave theory.

- The total maximum force F_m, where the sum of the inertia and drag components is maximum, can be estimated using Figures VI-5-131 to VI-5-134. These figures were also constructed using stream-function theory. Figure selection is based on the nondimensional parameter

$$W = \frac{C_M D}{C_D H} \qquad \text{(VI-5-299)}$$

and the drawn curves give values of φ_m corresponding to the known parameters H/gT^2 and d/gT^2.

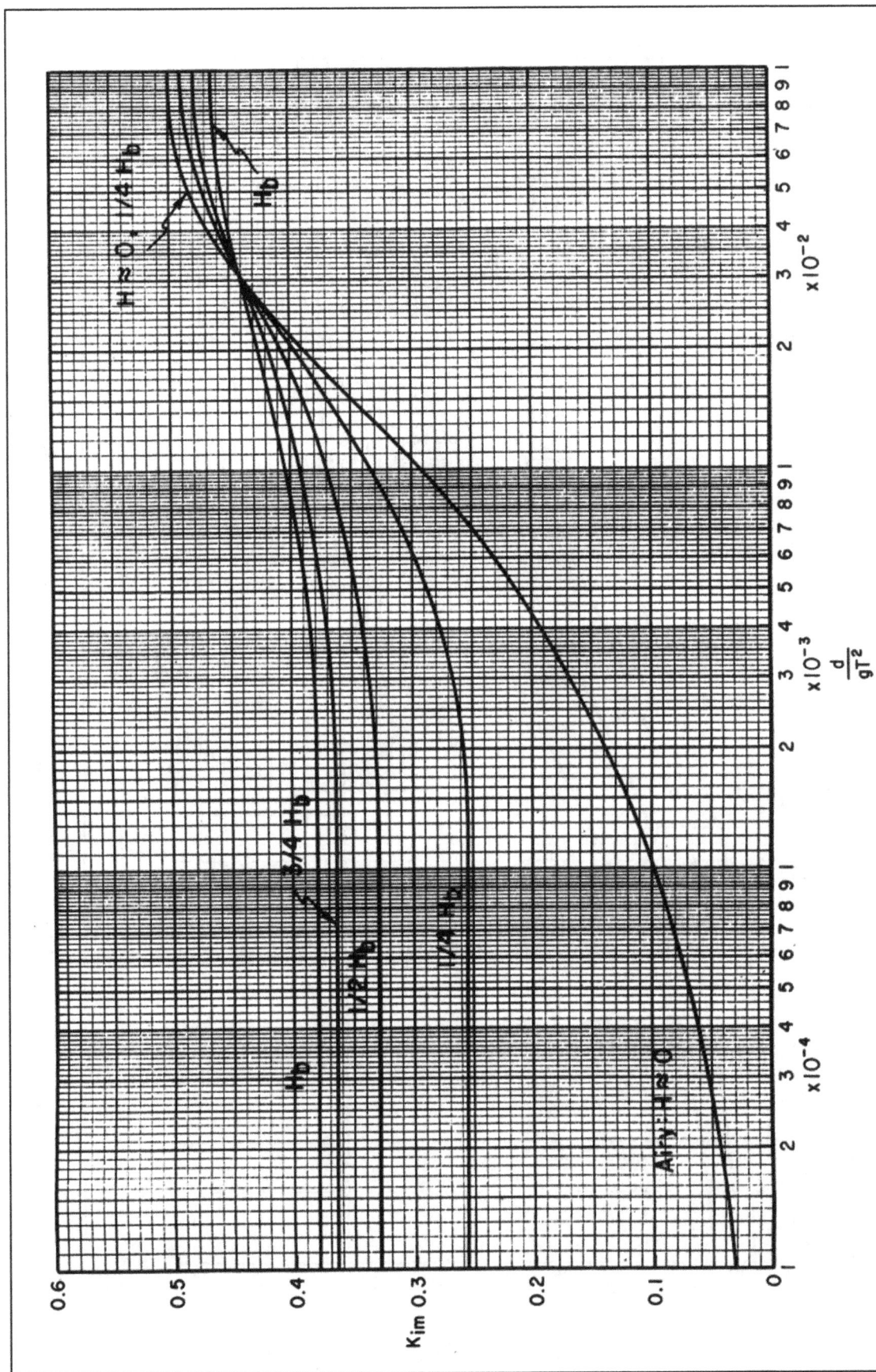

Figure VI-5-126. K_{im} versus relative depth, d/gT^2

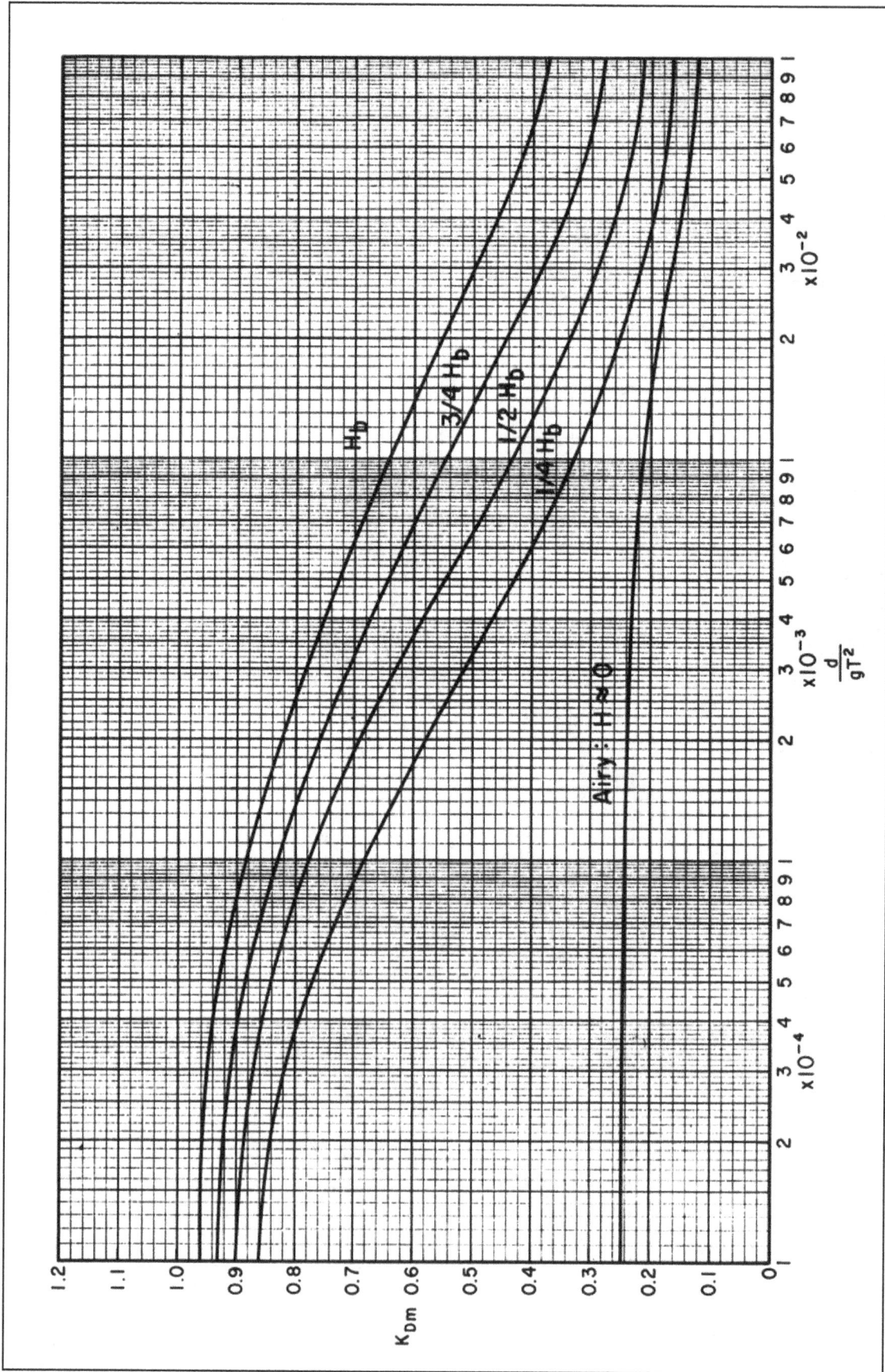

Figure VI-5-127. K_{Dm} versus relative depth, d/gT^2

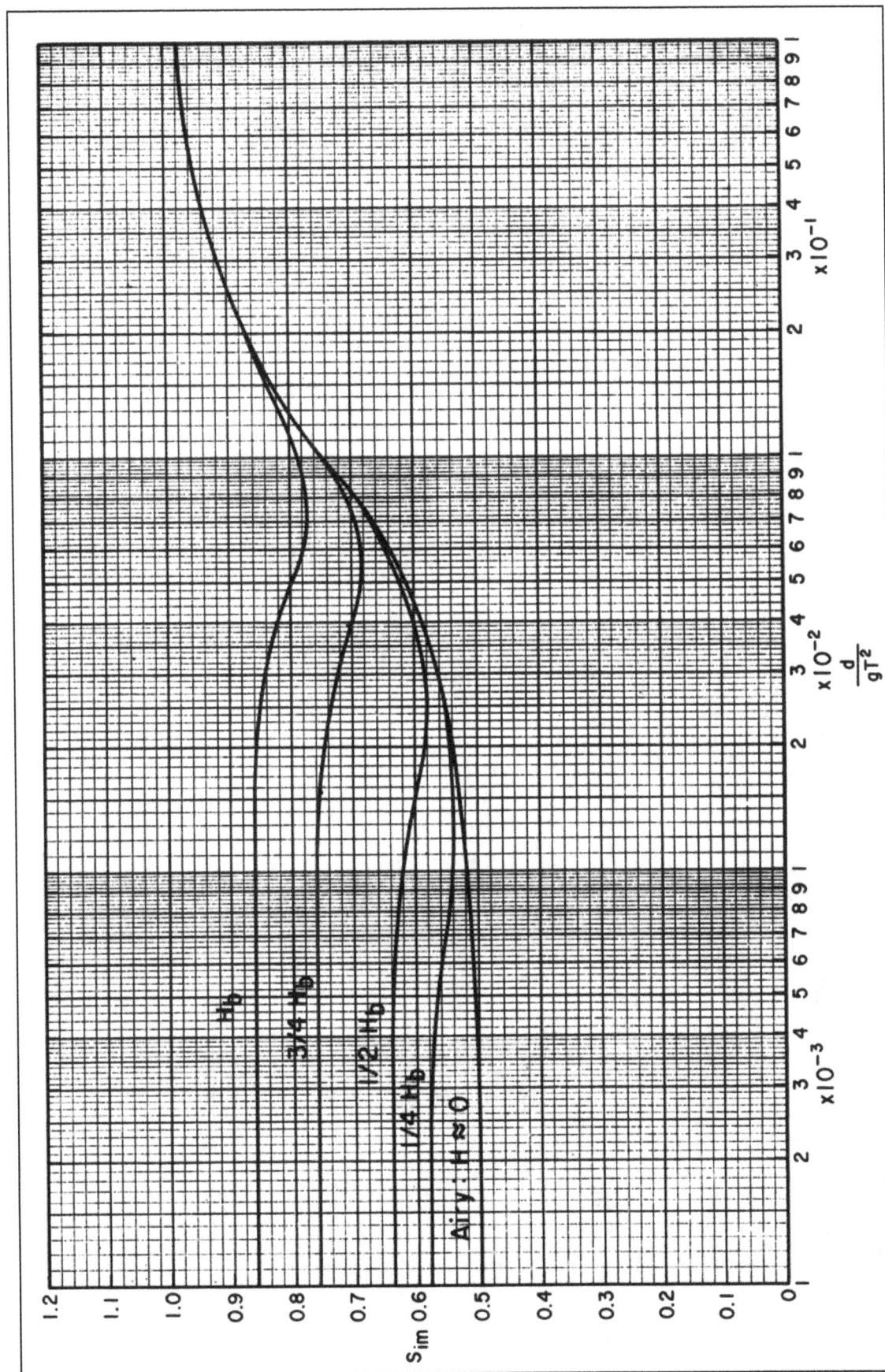

Figure VI-5-128. Inertia force moment arm S_{im} versus relative depth, d/gT^2

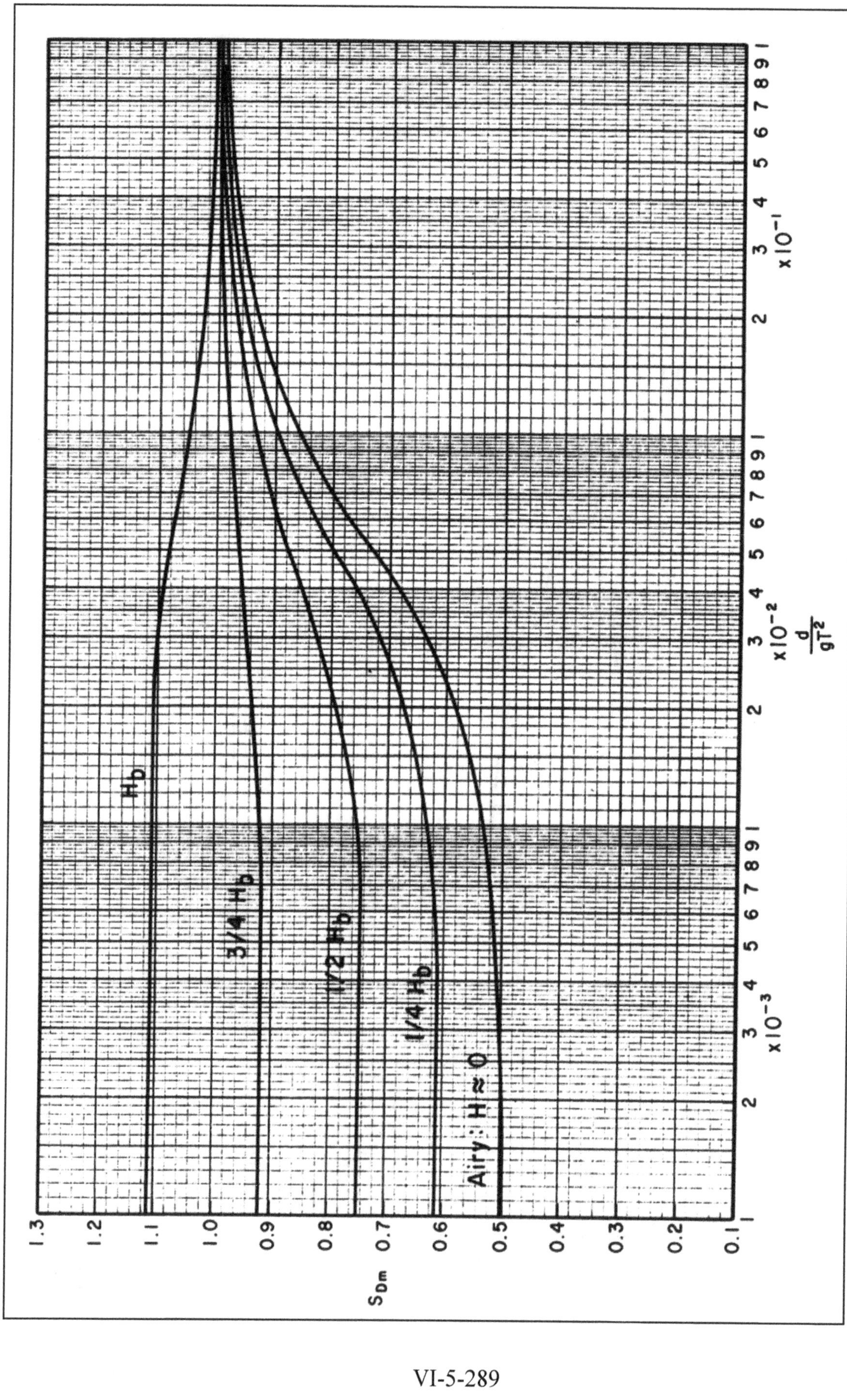

Figure VI-5-129. Drag force moment arm S_{Dm} versus relative depth, d/gT^2

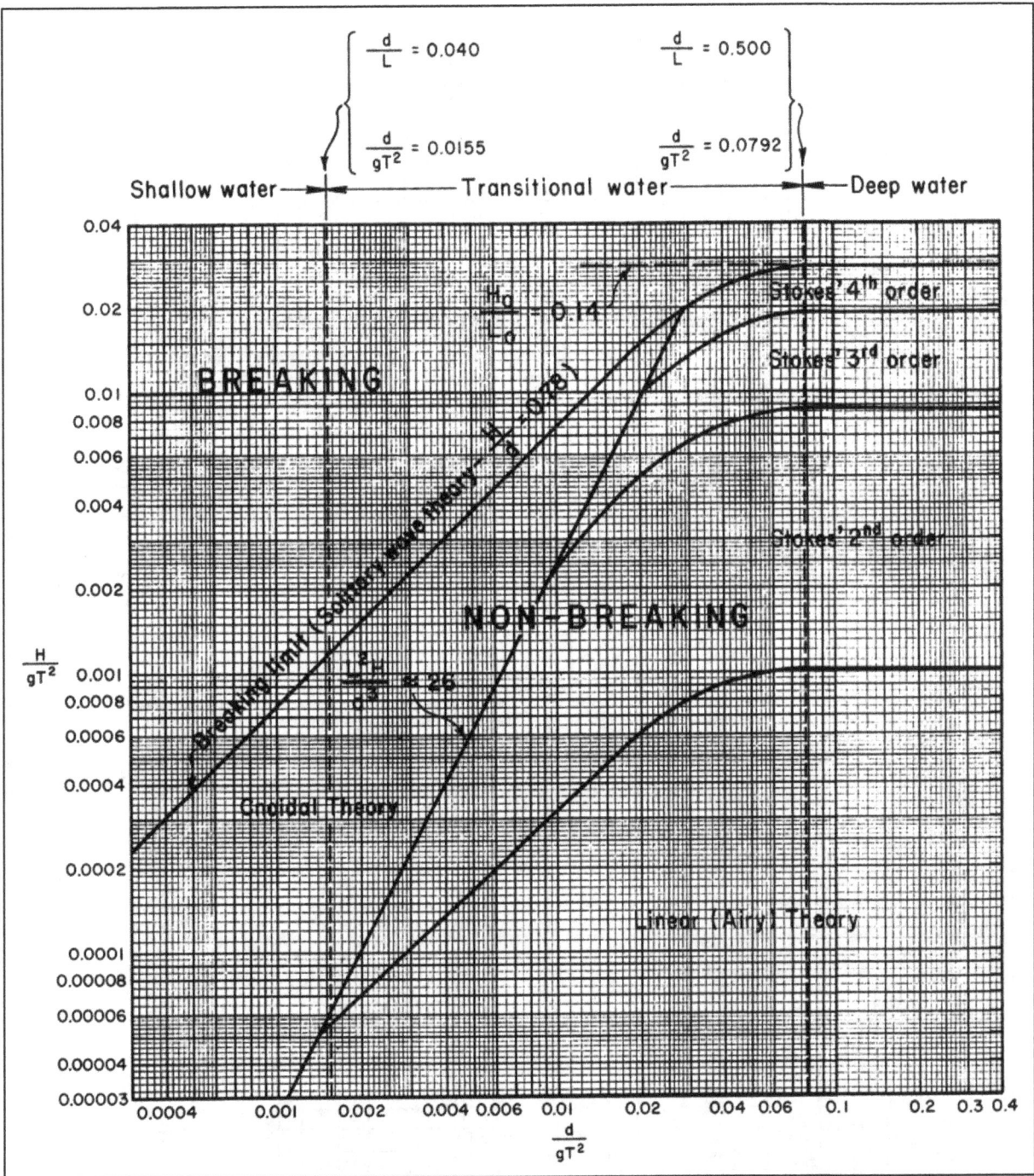

Figure VI-5-130. Breaking wave height and regions of validity of various wave theories

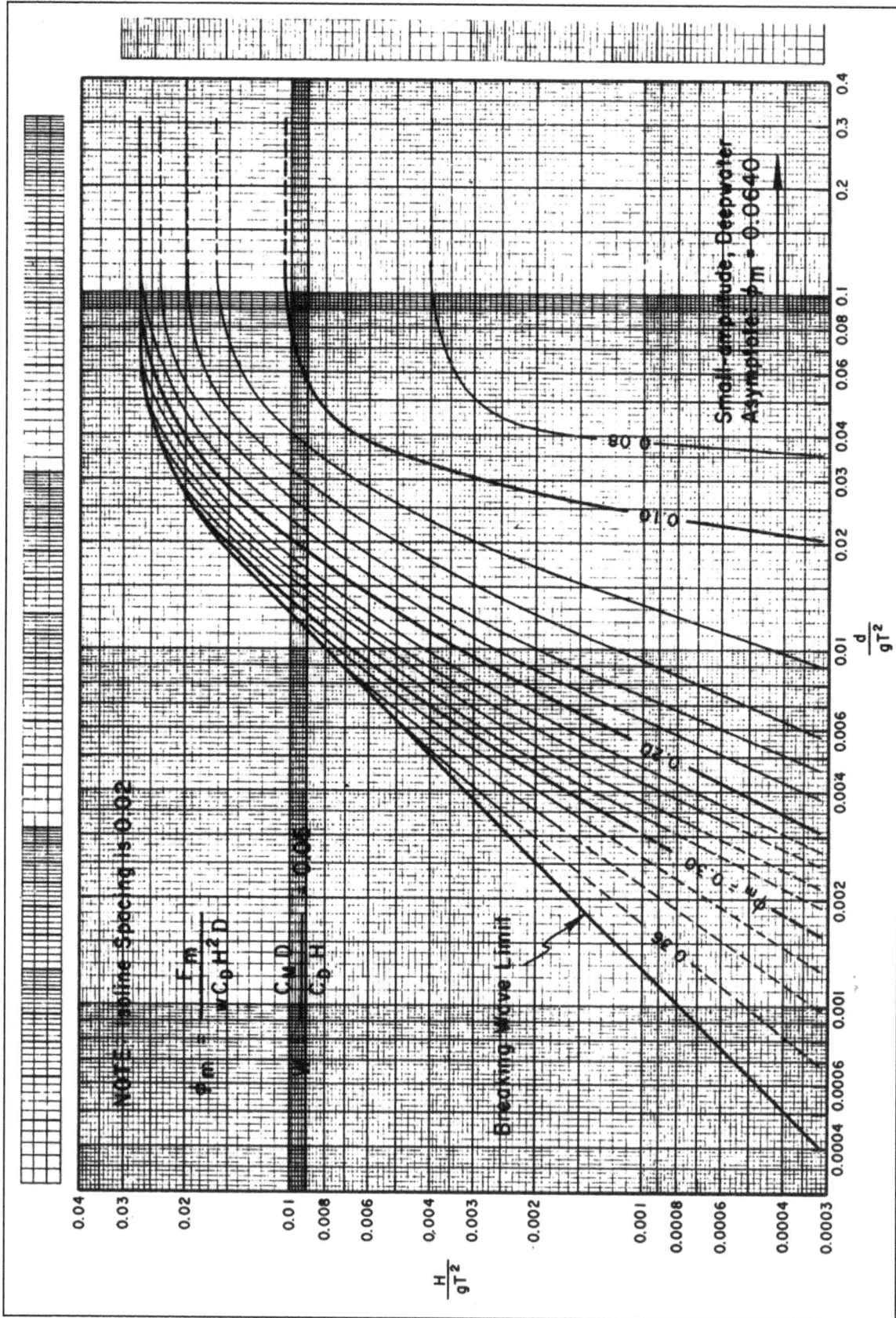

Figure VI-5-131. Isolines of φ_m versus H/gT^2 and d/gT^2 ($W = 0.05$)

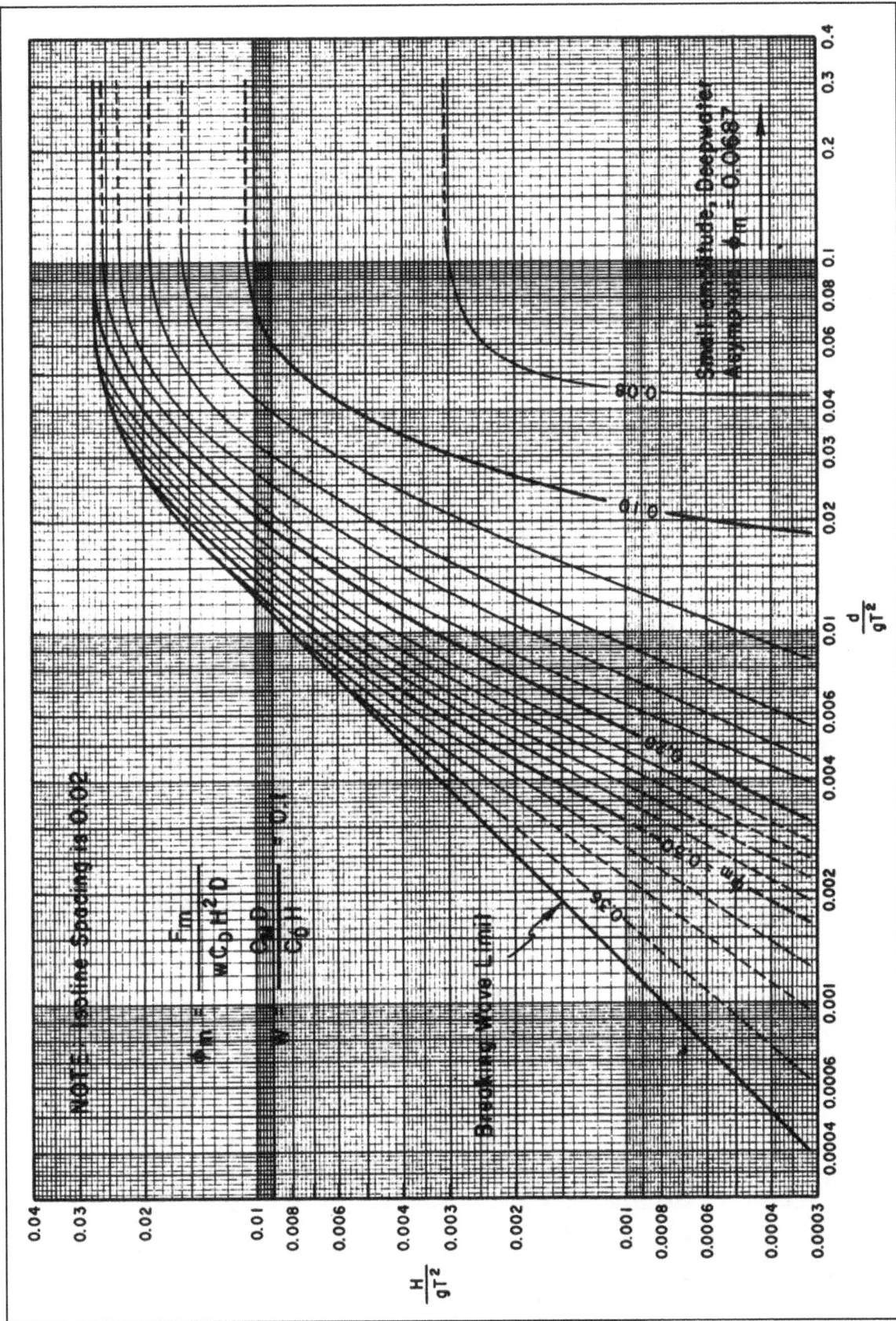

Figure VI-5-132. Isolines of φ_m versus H/gT^2 and d/gT^2 ($W = 0.10$)

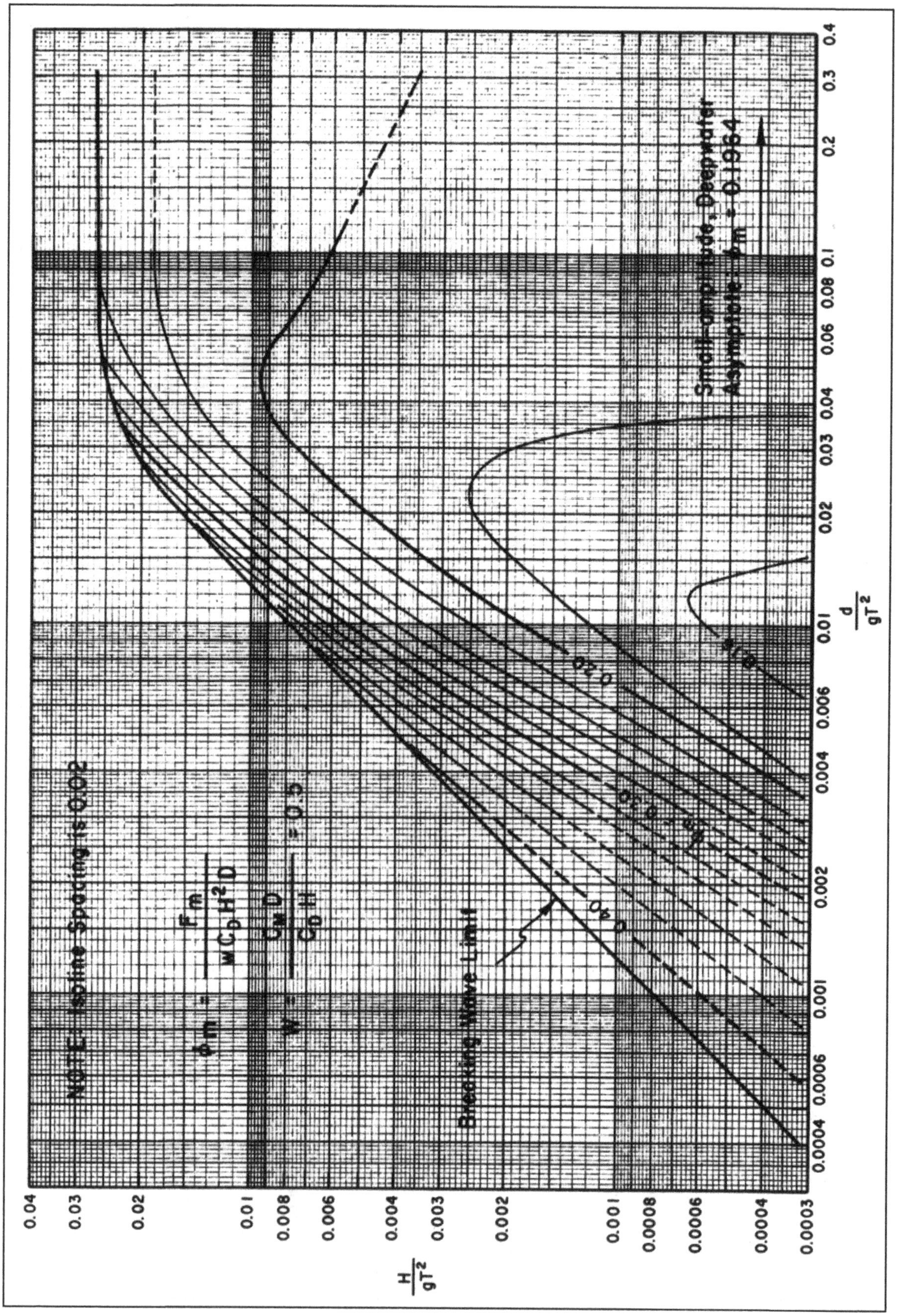

Figure VI-5-133. Isolines of φ_m versus H/gT^2 and d/gT^2 ($W = 0.50$)

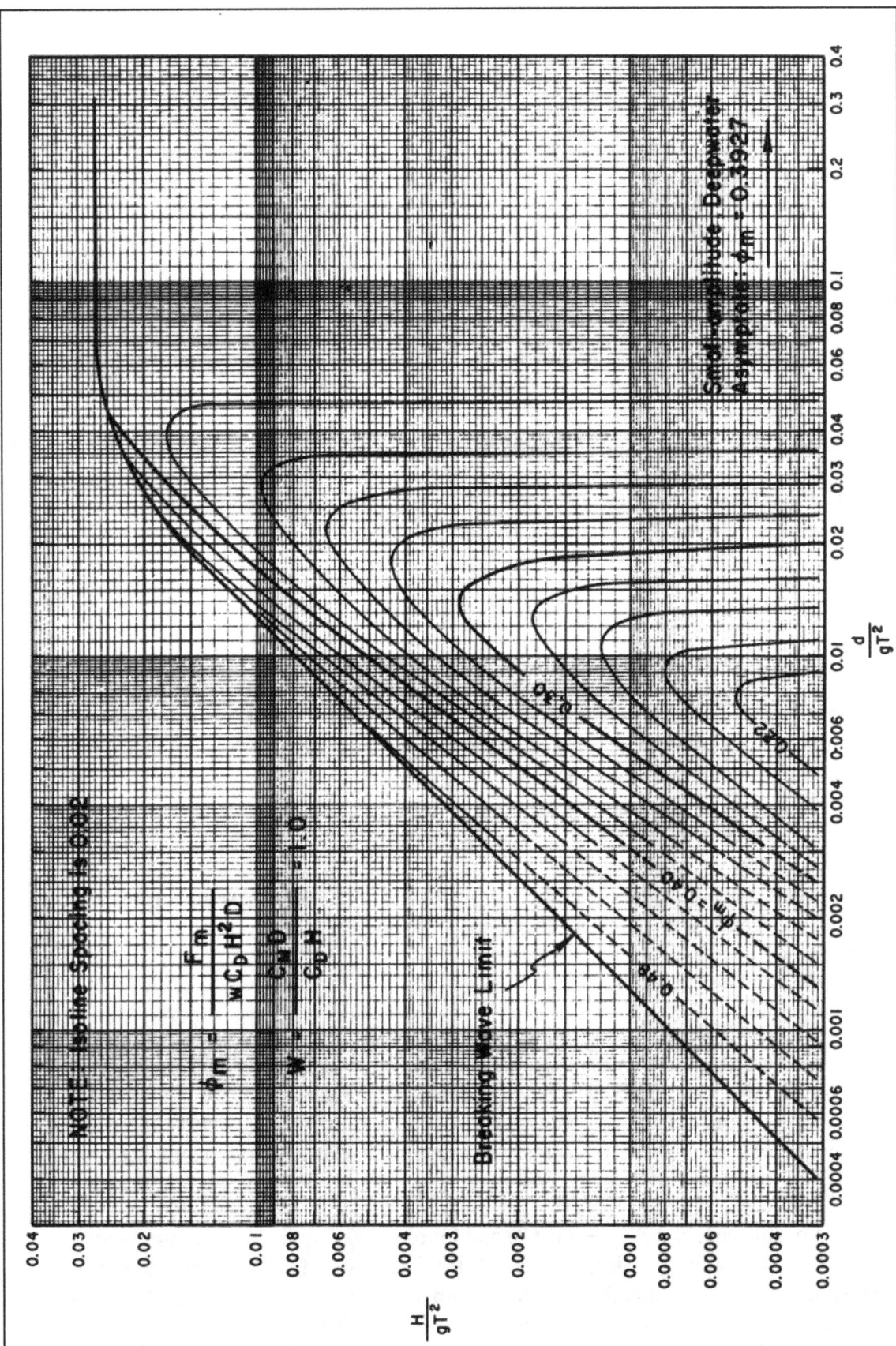

Figure VI-5-134. Isolines of φ_m versus H/gT^2 and d/gT^2 ($W = 1.0$)

• The maximum force is calculated as

$$F_m = \phi_m C_D \rho g H^2 D \tag{VI-5-300}$$

• Similarly, the total maximum moment M_m can be estimated using Figures VI-5-135 through VI-5-138 which were also constructed using stream-function theory. Choice of figure is based on the value of W given by Equation VI-5-299, and values for α_m are corresponding to the parameters H/gT^2 and d/gT^2. The moment about the mudline is given by

$$M_m = \alpha_m C_D \rho g H^2 D d \tag{VI-5-301}$$

• For both the total force and total moment calculations, the calculated value of W will likely lie between the values for which the figures are drawn. In this case, determine values of φ_m and α_m from the plots on either side of the W-value, then use linear interpolation to estimate values of φ_m and α_m for the calculated value of W.

• The maximum moment is calculated at the mudline, and the corresponding moment arm is the maximum moment divided by the maximum force, or

$$r_a = \frac{M_m}{F_m} \tag{VI-5-302}$$

• If the surrounding soil does not provide any lateral resistance, or if there has been scour around the pile, the effective moment arm must be increased and a new maximum total moment calculated. For example, if the scour depth beneath the surrounding bed is S_m, the modified maximum total moment will be

$$M_{m'} = (r_a + S_m) F_m \tag{VI-5-303}$$

• See Part VI-7, "Design of Specific Project Elements," for an example illustrating calculation of forces and moments on a vertical cylinder.

(3) Transverse forces due to eddy shedding.

(a) In addition to drag and inertia forces that act in the direction of wave advance, transverse forces may arise. Transverse forces are caused by vortex or eddy shedding on the downstream side of the pile. Eddies are shed alternately from each side of the pile resulting in a laterally oscillating force. Transverse forces act perpendicular to both wave direction and pile axis, and they are often termed lift forces because they are similar to aerodynamic lift acting on an airfoil.

(b) Laird, Johnson, and Walker (1960) and Laird (1962) studied transverse forces on rigid and flexible oscillating cylinders. In general, lift forces were found to depend on the dynamic response of the structure. For structures with a natural frequency of vibration about twice the wave frequency, a dynamic coupling between the structure motion and fluid motion occurs, resulting in large lift forces. Transverse forces have been observed 4.5 times greater than the drag force. However, for rigid structures a transverse force equal to the drag force is a reasonable upper limit. Larger transverse forces can occur where there is dynamic interaction between the waves and cylindrical pile. The design guidance in this section pertains only to rigid piles.

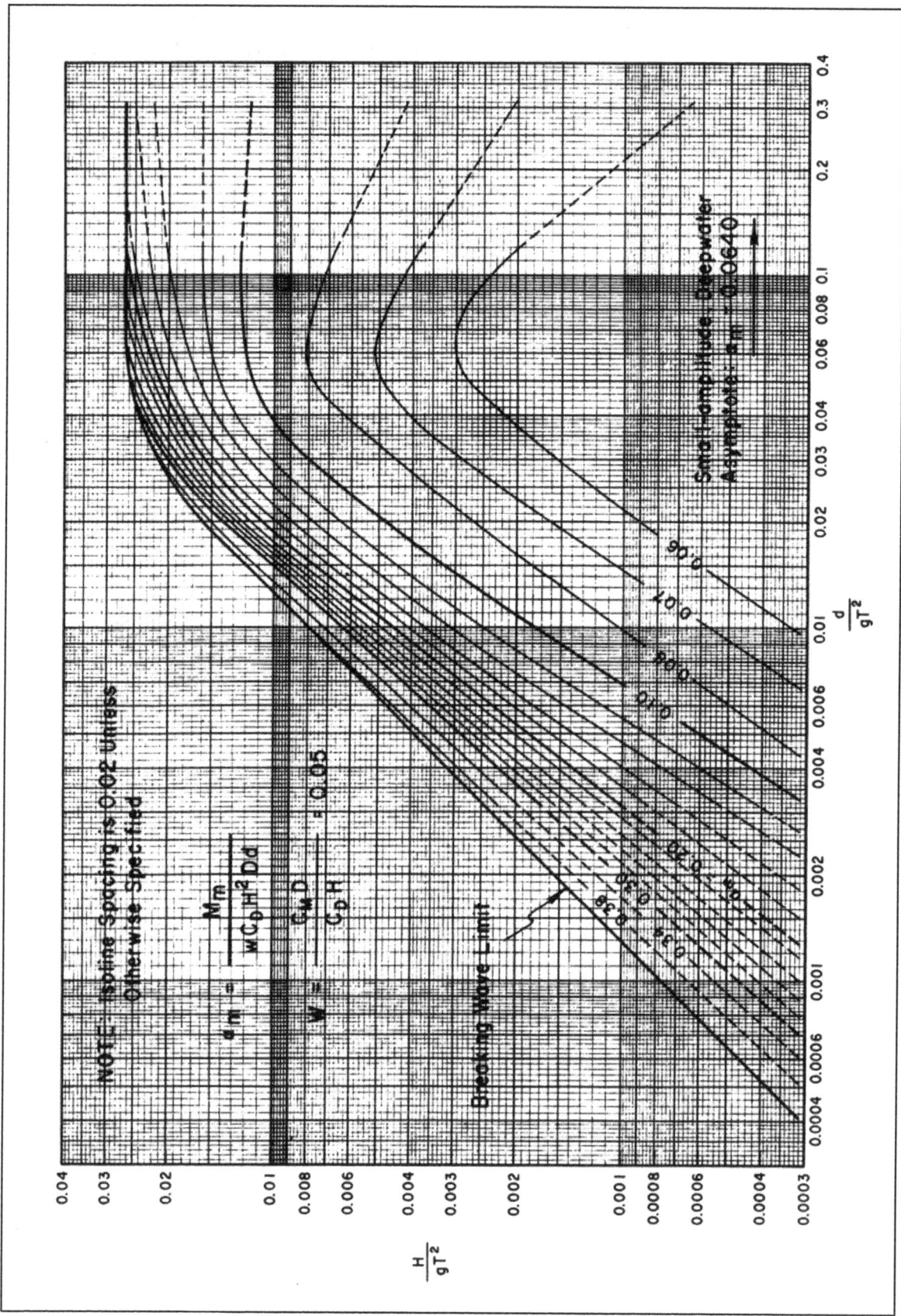

Figure VI-5-135. Isolines of α_m versus H/gT^2 and d/gT^2 ($W = 0.05$)

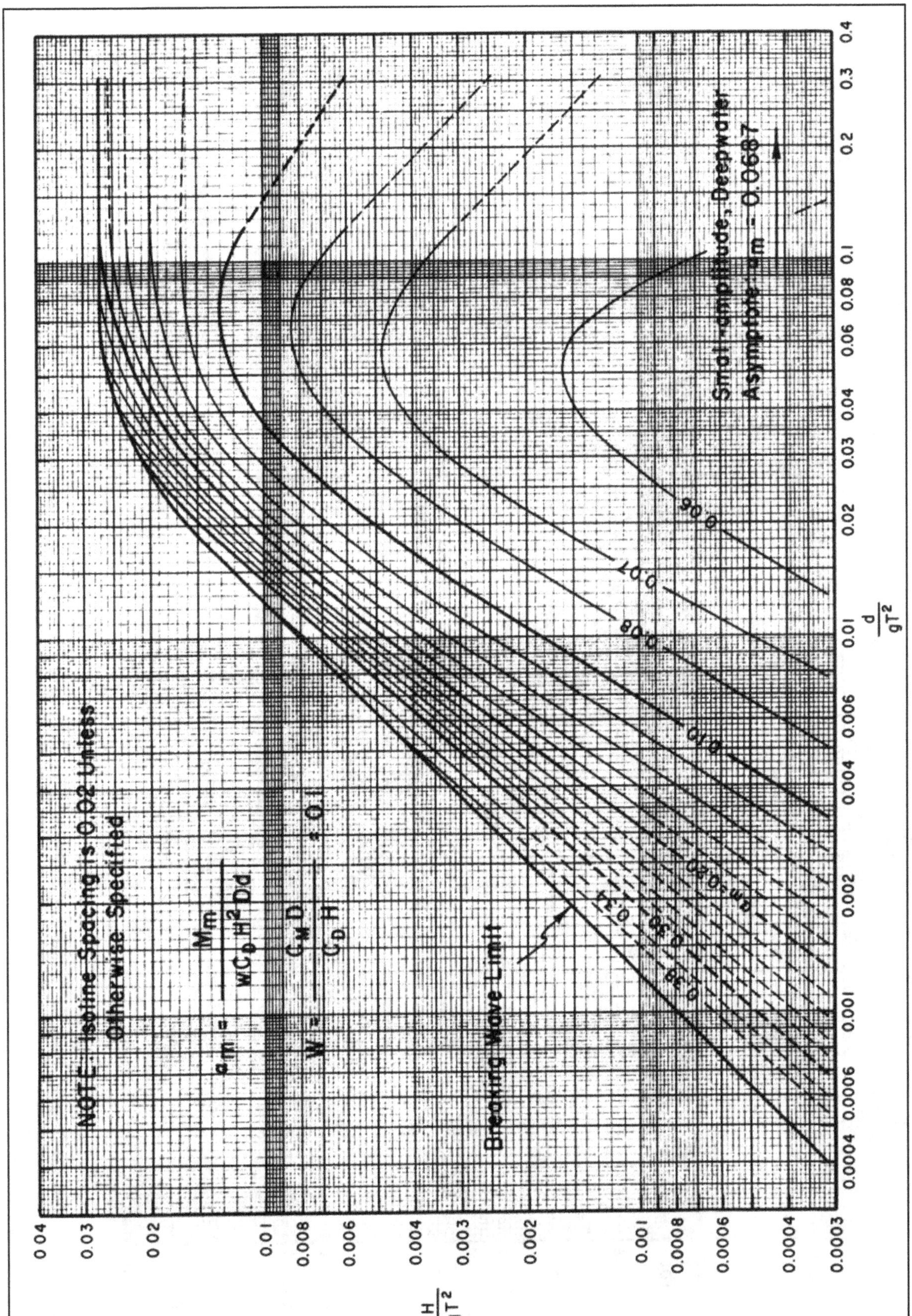

Figure VI-5-136. Isolines of α_m versus H/gT^2 and d/gT^2 ($W = 0.10$)

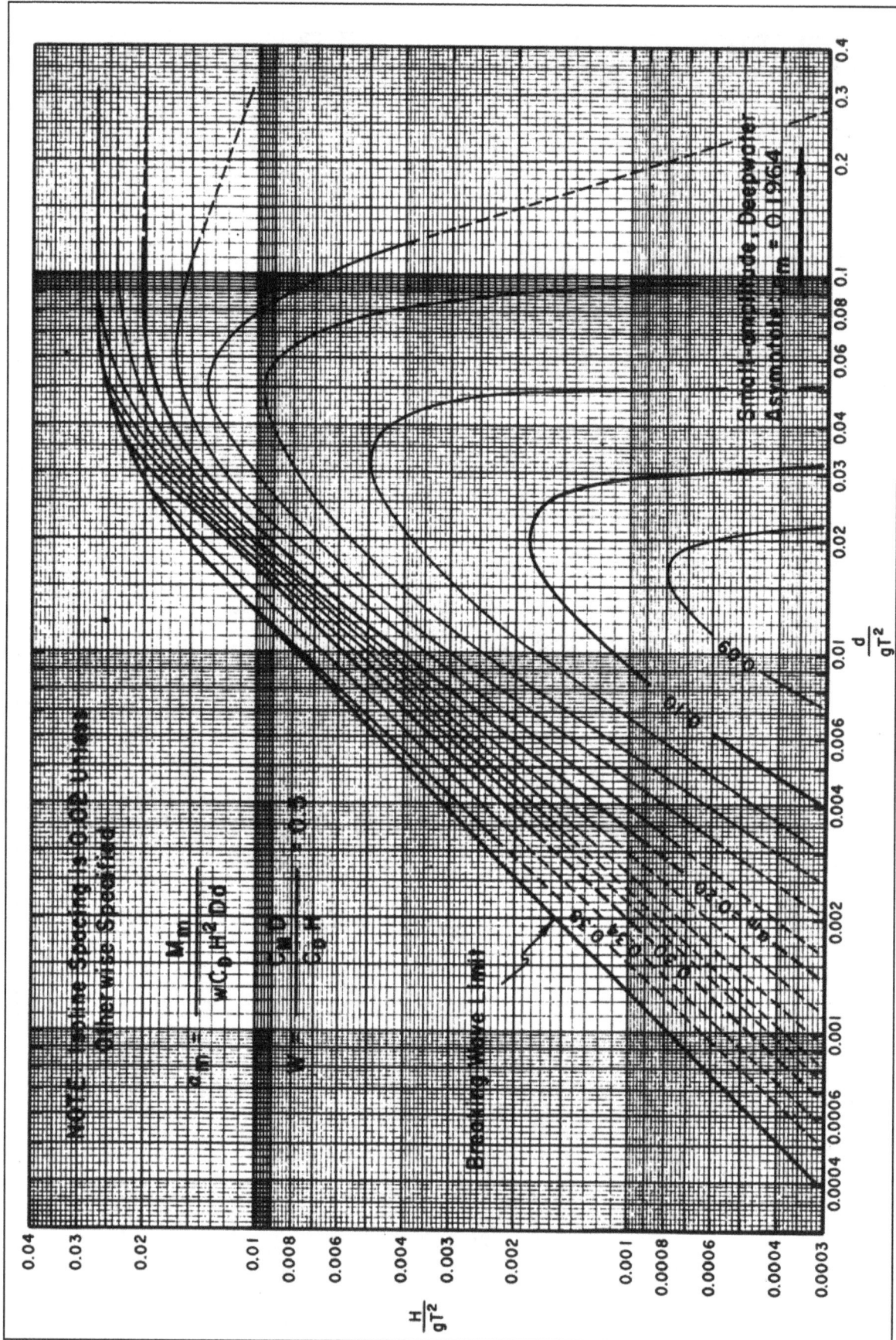

Figure VI-5-137. Isolines of α_m versus H/gT^2 and d/gT^2 ($W = 0.5$)

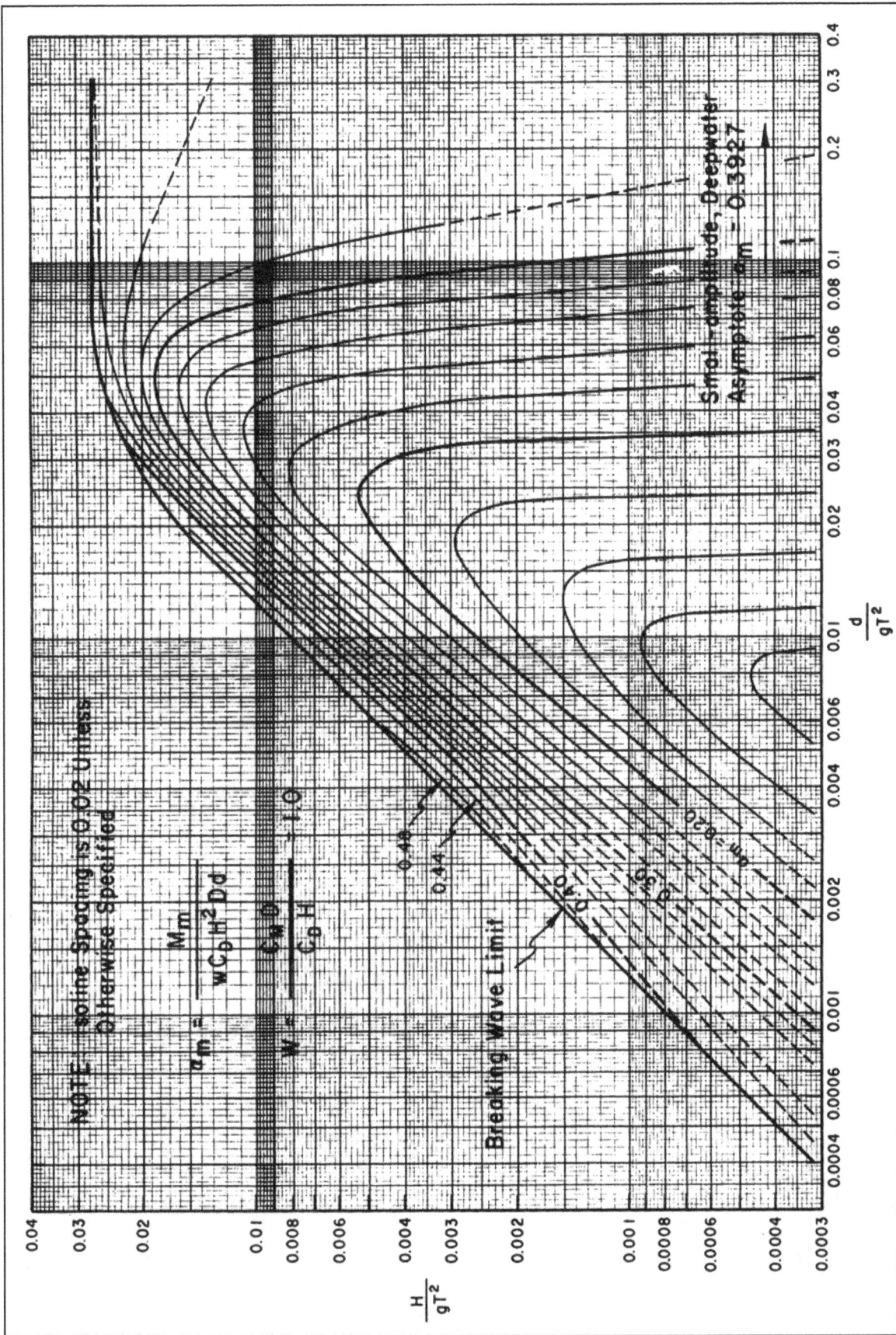

Figure VI-5-138. Isolines of α_m versus H/gT^2 and d/gT^2 ($W = 1.0$)

(c) Chang (1964) found in laboratory investigations that eddies are shed at a frequency that is twice the wave frequency. Two eddies are shed after passage of the wave crest (one on each side of the pile), and two are shed on the return flow after passage of the wave trough. The maximum lift force is proportional to the square of the horizontal wave-induced velocity in much the same way as the drag force. Consequently, for design estimates of the lift force the following equation can be applied.

$$F_L = F_{Lm} \cos 2\theta = C_L \frac{\rho g}{2} D H^2 K_{Dm} \cos 2\theta \qquad \text{(VI-5-304)}$$

where F_L is the time-varying transverse (lift) force, F_{Lm} is the maximum transverse force, θ is the wave phase angle ($\theta = 2\pi x/L - 2\pi t/T$), C_L is an empirical lift coefficient analogous to the drag coefficient in Equation VI-5-286, and K_{Dm} is the dimensionless parameter given in Figure VI-5-127. Chang found that C_L depends on the average Keulegan-Carpenter number given as

$$KC_{ave} = \frac{(u_{\max})_{ave} T}{D} \qquad \text{(VI-5-305)}$$

where $(u_{max})_{ave}$ is the maximum horizontal velocity averaged over the depth. When KC_{ave} is less than 3, no significant eddy shedding occurs and no lift forces are developed. As KC_{ave} increases, C_L increases until it is approximately equal to C_D for rigid piles. Consequently, it must be recognized that: the lift force can represent a major portion of the total force acting on a pile and therefore should not be neglected in the design of the pile.

c. Selection of hydrodynamic force coefficients C_D, C_M, and C_L.

(1) Sarpkaya (1976a, 1976b) conducted an extensive experimental investigation of the inertia, drag, and transverse forces acting on smooth and rough circular cylinders. The experiments were performed in an oscillating U-tube water tunnel for a range of Reynolds numbers up to 700,000 and Keulegan-Carpenter numbers up to 150. Relative roughness of the cylinders k/D varied between 0.002 and 0.02 (where k is the average height of the roughness element). Forces were measured on stationary cylinders, and the corresponding drag and inertia coefficients were determined using a technique of Fourier analysis and least-squares best fit of the Morison equation (Equation VI-5-281) to the measured forces.

(2) The results were presented as plots of the force coefficients versus Keulegan-Carpenter number

$$KC = \frac{u_m T}{D} \qquad \text{(VI-5-306)}$$

for given values of Reynolds number

$$R_e = \frac{u_m D}{\nu} \qquad \text{(VI-5-307)}$$

or the frequency parameter

$$\beta = \frac{R_e}{KC} = \frac{D^2}{v\,T} \tag{VI-5-308}$$

(3) In Equations VI-5-306 - VI-5-308 u_m is the maximum horizontal wave velocity, T is the wave period, D is the cylinder diameter, and v is the fluid kinematic viscosity.

(4) Figures VI-5-139 through VI-5-141 present Sarpkaya's (1976a, 1976b) experimental results for the force coefficients C_D, C_M, and C_L for smooth cylinders. In each figure the force coefficient is plotted versus Keulegan-Carpenter number for constant values of Reynolds number (dotted lines) and frequency parameter (solid lines). Drag and inertia force coefficients versus Reynolds number for rough cylinders are plotted on Figures VI-5-142 and VI-5-143, respectively, for selected values of relative roughness k/D. Sarpkaya cautioned that the force coefficients were developed for oscillatory flow with zero mean velocity, and it is possible that waves propagating on a uniform current may have different force coefficients.

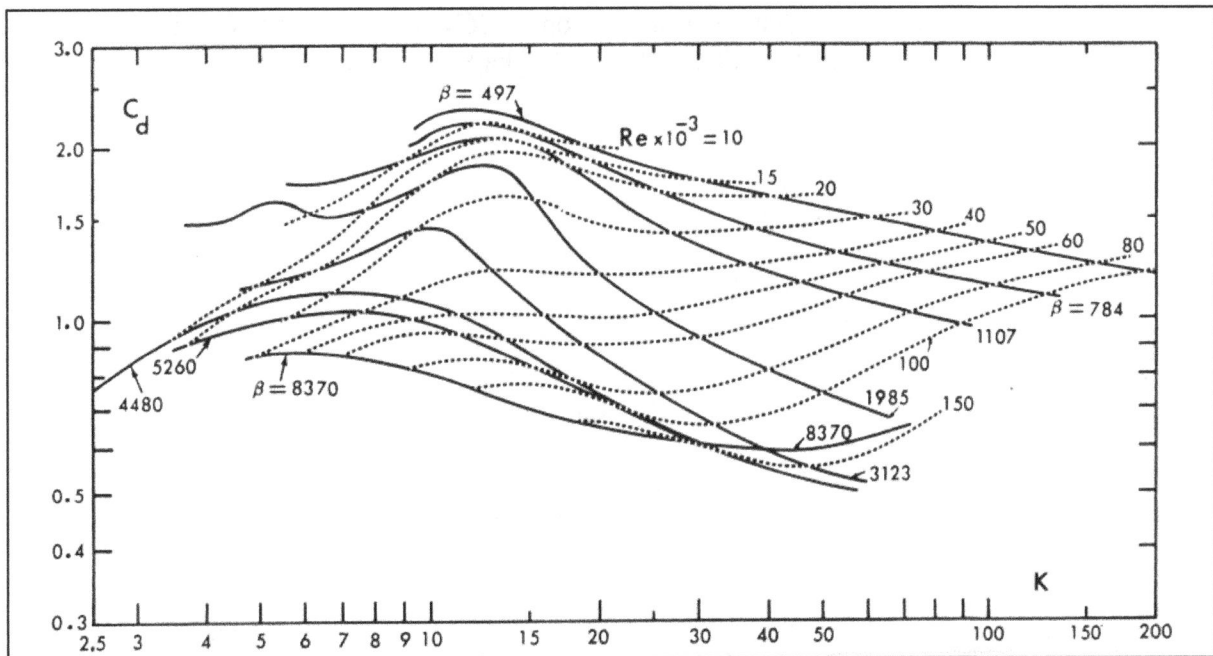

Figure VI-5-139. Drag coefficient C_D as a function of KC and constant values of R_e or β for smooth cylinders (from Sarpkaya 1976a)

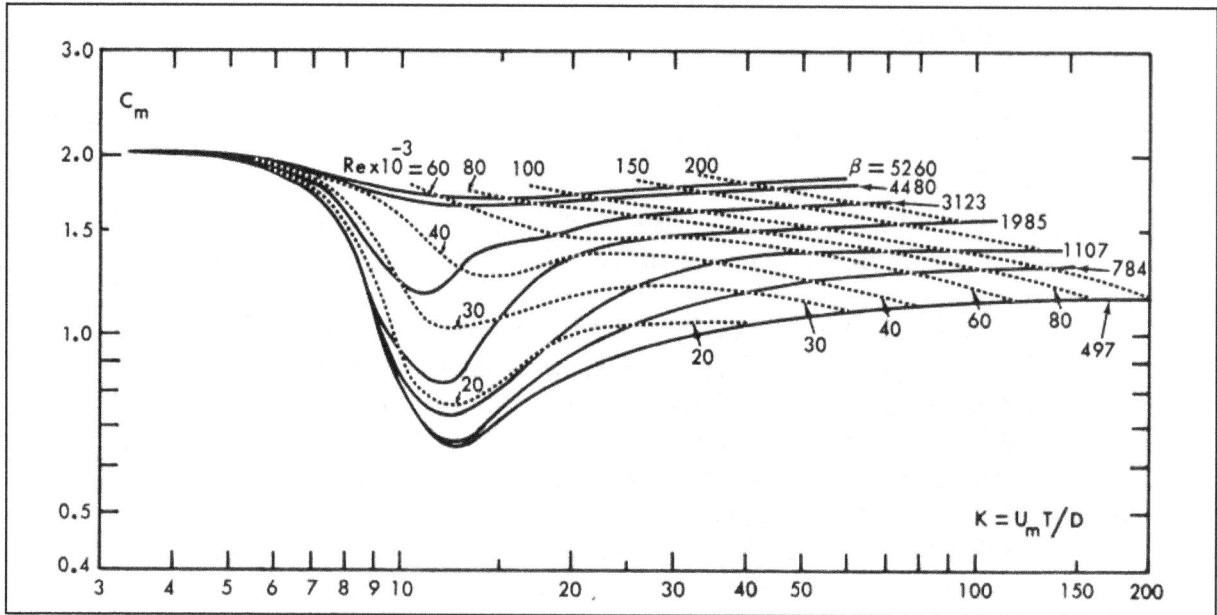

Figure VI-5-140. Inertia coefficient C_M as a function of KC and constant values of R_e or β for smooth cylinders (from Sarpkaya 1976a)

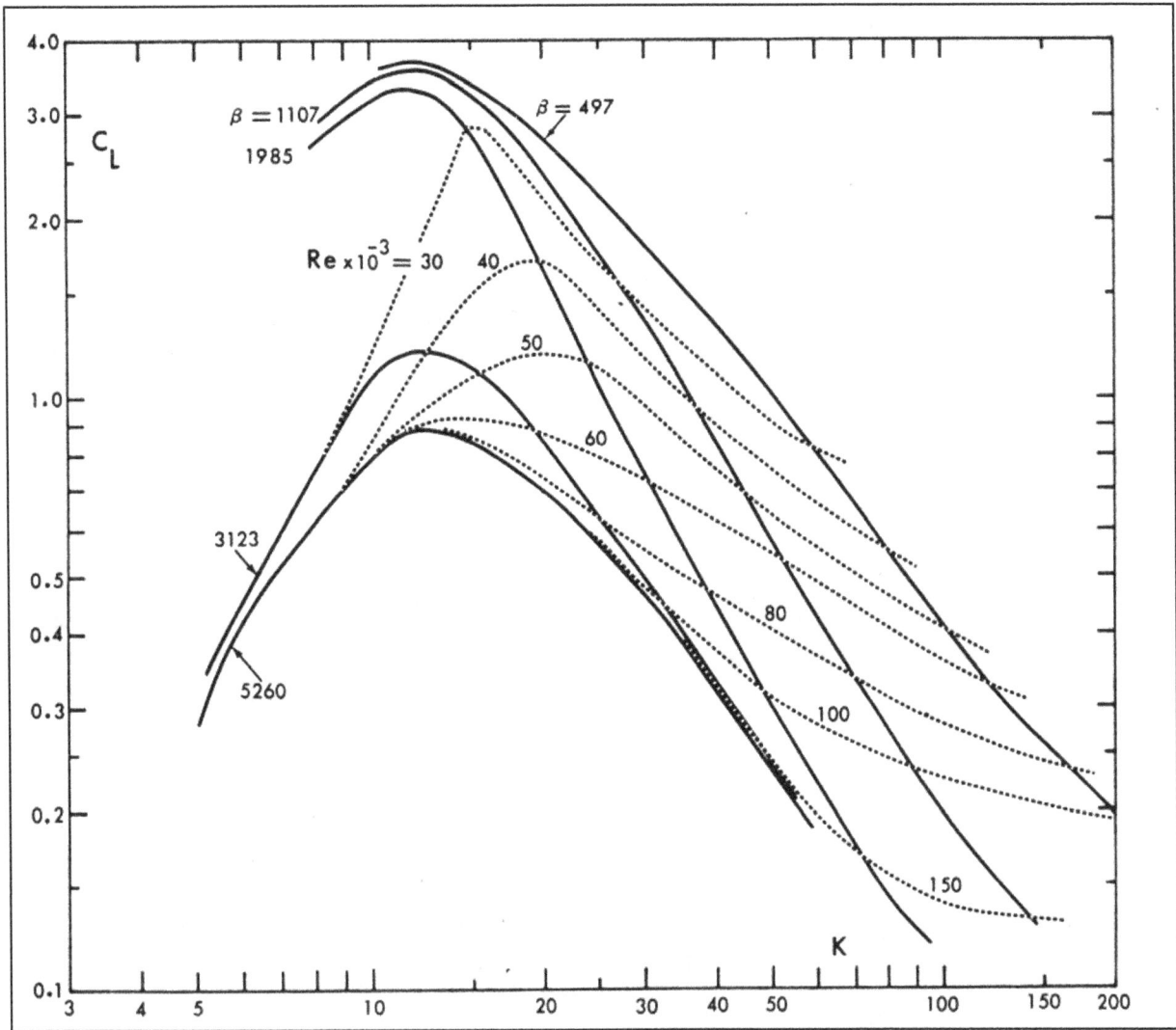

Figure VI-5-141. Lift coefficient C_L as a function of KC and constant values of R_e or β for smooth cylinders (from Sarpkaya 1976a)

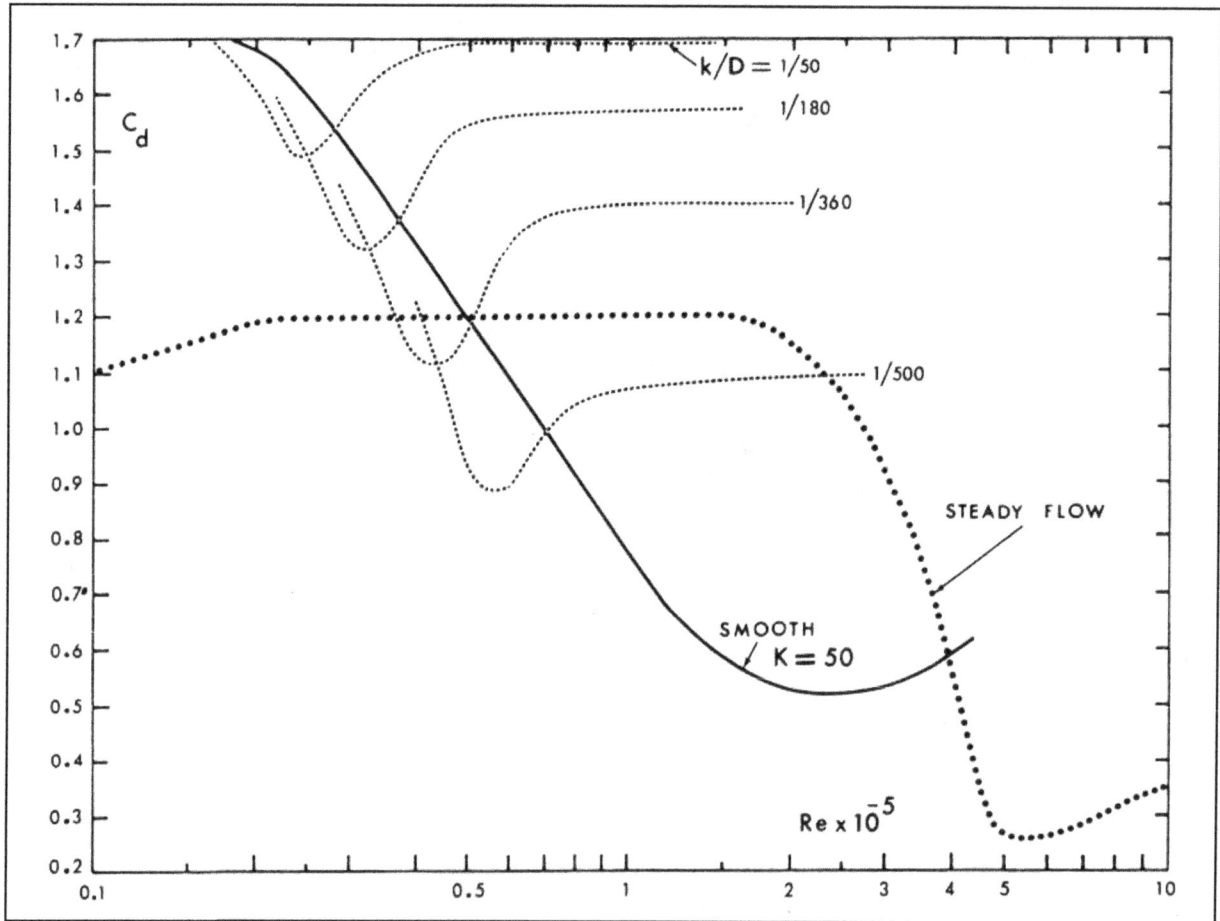

Figure VI-5-142. Drag coefficient C_D as a function of Reynolds number for rough cylinders
(from Sarpkaya 1976a)

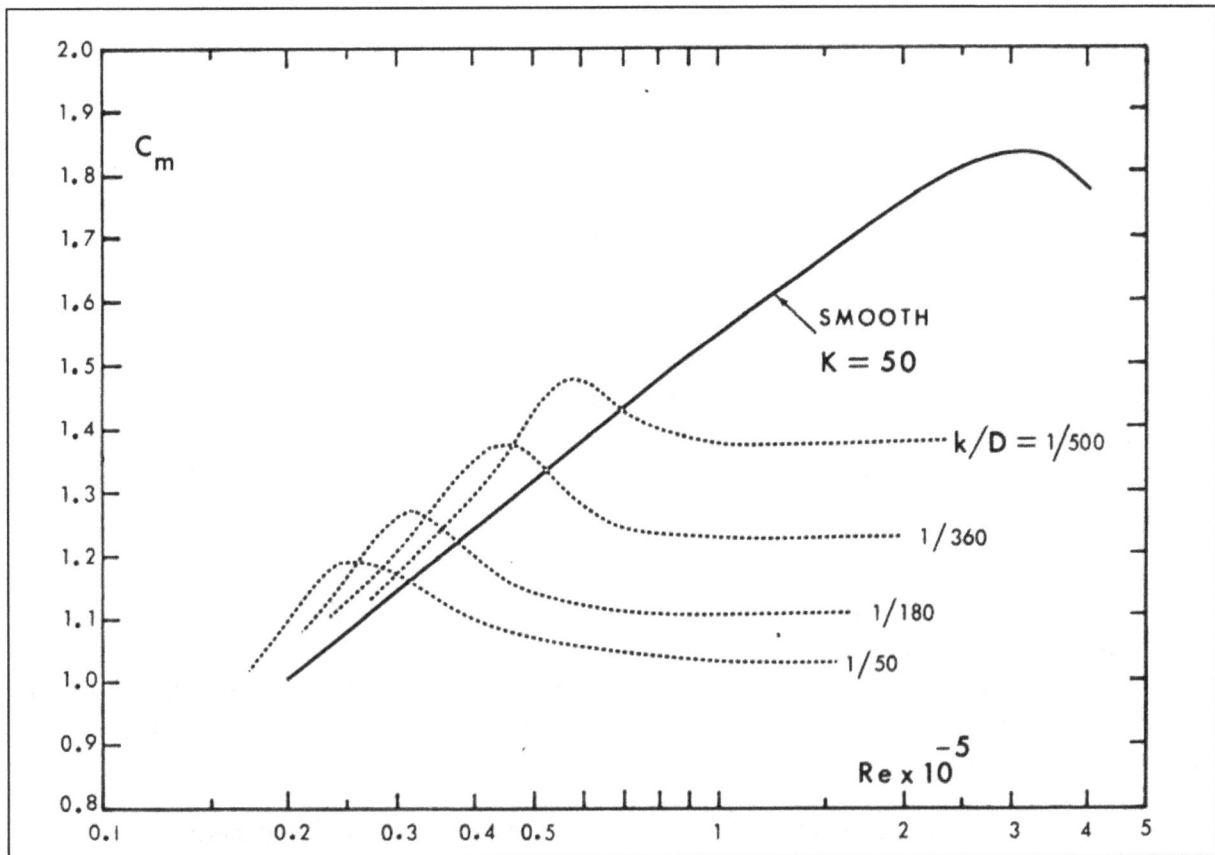

Figure VI-5-143. Inertia coefficient C_M as a function of Reynolds number for rough cylinders
(from Sarpkaya 1976a)

(5) The force coefficients given in Figures VI-5-139 through VI-5-143 should give reasonable force estimates when used with the design figures based on stream function theory given in the previous section. However, the design engineer should be aware of the limitations of assuming the force coefficients are constant over the water depth and throughout the wave cycle.

(6) Sarpkaya's experimental apparatus gave uniform values of Reynolds number and Keulegan-Carpenter number over the entire test pile. For a vertical pile exposed to waves, the maximum horizontal velocity will vary from its largest value at the sea surface to a somewhat smaller value near the bottom. Consequently, both R_e and KC will vary over the depth of the pile. For design purposes, it is reasonable to calculate R_e and KC based on the average value of u_m over the water depth in shallow water because the variation will not be too significant. In deeper water it may be wise to investigate the variation of force coefficients with depth to determine if using R_e and KC based on average u_m is appropriate.

(7) Sarpkaya=s experimental data do not cover the range of Reynolds numbers likely to be encountered with bigger waves and larger pile diameters. For larger calculated Reynolds numbers use the following guidance that has been repeated from the old *Shore Protection Manual* (1984).

$$C_D = \begin{cases} 1.2 - \dfrac{(R_e - 2(10)^5}{6(10)^5} & \text{for } 2(10)^5 < R_e - 5(10)^5 \\[2em] 0.7 & \text{for } 5(10)^5 < R_e \end{cases}$$

$$C_M = \begin{cases} 2.5 - \dfrac{R_e}{5(10)^5} & \text{for } 2.52(10)^5 < R_e - 5(10)^5 \\[2em] 1.5 & \text{for } 5(10)^5 < R_e \end{cases}$$

Bear in mind the above recommendations for higher Reynolds number are based on older experimental results, and more accurate estimates might be available from the offshore engineering literature for critical applications.

    d. Safety factors in pile design.

    (1) Before the pile is designed or the foundation analysis is performed, a safety factor is usually applied to calculated forces. Reasons for uncertainty to the design include approximations in applying the wave theory, estimated values for the force coefficients, potential loss of pile strength over time, and the probability that the design wave will be exceeded during the life of the structure.

    (2) The following recommendations for safety factors are offered as general rules of thumb. In situations where pile failure could lead to loss of life or catastrophic failure of supported infrastructure, safety factors should be increased.

    (a) When the design wave has low probability of occurrence, it is recommended that a safety factor of 1.5 be applied to calculated forces and moments that are to be used as the basis for structural and foundation design.

    (b) If the design wave is expected to occur frequently, such as in depth-limited situations, a safety factor of at least 2.0 should be applied to the calculated forces and moments.

    (3) In addition to the safety factor, changes occurring during the expected life of the pile should be considered in design. Such changes as scour about the pile base and added pile roughness due to marine growth may be important.

    (4) The design procedure presented in the previous sections is a static procedure; forces are calculated and applied to the structure statically. The dynamic nature of forces from wave action must be considered in the design of some offshore structures. When a structure's natural frequency of oscillation is such that a significant amount of energy in the wave spectrum is available at that frequency, the dynamics of the structure must be considered. In addition, stress reversals in structural members subjected to wave forces may cause failure by fatigue. If fatigue problems are anticipated, the safety factor should be increased or allowable stresses should be decreased. Evaluation of these considerations is beyond the scope of this manual.

(5) Corrosion and fouling of piles also require consideration in design. Corrosion decreases the strength of structural members. Consequently, corrosion rates over the useful life of an offshore structure must be estimated and the size of structural members increased accordingly.

(6) Fouling of a structural member by marine growth increases the roughness and effective diameter of the pile and also changes the values of the force coefficients. The increased diameter must be carried through the entire design procedure to determine forces on the fouled member.

e. Other considerations related to forces on slender cylindrical piles.

(1) Wave forces on pile groups. For a group of piles supporting a structure such as a platform or pier, the methods given in the previous sections can be used provided the piles are sufficiently separated so that flow around one pile does not influence the flow around adjacent piles. One approach is to assume waves are long crested and of permanent form. Given the relative orientation of the piles to each other and to the incoming wave, forces can be estimated on each pile at different times during the wave passage. Typically, the maximum force on individual piles occurs at different times unless all the piles are parallel to the wave crest. Therefore, numerous calculations throughout the wave passage are needed to determine the worst loading on the overall structure. Because the tops of the piles are connected by the superstructure, and the connections may provide some rigidity; it may be necessary to analyze the pile group as a frame.

(a) As the distance between piles becomes small relative to the wavelength, maximum forces and moments on pile groups may be conservatively estimated by summing the maximum forces and moments on each pile.

(b) The assumption that piles are unaffected by neighboring piles is not valid when the distance between piles is less than about three times the pile diameter. Chakrabarti (1991) presented design graphs giving maximum force on a pile in a linear pile group (piles aligned in a row) as a function of Keulegan-Carpenter number and relative separation distance S/D where S is the distance between center lines of adjacent piles. Graphs were provided for pile groups consisting of two, three and five piles with waves approaching parallel and perpendicular to the line of piles. Graphs were also given for estimating C_D, C_M, and C_L for pile groups of three and five piles exposed to waves parallel and perpendicular to the pile line.

(2) Wave forces on nonvertical piles. Forces and moments on nonvertical cylindrical piles can be estimated using Morison's equation (Equation VI-5-281) where the values for velocity u and acceleration du/dt are given as the velocity and acceleration components perpendicular to the pile. Calculations will need to be performed using an appropriate wave theory along with the force coefficients given in Part VI-5-7c, "Selection of hydrodynamic force coefficients C_D, C_M, and C_L." Do not use the curves provided in design Figures VI-5-126 through VI-5-129 and VI-5-131 through VI-5-134 because these figures are only for vertical piles. For nonvertical piles, the pile self weight (immersed and above water) will contribute to the overturning moment and must be included in the calculation.

(3) Broken wave forces. Forces resulting from action of broken waves on piles are much smaller than forces due to breaking waves. When pile-supported structures are constructed in the surf zone, lateral forces from the largest wave breaking on the pile should be used for design. Although breaking-wave forces in the surf zone are great per unit length of pile, the pile length actually subjected to wave action is usually short. Hence, the resulting total force and moment are small. Pile design in the surf zone is usually governed primarily by vertical loads acting along the pile axis.

VI-5-8. Other Forces and Interactions.

a. Impact forces. Impact force loading on coastal projects occurs when waves or solid objects collide with typically stationary coastal structure elements. Only solid body impacts are discussed in this section. Impact loads between shifting concrete armor units are discussed in Part VI-5-3c, "Structural integrity of concrete armor units."

(1) Certain coastal structures such as thin-walled flood barriers, sheet-pile bulkheads, mooring facilities, coastal buildings, or other infrastructure may be subject to impact damage by solid objects carried by waves, currents, or hurricane-force winds. During severe storms, high winds may propel small pleasure craft, barges, and floating debris which can cause significant horizontal impact loads on structures. Likewise, floating ice masses can also cause great impact loads. Impact loads are an important consideration in design of vessel moorings and fendering systems.

(2) Designing a structure to resist impact loads during extreme events is difficult because of uncertainty associated with impact speed and duration. In situations where impact damage by large floating objects could cause catastrophic loss, it may be prudent to limit adjacent water depth by constructing sloping rubble-mound protection fronting the structure or by placing submerged breakwaters seaward of the structure to ground large floating masses and eliminate the hazard.

(3) Impact forces are evaluated using impulse-momentum and energy considerations found in textbooks on fundamental dynamics. However, application of these principles to particular impact problems is difficult unless reliable estimates can be made of object mass (including added mass in water), the mass initial and final velocities, duration of impact loading, and distribution of impact force over time. In addition, some evaluation must be made on whether the collision of the floating object with a coastal structure results in purely elastic behavior in which momentum is conserved, purely plastic impact with all the kinetic energy of the impact being absorbed, or some combination of the two.

(4) Fendering systems in ports and harbors are designed to absorb low-velocity impacts by vessels during docking maneuvers and seiching motions. Design of fendering systems is adequately covered by numerous textbooks and design standards. Examples of typical design references in the coastal engineering literature include Quinn (1972) and Costa (1973). The modes of kinetic energy absorption by fendering systems were studied theoretically by Hayashi and Shirai (1963). Otani, Horii, and Ueda (1975) presented field measurements related to absorption of impact kinetic energy of 50 large tankers. They observed that most berthing velocities are generally below 6 cm/s, and that measured impact energy was substantially larger

than calculated using the design standards that existed at that time. Kuzmanovic and Sanchez (1992) discussed protective systems for bridge piers and pilings, and they gave procedures for accessing the equivalent static force acting on bridge piers due to vessel impacts.

    b. Ice forces. A description of ice loading and how it may impact various types of coastal structures in the context of site-specific design criteria is given in Part VI-3-5, "Ice." Other general references include Chen and Leidersdorf (1988); Gerwick (1990); and Leidersdorf, Gadd, and Vaudrey(1996). The following section presents methods for calculating ice forces under specific loading conditions.

    (1) Horizontal ice forces.

    (a) Solid ice forces.

    • Large horizontal forces can result when solid sheet ice, or large chunks of solid ice that have broken free, come in contact with vertical-front coastal structures. Most ice sheets are large enough that impact forces are limited by ice failure in the weakest mode permitted by the mechanics of interaction as the structure penetrates the ice, i.e., crushing, splitting, shear, or bending. For smaller ice blocks or wide structures, the maximum impact force may be limited by the kinetic energy available at the moment of impact (HQUSACE 1982). Ice impact calculations should be based on impulse-momentum considerations, but such calculations will be difficult because of uncertainty in estimating a value for ice block velocity.

    • Wind and water current drag acting on large floating blocks of ice press the ice blocks against structures creating large pressures at the points of contact. The force due to drag on a block of ice can be calculated for wind and water currents using the following formula (PIANC 1992)

$$F_d = C_{sf}\, \rho\, A\, (u - u_i)^2 \qquad\qquad\text{(VI-5-309)}$$

where

    C_{sf} = coefficient of skin friction between wind and ice or water and ice (see Table VI-5-89).

    ρ = fluid density (air or water)

    A = horizontal area of ice sheet

    u = fluid velocity (10 m above ice for air or 1 m below ice for water)

    u_i = velocity of ice in the direction of u

Table VI-5-89
Values of Skin Friction Coefficient, C_{sf} (PIANC 1992)

	Smooth Ice	Rough (Pack) Ice
Wind drag	0.001 - 0.002	0.002 - 0.003
Water drag	0.002 - 0.004	0.005 - 0.008

• Separate drag calculations should be performed for both wind and water currents with the results treated as vector forces on the ice mass. Because drag force is directly proportional to ice surface area, larger ice sheets will exert greater forces.

• Once an ice sheet has come to rest against a structure, u_i is zero, and the total drag force can be calculated. Intact ice sheets should be treated as solid bodies with the resultant loads vectorially distributed among the structure/ice contact points using force and moment balance. The total force may be somewhat uniformly distributed along a lineal vertical wall. However, if the ice block comes in contact at only a few discrete points, the contact pressure may be very large. In these cases, the calculated force due to drag may exceed the force necessary for local crushing of the ice, in which case the local crushing strength becomes the limiting force applied to the structure.

(b) Localized ice crushing forces.

• The limiting ice force on a vertical structure is determined by the crushing failure strength of the ice in compression. A theoretical expression for the horizontal ice crushing force was given in Korzhavin in a 1962 Russian publication (Ashton 1986) as

$$\frac{F_c}{bh_i} = m\,I\,k\,x\,\sigma_c \tag{VI-5-310}$$

where

F_c = horizontal crushing force

b = structure horizontal width or diameter

h_i = thickness of ice sheet

m = plan shape coefficient

I = indentation coefficient

k = contact coefficient

x = strain rate function

σ_c = ice compressive failure strength in crushing

• This formula is usually applied to piles and pier structures rather than long vertical walls. The plan shape coefficient, m, is 1.0 for flat surfaces, 0.9 for circular piles, and $0.85[sin(\beta/2)]^{1/2}$ for wedge-shaped structures having a wedge angle of β. The indentation coefficient, I, has been found experimentally to be a function of the aspect ratio, b/h_i, and it is usually presented in graphical form. The contact coefficient, k, is a function of ice velocity and width of structure, and it varies between values of 0.4 to 0.7 for ice velocities between 0.5 and 2.0 m/s. The strain rate coefficient is also a function of ice speed. Ashton (1986) provided further details about the theoretical development of Equation VI-5-310 and its associated coefficients.

• In a Russian publication, Afansev (Ashton 1986) combined the coefficients I, k, and x of Equation VI-5-310 into a single coefficient, C, giving the formula

$$\frac{F_c}{bh_i} = C\, m\, \sigma_c \qquad\qquad \text{(VI-5-311)}$$

• Afansev established the following empirical relationship for C based on model tests using laboratory-grown, saline ice.

$$
\begin{aligned}
C &= \left(5\frac{h_i}{b} + 1\right)^{1/2} \qquad && for\ 1 < \frac{b}{h_i} \\[2mm]
C &= 4.17 - 1.72\left(\frac{b}{h_i}\right) \qquad && for\ 0.1 < \frac{b}{h_i} < 1
\end{aligned}
\qquad \text{(VI-5-312)}
$$

• The lower formula in Equation VI-5-312 is a linear interpolation as recommended in Ashton (1986).

• In Equation VI-5-311 values of the shape coefficient are the same as given for the Korzhavin formula (Equation VI-5-310).

• The Canadian Standards Association Bridge Code (Canadian Standards Association 1978) recommended an even more simplified version of Equation VI-5-310 given by

$$\frac{F_c}{bh_i} = \sigma_c \qquad\qquad \text{(VI-5-313)}$$

using the range of values for sheet ice compressive crushing strength shown in Table VI-5-90. Equation VI-5-313 and the crushing strength values of Table VI-5-90 were also adopted by the American Association of State Highway and Transportation Officials.

• Use of Equation VI-5-313 implies that the product $C\, A\, m = 1$ in Equation VI-5-311, which corresponds to large values of b/h_i. This is a realistic assumption for large bridge piles and piers, but ice crushing forces on smaller diameter piles should be calculating using the appropriate strength values from Table VI-5-90 in Equation VI-5-311.

Table VI-5-90 Values of Effective Ice Crushing Strength, σ_c	
Ice Crushing Stress	Environmental Situation
0.7 MPa (100 psi)	Ice breakup occurs at melting temperatures and the ice moves in small pieces that are essentially disintegrated.
1.4 MPa (200 psi)	Ice breakup occurs at melting temperatures, but the ice moves in large pieces that are generally sound.
2.1 MPa (300 psi)	Ice breakup consists of an initial movement of the entire ice sheet or large sheets of sound ice impact piers.
2.8 MPa (400 psi)	Ice breakup occurs with an ice temperature significantly below the melting point and ice movement consists of large sheets.

(c) Thermal ice forces. Equations are available for predicting ice temperature based on an energy balance between the atmosphere and the ice sheet. However, the required parameters (air temperature, air vapor pressure, wind, and cloud cover) needed to calculate thermal expansion are difficult to estimate. Thermal strain is equal to the ice thermal expansion coefficient times the change in ice temperature. For restrained or partially restrained ice sheets a nonlinear, time dependent stress-strain law is used to predict thermal stresses (HQUSACE 1982). Because of stress relaxation due to creep, the rate of thermal change is an important factor; and even a thin snow cover can drastically reduce thermal stresses in ice sheets.

• A design rule-of-thumb for thermal expansion loads per unit horizontal length on dams and other rigid structures is 145 - 220 kN/m (10,000 - 15,000 lbs/ft) (HQUSACE 1982). Movable structures should allow for 73 kN/m (5,000 lbs/ft). These values are based on field measurements.

• Thermal expansion of water frozen between elements of a coastal structure can result in dislocation of individual elements or cracking of armor units making the protection vulnerable to wave attack.

(2) Ice forces on slopes.

(a) Ride-up of ice on slopes.

• When horizontally moving ice encounters a sloping structure, a component of the horizontal force pushes the ice up the slope. This action induces a bending failure in the ice sheet at loads less than required for ice crushing failure. Ashton (1986) showed the derivation of a simple two-dimensional theory for calculating the horizontal force exerted by ice on a sloping structure as illustrated in Figure VI-5-144. (Ashton also included discussion and analysis of the more complex case of ice ride-up on three-dimensional structures).

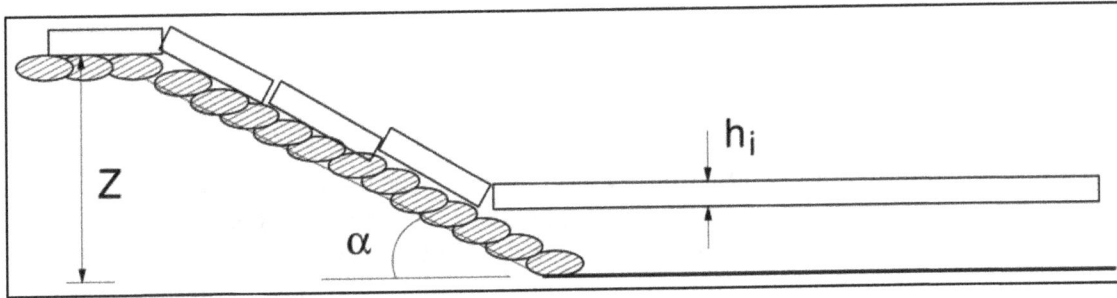

Figure VI-5-144. Ice riding up on structure slope

•   For the two-dimensional case the horizontal force per unit width of structure was given by the expression

$$\frac{F_h}{b} = C_1 \sigma_f \left(\frac{\rho_w \, g \, h_i^5}{E} \right)^{1/4} + C_2 Z \rho_i \, g \, h_i \qquad \text{(VI-5-314)}$$

with

$$C_1 = 0.68 \left[\frac{\sin \alpha + \mu \cos \alpha}{\cos \alpha - \mu \sin \alpha} \right] \qquad \text{(VI-5-315)}$$

and

F_h = total horizontal force

$$C_2 = \left[\frac{(\sin \alpha + \mu \cos \alpha)^2}{\cos \alpha - \mu \sin \alpha} + \frac{\sin \alpha + \mu \cos \alpha}{\tan \alpha} \right] \qquad \text{(VI-5-316)}$$

b = horizontal width of structure contact zone

h_i = ice sheet thickness

σ_f = flexural strength of ice (0.5 = 1.5 MPa)

ρ_w = water density

ρ_i = ice density (915 = 920 kg/m^3)

E = modulus of elasticity of ice (1,000 = 6,000 MPa)

Z = maximum vertical ice ride-up distance

g = gravitational acceleration

α = structure slope angle relative to horizontal

μ = structure slope friction factor (0.1 = 0.5)

• The first term in Equation VI-5-314 is interpreted as the force necessary to break the ice in bending, and the second term is the force that pushes the ice blocks up the sloping structure. The modulus of elasticity varies from 1,000 MPa for very salty water up to about 6,000 MPa for fresh water (Machemehl 1990). Ashton (1986) warned that this simple two-dimensional theory will be inadequate for narrow structures because the zone of ice failure will be wider than the structure.

• Low values of friction factor ($\mu = 0.1$) are associated with smooth slopes such as concrete or carefully layed block protection, whereas high values ($\mu = 0.5$) are applicable for randomly-placed stone armor, riprap, or filled geotextile bags. For slopes steeper than 1:1, the horizontal ice force increases rapidly for the higher friction factors, and there is a risk of the dominant failure mode being crushing or buckling rather than bending. Milder slopes with smooth surfaces are much more effective in reducing horizontal ice forces. Croasdale, Allyn, and Roggensack (1988) discussed several additional aspects related to ice ride-up on sloping structures.

• Quick "rough" estimates of horizontal forces on sloping structures can be made using a variation of Equation VI-5-313 as proposed in Ashton (1986), i.e.,

$$\frac{F_h}{b} = K_h h_i \sigma_c \qquad \text{(VI-5-317)}$$

where K_h is approximated from a curve given in Ashton (1986) by the formula

$$K_h = 1 - 0.654 f^{0.38} \qquad \text{(VI-5-318)}$$

with

$$f = \frac{1 - \mu \tan \alpha}{\mu + \tan \alpha} \qquad \text{(VI-5-319)}$$

and σ_c is the ice compressive strength as given in Table VI-5-90. As slope angle increases, K_h approaches a value of unity which represents failure by crushing. For decreasing slope angles, K_h decreases because of the increasing tendency of the ice to fail in bending. Values of K_h less than 0.2 should never to used in Equation VI-5-317.

(b) Adfreeze loads. When ice that is in contact with a coastal structure is stationary for a sufficient time, the ice will freeze to the structure or its elements. Adfreeze loads result if the ice then moves either horizontally by dynamic ice thrust or vertically due to changing water level. This is more of a problem in lakes with slowly varying water levels than in tidal waters.

• Little guidance is available on adfreeze stresses with adhesion strength varying between 140 kPa to 1050 kPa for freshwater ice (PIANC 1992). Adfreeze may dislodge individual armor stones on rubble-mound slopes creating a weakness in the armor layer. This can be prevented by using oversized stones or interlocking armor on the slope. A survey of riprap structures at Canadian hydropower reservoirs concluded that plucking of individual stones frozen to ice could be largely prevented by sizing the riprap median diameter (d_{50}) greater than the expected maximum winter ice thickness (Wuebben 1995).

(3) Vertical ice forces. Ice frozen to coastal structures can create vertical forces due to ice buoyancy effects when water level rises, or by ice weight when water level falls. These vertical forces will persist until the ice sheet fractures due to bending or the adfreeze force is exceeded.

(a) Cylindrical piles.

• In cases where the ice sheet freezes around a pile, forces will be exerted on the pile if the water level rises or falls. A rising water level will lift the ice sheet, and under certain conditions the uplift force on the pile may be sufficient to pull the pile free. Similarly, during falling water levels the weight of the ice sheet will exert a downward force on a pile which may be sufficient to buckle a slender pile.

• Kerr (1975) studied vertical loads on cylindrical piles and presented equations for calculating loads under the conservative assumption that the water level change is rapid enough to assure elastic ice behavior before failure. A closed-form solution to the governing equation was obtained in terms of Bessel functions, and Kerr presented a numerically evaluated solution in graphical form as shown on Figure VI-5-145. The graphical solution is dimensional, and it has the functional form of

$$P = f\ (a, h, E_{av}, \varDelta, v) \tag{VI-5-320}$$

where

$P =$ uplift force in metric tons (tonnes)

$a =$ pile radius (cm)

$h =$ ice plate thickness (cm)

$E_{av} =$ averaged Young's modulus for ice (kg/cm^2)

$\varDelta =$ water level rise (cm) up to the thickness of the ice

$v =$ Poisson's ratio

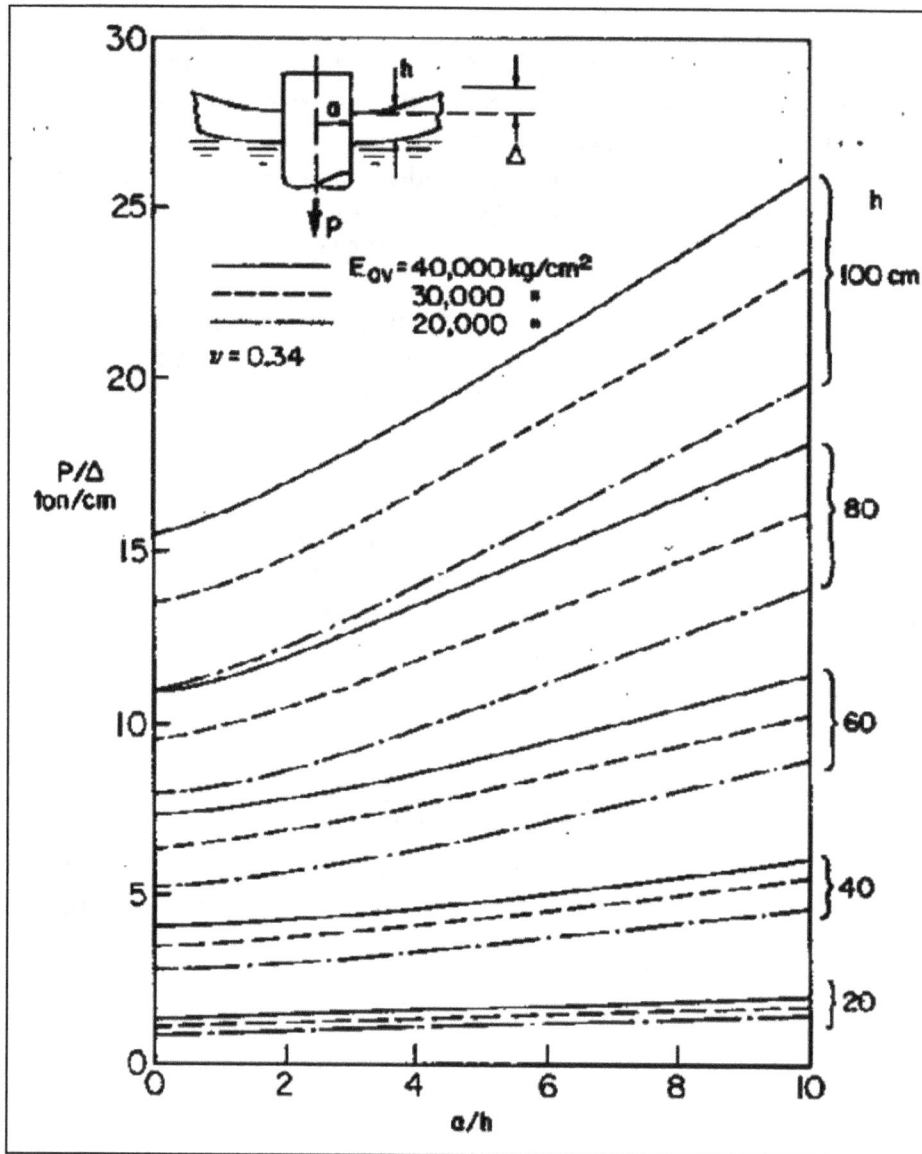

Figure VI-5-145. Vertical ice forces on a cylindrical pile (Kerr 1975)

• Kerr's solution gives estimates of the maximum vertical load assuming the ice sheet does not fail in shear or bending before the maximum load on the pile is reached. For example, the maximum uplift force on a pile with radius $a = 100$ cm surrounded by a 40-cm-thick ice sheet having an average Young's modulus of 30,000 kg/cm^2 would be estimated from Figure VI-5-145 using a value of $a/h = 2.5$ giving $P_{max} = 3.7\,\Delta$. The total maximum force for a 5-cm water level rise would be

$$P_{max} = 3.7\,\Delta = 3.7\,(5\,cm) = 18.5\,tonnes$$

• Kerr (1975) pointed out that the same analysis applied for falling water levels with only a sign change, thus Figure VI-5-145 can also be used decreasing water levels.

(b) Vertical walls.

• Uplift or downward forces per unit horizontal length caused by vertical movement of ice sheets frozen to vertical walls can be approximated using the following formula (PIANC 1992)

$$\frac{F_v}{b} = \rho_w \, g \, \Delta h \, L_c \qquad (VI\text{-}5\text{-}321)$$

where the characteristic length L_c is given as

$$L_c = \left[\frac{E h_i^3}{12 \rho_w g \left(1 - v^2\right)} \right]^{\frac{1}{4}} \qquad (VI\text{-}5\text{-}322)$$

and

$F_v =$ total vertical force acting on the wall

$b =$ horizontal length of wall

$\Delta h =$ change in water level

$\rho_w =$ density of water

$g =$ gravitational acceleration

$E =$ modulus of elasticity of ice

$h_i =$ ice thickness

$v =$ Poisson's ratio (0.31-0.35)

• As previously mentioned, the modulus of elasticity for ice varies with brine volume from about 1,000 MPa for very salty water to about 6,000 MPa for freshwater ice. For freshwater ice, L_c is typically between 15 to 20 times the ice thickness with a reasonable rule-of-thumb being $L_c = 17 \, h_i$.

(c) Sloping structures. The additional vertical load caused by the ride-up and piling of ice on sloping structures needs to be evaluated for the local conditions and specific type of structure. Ice piled up on the slope could initiate slumping of the armor layer on steeper slopes. During rising waters, individual armor stones or revetment units might be lifted out by adfreeze forces.

(4) Aspects of slope protection design.

(a) Much of our understanding of successful slope protection design in cold coastal regions stems from practical experience as documented in the technical literature. In general the design philosophy recognizes that little can be done to prevent ice contact with slope protection structures. Therefore, emphasis is placed on minimizing potential ice damage using a variety of techniques.

(b) Leidersdorf, Gadd, and McDougal (1990) reviewed the performance aspects of three types of slope protection used for coastal projects related to petroleum activities in the Beaufort Sea. For water depths less than 2 m, sacrificial beaches appeared to function well. In water depths ranging from 7 m to 15 m, gravel-filled geotextile bags were able to withstand the larger wave forces, but they were susceptible to ice damage and required regular maintenance. Linked concrete mat armor (Leidersdorf, Gadd, and McDougal 1988; Gadd and Leidersdorf 1990) withstood both wave and ice loads in depths up to about 14 m. Mats were recommended for projects with a lengthy service life so that high initial capital costs would be offset by lower maintenance expenses. Wuebben (1995) reviewed the effects of ice on riprap structures constructed along ice-prone waterways. This paper provided a good summary of successful riprap revetment design and construction practices based on actual field experience. Numerous useful references documenting ice effects on riprap are included in Wuebben's paper. The following rules-of-thumb for arctic slope protection were given in Chen and Leidersdorf (1988) and summarized in PIANC (1992).

• Cover layers and underlayers should be strong enough to withstand local penetration by thick ice sheets.

• Smooth slopes without protrusions will reduce loads and allow the ice to ride up more easily without plucking out individual armor elements. (However, wave runup will be greater.)

• Flexible cover layers consisting of graded riprap may help absorb impacts by smaller ice blocks during wave action without appreciable damage. Sand bags are effective for structures with intended short service lives.

• Mild structure slopes are essential because they reduce the risk of ice penetration into the slope. Maximum slope 15 deg is recommended in the zone of ice attack.

• Compound slopes with a nearly horizontal berm above the swl provide a platform for piled-up ice in regions which experience frequent ride-up of ice sheets.

• Maximum ice loads will not occur at the same time as maximum expected wave loads. Therefore, slope design can consider each load condition separately.

VI-5-9. <u>References</u>.

EM 1110-2-1601
Hydraulic Design of Flood Control Channels

EM 1110-2-1612
Ice Engineering

ACI 1989
American Concrete Institute. 1989 (revised annually). *Manual of Concrete Practice*, Part 3, Detroit, MI.

Ahrens 1981a
Ahrens, J. P. 1981a. "Irregular Wave Runup on Smooth Slopes," Technical Aid No. 81-17, U.S. Army Engineer Waterways Experiment Station, Coastal Engineering Research Center, Vicksburg, MS.

Ahrens 1981b
Ahrens, J. P. 1981b. "Design of Riprap Revetments for Protection Against Wave Attack," Technical Paper TP-81-5, U.S. Army Engineer Waterways Experiment Station, Coastal Engineering Research Center, Vicksburg, MS.

Ahrens 1984
Ahrens, J. P. 1984. "Reef Type Breakwaters," *Proceedings of the 19th International Coastal Engineering Conference*, American Society of Civil Engineers, Vol 3, pp 2648-2662.

Ahrens 1987
Ahrens, J. P. 1987. "Characteristics of Reef Breakwaters," Technical Report CERC-87-17, U.S. Army Engineer Waterways Experiment Station, Vicksburg, MS.

Ahrens and Heinbaugh 1988a
Ahrens, J. P., and Heinbaugh, M. S. 1988a. "Approximate Upper Limit of Irregular Wave Runup on Riprap," Technical Report CERC-88-5, U.S. Army Engineer Waterways Experiment Station, Coastal Engineering Research Center, Vicksburg, MS.

Ahrens and Heinbaugh 1988b
Ahrens, J. P., and Heinbaugh, M. S. 1988b. "Seawall Overtopping Model," *Proceedings of the 21st International Coastal Engineering Conference*, American Society of Civil Engineers, Vol 1, pp 795-806.

Ahrens, Viggosson, and Zirkle 1982
Ahrens, J. P., Viggosson, G., and Zirkle, K. P. 1982. "Stability and Wave Transmission Characteristics of Reef Breakwaters," Interim Report, U.S. Army Engineer Waterways Experiment Station, Coastal Engineering Research Center, Vicksburg, MS.

Allsop 1983
Allsop, N. W. 1983. "Low-Crest Breakwaters, Studies in Random Waves," *Proceedings of Coastal Structures '83*, American Society of Civil Engineers, pp 94-107.

Allsop 1990
Allsop, N. W. 1990. "Reflection Performance of Rock Armoured Slopes in Random Waves," *Proceedings of the 22nd International Coastal Engineering Conference*, American Society of Civil Engineers, Vol 2, pp 1460-1472.

Allsop 1995
Allsop, N. W. 1995. "Stability of Rock Armour and Riprap on Coastal Structures," *River, Coastal and Shoreline Protection: Erosion Control Using Riprap and Armourstone,* C. R. Thorne, S. R. Abt, F. B. Barends, S. T. Maynord, and R. W. Pilarczyk, eds., John Wiley and Sons, New York, pp 213-226.

Allsop and Hettiarachchi 1988
Allsop, N. W., and Hettiarachchi, S. S. 1988. "Reflections from Coastal Structures," *Proceedings of the 21st International Coastal Engineering Conference*, American Society of Civil Engineers, Vol 1, pp 782-794.

Allsop et al. 1985
Allsop, N. W., Hawkes, P. I., Jackson, F. A., and Franco, L. 1985. "Wave Run-Up on Steep Slopes - Model Tests Under Random Waves," Report No. SR2, Hydraulics Research Station, Wallingford, England.

Allsop, McBride, and Colombo 1994
Allsop, N. W., McBride, M. W., and Colombo, D. 1994. "The Reflection Performance of Vertical Walls and Low Reflection Alternatives: Results of Wave Flume Tests," *Proceedings of the 3rd MCS Project Workshop, MAS2-CT92-0047, Monolithic (Vertical) Coastal Structures,* De Voorst, The Netherlands.

Aminti and Franco 1988
Aminti, P., and Franco, L. 1988. "Wave Overtopping on Rubble Mound Breakwaters," *Proceedings of the 21st International Coastal Engineering Conference*, American Society of Civil Engineers, Vol 1, pp 770-781.

Andersen 1981
Andersen, K. H. 1981. Discussion of "Cyclic Simple Shear Behaviour of Fine Grained Soil," Dyvik, Zimme, and Schimelfenjg, *International Conference on Recent Advances in Geotechnical Earthquake Engineering and Soil Dynamics*, St. Louis, MO, pp. 920-921.

Andersen 1983
Andersen, K. H. 1983. "Strength and Deformation Properties of Clay Subjected to Cyclic Loading," Report 52412-8, Norwegian Geotechnical Institute.

Andersen 1988
Andersen, K. H. 1988. "Properties of Soil Clay Under Static and Cyclic Loading," *Proceedings of the International Conference on Engineering Problems of Regional Soils*, Beijing, China, pp 7-26.

Andersen 1991
Andersen, K. H. 1991. "Foundation Design of Offshore Gravity Structures," *Cyclic Loading of Soils*, M. P. O'Reilly and S. F. Brown, eds., Blackie and Son Ltd., Glasgow, Scotland.

Andersen and Høeg 1991
Andersen, K. H., and Høeg, K. 1991. "Deformation of Soils and Displacements of Structures Subjected to Combined Static and Cyclic Loads," *Proceedings 10th European Conference on Soil Mechanics and Foundation Engineering*, Florence, Italy, Vol 4, pp 1147-1158.

Andersen and Richards 1985
Andersen, M. G., and Richards, K. S. 1985. *Slope Stability*, John Wiley & Sons, New York, NY.

Anderson et al. 1976
Anderson, K. H., Brown, S. F., Foss, I., Pool, J. H., and Rosenbrand, F. W. 1976. "Effect of Cyclic Loading on Clay Behavior," *Proceedings of the Conference on Design and Construction of Offshore Structures*, Institution of Civil Engineers, London, UK, pp 75-79.

Andersen, Kleven, and Heien 1988
Andersen, K. H., Kleven, A., and Heien, D. 1988. "Cyclic Soil Data for Design of Gravity Structures," *Journal of Geotechnical Engineering*, American Society of Civil Engineers, Vol 114, No. 5, pp 517-539.

Andrews and Borgman 1981
Andrews, M. E., and Borgman, L. E. 1981. "Procedures for Studying Wave Grouping in Wave Records from the California Coastal Data Collection Program," Statistics Laboratory Report No. 141, University of Wyoming, Laramie, WY.

Anglin et al. 1990
Anglin, C. D., Scott, R. D., Turcke, D. J., and Turke, M. A. (1990). "The Development of Structural Design Criteria for Breakwater Armor Units," *Stresses in Concrete Armor Units*, American Society of Civil Engineers, pp 123-148.

Anglin et al. 1996
Anglin, C. D., Nairn, R. B., Cornett, A. M., Dunaszegi, L, Turnham, J, and Annandale, G. W. 1996. "Bridge Pier Scour Assessment for the Northumberland Strait Crossing," *Proceedings of the 25th International Coastal Engineering Conference*, American Society of Civil Engineers, Vol 2, pp 2230-2243.

Ashton 1986
Ashton, G. D. 1986. *River and Lake Ice Engineering*. Water Resources Publications, Littleton, CO.

Bagnold 1939
Bagnold, R. A. 1939. "Interim Report on Wave-Pressure Research," *Journal of Institution of Civil Engineers*, Vol 12, pp 202-226.

Bakker et al. 1994
Bakker, K. J., Verheij, H. J., and de Groot, M. B. 1994. "Design Relationship for Filters in Bed Protection," *Journal of Hydraulic Engineering*, American Society of Civil Engineers, Vol 120, No. 9, pp 1082-1088.

Barends 1988
Barends, F. B. 1988. Discussion of "Pore Pressure Response and Stability of Rubble-Mound Breakwaters," Simm and Hedges, *Proceedings of the Breakwaters '88 Conference: Design of Breakwaters,* Institution of Civil Engineers, Thomas Telford, London, UK, pp 85-88.

Barends et al. 1983
Barends, F. B., van der Kogel, H., Uijttewall, F. J., and Hagenaar, J. 1983. "West Breakwater - Sines: Dynamic-Geotechnical Stability of Breakwaters," *Proceedings of Coastal Structures '83*, American Society of Civil Engineers, pp 31-43.

Barnett and Wang 1988
Barnett, M. R., and Wang, H. 1988. "Effects of a Vertical Seawall on Profile Response," *Proceedings of the 21st International Coastal Engineering Conference*, American Society of Civil Engineers, Vol 2, pp 1493-1507.

Bass and Fulford 1992
Bass, G. P, and Fulford, E. T. 1992. "South Jetty Scour Hole Stabilization, Ocean City, Maryland," *Proceedings of Coastal Engineering Practice '92*, American Society of Civil Engineers, pp 583-597.

Battjes 1974
Battjes, J. A. 1974. "Computation of Set-Up, Longshore Currents, Run-Up, and Overtopping Due to Wind-Generated Waves," Report 74-2, Committee on Hydraulics, Department of Civil Engineering, Delft University of Technology, The Netherlands.

Battjes 1974b
Battjes, J. A. 1974b. "Surf Similarity," *Proceedings of the 14th International Coastal Engineering Conference*, American Society of Civil Engineers, Vol 1, pp 466-479.

Battjes 1982
Battjes, J. A. 1982. AEffects of Short-Crestedness on Wave Loads on Long Structures," *Applied Ocean Research*, Vol 4, No. 3, pp 165-172.

Bea et al. 1999
Bea, R. G., Xu, T., Stear, J., and Ramos, R. 1999. "Wave Forces on Decks of Offshore Platforms," *Journal of Waterway, Port, Coastal, and Ocean Division*, American Society of Civil Engineers, Vol 125, No. 3, pp 136-144.

Belfadhel, Lefebvre, and Rohan 1996
Belfadhel, M. B., Lefebvre, G., and Rohan, K. 1996. "Comparison and Evaluation of Different Riprap Stability Formulas Using Field Performance," *Journal of Waterway, Port, Coastal and Ocean Engineering*, American Society of Civil Engineers, Vol. 122, No. 1, pp 8-15.

Benoit and Teisson 1994
Benoit, M. and Teisson, C. 1994. "Laboratory Study of Breakwater Reflection - Effect of Wave Obliquity, Wave Steepness and Mound Slope," *International Symposium: Waves - Physical and Numerical Modelling,* Vol 2, pp 1021-1030.

Bijker and de Bruyn 1988
Bijker, E. W., and de Bruyn, C. A. 1988. "Erosion Around a Pile Due to Current and Breaking Waves," *Proceedings of the 21st International Coastal Engineering Conference*, American Society of Civil Engineers, Vol 2, pp 1368-1381.

Bishop 1955
Bishop, A. W. 1955. "The Use of the Slip Circle in the Stability Analysis of Slopes," *Geotechnique*, Vol V, No. 1, pp 1-17.

Bishop and Green 1965
Bishop, A. W., and Green, G.E. 1965. "The Influence of End Restraint on the Compression Strength of a Cohesionless Soil," *Geotechnique*, Vol 15, No. 3, pp. 243-266.

Bjerrum 1973
Bjerrum, L. 1973. "Problems of Soil Mechanics and Construction on Soft Clays," State-of-the-Art Report to Session IV, *Proceedings of the International Conference on Soil Mechanics and Foundation Engineering 3*, pp 111-159.

Bolton 1979
Bolton, M. D. 1979. *A Guide to Soil Mechanics.* Macmillan Press Ltd., London, UK.

Bolton 1986
Bolton, M. D. 1986. "The Strength and Dilatancy of Sands," *Geotechnique,* Vol 36, No. 1, pp 65-78.

Bottin, Chatham, and Carver 1976
Bottin, R., Chatham, C., and Carver, R. 1976. "Waianae Small-Boat Harbor, Oahu, Hawaii, Design for Wave Protection," TR H-76-8, U.S. Army Engineer Waterways Experiment Station, Vicksburg, MS.

Bradbury and Allsop 1988
Bradbury, A. P., and Allsop, N. W. 1988. "Hydraulic Effects of Breakwater Crown Walls," *Proceedings of the Breakwaters '88 Conference,* Institution of Civil Engineers, Thomas Telford Publishing, London, UK, pp 385-396.

Bradbury et al. 1988
Bradbury, A. P., Allsop, N. W., and Stephens, R. V. 1988. "Hydraulic Performance of Breakwater Crown Walls," Report No. SR146, Hydraulics Research, Wallingford, UK.

Breusers, Nicollet, and Shen 1977
Breusers, H. N. C., Nicollet, G., and Shen, H. W. 1977. "Local Scour Around Cylindrical Piers," *Journal of Hydraulic Research*, Vol 15, No. 3, pp 211-252.

Brinch Hansen 1961
Brinch Hansen, J. 1961. "A General Formula for Bearing Capacity," Bulletin No. 11, Danish Geotechnical Institute, Copenhagen, Denmark.

Brinch Hansen 1967
Brinch Hansen, J. 1967. "Støttemures Bæreevne (The Bearing Capacity of Retaining Walls)," B-undervisning og Forskning 67, Danmarks Ingeniørakademi, Copenhagen, Denmark, pp 307-320.

Brinch Hansen 1970
Brinch Hansen, J. 1970. "A Revised and Extended Formula for Bearing Capacity," Bulletin No. 28, Danish Geotechnical Institute, Copenhagen, Denmark.

Broderick 1983
Broderick, L. L. 1983. "Riprap Stability A Progress Report," *Proceedings of Coastal Structures '83*, American Society of Civil Engineers, pp 320-330.

Brorsen, Burcharth, and Larsen 1974
Brorsen, M., Burcharth, H. F., and Larsen, T. 1974. "Stability of Dolos Slopes," *Proceedings of the 14th International Coastal Engineering Conference*, American Society of Civil Engineers, Vol 3, pp 1691-1701.

Bruining 1994
Bruining, J. W. 1994. "Wave Forces on Vertical Breakwaters. Reliability of Design Formula," Delft Hydraulics Report H 1903, MAST II contract MAS2-CT92-0047.

Burcharth 1984
Burcharth, H. F. 1984. "Fatigue in Breakwater Concrete Armour Units," *Proceedings of the 19th International Coastal Engineering Conference*, American Society of Civil Engineers, Vol 3, pp 2592-2607.

Burcharth 1993
Burcharth, H. F. 1993. "The Design of Breakwaters," Department of Civil Engineering, Aalborg University, Denmark.

Burcharth 1993b
Burcharth, H. F. 1993b. "Structural Integrity and Hydraulic Stability of Dolos Armor Layers," Series Paper 9, Department of Civil Engineering, Aalborg University, Denmark.

Burcharth 1998
Burcharth, H. F. 1998. "Breakwater with Vertical and Inclined Concrete Walls: Identification and Evaluation of Design Tools," Report of Sub-group A, Working Group 28, PIANC PTCII.

Burcharth and Andersen 1995
Burcharth, H.F., and Andersen, O.H. 1995. "On the One-Dimensional Unsteady Porous Flow Equation," *Coastal Engineering,* Vol 24, pp. 233-257.

Burcharth and Brejnegaard-Nielsen 1986
Burcharth, H. F., and Brejnegaard-Nielsen, T. 1986. "The Influence of Waist Thickness of Dolosse on the Hydraulic Stability of Dolosse Armour," *Proceedings of the 20th International Coastal Engineering Conference*, American Society of Civil Engineers, Vol 2, pp 1783-1796.

Burcharth and Liu 1992
Burcharth, H. F., and Liu, Z. 1992. "Design of Dolos Armour Units," *Proceedings of the 23rd International Coastal Engineering Conference*, American Society of Civil Engineers, Vol 1, pp 1053-1066.

Burcharth and Liu 1995
Burcharth, H. F., and Liu, Z. 1995. "Design Formula for Dolos Breakage," *Proceedings of the Final Workshop, Rubble Mound Breakwater Failure Modes,* Sorrento, Italy.

Burcharth and Liu 1998
Burcharth, H. F., and Liu, Z. 1998. "Force Reduction of Short-Crested Non-Breaking Waves on Caissons," Section 4.3, Part 4, Class II Report of MAST II Project: PROVERBS, Technical University of Braunschweig, Germany, ed.

Burcharth and Thompson 1983
Burcharth, H. F., and Thompson, A. C. 1983. "Stability of Armour Units in Oscillatory Flow," *Proceedings of Coastal Structures '83*, American Society of Civil Engineers, pp 71-82.

Burcharth et al. 1991
Burcharth, H. F., Toschi, P. B., Turrio, E., Balestra, T, Noli, A., Franco, L., Betti, A., Mezzedimi, S. 1991. "Bosaso Harbour, Somalia. A New Hot-Climate Port Development," *Third International Conference on Coastal & Port Engineering in Developing Countries*, Vol 1, pp 649-666.

Burcharth et al. 1995a
Burcharth H. F., Frigaard, P., Uzcanga, J., Berenguer, J. M., Madrigal, B. G., and Villanueva, J. 1995a. "Design of the Ciervana Breakwater, Bilbao," *Proceedings of the Advances in Coastal Structures and Breakwaters Conference*, Institution of Civil Engineers, Thomas Telford, London, UK, pp 26-43.

Burcharth et al. 1995b
Burcharth, H. F., Jensen, M. S., Liu, Z., van der Meer, J. W., and D'Angremond, K. 1995b. "Design Formula for Tetrapod Breakage," *Proceedings of the Final Workshop, Rubble Mound Breakwater Failure Modes*, Sorrento, Italy.

Burcharth et al. 1998
Burcharth, H. F., Christensen, M., Jensen, T. and Frigaard, P. 1998. "Influence of Core Permeability on Accropode Armour Layer Stability," *Proceedings of International Conference on Coastlines, Structures, and Breakwaters '98*, Institution of Civil Engineers, London, UK, pp 34-45.

Burcharth, Liu, and Troch 1999
Burcharth, H.F., Liu, Z., and Troch, P. 1999. "Scaling of Core Material in Rubble Mound Breakwater Model Tests," *Proceedings of Fifth International Conference on Coastal and Port Engineering in Developing Countries*, Cape Town, South Africa, pp 1518-1528.

Burcharth et al. 2000
Burcharth, H. F., d'Angremond, K, van der Meer, J. W., and Liu, Z. 2000. "Empirical Formula for Breakage of Dolosse and Tetrapods," *Coastal Engineering*, Elsevier, Vol 40, No. 3, pp 183-206.

Calabrese and Allsop 1997
Calabrese, M., and Allsop, N. W. 1997. "Impact Loading on Vertical Walls in Directional Seas," Contribution to the 1st Task, MAST III Workshop (PROVERBS), Edinburgh.

Camfield 1990
Camfield, F. E. 1990. "Tsunami," *Handbook of Coastal and Ocean Engineering - Volume 1: Wave Phenomena and Coastal Structures*. John B. Herbich, ed., Gulf Publishing Company, Houston, TX, pp 591-634.

Camfield 1991
Camfield, F. E. 1991. "Wave Forces on Wall," *Journal of Waterway, Port, Coastal, and Ocean Engineering*, American Society of Civil Engineers, Vol 117, No. 1, pp 76-79.

Canadian Standards Association 1978
Canadian Standards Association. 1978. "Design of Highway Bridges," Standard CAN3-56-M78.

Carpenter and Powell 1998
Carpenter, K. E., and Powell, K. A. 1998. "Toe Scour at Vertical Walls: Mechanisms and Prediction Methods," Report No. SR506, HR Wallingford, UK.

Carstens 1976
Carstens, T. 1976. "Seabed Scour by Currents Near Platforms," Proceedings of the 3rd International Conference on Port and Ocean Engineering Under Arctic Conditions, pp 991-1006.

Carver 1980
Carver, R. D. 1980. "Effects of First Underlayer Weight on the Stability of Stone-Armored, Rubble-Mound Breakwater Trunks Subjected to Nonbreaking Waves with No Overtopping," Technical Report HL-80-1, U.S. Army Engineer Waterways Experiment Station, Vicksburg, MS.

Carver and Davidson 1977
Carver, R. D., and Davidson, D. D. 1977. "Dolos Armor Units Used on Rubble-Mound Breakwater Trunks Subjected to Nonbreaking Waves with No Overtopping," Technical Report H-77-19, U.S. Army Engineer Waterways Experiment Station, Vicksburg, MS.

Carver and Davidson 1983
Carver, R. D., and Davidson, D. D. 1983. "Jetty Stability Study, Oregon Inlet, North Carolina," Technical Report CERC-83-3, U.S. Army Engineer Waterways Experiment Station, Vicksburg, MS.

Carver and Heimbaugh 1989
Carver, R. D., and Heimbaugh, M. S. 1989. "Stability of Stone- and Dolos-Armored Rubble-Mound Breakwater Heads Subjected to Breaking and Nonbreaking Waves with No Overtopping," Technical Report CERC-89-4, U.S. Army Engineer Waterways Experiment Station, Coastal and Hydraulics Laboratory, Vicksburg, MS.

Carver and Markle, 1981
Carver, R. D., and Markle, D. G. 1981. "Stability of Rubble-Mound Breakwater; Maalea Harbor, Maui, Hawaii," Miscellaneous Paper HL-81-1, U.S. Army Engineer Waterways Experiment Station, Vicksburg, MS.

Chakrabarti 1991
Chakrabarti, S. K. 1991. "Wave Forces on Offshore Structures," *Handbook of Coastal and Ocean Engineering - Volume 2: Offshore Structures, Marine Foundations, Sediment Processes, and Modeling*. John B. Herbich, ed., Gulf Publishing Company, Houston, TX. pp 1-54.

Chang 1964
Chang, K. S. 1964. "Transverse Forces on Cylinders Due to Vortex Shedding in Waves," M.A. thesis, Department of Civil Engineering, Massachusetts Institute of Technology, Cambridge, MA.

Chen 1961
Chen, T. C. 1961. "Experimental Study on the Solitary Wave Reflection Along a Straight Sloped Wall at Oblique Angle of Incidence," Technical Memorandum No. 124, U.S. Army Corps of Engineers, Beach Erosion Board, Washington, DC.

Chen and Leidersdorf 1988
Chen, A. T., and Leidersdorf, C. B., ed. 1988. "Arctic Coastal Processes and Slope Protection Design," Technical Council on Cold Regions Engineering Monograph, American Society of Civil Engineers, New York, NY.

Chesnutt and Schiller 1971
Chesnutt, C. B., and Schiller, R. E. 1971. "Scour of Simulated Gulf Coast Sand Beaches Due to Wave Action in Front of Sea Walls and Dune Barriers," COE Report No. 139, *TAMU-SG-71-207*, Texas A&M University, College Station, TX.

Chiew 1990
Chiew, Y-M. 1990. "Mechanics of Local Scour Around Submarine Pipelines," *Journal of Hydraulic Engineering*, American Society of Civil Engineers, Vol 116, No. 4, pp 515-539.

Cété-Laboratoire Régional Nord - Pas de Calais 1990
Cété-Laboratoire Régional Nord - Pas de Calais. 1990. "Le Havre - Route du Mole Central - Essais de Frottement Beton Ondulé-Grave Concassée," Laboratoire Régional Nord - Pas de Calais.

Costa 1973
Costa, F. V. 1973. "Berthing Maneuvers of Large Ships," *Port Engineering*. Gulf Publishing Company, Houston, TX, pp 401-417.

Croasdale, Allyn, and Roggensack 1988
Croasdale, K. R., Allyn, N., and Roggensack, W. 1988. "Arctic Slope Protection: Considerations for Ice," *Arctic Coastal Processes and Slope Protection Design*. A. T. Chen and C. B. Leidersdorf, ed., American Society of Civil Engineers, pp 216-243.

Cross 1967
Cross, R. H. 1967. "Tsunami Surge Forces," *Journal of the Waterways and Harbors Division*, American Society of Civil Engineers, Vol 93, No. ww4, pp 201-231.

Centre for Civil Engineering Research and Codes 1995
Centre for Civil Engineering Research and Codes. 1995. "Manual on the Use of Rock in Hydraulic Engineering," CUR/RWS Publication 169, CUR, Gouda, The Netherlands.

Czerniak and Collins 1977
Czerniak, M. T., and Collins, J. I. 1977. "Design Considerations for a Tetrapod Breakwater," *Proceedings of Ports '77*, American Society of Civil Engineers, New York, pp 178-195.

d'Angremond, van der Meer, van Nes 1994
d'Angremond, K., van der Meer, J. W., and van Nes, C. P. 1994. "Stresses in Tetrapod Armour Units Induced by Wave Action," *Proceedings of the 24th International Coastal Engineering Conference*, American Society of Civil Engineers, Vol 2, pp 1713-1726.

Daemrich and Kahle 1985
Daemrich, K. F., and Kahle, W. 1985. "Shutzwirkung von Unterwasserwellen Brechern Unter Dem Einfluss Unregelmassiger Seeganswellen," Eigenverlag des Franzius-Instituts fur Wasserbau und Kusteningenieurswesen, Heft 61 (in German).

Daemen 1991
Daemen, I. F. 1991. "Wave Transmission at Low-Crested Breakwaters," M.S. thesis, Delft University of Technology, The Netherlands.

Davidson et al. 1994
Davidson, M. A., Bird, P. A., Bullock, G. N., and Huntley, D. A. 1994. "Wave Reflection: Field Measurements, Analysis and Theoretical Developments," *Proceedings of Coastal Dynamics '94*, American Society of Civil Engineers, pp 642-655.

Dean 1974
Dean, R. G. 1974. "Evaluation and Development of Water Wave Theories for Engineering Application," Special Report 1, Vols I and II, U.S. Army Coastal Engineering Research Center, Stock Nos. 008-022-00083-6 and 008-022-00084-6, U.S. Government Printing Office, Washington, DC.

de Best, Bijker, and Wichers 1971
de Best, A., Bijker, E. W., and Wichers, J. E. W. 1971. "Scouring of a Sand Bed in Front of a Vertical Breakwater," *Proceedings of the 1st International Conference on Port and Ocean Engineering Under Artic Conditions*, Vol II, pp 1077-1086.

de Gerloni et al. 1991
de Gerloni, M, Cris, E., Franco, L., and Passoni, G. 1991. "The Safety of Breakwaters Against Wave Overtopping," *Proceedings of the Coastal Structures and Breakwaters Conference*, Institution of Civil Engineers, Thomas Telford, London, UK, pp 335-342.

de Graauw, van der Meulen, and van der Does de Bye 1984
de Graauw, A. F., van der Meulen, T., and van der Does de Bye, M. R. 1984. "Granular Filters: Design Criteria," *Journal of Waterway, Port, Coastal, and Ocean Engineering*, American Society of Civil Engineers, Vol 110, No. 1, pp 80-96.

de Groot et al. 1996
de Groot, M. M., Andersen, K. H., Burcharth, H. F., Ibsen, L. B., Kortenhaus, A., Lundgren, H., Magda, W., Oumeraci, H., and Richwien, W. 1996. "Foundation Design of Caisson Breakwaters," Publication No. 198, Norwegian Geotechnical Institute, Oslo, Norway.

de Rouck 1991
de Rouck, J. 1991. "De Stabilitet Van Stortsteengolfbrekers. Algemeen Glijdingsevenwicht - Een Nieuw Deklaagelement," Hydraulic Laboratory, University of Leuven, Belgium.

de Waal and van der Meer 1992
de Waal, J. P., and van der Meer, J. W. 1992. "Wave Run-Up and Overtopping on Coastal Structures," *Proceedings of the 23rd International Coastal Engineering Conference*, American Society of Civil Engineers, Vol 2, pp 1758-1771.

Delft Hydraulics 1989
Delft Hydraulics. 1989. "Slopes of Loose Materials: Wave Run-Up on Statistically Stable Rock Slopes Under Wave Attack," Report M 1983 (in Dutch), Delft Hydraulics Laboratory, The Netherlands.

Eadie and Herbich 1986
Eadie, R. W., and Herbich, J. B. 1986. "Scour About a Single Cylindrical Pile Due to Combined Random Waves and a Current," *Proceedings of the 20th International Coastal Engineering Conference*, American Society of Civil Engineers, Vol 3, pp 1858-1870.

Eckert 1983
Eckert, J. W. 1983. "Design of Toe Protection for Coastal Structures," *Proceedings of Coastal Structures '83*, American Society of Civil Engineers, pp 331-341.

Edge et al. 1990
Edge, B. L., Crapps, D. K., Jones, J. S., and Dean, W. L. 1990. "Design and Installation of Scour Protection for the Acosta Bridge," *Proceedings of the 22nd International Coastal Engineering Conference,* American Society of Civil Engineers, Vol 3, pp 3268-3280.

Endoh and Takahashi 1994
Endoh, K., and Takahashi, S. 1994. "Numerically Modelling Personnel Danger on a Promenade Breakwater Due to Overtopping Waves," *Proceedings of the 24th International Coastal Engineering Confernece*, American Society of Civil Engineers, Vol 1, pp 1016-1029.

Fagerlund and Larsson 1979
Fagerlund, G., and Larsson, B. 1979. "Betongs Slaghallfasthed." (in Swedish), Swedish Cement and Concrete Research Institute, Institute of Technology, Stockholm, Sweden.

Fellenius 1936
Fellenius, W. 1936. "Calculation of the Stability of Earth Dams," *Proceeding of the 2nd International Conference on Large Dams*, Washington DC, Vol 4, pp 445-449.

Fowler 1992
Fowler, J. E. 1992. "Scour Problems and Methods for Prediction of Maximum Scour at Vertical Seawalls," Technical Report CERC-92-16, U.S. Army Engineer Waterways Experiment Station, Coastal Engineering Research Center, Vicksburg, MS.

Fowler 1993
Fowler, J. E. 1993. "Coastal Scour Problems and Methods for Prediction of Maximum Scour," Technical Report CERC-93-8, U.S. Army Engineer Waterways Experiment Station, Coastal Engineering Research Center, Vicksburg, MS.

Fox and McDonald 1985
Fox, R. W., and McDonald, A. T. 1985. *Introduction to Fluid Mechanics*. Third ed., John Wiley and Sons, New York, NY.

Franco and Franco 1999
Franco, C., and Franco, L. 1999. "Overtopping Formulas for Caisson Breakwaters with Nonbreaking 3D Waves," *Journal of Waterway, Port, Coastal, and Ocean Engineering*, American Society of Civil Engineers, Vol 125, No. 2, pp 98-108.

Franco, de Gerloni, and van der Meer 1994
Franco, L., de Gerloni, M., and van der Meer, J. W. 1994. "Wave Overtopping on Vertical and Composite Breakwaters," *Proceedings of the 24th International Coastal Engineering Conference*, American Society of Civil Engineers, Vol 1, pp 1030-1045.

Franco, van der Meer, and Franco 1996
Franco, C., van der Meer, J. W., and Franco, L. 1996. "Multidirectional Wave Loads on Vertical Breakwaters," *Proceedings of the 25th International Coastal Engineering Conference*, Vol 2, pp 2008-2021.

Fredsøe and Sumer 1997
Fredsøe, J., and Sumer, B. M. 1997. "Scour at the Round Head of a Rubble-Mound Breakwater," *Coastal Engineering*, Elsevier, Vol 29, No. 3, pp 231-262.

French 1969
French, J. A. 1969. "Wave Uplift Pressure on Horizontal Platforms," Report No. KH-R-19, W. M. Keck Laboratory of Hydraulics and Water Resources, California Institute of Technology, Pasadena, CA.

Froehlich 1988
Froehlich, D. C. 1988. "Analysis of Onsite Measurements of Scour at Piers," *Hydraulic Engineering- Proceedings of the 1988 National Conference*, American Society of Civil Engineers, pp 534-539.

Führböter, Sparboom, and Witte 1989
Führböter, A., Sparboom, U., and Witte, H. H. 1989. "Großer Wellenkanal Hannover: Versuchsergebnisse über den Wellenauflauf auf Glatten und Rauhen Deichböschungen mit der Neigung 1:6," Die Küßte. Archive for Research and Technology on the North Sea and Baltic Coast.

Fukuda, Uno, and Irie 1974
Fukuda, N., Uno, T., and Irie, I. 1974. "Field Observations of Wave Overtopping of Wave Absorbing Revetment," *Coastal Engineering in Japan*, Vol 17, pp 117-128.

Gadd and Leidersdorf 1990
Gadd, P. E., and Leidersdorf, C. B. 1990. "Recent Performance of Linked Concrete Mat Armor Under Wave and Ice Impact," *Proceedings of the 22nd International Conference on Coastal Engineering*, American Society of Civil Engineers, Vol 3, pp 2768-2781.

Gerwick 1990
Gerwick, B. C., Jr. 1990. "Ice Forces on Structures," *The Sea: Ocean Engineering Science*. B. Le Mehaute and D. M. Hanes, eds., Vol 9, Part B, pp 1263-1301.

Givler and Sorensen 1986
Givler, L. D., and Sorensen, R. M. 1986. "An Investigation of the Stability of Submerged Homogeneous Rubble-Mound Structures Under Wave Attack," Report IHL 110-86, H.R. IMBT Hydraulics, Lehigh University, Philadelphia, PA.

Goda 1969
Goda, Y. 1969. "Reanalysis of Laboratory Data on Wave Transmission Over Breakwaters," *Report of Port and Harbour Research Institute*, Japan, Vol 8, No. 3, pp 3-18.

Goda 1970
Goda, Y. 1970. "Estimation of the Rate of Irregular Overtopping of Seawalls," *Report of Port and Harbor Research Institute*, Vol 9, No 4, 1970 (in Japanese).

Goda 1974
Goda, Y. 1974. "New Wave Pressure Formulae for Composite Breakwaters," *Proceedings of the 14th International Coastal Engineering Conference*, Vol 3, pp 1702-1720.

Goda 1985
Goda, Y. 1985. *Random Seas and Design of Maritime Structures,* University of Tokyo Press, Tokyo, Japan.

Goda and Suzuki 1976
Goda, Y., and Suzuki, Y. 1976. "Estimation of Incident and Reflected Waves in Random Wave Experiments," *Proceedings of the 15th International Coastal Engineering Conference*, American Society of Civil Engineers, Vol 1, pp 828-845.

Hales 1980
Hales, L. Z. 1980. "Erosion Control of Scour During Construction," Technical Report HL-80-3, U.S. Army Engineer Waterways Experiment Station, Coastal Engineering Research Center, Vicksburg, MS.

Hall 1967
Hall, J. R. 1967. "Coupled Rocking and Sliding Oscillations of Rigid Circular Footings," *Proceedings of the International Symposion on Wave Propagation and Dynamic Properties of Earth Materials*, American Society of Civil Engineers, pp 139-167.

Hansbo 1994
Hansbo, S. 1994. *Foundation Engineering*. Elsevier. The Netherlands.

Hansen 1979
Hansen, B. 1979. "Definition and Use of Friction Angles," *Seventh European Conference on Soil Mechanics and Foundation Engineering*, Brighton, Vol 1, pp 173-176.

Hanzawa et al. 1996
Hanzawa, M., Sato, H., Takahashi, S., Shimosako, K., Takayama, T., and Tanimoto, K. 1996. "New Stability Formula for Wave-Dissipating Concrete Blocks Covering Horizontally Composite Breakwaters," *Proceedings of the 25th International Coastal Engineering Conference*, American Society of Civil Engineers, Vol 2, pp 1665-1678.

Hardin 1978
Hardin, B. O. 1978. "The Nature of Stress-Strain Behaviour of Soils," *Earthquake Engineering and Soil Dynamics*, American Society of Civil Engineers, Vol I, pp 3-90.

Hardin and Black 1968
Hardin, B. O., and Black, W. L. 1968. "Vibration Modulus of Normally Consolidated Clays," *Journal of the Soil Mechanics and Foundations Division*, American Society of Civil Engineers, Vol 94, No. SM2, pp 353-369.

Hardin and Drnevich 1972
Hardin, B. O., and Drnevich, V. P. 1972. "Shear Modulus and Damping in Soils: Design Equations and Curves," *Journal of the Soil Mechanics and Foundations Division*, American Society of Civil Engineers, Vol 98, No. SM7, pp 667-692.

Hayashi and Shirai 1963
Hayashi, T., and Shirai, M. 1963. "Force of Impact at the Moving Collision of a Ship with the Mooring Construction," *Coastal Engineering in Japan*, Japan Society of Civil Engineering, Vol 6, pp 9-19.

Helm-Petersen 1998
Helm-Petersen, J. 1998. "Estimation of Wave Disturbance in Harbours," Ph.D. diss., Department of Civil Engineering, Aalborg University, Denmark.

Herbich 1991
Herbich, J. B. 1991. "Scour Around Pipelines, Piles, and Seawalls," Handbook of Coastal and Ocean Engineering - Volume 2: Offshore Structures, Marine Foundations, Sediment Processes, and Modeling. John B. Herbich, ed., Gulf Publishing Company, Houston, TX, pp 867-958.

Herbich and Ko 1968
Herbich, J. B., and Ko, S. C. 1968. "Scour of Sand Beaches in Front of Seawalls," *Proceedings of the 11th International Coastal Engineering Conference*, American Society of Civil Engineers, Vol 1, pp 622-643.

Hjorth 1975
Hjorth, P. 1997. "Studies on the Nature of Local Scour," Department of Water Resources Engineering Bulletin, Series A, No. 46, University of Lund, Sweden.

Hoffmans and Verheij 1997
Hoffmans, G.J.C.M., and Verheij, H. J. 1997. *Scour Manual*. A. A. Balkema, Rotterdam, The Netherlands.

Holtzhausen and Zwamborn 1990
Holtzhausen, A. H., and Zwamborn, J. A. 1990. "Stability of Dolosse with Different Waist Thicknesses for Irregular Waves," *Proceedings of the 22nd International Coastal Engineering Conference*, American Society of Civil Engineers, Vol 2, pp 1805-1818.

Holtzhausen and Zwamborn 1991
Holtzhausen, A. H., and Zwamborn, J. A. 1991. "Stability of Accropode7 and Comparison with Dolosse," *Coastal Engineering*, Vol 15, pp 59-86.

Hsu 1991
Hsu, J. R. 1991. "Short-Crested Waves," *Handbook of Coastal and Ocean Engineering - Volume 1: Wave Phenomena and Coastal Structures*. John B. Herbich, ed., Gulf Publishing Company, Houston, TX, pp 95-174.

Hudson 1958
Hudson, R. Y. 1958. "Design of Quarry-Stone Cover Layers for Rubble-Mound Breakwaters; Hydraulic Laboratory Investigation," Research Report No. 2-2, U.S. Army Engineer Waterways Experiment Station, Vicksburg, MS.

Hudson 1959
Hudson, R. Y. 1959. "Laboratory Investigation of Rubble-Mound Breakwaters," *Journal of the Waterways and Harbors Division*, American Society of Civil Engineers, Vol 85, No. WW3, pp 93-121.

Hudson 1974
Hudson, R. Y. (editor). 1974. "Concrete Armor Units for Protection Against Wave Attack," Miscellaneous Paper H-74-2, U.S. Army Engineer Waterways Experiment Station, Vicksburg, MS.

Hughes 1992
Hughes, S. A. 1992. "Estimating Wave-Induced Bottom Velocities at Vertical Wall," *Journal of Waterway, Port, Coastal and Ocean Engineering*, American Society of Civil Engineers, Vol 118, No. 2, pp 175-192.

Hughes and Fowler 1991
Hughes, S. A., and Fowler, J. E. 1991. "Wave-Induced Scour Prediction at Vertical Walls," *Proceedings of Coastal Sediments '91*, American Society of Civil Engineers, Vol. 2, pp 1886-1900.

Hughes and Fowler 1995
Hughes, S. A., and Fowler, J. E. 1995. "Estimating Wave-Induced Kinematics at Sloping Structures," *Journal of Waterway, Port, Coastal and Ocean Engineering*, American Society of Civil Engineers, Vol 121, No. 4, pp 209-215.

Hughes and Kamphuis 1996
Hughes, S. A., and Kamphuis, J. W. 1996. "Scour at Coastal Inlet Structures," *Proceedings of the 25th International Coastal Engineering Conference*, American Society of Civil Engineers, New York, Vol 2, pp 2258-2271.

Hughes and Schwichtenberg 1998
Hughes, S. A., and Schwichtenberg, B. R. 1998. "Current-Induced Scour Along a Breakwater at Ventura Harbor, California - Experimental Study," *Coastal Engineering*, Elsevier, Vol 34, No. 1-2, pp 1-22.

Ibsen 1999
Ibsen, L. B. 1999. "Cyclic Fatique Model," Geotechnical Engineering Papers, Aalborg University.

Ibsen and Lade 1998
Ibsen, L. B., and Lade, P.V. 1998. The Role of the Characteristic Line in Static Soil Behaviour," *Localization and Bifurcation Theory for Soils and Rocks*. Adachi and Yashima, Balkema, ed., Rotterdam.

Iribarren 1938
Iribarren, R. 1938. "Una Formula Para el Cálculo de los Digues de Escollera," Technical Report HE 116-295, Fluid Mechanics Laboratory, University of California, Berkeley, CA (Translated by D. Heinrich in 1948).

Iribarren and Nogales 1954
Iribarren, R., Nogales, C. 1954. "Other Verifications of the Formula for Calculating Breakwater Embankments," PIANC Bulletin No. 39, Permanent International Association of Navigation Congresses.

Irie and Nadaoka 1984
Irie, I., and Nadaoka, K. 1984. "Laboratory Reproduction of Seabed Scour in Front of Breakwaters," *Proceedings of the 19th International Coastal Engineering Conference*, American Society of Civil Engineers, Vol 2, pp 1715-1731.

Ito, Tanimoto, and Yamamoto 1972
Ito, Y., Tanimoto, K., and Yamamoto, S. 1972. "Wave Height Distribution in the Region of Ray Crossings - Application of Numerical Analysis Method of Wave Propagation," Report of Port and Harbour Research Institute, Vol 11, No.3, pp 87-109 (in Japanese).

Jackson 1968
Jackson, R. A. 1968. "Design of Cover Layers for Rubble-Mound Breakwaters Subjected to Nonbreaking Waves," Research Report No. 2-11, U.S. Army Engineer Waterways Experiment Station, Vicksburg, MS.

Jacobsen 1967
Jacobsen, M. 1967. "The Undrained Shear Strength of Preconsolidated Boulder Clay," *Proceedings of Geotechnical Conference*, Oslo, Norway, Vol I, pp 119-122.

Janbu 1954a
Janbu, N. 1954. "Application of Composite Slip Surfaces for Stability Analysis," *Proceedings of the Conference on Stability*, Vol III, p. 43, Stockholm, Sweden.

Janbu 1954b
Janbu, N. 1954. "Stability Analysis of Slopes with Dimensionless Parameters," Ph.D. diss., Harvard University, Cambridge, MA.

Janbu 1973
Janbu, N. 1973. "The Generalized Procedure of Slices," *Embankment Dam Engineering*, Casagrande Volume, pp 47-86.

Jensen, 1983
Jensen, O. J. 1983. "Breakwater Superstructures," *Proceedings of Coastal Structures '83*, American Society of Civil Engineers, New York, pp 272-285.

Jensen 1984
Jensen, O. J. 1984. "A Monograph on Rubble Mound Breakwaters," Danish Hydraulic Institute, Denmark.

Johnson 1992
Johnson, P. A. 1992. "Reliability-Based Pier Scour Engineering," *Journal of Hydraulic Engineering*, American Society of Civil Engineers, Vol 118, No. 10, pp 1344-1358.

Johnson 1995
Johnson, P. A. 1995. "Comparison of Pier-Scour Equations Using Field Data," *Journal of Hydraulic Engineering*, American Society of Civil Engineers, Vol 121, No. 8, pp 626-630.

Juhl and van der Meer 1992
Juhl, J. and van der Meer, J. W. 1992. "Quasi-Static Forces on Vertical Structures. Re-Analysis of Data at the Danish Hydraulic Institute and Delft Hydraulics," Report MAST G6-S, Coastal Structures, Project 2.

Kawata and Tsuchiya 1988
Kawata, Y., and Tsuchiya, Y. 1988. "Local Scour Around Cylindrical Piles Due to Waves and Currents Combined," *Proceedings of the 21st International Coastal Engineering Conference*, American Society of Civil Engineers, Vol 2, pp 1310-1322.

Kerr 1975
Kerr, A. D. 1975. "Ice Forces on Structures Due to a Change of the Water Level," *Proceedings IAHR 3rd International Symposium on Ice Problems*, Interational Association for Hydraulic Research, pp 419-427.

Kjærnsli, Valstad, and Høeg 1992
Kjærnsli, B., Valstad, T., and Høeg, K. 1992. "Rockfill Dams," Norwegian Institute of Technology, Division of Hydraulic Engineering, Trondheim, Norway.

Klomb and Tonda 1995
Klomb, W. H. G., and Tonda, P. L. 1995. "Pipeline Cover Stability," *Proceedings of the 5th International Conference on Offshore Mechanics and Arctic Engineering*, American Society of Mechanical Engineering, Vol 2, pp 15-22.

Kobayashi and Oda 1994
Kobayashi, T., and Oda, K. 1994. "Experimental Study on Developing Process of Local Scour Around a Vertical Cylinder," *Proceedings of the 24th International Coastal Engineering Conference*, American Society of Civil Engineers, Vol 2, pp 1284-1297.

Kortenhaus and Oumeraci 1998
Kortenhaus, A., and Oumeraci, H. 1998. "Classification of Wave Loading on Monolithic Coastal Structures," *Proceedings of the 26th International Coastal Engineering Conference*, Vol 1, pp 867-880.

Kraus 1988
Kraus, N. C. 1988. "The Effects of Seawalls on the Beach: An Extended Literature Review," *Journal of Coastal Research*, The Coastal Education and Research Foundation, Special Issue No. 4, N. C. Kraus and O. H. Pilkey, eds., pp 1-28.

Kraus and McDougal 1996
Kraus, N. C., and McDougal, W. G. 1996. "The Effects of Seawalls on the Beach: Part I, An Updated Literature Review," *Journal of Coastal Research*, The Coastal Education and Research Foundation, Vol 12, No. 3, pp 691-701.

Kriebel, Sollitt, and Gerken 1998
Kriebel, D., Sollitt, C, and Gerken, W. 1998. "Wave Forces on a Vertical Wave Barrier," *Proceedings of the 26th International Coastal Engineering Conference*, American Society of Civil Engineers, Vol 2, pp 2069-2081.

Kuzmanovic and Sanchez 1992
Kuzmanovic, B. O., and Sanchez, M. R. 1992. "Design of Bridge Pier Pile Foundations for Ship Impact," *Journal of Structural Engineering*, American Society of Civil Engineers, Vol 118, No. 8, pp 2151-2167.

Laird 1962
Laird, A. D. K. 1962. "Water Forces on Flexible Oscillating Cylinders," *Journal of the Waterways and Harbor Division*, American Society of Civil Engineers, Vol 88, No. WW3, pp 125-137.

Laird, Johnson, and Walker 1960
Laird, A. D. K., Johnson, C. A., and Walker, R. W. 1960. "Water Eddy Forces on Oscillating Cylinders," *Journal of the Hydraulics Division*, American Society of Civil Engineers, Vol 86, No. HY9, pp 43-54.

Lambe and Whitman 1979
Lambe, T. W., and Whitman, R. V. 1979. *Soil Mechanics* (SI version). John Wiley and Sons.

Lamberti 1995
Lamberti, A. 1995. "Preliminary Results on Main Armour - Toe Berm Interaction," Final Proceedings of MAST II Project, *Rubble Mound Breakwater Failure Modes*, Vol 2, Aalborg University, ed.

Lee 1972
Lee, T. T. 1972. "Design of Filter System for Rubble-Mound Structures," *Proceedings of the 13th International Conference on Coastal Engineering*, American Society of Civil Engineers, Vol 3, pp 1917-1933.

Lee and Lai 1986
Lee, J. J., and Lai, C. P. 1986. "Wave Uplift on Platforms or Docks," *Proceedings of the 20th International Coastal Engineering Conference*, American Society of Civil Engineers, Vol 2, pp 2023-2034.

Leidersdorf, Gadd, and McDougal 1988
Leidersdorf, C. B., Gadd, P. E., and McDougal, W. G. 1988. "Articulated Concrete Mat Slope Protection," *Proceedings of the 21st International Conference on Coastal Engineering*, American Society of Civil Engineers, Vol 3, pp 2400-2415.

Leidersdorf, Gadd, and McDougal 1990
Leidersdorf, C. B., Gadd, P. E., and McDougal, W. G. 1990. "Arctic Slope Protection Methods," *Proceedings of the 22nd International Conference on Coastal Engineering*, American Society of Civil Engineers, Vol 2, pp 1687-1701.

Leidersdorf, Gadd, and Vaudrey 1996
Leidersdorf, C. B., Gadd, P. E., and Vaudrey, K. D. 1996. "Design Considerations for Coastal Protection Projects in Cold Regions," *Proceedings of the 25th International Conference on Coastal Engineering*, American Society of Civil Engineers, Vol 4, pp 4397-4410.

Leps 1970
Leps, T. M. 1970. "Review of Shearing Strength of Rockfill," *Journal of the Soil Mechanics and Foundation Division*, American Society of Civil Engineers, Vol 96, No. SM4, pp 1159-1170.

Lillevang 1977
Lillevang, O. V. 1977. "Breakwater Subjected to Heavy Overtopping; Concept Design, Construction, and Experience," *Proceedings of Ports '77*, American Society of Civil Engineers, New York, pp 61-93.

Lillycrop and Hughes 1993
Lillycrop, W. J., and Hughes, S. A. 1993. "Scour Hole Problems Experienced by the Corps of Engineers; Data Presentation and Summary," Miscellaneous Paper CERC-93-2, U.S. Army Engineer Waterways Experiment Station, Coastal Engineering Research Center, Vicksburg, MS.

Lin et al. 1986
Lin, M-C., Wu, C-T., Lu, Y-C., and Liang, N-K. 1986. "Effects of Short-Crested Waves on the Scouring Around the Breakwater," *Proceedings of the 20th International Coastal Engineering Conference*, American Society of Civil Engineers, Vol 3, pp 2050-2064.

Losada, Silva, and Losada 1997
Losada, I. J., Silva, R, and Losada, M. A. 1997. "Effects of Reflective Vertical Structures Permeability on Random Wave Kinematics," *Journal of Waterway, Port, Coastal, and Ocean Engineering*, American Society of Civil Engineers, Vol 123, No. 6, pp 347-353.

Lundgren 1969
Lundgren, H. 1969. "Wave Shock Forces: An Analysis of Deformations and Forces in the Wave and in the Foundation," *Proceedings of the Symposium on Research on Wave Action*, Delft Hydraulics Laboratory, Emmeloord, The Netherlands, Vol II, Paper 4.

Lysmer and Richart 1966
Lysmer, J., and Richart, F. E. 1966. "Dynamic Response of Footings to Vertical Loading," *Journal of Soil Mechanics and Foundations Division*, American Society of Civil Engineers, Vol 92, No. SM 1, pp 65-91.

MacCamy and Fuchs 1954
MacCamy, R. C., and Fuchs, R. A. 1954. "Wave Forces on Piles: A Diffraction Theory," Technical Memorandum TM-69, U.S. Army Corps of Engineers, Beach Erosion Board, Washington, DC.

Machemehl 1990
Machemehl, J. L. 1990. "Wave and Ice Forces on Articicial Islands and Arctic Structures," *Handbook of Coastal and Ocean Engineering - Volume 1: Wave Phenomena and Coastal Structures*. John B. Herbich, ed., Gulf Publishing Company, Houston, TX, pp 1081-1124.

Madrigal and Valdés 1995
Madrigal, B. G., and Valdés, J. M. 1995. "Study of Rubble Mound Foundation Stability," Proceedings of the Final Workshop, MAST II, MCS-Project.

Magoon et al. 1974
Magoon, O. T., Sloan, R. L., and Foote, G. L. 1974. "Damages to Coastal Structures," *Proceedings of the 14th International Coastal Engineering Conference*, American Society of Civil Engineers, Vol 3, pp 1655-1676.

Markle, 1982
Markle, D. G. 1982. "Kahului Breakwater Stability Study, Kahului, Maui, Hawaii," Technical Report HL-82-14, U.S. Army Engineer Waterways Experiment Station, Vicksburg, MS.

Markle 1989
Markle, D. G. 1989. "Stability of Toe Berm Armor Stone and Toe Buttressing Stone on Rubble-Mound Breakwaters and Jetties; Physical Model Investigation," Technical Report REMR-CO-12, U.S. Army Engineer Waterways Experiment Station, Vicksburg, MS.

Maynord 1998
Maynord, S. T. 1998. "Corps of Engineers Riprap Design For Bank Stabilization," *Proceedings of the International Water Resources Engineering Conference*, American Society of Civil Engineers, pp 471-476.

McConnell 1998
McConnell, K. 1998. *Revetment Systems Against Wave Attack - A Design Manual,* Thomas Telford Publishing, London, UK.

Melby 1990
Melby, J. A. 1990. "An Overview of the Crescent City Dolos Design Procedure," *Stresses in Concrete Armor Units*, American Society of Civil Engineers, pp 312-326.

Melby 1993
Melby, J. A. 1993. "Dolos Design Procedure Based on Crescent City Prototype Data," Technical Report CERC-93-10, U.S. Army Engineer Waterways Experiment Station, Coastal Engineering Research Center, Vicksburg, MS.

Melby 1999
Melby, J. A. 1999. "Damage Progression on Rubble-Mound Breakwaters," Technical Report CHL-99-17, U.S. Army Engineer Waterways Experiment Station, Coastal and Hydraulics Laboratory, Vicksburg, MS.

Melby and Kobayashi 1998a
Melby, J. A., and Kobayashi, N. 1998a. "Progression and Variability of Damage on Rubble-Mound Breakwaters," *Journal of Waterway, Port, Coastal, and Ocean Engineering*, American Society of Civil Engineers, New York, Vol 124, No. 6, pp 286-294.

Melby and Kobayashi 1998b
Melby, J. A., and Kobayashi, N. 1998b. "Damage Progression on Breakwaters," *Proceedings of the 26th International Coastal Engineering Conference*, American Society of Civil Engineers, Vol 2, pp 1884-1897.

Melby and Mlaker 1997
Melby, J. A., and Mlaker, P. R. 1997. "Reliability Assessment of Breakwaters," Technical Report CHL-97-9, U.S. Army Engineer Waterways Experiment Station, Vicksburg, MS.

Melby and Turk 1992
Melby, J. A., and Turk, G. F. 1992. "Dolos Design Using Reliability Methods," *Proceedings of the 23rd International Coastal Engineering Conference*, American Society of Civil Engineers, Vol 2, pp 1385-1399.

Melby and Turk 1994
Melby, J. A., and Turk, G. F. 1994. "The CORE-LOC: Optimized Concrete Armor," *Proceedings of the 24th International Coastal Engineering Conference*, American Society of Civil Engineers, Vol 2, pp 1426-1438.

Mettam 1976
Mettam, J. D. 1976. "Design of Main Breakwater at Sines Harbor," *Proceedings of the 15th International Coastal Engineering Conference*, American Society of Civil Engineers, Vol 3, pp 2499-2518.

Meyerhof 1951
Meyerhof, G. G. 1951. "The Ultimate Bearing Capacity of Foundations," *Geotechnique*, Vol 2, pp 301-332.

Meyerhof 1953
Meyerhof, G. G. 1953. "The Bearing Capacity of Foundations Under Eccentric and Inclined Loads," *Proceedings of the 3rd International Conference on Soil Mechanics*, Zurich, Vol 1, pp 440-445.

Meyerhof 1963
Meyerhof, G. G. 1963. "Some Recent Research on the Bearing Capacity of Foundations," *Canadian Geotechnique* I., Vol 1, No. 1, pp 16-26.

Morgenstern and Price 1965
Morgenstern, N. R., and Price, V. E. 1965. "The Analysis of the Stability of Slip Surfaces," *Geotechnique*, Vol 15, No 1, pp 79-93.

Morison et al. 1950
Morison, J. R., O'Brien, M. P., Johnson, J. W., and Shaff, S. A. 1950. "The Force Exerted by Surface Waves on Piles," *Petroleum Transactions*, AIME, Vol 189, pp 149-154.

Newark 1942
Newark, N. M. 1942. "Influence Charts for Computation of Stresses in Elastic Foundations," Bulletin No. 45, University of Illinois, Urbana-Champaign, IL.

Norwegian Geotechnical Institute 1992
Norwegian Geotechnical Institute. 1992. Publication No. 185, Norwegian Geotechnical Institute, Oslo, Norway.

Niedoroda and Dalton 1986
Niederoda, A. W. and Dalton, C. 1986. "A Review of the Fluid Mechanics of Ocean Scour," *Ocean Engineering*, Vol 9, No. 2, pp.159-170.

O'Donoghue and Goldsworthy 1995
O'Donoghue, T., and Goldsworthy, C. J. 1995. "Random Wave Kinematics in Front of Sea Wall," *Proceedings of Breakwaters '95*, Institution of Civil Engineering, London, UK, pp 1-12.

Otani, Horii, and Ueda 1975
Otani, H., Horii, O., and Ueda, S. 1975. "Field Study on the Impact Energy of Large Tankers," *Coastal Engineering in Japan*, Japan Society of Civil Engineering, Vol 18, pp 185-194.

Oumeraci 1991
Oumeraci, H. 1991. "Dynamic Loading and Response of Caisson Breakwaters - Results of Large-Scale Model Tests," MAST G6-S, Project 2 Report, Franzius Institute, University of Hannover, Hannover, Germany.

Oumeraci 1994
Oumeraci, H. 1994. "Scour in Front of Vertical Breakwaters - Review of Problems," *Proceedings of International Workshop on Wave Barriers in Deepwaters*, Port and Harbor Research Institute, Japan, pp 281-307.

Owen 1980
Owen, M. W. 1980. "Design of Seawalls Allowing for Wave Overtopping," Report No. 924, Hydraulics Research Station, Wallingford, UK.

Owen 1982
Owen, M. W. 1982. "The Hydraulic Design of Seawall Profiles," *Proceedings of the Coastal Protection Conference*, Institution of Civil Engineers, Thomas Telford Publishing, London, UK, pp 185-192.

Pedersen 1996
Pedersen, J. 1996. "Experimental Study of Wave Forces and Wave Overtopping on Breakwater Crown Walls," Series paper 12, Hydraulics & Coastal Engineering Laboratory, Department of Civil Engineering, Aalborg University, Denmark.

Pedersen 1997
Pedersen, J. 1997. "Dynamic Response of Caisson Breakwaters Subjected to Impulsive Wave Loading - Design Diagrams for Dynamic Load Factors," *Proceedings*, Annex 2 C, 1st overall Workshop, EU-MASTIII project PROVERBS, Las Palmas, Gran Canaria, Spain.

Pedersen and Burcharth 1992
Pedersen, J., and Burcharth, H. F. 1992. "Wave Forces on Crown Walls," *Proceedings of the 23rd International Coastal Engineering Conference*, American Society of Civil Engineers, Vol 2, pp 1489-1502.

Perroud 1957
Perroud, P. H. 1957. "The Solitary Wave Reflection Along a Straight Vertical Wall at Oblique Incidence," Ph.D. diss., University of California, Berkeley, CA.

Permanent International Association of Navigation Congresses 1992
Permanent International Association of Navigation Congresses. 1992. "Guidelines for the Design and Construction of Flexible Revetments Incorporating Geotextiles in Marine Environment," Report of Working Group No. 21 of the Permanent Technical Committee II, Supplement to Bulletins No. 78/79.

Pilarczyk 1990
Pilarczyk, Krystian, W. 1990. "Design of Seawalls and Dikes - Including Overview of Revetments," *Coastal Protection*, K. Pilarczyk, ed., A.A. Balkema Publishers, Rotterdam, The Netherlands.

Postma 1989
Postma, G. M. 1989. "Wave Reflection from Rock Slopes Under Random Wave Attack," M.S. thesis, Delft University of Technology, The Netherlands.

Powell 1987
Powell, K. A. 1987. "Toe Scour at Sea Walls Subject to Wave Action," Report SR 119, Hydraulic Research Limited, Wallingford, UK.

Powell and Allsop 1985
Powell, K. A., and Allsop, N. W. 1985. "Low-Crest Breakwaters, Hydraulic Performance and Stability," Report No. SR 57, Hydraulics Research Station, Wallingford, England.

Price 1979
Price, A. W. 1979. "Static Stability of Rubble Mound Breakwaters," *Dock and Harbour Authority*, Vol LX, No. 702, pp 2-7.

Quinn 1972
Quinn, A. D. 1972. *Design and Construction of Ports and Marine Structures*. McGraw-Hill, New York.

Rance 1980
Rance, P. J. 1980. "The Potential for Scour Around Large Objects," *Scour Prevention Techniques Around Offshore Structures*, London Seminar, Society for Underwater Technology, pp 41-53.

Richardson and Davis 1995
Richardson, E. V., and Davis, S. R. 1995. "Evaluating Scour at Bridges," *Hydraulic Engineering Circular No. 18*, FHWA-IP-90-017, Federal Highway Administration, Washington, DC.

Richardson and Whitman 1964
Richardson, A. M., Jr., and Whitman, R. V. 1964. "Effect of Strain Rate Upon Undrained Shear Resistance of Saturated Remolded Fat Clay," *Geotechnique*, Vol 13, No. 4, pp 310-346.

Rouse 1950
Rouse, H., ed. 1950. *Engineering Hydraulics*, John Wiley and Sons, New York, NY.

Rowe and Barden 1964
Rowe, P.W., and Barden, L. 1964. "Importance of Free Ends in Triaxial Testing," *Journal of the Soil Mechanics and Foundation Division*, American Society of Civil Engineers, Vol 90, No. SM1, pp 1-17.

Sainflou 1928
Sainflou, M. 1928. "Treatise on Vertical Breakwaters," *Annals des Ponts et Chaussee*, Paris, France (Translated by W. J. Yardoff, U.S. Army Corps of Engineers.)

Sakaiyama and Kajima 1997
Sakaiyama, T., and Kajima, R. 1997. "Stability of Armor Units at Concave Section of Man-Made Island," *Proceedings of the Combined Australasian Coastal Engineering and Ports Conference - Pacific Coasts and Ports '97,* Center for Advanced Engineering, Vol 1, pp 213-218.

Sarpkaya 1976a
Sarpkaya, T. 1976a. "Vortex Shedding and Resistance in Harmonic Flow About Smooth and Rough Circular Cylinders at High Reynolds Numbers," Report No. NPS-59SL76021, Naval Postgraduate School, Monterey, CA.

Sarpkaya 1976b
Sarpkaya, T. 1976b. "In-Line and Transverse Forces on Cylinders in Oscillatory Flow at High Reynolds Number," *Proceedings of the Eighth Offshore Technology Conference*, Houston, TX, Paper No. OTC 2533, Vol 2, pp 95-108.

Sarpkaya and Isaacson 1981
Sarpkaya, T., and Isaacson, M. 1981. *Mechanics of Wave Forces on Offshore Structures,* Van Nostrand Reinhold.

Sato, Tanaka, and Irie 1968
Sato, S., Tanaka, N., and Irie, I. 1968. "Study on Scouring at the Foot of Coastal Structures," *Proceedings of the 11th International Coastal Engineering Conference*, American Society of Civil Engineers, Vol 1, pp 579-598.

Seed and Idriss 1970
Seed H. B., and Idriss, I. M. 1970. "Soil Moduli and Damping Factors for Dynamic Response Analysis," Report No. EERC 75-29, Earthquake Engineering Research Center, University of California, Berkeley, CA.

Seed et al. 1986
Seed H. B., Wong, R. T., Idriss, I. M., and Tokimatsu, K. 1986. "Moduli and Damping Factors for Dynamic Analyses of Cohesive Soils," *Journal of Geotechnical Engineering*, American Society of Civil Engineers, Vol 112, No. GT 11, pp 1016-1032.

Seelig 1980
Seelig, W. N. 1980. "Two-Dimensional Tests of Wave Transmission and Reflection Characteristics of Laboratory Breakwaters," Technical Report No. 80-1, U.S. Army Engineer Waterways Experiment Station, Vicksburg, MS.

Seelig 1983
Seelig, W. N. 1983. "Wave Reflection from Coastal Structures," *Proceedings of Coastal Structures '83*, American Society of Civil Engineers, pp 961-973.

Shore Protection Manual 1977
Shore Protection Manual. 1977. 3rd ed., U.S. Army Engineer Waterways Experiment Station, U.S. Government Printing Office, Washington, DC.

Shore Protection Manual 1984
Shore Protection Manual. 1984. 4th ed., U.S. Army Engineer Waterways Experiment Station, U.S. Government Printing Office, Washington, DC.

Silvester 1991
Silvester, R. 1991. "Scour Around Breakwaters and Submerged Structures," *Handbook of Coastal and Ocean Engineering - Volume 2: Offshore Structures, Marine Foundations, Sediment Processes, and Modeling*. John B. Herbich, ed., Gulf Publishing Company, Houston, TX, pp 959-996.

Skjelbriea et al. 1960
Skjelbriea, L., Hendrickson, J. A., Grogg, W., and Webb, L. M. 1960. "Loading on Cylindrical Pilings Due to the Action of Ocean Waves," Contract NBy-3196, 4 Volumes, U.S. Naval Civil Engineering Laboratory.

Smith 1999
Smith, E. R. 1999. "Toe Stability of Rubble-Mound Structures in a Breaking Wave and Ebb Flow Environment," Technical Report REMR-CO-20, U.S. Army Engineer Waterways Experiment Station, Coastal and Hydraulics Laboratory, Vicksburg, MS.

Smith, Seijffert, and van deer Meer 1994
Smith, G. M., Seijffert, J. W., and van der Meer, J. W. 1994. "Erosion and Overtopping of a Grass Dike: Large Scale Model Tests," *Proceedings of the 24th International Coastal Engineering Conference*, American Society of Civil Engineers, Vol 3, pp 2639-2652.

Smits, Anderson, and Gudehus 1978
Smits, F. P., Andersen, K. H., and Gudehus, G. 1978. "Pore Pressure Generation," *International Symposium on Soil Mechanics Research and Foundation Design for the Oosterschelde Storm Surge Barrier*, Vol 1, paper II-3.

Song and Schiller 1973
Song, W. O., and Schiller, R. E. 1973. "Experimental Studies of Beach Scour," COE Report No. 166, Texas A&M University, College Station, TX.

Steenfelt 1992
Steenfelt, J. S. 1992. "Strength and Dilatancy Revisited," Miscellaneous Papers in Civil Engineering, 35th Anniversary of the Danish Engineering Academy, Lyngby, Denmark, pp 157-188. (Also published by the Danish Geotechnical Institute, Lyngby, Denmark).

Steenfelt and Foged 1994
Steenfelt, J. S., and Foged, N. 1994. "Dilational Behaviour of Crushed Stone," *Proceedings of the 7th International IAEG Congress*, Lisbon, Portugal, Vol V, pp 3357-3364.

Steinbrenner 1936
Steinbrenner, W. 1936. "A Rational Method for Determination of the Vertical Normal Stresses Under Foundations," *Proceedings of the International Conference on Soil Mechanics and Foundation Engineering*, Cambridge, MA, Vol 2, pp 142-143.

Stückrath 1996
Stückrath, T. 1996. "Recommendation for the Construction of Breakwaters With Vertical and Inclined Concrete Walls," Report of Sub-Group C, Working Group 28, PIANC PTC II.

Sumer and Fredsøe 1990
Sumer, B. M., and Fredsøe, J. 1990. "Scour Below Pipelines in Waves," *Journal of Waterway, Port, Coastal, and Ocean Engineering*, American Society of Civil Engineers, Vol 116, No. 3, pp 307-323.

Sumer and Fredsøe 1991
Sumer, B. M., and Fredsøe, J. 1991. "Onset of Scour Below a Pipeline Exposed to Waves," *International Journal of Offshore and Polar Engineering*, Vol 1, No. 3, pp 189-194.

Sumer and Fredsøe 1992
Sumer, B. M., and Fredsøe, J. 1992. "A Review of Wave/Current-Induced Scour Around Pipelines," *Proceedings of the 23rd International Coastal Engineering Conference*, American Society of Civil Engineers, Vol 3, pp 2839-2852.

Sumer and Fredsøe 1996
Sumer, B. M., and Fredsøe, J. 1996. "Scour Around Pipelines in Combined Waves and Currents," *Proceedings of the 7th International Conference on Offshore Mechanics and Arctic Engineering*, American Society of Mechanical Engineering, Vol 5, pp 595-602.

Sumer and Fredsøe 1997
Sumer, B. M., and Fredsøe, J. 1997. "Scour at the Head of a Vertical-Wall Breakwater," *Coastal Engineering*, Elsevier, Vol 29, No. 3, pp 201-230.

Sumer and Fredsøe 1998a
Sumer, B. M., and Fredsøe, J. 1998a. "Wave Scour Around Structures," *Advances in Coastal Engineering*, Phillip Liu, ed., Vol 4, World Scientific, Singapore.

Sumer and Fredsøe 1998b
Sumer, B. M., and Fredsøe, J. 1998b. "Wave Scour Around Group of Vertical Piles," *Journal of Waterway, Port, Coastal, and Ocean Engineering Division*, American Society of Civil Engineers, Vol 124, No. 5, pp 248-256.

Sumer, Christiansen, and Fredsøe 1992a
Sumer, B. M., Christiansen, N., and Fredsøe, J. 1992a. "Scour Around Vertical Pile in Waves," *Journal of Waterway, Port, Coastal, and Ocean Engineering Division*, American Society of Civil Engineers, Vol 118, No. 1, pp 15-31.

Sumer, Christiansen, and Fredsøe 1992b
Sumer, B. M., Christiansen, N., and Fredsøe, J. 1992b. "Time Scale of Scour Around a Vertical Pile," *Proceedings of the 2nd International Offshore and Polar Engineering Conference*, International Society of Offshore and Polar Engineers, Vol 3, pp 308-315.

Sumer, Christiansen, and Fredsøe 1993
Sumer, B. M., Christiansen, N., and Fredsøe, J. 1993. "Influence of Cross Section on Wave Scour Around Piles," *Journal of Waterway, Port, Coastal, and Ocean Engineering Division*, American Society of Civil Engineers, Vol 119, No. 5, pp 477-495.

Sutherland and O'Donoghue 1997
Sutherland, J., and O'Donoghue, T. 1997. "CRF Study of Wave Kinematics in Front of Coastal Structures," *Proceedings of Coastal Dynamics '97*, American Society of Civil Engineers, pp 694-703.

Sutherland and O'Donoghue 1998a
Sutherland, J., and O'Donoghue, T. 1998a. "Wave Phase Shift at Coastal Structures," *Journal of Waterway, Port, Coastal and Ocean Engineering*, American Society of Civil Engineers, Vol 124, No. 2, pp 90-98.

Sutherland and O'Donoghue 1998b
Sutherland, J., and O'Donoghue, T. 1998b. "Characteristics of Wave Reflection Spectra," *Journal of Waterway, Port, Coastal and Ocean Engineering*, American Society of Civil Engineers, Vol 124, No. 6, pp 303-311.

Svee 1962
Svee, R. 1962. "Formulas for Design of Rubble Mound Breakwaters," *Journal of the Waterways and Harbours Division,* American Society of Civil Engineers, Vol 88, No. WW2, pp 11-21.

Tait and Mills 1980
Tait, R. B., and Mills, R. D. 1980. "An Investigation into the Material Limitations of Breakwater Dolosse," ECOR Newsletter, No. 12.

Takahashi 1996
Takahashi, S., 1996. "Design of Vertical Breakwaters," Reference Documents, No. 34, Port and Harbour Research Institute, Japan.

Takahashi and Hosoyamada 1994
Takahashi, S., and Hosoyamada, S. 1994. "Hydrodynamic Characteristics of Sloping Top Caissions," *Proceedings of International Conference on Hydro-Technical Engineering for Port and Harbour Construction*, Port and Harbour Research Institute, Japan, Vol 1, pp 733-746.

Takahashi, Tanimoto, and Shimosako 1990
Takahashi., S., Tanimoto, K., and Shimosako, K. 1990. "Wave and Block Forces on a Caisson Covered With Wave Dissipating Blocks," Report of Port and Harbour Research Institute, Yokosuka, Japan, Vol 30, No. 4, pp 3-34 (in Japanese).

Takahashi, Shimosako, and Sasaki 1991
Takahashi, S., Shimosako, K., and Sasaki, H. 1991. "Experimental Study on Wave Forces Acting on Perforated Wall Caisson Breakwaters," Report of Port and Harbour Research Institute, Yokosuka, Japan, Vol 30, No. 4, pp 3-34 (in Japanese).

Takahashi, Tanimoto, and Shimosako 1994a
Takahashi, S., Tanimoto, K., and Shimosako, K. 1994a. "A Proposal of Impulsive Pressure Coefficient for Design of Composite Breakwaters," *Proceedings of the International Conference on Hydro-Technical Engineering for Port and Harbor Construction,* Port and Harbour Research Institute, Yokosuka, Japan, pp 489-504.

Takahashi, Tanimoto, and Shimosako 1994b
Takahashi, S., Tanimoto, K., and Shimosako, K. 1994. "Wave Pressure on Perforated Wall Caissons," *Proceeding of International Conference on Hydro-Technical Engineering for Port and Harbor Construction*, Port and Harbour Research Institute, Yokosuka, Japan, pp 747-764.

Takayama 1992
Takayama, T. 1992. "Estimation of Sliding Failure Probability of Present Breakwaters for Probabilistic Design," Port and Harbour Research Institute, Vol 31, No.5, pp 80-96.

Tanimoto and Kimura 1985
Tanimoto, K., and Kimura, K. 1985. "A Hydraulic Experimental Study on Trapezoidal Caisson Breakwaters," Technical Note No. 528, Port and Harbour Research Institute, Yokosuka, Japan (in Japanese).

Tanimoto, et al. 1976
Tanimoto, K., Moto, K., Ishizuka, S., and Goda Y. 1976. "An Investigation on Design Wave Force Formulae of Composite-Type Breakwaters," *Proceedings of the 23rd Japanese Conference on Coastal Engineering*, pp 11-16 (in Japanese).

Tanimoto, Takahashi, and Kitatani 1981
Tanimoto, K., Takahashi, K., and Kitatani, T. 1981. "Experiment Study of Impact Breaking Wave Forces on a Vertical Wall Caisson of Composite Breakwater," Report of Port and Harbour Research Institute, Vol 20, No. 2, pp 3-39 (in Japanese).

Tanimoto, Yagyu, and Goda 1982
Tanimoto, T., Yagyu, T., and Goda, Y. 1982. "Irregular Wave Tests for Composite Breakwater Foundations," *Proceedings of the 18th International Coastal Engineering Conference*, American Society of Civil Engineers, Vol 3, pp 2144-2163.

Tanimoto, Haranaka, and Yamazaki 1985
Tanimoto, K, Haranaka, S., and Yamazaki, K. 1985. "Experimental Study of Wave Dissipating Concrete Blocks Against Irregular Waves," *Report of the Port and Harbour Research Institute*, Vol 24, No. 2, pp 85-121 (in Japanese).

Tanimoto, Takahashi, and Kimura 1987
Tanimoto, K., Takahashi, S., and Kimura, K. 1987. "Structures and Hydraulic Characteristics of Breakwaters - The State of the Art of Breakwater Design in Japan," *Report of the Port and Harbour Research Institute*, Japan, Vol. 26, No. 5, pp 11-15.

Taylor 1958
Taylor, D. W. 1958. *Fundamentals of Soil Mechanics*, John Wiley & Sons, New York, NY.

Tepfers and Kutti 1979
Tepfers, R., and Kutti, T. 1979. "Fatigue Strength of Plain Ordinary and Lightweight Concrete," *ACI Journal*, May 1979, pp 635-652.

Terzaghi and Peck 1944
Terzaghi, K., and Peck, R. B. 1944. *Soil Mechanics in Engineering Practice*, John Wiley & Sons, New York, NY.

Thomsen, Wohlt, and Harrison 1972
Thomsen, A. L., Wohlt, P. E., and Harrison, A. S. 1972. "Riprap Stability on Earth Embankments Tested in Large- and Small-Scale Wave Tanks," TM-37, U.S. Army Corps of Engineers, Coastal Engineering Research Center, Washington, DC.

Tomlinson 1980
Tomlinson, M. J. 1980. *Foundation Design and Construction.* Pitman Publishing Ltd., London.

Tsuruta and Goda 1968
Tsuruta, S., and Goda, Y. 1968. "Expected Discharge of Irregular Wave Overtopping," *Proceedings of the 11th International Coastal Engineering Conference*, American Society of Civil Engineers, Vol 2, pp 833-852.

Turk and Melby 1997
Turk, G. F., and Melby, J. A. 1997. "CORE-LOC[7] Concrete Armor Units: Technical Guidelines," Miscellaneous Paper CHL-97-6, U.S. Army Engineer Waterways Experiment Station, Coastal and Hydraulics Laboratory, Vicksburg, MS.

van der Meer 1988
van der Meer, J. W. 1988. "Rock Slopes and Gravel Beaches Under Wave Attack," Ph.D. diss., Delft University of Technology, The Netherlands. (Also Delft Hydraulics Publication No. 396)

van der Meer 1988b
van der Meer, J. W. 1988b. "Stability of Cubes, Tetrapodes and Accropode," *Proceedings of the Breakwaters '88 Conference; Design of Breakwaters*, Institution of Civil Engineers, Thomas Telford, London, UK, pp 71-80.

van der Meer 1990
van der Meer, J. W. 1990. "Low-Crested and Reef Breakwaters," Report H198/Q638, Delft Hydraulics Laboratory, The Netherlands.

van der Meer 1991
van der Meer, J. W. 1991. "Stability and Transmission at Low-Crested Structures," Delft Hydraulics Publication No. 453, Delft Hydraulics Laboratory, The Netherlands.

van der Meer 1997
van der Meer, J. W. 1997. Discussion of "Comparison and Evaluation of Different Riprap Stability Formulas Using Field Performance," Belfadhel et al., *Journal of Waterway, Port, Coastal and Ocean Engineering*, American Society of Civil Engineers, Vol. 123, No. 3, pp 147-148.

van der Meer and d'Angremond 1991
van der Meer, J. W., and d'Angremond, K. 1991. "Wave Transmission at Low Crested Structures," *Proceedings of the Coastal Structures and Breakwaters Conference*, Institution of Civil Engineers, Thomas Telford Publishing, London, UK, pp 25-41.

van der Meer and Janssen 1995
van der Meer, J. W., and Janssen, W. 1995. "Wave Run-Up and Wave Overtopping at Dikes," In *Wave Forces on Inclined and Vertical Wall Structures*, Kobayashi and Demirbilek, eds., American Society of Civil Engineers, pp 1-27.

van der Meer and Stam 1992
van der Meer, J. W., and Stam C. M. 1992. "Wave Run-Up on Smooth and Rock Slopes of Coastal Structures," *Journal of Waterway, Port, Coastal and Ocean Engineering*, American Society of Civil Engineers, Vol. 118, No. 5, pp 534-550.

van der Meer, d=Agremond, and Juhl 1994
van der Meer, J. W., d=Angremond, K., and Juhl, J. 1994. "Probabilistic Calculation of Wave Forces on Vertical Structures," *Proceedings of the 24th International Coastal Engineering Conference*, American Society of Civil Engineers, Vol 2, pp 1754-1769.

van der Meer, d=Angremond, and Gerding 1995
van der Meer, J. W., d'Angremond, K., and Gerding, E. 1995. "Toe Structure Stability of Rubble Mound Breakwaters," *Proceedings of the Advances in Coastal Structures and Breakwaters Conference*, Institution of Civil Engineers, Thomas Telford Publishing, London, UK, pp 308-321.

van Oorschot and d'Angremond 1968
van Oorschot, J. H., and d'Angremond, K. 1968. "The Effect of Wave Energy Spectra on Wave Run-Up," *Proceedings of the 11th International Coastal Engineering Conference*, American Society of Civil Engineers, Vol 2, pp 886-900.

Vidal et al. 1992
Vidal, C., Losada, M. A., Medina, R., Mansard, E. P., and Gomez-Pina, G. 1992. "A Universal Analysis for the Stability of Both Low-Crested and Submerged Breakwaters," *Proceedings of the 23rd International Coastal Engineering Conference*, American Society of Civil Engineers, Vol 2, pp 1679-1692.

Wang and Herbich 1983
Wang, R. K., and Herbich, J. B. 1983. "Combined Current and Wave Produced Scour Around a Single Pile," Report No. COE 269, Texas Engineering Experiment Station, Texas A&M University, College Station, TX.

Wiegel 1964
Wiegel, R. L. 1964. *Oceanographical Engineering*, Prentice-Hall, New Jersey.

Wuebben 1995
Wuebben, J. L. 1995. "Ice Effects on Riprap," *River, Coastal and Shoreline Protection: Erosion Control Using Riprap and Armourstone*. C. R. Thorne, S. R. Abt, F. B. Barends, S. T. Maynord, and R. W. Pilarczyk, eds., John Wiley & Sons, Ltd., Chichester, UK, pp 513-530.

Xie 1981
Xie, S.-L. 1981. "Scouring Patterns in Front of Vertical Breakwaters and their Influences on the Stability of the Foundations of the Breakwaters," Department of Civil Engineering, Delft University of Technology, Delft, The Netherlands.

Xie 1985
Xie, S.-L. 1985. "Scouring Patterns in Front of Vertical Breakwaters," *Acta Oceanologica Sinica*, Vol 4, No. 1, pp 153-164.

Zielinski, Reinhardt, and Körmeling 1981
Zielinski, A. J., Reinhardt, H. W., and Körmeling, H. A. 1981. "Experiments on Concrete Under Repeated Uniaxial Impact Tensile Loading," RILEM, Materials and Structures, Vol 14, No. 81.

Zwamborn and Phelp 1990
Zwamborn, J. A., and Phelp, D. 1990. "Structural Tests on Dolosse," *Proceedings of Stresses in Concrete Armor Units*, American Society of Civil Engineers, pp 40-60.

VI-5-10. <u>Acknowledgements</u>.

Authors: Dr. Hans F.Burcharth, Department of Civil Engineering, Aalborg University, Aalbort, Denmark; and Dr. Steven A. Hughes, Coastal and Hydraulics Laboratory (CHL), U.S. Army Engineer Research and Development Center, Vicksburg, MS.

Reviewers:
H. Lee Butler, CHL (retired)
Dr. David R. Basco, Department of Civil Engineering, Old Dominion University, Norfolk, VA
Students Han Ligteringen, Delft University of Technology, The Netherlands; Jorgen Fredsøe and Mutly Sumer, Institute of Hydrodynamics and Hydraulic Engineering, Technical University of Denmark, Lyngby, Denmark.

VI-5-11. <u>Symbols</u>.

α (alpha)

α	Angle a surface-piercing sloped plane forms with the horizontal [deg]
α	Angle of wave approach [deg]
α	Tangent of seaward armor slope

β (beta)

β	Angle of incidence of waves [deg]
β	Concave angle at vertical walls [radians]
β	Frequency parameter [dimensionless]

γ (gamma)

γ	Load factor [dimensionless]
γ_β	Factor for influence of angle of incidence β of the waves [dimensionless]
γ_b	Reduction factor for influence of a berm [dimensionless]
γ_h	Reduction factor for influence of shallow-water conditions where the wave height distribution deviates form the Rayleigh distribution [dimensionless]
γ_r	Reduction factor for influence of surface roughness [dimensionless]
γ_w	Specific weight of water or salt water [force/length^3]
$\_!$	Average effective weight of soil from base to depth B under base level [force]

δ (delta)

δ	Logarithmic decrement
δ_0	Vertical shift in the wave crest and wave trough at the wall [length]

Δ (delta)

Δ	$(= \rho_s / \rho_w) - 1$
Δ	Water level rise up to the thickness of the ice [length]
$\Delta\tau_s$	Change in the average shear stress due to the submerged weight of the structure [force/length^2]

ε (epsilon)

ε_i	Random wave phase angle of the ith incident wave component [deg]
$\dot{\varepsilon}_1, \dot{\varepsilon}_3$	Strain rates in principal stress directions 1 and 3
$\dot{\varepsilon}_{vol}$	Volume strain rate

η (eta)

η	Sea surface elevation adjacent to a reflective structure [length]
η^2_{rms}	Root-mean-squared sea surface elevation [length^2]

θ (theta)

θ	Angle of wave incidence [deg]
θ	Bottom slope [deg]

θ	Channel side wall slope [degrees]
θ	Wave phase angle ($= 2\pi x/L - 2\pi t/T$) [radians]
θ_i	Reflection phase angle of the ith incident wave component [deg]
κ *(kappa)*	
κ	von Karman constant ($= 0.4$) [dimensionless]
λ *(lamda)*	
$\lambda_{1,2,3}$	Modification factors depending on the structure type [dimensionless]
μ *(mu)*	
μ	Dynamic friction coefficient corresponding to caisson displacement S [dimensionless]
μ	Friction coefficient for the base plate against the rubble stones [dimensionless]
μ	Structure slope friction factor [dimensionless]
v *(nu)*	
v	Kinematic viscosity [length^2/time]
v	Poisson= ratio [dimensionless]
v_*	Shear velocity
ξ *(xi)*	
ξ	Principal stress reduction factor
ξ_0	Surf similarity parameter for regular waves (Equation VI-5-1)
ξ_{0m}	Surf similarity parameter for irregular waves (Equation VI-5-2)
ξ_{0p}	Surf similarity parameter for irregular waves (Equation VI-5-2)
ξ_{eq}	Breaking wave surf similarity parameter
ρ *(rho)*	
ρ	Bulk density [force/length^3]
ρ_a	Mass density of armor units [force/length^3]
ρ_c	Mass density of the structure [force/length^3]
ρ_i	Ice density [force/length^3]
ρ_s	Mass density of armor units [force/length^3]
ρ_w	Mass density of water (salt water = 1,025 kg/m^3 or 2.0 slugs/ft^3; fresh water = 1,000 kg/m^3 or 1.94 slugs/ft^3) [force-time^2/length^4]
σ *(sigma)*	
σ	Normal stress on a section through a soil element [force/length^2]
σ	Spreading of short-crested waves
σ_l	Principal stress [force/length^2]
σ_c	Ice compressive failure strength in crushing [force/length^2]
σ_c	Standard deviation of the average non-dimensional cover armor depth
σ_e	Standard deviation of the average non-dimensional eroded armor depth

σ_f	Flexural strength of ice
$\sigma_f\prime$	Effective stress at failure [force/length^2]
σ_i	Angular wave frequency of the ith incident wave component [time^{-1}]
$\sigma_n\prime$	Normal stress on failure plane [force/length^2]
σ_S	Standard deviation of average damage
τ (tau)	
τ	Shear stress on a section through a soil element [force/length^2]
τ_0	Shear stress acting on the bed [force/length^2]
Ψ (psi)	
Ψ	Angle of dilation [degrees]
Ψ	Shields parameter
Ω (omega)	
Ω	Dynamic load factor [dimensionless]
ν	
ν	Angle of internal friction of the soil [degrees]
ν	Angle of repose of the armor [degrees]
ν	Strength factor [dimensionless]
$\nu\prime$	Angle of friction in granular material [degrees]
$\nu\prime_{crit}$	Critical angle of friction [degrees]
$\nu\prime_s$	Effective secant angle of friction
$\nu\prime_t$	Effective tangent angle of friction
A	
a	Pile radius [length]
a_i	Amplitude of the ith incident wave component [length]
A	Area of structure slope [length^2]
A	Horizontal area of ice sheet [length^2]
A_c	Berm crest height [length]
A_e	Area of eroded armor layer [length^2]
A_n	Projected area of solid body normal to the flow direction [length^2]
A_s	Total area of steel intersecting the crack [length^2]
A_t	Area of initial cross section of structure [length^2]
A_z	Projected area of solid body in the horizontal plane [length^2]
B	
b	Pile width [length]
b	Structure horizontal width or diameter [length]

B	Berm width [length]
B	Diameter of the vertical breakwater circular head [length]
B	Function of Reynolds number (= 8.5 for fully rough, turbulent flow)
B	Horizontal width of the barrier [length]
B	Relative breakage
B	Width of footing [length]
B	Width of structure crest [length]
C	
c	Infiltration factor [dimensionless]
c	Shear strength of soil [force/length^2]
c_u	Undrained shear strength [force/length^2]
$c!$	Cohesion intercept
C	Damage parameter for structure armor layer [dimensionless]
C	Dolos fluke length [length]
C_0	Zero-damage cover layer thickness [length]
C_D	Drag hydrodynamic force coefficient [force/length]
C_L	Empirical lift coefficient
C_M	Inertia or mass hydrodynamic force coefficient [force/length]
C_r	Bulk reflection coefficient [dimensionless]
C_{ri}	Reflection coefficient of the ith incident wave component [dimensionless]
C_s	Stability coefficient for incipient motion [dimensionless]
C_{sf}	Coefficient of skin friction between wind and ice or water and ice [dimensionless]
C_t	Wave transmission coefficient [dimensionless]
C_{to}	Wave transmission coefficient for overtopping [dimensionless]
C_{tp}	Wave transmission coefficient for wave penetration [dimension]
C_u	Uncertainty factor [dimensionless]
C_U	Laboratory derived slamming coefficient [dimensionless]
C_V	Coefficient of consolidation
D	
d	Grain diameter [length]
d_B	Berm horizontal surface above the still-water line [length]
d_c	Depth of armor cover [length]
d_c	Elevation of the lower edge of the sloping face relative to the SWL [length]
d_e	depth of sheet-pile penetration below the seabed [length]
d_e	Depth of eroded armor layer [length]

d_i	Depth at the toe of the sloping structure [length]
d_s	Water depth at the structure toe [length]
D	Cylindrical pile diameter [length]
D	Damping ratio [dimensionless]
D	Minimum depth of footing below soil surface [length]
D	Pipe diameter [length]
D	Sphere diameter [length]
D_e	Equivalent pile diameter [length]
D_H	Distance between centroids of two adjacent units on the same horizontal row [length]
D_n	Cube length [length]
D_{n50}	Median of nominal diameter of rocks for design conditions [length]
D_r	Relative density of soils [percent]
D_{swl}	Vertical distance from SWL to location of stressed dolos [length]
D_U	Distance between the centroids of units upslope in the plane of the structure slope [length]
E	
e	In-place void ratio [dimensionless]
e_{cr}	Critical embedment [length]
e_{max}	Void ratio of soil in loosest condition [dimensionless]
e_{min}	Void ratio of soil in most dense condition [dimensionless]
E	Damage parameter for structure armor layer [dimensionless]
E	Modulus of elasticity of ice
E	Young=s modulus
E_{av}	Averaged Young=s modulus for ice [force/length^2]
E_d	Dissipated wave energy in one wavelength per unit crest width [length-force/length^2]
E_i	Incident wave energy in one wavelength per unit crest width [length-force/length^2]
E_r	Reflected wave energy in one wavelength per unit crest width [length-force/length^2]
E_t	Transmitted wave energy in one wavelength per unit crest width [length-force/length^2]
F	
f_c	Concrete compressive strength [force/length^2]
f_c	Height of wall not protected by the armor layer [length]
f_{ct}	Concrete splitting tensile strength [force/length^2]
f_D	Drag force per unit length of pile [force/length]

f_i	Inertial force per unit length of pile [force/length]
f_i	Reduction factor [dimensionless]
f_T	Concrete static tensile strength [force/length]
f_y	Yield strength of the steel
F	Safety factor [dimensionless]
F_0	Significant force per unit width for a vertical wall [force/length]
F_B	Buoyancy force [force]
F_c	Horizontal crushing force [force]
F_D	Drag force [force]
F_G	Gravitational force [force]
F_G	Reduced weight of the vertical structure due to buoyancy [force]
F_H	Wave induced horizontal force [force]
F_I	Inertia force [force]
F_L	Lift force [force]
F_L	Time-varying transverse (lift) force
F_{Lm}	Maximum transverse force [force]
F_{m0}	Significant force per unit width of barrier [force/length]
F_r	Flow Froude number [dimensionless]
F_U	Wave induced uplift force [force]
F_v	Total vertical force acting on the wall [force]
F_W	Irregular wave loading [force]
G	
g	Gravitational acceleration [length/time^2]
G	Berm width [length]
G	Factor dependent on the armor layer gradation [dimensionless]
G	Shear modulus
H	
h	Ice plate thickness [length]
h	Pre-scour water depth at the vertical wall [length]
h	Water depth [length]
h_b	Water depth at a distance of $5H_s$ seaward of the breakwater front wall [length]
h_b	Water depth at top of toe berm [length]
h_c	Equilibrium height of the structure [length]
h_i	Thickness of ice sheet [length]
h_s	Water depth in front of structure [length]
$h!$	Height of wall protected by the armor layer [length]

$h!$	Submerged height of the wall from the toe to the still water line [length]
$h!_c$	Initial height of structure over seabed level [length]
Δh	Change in water level [length]
H	Characteristic wave height [length]
H	Drainage distance [length]
H_0	Deepwater wave height [length]
H_b	Breaking wave height [length]
H_c	Wave height in the corner [length]
H_i	Incident wave height [length]
H_I	Incident wave height [length]
H_{m0}	Zeroth-moment wave height [length]
H_S	Significant wave height [length]
H_{sr}	Significant reflected wave height [length]
H_w	Wave or surge height at the wall [length]
I	
i	Hydraulic gradient [length/length]
I	Indentation coefficient [dimensionless]
I_p	Plasticity index
K	
k	Contact coefficient [dimensionless]
k	Permeability coefficient
k	Wave number (= $2\pi/L = 2\pi/CT$) [length^{-1}]
k_i	Wave number of the ith incident wave component [length^{-1}]
k_M	Moment contribution factor
k_p	Wave number associated with the spectral peak by linear wave theory [length^{-1}]
k_p	Wave number associated with the spectral peak period T_p [length^{-1}]
k_s	Boundary or bed roughness
k_T	Torque contribution factor [dimensionless]
k_x	Stiffness coefficient
k_v	Stiffness coefficient
k_Δ	Layer coefficient (Table VI-5-51)
K	Bulk modulus
K	Coefficient of lateral stress [dimensionless]
K	Factor to account for blankets plaved on sloping channel side walls [dimensionless]
K_1	Pile shape factor [dimensionless]
K_2	Pile orientation factor [dimensionless]

KC	Keulegan-Carpenter number [dimensionless]
K_D	Stability coefficient
K_o	Coefficient of lateral stress at rest [dimensionless]
L	
l_e	Upslope eroded length [length]
L	Damage parameter for structure armor layer [dimensionless]
L	Length of footing [length]
L	Local wave length [length]
L_0	Deepwater wave length (= $gT^2/2\pi$) [length]
L_{0m}	Deepwater wave length corresponding to mean wave period [length]
L_{0p}	Deepwater wave length corresponding to the peak of the wave spectrum [length]
L_p	Local wavelength associated with the peak spectral period T_p [length]
M	
m	Plan shape coefficient [dimensionless]
m	Total mass [force]
m_0	Area beneath the measured force spectrum [length^2]
M	Armor unit mass [force]
M	Constrained modulus
M_{50}	Medium mass of rocks; mass of Core-Loc armor unit (= ρ_s $(D_{n50})^3$) [force]
M_{cr}	Critical strength of concrete in moment
M_d	Overturning moment per unit horizontal length about the toe of the wall due to the dynamic pressure [length-force/length]
M_{FG}	Stabilizing moment around the heel by buoyancy-reduced weight of the caisson [length-force]
M_{FH}	Antistabilizing moment by wave induced horizontal force [length-force]
M_{FU}	Antistabilizing moment by wave induced uplift force [length-force]
M_{max}	Maximum wave-load-induced moment around the center of gravity [length-force]
M_s	Hydrostatic overturning moment per unit width [length-force/length]
M_s	Stabilizing moment due to friction and cohesion [length-force]
N	
n	Model scale factor [dimensionless]
n	Porosity
$\underline{n_z}$	Normal unit velocity in the positive z-direction [length/time]
N_{0w}	Number of overtopping waves
N_a	Total number of armor layer units or number of rocks in the mound
N_f	Number of cycles to failure

N_{od}	Number of units displaced out of the armor layer
N_{or}	Number of rocking units
N_S	Stability parameter [dimensionless]
N_w	Number of incoming waves
N_z	Number of waves
N^*_S	Spectral stability number [dimensionless]
P	
p	Porosity of the armor layer [dimensionless]
p_a	Atmospheric pressure [force/length^2]
p_s	Pressure on solid body surface due to moving fluid [force/length^2]
$p_{1,2,3}$	Wave pressure at the SWL corresponding to wave crest, at the base, at the SWL, corresponding to wave trough [force/length^2]
$p!$	Mean effective stress [force/length^2]
$p!_o$	Vertical effective overburden pressure [force/length^2]
P	Notational permeability parameter (Figure VI-5-11)
P	Uplift force in metric tons
P_{ow}	Probability of overtopping per incoming wave
P_s	Hydrostatic pressure [force/length^2]
Q	
q	Average overtopping discharge per unit length of structure [length^3/time/length]
$q!$	Effective overburden pressure [force/length^2]
Q	Dimensionless average discharge per meter (Equations VI-5-20 and VI-5-21)
Q_n	Nominal load [force]
R	
r	Dolos waist ratio [dimensionless]
r	Thickness of armor cover or under layer [length]
R_a	Maximum vertical runup height [length]
R_c	Crest freeboard (Figure VI-5-13) [length]
R_d	Force per unit horizontal length of wall [force/length]
R_d	Minimum rundown or water-surface elevation measured vertically from the still-water level [length]
R_e	Reynolds number [dimensionless]
R_h	Distance to the center of the section [length]
R_n	Nominal strength
R_s	Hydrostatic force per unit horizontal width of wall [force/length^2/length]
R_u	Maximum runup or water-surface elevation measured vertically from the still-water level [length]

$R_{ui\%}$	Runup level exceeded by i percent of the incident waves (Equation VI-5-3) [length]
R_{us}	Significant runup level [length]
S	
s_0	Deepwater wave steepness (=H_0/L_0) [dimensionless]
s_{0m}	Deepwater mean wave steepness (=H_s/L_{0m})
s_{0p}	Deepwater wave steepness corresponding to the peak of the wave spectrum [dimensionless]
s_m	Wave steepness (= H_s / L_{0m})
s_p	Local wave steepness [dimensionless]
s_t	Settlement (decrease in layer thickness) at time t [length]
s_4	Final settlement reached when the soil skeleton is fully carrying the load [length]
S	Caisson displacement [length]
S	Horizontal seismic inertia force
S	Relative eroded area or damage parameter for structure armor layer [dimensionless]
S_f	Safety factor at allow for debris impacts or other unknowns [dimensionless]
S_m	Maximum scour depth [length]
S_M	Section moduli for flexure
S_t	Cohesive soil sensitivity (ratio between the undrained shear strength of a specimen in undisturbed and in remoulded states)
S_T	Section moduli for torsion
T	
t	Time at end of storm n
t_n	Time at start of storm n
T	Wave period [time]
T_{0m}	Wave period associated with the spectral peak in deep water [time]
T_{0p}	Wave period associated with the spectral peak in deep water [time]
T_{cr}	Critical strength of concrete in torque
T_m	Average or mean wave period [time]
T_p	Wave period corresponding to the peak of the wave spectrum [time]
T_s	Strength contribution from the torsional steel reinforcement
U	
u	Horizontal component of the wave orbital velocity [length/time]
u	Magnitude of flow velocity [length/time]
u	Pore pressure [force/length^2]
u_i	Velocity of ice [length/time]

u_p	Pore water pressure [force/length^2]	
u_s	Water pressure along the surface of the slope [force/length^2]	
u^2_{rms}	Root-mean-squared horizontal wave velocity [length^2]	
U	Current magnitude [length/time]	
U	Degree of consolidation [dimensionless]	
U_c	Critical depth-averaged flow velocity [length/time]	
U_m	Maximum wave orbital velocity at the bed [length/time]	
V		
v	Bulk flow velocity [length/time]	
V	Overtopping volume per wave per unit width [length^3/length]	
V	Total volume [length^3]	
V_p	Volume of voids [length^3]	
V_s	Volume of solids [length^3]	
W		
w	Barrier penetration depth [length]	
w	Vertical component of flow velocity at level of object [length/time]	
w_a	Specific weight of armor material [force/length^3]	
W	Stone or armor weight [force]	
W	Total weight of the slice including surface load [force]	
W	Width of scour apron [length]	
W_T	Total weight of riprap [force]	
X		
x	Horizontal coordinate with positive toward the structure and $x = 0$ located at the structure toe [length]	
x	Strain rate function	
Y		
y	Elevation above the bed [length]	
Z		
z	Vertical coordinate with $z = 0$ at the SWL and $z = -h$ at bottom [length]	
Z	Maximum vertical ice ride-up distance [length]	

CHAPTER 6

Reliability Based Design of Coastal Structures

TABLE OF CONTENTS

List of Figures

List of Tables

CHAPTER VI-6

Reliability Based Design of Coastal Structures

VI-6-1. Introduction.

a. Conventional design practice for coastal structures is deterministic in nature and is based on the concept of a design load which should not exceed the resistance (carrying capacity) of the structure. The design load is usually defined on a probabilistic basis as a characteristic value of the load, for example the expectation (mean) value of the 100-year return period event. However, this selection is often made without consideration of the involved uncertainties. In most cases the resistance is defined in terms of the load that causes a certain design impact or damage to the structure, and it is not given as an ultimate force or deformation. This is because most of the available design formulae only give the relationship between wave characteristics and some structural response, such as runup, overtopping, armor layer damage, etc. An example is the Hudson formula for armor layer stability.

b. Almost all coastal structure design formulae are semiempirical and based mainly on central fitting to model test results. The often considerable scatter in test results is not considered in general because the formulae normally express only the mean values. Consequently, the applied characteristic value of the resistance is then the mean value and not a lower fractile as is usually the case in other civil engineering fields. The only contribution to a safety margin in the design is inherent in the choice of the return period for the design load. (The exception is when the design curve is fitted to the conservative side of the data envelope to give a built-in safety margin.) It is now more common to choose the return period with due consideration of the encounter probability, i.e., the probability that the design load value is exceeded during the structure lifetime. This is an important step towards a consistent probabilistic approach.

c. In addition to design load probability, a safety factor (as given in some national standards) might be applied as well, in which case the method is classified as a Level I (deterministic/quasi-probabilistic) method. However, this approach does not allow determination of the reliability (or the failure probability) of the design; and consequently, it is not possible to optimize structure design or avoid overdesign of a structure. In order to overcome this problem, more advanced probabilistic methods must be applied where the uncertainties (the stochastic properties) of the involved loading and strength variables are considered.

d. Methods where the actual distribution functions for the variables are taken into account are denoted as Level III methods. Level II methods generally transform correlated and non-normally distributed variables into uncorrelated and standard normal distributed variables, and reliability indices are used as measures of the structural reliability. Both Level II and III methods are discussed in the following sections. Also described is an advanced partial coefficient system which takes into account the stochastic properties of the variables and makes it possible to design a structure for a specific failure probability level.

VI-6-2. Failure Modes and Failure Functions.

a. Evaluation of structural safety is always related to the structural response as defined by the failure modes. Failure modes for various structures are presented in Part VI-2-4, "Failure Modes of Typical Structure Types."

b. Each failure mode must be described by a formula, and the interaction (correlation) between the failure modes must be known. As an illustrative example consider only one failure mode, "hydraulic stability of the main armor layer" described by the Hudson formula

$$D_n^3 = \frac{H_s^3}{K_D \Delta^3 \cot \alpha} \qquad \text{(VI-6-1)}$$

where

D_n = nominal block diameter

$\Delta = \rho_s / \rho_w - 1$

ρ_s = block density

ρ_w = water density

α = armor slope angle

H_s = significant wave height

K_D = coefficient signifying the degree of damage (movements of the blocks)

c. The formula can be split into load variables X_i^{load} and resistance variables, X_i^{res}. Whether a parameter is a load or a resistance parameter can be seen from the failure function. If a larger value of a parameter results in a safer structure, it is a resistance parameter; and if a larger value results in a less safe structure, it is a load parameter.

d. According to this definition one specific parameter can in one formula act as a load parameter while in another formula the same parameter can act as a resistance parameter. An example is the wave steepness parameter in the van der Meer formulas for rock, which is a load parameter in the case of surging waves, but a resistance parameter in the case of plunging waves. The only load variable in Equation VI-6-1 is H_s while the others are resistance variables.

e. Equation VI-6-1 is formulated as a failure function (performance function)

$$g = A \cdot \Delta \cdot D_n \left(K_D \cot \alpha \right)^{1/3} - H_s \begin{cases} <0 \ \textit{failure} \\ =0 \ \textit{limit state (failure)} \\ >0 \ \textit{no failure (safe region)} \end{cases} \qquad \text{(VI-6-2)}$$

f. All the involved parameters are regarded as stochastic variables, X_i, except K_D, which signifies the failure, i.e., a specific damage level chosen by the designer. The factor A in Equation VI-6-2 is also a stochastic variable signifying the uncertainty of the formula. In this case the mean value of A is 1.0.

g. In general Equation VI-6-2 is formulated as

$$g = R - S \qquad\qquad\qquad\qquad\qquad\qquad\qquad \text{(VI-6-3)}$$

where R stands for resistance and S for loading. Usually R and S are functions of many random variables, i.e.,

$$R = R(X_1^{res}, X_2^{res}, \ldots, X_m^{res}) \quad and \quad S = S(X_{m+1}^{load}, \ldots, X_n^{load}) \quad or \quad g = g(\overline{X})$$

The limit state is given by

$$g = 0 \qquad\qquad\qquad\qquad\qquad\qquad\qquad\qquad \text{(VI-6-4)}$$

which is denoted the *limit state equation* and defines the so-called *failure surface* which separates the safe region from the failure region.

h. In principle, R is a variable representing the variations in resistance between nominally identical structures, whereas S represents the maximum load effects within a period of time, for instance T successive years. The distributions of R and S are both assumed independent of time. The *probability of failure*, P_f, during any reference period of duration T years is then given by

$$P_f = \text{Prob}(g \le 0) \qquad\qquad\qquad\qquad\qquad \text{(VI-6-5)}$$

i. The reliability R_f is defined as

$$R_f = 1 - P_f \qquad\qquad\qquad\qquad\qquad\qquad \text{(VI-6-6)}$$

VI-6-3. Single Failure Modes Probability Analysis.

a. Level III methods.

(1) A simple method (in principle) of estimating P_f is the Monte Carlo method where a very large number of realizations x of the variables X are simulated. P_f is then approximated by the proportion of the simulations where $g \# 0$. The reliability of the Monte Carlo method depends on a realistic assessment of the distribution functions for the variables X and their correlations.

(2) Given $f_{\overline{x}}$ as the joint probability density function (*jpdf*) of the vector $\overline{X} = (X_1, X_2, \ldots, X_n)$, then Equation VI-6-5 can be expressed by

$$P_f = \int_{R \le S} f_{\bar{X}}(\bar{x}) d\bar{x} \qquad (\text{VI-6-7})$$

(3) Note that the symbol x is used for values of the random variable X. If only two variables R and S are considered then Equation VI-6-7 reduces to

$$P_f = \int_{R \le S} f_{(R,S)}(r,s)\, dr\, ds \qquad (\text{VI-6-8})$$

which is conceptually illustrated in Figure VI-6-1. If more than two variables are involved it is not possible to describe the *jpdf* as a surface but requires an imaginary multidimensional description.

(4) Figure VI-6-1 also shows the so-called design point which is the point of failure surface where the joint probability density function attains the maximum value, i.e., the most probable point of failure.

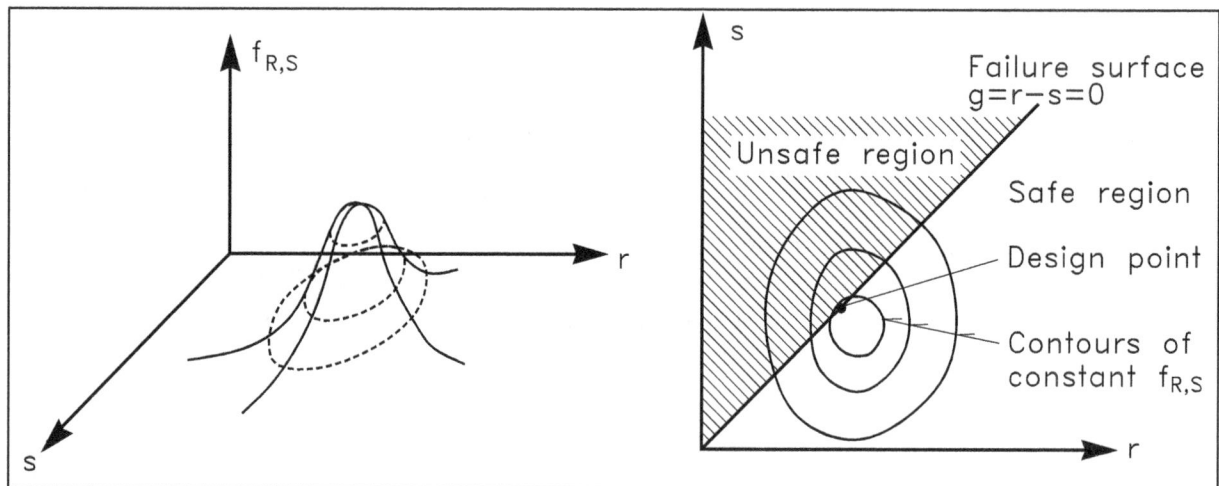

Figure VI-6-1. Illustration of the two-dimensional joint probability density function for loading and strength

(5) Unfortunately, the *jpdf* is seldom known. However, the variables can often be assumed independent (noncorrelated) in which case Equation VI-6-7 is given by the n-fold integral

$$P_f = \iiint_{R \le S} \cdots \int f_{X_1}(x_1) \ldots f_{X_n}(x_n)\, d\,x_1 \ldots d\,x_n \qquad (\text{VI-6-9})$$

where f_{Xi} are the marginal probability density function of the variables X_i. The amount of calculations involved in the multidimensional integration Equation VI-6-9 is enormous if the number of variables, n, is larger than 5.

(6) If only two independent variables are considered, e.g., R and S, then Equation VI-6-9 simplifies to

$$P_f = \iint_{R \leq S} f_R(r) f_S(s) \, dr \, ds \qquad \text{(VI-6-10)}$$

which by partial integration can be reduced to a single integral

$$P_f = \int_0^\infty F_R(x) f_S(x) \, dx \qquad \text{(VI-6-11)}$$

where F_R is the cumulative distribution function for R. Formally the lower integration limit should be -4, but it is replaced by 0 since, in general, negative strength is not meaningful.

(7) Equation VI-6-11 represents the product of the probabilities of two independent events, namely the probability that S lies in the range x, $x+dx$ (i.e., $f_S(x) \, dx$) and the probability that $R \# x$ (i.e., $F_R(x)$), as shown in Figure VI-6-2.

b. Level II methods. This section gives a short introduction to reliability calculations at Level II. Only the so-called first-order reliability method (FORM), where the failure surface is approximated by a tangent hyberplane at some point, is presented. A more accurate method is the second-order reliability method (SORM), which uses a quadratic approximation to the failure surface.

(1) Linear failure functions of normally-distributed random variables.

(a) Assume the loading $S(x)$ and the resistance $R(x)$ for a single failure mode to be statistically independent and with density functions as illustrated in Figure VI-6-2. The failure function is given by Equation VI-6-3 and the probability of failure is expressed by Equation VI-6-10 or Equation VI-6-11.

(b) However, in many cases these functions are not known, but under certain assumptions the functions might be estimated using only the mean values and standard deviations. If S and R are assumed to be independent normally distributed variables with known means and standard deviations, then the linear failure function $g = R - S$ is normally distributed with mean value,

$$\mu_g = \mu_R - \mu_S \qquad \text{(VI-6-12)}$$

and standard deviation

$$\sigma_g = \sqrt{(\sigma_R^2 + \sigma_S^2)} \qquad \text{(VI-6-13)}$$

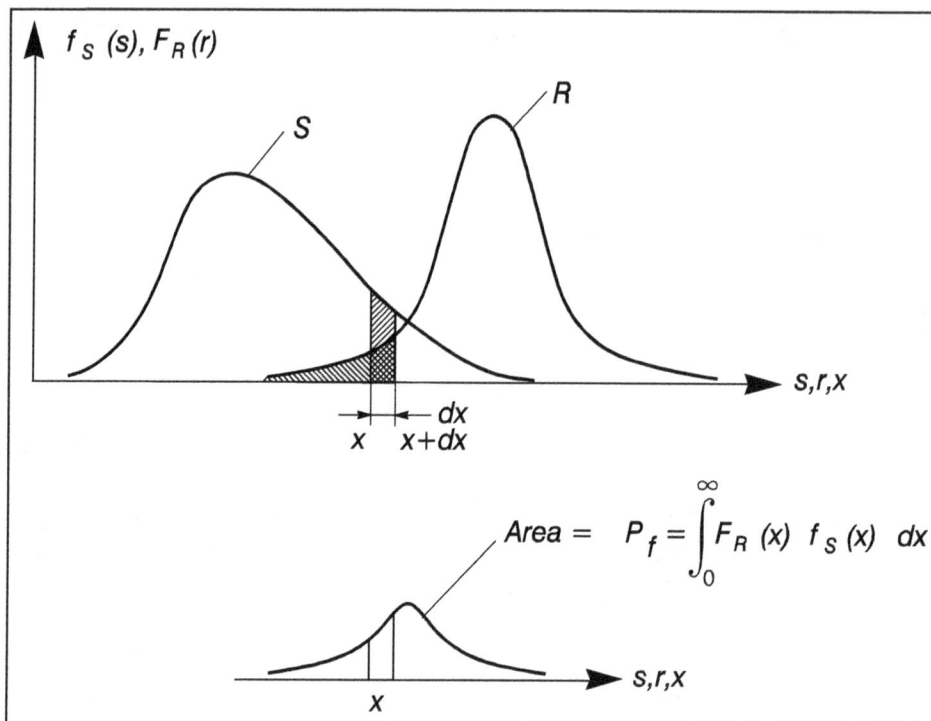

Figure VI-6-2. Illustration of failure probability in case of two independent variables, S and R

The quantity $(g - \mu_g) / \sigma_g$ will be unit standard normal, and consequently,

$$P_f = prob\,(\,g \le 0\,) = \int_{-\infty}^{0} f_g(x)\,dx = \Phi\left(\frac{0 - \mu_g}{\sigma_g}\right) = \Phi\,(-\beta) \qquad\text{(VI-6-14)}$$

where

$$\beta = \frac{\mu_g}{\sigma_g} \qquad\text{(VI-6-15)}$$

is a measure of the probability of failure referred to as the reliability index (Cornell 1969). Figure VI-6-3 illustrates β and the reliability index. Note that β is the inverse of the coefficient of variation, and it is the distance (in terms of number of standard deviations) from the most probable value of g (in this case the mean) to the failure surface, $g = 0$.

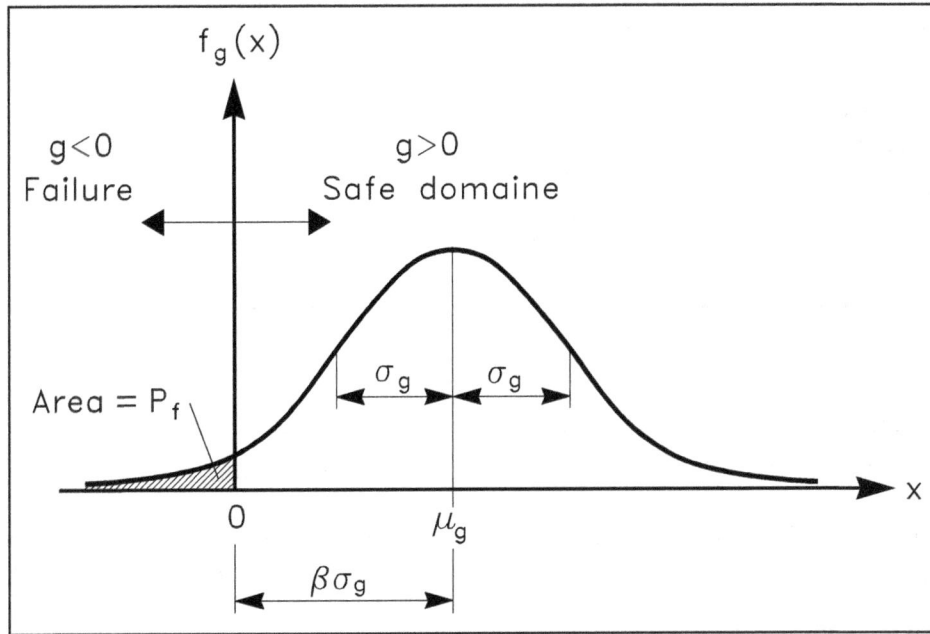

Figure VI-6-3. Illustration of reliability index

(c) If R and S are normally distributed and "correlated," then Equation VI-6-14 still holds, but σ is given by

$$\sigma_g = \sqrt{(\sigma_R^2 + \sigma_S^2 + 2\,\rho_{RS}\,\sigma_R\,\sigma_S\,)} \qquad \text{(VI-6-16)}$$

where ρ_{RS} is the *correlation coefficient*

$$\rho_{RS} = \frac{Cov\,(R,S)}{\sigma_R\,\sigma_S} = \frac{E\,[(R-\mu_R)\,(S-\mu_S)]}{\sigma_R\,\sigma_S} \qquad \text{(VI-6-17)}$$

R and S are said to be uncorrelated if $\rho_{RS} = 0$.

(d) In addition to the illustration of β in Figure VI-6-3, a simple geometrical interpretation of β can be given in the case of a linear failure function $g = R - S$ of the independent variables R and S by a transformation into a normalized coordinate system of the random variables $RN = (R - \mu_R)\,/\sigma_R$ and $SN = (S - \mu_S)\,/\sigma_S$, as shown in Figure VI-6-4.

(e) With these variables the failure surface $g = 0$ is linear and given by

$$R'\,\sigma_R - S'\,\sigma_S + \mu_R - \mu_S = 0 \qquad \text{(VI-6-18)}$$

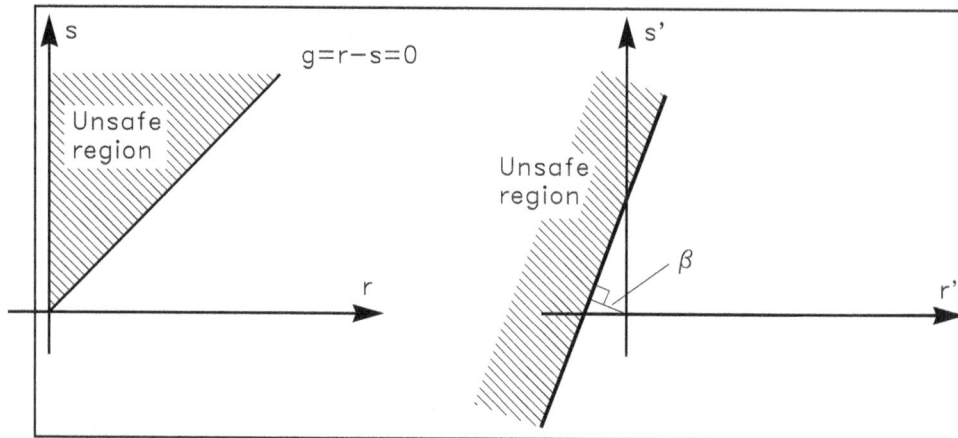

Figure VI-6-4. Illustration of β in normalized coordinate system

(f) By geometrical considerations it can be shown that the shortest distance from the origin to this linear failure surface is equal to in which Equations VI-6-12 and VI-6-13 are used.

$$\beta = \frac{\mu_g}{\sigma_g} = \frac{\mu_R - \mu_S}{\sqrt{\sigma_R^2 + \sigma_S^2}} \qquad \text{(VI-6-19)}$$

(2) Nonlinear failure functions of normally-distributed random variables.

(a) If the failure function $g = g\,(\overline{X})$ is *nonlinear*, then approximate values for μ_g and σ_g can be obtained by using a linearized failure function. Linearization is generally performed by retaining only the linear terms of a Taylor-series expansion about some point. However, the values of μ_g and σ_g, and thus the value of β, depend on the choice of linearization point. Moreover, the value of β defined by Equation VI-6-15 will change when different, but functionally equivalent, nonlinear failure functions are used.

(b) To overcome these problems, a transformation of the basic variables $\overline{X} = (X_1, X_2, \ldots, X_n)$ into a new set of normalized variables $\overline{Z} = (Z_1, Z_2, \ldots, Z_n)$ is performed. For uncorrelated normally distributed basic variables $\overline{X}$ the transformation is

$$Z_i = \frac{X_i - \mu_{X_i}}{\sigma_{X_i}} \qquad \text{(VI-6-20)}$$

in which case $\mu_{Zi} = 0$ and $\sigma_{Zi} = 1$. By this linear transformation the failure surface g = 0 in the x-coordinate system is mapped into a failure surface in the z-coordinate system which also divides the space into a safe region and a failure region as illustrated in Figure VI-6-5.

(c) Figure VI-6-5 introduces the Hasofer and Lind reliability index β_{HL} which is defined as the distance from the origin to the nearest point, D, of the failure surface in the z-coordinate system (Hasofer and Lind 1974). This point is called the design point. The coordinates of the design point in the original x-coordinate system are the most probable values of the variables $\overline{X}$ at failure. β_{HL} can be formulated as

$$\beta_{HL} = \min_{g(\overline{z})=0} \left(\sum_{i=1}^{n} z_i^2 \right)^{1/2} \tag{VI-6-21}$$

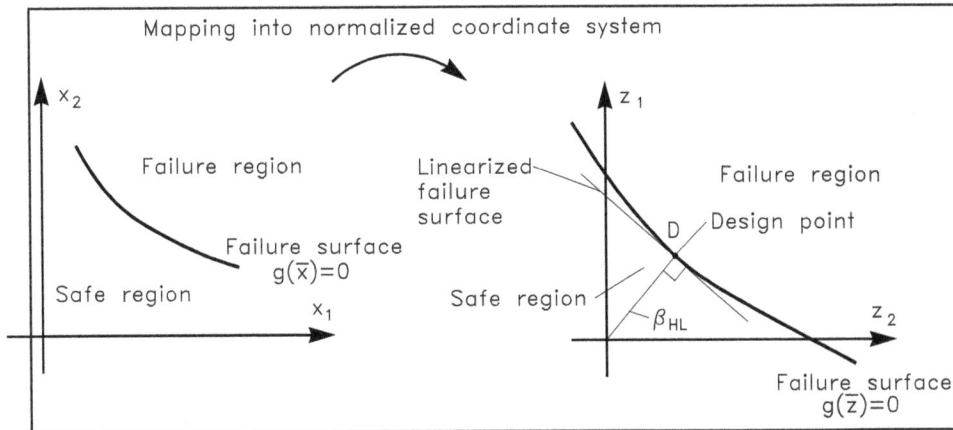

Figure VI-6-5. Definition of the Hasofer and Lind reliability index, β_{HL}

(d) The special feature of β_{HL}, as opposed to β, is that β_{HL} is related to the failure "surface" $g(\overline{z})=0$ which is invariant to the failure function because equivalent failure functions result in the same failure surface.

(e) The calculation of β_{HL} and the design point coordinates can be undertaken in a number of different ways. An iterative method must be used when the failure surface is nonlinear. A widely used method of calculating β_{HL} is

- Step 1. Select some trial coordinates of the design point in the z-coordinate system

$$\overline{z}^d = (z_1^d, z_2^d, \ldots, z_n^d)$$

- Step 2. Calculate α_i $i = 1, 2, \ldots, n$ by

$$\alpha_i = \left. \frac{\partial g}{\partial z_i} \right|_{\overline{z} = \overline{z}^d}$$

- Step 3. Determine a better estimate of $\bar{z}^d$ by

$$z_i^d = \alpha_i \frac{\sum_{i=1}^{n}\left(\alpha_i z_i^d\right) - g\big|_{\bar{z}=\bar{z}^d}}{\sum_{i=1}^{n}\left(\alpha_i\right)^2}$$

- Step 4. Repeat Steps 2 and 3 to achieve convergence

- Step 5. Evaluate β_{HL} by

$$\beta_{HL} = \left(\sum_{i=1}^{n}\left(z_i^d\right)^2\right)^{1/2}$$

The method is based on the assumption of the existence of only one minimum. However, several "local" minima might exist. In order to avoid convergence toward a local minima (and thereby overestimation of β_{HL} and the reliability) several different sets of trial coordinates might be tried.

(3) Nonlinear failure functions of non-normal random variables.

(a) It is not always a reasonable assumption to consider the random variables normally distributed. For example, parameters characterizing the sea state in long-term wave statistics, such as H_s, will in general follow extreme distributions (e.g., Gumbel and Weibull). These distributions are quite different from the normal distribution and cannot be described using only the mean value and the standard deviation.

(b) For such cases it is still possible to use the reliability index β_{HL}, but an extra transformation of the non-normal basic variables into normal basic variables must be performed before β_{HL} can be determined as previously described.

(c) A commonly used transformation is based on the substitution of the non-normal distribution of the basic variable X_i by a normal distribution in such a way that the density and distribution functions f_{Xi} and F_{Xi} are unchanged at the design point.

(d) If the design point is given by $x_1^d, x_2^d, \ldots, x_n^d$, then the transformation reads

$$F_{X_i}(x_i^d) = \Phi\left(\frac{x_i^d - \mu_{X_i}}{\sigma_{X_i}}\right)$$

$$f_{X_i}(x_i^d) = \frac{1}{\sigma_{X_i}} \varphi\left(\frac{x_i^d - \mu_{X_i}}{\sigma_{X_i}}\right)$$

(VI-6-22)

where μN_{Xi} and σN_{Xi} are the mean and standard deviation of the approximate (fitted) normal distribution.

(e) Equation VI-6-22 yields

$$\sigma_{X_i} = \frac{\varphi\left(\Phi^{-1}\left(F_{X_i}(x_i^d)\right)\right)}{f_{X_i}(x_i^d)}$$ (VI-6-23)

$$\mu_{X_i} = x_i^d - \Phi^{-1}\left(F_{X_i}(x_i^d)\right)\sigma_{X_i}$$

(f) Equation VI-6-22 can also be written

$$F_{X_i}(x_i^d) = \Phi\left(\frac{x_i^d - \mu_{X_i}}{\sigma_{X_i}}\right) = \Phi(z_i^d) = \Phi(\beta_{HL}\,\alpha_i)$$

(g) Solving with respect to x_i^d gives

$$x_i^d = F_{X_i}^{-1}[\Phi(\beta_{HL}\,\alpha_i)]$$ (VI-6-24)

(h) The iterative method presented above for calculation of β_{HL} can still be used if for each step of iteration the values of μN_{Xi} and σN_{Xi} given by Equation VI-6-24 are calculated for those variables where the transformation (Equation VI-6-22) has been used. For correlated random variables the transformation into noncorrelated variables is used before normalization.

(4) Time-variant random variables. The failure functions within breakwater engineering are generally of the form

$$g = f_1(\overline{R}) - f_2(H_s, W, T_m)$$ (VI-6-25)

where $\overline{R}$ represents the resistance variables and H_s, W, and T_m are the load variables signifying the wave height, the water level, and the wave period. The random variables are in general time-variant.

(a) Discussion of load variables:

• The most important load parameter in breakwater engineering is the wave height. It is a time-varying quantity which is best modeled as a stochastic process. Distinction is made between short-term and long-term statistics of the wave heights. Short-term statistics deal with the distribution of the wave height H during a stationary sequence of a storm, i.e., during a period of constant H_s (or any other characteristic wave height). The short-term wave height distribution follows the Rayleigh distribution for deepwater waves and some truncated distribution in the case of shallow-water waves.

• Long-term statistics deal with the distribution of the storms which are then characterized by the maximum value of H_s occurring in each storm. The storm history is given as the sample $(H_{s1}, H_{s2},..., H_{sn})$ covering a period of observation, Y. Extreme-value distributions like the Gumbel and Weibull distributions are then fitted to the data sample. For strongly depth-limited wave conditions a normal distribution with mean value as a function of water depth might be appropriate.

- The true distribution of H_s can be approximated by the distribution of the maximum value over T years, which is denoted as the distribution of H_s^T. The calculated failure probability then refers to the period T (which in practice might be the lifetime of the structure) if distribution functions of the other variables in Equation VI-6-25 are assumed to be unchanged during the period T.

- As an example, consider a sample of n independent storms, i.e., H_{s1}, H_{s2}, ... , H_{sn}, obtained within Y years of observation. Assume that H_s follows a Gumbel distribution given by

$$F(H_s) = \exp\left[-e^{-\alpha(H_s - \beta)}\right]$$

(VI-6-26)

which is the distribution of H_s over a period of Y years with average time span between observations of Y/n.

- The distribution parameters α and β can be estimated from the data using techniques such as the maximum likelihood method or the methods of moments. Moreover, the standard deviations of α and β, signifying the statistical uncertainty due to limited sample size, can also be estimated.

- The sampling intensity is $\lambda = n/Y$. Within a T-year reference period the number of data will be λT. The probability of the maximum value of H_s within the period T is then

$$F(H_s^T) = [F(H_s)]^{\lambda T} = \left[\exp\left(-e^{-\alpha(H_s - \beta)}\right)\right]^{\lambda T}$$

(VI-6-27)

- The expectation (mean) value of H_s^T is given by

$$\mu_{H_s^T} = \beta - \frac{1}{\alpha}\ln\left[-\ln\left(1 - \frac{1}{\lambda T}\right)\right]$$

(VI-6-28)

and the standard deviation of H_s^T (from maximum likelihood estimates) is

$$\sigma_{H_s^T} = \left[\frac{1}{n\alpha^2}\left\{1.109 + 0.514\left(-\ln\left[-\ln\left(1 - \frac{1}{\lambda T}\right)\right]\right) + \right.\right.$$
$$\left.\left. + 0.608\left(-\ln\left[-\ln\left(1 - \frac{1}{\lambda T}\right)\right]\right)^2\right\}\right]^{1/2}$$

(VI-6-29)

- Equation VI-6-29 includes the statistical uncertainty due to limited sample size. Some uncertainty is related to the estimation of the sample values H_{s1}, H_{s2}, ... , H_{sn} arising from measurement errors, errors in hindcast models, etc. This uncertainty corresponds to a coefficient of variation σ_{Hs}/μ_{Hs} on the order of 5 - 20 percent. The effect of this might be implemented in the calculations by considering a total standard deviation of

$$\sigma = \sqrt{\sigma_{H_s^T}^2 + \sigma_{H_s}^2} \qquad\qquad\text{(VI-6-30)}$$

- In Level II calculations, Equation VI-6-27 is normalized around the design point, and Equations VI-6-28 and VI-6-29 or VI-6-30 are used for the mean and the standard deviation.

- Instead of substituting H_s in Equation VI-6-25 with H_s^T, the following procedure might be used: Set T in Equations VI-6-27 to VI-6-29 to be 1 year. The outcome of the calculations will then be the probability of failure in a 1-year period, P_f *(1 year)*. If the failure events of each year are assumed independent for all variables then the failure probability in T years is

$$P_f \,(T\ years) = 1-[1-P_f \,(1\ year)]^T \qquad\qquad\text{(VI-6-31)}$$

- This assumption simplifies the probability estimation somewhat, and for some structures it is reasonable to assume failure events are independent, e.g., rubble-mound stone armor stability. However, for some resistance variables, such as concrete strength, it is unrealistic to assume the events of each year are independent. The calculated values of the failure probability in T-years using $H_s^{1\ year}$ and H_s^T will be different. The difference will be very small if the variability of H_s is much larger than the variability of other variables.

- The water level W is also an important parameter because it influences the structure freeboard and limits wave heights in shallow-water situations. Consequently, for the general case it is necessary to consider the joint distribution of H_s, W, and T_m. However, for deepwater waves W is often almost independent (except for barometric effects) of H_s and T_m and can be approximated as a noncorrelated variable that might be represented by a normal distribution with a certain standard deviation. The distribution of W is assumed independent of the length of the reference period T. In shallow water, W will be correlated with H_s due to storm surge effects.

- The *wave period* T_m is correlated to H_s. As a minimum the mean value and the standard deviation of T_m and the correlation of T_m with H_s should be known in order to perform a Level II analysis. However, the linear correlation coefficient is not very meaningful because it gives an insufficient description when the parameters are non-normally distributed. Alternatively the following approach might be used: From a scatter diagram of H_s and T_m a relationship of the form $T_m = A\,f(H_s)$ is established in which the parameter A follows a normal distribution (or some other distribution) with mean value $\mu_A = 1$ and a standard deviation σ_A which signifies the scatter. T_m can then be replaced by the variable A in Equation VI-6-25. The variable A is assumed independent of all other parameters.

- Generally, the best procedure for coping with the correlations between H_s, W, and T_m is to work on the conditional distributions. Assume the distribution of the maximum value of H_s within the period T is given as $F_1(H_s^T)$. Furthermore, assume the conditional distributions $F_2(W|H_s^T)$ and $F_3(T_m|H_s^T)$ are known. Let Z_1, Z_2 and Z_3 be independent standard normal variables and

$$\Phi(z_1) = F_1(H_s^T)$$
$$\Phi(z_2) = F_2(W|H_s^T)$$
$$\Phi(z_3) = F_3(T_m|H_s^T)$$

- The inverse relationships are given by

$$H_s^T = F_1^{-1}[\Phi(z_1)]$$
$$W = F_2^{-1}[\Phi(z_2)|H_s^T]$$
$$T_m = F_3^{-1}[\Phi(z_3)|H_s^T]$$

- By converting the resistance variables $\overline{R}$ into standard normal variable $\overline{Z}_o$, i.e., the resistance term is written $f_1(\overline{R}) = f_3(\overline{Z}_o)$, then the failure function Equation VI-6-25 becomes

$$g = f_3(\overline{z}_o) - f_2\left(F_1^{-1}[\Phi(z_1)]\ ,\ F_2^{-1}[\Phi(z_2)|H_s^T]\ ,\ F_3^{-1}[\Phi(z_3)|H_s^T]\right) = 0$$

- Because g now comprises only independent standard normal variables, the usual iteration methods for calculating β_{HL} can be applied.

(b) Discussion of resistance parameters:

- The service life of coastal structures spans anywhere between 20 to 100 years. Over periods of this length a decrease in the structural resistance is to be expected because of various types of material deterioration. Chemical reaction, thermal effect, and repeated loads (fatigue load) can cause deterioration of concrete and natural stone leading to disintegration and rounding of elements. Also the resistance against displacements of armor layers made of randomly placed armor units will decrease with the number of waves (i.e., with time) due to the stochastic nature of the resistance. Consequently, for armor layers this means a reduction over time of the D_n and K_D parameters in the Hudson equation.

- Although material effects can greatly influence reliability in some cases, they are not easy to include in reliability calculations. The main difficulty is the assessment of the variation with time which depends greatly on the intrinsic characteristics of the placed rock and concrete. At this time only fairly primitive methods are available for assessment of the relevant material characteristics. In addition, the variation with time depends very much on the load-history which can be difficult to estimate for the relevant period of structural life.

- Figure VI-6-6 illustrates an example situation representing the tensile strength of concrete armor units where a resistance parameter $R(t)$ decreases with time t. $R(t)$ is assumed to be a deterministic function. The load $S(t)$ (the tensile stress caused by wave action) is assumed to be a stationary process. The probability of failure, $P(S > R)$, within a period T is

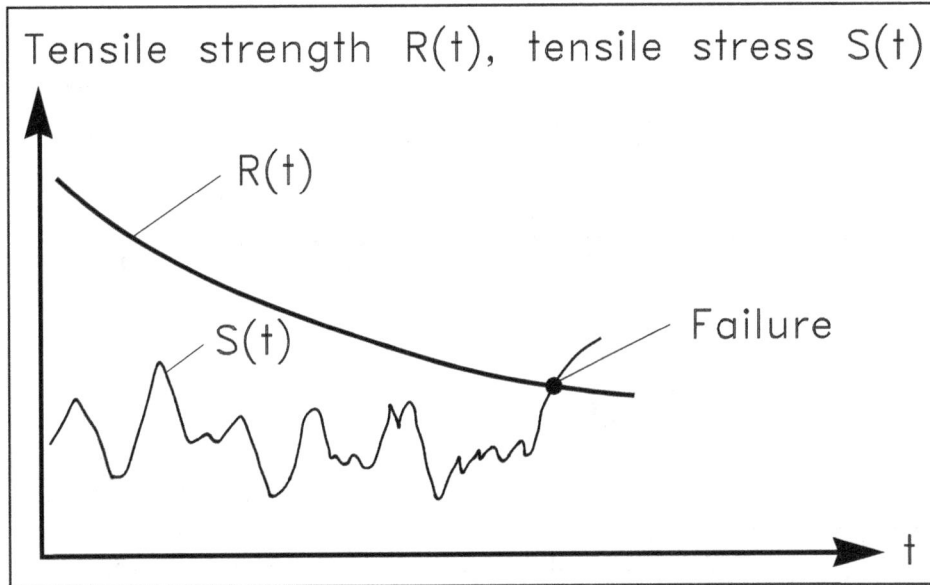

Figure VI-6-6. Illustration of a first-passage problem

$$P_f(T) = 1 - \exp\left[-\int_0^T v^+[R(t)]\,dt\right]$$
(VI-6-32)

where $v^+[R(t)]$ is the mean-upcrossing rate (number of upcrossings per unit time) of the level $R(t)$ by the process $S(t)$ at time t. v^+ can be computed by Rice's formula

$$v^+[R(t)] = \int_{\dot{R}}^{\infty}(\dot{S}-\dot{R})\,f_{S\dot{S}}[R(t),\dot{S}]\,d\dot{S}$$

in which $f_{S\dot{S}}$ is the joint density function for $S(t)$ and $\dot{S}(t)$. Implementation of time-variant variables into Level II analyses is rather complicated. For further explanation, see Wen and Chen (1987).

VI-6-4. <u>Failure Probability Analysis of Failure Mode Systems</u>.

a. A coastal structure can be regarded as a system of components which can either function or fail. Due to interactions between the components, failure of one component may impose failure of another component and even lead to failure of the system. A so-called fault tree is often used to clarify the relationships between the failure modes.

b. A fault tree describes the relationships between the failure of the system (e.g., excessive wave transmission over a breakwater protecting a harbor) and the events leading to this failure. Figure VI-6-7 shows a simplified example based on some of the failure modes of a rubble-mound breakwater.

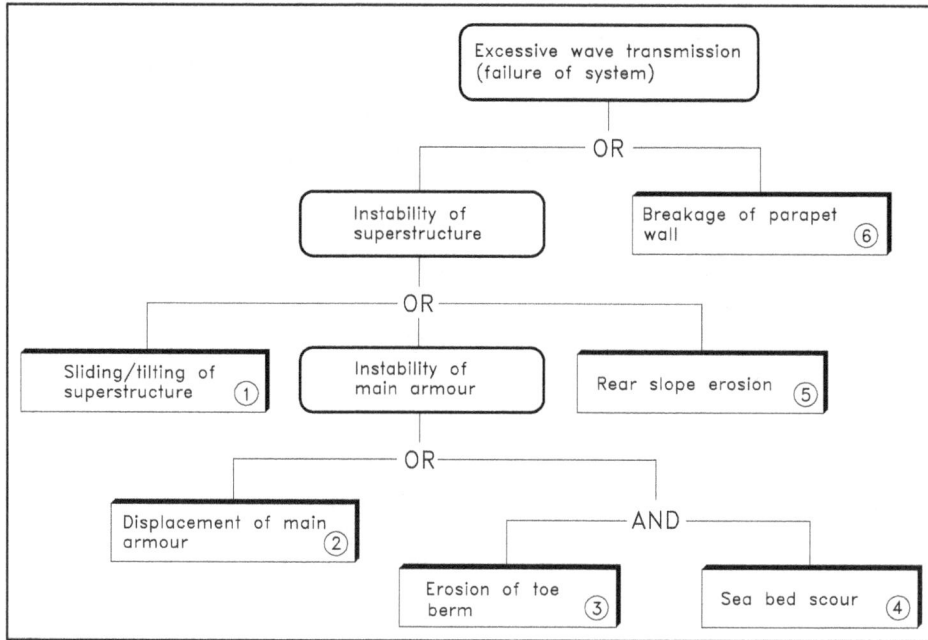

Figure VI-6-7. Example of simplified fault tree for a breakwater

c. A fault tree is a simplification and a systematization of the more complete so-called cause-consequence diagram that indicates the causes of partial failures as well as the interactions between the failure modes. An example is shown in Figure VI-6-8.

d. The failure probability of the system (for example, the probability of excessive wave transmission in Figure VI-6-7) depends on the failure probability of the single failure modes and on the correlation and linking of the failure modes. The failure probability of a single failure mode can be estimated by the methods described in Part VI-6-3. Two factors contribute to the correlation, namely physical interaction, such as sliding of main armor caused by erosion of a supporting toe berm, and correlation through common parameters like H_s. The correlations caused by physical interactions are not yet quantified. Consequently, only the common-parameter-correlation can be dealt with in a quantitative way. However, it is possible to calculate upper and lower bounds for the failure probability of the system.

e. A system can be split into two types of fundamental systems, namely series systems and parallel systems as illustrated by Figure VI-6-9.

(1) Series systems.

(a) In a series system, failure occurs if any of the elements $i = 1, 2, \ldots, n$ fails. The upper and lower bounds of the failure probability of the system, P_{fS} are

$$\textit{Upper bound} \quad P_{f\,S}^{U} = 1 - (1 - P_{f1})\,(1 - P_{f\,2}) \ldots (1 - P_{f\,n}) \tag{VI-6-33}$$

$$\textit{Lower bound} \quad P_{f\,S}^{L} = \max[\,P_{fi}\,] \tag{VI-6-34}$$

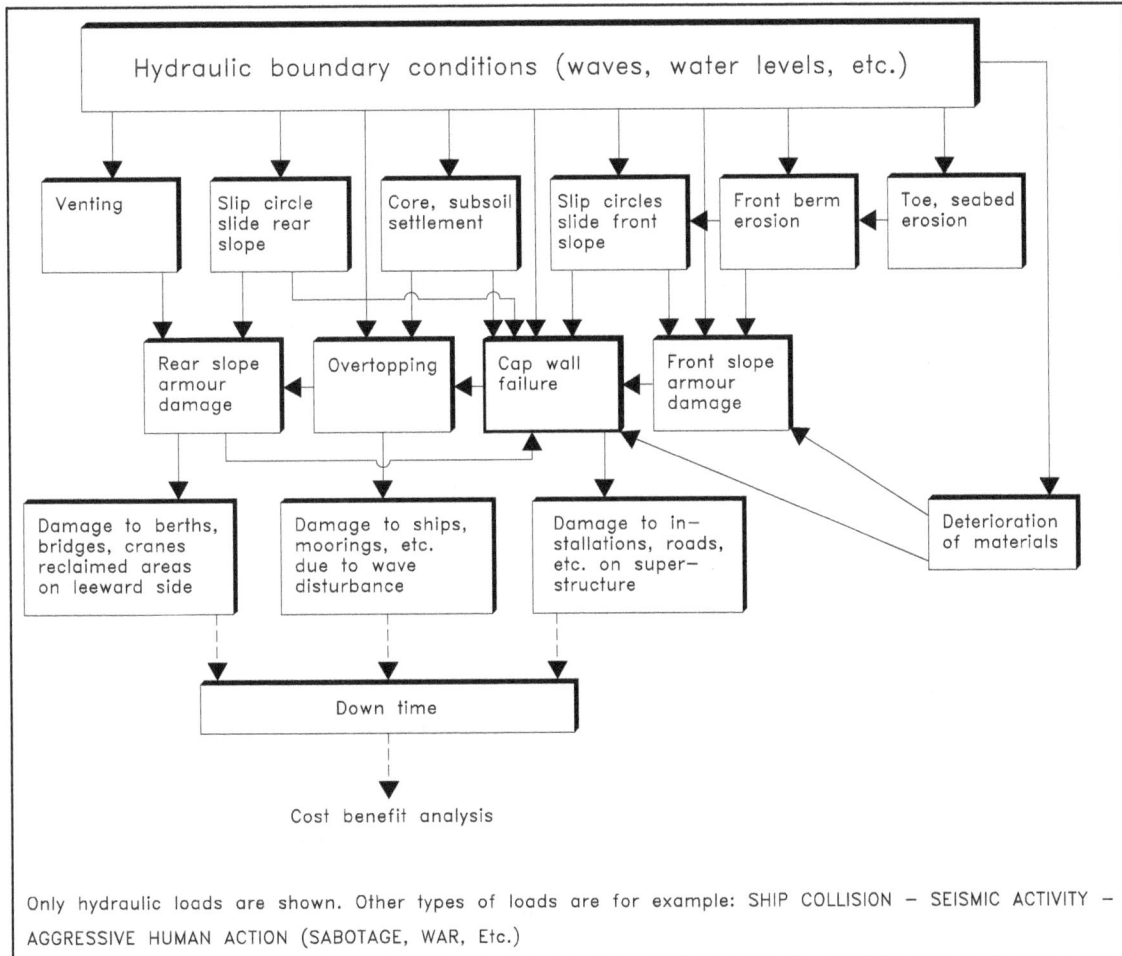

Figure VI-6-8. Example of cause-consequence diagram for a rubble-mound breakwater

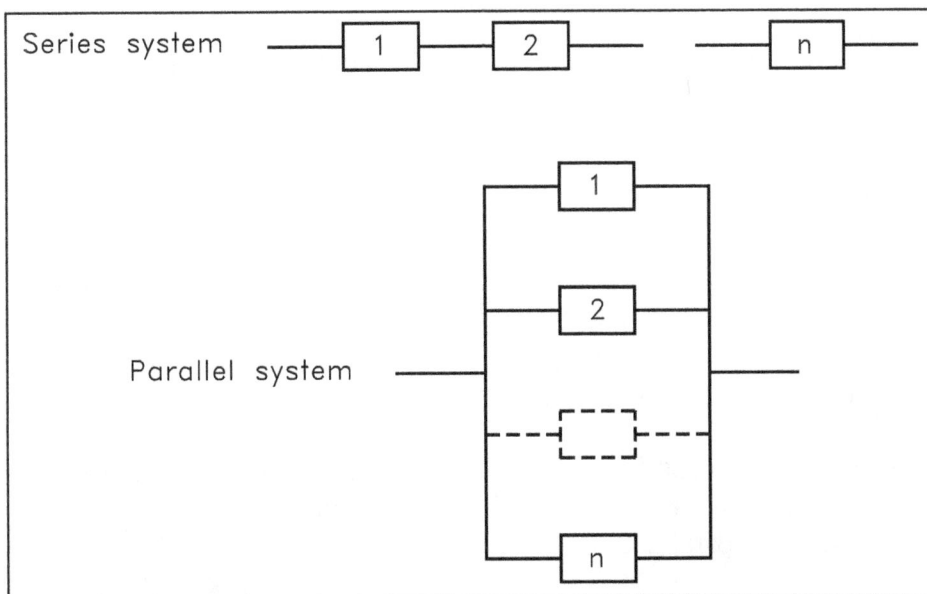

Figure VI-6-9. Series and parallel systems

where max $[P_{fi}]$ is the largest failure probability among all elements. The upper bound corresponds to no correlation between the failure modes and the lower bound to full correlation.

Equation VI-6-33 is sometimes approximated by $P_{fS}^U = \sum_{i=1}^{n} P_{fi}$ which is applicable only for small P_{fi} because P_{fS}^U should not be larger than 1.

   (b) The OR-gates in a fault tree correspond to series components. Series components are dominant in breakwater fault trees. In fact, the AND-*gate* shown in Figure VI-6-7 is included for illustration purposes, and in reality it should be an OR-gate.

   (2) Parallel systems.

   (a) A parallel system fails only if all the elements fail.

Upper bound $P_{fS}^U = \min [\, P_{fi} \,]$ (VI-6-35)

Lower bound $P_{fS}^L = P_{f1} \bullet P_{f2} \cdots P_{fn}$ (VI-6-36)

   (b) The upper bound corresponds to full correlation between the failure modes, and the lower bound corresponds to no correlation.

   • The AND-gates in a fault tree represent parallel components. To calculate upper and lower failure probability bounds for a system, it is convenient to decompose the overall system into series and parallel systems. Figure VI-6-10 shows a decomposition of the fault tree (Figure VI-6-7).

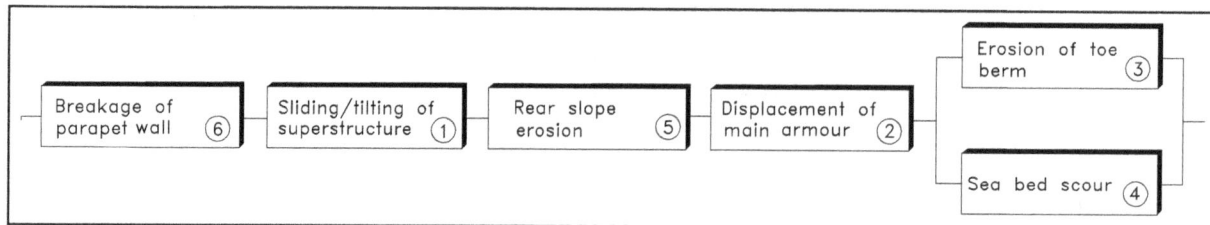

Figure VI-6-10. Decomposition of the fault tree into series and parallel systems

   • To obtain correct P_{fS}-values it is very important that the fault tree represents precisely the real physics of the failure development. This is illustrated by Example VI-6-2 where a fault tree alternative to Figure VI-6-7 is analyzed. In Example VI-6-2 the same failure mode probabilities as given in Example VI-6-1 are used.

   • The real failure probability of the system P_{fS} will always be in between P_{fS}^U and P_{fS}^L because some correlation exists between the failure modes due to the common loading represented by the sea state parameters, e.g., H_s.

EXAMPLE PROBLEM VI-6-1

The Level II analysis of the single failure modes for a specific breakwater schematized in Figure VI-6-10 revealed the following probabilities of failure in a 1-year period

i	1	2	3	4	5	6
$P_{fi}\%$	3	6	4	3	0.5	1

Note that these P_{fi}-values cannot be used in general because they relate to a specific structure. However, they are typical for conventionally designed breakwaters with respect to order of magnitude and large variations.

The simple failure probability bounds for the system are given by Equations VI-6-33, VI-6-34, VI-6-35, and VI-6-36:

Upper bound (no correlation):

$$P_{fs}^{U} = 1 - (1 - P_{f6})\,(1 - P_{f1})\,(1 - P_{f5})\,(1 - P_{f2})\,(1 - \min\,[P_{f3},\, P_{f4}]) = 12.9\%$$

or alternately for small values of P_{fi}

$$P_{fs}^{U} = P_{f6} + P_{f1} + P_{f5} + P_{f2} + \min\,[P_{f3},\, P_{f4}]) = 13.5\%$$

Lower bound (full correlation):

$$P_{fs}^{L} = \max\,[P_{f6}, P_{f1}, P_{f5}, P_{f2}, (P_{f3} \bullet P_{f4})] = 6\%$$

The simple bounds corresponding to T-years structural life might be approximated by the use of Equation VI-6-31[1]

	Structure life in years		
	20	50	100
$P_{fs}^{U}\%$	94	100	100
$P_{fs}^{L}\%$[1]	71	95	100

[1] It is very important to notice that the use of Equation VI-6-31, which assumes independent failure events from one year to another, can be misleading. This will be the case if some of the parameters which contribute significantly to the failure probability are time-invariant, i.e., are not changed from year to year. An example would be the parameter signifying a large uncertainty of a failure mode formula, such as the parameter A in Equation VI-6-2. If all parameters were time-invariant then the correct lower bound would be

$$P_{fs}^{L} = \max_{i=1-n}[P_{fi}]$$

independent of T, i.e., 6% for all T in the example. It follows that use of Equation VI-6-31 results in values of P_{fs}^{L} for $T > 1$ year that are too large.

EXAMPLE PROBLEM VI-6-2

Figure VI-6-11 shows a fault tree that differs from the fault tree in Figure VI-6-7. In Figure VI-6-11 only failure mode 6 can directly cause system failure, whereas in Figure VI-6-7 each of the failure modes 6, 5, 1, 2 and (3+4) can cause system failure.

The decomposition of the fault tree is shown in two steps in Figure VI-6-12. Note that the same failure mode can appear more than once in the decomposed system.

The simple bounds for the system are given by Equations VI-6-33, VI-6-34, VI-6-35, and VI-6-36:

Upper bound:

$$P_{fS}^U = 1 - (1 - P_{f6})\,(1 - \min\,[\,P_{f1}, P_{f5}\,])\,[\,P_{f1}, P_{f2}, P_{f3}, P_{f4}\,] = 4.5\%$$

or for smaller values of P_{fi}

$$P_{fS}^U = P_{f6} + \min\,[\,P_{f1}, P_{f5}\,] + \min\,[\,P_{f1}, P_{f2}, P_{f3}, P_{f4}\,] = 4.5\%$$

Lower bound:

$$P_{fS}^L = \max\,[\,P_{f6}, (P_{f1} \cdot P_{f5}), (P_{f1} \cdot P_{f2} \cdot P_{f3} \cdot P_{f4})\,] = 1\%$$

Using the same P_{fi}-values and procedure as given in Example VI-6-1 the following system failure probabilities are obtained

	Structure life in years		
	20	50	100
$P_{fs}^U\,\%$	60	90	99
$P_{fs}^L\,\%^1$	18	39	63

These values are quite different from the values of Example VI-6-1 which emphasizes the importance of a correct fault tree.

^{1} See note in Example VI-6-1.

- It would be possible to estimate P_{fS} if the physical interactions between the various failure modes were known and described by formulae, and if the correlations between the involved parameters were known. However, the procedure for determining such correlations are complicated and are not yet fully developed for practical use.

- The probability of failure cannot in itself be used as the basis for an optimization of a design. Optimization must be related to a kind of measure (scale), which for most structures is the economy, but can include other measures such as loss of human life.

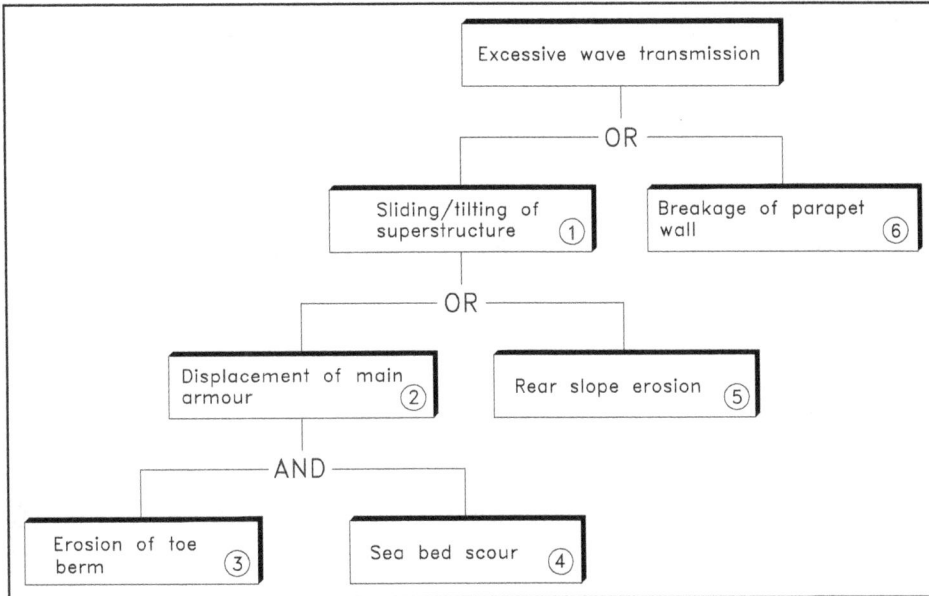

Figure VI-6-11. Example of simplified fault tree for a breakwater

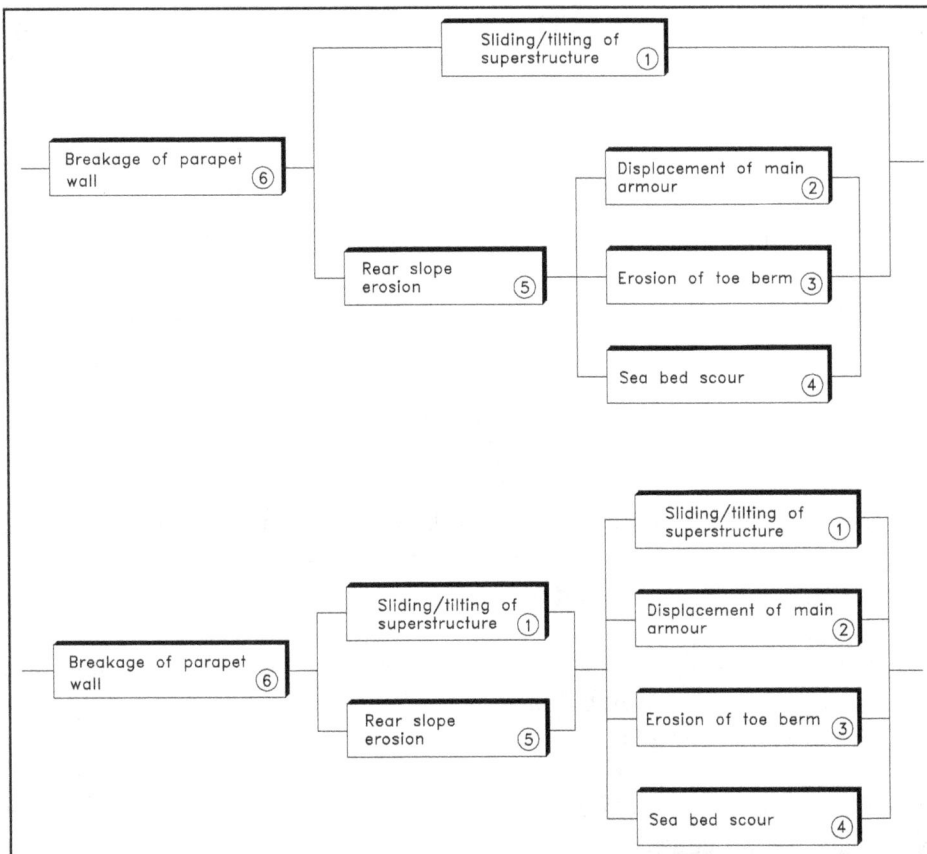

Figure VI-6-12. Decomposition of the fault tree into series and parallel systems

• The so-called risk, defined as the product of the probability of failure and the economic consequences, is used in optimization considerations. The economic consequences must cover all kinds of expenses related to the failure in question, i.e., cost of replacement, downtime costs, etc.

VI-6-5. <u>Parameter Uncertainties in Determining the Reliability of Structures</u>. Calculation of reliability or failure probability of a structure is based on formulae describing the structure's response to loads and on information about the uncertainties related to the formulae and relevant parameters. Basically, uncertainty is best given by a probability distribution; but because the true distribution is rarely known, it is common to assume a normal distribution and a related coefficient of variation, defined as

$$\sigma' = \frac{\sigma}{\mu} = \frac{standard\ deviation}{mean\ value} \qquad\qquad \text{(VI-6-37)}$$

as the measure of the uncertainty. The term "uncertainty" is used in this chapter as a general term referring to errors, to randomness, and to lack of knowledge.

    a.   Uncertainty related to failure mode formulae. The uncertainty associated with a formula can be considerable. This is clearly seen from many diagrams presenting the formula as a smooth curve shrouded by a wide scattered cloud of data points (usually from experiments) that are the basis for the curve fitting. Coefficients of variation of 15 - 20 percent or even larger are quite normal. The range of validity and the related coefficient of variation should always be considered when using a design formula.

    b.   Uncertainty related to environmental parameters. The sources of uncertainty contributing to the total uncertainties in environmental design values are categorized as follows:

    (1) Errors related to instrument response (e.g., from accelerometer buoy and visual observations).

    (2) Variability and errors due to different and imperfect calculations methods (e.g., wave hindcast models, algorithms for time-series analysis).

    (3) Statistical sampling uncertainties due to short-term randomness of the variables (variability within a stochastic process, e.g., two 20-min. records from a stationary storm will give two different values of the significant wave height)

    (4) Choice of theoretical distribution as a representative of the unknown long-term distribution (e.g., a Weibull and a Gumbel distribution might fit a data set equally well but can provide quite different values for a 200-year event).

    (5) Statistical uncertainties related to extrapolation from short samples of data sets to events of low probability of occurrence.

    (6) Statistical vagaries of the elements.

    (a) Distinction must be made between short-term sea state statistics and long-term (extreme) sea statistics. Short-term statistics are related to the stationary conditions during a sea state, e.g., wave height distribution within a storm of constant significant wave height, H_s. Long-term statistics deal with the extreme events, e.g., the distribution of H_s over many storms.

(b) Related to the short-term sea state statistics the following aspects must be considered:

- The distribution for individual wave heights in a record in deepwater and shallow-water conditions, i.e., Rayleigh distribution and some truncated distributions, respectively.

- Variability due to short samples of single peak spectra waves in deep and shallow water based on theory and physical simulations.

- Variability due to different spectral analysis techniques, i.e., different algorithms, smoothing and filter limits.

- Errors in instrument response and influence of measurement location. For example, floating accelerometer buoys tend to underestimate the height of steep waves. Characteristics of shallow-water waves can vary considerably in areas with complex seabed topography. Wave recordings at positions with depth-limited breaking waves cannot produce reliable estimates of the deepwater waves.

- Imperfection of deep and shallow-water numerical hindcast models and quality of wind input data.

(c) Estimates of overall uncertainties for short-term sea state parameters (first three items) are presented in Table VI-6-1 for use when more precise site specific information is not available.

(d) Evaluation of the uncertainties related to the long-term sea state statistics, and use of these estimates for design, involves the following considerations:

- The encounter probability.

- Estimation of the standard deviation of a return-period event for a given extreme distribution.

- Estimation of extreme distributions by fitting to data sets consisting of uncorrelated values of H_s from

  - Frequent measurements of H_s equally spaced in time.

  - Identification of the largest H_s in each year (annual series).

  - Maximum values of H_s for a number of storms exceeding a certain threshold value of H_s using peak over threshold (POT) analysis.

The methods of fitting are the maximum likelihood method, the method of moments, the least square method, and visual graphical fit.

- Uncertainty on extreme distribution parameters due to limited data sample size.

- Influence on the extreme value of H_s on the choice of threshold value in the POT analysis. (The threshold level should exclude all waves which do not belong to the statistical population of interest).

- Errors due to lack of knowledge about the true extreme distribution. Different theoretical distributions might fit a data set equally well, but might provide quite different return period values of H_s. (The error can be estimated only empirically by comparing results from fits to different theoretical distributions).

Table VI-6-1

Typical Variational Coefficients $\sigma N = \sigma/\mu$ (standard deviation over mean value) for Measured and Calculated Sea State Parameters (Burcharth 1992)

Parameter	Methods of Determination	Estimated Typical Values ΣN	Bias	Comments
Significant wave height, OFFSHORE	Accelerometer buoy, pressure cell, vertical radar	0.05 - 0.1	–0	
	Horizontal radar	0.15	–0	
	Hindcast numerical models	0.1 - 0.2	0 - 0.1	Very dependent on quality of weather maps
	Hindcast, SMB method	0.15 - 0.2	?	Valid only for storm conditions in restricted sea basins
Significant wave height NEARSHORE determined from offshore significant wave height accounting for shallow-water effects	Visual observations from ships	0.2	0.05	
	Numerical models	0.1 - 0.20	0.1	σN can be much larger in some cases
	Manual calculations	0.15 - 0.35		
Mean wave period offshore on condition of fixed significant wave height	Accelerometer buoy	0.02 - 0.05	–0	
	Estimates from ampl. Spectra	0.15	–0	
	Hindcast, numerical models	0.1 - 0.2	–0	

Duration of sea state with significant wave height exceeding a specific level	Direct measurements	0.02	−0
	Hindcast numerical models	0.05 - 0.1	−0
Spectral peak frequency offshore	Measurements	0.05 - 0.15	−0
	Hindcast numerical models	0.1 - 0.2	−0
Spectral peakedness offshore	Measurements and hindcast numerical models	0.4	−0
Mean direction of wave propagation offshore	Pitch-roll buoy	5 degrees	
	Measurements η, u, v or p, u, v [1]	10 degrees	
	Hindcast numerical models	15 - 30 degrees	
Astronomical tides	Prediction from constants	0.001 - 0.07	−0
Storm surge	Numerical models	0.1 - 0.25	∀0.1

[1] Two horizontal velocity components and water-level elevation or pressure.

- Errors due to applied plotting formulae in the case of graphical fitting. Depending on the applied plotting formulae quite different extreme estimates can be obtained. The error can only be empirically estimated.

- Climatological changes.

- Physical limitations in extrapolation to events of low probability. The most important example might be limitations in wave heights due to limited water depths and fetch restrictions.

- The effect of measurement error on the uncertainty related to an extreme event.

(e) It is beyond the scope of this chapter to discuss in more detail the mentioned uncertainty aspects related to the environmental parameters. Additional information is given in Burcharth (1992).

c. Uncertainty related to structural parameters. The uncertainties related to material parameters (such as density) and geometrical parameters (such as slope angle and size of structural elements) are generally much smaller than the uncertainties related to the environmental parameters and to the design formulae.

VI-6-6. Partial Safety Factor System for Implementing Reliability in Design.

    a. Introduction to partial safety factors.

    (1) The objective of using partial safety factors in design is to assure a certain reliability of the structures. This section presents the partial safety factors developed by the Permanent International Association of Navigation Congresses (PIANC) PTCII Working Group 12 (Analysis of Rubble-Mound Breakwaters) and Working Group 28 (Breakwaters with Vertical and Inclined Concrete Walls), Burcharth (1991) and Burcharth and Sørensen (1999).

    (2) The partial safety factors, γ_i, are related to characteristic values of the stochastic variables, $X_{i,ch}$. In conventional civil engineering codes the characteristic values of loads and other action parameters are often chosen to be an upper fractile (e.g., 5 percent), while the characteristic values of material strength parameters are chosen to be a lower fractile. The values of the partial safety factors are uniquely related to the applied definition of the characteristic values.

    (3) The partial safety factors, γ_i, are usually larger than or equal to 1. Consequently, if we define the variables as either load variables X_i^{load} (for example H_s) or resistance variables X_i^{res} (for example the block volume) then the related partial safety factors should be applied as follows to obtain the design values:

$$X_i^{design} = \gamma_i^{load} \bullet X_{i,ch}^{load}$$
$$X_i^{design} = \frac{X_{i,ch}^{res}}{\gamma_i^{res}} \tag{VI-6-38}$$

    (4) The magnitude of γ_i reflects both the uncertainty of the related parameter X_i, and the relative importance of X_i in the failure function. A large value, e.g., $\gamma_{Hs} = 1.4$, indicates a relatively large sensitivity of the failure probability to the significant wave height, H_s. On the other hand, $\gamma \square 1$ indicates little or negligible sensitivity, in which case the partial coefficient should be omitted. Bear in mind that the magnitude of γ_i is not (in a mathematical sense) a stringent measure of the sensitivity of the failure probability of the parameter, X_i.

    (5) As an example, when partial safety factors are applied to the characteristic values of the parameters in Equation VI-6-2, a design equation is obtained, i.e., the definition of how to apply the coefficients. The partial safety factors can be related either to each parameter or to combinations of the parameters (overall coefficients). The design equation obtained when partial safety factors are applied to each parameter is given by

$$G = \frac{A_{ch}}{\gamma_A} \frac{\Delta_{ch}}{\gamma_\Delta} \frac{D_{n,ch}}{\gamma_{D_n}} \left(K_D \frac{\cot \alpha_{ch}}{\gamma_{\cot \alpha}} \right)^{1/3} - \gamma_{H_s} H_{s,ch} \geq 0$$

$$\text{or} \tag{VI-6-39}$$

$$D_{n,ch} \geq \gamma_A \gamma_\Delta \gamma_{D_n} \gamma_{\cot \alpha}^{1/3} \gamma_{H_s} \frac{H_{s,ch}}{A_{ch} \Delta_{ch} K_D \cot \alpha_{ch}}$$

(6) If the partial safety factors are applied to combinations of parameters, there may be only γ_{Hs} and an overall coefficient γ_Z related to the first term on the right-hand side of Equation VI-6-39. The design equation would then become

$$G = \frac{A_{ch}}{\gamma_Z} \Delta_{ch} D_{n,ch} \left(K_D \cot \alpha \right)^{\frac{1}{3}} - \gamma_{H_s} H_{s,ch} \geq 0$$

or

$$D_{n,ch} \geq \gamma_Z \gamma_{H_s} \frac{H_{s,ch}}{A_{ch} \Delta_{ch} \left(K_D \cot \alpha_{ch} \right)^{\frac{1}{3}}}$$

(VI-6-40)

(7) Equations VI-6-39 and VI-6-40 express two different "code formats." By comparing the two equations it is seen that the product of the partial coefficients is independent of the chosen format if the other parameters are equal. A goal is to have a system which is as simple as possible, i.e., with as few partial safety factors as possible, but without invalidating the accuracy of the design equation beyond acceptable limits. Fortunately, it is often possible to use overall coefficients, such as γ_A in Equation VI-6-40, without losing significant accuracy within the realistic range of parameter value combinations. This is the case for the partial safety factors system presented in this chapter where only two partial safety factors, γ_{Hs} and γ_Z, are used in each design formula.

(8) Usually several failure modes are relevant to a particular design. The relationship between the failure modes are characterized either as series systems or parallel systems. A fault tree can be used to illustrate the complete system. The partial safety factors for failure modes associated with a system having a failure probability, P_f, are different from the partial safety factors for single failure modes having the same failure probability, P_f. Therefore, partial safety factors for single failure modes and multifailure mode systems must be treated separately.

b.   Uncertainties and statistical models. Uncertainties in relation to rubble-mound breakwaters can be divided in uncertainties related to the following three groups:

(1) Load uncertainties (wave modeling).

(2) Soil strength uncertainties (modeling of soil strength parameters).

(3) Model uncertainties (both wave load models and models for bearing capacity of the foundation).

(a) Wave modeling.

- For calibration of partial safety factors the maximum significant wave height in T years is denoted as $F_{H_s^T}$, and it is modeled (for example) by the extreme Weibull distribution, given as

$$F_{H_S^T}(H_s) = \left\{ 1 - \exp\left[-\left(\frac{H_s - H_{\bar{s}}}{\beta} \right)^{\alpha} \right] \right\}^{\lambda T} \qquad (VI-6-41)$$

where λ is the number of observations per year, $H_S N$ is the threshold level, and α and β are the Weibull distribution parameters.

- For calibration of the PIANC partial safety factor system, wave data from four quite different geographical locations were selected as presented in Table VI-6-2. In Table VI-6-2, N is the number of data samples and h is the water depth in meters.

Table VI-6-2
Wave Data from Different Locations Fitted to a Weibull Distribution
(β, $H_S N$ and h are in meters)

	N	λ	α	β (m)	$H_S N$ (m)	h (m)
Bilbao	50	4.17	1.39	1.06	4.9	25
Sines	15	1.25	1.78	2.53	7.1	25
Tripoli	15	0.75	1.83	3.24	2.9	25
Fallonica	46	5.94	1.14	0.58	2.7	10

- The wave data from Bilbao, Sines and Tripoli correspond to deepwater waves, whereas the wave data from Fallonica corresponds to shallow-water waves. To model the statistical uncertainty, α and β are modeled as independent and normally distributed.

- The model uncertainty related to the quality of the measured wave data is modeled by a multiplicative stochastic variable F_{Hs} which is assumed to be normally distributed with expected value 1 and standard deviation $\sigma_{F_{Hs}}$. High quality and low quality wave data could be represented by $\sigma_{F_{Hs}} = 0.05$ and 0.2, corresponding to accelerometer buoy and fetch diagram estimates, respectively, as given by Table VI-6-1.

(b) Soil strength modeling.

- Statistical modeling of the soil strength (sand and/or clay) is generally difficult, and only few models are available in the literature that can be used for practical reliability calculations. In general the material characteristics of the soil have to be modeled as a stochastic field. The parameters describing the stochastic field have to be determined on the basis of the measurements which are usually performed to characterize the soil characteristics. Because these measurements are only performed in a few locations, statistical uncertainty due to the sparse data is introduced, and this uncertainty must be included in the statistical model. Furthermore, the uncertainty in the determination of the soil properties and the measurement uncertainty must also be included in the statistical model.

- Because breakwaters are composed of loose material in frictional contact, and it is assumed that the foundation failure modes are developed in the core; only statistical models for the effective friction angle and the angle of dilation are needed. Usually these angles are modeled by normal or lognormal distributions.

- The bearing capacities related to the geotechnical failure modes are estimated using the upper bound theorem of classical plasticity theory where an associated flow rule is assumed. However, the friction angle and the dilation angle for the rubble-mound material and the sand subsoil are usually different. Therefore, in order to use the theory based on an associated flow rule, the following reduced effective friction angle v_d is used (Hansen 1979):

$$\tan \varphi_d = \frac{\sin \varphi' \cos \psi}{1 - \sin \varphi' \sin \psi} \tag{VI-6-42}$$

where v_N is the effective friction angle and ψ is the dilation angle.

(c) Model uncertainties.

- In general, model uncertainties related to a given mathematical model can be evaluated on the basis of:

  – Comparisons between experimental tests/measurements and numerical model calculations.

  – Comparisons between numerical calculations with the given mathematical model and a more advanced/complex model.

  – Expert opinions.

  – Information from the literature.

- Many laboratory experiments have been performed for most of the failure modes related to hydraulic instability of the armor layer. Based on these experiments the model uncertainty can be estimated. Model uncertainty connected with extrapolation from laboratory to a real structure can be judged on the basis of expert opinions, information from the literature, and observations of similar existing structures.

- For soil strength models no similar measurements models are available. However, if "simple" rotation and translation failure models based on the upper bound theorem of plasticity theory are used, then these can be evaluated by comparison with results from more refined numerical calculations using nonlinear finite element programs. Estimates of the model uncertainties can thus be obtained.

   c. Format for partial safety factors.

   (1) The PIANC partial safety factors are calibrated with the following input:

   (a) Design lifetime T_L (= 20, 50 or 100 years).

   (b) Acceptable probability of failure P_f (= 0.01, 0.05, 0.10, 0.20, or 0.40).

   (c) Coefficient of variation σ_{FH_s} = (0.05 and 0.20).

   (d) Deep or shallow-water conditions.

   (e) Wave loads determined with or without hydraulic model tests.

   (2) The partial safety factors are as follows:

   (a) A load partial safety factor γ_P to be applied to the mean value of the permanent load (= 1).

   (b) A load partial safety factor γ_H to be applied to $\hat{H}_s^{T_L}$ (the central estimate of the significant wave height which, in average, is exceeded once every T_L years).

   (c) A partial safety factor to be used to the combination of the mean values of the resistance variables as shown in the design equation. γ_Z is to be used with friction materials in rubble-mound and/or subsoils (tangent to the mean value of the friction angle is divided by γ_Z).

   (d) A partial safety factor γ_C to be used with the mean value of the undrained shear strength of clay materials in the subsoil (the mean value of the undrained shear strength is divided by γ_C).

   d. Tables of partial safety factors.

   (1) Partial safety factors are presented in Table VI-6-3.

   (2) In the case of vertical walls, wave forces are calculated from the Goda formula. Furthermore, the following factors are used to compensate for the positive bias inherent in the Goda formula (see Table VI-5-55):

$\hat{U}_{Hor.Force}$ = 0.90, bias factor to be applied to the Goda horizontal wave force

$\hat{U}_{Ver.Force}$ = 0.77, bias factor to be applied to the Goda vertical wave force

$\hat{U}_{Hor.Moment}$ = 0.81, bias factor to be applied to the moment from the Goda horizontal wave forces around the shoreward heel of the base plate

$\hat{U}_{Ver.Moment}$ = 0.72, bias factor to be applied to the moment from the Goda vertical wave forces around the shoreward heel of the base plate

A carrot symbol ($^\wedge$) over the variable indicates a mean value.

Table VI-6-3 Partial Safety Factor Tables			
Structure	Failure	Armor	Table(s)
Rubble-mound structures	Armor stability	Rocks	VI-6-4 - VI-6-6
		Cubes	VI-6-7
		Tetrapods	VI-6-8
		Dolosse	VI-6-9 & VI-6-10
		Hollowed Cubes	VI-6-11 & VI-6-12
	Toe berm		VI-6-13
	Breakage	Dolosse	VI-6-14 & VI-6-15
		Tetrapods	VI-6-16
	Runup	Rock	VI-6-17
		Hollowed Cubes	VI-6-18
		Dolosse	VI-6-19
	Scour		VI-6-20 & VI-6-21
Vertical-wall caisson structures	Foundation: sand subsoil		VI-6-22
	Foundation: clay subsoil		VI-6-23
	Sliding failure		VI-6-24
	Overturning failure		VI-6-25
	Scour		VI-6-26
	Toe berm		VI-6-27

(3) Part VI-7, "Example Problems," contains worked design examples for the most common coastal structures. Some of these examples include a reliability analysis based on the information contained in Tables VI-6-4 to VI-6-27 either as part of the design or as an alternative to deterministic methods based on a single return period of occurrence. The Part VI-7 examples provide coastal engineers with guidance on selection of partial safety factors γ_{Hs} and γ_Z for various levels of P_f and $\sigma_{F_{H's}}$.

Table VI-6-4
Partial Safety Factors for Stability Failure of Rock Armor, Hudson Formula, Design Without
Model Tests

Design equation (cf. Table VI-5-22)

$$G = \frac{1}{\gamma_Z} \hat{\Delta} \hat{D}_n (K_D \, c\hat{o}t \, \alpha)^{1/3} - \gamma_H \hat{H}_S^T \qquad (VI-6-43)$$

P_f	$\sigma'_{FH_S} = 0.05$		$\sigma'_{FH_S} = 0.2$	
	γ_H	γ_Z	γ_H	γ_Z
0.01	1.7	1.04	2.0	1.00
0.05	1.4	1.06	1.6	1.02
0.10	1.3	1.04	1.4	1.06
0.20	1.2	1.02	1.3	1.00
0.40	1.0	1.08	1.1	1.00

Table VI-6-5
Partial Safety Factors for Stability Failure of Rock Armor, Plunging Waves, van der Meer
Formula, Design Without Model Tests

Design equation (cf. Table VI-5-23)

$$G = \frac{1}{\gamma_Z} 6.2 \, \hat{S}^{0.2} \, \hat{P}^{0.18} \, \hat{\Delta} \hat{D}_n \hat{f} \, (c\hat{o}t \, \alpha)^{0.5} \, (\hat{s}_{om})^{0.25} \, \hat{N}_z^{-0.1} - \gamma_H \hat{H}_S^T \qquad (VI-6-44)$$

where the factor f models the effect of low crested breakwaters:

$$\hat{f} = \frac{1}{1.25 - 4.8 \frac{R_c}{\hat{H}_S^T} \sqrt{\frac{\hat{s}_{op}}{2\pi}}}$$

P_f	$\sigma'_{FH_S} = 0.05$		$\sigma'_{FH_S} = 0.2$	
	γ_H	γ_Z	γ_H	γ_Z
0.01	1.6	1.04	1.9	1.00
0.05	1.4	1.02	1.5	1.06
0.10	1.3	1.00	1.3	1.10
0.20	1.2	1.00	1.2	1.06
0.40	1.0	1.08	1.0	1.10

Table VI-6-6
Partial Safety Factors for Stability Failure of Rock Armor, Surging Waves, van der Meer
Formula, Design Without Model Tests

Design equation (cf. Table VI-5-23)

$$G = \frac{1}{\gamma_Z} \hat{S}^{0.2} \hat{P}^{-0.13} \hat{\Delta} \hat{D}_n \hat{f} (\cot \alpha)^{(0.5-P)} (\hat{s}_{om})^{-0.5P} \hat{N}_z^{-0.1} - \gamma_H \hat{H}_S^T \qquad (VI-6-45)$$

where

$$\hat{f} = \frac{1}{1.25 - 4.8 \frac{R_c}{\hat{H}_S^T} \sqrt{\frac{\hat{s}_{op}}{2\pi}}}$$

	$\sigma'_{FH_S} = 0.05$		$\sigma'_{FH_S} = 0.2$	
P_f	γ_H	γ_Z	γ_H	γ_Z
0.01	1.7	1.00	1.9	1.02
0.05	1.3	1.10	1.6	1.00
0.10	1.3	1.02	1.4	1.04
0.20	1.1	1.10	1.2	1.08
0.40	1.0	1.08	1.1	1.00

Table VI-6-7
Partial Safety Factors for Stability Failure of Cube Block Armor, van der Meer Formula,
Design Without Model Tests

Design equation (cf. Table VI-5-29)

$$G = \frac{1}{\gamma_Z} \left(6.7 \frac{(\hat{N}_{od})^{0.4}}{(\hat{N}_z)^{0.3}} + 1.0 \right) (\hat{s}_{om})^{-0.1} \hat{\Delta} \hat{D}_n - \gamma_H \hat{H}_S^T \qquad (VI-6-46)$$

	$\sigma'_{FH_S} = 0.05$		$\sigma'_{FH_S} = 0.2$	
P_f	γ_H	γ_Z	γ_H	γ_Z
0.01	1.5	1.10	1.8	1.04
0.05	1.3	1.08	1.5	1.04
0.10	1.3	1.00	1.4	1.02
0.20	1.2	1.00	1.2	1.06
0.40	1.0	1.08	1.0	1.10

Table VI-6-8
Partial Safety Factors for Stability Failure of Tetrapods, van der Meer Formula, Design Without Model Tests

Design equation (cf. Table VI-5-30)

$$G = \frac{1}{\gamma_Z}\left(3.75\frac{(\hat{N}_{od})^{0.5}}{(\hat{N}_z)^{0.25}} + 0.85\right)(\hat{s}_{om})^{-0.2}\hat{\Delta}\hat{D}_n - \gamma_H\hat{H}_S^T \qquad (VI-6-47)$$

	$\sigma'_{FH_S} = 0.05$		$\sigma'_{FH_S} = 0.2$	
P_f	γ_H	γ_Z	γ_H	γ_Z
0.01	1.7	1.02	1.9	1.04
0.05	1.4	1.06	1.5	1.08
0.10	1.3	1.04	1.4	1.04
0.20	1.2	1.02	1.3	1.00
0.40	1.0	1.08	1.1	1.00

Table VI-6-9
Partial Safety Factors for Stability Failure of Dolosse, Without Superstructure, Burcharth Formula, Design Without Model Tests

Design equation (cf. Table VI-5-31)

$$G = \frac{1}{\gamma_Z}\hat{\Delta}\hat{D}_n(47 - 72\hat{r})\,\hat{\varphi}\,\hat{D}^{1/3}\,\hat{N}_z^{-0.1} - \gamma_H\hat{H}_S^T \qquad (VI-6-48)$$

	$\sigma'_{FH_S} = 0.05$		$\sigma'_{FH_S} = 0.2$	
P_f	γ_H	γ_Z	γ_H	γ_Z
0.01	2.1	1.08	2.4	1.02
0.05	1.7	1.00	1.7	1.08
0.10	1.5	1.00	1.6	1.00
0.20	1.3	1.00	1.3	1.04
0.40	1.0	1.10	1.1	1.02

Table VI-6-10
Partial Safety Factors for Stability Failure of Dolosse, With Superstructure, Burcharth and Liu (1995a), Design Without Model Tests

Design equation

$$G = \frac{1}{\gamma_Z} \hat{\Delta} \hat{D}_n (43 - 66\hat{r}) \, \hat{\varphi} \, \hat{D}^{1/3} \, \hat{N}_z^{-0.1} - \gamma_H \hat{H}_S^T \qquad \text{(VI}-6-49)$$

P_f	$\sigma'_{FH_S} = 0.05$		$\sigma'_{FH_S} = 0.2$	
	γ_H	γ_Z	γ_H	γ_Z
0.01	1.9	1.10	2.2	1.04
0.05	1.6	1.02	1.7	1.04
0.10	1.4	1.04	1.5	1.04
0.20	1.2	1.06	1.3	1.04
0.40	1.0	1.10	1.1	1.02

H_s^T	Significant wave height with return period T
ρ_s	Mass density of concrete
ρ_w	Mass density of water
Δ	$(\rho_s/\rho_w) - 1$
D_n	Equivalent cube length, i.e., length of cube with the same volume as Dolosse
r	Dolos waist ratio
φ	Packing density
D	Relative number of units within levels SWL $\pm$ 6.5 Dn displaced one Dolos height h, or more (e.g., for 2% displacement insert $D = 0.02$)
N_z	Number of waves. For $N_z \geq 3000$ use $N_z = 3000$

Table VI-6-11
Partial Safety Factors for Stability Failure of Trunk of Hollowed Cubes, Slope 1:1.5 and 1:2,
Berenguer and Baonza (1995), Design Without Model Tests

Design equation

$$G = \frac{1}{\gamma_Z}\hat{\zeta}_p^{-0.1}(3.3 + 0.7\hat{N}_{0d}^{0.4})\hat{\Delta}\hat{D}_n - \gamma_H\hat{H}_S^T \qquad (VI-6-50)$$

where

$$\hat{\zeta}_p = (\hat{cot}\ \alpha)^{-1}(\hat{s}_{op})^{-0.5}$$

P_f	$\sigma'_{FH_S} = 0.05$		$\sigma'_{FH_S} = 0.2$	
	γ_H	γ_Z	γ_H	γ_Z
0.01	3.5	1.10	3.5	1.10
0.05	2.3	1.08	2.5	1.02
0.10	1.8	1.06	1.9	1.04
0.20	1.4	1.06	1.5	1.02
0.40	1.1	1.04	1.1	1.04

H_s^T	Significant wave height with return period T
ρ_s	Mass density of concrete
ρ_w	Mass density of water
Δ	$(\rho_s/\rho_w) - 1$
D_n	Equivalent cube length, i.e., length of cube with the same volume as Dolosse
N_{od}	Number of displaced units within a strip width of one equivalent cube length D_n

Table VI-6-12
Partial Safety Factors for Stability Failure of Roundhead of Hollowed Cubes, Slope 1:1.5 and 1:2, Berenguer and Baonza (1995), Design Without Model Tests

Design equation

$$G = \frac{1}{\gamma_Z}(1.8 + 6.6\hat{D}^{0.33}\hat{\zeta}_p^{-0.1})\hat{\Delta}\hat{D}_n - \gamma_H\hat{H}_S^T \qquad (VI-6-51)$$

where

$$\hat{\zeta}_p = (\hat{cot}\,\alpha)^{-1}(\hat{s}_{op})^{-0.5}$$

P_f	$\sigma'_{FH_S} = 0.05$		$\sigma'_{FH_S} = 0.2$	
	γ_H	γ_Z	γ_H	γ_Z
0.01	1.8	1.00	1.9	1.06
0.05	1.5	1.00	1.5	1.10
0.10	1.3	1.06	1.4	1.06
0.20	1.2	1.02	1.3	1.00
0.40	1.0	1.08	1.1	1.00

H_s^T Significant wave height with return period T

ρ_s Mass density of concrete

ρ_w Mass density of water

Δ $(\rho_s/\rho_w) - 1$

D_n Equivalent cube length, i.e., length of cube with the same volume as Dolosse

D Relative number of displaced units

s_{op} Wave steepness, Hs/L_{op}

L_{op} Deepwater wave length corresponding to peak wave period

Table VI-6-13
Partial Safety Factors for Stability Failure of Toe Berm, Parallelepiped Concrete Blocks and Rocks., Burcharth Formula, Design Without Model Tests

Design equation (cf. Table VI-5-47)

$$G = \frac{1}{\gamma_Z} \left(0.4 \frac{\hat{h}_b}{\hat{\Delta}\hat{D}_{n50}} + 1.6 \right) (\hat{N}_{od})^{0.15} \hat{\Delta}\hat{D}_{n50} - \gamma_H \hat{H}_S^T \qquad (VI-6-52)$$

P_f	$\sigma'_{FH_S} = 0.05$		$\sigma'_{FH_S} = 0.2$	
	γ_H	γ_Z	γ_H	γ_Z
0.01	1.6	1.06	1.8	1.06
0.05	1.3	1.10	1.5	1.06
0.10	1.3	1.02	1.4	1.04
0.20	1.1	1.10	1.2	1.08
0.40	1.0	1.08	1.0	1.10

Table VI-6-14
Partial Safety Factors for Trunk Dolos Breakage, Burcharth Formula, Design Without Model Tests

Design equation (cf. Table VI-5-40)

$$G = \frac{1}{\gamma_Z} B - C_0 \hat{M}^{C_1} \hat{f}_T^{C_2} (\gamma_H \hat{H}_S^T)^{C_3} \qquad (VI-6-53)$$

P_f	$\sigma'_{FH_S} = 0.05$		$\sigma'_{FH_S} = 0.2$	
	γ_H	γ_Z	γ_H	γ_Z
0.01	1.9	1.00	2.1	1.00
0.05	1.5	1.04	1.6	1.10
0.10	1.4	1.00	1.5	1.00
0.20	1.2	1.10	1.3	1.00
0.40	1.1	1.00	1.1	1.02

Table VI-6-15
Partial Safety Factors for Roundhead Dolos Breakage, Burcharth Formula, Design Without Model Tests

Design equation (cf. Table VI-5-40)

$$G = \frac{1}{\gamma_Z} B - 0.025 \, \hat{M}^{-0.65} \, \hat{f_T}^{-0.66} \, (\gamma_H \hat{H}_S^T)^{2.42}$$

$$(VI-6-54)$$

P_f	$\sigma'_{FH_S} = 0.05$		$\sigma'_{FH_S} = 0.2$	
	γ_H	γ_Z	γ_H	γ_Z
0.01	1.8	1.02	2.0	1.00
0.05	1.4	1.10	1.6	1.00
0.10	1.3	1.06	1.4	1.08
0.20	1.2	1.02	1.3	1.00
0.40	1.1	1.00	1.1	1.00

Table VI-6-16
Partial Safety Factors for Trunk Tetrapod Breakage, Burcharth Formula, Design Without Model Tests

Design equation (cf. Table VI-5-40)

$$G = \frac{1}{\gamma_Z} B - 3.39(10)^{-3} \, \hat{M}^{-0.79} \, \hat{f_T}^{-2.73} \, (\gamma_H \hat{H}_S^T)^{3.84}$$

$$(VI-6-55)$$

P_f	$\sigma'_{FH_S} = 0.05$		$\sigma'_{FH_S} = 0.2$	
	γ_H	γ_Z	γ_H	γ_Z
0.01	1.9	1.10	2.1	1.06
0.05	1.6	1.00	1.7	1.00
0.10	1.4	1.04	1.5	1.04
0.20	1.2	1.10	1.3	1.06
0.40	1.1	1.00	1.1	1.04

Table VI-6-17
Partial Safety Factors for Runup, Rock Armored Slopes, De Waal and van der Meer (1992),
Design Without Model Tests

Design equation

$$\text{For} \quad \zeta_m = (\cot \alpha)^{-1}(s_{om})^{-0.5} \le 1.5 \quad : \quad R_u/H_s = a\,\zeta_m$$

$$G = \frac{1}{\gamma_Z}\hat{R}_u\,\hat{a}^{-1}\,(cot\,\hat{\alpha})(\hat{s}_{om})^{0.5} - \gamma_H\hat{H}_s^T \qquad\qquad (VI-6-56)$$

P_f	$\sigma'_{FH_S} = 0.05$		$\sigma'_{FH_S} = 0.2$	
	γ_H	γ_Z	γ_H	γ_Z
0.01	1.7	1.04	2.0	1.00
0.05	1.4	1.06	1.6	1.02
0.10	1.3	1.04	1.4	1.06
0.20	1.2	1.02	1.3	1.00
0.40	1.0	1.08	1.1	1.00

$$\text{For} \quad \zeta_m = (\cot \alpha)^{-1}(s_{om})^{-0.5} > 1.5 \quad : \quad R_u/H_s = b\,(\zeta_m)^c$$

$$G = \frac{1}{\gamma_Z}\hat{R}_u\,\hat{b}^{-1}[\cot\hat{\alpha}\,(\hat{s}_{om})^{0.5}]^{\hat{c}} - \gamma_H\hat{H}_s^T \qquad\qquad (VI-6-57)$$

P_f	$\sigma'_{FH_S} = 0.05$		$\sigma'_{FH_S} = 0.2$	
	γ_H	γ_Z	γ_H	γ_Z
0.01	1.5	1.08	1.8	1.02
0.05	1.3	1.06	1.4	1.10
0.10	1.2	1.06	1.3	1.08
0.20	1.1	1.08	1.2	1.06
0.40	1.0	1.06	1.0	1.10

For permeable structures, $P > 0.4$, the upper limit of R_u is given by $R_u/H_s = d$

α	Slope angle
s_{om}	Wave steepness, Hs/L_{om}
L_{om}	Deepwater wave length corresponding to mean wave period
R_u	Wave runup
H_s^T	Significant wave height with return period T
P	Notational permeability, cf. Figure VI-5-11

Values of a, b, c, and d coefficients.

exceedence probability (%)	a	b	c	d
0.1	1.12	1.34	0.55	2.58
2	0.96	1.17	0.46	1.97
Significant	0.72	0.88	0.41	1.35

Table VI-6-18
Partial Safety Factors for Runup, Hollowed Cubes, Slopes 1:1.5 and 1:2, Berenguer and
Baonza (1995), Design Without Model Tests

Design equation

$$G = \frac{1}{\gamma_z}\hat{R}_u - \gamma_H \hat{H}_S^T \, (0.78 + 0.17\,\hat{\zeta}_p) \qquad\qquad (VI-6-58)$$

where

$$\hat{\zeta}_p = (\cot\,\alpha)^{-1}(\hat{s}_{op})^{-0.5}$$

P_f	$\sigma'_{FH_S} = 0.05$		$\sigma'_{FH_S} = 0.2$	
	γ_H	γ_Z	γ_H	γ_Z
0.01	1.8	1.02	2.0	1.04
0.05	1.4	1.10	1.7	1.00
0.10	1.3	1.08	1.5	1.02
0.20	1.2	1.06	1.3	1.02
0.40	1.0	1.10	1.1	1.02

α	Slope angle
s_{op}	Wave steepness, Hs/L_{op}
L_{op}	Deepwater wavelength corresponding to peak wave period
R_u	Wave runup
H_s^T	Significant wave height with return period T

Table VI-6-19
Partial Safety Factors for Runup, Dolosse, Slopes 1:1.5, Burcharth and Liu (1995b), Design
Without Model Tests

Design equation

$$G = \frac{1}{\gamma_z} \hat{R}_u - \gamma_H \hat{H}_S^T \left(0.75 + 0.11\, \hat{\zeta}_p\right) \qquad\qquad (VI-6-59)$$

where

$$\hat{\zeta}_p = \left(\cot\alpha\right)^{-1} \left(\hat{s}_{op}\right)^{-0.5}$$

P_f	$\sigma'_{FH_S} = 0.05$		$\sigma'_{FH_S} = 0.2$	
	γ_H	γ_Z	γ_H	γ_Z
0.01	1.5	1.10	1.8	1.04
0.05	1.4	1.00	1.5	1.04
0.10	1.3	1.00	1.4	1.02
0.20	1.2	1.00	1.2	1.06
0.40	1.0	1.08	1.0	1.10

α Slope angle
s_{op} Wave steepness, Hs/L_{op}
L_{op} Deepwater wavelength corresponding to peak wave period
R_u Wave runup
H_s^T Significant wave height with return period T

Table VI-6-20
Partial Safety Factors for Steady Stream Scour Depth in Sand at Conical Roundheads, Fredsøe
and Sumer (1997), Design Without Model Tests

Design equation, cf. Eqn. VI-5-262

$$G = \frac{1}{\gamma_z} \frac{\hat{S}}{\hat{B}} - 0.04 \left(1 - \frac{1}{\exp[4(\hat{KC} - 0.05)]} \right) \qquad (VI-6-60)$$

where

$$KC = \frac{U_m T_p}{B}$$

In calculation of U_m (maximum wave orbital velocity at the bed with no structure) use
the wave height $\gamma_H \hat{H}_s^T$

P_f	$\sigma'_{FH_S} = 0.05$		$\sigma'_{FH_S} = 0.2$	
	γ_H	γ_Z	γ_H	γ_Z
0.01	1.7	1.10	1.9	1.10
0.05	1.4	1.10	1.6	1.08
0.10	1.3	1.10	1.4	1.10
0.20	1.2	1.06	1.2	1.10
0.40	1.0	1.10	1.1	1.02

Table VI-6-21
Partial Safety Factors for Scour Depth in Sand at Conical Roundheads in Breaking Wave
Conditions, Fredsøe and Sumer (1997), Design Without Model Tests

Design equation, cf. Eqn. VI-5-264

$$G = \frac{1}{\gamma_z} \hat{S} - 0.01 \left(\frac{\hat{T}_p \sqrt{g \gamma_H \hat{H}_S^T}}{\hat{h}} \right)^{1.5} \qquad (VI-6-61)$$

P_f	$\sigma'_{FH_S} = 0.05$		$\sigma'_{FH_S} = 0.2$	
	γ_H	γ_Z	γ_H	γ_Z
0.01	1.6	1.08	1.8	1.10
0.05	1.4	1.02	1.5	1.10
0.10	1.3	1.02	1.4	1.06
0.20	1.2	1.00	1.3	1.00
0.40	1.1	1.00	1.1	1.00

Table VI-6-22
Partial Safety Factors for Foundation Failure of Vertical Wall Caissons - Sand Subsoil

---

Design equation

$$G = G(\gamma_H \hat{H}_S^T, \hat{\rho}_c, \hat{U}_{Hor.Force}, \hat{U}_{Ver.Force}, \hat{U}_{Hor.Moment}, \hat{U}_{Ver.Moment},$$

$$\hat{\zeta}, \frac{1}{\gamma_Z}\hat{tan}\,\varphi_{d_1}, \frac{1}{\gamma_Z}\hat{tan}\,\varphi_{d_2}, B) \qquad\qquad (VI-6-62)$$

Deep water. Design without model tests. γ_Z is used for both rubble-mound and sand subsoil.

P_f	$\sigma'_{FH_S} = 0.05$		$\sigma'_{FH_S} = 0.2$	
	γ_H	γ_Z	γ_H	γ_Z
0.01	1.4	1.3	1.4	1.3
0.05	1.3	1.2	1.3	1.2
0.10	1.2	1.2	1.2	1.2
0.20	1.1	1.1	1.1	1.2
0.40	1.1	1.0	1.1	1.0

Deep water. Wave load determined by model tests. γ_Z is used for both rubble-mound and sand subsoil.

P_f	$\sigma'_{FH_S} = 0.05$		$\sigma'_{FH_S} = 0.2$	
	γ_H	γ_Z	γ_H	γ_Z
0.01	1.3	1.2	1.4	1.2
0.05	1.3	1.1	1.4	1.1
0.10	1.2	1.1	1.3	1.1
0.20	1.1	1.1	1.1	1.1
0.40	1.1	1.0	1.1	1.0

Shallow water. Design without model tests. γ_Z is used for both rubble-mound and sand subsoil.

P_f	$\sigma'_{FH_S} = 0.05$		$\sigma'_{FH_S} = 0.2$	
	γ_H	γ_Z	γ_H	γ_Z
0.01	1.3	1.4	1.3	1.4
0.05	1.2	1.3	1.3	1.3
0.10	1.2	1.2	1.2	1.2
0.20	1.1	1.1	1.1	1.2
0.40	1.1	1.0	1.1	1.0

Shallow water. Wave load determined by model tests. γ_Z is used for both rubble-mound and sand subsoil.

P_f	$\sigma'_{FH_S} = 0.05$		$\sigma'_{FH_S} = 0.2$	
	γ_H	γ_Z	γ_H	γ_Z
0.01	1.3	1.2	1.4	1.2
0.05	1.3	1.1	1.4	1.1
0.10	1.2	1.1	1.3	1.1
0.20	1.1	1.1	1.1	1.1
0.40	1.1	1.0	1.1	1.0

H_s^T	Significant wave height with return period T
B	Width of caisson
$\hat{U}_{Hor.Force}$	0.90, bias factor to be applied to the Goda horizontal wave force
$\hat{U}_{Ver.Force}$	0.77, bias factor to be applied to the Goda vertical wave force
$\hat{U}_{Hor.Moment}$	0.81, bias factor to be applied to the moment from the Goda horizontal wave forces around the shoreward heel of the base plate
$\hat{U}_{Ver.Moment}$	0.72, bias factor to be applied to the moment from the Goda vertical wave forces around the shoreward heel of the base plate
φ_d	$\frac{\sin\varphi' \cos\psi}{1-\sin\varphi' \sin\psi}$
φ'	Effective friction angle of friction material (sand or rubble stone)
ψ	Dilation angle of friction material (sand or rubble stone)
ρ_c	Mass density of caisson

Table VI-6-23
Partial Safety Factors for Foundation Failure of Vertical Wall Caissons - Clay Subsoil

Design equation

$$G = G(\gamma_H \hat{H}_S^T, \hat{\rho}_c, \hat{U}_{Hor.Force}, \hat{U}_{Ver.Force}, \hat{U}_{Hor.Moment}, \hat{U}_{Ver.Moment},$$

$$\hat{\zeta}, \frac{1}{\gamma_Z} \tan \varphi_{d_1}, \frac{1}{\gamma_c} \hat{c}_u, B) \qquad (VI-6-63)$$

Deep water. Design without model tests. γ_Z is used for rubble-mound and γ_C clay subsoil.

P_f	$\sigma'_{FH_S} = 0.05$			$\sigma'_{FH_S} = 0.2$		
	γ_H	γ_Z	γ_C	γ_H	γ_Z	γ_C
0.01	1.3	1.5	1.6	1.4	1.5	1.6
0.05	1.2	1.4	1.5	1.3	1.4	1.5
0.10	1.1	1.3	1.5	1.2	1.3	1.5
0.20	1.0	1.3	1.4	1.0	1.3	1.5
0.40	1.0	1.1	1.1	1.0	1.1	1.2

Deep water. Wave load determined by model tests. γ_Z is used for rubble-mound and γ_C clay subsoil.

P_f	$\sigma'_{FH_S} = 0.05$			$\sigma'_{FH_S} = 0.2$		
	γ_H	γ_Z	γ_C	γ_H	γ_Z	γ_C
0.01	1.2	1.5	1.6	1.3	1.5	1.6
0.05	1.1	1.3	1.5	1.2	1.3	1.5
0.10	1.0	1.3	1.5	1.1	1.3	1.4
0.20	1.0	1.2	1.3	1.0	1.3	1.3
0.40	1.0	1.1	1.1	1.0	1.1	1.1

Shallow water. Design without model tests. γ_Z is used for rubble-mound and γ_C for clay subsoil.

P_f	$\sigma'_{FH_S} = 0.05$			$\sigma'_{FH_S} = 0.2$		
	γ_H	γ_Z	γ_C	γ_H	γ_Z	γ_C
0.01	1.2	1.5	1.6	1.3	1.5	1.6
0.05	1.1	1.4	1.5	1.2	1.4	1.5
0.10	1.1	1.3	1.3	1.2	1.3	1.3
0.20	1.0	1.3	1.3	1.1	1.2	1.3
0.40	1.0	1.1	1.1	1.1	1.1	1.1

Shallow water. Wave load determined by model tests. γ_Z is used for rubble-mound and γ_C for clay subsoil.

P_f	$\sigma'_{FH_S} = 0.05$			$\sigma'_{FH_S} = 0.2$		
	γ_H	γ_Z	γ_C	γ_H	γ_Z	γ_C
0.01	1.2	1.3	1.4	1.3	1.3	1.4
0.05	1.1	1.2	1.4	1.2	1.2	1.4
0.10	1.1	1.2	1.3	1.1	1.2	1.3
0.20	1.0	1.1	1.3	1.1	1.1	1.2
0.40	1.0	1.0	1.0	1.1	1.0	1.0

H_s^T	Significant wave height with return period T
B	Width of caisson
$\hat{U}_{Hor.Force}$	0.90, bias factor to be applied to the Goda horizontal wave force
$\hat{U}_{Ver.Force}$	0.77, bias factor to be applied to the Goda vertical wave force
$\hat{U}_{Hor.Moment}$	0.81, bias factor to be applied to the moment from the Goda horizontal wave forces around the shoreward heel of the base plate
$\hat{U}_{Ver.Moment}$	0.72, bias factor to be applied to the moment from the Goda vertical wave forces around the shoreward heel of the base plate
φ_d	$= \frac{\sin\varphi' \cos \psi}{1-\sin \varphi' \sin \psi}$
φ'	Effective friction angle of friction material (sand or rubble stone)
ψ	Dilation angle of friction material (sand or rubble stone)
ρ_c	Mass density of caisson
c_u	Undrained shear strength of clay

Table VI-6-24
Partial Safety Factors for Sliding Failure of Vertical Wall Caissons

Design equation

$$G = G(\gamma_H \hat{H}_s^T, \hat{\rho}_c, \hat{U}_{Hor.Force}, \hat{U}_{Ver.Force}, \hat{\zeta}, \frac{1}{\gamma_Z}\hat{f}, B)$$

$$= (\hat{F}_G - \hat{U}_{Ver.Force}\hat{F}_U)\frac{1}{\gamma_Z}\hat{f} - \hat{U}_{Hor.Force}\hat{F}_H \qquad (VI-6-64)$$

In calculation of $\hat{F}_U$ and $\hat{F}_H$ use wave height $\gamma_H \hat{H}_s^T$.

Deep water. Design without model tests.

P_f	$\sigma'_{FH_S} = 0.05$		$\sigma'_{FH_S} = 0.2$	
	γ_H	γ_Z	γ_H	γ_Z
0.01	1.4	1.7	1.5	1.7
0.05	1.3	1.4	1.4	1.4
0.10	1.3	1.2	1.4	1.3
0.20	1.2	1.2	1.3	1.2
0.40	1.1	1.0	1.1	1.1

Deep water. Wave load determined by model tests.

P_f	$\sigma'_{FH_S} = 0.05$		$\sigma'_{FH_S} = 0.2$	
	γ_H	γ_Z	γ_H	γ_Z
0.01	1.3	1.5	1.4	1.5
0.05	1.2	1.4	1.3	1.4
0.10	1.2	1.2	1.3	1.2
0.20	1.1	1.2	1.2	1.2
0.40	1.0	1.2	1.1	1.0

Shallow water. Design without model tests.

P_f	$\sigma'_{FH_S} = 0.05$		$\sigma'_{FH_S} = 0.2$	
	γ_H	γ_Z	γ_H	γ_Z
0.01	1.3	1.9	1.4	1.9
0.05	1.2	1.6	1.3	1.6
0.10	1.2	1.4	1.3	1.4
0.20	1.1	1.3	1.2	1.3
0.40	1.0	1.2	1.0	1.2

Shallow water. Wave load determined by model tests.

P_f	$\sigma'_{FH_S} = 0.05$		$\sigma'_{FH_S} = 0.2$	
	γ_H	γ_Z	γ_H	γ_Z
0.01	1.2	1.6	1.3	1.6
0.05	1.1	1.5	1.2	1.5
0.10	1.1	1.3	1.2	1.3
0.20	1.1	1.2	1.1	1.2
0.40	1.0	1.1	1.0	1.1

H_s^T	Significant wave height with return period T
B	Width of caisson
$\hat{U}_{Hor.Force}$	0.90, bias factor to be applied to the Goda horizontal wave force
$\hat{U}_{Ver.Force}$	0.77, bias factor to be applied to the Goda vertical wave force
$\hat{U}_{Hor.Moment}$	0.81, bias factor to be applied to the moment from the Goda horizontal wave forces around the shoreward heel of the base plate
$\hat{U}_{Ver.Moment}$	0.72, bias factor to be applied to the moment from the Goda vertical wave forces around the shoreward heel of the base plate
ρ_c	Mass density of caisson
F_G	Buoyancy reduced weight of caisson
F_H	Horizontal wave force calculated by the Goda formula
F_U	Wave induced uplift force calculated by the Goda formula
f	Friction coefficient

Table VI-6-25
Partial Safety Factors for Overturning Failure of Vertical Caissons

Design equation

$$G = G(\gamma_H \hat{H}_S^T, \hat{\rho}_c, \hat{U}_{Hor.Moment}, \hat{U}_{Ver.Moment}, \hat{\zeta}, B)$$
$$= (\hat{M}_G - \hat{U}_{Ver.Moment} M_U) - \hat{U}_{Hor.Moment} M_H \qquad (VI-6-65)$$

In calculation of $\hat{M}_U$ and $\hat{M}_H$ use wave height $\gamma_H \hat{H}_s^T$.

Design without model tests.

P_f	$\sigma'_{FH_S} = 0.05$ γ_H	$\sigma'_{FH_S} = 0.2$ γ_H
0.01	-	-
0.05	2.7	-
0.10	2.0	2.5
0.20	1.6	1.7
0.40	1.2	1.2

Wave load determined by model tests.

P_f	$\sigma'_{FH_S} = 0.05$ γ_H	$\sigma'_{FH_S} = 0.2$ γ_H
0.01	2.1	2.3
0.05	1.7	1.9
0.10	1.4	1.6
0.20	1.3	1.4
0.40	1.1	1.2

H_s^T	Significant wave height with return period T
B	Width of caisson
$\hat{U}_{Hor.Force}$	0.90, bias factor to be applied to the Goda horizontal wave force
$\hat{U}_{Ver.Force}$	0.77, bias factor to be applied to the Goda vertical wave force
$\hat{U}_{Hor.Moment}$	0.81, bias factor to be applied to the moment from the Goda horizontal wave forces around the shoreward heel of the base plate
$\hat{U}_{Ver.Moment}$	0.72, bias factor to be applied to the moment from the Goda vertical wave forces around the shoreward heel of the base plate
ρ_c	Mass density of caisson
M_G	Moment of F_G around heel of caisson
M_H	Moment of F_H around heel of caisson
M_U	Moment of F_U around heel of caisson
F_G	Buoyancy reduced weight of caisson
F_H	Horizontal wave force calculated by the Goda formula
F_U	Wave induced uplift force calculated by the Goda formula

Table VI-6-26
Partial Safety Factors for Scour at Circular Vertical Wall Roundheads, Sumer and Fredsøe (1997), Design Without Model Tests

Design equation, cf. Equation VI-5-257

$$G = \frac{1}{\gamma_Z} \frac{\hat{S}}{\hat{B}} - 0.5 \left[1 - \exp(-0.175 \left[\hat{KC} - 1\right])\right] \qquad (VI-6-66)$$

where

$$KC = \frac{U_m T_p}{B}$$

In calculation of U_m (maximum wave orbital velocity at the bed with no structure) use the wave height $\gamma_H \hat{H}_s^T$.

Deep water.

P_f	$\sigma'_{FH_S} = 0.05$		$\sigma'_{FH_S} = 0.2$	
	γ_H	γ_Z	γ_H	γ_Z
0.01	2.0	2.4	2.0	2.4
0.05	2.0	2.0	2.0	2.0
0.10	2.0	1.8	2.0	1.8
0.20	2.0	1.5	2.0	1.5
0.40	2.0	1.2	2.0	1.2

Shallow water.

P_f	$\sigma'_{FH_S} = 0.05$		$\sigma'_{FH_S} = 0.2$	
	γ_H	γ_Z	γ_H	γ_Z
0.01	2.0	2.4	2.0	2.4
0.05	2.0	2.0	2.0	2.0
0.10	2.0	1.8	2.0	1.8
0.20	2.0	1.5	2.0	1.5
0.40	2.0	1.2	2.0	1.2

Table VI-6-27
Partial Safety Factors for Toe Berm Rock Armor Failure in Front of Vertical Wall Caissons, Design Without Model Tests

Design equation (cf. Table VI-5-48)

$$G = \frac{1}{\gamma_Z} \hat{\Delta} \hat{D}_n \left(5.8 \frac{\hat{h}_b}{\hat{h}_s} - 0.60\right) (N_{od})^{0.19} - \gamma_H \hat{H}_S^T \qquad (VI-6-67)$$

Deep water.

P_f	$\sigma'_{FH_S} = 0.05$		$\sigma'_{FH_S} = 0.2$	
	γ_H	γ_Z	γ_H	γ_Z
0.01	1.6	1.3	1.7	1.3
0.05	1.4	1.2	1.5	1.2
0.10	1.3	1.2	1.4	1.2
0.20	1.2	1.1	1.3	1.1
0.40	1.1	1.0	1.2	1.0

Shallow water.

P_f	$\sigma'_{FH_S} = 0.05$		$\sigma'_{FH_S} = 0.2$	
	γ_H	γ_Z	γ_H	γ_Z
0.01	1.5	1.5	1.6	1.5
0.05	1.3	1.3	1.4	1.3
0.10	1.2	1.2	1.3	1.2
0.20	1.1	1.2	1.2	1.2
0.40	1.1	1.0	1.2	1.0

VI-6-7. <u>References</u>.

Berengu☐r and Baonza 1995
Berengu☐r, J. M., and Baonza, A. 1995. "Hollowed Cube Research," Final Proceedings,
EU-MAST 2 Project MAS 2- CT92 - 0042, Rubble-Mound Breakwater Failure Modes, Aalborg
University, Denmark.

Burcharth 1991
Burcharth, H. F. 1991. "Introduction of Partial Coefficient in the Design of Rubble Mound
Breakwaters," *Proceedings of the Conference on Coastal Structures and Breakwaters*, Institution
of Civil Engineers, London, Thomas Telford Publishing, London, UK, pp 543-565.

Burcharth 1992
Burcharth, H. F. 1992. "Uncertainty Related to Environmental Data and Estimated Extreme
Events," Final Report of PIANC Working Group 12, Group B, June 1992.

Burcharth and Liu 1995a
Burcharth, H. F., and Liu, Z. 1995a. "2-D Model Tests of Dolos Breakwater," Final Proceedings,
EU-MAST 2 project MAS 2 - CT92 - 0042, Rubble Mound Breakwater Failure Modes, Aalborg
University, Denmark.

Burcharth and Liu 1995b
Burcharth, H. F., and Liu, Z. 1995b. "Design Formulae for Hydraulic Stability and Structural
Integrity of Dolos Breakwater Roundheads," Final Proceedings, EU-MAST 2 project MAS 2 -
CT92 -0042, Rubble Mound Breakwater Failure Modes, Aalborg University, Denmark.

Burcharth and Sørensen 1999
Burcharth, H. F., and Sørensen, J. D. 1999. "The PIANC Safety Factor System for Breakwaters,"
Proceedings of Coastal Structures' 99, J. J. Losada and A. A. Balkema, eds., Vol 2, pp 1125-
1144.

Cornell 1969
Cornell, C. A. 1969. "A Reliability-Based Structural Code," *ACI-Journal*, Vol 66, pp 974-985.

De Waal and van der Meer 1992
De Waal, J. P., and van der Meer, J. W. 1992. "Wave Run-Up and Overtopping of Coastal
Structures," *Proceedings of the 23rd International Coastal Engineering Conference*, Vol 2, pp
1758-1771.

Fredsøe and Sumer 1997
Fredsøe, J., and Sumer, B. M. 1997. "Scour at the Roundhead of a Rubble Mound Breakwater,"
Coastal Engineering, Elsevier, Vol. 29, No 3-4, pp 231-262.

Hasofer and Lind 1974
Hasofer, A. M., and Lind, N. C. 1974. "Exact and Invariant Second-Moment Code Format,"
Journal of Engineering Mechanics Division, American Society of Civil Engineers, Vol 100, No.
EM1, pp 111-121.

Hansen 1979
Hansen, B. 1979. "Definition and Use of Friction Angles," Proceedings of the Seventh European Conference on Soil Mechanics and Foundation Engineering, Brighton, UK.

Sumer and Fredsøe 1997
Sumer, B. M., and Fredsøe, J. 1997. "Scour at the Head of a Vertical-Wall Breakwater," *Coastal Engineering*, Elsevier, Vol 29, No. 3-4, pp 201-230.

Wen and Chen 1987
Wen, Y. K., and Chen, H. C. 1987. "On Fast Integration for Time Variant Structure Reliability," *Probabilistic Engineering Mechanics*, Vol 2, pp 156-162.

VI-6-8. Acknowledgments.

Author: Dr. Hans F. Burcharth, Department of Civil Engineering, Aalborg University, Aalborg, Denmark.

Reviewers: H. Lee Butler (CHL retired), Dr. Steven A. Hughes, and Jeffrey A. Melby, Coastal and Hydraulics Laboratory (CHL), U.S. Army Engineer Research and Development Center (ERDC), Vicksburg, MS.

VI-6-9. Symbols.

α — Weibull distribution parameter

α — Armor slope angle

β — Weibull distribution parameter

β_{HL} — Hasofer and Lind reliability index

β — Reliability index

γ_i — Partial safety factors

λ — Number of observations per year

μ — Mean value

ρ_c — Mass density of caisson [force/length^3]

ρ_w — Mass density of water (salt water = 1,025 kg/m^3 or 2.0 slugs/ft^3; fresh water = 1,000 kg/m^3 or 1.94 slugs/ft^3) [force-time^2/length^4]

ρ_s — Mass density of concrete [force/length^3]

ρ_s — Block density

$\sigma!_{FHs}$ — Coefficient of variation

σ — Standard deviation

Ψ — Angle of dilation [degrees]

ν	Packing density
$\nu!$	Angle of friction in granular material [degrees]
B	Width of caisson [length]
c_u	Undrained shear strength of clay [force/length^2]
D	Relative number of units
D_n	Nominal block diameter [length]
f	Friction coefficient [dimensionless]
F_U	Wave induced uplift force [force]
F_H	Horizontal wave force [force]
F_G	Buoyancy reduced weight of caisson [force]
g	Failure function
H^T_s	Significant wave height with return period T
h	Water depth [length]
$h_S!$	Threshold level
H_S	Significant wave height [length]
K_D	Coefficient signifying the degree of damage [dimensionless]
L_{om}	Deepwater wave length corresponding to mean wave period [length]
L_{op}	Deepwater wave length corresponding to the peak wave period [length]
M_G	Moment around the heel of caisson by buoyancy-reduced weight of the caisson [length-force]
M_H	Moment around the heel of caisson by wave induced horizontal force [length-force]
M_U	Moment around the heel of caisson by wave induced uplift force [length-force]
N_{od}	Number of units displaced out of the armor layer
N_z	Number of waves
N	Number of data points
P	Notational permeability parameter (Figure VI-5-11)
P_f	Probability of failure
R_u	Maximum runup or water-surface elevation measured vertically from the still-water level [length]
r	Dolos waist ratio [dimensionless]

R_f Reliability

R Variable representing the variations in resistance between nominally identical structures

s_{om} Wave steepness ($=H_s/L_{om}$)

s_{op} Deepwater wave steepness [dimensionless]

S Represents the maximum load effects within a period of time

T_L Design lifetime [years]

T_m Wave period [time]

U_m Maximum wave orbital velocity at the bed with no structure [length/time]

W Water level [length]

CHAPTER 7

Example Problems

TABLE OF CONTENTS

List of Figures

CHAPTER VI-7

Example Problems

VI-7-1. <u>Introduction</u>.

a. "Only the application makes the rod into a lever" is the famous remark of the philosopher Ludwig Wittgenstein (Pitcher 1964). All engineers remember their university days (and nights) doing homework problems that turned the lectures (rod) into useful information and tools (lever) by the application of the materials presented. Those textbooks with many example problems (and answers to the homework problems) always rate as the best.

b. The *Coastal Engineering Manual* (CEM) is divided into six parts. The first four parts mainly cover the science surrounding the subject while the remaining Parts V and VI summarize the latest engineering knowledge, studies, designs, and constructions. Part VI-7 has been set aside for example problems. This chapter includes wave runup, wave overtopping, armor-layer stability, and forces on vertical-front structures.

c. The single, most important coastal engineering advance has been the use of irregular water-wave spectra in the analytical treatment, physical (laboratory) experiments, and numerical model simulations to study wave runup, overtopping, and armor-layer stability. Coastal engineers must adopt this new technology quickly to prepare more cost-effective and safe designs in the future.

d. Throughout the example problems chapter, several references to the *Shore Protection Manual* (1984) are made. By referencing the older document, an attempt has been made to identify differences in engineering practice between the older *Shore Protection Manual* and the newer *Coastal Engineering Manual*.

VI-7-2. Wave Runup.

EXAMPLE PROBLEM VI-7-1

FIND: The surf-similarity parameter (also called the Iribarren number) for use in wave runup and wave overtopping calculations for long-crested, irregular waves on impermeable (without water penetration) and permeable slopes.

GIVEN: An impermeable structure has a smooth slope of 1 on 2.5 and is subjected to a design significant wave, H_s = 2.0 m (6.6 ft) measured at a gauge located in a depth, d = 4.5 m (14.8 ft). Design wave peak period is T_p = 8 s. Water depth at structure toe at high water is d_{toe} = 3.0 m (9.8 ft). (Assume no change in the refraction coefficient between the structure and the wave gauge.)

SOLUTION: The surf-similarity parameter for irregular waves depends on the wave steepness and structure slope. Two definitions are given in Equation VI-5-2 formulated with either the peak wave period, T_p or the mean wave period, T_m; but both use the significant wave height at the toe of the structure.

(Sheet 1 of 5)

EXAMPLE PROBLEM VI-7-1 (Continued)

Ideally, a spectral wave model would be used to shoal the irregular wave H_s to the structure toe. However, for purposes of illustration, it is assumed that H_s will shoal according to linear wave theory. For swell-type spectra this is reasonable assumption, but linear shoaling overestimates shoaling of fully saturated storm spectra.

Item 1. Linear, regular wave shoaling (illustrated by several of the available methods).

(a) Deep water.

First calculate the deep water, unrefracted wave height, H_o' from where measured back out to deep water. Using the depth where waves measured, and assuming $T = T_p = 8$ s and $H = H_s$ gives

$$\frac{d}{L_o} = \frac{2\pi d}{gT^2} = \frac{2\pi(4.5 \text{ m})}{(9.81 \text{ m/s}^2)(8 \text{ s})^2} = 0.0450$$

(1) From the *Shore Protection Manual* (1984), Table C-1, Appendix C for $d/L_o = 0.0450$.

$$\frac{H}{H_o'} = 1.042, \text{ the shoaling coefficient, } k_s$$

Therefore,

$$H_o' = \frac{2.0 \text{ m}}{1.042} = 1.92 \text{ m} (6.3 \text{ ft})$$

(2) or, using ACES (Leenknecht et al. 1992), Snell's Law, crest angle = 0.0°

$$H_o' = 1.92 \text{ m} (6.3 \text{ ft})$$

(3) or, using explicit approximations (e.g., Nielsen 1984)

$$\frac{C_g}{C_o} = \sqrt{k_o d}\left[1 - \frac{1}{2}k_o d + \frac{7}{72}(k_o d)^2\right]$$

$$k_o = \frac{2\pi}{L_o} = \frac{4\pi^2}{gT^2}$$

$$K_s = \sqrt{0.5\, C_o / C_g}$$

(Sheet 2 of 5)

EXAMPLE PROBLEM VI-7-1 (Continued)

gives

$$k_o d = 0.28296$$

$$\frac{C_g}{C_o} = 0.46082$$

$$K_s = 1.0416$$

$$H_o{}' = 1.920 \ m \ (6.3 \ ft)$$

(b) Toe of structure

Next, shoal the deepwater wave to a depth, $d = 3.0$ m (9.8 ft) at the toe of the structure

(1) From the *Shore Protection Manual* (1984), Table C-1, Appendix C for

$$\frac{d_{toe}}{L_o} = \frac{2\pi d_{toe}}{gT^2} = \frac{2\pi (3.0 \ m)}{(9.81 \ m/s^2)(8 \ s)^2} = 0.030023$$

$$K_s = \frac{H_{toe}}{H_o{}'} = 1.125$$

$$H_{toe} = 1.92 \ m \ (1.125) = 2.16 \ m (7.09 \ ft)$$

(2) From ACES, Snell's law, crest angle = 0.0°.

$$H_{toe} = 2.161 \ (7.09 \ ft)$$

(3) From explicit approximations

$$k_o d = 0.18864$$

$$\frac{C_g}{C_o} = 0.39486$$

$$K_s = 1.1253$$

$$H_{toe} = 2.161 \ m \ (7.09 \ ft)$$

(Sheet 3 of 5)

EXAMPLE PROBLEM VI-7-1 (Continued)

<u>Item 2</u>. Deepwater wave steepness, s_{op}

$$\left(s_{op}\right)_{toe} \equiv \frac{\left(H_s\right)_{toe}}{L_{op}} = \frac{2\pi H_{toe}}{gT_p^2} = \frac{2\pi\left(2.16\text{ m}\right)}{9.81\text{ m/s}^2\left(8\text{ s}\right)^2}$$

$$s_{op} = 0.02162$$

<u>Item 3</u>. Surf-similarity parameter, ξ_{op}

Finally, the surf-similarity parameter, ξ_{op} as defined by Equation VI-5-2 gives

$$\xi_{op} \equiv \frac{\tan\alpha}{\sqrt{s_{op}}} = \frac{1/2.5}{\sqrt{0.02162}} = \frac{0.4}{0.1469}$$

Therefore,

$$\xi_{op} = 2.72$$

Note that the subscript notation means using the deepwater wavelength, L_o, and the peak wave period, T_p, to calculate ξ_{op}.

<u>Item 4</u>. Surf-similarity parameter, ξ_{om}

The mean wave period, T_m, requires knowledge of variations in the width of the wave spectrum. From Section VI-5-2-a-(3)-(b) for the theoretical spectrums

JONSWAP spectra $\qquad\qquad T_m/T_p = 0.79$ to 0.87

or

PIERSON-MOSKOWITZ spectra $\qquad T_m/T_p = 0.71$ to 0.82

Therefore, assuming $T_m/T_p = 0.76$, gives $T_m = 6.1$ s, hence

$$s_{om} = \frac{H_{toe}}{\frac{g}{2\pi}T_m^2} = \frac{\left(2\pi\right)\left(2.16\text{ m}\right)}{9.81\text{ m/s}^2\left(6.1\text{ s}\right)^2} = 0.03718$$

(Sheet 4 of 5)

EXAMPLE PROBLEM VI-7-1 (Continued)

Therefore,

$$\xi_{om} \equiv \frac{\tan \alpha}{\sqrt{s_{om}}} = \frac{1/2.5}{\sqrt{0.03718}} = \frac{0.4}{0.1928}$$

$$\xi_{om} = 2.07$$

Both ξ_{op} and ξ_{om} are employed in wave runup and overtopping formulations.

DISCUSSION: In general (for either ξ_{op} or ξ_{om})

$$\xi_o \equiv \frac{\tan \alpha}{\sqrt{s_o}}$$

$$s_o \equiv \frac{H_s}{L_o} = \frac{2\pi \, H_s}{gT^2}$$

$$\xi_o \equiv \frac{(\tan \alpha) \, T}{\left(\dfrac{2\pi}{g}\right)^{1/2} (H_s)^{1/2}}$$

or

$$\xi_o \equiv \frac{T}{(K)\cot \alpha \, (H_s)^{1/2}}$$

where k is a constant.

(1) As T increases, ξ_o increases

(2) As cot α increases (flatter slope), ξ_o decreases

(3) As H_s increases, ξ_o decreases, nonlinearly

(Sheet 5 of 5)

EXAMPLE PROBLEM VI-7-2

FIND:

(a) The height above the still-water level (SWL) to which a new revetment must be built to prevent wave overtopping by the design wave. The structure is to be impermeable.

(b) The reduction in required structure height if uniform-sized armor stone is placed on the slope.

GIVEN: An impermeable structure has a smooth slope of 1 on 2.5 and is subjected to a design, significant wave $H_s = 2.0$ m (6.6 ft) measured at a gauge located in a depth $d = 4.5$ m (14.8 ft). Design wave peak period is $T_p = 8$ s. Water depth at structure toe at high water is $d_{toe} = 3.0$ m (9.8 ft).

SOLUTION: From Example Problem VI-7-1, linear wave theory estimates the wave height due to wave shoaling as

$$\left(H_s\right)_{toe} = 2.16 \text{ m } (7.1 \text{ ft})$$

and the surf-similarity parameter as

$$\xi_{op} = 2.72$$

To prevent wave overtopping, the wave runup value at the 2 percent probability of exceedance level is calculated. Figure VI-5-3 displays the considerable scatter in the data for smooth slopes, irregular, long-crested, head-on waves and Table VI-5-2 gives the coefficients for use in Equation VI-5-3 when

$$\xi_{op} > 2.5$$

Namely

$$\frac{R_{u2\%}}{\left(H_s\right)_{toe}} = -0.2\xi_{op} + 4.5$$

$$= -0.2(2.72) + 4.5 = -0.544 + 4.5 = 3.956$$

$$R_{u2}\% = 3.956 \ (2.16 \text{ m}) = 8.545 \text{ m } (28.0 \text{ ft})$$

(Sheet 1 of 4)

EXAMPLE PROBLEM VI-7-2 (Continued)

So that,

(a) Smooth slopes

(1) To prevent overtopping; $R_{u2\%} = 8.55$ m (28.1 ft)
(Note that $\gamma_r = \gamma_b = \gamma_h = \gamma_\beta = 1.0$ are taken in Equation VI-5-3 for smooth, no berm, Rayleigh distribution, and zero incidence angle conditions, respectively.)
Another set of runup data for smooth slopes is presented in Figure VI-5-5 and Equation VI-5-6 (from de Waal and van der Meer 1992). When

$$\xi_{op} > 2.0$$

namely

$$\frac{R_{u2\%}}{\left(H_s\right)_{toe}} = 3.0$$

hence,

(2) To prevent overtopping; $R_{u2\%} = 3.0 \ (2.18 \text{ m}) = 6.48$ m (21.3 ft)
Note that the data in Figure VI-5-5 is for slopes milder than 1 on 2.5, and thus may not be appropriate for this example.

(b) Rough slopes
The surface roughness reduction factor γ_r for Equation VI-5-3 is given in Table VI-5-3 and lies in the range, $\gamma_r = 0.5\text{-}0.6$ for one or more layers of rock.

Use $\gamma_r = 0.55$ to get

(1) Equation VI-5-3 $\dfrac{R_{u2\%}}{\left(H_s\right)_{toe}} = 3.956 \ \gamma_r = 3.956\,(0.55) = 2.176$

hence

$$R_{u2\%} = 2.176 \ (2.16 \text{ m}) = 4.70 \ (15.4 \text{ ft})$$

(Sheet 2 of 4)

EXAMPLE PROBLEM VI-7-2 (Continued)

and

(2) Equation VI-5-6 $\dfrac{R_{u2\%}}{(H_s)_{toe}} = 3.0$ $\gamma_r = 3.0(0.55) = 1.65$

$R_{u2\%}$ = 1.65 (2.16 m) = 3.56 m (11.7 ft)

(3) The Delft Hydraulics test program (Table VI-5-4) also provided data for impermeable rock slopes. Here, the surf-similarity parameter based on mean wave period, ξ_{om}, is employed to develop design Equation VI-5-12 with coefficients in Table VI-5-5 for a wide range of exceedance probabilities. The ξ_{om}-value for this example ($\xi_{om} = 2.07$) was estimated in Example Problem VI-7-1.

When $\xi_{om} > 1.5$

$$\frac{R_{u2\%}}{(H_s)_{toe}} = 1.17(\xi_{om})^{0.46}$$
$$= 1.17(2.07)^{0.46} = 1.635$$

Therefore, $R_{u2\%}$ = 1.635 (2.16 m) = 3.353 m (11.6 ft) using coefficients for B and C at the 2% exceedance probability level. This result is very close to that in the preceding Equation VI-5-6 (2) taking $\gamma_r = 0.55$.

(4) Partial safety factors, γ_H and γ_z and $\hat{R}_{u2\%}$. The Delft Hydraulics data set has been analyzed for partial safety factors as discussed in Part VI-6 and presented in Table VI-6-17.

Assume the annual failure probability $P_f = 0.10$ (90% reliability). For relatively low uncertainty in knowledge of the wave height ($\sigma'_{FHS} = 0.05$) the values associated with Equation VI-6-57 yield

$\gamma_H \gamma_z = 1.2\ (1.06) = 1.272$

(Sheet 3 of 4)

EXAMPLE PROBLEM VI-7-2 (Concluded)

and a probabilistic estimate, $\hat{R}_{u2\%}$ = 1.272 (3.53 m) = 4.49 m (14.7 ft). If the uncertainty is higher regards wave height (σ'_{FHS} = 0.2) then

$$\gamma_H \gamma_z = 1.3 (1.08) = 1.404$$

and a probabilistic estimate, $\hat{R}_{u2\%}$ = 1.404 (3.53 m) = 4.96 m (16.3 ft).

The range of $\hat{R}_{u2\%}$ = 4.5-5.0 m (14.8-16.4 ft) brackets the estimate of $\hat{R}_{u2\%}$ = 4.7 m (15.4 ft) as found from Equation VI-5-3. The higher estimate of the $\hat{R}_{u2\%}$ value found from Equation VI-5-3 could also be explained as being reliable at the 90-percent annual level.

DISCUSSION: As seen in Figure VI-5-3, at ξ_{op} = 2.5, $R_{u2\%}/H_s$ reaches a maximum value. Solving for the variables involved in ξ_{op} gives approximately

Metric system

$$\frac{(\tan \alpha) T_p}{(H_{toe})^{1/2}} = 2.0$$

English system

$$\frac{(\tan \alpha) T_p}{(H_{toe})^{1/2}} = 1.10$$

For the preceding example, keeping $\tan$ = 0.4, H_{toe} = 2.16 m gives

$$(T_p) = 7.35 \text{ s for maximum runup}$$

For the preceding example, keeping T_p = 8.0 s, H_{toe} = 2.16 m gives

$$\tan \alpha = 0.36, \cot \alpha = 2.7 \text{ for maximum runup.}$$

Using a steeper or flatter slope will reduce the wave runup, all else being equal.

(Sheet 4 of 4)

EXAMPLE PROBLEM VI-7-3

FIND: The height above the SWL to which a rock-armored structure (permeable) should be built to prevent wave overtopping by the design wave.

GIVEN: The same information for Example Problem VI-7-2 as summarized as follows, but now for a permeable breakwater (jetty) structure

$$
\begin{aligned}
\text{slope} &= 1\!:\!2.5 \\
H_s &= 2.0 \text{ m (6.6 ft)} \\
\text{measured at } d &= 4.5 \text{ m (14.8 ft)} \\
T_p &= 8 \text{ s} \\
d_{toe} &= 3.0 \text{ m (9.8 ft)} \\
\xi_{om} &= 2.07 \\
(H_s)_{toe} &= 2.16 \text{ m}
\end{aligned}
$$

SOLUTION: Core permeability may significantly influence wave runup. Notational permeability coefficients are defined in Figure VI-5-11. The previous Example Problem VI-7-2 was for $P = 0.1$ defined as impermeable. Test results shown in Figures VI-5-12 are with $P = 0.1$ and $P = 0.5$ and clearly reveal the runup reduction when $\xi_{om} > 3$ for permeable structures. Equation VI-5-13 has been developed as the central fit to the permeable data with coefficients again found in Table VI-5-5. For $R_{u2\%}$, $B = 1.17$, $C = 0.46$, and $D = 1.97$. Selection of the appropriate equation requires calculation of

$$
(D/B)^{1/C} = \left(\frac{1.97}{1.17}\right)^{\frac{1}{0.46}} = (1.68)^{2.17} = 3.10
$$

Because $1.5 < \xi_{om} < (D/B)^{1/C}$, use the equation

$$
\frac{R_{u2\%}}{\left(H_s\right)_{toe}} = B\left(\xi_{om}\right)^C
$$

$$
= 1.17(2.07)^{0.46} = 1.635
$$

Therefore, $R_{u2\%} = 1.635 \,(2.16 \text{ m}) = 3.53 \text{ m (11.6 ft)}$, and this is a similar result as for $P = 0.1$, impermeable slopes.

For the 2% runup exceedence level a value of

$$
\xi_{om} \equiv \frac{\tan \alpha}{\sqrt{s_{om}}} \geq 3.10
$$

(Sheet 1 of 2)

EXAMPLE PROBLEM VI-7-3 (Concluded)

is the point where the permeable core begins to reduce wave runup. Longer period waves will increase ξ_{om}, but the runup remains constant because of the structure permeability. At this limit,

$$\frac{R_{u2\%}}{(H_s)_{toe}} = D = 1.97 \text{ (see Table VI-5-5)}$$

Therefore,

$$(R_{u2\%})\max = 1.97\,(2.16 \text{ m}) = 4.25 \text{ m (13.9 ft) for } T_m \geq 9.1 \sec\left(T_p \approx 12s\right)$$

for a slope of 1 to 2.5.

(Sheet 2 of 2)

EXAMPLE PROBLEM VI-7-4

FIND: The height above the still-water level to which a revetment must be built to prevent wave overtopping by the design wave (same as Example Problem VI-7-2) but for the following conditions:

(a) Statistical distributions of wave runup

(b) Influence of shallow water on wave runup

(c) Influence of wave angle and directional spreading on wave runup

GIVEN: Same conditions as Example Problem VI-7-2 for smooth slope

SOLUTION: Equation VI-5-3 holds in general for any $R_{ui\%}$ defined as the runup level exceeded by $i\%$ of the incident waves. Coefficients A and C depend on both ξ_{op} and i for Rayleigh distributed wave heights.

(a) Statistical distributions

(1) Significant runup. Figure VI-5-4 displays the data scatter and Table VI-5-2 provides coefficients to calculate the significant wave runup, R_{us}. Again from Example Problem VI-7-1

$$\left(H_s\right)_{toe} = 2.16 \text{ m (7.1 ft)}$$

and

$$\xi_{op} = 2.72$$

For R_{us} in the range $2 < \xi_{op} < 9$

$$\frac{R_{us}}{\left(H_s\right)_{toe}} = -0.25\,\xi_{op} + 3.0$$

$$= -0.25(2.72) + 3.0 = -0.68 + 3.0 = 2.32$$

therefore,

$$R_{us} = 2.32 \ (2.16 \text{ m}) = 5.01 \text{ m (16.4 ft)}$$

(Sheet 1 of 5)

EXAMPLE PROBLEM VI-7-4 (Continued)

The *Shore Protection Manual* (1984) calculated a runup value of 5.6 m (18.4 ft) for Example Problem No. 4 for the same data taking the design wave as the significant wave height. The Rayleigh distribution for wave heights gives the following relationships for extreme events

$$\frac{H_{0.1}}{H_{0.135}} = \frac{H_{0.1}}{H_s} = 1.072$$

$$\frac{H_{0.02}}{H_s} = 1.398$$

$$\frac{H_{0.01}}{H_s} = 1.516$$

If the wave runup also followed a Rayleigh distribution, then it might be expected that

$$\frac{R_{u0.2}}{R_{us}} = 1.398$$

gives

$$R_{u0.2} = 1.398 \ (5.01 \text{ m}) = 7.0 \text{ m} \ (23.0 \text{ ft})$$

This result is much lower than $R_{u2\%} = 8.55$ m (28.1 ft) calculated in Example Problem-7-2 for the smooth slope. In general, values for R_{us} and $R_{u2\%}$ calculated from Equation VI-5-3 and coefficients in Table VI-5-2 do not follow a Rayleigh distribution for wave runup.

(2) Statistical distribution of runup on permeable slopes

For the following restrictions: (1) Rayleigh distributed wave heights
                                    (2) Permeable, rock armored slopes
                                    (3) Slope, cot $\alpha > 2$

Equation VI-5-15 says

$$R_{up\%} = B \ (-ln \ p)^{1/C}$$

where
    $R_{up\%}$ = runup level exceeded by p% of runup
and
        B, C are calculated from Equation VI-5-16 and 17, respectively

_____
[1] See discussion, p. VI-7-16.

(Sheet 2 of 5)

EXAMPLE PROBLEM VI-7-4 (Continued)

From Example Problem VI-7-1

$$\xi_{om} = 2.07 \ (S_{om} = 0.03718)$$

and using the values for the permeable slope in Example Problem VI-7-3

$$P = 0.5$$
$$\tan \alpha = 0.4$$

Equation VI-5-18 gives

$$\xi_{omc} = \left(5.77 \ P^{0.3} \sqrt{\tan \alpha}\right)^{[1/(P+0.75)]}$$

$$= \left(5.77(0.5)^{0.3} \sqrt{0.4}\right)^{[1/(1.25)]}$$

$$= (2.964)^{0.8}$$

$$= 2.385$$

Because

$$\xi_{om} < \xi_{omc}$$

The value of C in Equation VI-5-17 is given for plunging waves as

$$C = 3.0\left(\xi_{om}\right)^{-3/4}$$

$$= 3.0(2.07)^{-3/4} = 1.738$$

and

$$\frac{1}{C} = 0.5754$$

(NOTE: When $C = 2$, Equation VI-5-15 becomes the Rayleigh distribution)

The scale parameter from Equation VI-15-16 becomes

$$B = H_s \left[0.4\left(s_{om}\right)^{-1/4} \left(\cot \alpha\right)^{-0.2}\right]$$

or

(Sheet 3 of 5)

EXAMPLE PROBLEM VI-7-4 (Continued)

$$= 2.16 \text{ m} \left[0.4(0.03718)^{-1/4} (2.5)^{-0.2} \right]$$

$$= 2.16 \text{ m} \left[0.4(2.2773)(0.83255) \right]$$

$$= 2.16 \text{ m} \left[0.7584 \right]$$

$$B = 1.638 \text{ m}$$

Now check previous results using the above values for B and C in Equation VI-5-15

$$R_{u2\%} = 1.638 \text{ m} \left[-\ln(0.02) \right]^{1/1.738}$$

$$R_{u2\%} = 1.638 \text{ m} \, (3.912)^{0.5754}$$

$$= 1.638 \text{ m} \, (2.192)$$

$$R_{u2\%} = 3.59 \text{ m (11.8 ft)} \left[(\text{From Example Problem VI-7-3, } R_{u2\%} = 3.53 \text{ m (11.6 ft)}) \right]$$

and

$$R_{us} = 1.638 \text{ m} \left[-\ln(0.135) \right]^{0.5754}$$

$$R_{us} = 1.638 \text{ m} \, (2.002)^{0.5754}$$

$$= 1.638 \text{ m} \, (1.491)$$

$$R_{us} = 2.44 \text{ m (8.0 ft)}$$

now

$$\frac{R_{u2\%}}{R_{us}} = \frac{3.59 \text{ m}}{2.44 \text{ m}} = 1.47$$

which does not give the same ratio as the Rayleigh distribution for wave heights where $H_{2\%} = 1.398 \, H_s$.

(Sheet 4 of 5)

EXAMPLE PROBLEM VI-7-4 (Concluded)

At the 1 percent level

$$R_{u1\%} = 1.638 \text{ m} \left[-\ln(0.01) \right]^{0.5754}$$
$$= 1.638 \text{ m} \left(4.605 \right)^{0.5754}$$
$$= 1.638 \text{ m} \left(2.4079 \right) = 3.94 \text{ m} \ (12.9 \text{ ft})$$

For design, wave runup values calculated at the 2 percent exceedance probability level are considered a reasonable upper limit ". . . to prevent wave overtopping."

(b) Influence of shallow water on wave runup

Assuming the breaker index for shallow-water wave breaking is given by the ratio

$$\left(\frac{H}{d} \right)_b = 0.78$$

then

$$H_b = 0.78 \ d = 0.78 \ (3.0) = 2.34 \text{ m} \ (7.7 \text{ ft})$$

Therefore, because $H_s = 2.16 \text{ m} < H_b$, no breaking occurs. Therefore, assuming $_h = 1.0$ is justified.

Note that if the design water depth at the structure toe dropped to 2.8 m (9.2 ft), then breaking begins. Equation VI-5-10 can only be applied where $H_{2\%}$ and H_s are known from field data or numerical model results.

(c) Influence of wave angle and directional spreading

As seen in Equation VI-5-11, the previous results hold for wave angles, β less than 10 deg from normal incidence of long-crested swell-type, wave spectrums. Angles of incidence larger than 10 deg will reduce the wave runup ($\gamma_\beta < 1.0$).

DISCUSSION: For other conditions, the statistical distribution of the wave runup has not been analyzed.

(Sheet 5 of 5)

EXAMPLE PROBLEM VI-7-5

FIND: Determine the wave runup at the 2 percent exceedance probability level for a composite slope shown in the following.

GIVEN: A smooth-faced breakwater of composite slope (m) shown with water depth, d_{toe} = 1.2 m (3.9 ft) is subjected to a significant wave height in deep water H_o' = 1.5 m (4.9 ft) and T_p = 8 s. The offshore slope is 1:20.

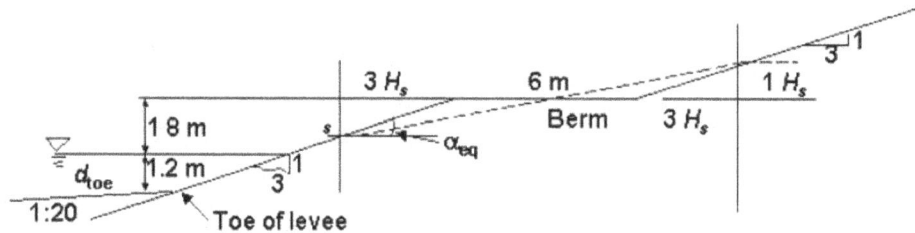

Figure VI-7-1. Smooth faced levee

SOLUTION:
(1) Wave height, (H_s) toe

(a) SPM (1984)

$$\frac{H_o'}{gT^2} = \frac{1.5 \text{ m}}{9.81 \text{ m/s}^2 (8 \text{ s})^2} = 0.0024$$

From Figure 7-3 (*Shore Protection Manual* 1984), at m = 0.05

$$\frac{H_b}{H_o'} = 1.46 \text{ or } H_b = 1.46 \ (1.5 \text{ m}) = 2.19 \text{ m (7.19 ft)}$$

From Figure 7-2, *Shore Protection Manual* (1984), for m = 0:05 and

$$\frac{H_b}{gT^2} = \frac{2.19 \text{ m}}{9.81 \text{ m/s}^2 (8 \text{ s})^2} = 0.0035$$

(Sheet 1 of 3)

EXAMPLE PROBLEM VI-7-5 (Continued)

$$\frac{d_b}{H_b} = 0.93$$

$d_b = 0.93\ (2.19\ \text{m}) = 2.04\ \text{m}\ (6.68\ \text{ft})$

and occurs about 17 m (56 ft) in front of toe

(b) ACES (Leenknecht et al. 1992)
Goda method not applicable for $d < 3.048$ m (10 ft)

Linear theory/Snell's law - Wave broken, $H_b = 2.41$ m, $d_b = 2.33$ m gives

$$\frac{d_b}{H_b} = \frac{2.33}{2.41} = 0.97 \qquad \text{(Checks okay)}$$

(c) Assume wave energy decay continues from $d_b = 2.1 - 2.3$ m to toe of levee, hence,

$$\left(H\right)_{toe} = \frac{d_b}{0.93} = \frac{1.2}{0.93} = 1.29\ \text{m}\ (4.2\ \text{ft})$$

Use $(H_s)_{toe} = 1.29$ m at toe of levee

(2) Berm influence factor, γ_b

(a) Breaking wave surf similarity parameter based on an equivalent slope, ξ_{eq}.
(See Figure VI-7-1)

equivalent structure slope $\alpha_{eq} = \tan^{-1}\left[\dfrac{1.29}{3+3(1.29)}\right] = \tan^{-1} 0.188 = 10.63°\ (1\text{:}5.3\ \text{slope})$

average slope

$$\alpha = \tan^{-1} 0.333 = 18.43°$$

$$S_{op} \equiv \frac{\left(H_s\right)_{toe}}{L_{op}} = \frac{2\pi\left(H_s\right)_{toe}}{gT_p^2} = \frac{2\pi(1.29)}{9.81\ \text{m/s}^2(8\ \text{s})^2} = 0.01291$$

(Sheet 2 of 3)

EXAMPLE PROBLEM VI-7-5 (Concluded)

therefore

$$\xi_{eq} = \frac{\tan \alpha_{eq}}{\sqrt{s_{op}}} = \frac{0.188}{\sqrt{0.01291}} = 1.65$$

$$\xi_{op} = \frac{\tan \alpha}{\sqrt{s_{op}}} = \frac{0.333}{\sqrt{0.01291}} = 2.93$$

From Equation VI-5-8

$$\gamma_b = \frac{\xi_{eq}}{\xi_{op}} = \frac{1.65}{2.93} = 0.56$$

Since $0.6 < \gamma_b$ 1.0, use $\gamma_b = 0.6$

For $\xi_{eq} \leq 2$ from Equation VI-5-7

$$\frac{R_{u2\%}}{(H_s)_{toe}} = 1.5 \xi_{op} \gamma_r \gamma_b \gamma_h \gamma_\beta$$

Take,

$\gamma_r = 1.0$ (smooth)

$\gamma_b = 0.6$ berm influence

$\gamma_h = 0.9$ (since wave breaking begins 19 m (59 ft) from toe, assume some reduction in γ_h

$\gamma_\beta = 1.0$ ($\theta = 0°$)

gives

$$\frac{R_{u2\%}}{(H_s)_{toe}} = 1.5(2.93)(1.0)(0.6)(0.9)(1.0)$$

$$= 2.37$$

$$R_{u2\%} = 2.37(1.29 \text{ m}) = 3.06 \text{ m}(10.0 \text{ ft})$$

[NOTE: By composite method, *Shore Protection Manual* (1984) gave $R_{us} = 1.8$ m (5.9 ft). Assuming Rayleigh distribution

$R_{u2\%}/R_{us} = 1.4$ m (4.6 ft) so $R_{u2\%} = 2.5$ m (8.2 ft)].

(Sheet 3 of 3)

VI-7-3. Wave Overtopping.

EXAMPLE PROBLEM VI-7-6

FIND: Estimate the average overtopping discharge rate for the given wave, water level, and structure geometry.

GIVEN: An impermeable structure with a smooth slope of 1-on-2.5 (tan α = 0.4) is subjected to waves having a deepwater, significant height $H_o' = 1.5$ m (4.9 ft) and a period $T = 8$ s. Water depth at the structure toe is $d_{toe} = 3.0$ m (9.8 ft) relative to design still-water level (SWL). The crest elevation, R_L is 1.5 m (4.9 ft) above the design, SWL. Onshore winds of 18 m/s (35 knots) are assumed.

SOLUTION: Table VI-5-7 lists two models applicable for this example, to determine the average overtopping discharge rate, q (cu m/s per meter) from two formulas, namely Owen (1980, 1982) and van der Meer and Janssen (1995). Both require knowledge of the wave height, H_s at the toe of the structure.

Assume wave direction is shore normal to the structure.

(1) Wave height, H_s at structure toe

(a) Linear wave theory

$$L_o = \frac{g}{2\pi}T^2 = \frac{9.81 \text{ m/s}^2}{2\pi}(8 \text{ s})^2 = 99.92 \text{ m (327.8 ft)}$$

$$\frac{d}{L_o} = \frac{3.0}{99.92} = 0.0300$$

$$\frac{H}{H_o'} = 1.125$$

(Table C-1, Shore Protection Manual 1984)

Assume $H = H_s = 1.125(1.5 \text{ m}) = 1.69$ m (5.5 ft) (nonbreaking)

(b) Irregular wave, Goda method, see ACES (Leenknecht et al. 1992)

$H_s = 1.6$ m (5.2 ft) (Checks okay)

(Sheet 1 of 4)

EXAMPLE PROBLEM VI-7-6 (Continued)

(c) Use $(H_s)_{toe} = 1.69$ m (5.5 ft) (conservative)

(2) Table VI-5-8, Owen (1980, 1982)

Using Equation VI-5-22

$$\frac{q}{gH_sT_{om}} = a\exp\left(-b\frac{R_c}{H_s}\sqrt{\frac{s_{om}}{2\pi}}\frac{1}{\gamma_r}\right)$$

requires knowledge of T_{om}. As discussed in VI-5-2-a-(3)-(b), the relation between T_m and T_p can be estimated from

JONSWAP spectra $\qquad T_m/T_p = 0.79 - 0.87$

or in deep water

Pierson-Moskowitz spectra $\quad T_m/T_p = 0.71 - 0.82$

here, take $T_m = 0.8\ T_p$ so that $T_m = 6.4$ sec $= T_{om}$

Therefore,

$$L_{om} = \frac{g}{2\pi}T_m^2 = \frac{9.81\ \text{m/s}^2}{2\pi}(6.4\ \text{s})^2 = 63.9\ \text{m (210 ft)}$$

and

$$s_{om} = \frac{H_s}{L_{om}} = \frac{1.69\ \text{m}}{63.9\ \text{m}} = 0.02645$$

Now from the coefficients table for smooth slopes shown in Table VI-5-8

Slope	a	b
1:2.0	0.0130	22
1:2.5	0.0145	27 ← by linear interpolation
1:3.0	0.0160	32

Therefore, with $\gamma_r = 1.0$ (smooth slope)

$$\frac{q}{(9.81\ \text{m/s}^2)(1.69\ \text{m})(6.4\ \text{s})} = 0.0145\exp\left(-27\frac{1.5\ \text{m}}{1.69\ \text{m}}\sqrt{\frac{0.02645}{2\pi}}\frac{1}{1.0}\right)$$

$$= 0.0145\exp(-1.5549)$$

$$= 0.0145(0.21122) = 0.003063$$

(Sheet 2 of 4)

EXAMPLE PROBLEM VI-7-6 (Continued)

or

$$q = \left(9.81 \text{ m/s}^2\right)\left(1.69 \text{ m}\right)\left(6.4 \text{ s}\right)\left(0.003063\right)$$

$$= \left(106.10\right)\left(0.003063\right)$$

$$= 0.325 \text{ m}^3/\text{s per meter width} \ (3.5 \text{ ft}^3/\text{s per foot width})$$

(3) Table VI-5-11, van der Meer and Janssen (1995)

The data used to develop Equation VI-5-24 are shown in Figure VI-5-15 (top plot) for $\xi_{op} < 2$. This is a comprehensive data set showing the 95 percent confidence bands for the data.

Using $T_p = 8$ s gives:

$$s_{op} = \frac{H_s}{L_{op}} = \frac{1.69 \text{ m}}{99.92 \text{ m}} = 0.01691$$

so that

$$\xi_{op} = \frac{\tan \alpha}{\sqrt{s_{op}}} = \frac{0.4}{\sqrt{0.01691}} = \frac{0.4}{0.13} = 3.08$$

Therefore:

$$\xi_{op} > 2$$

so that Equation VI-5-25 (see bottom plot of Figure VI-5-15) governs.

Using Equation VI-5-25

$$\frac{q}{\sqrt{9.81 \text{ m/s}^2 \left(1.69 \text{ m}\right)^3}} = 0.2 \exp\left(-2.6 \frac{1.5 \text{ m}}{1.69 \text{ m}} \frac{1}{\left(1.0\right)\left(1.0\right)\left(1.0\right)\left(1.0\right)}\right)$$

with all the reduction factors as unity.

(Sheet 3 of 4)

EXAMPLE PROBLEM VI-7-6 (Concluded)

or

$$q = \sqrt{(9.81 \text{ m/s}^2)(1.69 \text{ m})^3} \cdot 0.2\exp(-2.30769)$$

$$q = 6.88 \text{ m}^3/\text{s} \ (0.0199)$$

$$q = 0.137 \text{ m}^3/\text{s per meter width } (1.47 \text{ ft}^3/\text{s per foot width})$$

This result is considerably lower than that found from Table VI-5-8 by Owen (1980, 1982). However, a check of this result by examining the data scatter in the lower plot of Figure VI-5-15 provides some insight.

For a value on the horizontal axis of

$$\frac{R_c}{H_s} \frac{1}{\gamma_r\gamma_b\gamma_h\gamma_\beta} = \frac{1.5 \text{ m}}{1.69 \text{ m}} \frac{1}{1.0} = 0.89$$

the range covered on the vertical axis by the data is about

$$\frac{q}{\sqrt{gH_s^3}} = \begin{cases} 5(10^{-2}) \\ 1(10^{-2}) \end{cases} 0.0199 \approx 2(10^{-2}) \text{ (mean)}$$

Therefore

$$q = \begin{cases} 0.344 \\ 0.136 \leftarrow \text{ mean, m}^3/\text{s per meter} \\ 0.069 \end{cases} \quad \begin{cases} 3.7 \\ 1.46 \leftarrow \text{ mean, ft}^3/\text{s per foot} \\ 0.74 \end{cases}$$

The range of q at the 95 percent confidence level is about 0.07 to 0.34 m^3/s per meter. The result from Table VI-5-8 with $q = 0.32$ m^3/sec per meter (3.44 ft^3/sec per foot) now seems reasonable.

This example problem is identical to Example Problem 8 in Chapter 7 of the *Shore Protection Manual* (1984) where the average overtopping rate $\bar{Q} = 0.3$ m^3/s per meter (3.23 ft^3/s per foot) was found. The *Shore Protection Manual* result included a factor for wind that is not included. Because of the range of variability in the time-average overtopping discharge rate, the rate of

$$q = 0.3 \text{ m}^3/\text{s per meter width}$$

indicates a potential danger for vehicles, pedestrians, and the safety of structures as illustrated in Table VI-5-6. Therefore, raising the crest elevation should be considered.

(Sheet 4 of 4)

EXAMPLE PROBLEM VI-7-7

FIND:
a. Estimate the overtopping volumes of individual waves, and overtopping distributions for the given wave, water level, and structure geometry.

b. What effect does the structure permeability have on the results?

GIVEN: The identical conditions of Example Problem VI-7-6 (see sketch)

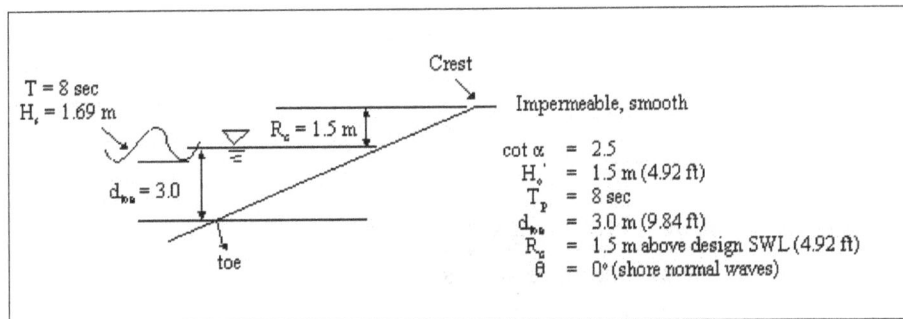

Figure VI-7-2. Overtopping of an impermeable structure

SOLUTION: From Example Problem VI-7-6

$$q = 0.3 \text{ m}^3/\text{s per meter (3.23 ft}^3/\text{s per foot)}$$

is the *average* overtopping discharge rate for waves with $(H_s)_{toe} = 1.69$ m (5.54 ft) and $T_m = 6.4$ s.

Equation VI-5-30 (or VI-5-31) with coefficient B (Equation VI-5-32) depend on P_{ow}, probability of overtopping per incoming wave.

(1) Rayleigh distribution for runup on smooth, impermeable slopes.

Assuming the runup levels follow a Raleigh distribution, Equation VI-5-33 gives for the probability of overtopping per incoming wave,

$$P_{ow} = \exp\left[-\left(\frac{R_c}{c\,H_s}\right)^2\right]$$

(Sheet 1 of 4)

EXAMPLE PROBLEM VI-7-7 (Continued)

and Equation VI-5-34 gives

$$c = 0.81\ \xi_{eq}\ \gamma_r\ \gamma_h\ \gamma_\beta$$

From Example Problem VI-7-6, taking

$$\xi_{eq} = \xi_{op} = 3.07 \qquad \text{(i.e., } \gamma_b = 1, \text{ no berm)}$$

and all other reduction factors of unity, gives

$$c = 0.81\ (3.07)\ (1.0)\ (1.0)\ (1.0) = 2.49$$

and

$$P_{ow} = \exp\left[-\left(\frac{1.5\text{ m}}{2.49(1.69\text{ m})}\right)^2\right]$$

$$P_{ow} = 0.88 = \frac{\text{number of overtopping waves}}{\text{number of incoming waves}}$$

This large percentage is due to the relatively low, crest elevation of the structure.

(2) Other distributions

As shown in Example Problem VI-7-4, the relation between

$$\frac{R_{u2\%}}{R_{us}} = \frac{8.55}{5.01} = 1.71$$

for smooth, impermeable slopes is much different than the Rayleigh distribution for wave heights where $H_{0.02}/H_s = 1.398$. Until further research is conducted, however, it must be assumed that wave runup on smooth, impermeable slopes can be approximated by the Rayleigh distribution.

(a) Overtopping volumes of individual wave, V

From Equation VI-5-32

$$B = 0.84\frac{T_m q}{P_{ow}} = 0.84\frac{(6.4\text{ s})(0.3\text{ m}^3/\text{s/m})}{0.88}$$

$$B = 1.833\text{ m}^3/\text{m}$$

(Sheet 2 of 4)

EXAMPLE PROBLEM VI-7-7 (Continued)

as the scale factor for the one-parameter, Weibull distribution given by Equation VI-5-31, so that

$$V = (1.833 \text{ m}^3/\text{m}) \, [-\ln (p_>)]^{4/3}$$

where $p_>$ is the probability of an individual wave overtopping volume (per unit width) exceeding the specified overtopping volume, V (per unit width), for some representative probabilities of exceedance for individual waves,

$P >$	V (m3/m)	V (ft3/ft)
0.5	1.12	12.1
0.135	4.62	49.7
0.10	5.57	60.0
0.05	7.92	85.3
0.02	11.29	121.5
0.01	14.04	151.1
0.001	24.11	259.5

The maximum overtopping volume per unit width, V_{max} produced by one wave can be estimated from Equation VI-5-35 with $B = 1.833 \text{ m}^3/\text{m}$, i.e., $V_{max} = 1.833 \, (\ln N_{ow})^{4/3}$ which depends on storm duration, t.

Assuming $T_m = 6.4$ s over the storm duration, t, and $P_{ow} = 0.88$

t hr	N_w	N_{ow}	V_{max} m³/m	ft³/ft	Remarks
1	563	495	20.9	225	
2	1125	990	24.7	266	Similar to $P> = 0.001$
5	2813	2475	28.4	306	
10	5630	4954	31.8	342	
15	8438	7425	33.8	364	
20	11250	9900	35.3	380	
24	13500	11880	36.3	391	Gives $P> = 0.0001$

(Sheet 3 of 4)

EXAMPLE PROBLEM VI-7-7 (Concluded)

Storm surge hydrographs with varying design water levels and accompanying wave condition variability during the storm will modify these results, considerably.

(b) Effect of structure permeability

No data exist for permeable, straight and bermed slopes as summarized in Table VI-5-7, to estimate average wave overtopping discharge rates. However, as shown in Example VI-7-2 for a rough, impermeable slope

$R_{u2\%}$ = 4.70 m (15.4 ft) (Equation VI-5-3) γ = 0.55

= 3.56 m (11.7 ft) (Equation VI-5-6) γ = 0.55

= 3.53 m (11.6 ft) (Equation VI-5-12) (Table VI-5-5) P = 0.1

using various models. And, as shown in Example VI-7-3 for rock-armored, permeable-slopes (P = 0.5).

$R_{u2\%}$ = 3.53 m (11.6 ft) (Equation VI-5-3) (Table VI-5-5)

The runup elevation at the 2% exceedance level for this example is about the same for permeable and impermeable slopes.

Statistical distribution for wave runup on rock-armored, permeable slopes are discussed in Part VI-5-2-b.(4)(b) and best-fit, by a two-parameter Weibull distribution (Equation VI-5-14). Structure permeability absorbs the higher frequency runup components to modify the distribution from that given by the Rayleigh distribution.

Research is needed for the probability distribution of wave overtopping per incoming waves on permeable slopes.

(Sheet 4 of 4)

VI-7-4. Armor Layer Stability.

EXAMPLE PROBLEM VI-7-8

FIND: The weight of uniform-sized armor stone placed on an impermeable revetment slope with nonovertopping waves.

GIVEN: An impermeable structure (revetment) on a freshwater shore has a slope of 1 on 2.5 and is subjected to a design, significant wave height, $H_s = 2.0$ m (6.6 ft) measured at a gauge located in a depth, $d = 4.5$ m (14.8 ft). Design wave peak period, $T_p = 8$ s. Design depth at structure toe at high water is $d_{toe} = 3.0$ m (9.8 ft).

SOLUTION: From Example Problem VI-7-1, linear wave theory gives the wave height due to shoaling at the structure toe as:

$(H_s)_{toe} = 2.16$ m (7.0 ft)

ASSUMPTIONS: (See Tables VI-5-22 and VI-5-23)

1. Fresh water, $\rho_w = 1,000$ kg/m^3
2. Rock, $\rho_s = 2,650$ kg/m^3
3. Two layers, $n = 2$, random placement
4. Quarry stone, rough angular
5. No damage criteria (see Table VI-5-21 for damage values, D and S)

Item 1. Hudson (1974), *Shore Protection Manual* (1984)
Use $H = H_{0.1} = 1.27 H_s$ for the Rayleigh distributed wave heights and related K_D-values for stability coefficient. These recommendations of *Shore Protection Manual* (1984) introduce a factor of safety compared to that recommended in the *Shore Protection Manual* (1977). The no-damage range is $D = 0$-5 percent.

From Equation VI-5-67, rearranged for the median rock mass

$$M_{50} = \frac{\rho_s H^3}{K_D \left(\rho_s / \rho_w - 1\right)^3 \cot \alpha}$$

Noting
$W_{50} = M_{50}g$

(Sheet 1 of 6)

EXAMPLE PROBLEM VI-7-8 (Continued)

this equation becomes

$$W_{50} = \frac{\rho_s g H^3}{K_D \left(\rho_s / \rho_w - 1\right)^3 \cot \alpha}$$

By definition

$\gamma_s = \rho_s \, g$ = unit weight of rock

and

$\rho_s/\rho_w = s$, the specific gravity for rock

so

$$W_{50} = \frac{\gamma_s H^3}{K_D \left(s - 1\right)^3 \cot \alpha}$$

which is the more familiar form of the Hudson formula. K_D is the Hudson stability coefficient.

At the toe, $\dfrac{H_s}{d} = \dfrac{2.16 \text{ m}}{3.0 \text{ m}} = 0.7$, and the wave condition is close to breaking for shallow water.

If H_s is assumed to be equivalent to the energy-based significant wave height, H_{mo}, then the maximum depth-limited H_{mo} 0.6 d. Therefore, the maximum breaking wave at the structure toe would be the maximum monochromatic breaking wave.

If H_s is taken equal to $H_{1/3}$, then $H_s > H_{mo}$ near the point where a significant portion of waves in the distribution are breaking. In this case, calculate $H_{0.1}$ to see if it is greater than the maximum breaking wave at the structure toe, then use the lesser of the two.

In summary, determination of wave breaking depends on which definition of significant wave height (H_{mo} or $H_{1/3}$) is used to transform the waves to the toe of the structure.

For this example, assume $H_s = H_{mo}$. Therefore, linear shoaling has given an unrealistically large estimate of H_s. So assume $H_s = 0.6 \, d = 0.6 \, (3.0 \text{ m}) = 1.8 \text{ m}$ (5.9 ft).

The maximum breaking wave height at this depth (assuming a horizontal approach slope) is

$$H_b = 0.78 \, d = 0.78 \, (3.0 \text{ m}) = 2.34 \text{ m} \ (7.7 \text{ ft})$$

(Sheet 2 of 6)

EXAMPLE PROBLEM VI-7-8 (Continued)

For breaking waves on randomly-placed, rough angular stone, use $K_D = 2.0$ in Hudson's equation

$$W_{50} = \frac{\left(2,650 \ kg/m^3\right)\left(9.81 \ m/s^2\right)\left(2.34 \ m\right)^3}{2.0\left(2.65-1\right)^3\left(2.5\right)} = 14,830 \ \text{N} \ (3,334 \ \text{lb})$$

The equivalent cube length is given by

$$D_{n50} = \left(\frac{W_{50}}{\rho_s g}\right)^{1/3} = \left[\frac{14,830 \ \text{kg - m/s}^2}{\left(2,650 \ \text{kg/m}^3\right)\left(9.81 \ \text{m/s}^2\right)}\right] = 0.83 \ \text{m} \ \left(2.7 \ \text{ft}\right)$$

Item 2. Van der Meer (1988), Table VI-5-23
Additional assumptions and data input are required. See Table VI-5-23.

(a) Notational permeability coefficient, P.
As shown on Figure VI-5-11 for impermeable, rock revetments,
   $P = 0.1$

(b) Number of waves, N_z
This value depends on the length of the storm and average wave period during the storm. For example, a 13-14 hr storm with average wave period, $T_m = 6.6$ s would produce about 7,500 waves. When $N_z > 7,500$, the equilibrium damage criteria is obtained.
   Let $N_z = 7,500$

(c) Relative eroded (damage) area, S. This variable is defined by Equation VI-5-60.

$$S \equiv \frac{A_e}{D_{n50}^2}$$

where A_e is the eroded cross-section area around the SWL. Thus, S is a dimensionless damage parameter, independent of slope length. Table VI-5-21 presents damage levels (initial, intermediate, failure) for a two-layer armor layer ($n=2$). For a slope 1:2.5 (interpolate between 1:2 and 1:3).

   Initial damage, $S = 2$
   Intermediate damage, $S = 5.0\text{-}7.5$
   Failure, $S \geq 10$

(Sheet 3 of 6)

EXAMPLE PROBLEM VI-7-8 (Continued)

Hence, for initial or no damage condition, use

$S = 2$ (nominal value)

For irregular waves striking the revetment at 90 deg (normal), the applicable formulas of van der Meer (1988) are found in Table VI-5-23. Two cases exist depending on whether the waves are (1) plunging or (2) surging against the revetment slope.

(1) Plunging waves: $\xi_m < \xi_{mc}$ (Equation VI-5-58)

Recall Example VI-7-1 where the surf-similarity parameter, ξ_m was defined and discussed. Here, it is determined for a mean wave period, T_m using the wave height at the toe, H_{toe}. Therefore, using $H_s = 1.8$ m and $T_m = 6.6$ s,

$$S_{om} = \frac{H_{toe}}{\frac{g}{2\pi}(T_m)^2} = \frac{1.8\ \text{m}(2\pi)}{9.81\ \text{m/s}^2 (6.6\ \text{s})^2} = 0.02647$$

Here, it is assumed that $T_m = 0.82\ T_p$, which is an average relation and slightly different than that employed for Example Problem VI-7-1.

Now it is found that

$$\xi_m = \frac{\tan\alpha}{\sqrt{S_{om}}} = \frac{1/2.5}{\sqrt{0.02647}} = 2.46$$

As discussed in Table VI-5-23, if $\square_m < \square_{mc}$ where

$$\xi_{mc} \equiv \left[6.2 P^{0.31} (\tan\alpha)^{0.5} \right]^{1/(P+0.5)}$$

then, the plunging waves Equation VI-5-68 is applicable. Hence,

$$\xi_{mc} = \left[6.2(0.1)^{0.31} (0.4)^{0.5} \right]^{1/0.6}$$

gives

$$\xi_{mc} = 2.97$$

(Sheet 4 of 6)

EXAMPLE PROBLEM VI-7-8 (Continued)

Therefore, $\xi_m < \xi_{mc}$ so that the plunging wave conditions apply. It is convenient to apply the stability paramater, N_z form (see Equation VI-5-58) to give

$$\frac{H_s}{\Delta D_n} = 6.2\, S^{0.2} P^{0.18} N_z^{-0.1} \xi_m^{-0.5}$$

from Table VI-5-23. Or,

$$\frac{H_s}{\Delta D_n} = 6.2(2)^{0.2}(0.1)^{0.18}(7500)^{-0.1}(2.46)^{-0.5} = 1.23$$

which gives for $H = H_s = 1.8$ m, $\Delta = (s-1) = 1.65$

$$D_{n50} = \frac{H_s}{1.23\Delta} = \frac{1.8\text{ m}}{1.23(1.65)} = 0.89 \text{ m (2.9 ft)}$$

and

$$W_{50} = \rho_s g\,(D_{n50})^3 = (2{,}650 \text{ kg/m}^3)(9.81 \text{ m/s}^2)(0.89 \text{ m})^3 = 18{,}327 \text{ N (4,120 lb)}$$

The stability number is $N_s = 1.23$. Statically stable breakwaters have this stability parameter in the range 1-4 for $H_o T_o < 100$ (van der Meer 1990).

DISCUSSION: In summary, for the breaking wave, storm, and damage conditions, i.e.,

$H_s = 1.8$ m
$H_b = 2.34$ m
$T_p = 8$ s
$S = 2$
$N_z = 7{,}500$ (13-14 hr storm)

Hudson (1974) $W_{50} = 14{,}830$ N (3,334 lb) breaking wave
van der Meer (1988) $W_{50} = 18{,}327$ N (4,120 lb) plunging wave

and it can be said that both methods give simular results.

(Sheet 5 of 6)

EXAMPLE PROBLEM VI-7-8 (Concluded)

The Hudson (1974) formula and therefore, the *Shore Protection Manual* (1984) method limitations include:

no wave period effects
no storm duration effects
damage level limited to range 0-5%

and others as discussed in subsequent examples. The wave period effects have long been discussed as an important missing element in the Hudson (1974) formulation. For example, as shown in Example VI-7-1, as T increases, the surf similarity parameter increases. If the period in the preceding example was increased, the following results would be obtained from the van der Meer (1988) formulation for plunging waves (Table VI-5-23).

Period, s		W_{50}, N	Remarks
T_p	T_m		
9.0	7.38	21, 478	Plunging Waves formula okay
10.0	8.2	23, 865	Use Surging Wave formula, Equation VI-5-69
11.0	9.02	23, 194	Use Surging Wave formula, Equation VI-5-69

Example VI-7-9 demonstrates the practical importance of wave period on armor layer stability.

(Sheet 6 of 6)

EXAMPLE PROBLEM VI-7-9

FIND:
1. The design wave height for a stable, uniform-sized armor stone placed on an impermeable revetment slope with non-overtopping waves.

2. Study the evolution in armor stability design since the 1960's including such factors as alteration in coefficients, wave period, and partial safety factors for design.

GIVEN: In the early 1960's, the Chesapeake Bay Bridge Tunnel (CBBT) islands were constructed with 10 ton (U.S. units) armor stones on a 1:2 slope (single layer) as a revetment for storm protection. The CBBT revetments have been relatively stable and survived many northeasters and hurricanes.

On 31 October 1991, the famous Halloween storm caused severe damage to the revetment. (This storm has been the subject of a best selling novel "The Perfect Storm," Junger (1997) and a Hollywood movie "The Storm of the Century"). Hydrographic surveys determined the extent of damage as discussed in Example Problem VI-7-10.

Wave conditions measured at the U.S. Army Engineers Field Research Facility (FRF) located 65 miles south in 8 m (26.2 ft) water depth were $H_s = 4.6$ m (15.1 ft), $T_p = 22$ sec. At the Virginia Beach wave gauge VA001 also located in 8 m depth, $H_s = 2.6$ m (8.53 ft) and T_p about 23 sec under peak conditions. These waves came from 90 deg (True North) direction and lasted about 12 hours. The measured storm surge at Hampton Roads tide gauge (Sewells Pt.) was 0.85 m (2.8 ft).

ASSUMPTIONS: (See Tables VI-5-22 and VI-5-23)
      1. Sea water, $\rho_w = 1,030$ kg/m^3.
      2. Rock, $\rho_s = 2,650$ kg/m^3.
      3. One layer, $n = 1$, rough angular, random placement.
      4. No-damage criteria (see Table VI-5-21 for D and S damage values).

SOLUTION:
<u>Item 1.</u> Hudson (1974), SPM (1977)
Estimate the stable design wave height $H = H_s$. From Equation VI-5-67.

$$\frac{H}{\Delta D_{n50}} = \left(K_D \cot \alpha \right)^{1/3}$$

Knowing $W_{50} = (\rho_s g)(D_{n50})^3 = 10$ tons $= 20,000$ lbs (89,000 N), and
      $\rho_s g = \gamma_s = (5.14$ slugs/ft$^3)$ $(32.2$ ft/s$^2) = 165.6$ lb/ft^3 (26,000 N/m^3)

(Sheet 1 of 7)

EXAMPLE PROBLEM VI-7-9 (Continued)

The equivalent cube length is given as:

$$D_{n50} = \left(\frac{20,000 \text{ lbs}}{165.6 \text{ lb/ft}^3}\right)^{1/3} = 4.94 \text{ ft } (1.51 \text{ m})$$

Now, considering only wave breaking events on the revetment, as seen in Table VI-5-22, K_D values employed in 1977 were $K_D = 3.5$ for randomly-placed, rough, angular stone. Rearranging Equation VI-5-67

$$H_b = \Delta D_{n50} \left(K_D \cot \alpha\right)^{1/3} = \left(\frac{\rho_s}{\rho_w} - 1\right)(4.94 \text{ m})(3.5 \cdot 2.0)^{1/3}$$

$$H_b = (1.57)(4.94 \text{ m})(1.913)$$

$$H_b = 14.8 \text{ ft } (4.5 \text{ m})$$

If the stones were smooth and rounded, $K_D = 2.1$ giving $H_b = 12.5$ ft (3.8 m). These 10-ton stones would be less stable.

Item 2. Hudson (1974), SPM (1984)
The SPM (1984) took $H_{1/10} = 1.27 H_s$ from the Rayleigh Distribution for a non-breaking conditions and reduced the Hudson coefficients as a result of additional testing using irregular waves. For breaking wave conditions, use H_b as the wave height.

$$H_b = \Delta D_{n50} \left(K_D \cdot \cot \alpha\right)^{1/3}$$

$$H_b = 1.57(4.94)(2.0 \cdot 2.0)^{1/3}$$

$$H_b = 12.3 \text{ ft } (3.76 \text{ m})$$

so that $H_b = 12.3$ ft (3.76 m) for the stable conditions

Note also that for smooth stones ($K_D = 1.2$) gives $H_b = 10.4$ ft (3.17 m).

These results should be interpreted to demonstrate that for a given armor stone weight the design wave height for the stable, no-damage condition has decreased by about 22 percent using the SPM (1984) for breaking waves. Assuming the breaking wave height is approximately equal to $H_{1/10}$, the corresponding significant wave height is

$$H_s = H_b / 1.27 = 12.3 \text{ ft}/1.27 = 9.7 \text{ ft } (2.96 \text{ m})$$

(Sheet 2 of 7)

EXAMPLE PROBLEM VI-7-9 (Continued)

Wave heights measured at Duck, NC (H_s = 4.6 m) and at Virginia Beach, VA, during the storm event exceeded the design wave height, so it reasonable to assume waves at the CBBT site also exceed the design wave height.

The Halloween Storm event was unique to the Atlantic Ocean, East Coast for the very long period swell waves (T_p > 20 sec) generated and recorded. Wave period is not a variable in the Hudson formula.

Item 3. van der Meer (1988), Table VI-5-23

Now consider for irregular, head-on waves on rock, non-overtopping slopes, the formulas of van der Meer (1988) as shown in Table VI-5-23.

Assume as additional, needed variables

P = 0.1 for impermeable, rock revetments

$$N_z = \frac{t}{T_m} = \frac{12 \text{ hrs (3,600 s/hr)}}{18.3 \text{ sec}} = 2,360 \text{ waves}$$

S = 2 (nominal value) for the initial, no-damage condition

T_p = 22.0 sec (T_m = 18.3 sec)

(a) Determine which stability equation is applicable.

Because of the very long wave period, T_p = 22 sec giving $T_m \approx 18.3$ sec, the surf-similarity parameter, ξ_m given by

$$\xi_m = \frac{\tan \alpha}{\sqrt{S_m}}$$

with

$$s_m = \frac{H_{toe}}{\frac{g}{2\pi}(T_m)^2}$$

gives a relatively large value of ξ_m. For example, taking H_s = 5.75 ft (1.75 m) gives ξ_m = 8.63. But the critical ξ_{mc} is found from Table VI-5-23.

(Sheet 3 of 7)

EXAMPLE PROBLEM VI-7-9 (Continued)

$$\xi_{mc} = \left(6.2\, P^{0.31} \left(\tan\alpha\right)^{0.5}\right)^{1/(P+0.5)} = \left(6.2\ (0.1)^{0.31}\ (0.5)^{0.5}\right)^{1/0.6}$$

$$\xi_{mc} = 3.57$$

Therefore, since $\xi_m > \xi_{mc}$, Table VI-5-23 requires that the Surging Waves, Equation VI-5-69 be employed.

(b) Use Surging Waves, Equation VI-5-69 ($\xi_m > \xi_{mc}$)

The stability parameter, N_s form in Table VI-5-23 is

$$\frac{H_s}{\Delta D_{n50}} = 1.0\, S^{0.2} P^{-0.13} N_z^{-0.1} \left(\cot\alpha\right)^{0.5} \xi_m^P$$

giving

$$= 1.0 (2)^{0.2} (0.1)^{-0.13} (2360)^{-0.1} (2.0)^{0.5} \left(\xi_m\right)^{0.1} = 1.008 \left(\xi_m\right)^{0.1}$$

Substituting $D_{n50} = 4.94$ ft (1.51 m) and $\Delta = 1.57$ and expanding ξ_m yields

$$H_s = 1.57\ (4.94\ \text{ft})(1.008) \left[\frac{\tan\alpha}{\sqrt{\dfrac{2\pi\, H_s}{g\, T_m^{\,2}}}}\right]^{0.1}$$

$$H_s = 7.82 \left[\sqrt{\frac{2\pi}{32.2\ \text{ft/s}^2\, (18.3\ \text{s})^2}}\right]^{0.1} \left[\frac{1}{H_s}\right]^{0.05}$$

$$H_s^{1.05} = 10.58$$

$$H_s = (10.58)^{1/1.05} = 9.5\ \text{ft (2.9 m)}$$

which is comparable to the value estimated by the Hudson equation.

(Sheet 4 of 7)

EXAMPLE PROBLEM VI-7-9 (Continued)

The stability number, $N_s = 1.22$, i.e., $N_s > 1$ for stable conditions.

Now for this same wave height, $H_s = 9.5$ ft (2.90 m), what armor layer weight, W_{50} is required for shorter wave periods, T_p to remain stable?

Wave period (sec)		Weight W_{50}		Remarks
T_p	T_M	lbs	(kN)	
20.0	16.7	20,770	(92.4)	OK - Surging Equation VI-5-69
15.0	12.5	22,640	(100.7)	OK - Surging Equation VI-5-69
12.0	10.0	24,206	(107.7)	OK - Surging Equation VI-5-69
10.0	8.3	19,140	(85.1)	Use Plunging Equation VI-5-68

Now we see that for surging waves, <u>lowering</u> the wave period <u>increases</u> the stone weight, W_{50}, for stability up to some point where the conditions for the Surging Wave Equation are no longer applicable. This is the opposite trend as shown in Example VI-7-8 for the case where the Plunging Wave Equation was applicable. In general, each equation is only applicable for the special conditions.

$\xi_m < \xi_{mc}$ Use Plunging Waves, Equation VI-5-68

$\xi_m > \xi_{mc}$ Use Surging Waves, Equation VI-5-69

and the wave period, T_p is an important variable in the equation for ξ_m.

All the above does not address the need for some safety factors in applying the van der Meer formulas for design.

<u>Item 4.</u> Partial Safety Factors (VI-6-6)

The theory behind the inclusion of partial safety factors for the stable design of armor stone is found in VI-6-6. In general, the safety factors increase:
1. as our knowledge of the wave height conditions decreases, and
2. as our desire for a risk free, low failure probability increases.

Table VI-6-6 presents the Partial Safety Factors ranging up to 1.9 for Surging Wave conditions on non-overtopping slopes using the van der Meer, 1988 Equation VI-6-45.

(Sheet 5 of 7)

EXAMPLE PROBLEM VI-7-9 (Continued)

Here, we consider the influence of the partial safety factors on the wave heights for a stable armor stone weight, 10 tons on a 1:2 slope for this example. In all cases, also take $P = 0.1$, $S = 2$, and $N_z = 2360$ waves for $T_p = 22.0$ sec. Consider two cases:

(a) Excellent Knowledge of Wave Conditions ($\square = 0.05$) at site.

Failure Probability	P_f	γ_H	γ_z	H_s, ft (m)
Low	0.01	1.7	1.00	5.73 (1.75)
Medium	0.10	1.3	1.02	7.25 (2.21)
High	0.40	1.0	1.08	8.85 (2.70)
No. S.F.	?	1.0	1.0	9.5 (2.90)

Decreasing the degree of risk of failure (i.e., including safety factors) means lowering the wave height design conditions for the same armor stone weight and revetment slope.

(b) Relatively Poor Knowledge of Wave Conditions ($\sigma = 0.2$) at site

Failure Probability	P_f	γ_H	γ_z	H_s ft (m)
Low	0.01	1.9	1.02	5.05 (1.54)
Medium	0.10	1.4	1.04	6.65 (2.03)
High	0.40	1.1	1.00	8.70 (2.65)

To ensure a low failure probability means using 10-ton armor stone on a 1:2 slope in regions with long period waves but wave heights only in the 5-6 ft range.

Clearly, since these armor stones were severly damaged in the 31 October 1991 storm, the storm wave heights must have been greater than all those calculated above.

(Sheet 6 of 7)

EXAMPLE PROBLEM VI-7-9 (Concluded)

5. SUMMARY:
Given:

$$W_{50} = 10 \text{ tons} = 20,000 \text{ lbs } (89,000 \text{ N})$$
$$D_{n50} = 4.94 \text{ ft } (1.51 \text{ m})$$
$$P = 0.1$$
$$S = 2$$
$$N_z = 2,360$$
$$T_p = 22 \text{ sec } (T_m = 18.3 \text{ sec})$$

Armor-Layer Stability Formula	Stable, Significant Wave Height, H_s		Remarks
	feet	meters	
Hudson (1974) SPM (1977)	14.9	4.5	No period effects, breaking waves No safety factor
Hudson (1974) SPM (1984)	9.7	2.32	No period effects, breaking waves Revised coefficients, conservative
van der Meer (1988) (no safety factor)	9.5	2.90	$T_p = 22.0$ sec, surging waves No safety factor
van der Meer (1988) (with safety factor)	5-9	1.5-2.7	$T_p = 22.0$ sec, surging waves Includes partial safety factors, Part VI, Chapter 6

With such long wave periods, it is possible that waves did not break on the revetment, and Hudson's equation could be applied with nonbreaking wave K_Ds.

Clearly a wide range of wave heights are possible based on these formulas. Example VI-7-10 considers the damage experience by the CBBT island revetments to determine the design wave conditions. Example VI-7-11 considers the size (weight) of the armor stones for repair.

(Sheet 7 of 7)

EXAMPLE PROBLEM VI-7-10

FIND: The damage curve relationship for wave energy above the design wave height for uniform-sized armor stone placed on an impermeable revetment slope with nonovertopping waves.

GIVEN: The same condition as found in Example VI-7-9 for the Cheasapeake Bay Bridge Tunnel (CBBT) island revetments with 1:2 sloped revetment, $T_p = 22.0$ s, but uncertain knowledge of the wave height, H_s.

SOLUTION: Method 1. Based on van der Meer (1988)

Table 7-9 (in the *Shore Protection Manual* (1984), (Volume II, p. 7-211) presented the following generic, H/H_d - vs - damage D in percent relationships for rough/quarrystone revetments. (Two layers, random placed, nonbreaking waves, minor overtopping.) This table was not well supported by data, so it was not included in the *Coastal Engineering Manual*. The value of H depends on what level $H_{D=0}$ is used for a stable design.

$H/H_{D=0}$	Damage. D Percent
1.0	0-5
1.08	5-10
1.19	10-15
1.27	15-20
1.37	20-30
1.47	30-40
1.56	40-50

(Sheet 1 of 3)

EXAMPLE PROBLEM VI-7-10 (Continued)

Now apply van der Meer's (1988) Equation VI-5-69 for surging waves with $H_s(S=2) = 9.5$ ft (2.90 m), and vary the significant wave height, H_s to calculate S, the damage level. We keep $D_{n50} = 4.95$ ft (1.51 m) and $W_{50} = 10$ tons (20,000 lb). This assumes adequate depth exists at the structure toe to support the increased significant wave heights without depth-limited breaking.

H_s feet	(m)	$H/H_s(S=2)$	S	Relative Damage Level	Remarks
9.5	(2.90)	1.00	2.018	1.009	Slight rounding error, $S = 2$
10.0	(3.05)	1.05	2.643	1.32	
11.0	(3.35)	1.16	4.36	2.18	Intermediate damage level, $S = 4 - 6$
11.5	(3.51)	1.21	5.51	2.76	
12.0	(3.66)	1.26	6.89	3.45	
12.5	(3.81)	1.32	8.54	4.27	
13.0	(3.96)	1.37	10.5	5.25	Failure, $S = 8$ m, armor layer damaged, underlayer exposed to direct wave attack
14.0	(4.27)	1.47	15.5	7.75	
15.0	(4.57)	1.58	22.2	11.1	
16.0	(4.88)	1.68	31.2	15.6	

From Table VI-5-21, van der Meer (1988) gives the following guidelines for 1:2 slopes.

Initial damage	$S = 2$	Initial damage - no displacement
Intermediate damage	$S = 4 - 6$	Units displaced but without underlayer exposure
Failure (of armor layer	$S = 8$	The underlayer is exposed to direct wave attack

Values of $H/H_s (S = 2)$ - vs - S are plotted in Figure VI-7-3. Also, approximate percentage damage, D scales from *Shore Protection Manual* (1984) are constructed for comparison.

(Sheet 2 of 3)

EXAMPLE PROBLEM VI-7-10 (Concluded)

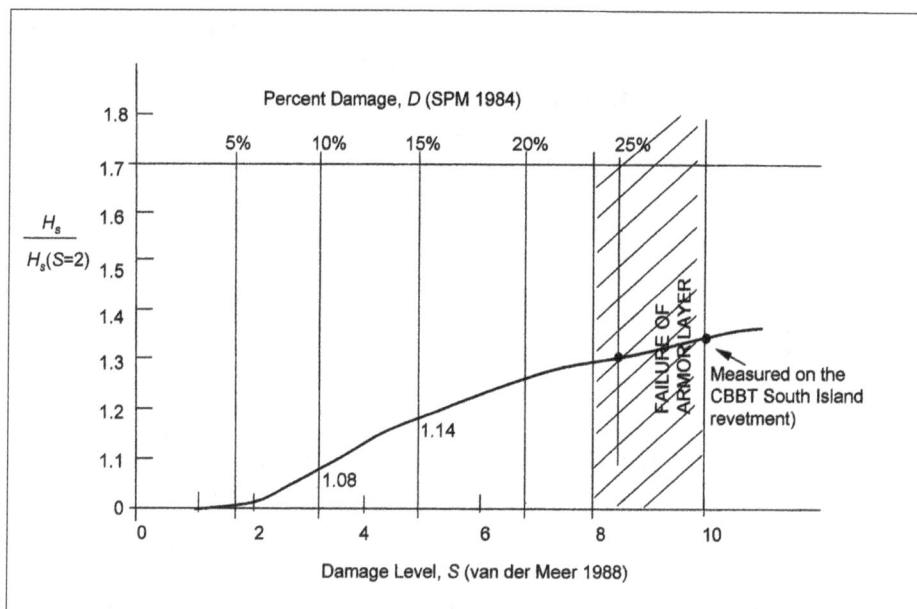

Figure VI-7-3. Percent damage curve for CBBT South Island revetment

Method 2. Damage measurements on CBBT Islands

Damage profile surveys taken by the engineering staff, CBBT District have been analyzed to learn that $S = 10$ from the 31 October 1991 Halloween Storm northeaster. Some underlayers were exposed and this level of the damage parameter is consistent with the criteria for "failure" as shown in Table VI-5-21.

This damage occurred on South Island on a curved section where the armor stones are more vulnerable, i.e., on the head of the structure rather than within the trunk section. Table VI-5-37 presents a method to estimate stability of rock breakwaters as proposed by Carver and Heimbaugh (1989).

From the analysis for $S = 10$, the ratio $H/H_s(S = 2) = 1.35$ giving a significant wave height of

$$H_s = 1.35 \ (9.5 \ \text{ft}) = 12.8 \ \text{ft} \ (3.9 \ \text{m})$$

necessary to produce this level of damage using the van der Meer (1988) formulation for surging waves. As shown in Example Problem VI-7-9, wave heights were measured at Duck, North Carolina, as 15.1 ft (4.6 m). These wave conditions are possible at the Chesapeake Bay entrance with relatively deep ($d = 12$ m (39.4 ft)) water.

Example VI-7-11 considers what value of H_s should be used to determine the size and weight of armor stone for repairs of the CBBT island.

(Sheet 3 of 3)

EXAMPLE PROBLEM VI-7-11

FIND: The weight of armor stone to repair the damage to the CBBT Island revetment.

GIVEN:
The results of Example Problems VI-7-9 and VI-7-10
Wave Information Study (WIS) hindcast information for nearby locations
Other extreme wave condition measurements/criteria
Appropriate partial safety factors for wave conditions

Method 1. WIS Station 2059 (Brooks and Brandon 1995)

$d = 14$ m (46.0 ft)	Lat. 37.00°N - Long. 75.75°W					
	Spectral Significant Wave Height, H_{mo} (m)					
Extreme Prob. Dist. Fisher-Tippett	Recurrence Interval, T_r, years					
	2	5	10	20	25	50
Type I	5.87	6.67	7.22	7.76	7.93	8.46
Type II	5.87	6.98	7.90	8.90	9.25	10.41

Method 2. Virginia Beach Hurricane Protection Project (U.S. Army Corps of Engineers)

1% chance storm each year (100-year recurrence interval)

$d = 30$ ft (9.1 m)
$H_s = 15.8$ ft (4.8m)
$T_p = 13.7$ sec

Storm surge elevation = 8.7 ft (2.65 m) above NGVD (1929)

SOLUTION: As shown in Example Problem VI-7-10, when $H_s = 12.8$ ft (3.90 m), the damage level parameter, S, in the van der Meer (1988) for surging waves was about 10 and this was also the average damage level measured by survey. For redesign, consider the following four cases:

a. Wave height, $H_s = 13.0$ ft (3.96 m)
   Wave period, $T_p = 22.0$ s

b. Consider what effect different wave periods will have on the armor stone size, keeping $H_s = 13.0$ ft.

(Sheet 1 of 4)

EXAMPLE PROBLEM VI-7-11 (Continued)

c. Consider what effect storm duration up to N_{max} = 7,500 waves will have on the stone size for H_s = 13.0 ft (3.96 m) and the critical periods.

d. Consider what effect some increase in allowable damage level, S, has on these results.

Case a. From Equation VI-6-45 in Table VI-6-6, for H_s = 13.0 ft and T_p = 22.0 s (T_m = 18.3 s).

$\qquad W_{50}$ = 54, 236 lb (241,250 N)
$\qquad D_{n50}$ = 6.9 ft (2.10 m)
with no safety factor, i.e., $\gamma_H = \gamma_z$ = 1.0

Assuming our knowledge of wave conditions is fairly good (σ_w = 0.05) and using a failure probability (P_f) of 0.10 gives γ_H = 1.3 and γ_z = 1.02. Using these partial safety factors in Equation VI-6-45 gives

$\qquad W_{50}$ = 126,490 lb (562,650 N)
$\qquad D_{n50}$ = 9.1 ft (2.77 m)

This is not a practical size in the quarry and for construction.

Case b. From Equation VI-6-45 for H_s = 13.0 ft and other wave periods, T_p with no partial safety factors.

Mean Wave Period, T_m s	Peak Wave Period, T_p s	Armor Layer		van der Meer Formula S = Surging P = Plunging	N_z = 2,360 Storm Duration t, hr	Remarks
		Stone wt, W_{50}, lb	Stone diam, D_n, ft (m)			
18.3	22	54,340	6.90 (2.10)	S	12.0	The Hudson
16.7	20	55,852	6.96 (2.12)	S	10.4	formula does
14.9	18	57,796	7.04 (2.15)	S	9.8	not consider
13.3	16	59,799	7.12 (2.17)	S	8.7	wave period
11.6	14	62,304	7.22 (2.20)	S	7.6	
10.0	12	51,585	6.78 (2.07)	P	6.6	
8.3	10	39,007	6.18 (1.88)	P	5.4	

(Sheet 2 of 4)

EXAMPLE PROBLEM VI-7-11 (Continued)

Lowering the wave period with $H_s = 13.0$ ft (3.96 m) slightly *increases* the armor stone weight by 15 percent for T_p from 22 - 14 s. The surging equations govern.

Case c. As noted in Table VI-5-23 for $N_z = 7,500$ waves, the equilibrium damage level is approximately reached.

Wave Period, T_p s	Armor Stone Weight, W50 lbs				Remarks
	$N_z =$ 2,360 (t = hr)	$N_z =$ 3,500 (t = hr)	$N_z =$ 5,000 (t = hr)	$N_z =$ 7,500 (t = hr)	
22	54.340 (12.0)	61,159 (17.8)	68,066 (25.4)	76,871 (38.1)	The very long durations are not physically realistic. Long durations at long periods also not realistic
18	57,796 (9.8)	65,049 (14.5)	72,396 (20.7)	81,760 (31.0)	
14	62.304 (7.6)	70,123 (11.3)	78,042 (16.1)	88,137 (24.2)	

Again, as expected, increasing the storm duration increases the weight of the armor stone required for stable weight at the $S = 2$ level. Note however, that some storm durations ($t >$ 18 hr) are physically unrealistic for sustained, long period waves and wave heights, $Hs = 13.0$ ft (3.96 m).

Case d. Consider realistic wave periods, $T_p = 14-22$ s for storms lasting 8 to 18 hr with $H_s = 13.0$ ft (3.96 m). Now vary the damage level allowable to the intermediate range, $S = 4 - 6$ (say $S = 5$) for rock on slopes with cot = 2:

Damage Parameter S	Armor Stone Weight, W_{50} lb					
	$N_z = 2,360$			$N_z = 3,500$		
	$T_p = 22$ s	$T_p = 18$ s	$T_p = 14$ s	$T_p = 22$ s	$T_p = 18$ s	$T_p = 14$ s
2	54,340	57,796	62,304	61,159	65,049	70,123
3	42,605	45,315	48,849	47,952	51,002	54,980
4	35,851	38,131	41,105	40,350	42,916	46,264
5	31,358	33,353	35,954	35,294	37,539	40,467
6	28,109	29,897	32,229	31,637	33,649	36,273
7	25,626	27,256	29,381	28,842	30,676	33,069

(Sheet 3 of 4)

EXAMPLE PROBLEM VI-7-11 (Concluded)

Allowing the damage level to rise up to $S = 6$ in effect means using roughly one-half the weight of armor stone needed when $S = 2$. This is a tremendous reduction and cost savings for initial construction costs and repair costs in a balanced design.

(Note that using the Hudson (1977) formula and *Shore Protection Manual* (1984) methodology $H_{1/10} = 1.27 H_s$, $K_D = 2.0$, gives $W_{50} = 48,036$ lb. This is roughly equivalent to that previously obtained for $T_p = 14$ s, $S = 3$ with an 11+ hr storm event.)

The CBBT island revetment was repaired in August 1994. The W_{50} was 13.5 tons (27,000 lb) with allowable range 12-15 tons. No stones $W_{50} < 12$ tons were allowed. As already demonstrated, this repair stone weight, $W_{50} = 13.5$ tons (35 percent increase in weight) is on the order required but with some damage expected in the future. All of the preceding is with no partial safety factors in the design.

Alternatives for repair would be to use artifically manufactured concrete cubes (Table VI-5-29), tetrapods (Table VI-5-30) or the Corps of Engineers' new Core-Loc® design (Table VI-5-34). These units have greater interlocking abilities and are stable with less weight. A detailed cost analysis is necessary to justify the additional repair expense.

SUMMARY:
Item 1. All of the preceding was calculated keeping $H_s = 13.0$ ft (3.96 m) for design. Wave period, storm duration, and allowable damage level are all additional, important factors, but are not considered in the Hudson formula (1977) nor in the *Shore Protection Manual* (1984).

Item 2. As shown in Table VI-6-6 for Equation VI-6-45 surging waves and van der Meer (1988) formulation, not including any partial safety factors (i.e., taking $\gamma_H = \gamma_F = 1.0$) implies:

a. Our knowledge of wave height conditions is good ($\sigma_w = 0.05$) and

b. A damage probability, $P_f > 40$ percent is expected sometime during the lifetime of the structures. This is acceptable if the damage can be repaired (economically) and if the additional risk is understood.

Item 3. Using artifically manufactured units (concrete cubes, tetrapods, Core-Loc®, etc.) can greatly reduce the level of risk by allowing P_f to decrease (say $P_f = 0.05$) including the appropriate γ_H and γ_F factors and repairing the damage on the CBBT island revetments. This will be shown in Example Problem VI-7-15.

(Sheet 4 of 4)

EXAMPLE PROBLEM VI-7-12

FIND: The weight of armor stone placed as a permeable, nearshore breakwater with overtopped wave conditions.

GIVEN: A permeable structure (nearshore, detached breakwater) has a slope of 1 on 2.5 and is subject to a design, significant wave height, H_s of 2.0 m (6.56 ft) measured at a gauge located in a depth, $d = 4.5$ m (14.8 ft). Design wave peak period, $T_p = 8$ s. Design water depth at the structure toe, $d_{toe} = 3.0$ m (9.8 ft). From Example Problem, VI-7-2 for these conditions, runup, $R_{U2}\%$ is 3.5-4.5 m above the SWL, hence some wave overtopping occurs.

SOLUTION: From Example Problem VI-7-1, linear wave theory shoaling to the structure toe gave: $(H_s)_{toe} = 2.16$ m (7.09 ft). However, it was noted in Example Problem VI-7-8 that this value of H_s exceeded the depth-limited energy-based wave height. So a value of $(H_s)_{toe} = 0.6$ d $= 0.6$ (3.0 m) $= 1.8$ m (5.9 ft) was used.

ASSUMPTIONS: (See Tables VI-5-22, VI-5-24, and VI-5-25 for conventional, two-layer armor stone designs. Also see Figure VI-5-11 for notational permeability coefficients.)

1. Fresh water $\rho_w = 1,000$ kg/m^3
2. Rock $\rho_r = 2,650$ kg/m^3
3. Two-layers $n = 2$, random placement
4. Quarrystone, rough, angular
5. No-damage criteria (see Table VI-5-21 for valves, D and S)

Item 1. Hudson (1974), *Shore Protection Manual* (1984)

See Example Problem VI-7-8 for results which do not change for rubble-mound revetment or nearshore breakwaters:

1. Nonbreaking waves $K_D = 4$, $H_{1/10} = 1.27$ (1.8 m) $= 2.28$ m (7.5 ft)
   $W_{50} = 6,859$ N (1,542 lb)
   $D_{n50} = 0.64$ m (2.1 ft)

2. Breaking waves, $K_D = 2$, $H_b = 2.34$ m (7.7 ft)
   $W_{50} = 14,830$ N (3,334 lb)
   $D_{n50} = 0.83$ m (2.7 ft)

(Sheet 1 of 6)

EXAMPLE PROBLEM VI-7-12 (Continued)

The Hudson formula was originally developed for nonovertopped slopes, but has been often applied to cases with moderate to substantial wave overtopping. In these cases, stone weights estimated with the Hudson equation will be conservative.

Item 2. Van der Meer (1991) Table VI-5-24.

Van der Meer (1991) developed an overtopping reduction factor, f_i given by Equation VI-5-71 in Table VI-5-24 to modify the original van der Meer (1988) stability formulas (Equations VI-5-68 and VI-5-69. The calculated D_{n50} value is reduced by f_i, and the relative freeboard R_c/H_s plays an important role in the factor, f_i. Here, R_c is the same as defined for overtopping depicted in Figure VI-5-14.

1. Nonovertopping conditions - impermeable revetment
    Recall from Example Problem VI-7-8 when

$P = 0.1$ (impermeable)
N_z 7,500 (t = 16-17 hr)
$S = 2$ (no damage condition)
$\xi_m < \xi_{mc}$

then for plunging wave conditions

$D_{n50} = 0.89$ m (2.9 ft)
$W_{50} = 18,327$ N (4,120 lb)

2. Overtopping conditions

The van der Meer (1991) equations can be written

$$\frac{f_i H_s}{\Delta D_{n50}} = 6.2\, S^{0.2} P^{0.18} N_z^{-0.1} \xi_m^{-0.5} \qquad \text{plunging } (\xi_m < \xi_{mc})$$

and

$$\frac{f_i H_s}{\Delta D_{n50}} = 1.0\, S^{-0.2} P^{-0.13} N_z^{-0.1} \left(\cot \alpha\right)^{0.5} \xi_m^{P} \qquad \text{surging } (\xi_m > \xi_{mc})$$

(Sheet 2 of 6)

EXAMPLE PROBLEM VI-7-12 (Continued)

where:

$$f_i = \left(1.25 - 4.8 \frac{R_c}{H_s}\sqrt{\frac{s_{op}}{2\pi}}\right)^{-1}$$

(VI-5-71)

within limits

$$0 < \frac{R_c}{H_s}\sqrt{\frac{s_{op}}{2\pi}} < 0.052$$

Note that now, the peak period wave steepness, s_{op}, is employed. It is convenient to set up a spreadsheet solution to investigate how D_{n50} and W_{50} vary with relative freeboard, R_c/H_s. First consider the case (unlikely) for an impermeable, nearshore breakwater design.

(a) Impermeable, $P = 0.1$

Relative Freeboard, R_c/H_s	R_c		D_{n50}		W_{50}		f_i	Remarks
	m	(ft)	m	(ft)	N	(tons)		
1.0	1.8	(5.9)	0.89	(2.9)	18,327	(2.06)	1.000	Exceeds Limit 0.052
0.85	1.53	(5.02)	0.86	(2.82)	16,556	(1.86)	0.969	
0.75	1.35	(4.43)	0.84	(2.75)	15,377	(1.73)	0.946	
0.50	0.90	(2.95)	0.79	(2.60)	12,883	(1.45)	0.892	
0.0	0	(0)	0.71	(2.33)	9,304	(1.05)	0.800	Limit Value = 0

As the relative freeboard, R_c/H_s decreases, more wave overtopping occurs, and the stable armor-layer weight also decreases, over the limiting factor range $0.8 < f_i < 1.0$.

(b) Permeable, $P = 0.4$ or 0.5

The primary application of the original van der Meer (1988) formulation with modification by the reduction factor, f_i, is for permeable structures such as nearshore breakwaters. The following tables illustrate application to permeable structures using the same wave and structure parameters.

(Sheet 3 of 6)

EXAMPLE PROBLEM VI-7-12 (Continued)

$P = 0.4$ (see Figure VI-5-11)

Relative Freeboard, R_c/H_s	R_c		D_{n50}		W_{50}		f_i	Limit Parameter	Remarks
	m	(ft)	m	(ft)	N	(tons)			
1.0	NOT APPLICABLE							>0.052	
0.85	1.53	(5.02)	0.67	(2.20)	7,830	(0.88)	0.969		Going from $P = 0.1$ (impermeable to $P = 0.4$ (permeable) produces a 50% or one-half lower weight requirement
0.80	1.44	(4.73)	0.66	(2.17)	7,545	(0.85)	0.957		
0.75	1.35	(4.43)	0.65	(2.15)	7,274	(0.82)	0.946		
0.50	0.90	(2.95)	0.62	(2.02)	6,094	(0.68)	0.891		
0.25	0.45	(1.48)	0.58	(1.91)	5,156	(0.58)	0.843		
0.10	0.18	(0.59)	0.56	(1.85)	4,684	(0.53)	0.817		
0.05	0.09	(0.30)	0.56	(1.85)	4,540	(0.51)	0.808		
0	0	(0)	0.55	(1.82)	4,400	(0.49)	0.800	0	

$P = 0.5$ (Permeable, D core = 0.3 D armor

Relative Freeboard, R_c/H_s	R_c		D_{n50}		W_{50}		f_i	Limit Parameter	Remarks
1.0	NOT APPLICABLE							>0.052	
0.85	1.53	(5.02)	0.64	(2.11)	6,942	(0.78)	0.969		Going from $P = 0.4$ to $P = 0.5$ gives a 11.3% drop in W_{50}
0.80	1.44	(4.73)	0.66	(2.09)	6,689	(0.75)	0.957		
0.75	1.35	(4.43)	0.63	(2.06)	6,448	(0.72)	0.946		
0.50	0.90	(2.95)	0.59	(1.94)	5,402	(0.61)	0.891		
0.25	0.45	(1.48)	0.56	(1.84)	4,570	(0.51)	0.843		
0.10	0.18	(0.59)	0.54	(1.78)	4,152	(0.47)	0.817		
0.05	0.09	(0.30)	0.53	(1.76)	4,024	(0.45)	0.808		
0	0	(0)	0.53	(1.76)	3,901	(0.44)	0.800	0	

The value of $P = 0.6$ is reserved for permeable breakwaters built with no core and homogeneous sized units as discussed in Example Problem VI-7-13.

(Sheet 4 of 6)

EXAMPLE PROBLEM VI-7-12 (Continued)

Note the significant (50% or more) reduction in the stable armor weight requirements due to the permeability, P, of the typical nearshore breakwater designs with a core.

SUMMARY:

Item 1. The Hudson formula is not applicable for wave overtopping conditions because it gives conservative results.

Item 2. Use the van der Meer (1991) formula to determine D_{n50}, then reduce D_{n50} by the factor, f_i.

Item 3. The reduction factor lies in the range $0.8 < f_i < 1.0$ where

$$f_i = 0.8 \quad \text{at } R_c / H_s = 0 \quad\quad \text{zero freeboard}$$

and

$$f_i = 1.0 \quad \text{at } \frac{R_c}{H_s}\sqrt{\frac{s_{op}}{2\pi}} = 0.052 \quad \text{limit}$$

Item 4. At limit of zero freeboard, $R_c = 0$, $f_i = 0.8$.

$$W_{50} = D_{n50}^3$$

Weight reduction $= (0.8)^3 = 0.512$ or almost a 50% drop.

Item 5. At limit $f_i = 1.0$ with

$$s_{op} = \frac{H_s}{L_{op}} = \frac{H_s}{\dfrac{g}{2\pi}T_p^2}$$

the limit for Equation VI-5-71 is

$$\frac{R_c}{H_s}\left(\frac{H_s}{\dfrac{g}{2\pi}T_p^2 2\pi}\right)^{1/2} < 0.052$$

$$\frac{R_c}{H_s}\left(\frac{H_s}{gT_p^2}\right)^{1/2} < 0.052$$

(Sheet 5 of 6)

EXAMPLE PROBLEM VI-7-12 (Concluded)

Thus, for a range of practical wave heights $1.0 < H_s < 10.0$ m and peak periods, $3 < T_p < 21$ s, a table can be prepared to calculate the maximum values of R_c/H_s at the 0.052 limit. This gives a practical range of $(R_c/H_s)_{max}$ values as shown in the table below.

| H_s | | $(R_c/H_s)_{max}$ Values at Limit = 0.052 | | | | | | |
| | | Peak Period, Tp | | | | | | |
m	(ft)	3	6	9	12	15	18	21
1.0	(3.28)	0.49	0.98	1.49	1.95	2.44	2.93	3.42
2.0	(6.56)	0.35	0.69	1.04	1.38	1.73	2.07	2.42
3.0	(9.84)	0.28	0.56	0.85	1.13	1.41	1.69	1.97
4.0	(13.1)	0.24	0.49	0.73	0.98	1.22	1.47	1.71
5.0	(16.4)	0.22	0.44	0.66	0.88	1.09	1.31	1.53
7.0	(23.0)	0.19	0.37	0.55	0.74	0.92	1.11	1.29
10.0	(32.8)	0.15	0.31	0.46	0.61	0.77	0.93	1.08

For (R_c/H_s) values greater than in the table, nonovertopping conditions prevail.

(Sheet 6 of 6)

EXAMPLE PROBLEM VI-7-13

FIND: The weight of armor stone placed as a permeable, nearshore breakwater with submerged water-level conditions.

GIVEN: The same data as for Example Problem VI-7-12 except now the design water depth, d_{toe} increases to submerge the structure. Assume H_s does not change as water level increases.

$$\cot \alpha = 2.5$$
$$(H_s)_{toe} = 1.8 \text{ m (5.9 ft)}$$
$$T_p = 8 \text{ s}$$
$$\rho_w = 1{,}000 \text{ kg/m}^3 \text{ (fresh water)}$$
$$\rho_r = 2{,}650 \text{ kg/m}^3 \text{ (rock)}$$

n = 2 layers, random placement quarrystone, rough angular, no-damage criteria, $S = 2$

$$d_{toe} = \text{varies}$$

SOLUTION:

Method 1. van der Meer (1991) Table VI-5-25

For irregular, head-on waves and data for $\cot \alpha = 1.5$, 2 slopes, van der Meer (1991) developed the formula

$$h_c'/h = (2.1 + 0.1\, S) \exp(-0.14\, N_s^*) \text{ (Equation VI-5-72)}$$

where

h_c' = crest height of structure above sea level
h = water depth
$h - h_c'$ = water depth over the structure crest
S = relative eroded area, (damage level)

and

N_s^* = spectral stability number

$$= \frac{H_s}{\Delta D_{n50}} s_p^{-1/3}$$

(Sheet 1 of 3)

EXAMPLE PROBLEM VI-7-13 (Continued)

with

$$s_p = H_s/L_p$$

where L_p is the local wavelength based on peak spectral period, T_p.

To determine the stable, armor stone diameter and weight, Equation VI-5-72 is solved first for a given h_c'/h ratio and s_p value to calculate N_s^*. Then rearrange N_s^* to solve for D_{n50}, i.e.,

$$N_s^* = -\frac{1}{0.14}\ln\left[\frac{(h_c')/h}{(2.1+0.1S)}\right]$$

A spreadsheet solution aids the calculation process. Note that the slope does not enter into the calculation and the empirical formula VI-5-72 has only been developed for two slopes.

ASSUMPTIONS:
1. Assume VI-5-72 also applicable when cot α = 2.5.
2. Using $S = 2$ gives the no-damage results.
3. Assume crest height, h_c' = 3.0 m above seabed.

h_c'/h	h/h_c'	Water Depth, h		Diameter, D_{n50}		Weight, W_{50}		N_s^*	Remarks
		m	(ft)	m	(ft)	N	(tons)		
1.1		Overtopped, not submerged, use Equation VI-5-71, h_c' = 3.0							
1.000	1.0	3.0	9.8	0.52	1.72	3,742	0.42	5.95	$R_c = 0$
0.909	1.1	3.3	10.8	0.48	1.57	2,826	0.32	6.63	
0.833	1.2	3.6	11.8	0.44	1.45	2,249	0.25	7.25	
0.769	1.3	3.9	12.8	0.41	1.36	1,858	0.21	7.82	
0.714	1.4	4.2	13.8	0.39	1.29	1,579	0.18	8.35	
0.667	1.5	4.5	14.8	0.38	1.23	1,372	0.15	8.85	
0.625	1.6	4.8	15.7	0.36	1.18	1,212	0.14	9.31	
0.588	1.7	5.1	16.7	0.35	1.14	1,087	0.12	9.74	
0.556	1.8	5.4	17.7	0.34	1.10	985	0.11	10.15	
0.526	1.9	5.7	18.7	0.33	1.07	902	0.10	10.53	
0.500	2.0	6.0	19.7	0.32	1.04	832	0.09	10.90	

The inverse of the h_c'/h ratio in Equation VI-5-72 is the relative submergence ratio $h/h_c' > 1.0$. At $h/h_c' = 1.0$ the water level is at the structure crest and this condition is equal to $R_c = 0$ as relative freeboard, R_c/H_s in Equation VI-5-71 when $f_i = 0.8$ (see previous Example Problem VI-7-12). Because the given data are the same for both Example Problem VI-7-12 and this problem VI-7-13, then the rock size D_{n50} and weight W_{50} should coincide at these extreme limits of these equations.

(Sheet 2 of 3)

EXAMPLE PROBLEM VI-7-13 (Concluded)

From Equation VI-5-71, at the limit when $R_c/H_s = 0$ (zero freeboard) the following values were obtained in Example Problem VI-7-12. These are compared to the W_{50} of 3,742 N (0.42 tons) found in this example for $h_c'/h = 1.0$.

P	D_{n50} (m)	(ft)	W_{50}, (N)	(tons)	Remarks
0.1	0.71	2.33	9,304	(1.05)	Substantially higher
0.4	0.55	1.82	4,400	(0.49)	Little higher
0.5	0.53	1.76	3,901	(0.44)	nearly the same

Therefore, it can be concluded that Equation VI-5-72 applies to permeable structures and may not be appropriate for submerged, impermeable structures.

(Sheet 3 of 3)

EXAMPLE PROBLEM VI-7-14

FIND: Select the armor stone size to withstand the design water level and wave conditions for a rubble-mound, nearshore breakwater constructed of homogeneous units with no core, i.e., permeable.

GIVEN: Breakwater crest elevations +4.0 ft, +5.0 ft, +6.0 ft (MLW) (1.22, 1.52, 1.83 m)

Design Water Levels			Design Wave Conditions			
Recurrence Interval, T_r, years	Storm Surge, S ft		Wave Height, H_s		Wave Period, T	Remarks
	(MLW)	(m)	ft	(m)	(sec)	
1	3.6	(1.10)	5.0	(1.52)	5.6	Storm surge and waves at structure toe
10	4.5	(1.37)	5.6	(1.71)	6.5	
25	5.2	(1.58)	6.5	(1.98)	7.6	
50	6.0	(1.83)	7.2	(2.19)	9.7	

(See Figure VI-7-4)

Tidal range, MTR = 1.0 ft (0.3 m)
Elevation at BW location = -3.5 ft (MLW) = Z (-1.07 m)
Vertical datum, MLW = 0.0 ft
Assume crest width, B = 5.0 ft (1.52 m)

Note: Values of H_s determined by numerical model. In some cases ratio of H_s to design water depth exceeds 0.6, which is the depth-limit value to use when more accurate methods are not available.

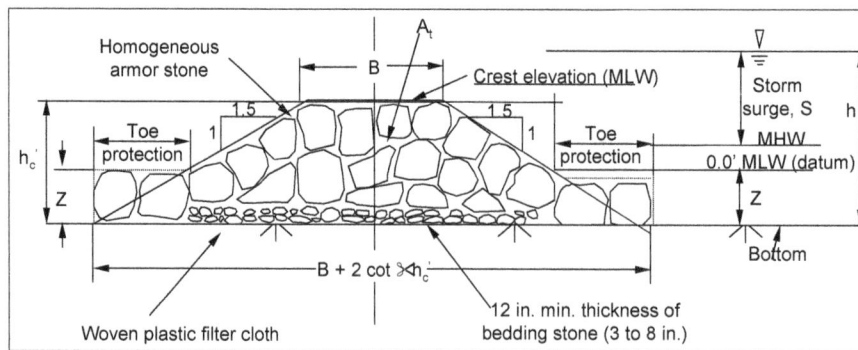

Figure VI-7-4. Homogeneous breakwater cross section

(Sheet 1 of 5)

EXAMPLE PROBLEM VI-7-14 (Continued)

Overtopping and submergence

Because of storm surge, some conditions will produce complete submergence so the breakwater acts as a low-crested reef. Table VI-5-27 gives Equation VI-5-73 as proposed by van der Meer (1990) using data by Ahrens (1987) and van der Meer (1990).

$$h_c = \sqrt{\frac{A_t}{\exp\left(aN_s^*\right)}}$$

where $h_c = h_c{'}$ is the no-damage condition. The equilbrium profile changes when wave energy reshapes the cross section to give a lower crest elevation height, h_c (damaged). The equilibrium area, A_t remains unchanged. The stability number is defined as

$$N_s^* = \frac{H_s}{\Delta D_{n50}} s_p^{-1/3}$$

where s_p is the wave steepness based on T_p and local wavelength.

Also, the parameter, a, is given by

$$a = -0.028 + 0.045 \frac{A_t}{\left(h_c{'}\right)^2} + 0.034 \frac{h_c{'}}{h} - \left(6 \times 10^{-9}\right) \frac{A_t^2}{D_{n50}^4}$$

Note that the stable stone size, D_{n50}, hence stable weight, W_{50}, is found by:

1. Specifying the reef breakwater dimensions, $h_c{'}$, B, and cot α, so the cross-sectional area A_t can be calculated.

2. Calculating a for a given water depth, h. (The formula for a is truncated, omitting the small last term.)

3. Calculating N_s^* when $h_c = h_c{'}$.

4. Calculating $s_p = H_s/L_p$, the local wave steepness.

5. Calculating D_n, i.e.

$$D_{n50} = \frac{H_s}{\Delta N_s^*} s_p^{-1/3}$$

(Sheet 2 of 5)

EXAMPLE PROBLEM VI-7-14 (Continued)

6.   and finally, calculating the homogeneous stone weight, W, from

$$W = \rho_s g \, (D_n)^3$$

Again, a spreadsheet solution aids the calculation process. Figure VI-7-5 illustrates the possible cases for design calculation. Calculations on the following table assumed rock specific weight of 165 lb/ft^3 and $\Delta = 1.58$ (salt water).

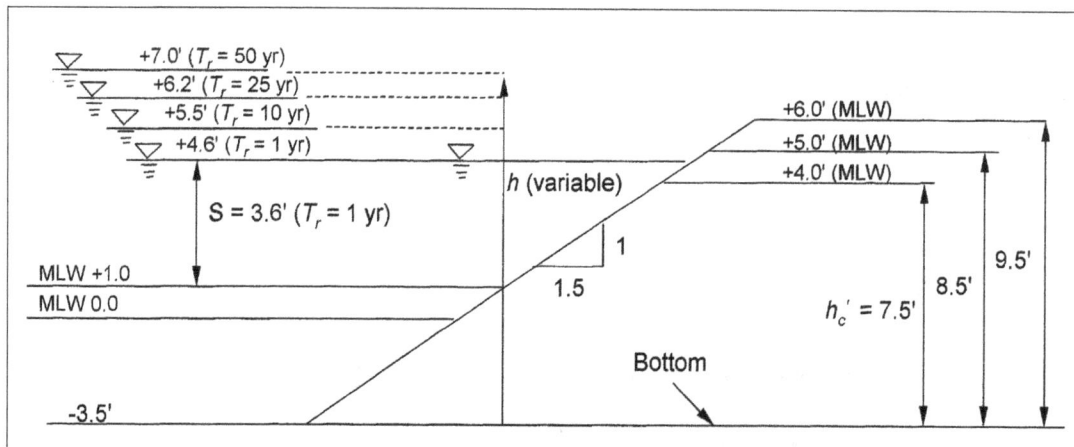

Figure VI-7-5. Variations in design water level and breakwater crest elevations

As expected, the rarer events with low exceedance probabilities each year (higher recurrence intervals) with larger wave heights require larger stones for stability. The crest elevation selected depends upon economics, allowable wave transmission, and resulting shoreline adjustment in the lee of each nearshore breakwater structure.

(Sheet 3 of 5)

EXAMPLE PROBLEM VI-7-14 (Continued)

Case 1. Crest Elevation +4.0 ft (MLW) $h_c^* = 7.5$ ft						
Recurrence Interval T_r, years	Water Depth h ft	Wave Conditions		Reef (Homogenous)		Remarks
		H_s ft	T_p, sec (T_m)	D_n ft	W lbs	
1.0	8.1	5.0	5.6	1.07	199	Sub
10.0	9.0	5.6	6.5	1.19	281	Sub
25.0	9.7	6.5	7.6	1.38	436	Sub
50.0	10.5	7.2	9.7	1.60	674	Sub
Case 2. Crest Elevation +5.0 ft (MLW) $h_c' = 8.5$ ft						
1.0	8.1	5.0	5.6	1.13	236	Overtop
10.0	9.0	5.6	6.5	1.26	328	Sub
25.0	9.7	6.5	7.6	1.45	504	Sub
50.0	10.5	7.2	9.7	1.67	773	Sub
Case 3. Crest Elevation +6.0 ft (MLW) $h_c' = 9.5$ ft						
1.0	8.1	5.0	5.6	1.19	278	Overtop
10.0	9.0	5.6	6.5	1.32	383	Overtop
25.0	9.7	6.5	7.6	1.52	584	Sub
50.0	10.5	7.2	9.7	1.75	889	Sub

This example problem was taken from Appendix A "Case Design Example of a Detached Breakwater Project" in Chasten et al. (1993). It is a real, constructed project for the community of Bay Ridge, Anne Arundel County, Maryland, on the western shore of the Chesapeake Bay near Annapolis. Example Problem V-3-1 in the cited report described the functional design of the breakwater layout (spacing distance offshore, length, etc.). A crest elevation of +4.0 ft (MLW) was selected for design. Structural stability design conditions were also given for the homogenous sized stone. Design conditions selected were:

$$T_r = 25 \text{ years}$$
$$H_s = 6.5 \text{ ft}$$
$$T_p = 7.6 \text{ s}$$
$$DSWL = +5.2 \text{ ft (MLW)}$$
$$h = 9.7 \text{ ft (MLW)}$$

The Hudson (1974) formula and *Shore Protection Manual* (1984) coefficients were used in the analysis. The breaking wave Hudson coefficient, $K_D = 2.0$ gives

$$D_n = 2.85 \text{ ft}$$
$$W = 3,830 \text{ lb}$$

(Sheet 4 of 5)

EXAMPLE PROBLEM VI-7-14 (Concluded)

The range of stone sizes accepted was 2,500-4,500 lb. The cross-sectional sketch of Figure VI-7-4 was taken from Chasten et al. (1993) Appendix A. It shows the large stone placed at the bottom for toe protection. The crest width B was 9.0 ft using three stone widths.

Use of the Hudson formula for Case 1 with completely submerged conditions gives stone weights over eight times heavier than required for these design conditions (436 lb – vs. – 3,830 lb). Even for the 50-year recurrence interval conditions, the Hudson formula results are roughly five times heavier. The crest width, based on three stone widths could also be reduced (6 ft – vs. – 9 ft which translates into less stone volume required and less costs).

SUMMARY: The Hudson formula was not meant to be applied for these conditions and is far too conservative.

The *Coastal Engineering Manual* presents stone stability formulations that give practical information for a range of water depths and waves impacting nearshore breakwaters.

(Sheet 5 of 5)

EXAMPLE PROBLEM VI-7-15

FIND: The weight and type (natural or artificial, concrete units) for repair of the north jetty at Barnegat Inlet, New Jersey.

GIVEN: The "Information Report on Effects of the North Jetty Repair on Navigation," especially Chapter 4, 'Hydraulic Analysis' (U.S. Army Engineer District, Philadelphia, 2001). Bathymetry from field surveys and numerical modeling of storm surge, water levels and wave conditions produced parameters with a range of frequency distributions. These permitted "sensitivity testing" for combined storm surge and wave conditions.

The design criteria specified less than 5 percent armor displacement or less than 0.5-ft crest elevation reduction for "storm conditions" at the 2 percent exceedance probability level in any one year, i.e., the 50-year recurrence interval. These criteria resulted in:

    Storm surge   =   +8.1 ft (MLLW)
    Wave height, H_{m0} = 13.8 ft (near structure head and toe)
    Seabed elevation =   -24ft (MLLW) (near structure head)

The raised jetty crest elevation was to be at +7.8 ft (MLLW).

SOLUTION:
Additional Assumptions: Sea water, ρ = 1.99 slugs/ft^3, γ = 64 lb/ft^3
                         Rock, ρ_s = 5.14 slugs/ft^3

$$\frac{\rho_s}{\rho} = 2.58$$

Use English System of units as given in USACE (2002).

Item 1. Natural Stone Armor Layer – Trunk

Assume quarrystone is available for all sizes required and construction equipment available for placement.

a. Hudson (1974) – Nonovertopped slopes

(1) Assume rough, angular, two layers
(2) Use no partial safety factors (Table VI-6-4)
(3) Assume nonbreaking wave conditions (H_{m0}/h) <0.6
(4) Use K_D = 4.0
(5) Use $H_{0.1}$ = 1.27 H_{m0} for design wave
(6) Use Eq. VI-6-43,
(7) $W_{50} = \rho_s\, g\, D_{n50}^{3}$

(Sheet 1 of 7)

EXAMPLE PROBLEM VI-7-15 (Continued)

cot α	W_{50} (lb)
1.5	37,646
2.0	28,235
2.5	22,588

No wave period, structure permeability, or storm duration effects are considered in the Hudson method. Damage is assumed to be between 0-5 percent.

b. van der Meer (1988) – Nonovertopped slopes

Additional assumptions include:
(1) Rock jetty, notational permeability, $P = 0.4$
(2) Nominal damage level, $S = 2$ (investigate W for $S = 5$)
(3) Number of waves, $N_z = 2,500$ (investigate W for $N_z = 5,000, 7,500$)
(4) Wave period, $T_p = 14$ s (investigate W for $T_p = 11$ s, 9 s)
(5) Use H_{m0} for design wave

The wave periods are taken from Table 4-11 for storm conditions as investigated in the report (Corps 2002). The results are found as follows when surging or plunging formulas are applicable, again with no partial safety factors. Mean period, T_m, was assumed to be 0.83 T_p.

T_p	cot α	W (lb)	Type	Applicable Eq.
14.0 sec	1.5	45,080	Surging	VI-6-45
	2.0	34,788	Plunging	VI-6-44
	2.5	24,892	Plunging	VI-6-44

Effects of wave period

T_p	cot α	W (lb)	Type	Applicable Eq.
11.0 sec	1.5	37,402	Plunging	VI-6-44
	2.0	24,291	Plunging	
	2.5	17,381	Plunging	
9.0 sec	1.5	27,680	Plunging	VI-6-44
	2.0	17,977	Plunging	
	2.5	12,863	Plunging	

It is clear that the longer period, $T_p = 14$ s ($T_m = 11.6$ s) requires heavier natural stones from the quarry. All are plunging type, except for cot $\alpha = 1.5$ and $T_p = 14$ s. But the storm surge will cause wave overtopping of the relatively low jetty crest.

(Sheet 2 of 7)

EXAMPLE PROBLEM VI-7-15 (Continued)

c. van der Meer (1991) – Overtopping slopes
Now the relative freeboard (water depth – jetty crest) must be considered. This will vary as
the total water level (storm surge, hydrograph plus local tidal range) changes during a storm
event. Consider a freeboard range, 8 ft, 5 ft, 3 ft, 1 ft, 0 ft and Equation VI-5-71 used to
reduce the values determined from van der Meer's Equations VI-6-44 and VI-6-45. To begin,
assume design wave height, H_{m0} = 13.8 ft remains constant as the storm surge increases, and
T_p = 14 s.

cot α	Stone Weights, W for Freeboard, R_c (ft)					Type
	8 ft	5 ft	3 ft	1 ft	0 ft	
1.5	32,057	28,209	25,980	23,890	23,056	S
2.0	24,817	21,838	20,113	18,564	17,849	P
2.5	17,739	15,610	14,377	13,270	12,759	P

The results reveal how overtopping reduces the stone weight requirements as formulated by
van der Meer (1991). The largest freeboard possible is actually +7.8 ft at MLLW with no
storm surge present. At the other extreme, with the design storm surge condition of +8.1 ft
(MLLW), the structure actually becomes submerged.

d. van der Meer (1991) – Submerged conditions
The crest height, h_c' = 7.8 ft + 24 ft water depth = 31.8 ft (from sea bottom).
The design storm surge water depth, h = 8.1 ft + 24 ft water depth = 32.1 ft (from sea
bottom).

The van de Meer (1991) formulation, Equation VI-5-72, uses the relative submergence, h_c'/h,
and all results are independent of structure slope, α.

The required stone weights decrease slightly with minor increases in submergence as
illustrated as follows for H_{m0} = 13.8 ft, T_p = 14.0 s.

cot α	Stone Weights, W (lb) for Submergence Levels $h - h_c'$ (ft)				
	0 ft	0.1 ft	0.2 ft	0.3 ft	0.5 ft
1.5	und	16,266	16,109	15,953	15,801
2.0	und	16,266	16,109	15,953	15,801
2.5	und	16,266	16,109	15,953	15,801

The design case is for 0.3-ft submergence = $h_{design} – h_c$ = 32.1 - 31.8 ft.

e. Effect of storm duration, t (hr)
All of the preceding results are for N_z = 2,500 waves and for T_p = 14 s which is roughly,
T_m = 11.6 and t = 8-hr storms. It is instructive to consider increased storm durations shown as
follows. Armor weights are in pounds

(Sheet 3 of 7)

EXAMPLE PROBLEM VI-7-15 (Continued)

cot α	$N_z = 2,500$ $t = 8$ hr	$N_z = 5,000$ $t = 16$ hr	$N_z = 7,500$ $t = 24$ hr	Type
1.5	45,080	55,500	62,679	S
2.0	34,788	42,829	48,369	P
2.5	24,892	30,646	34,610	P

These results return to the basic, nonovertopping formulations (Equations VI-6-44 and VI-6-45), for illustration purposes only. The longer storms may produce increased levels of the related damage parameter, S, as illustrated as follows.

The longest storm duration is recommended at $N_z = 7,500$ waves, beyond which, no increased effects are found.

f. Effect of damage parameter, S

All the preceding results are for $S = 2.0$, which is the nominal, almost zero-level damage parameter recommended. Increasing the damage parameter S to 5.0 will reduce the stable stone weight required, as illustrated as follows. This level is more consistent with longer storms with more waves. Armor weghts are given in pounds, and Eqs. VI-6-44 and VI-6-45 were used.

cot α	$N_z = 2,500$ $t = 8$ hr	$N_z = 5,000$ $t = 16$ hr	$N_z = 7,500$ $t = 24$ hr	Type
1.5	26,015	32,028	36,171	S
2.0	20,075	24,716	27,913	P
2.5	14,365	17,685	19,973	P

Again, these results return to the basic, nonovertopping formulations, for illustrative purposes.

The nominal stone diameter for cot α = 1.5, $N_z = 7,500$ waves, $S = 5.0$, and under surging conditions with $W_{50} = 36,171$ lb is:

$D_{nom} = 6.02$ ft

(Sheet 4 of 7)

EXAMPLE PROBLEM VI-7-15 (Continued)

g. Discussion/Summary

Point 1. The time phase relationship between storm surge hydrograph (including local tidal range) and the wave energy variation is important because the required stable stone weight varies considerably under nonovertopping, overtopping, and submerged conditions. Design for the combination that gives the largest armor stone weight.

Point 2. Increased storm duration increases armor weight requirements. A compromise is to accept more damage by increasing the nominal damage parameter from $S = 2$ to $S = 5$.

Point 3. Many factors not considered in the Hudson formulation can be included in the van der Meer formulation to give design insight regarding their importance.

Point 4. Partial safety factors can also be included to increase the stable stone weights required and increase the corresponding reliability. These should be studied for overtopped to submerged conditions.

Point 5. For design, consider for TRUNK SECTION ONLY the following values.

cot α	W, lb	W, tons	Remarks
1.5	36,000	18	These recommended values include overtopping,
2.0	28,000	14	$S = 5\%$, long storms, and are similar to Hudson (1974)
2.5	20,000	10	

Point 6. Physical model studies should be conducted for verification. This is especially true for the head section.

Point 7. Artificial, concrete armor units, e.g., Core-Loc®, must also be considered, as discussed as follows.

Item 2. Artificial, concrete armor units

a. Core-Loc® Non – or marginally-overtopped slopes
These units are discussed in Table VI-5-34 and in Melby and Turk (1994). Hereafter, referred to as Core-Loc, they have been under development since July 1992 by the U.S. Army Corps of Engineers. They incorporate all the following features that were previously distinct weaknesses of existing armor unit shapes (Melby and Turk 1994):

(1) High hydraulic stability when placed in a single-unit thickness at any slope angle

(2) Reserve stability for wave conditions that exceed the design event

(Sheet 5 of 7)

EXAMPLE PROBLEM VI-7-15 (Continued)

(3) No tendency for units to rock on slope

(4) Continued stability when broken or following renesting resulting from local instability

(5) Efficient combination of porosity and slope roughness to dissipate the maximum wave energy (and reduce wave runup)

(6) Maximum performance with a minimum concrete armor unit volume

(7) Hydraulically stable when placed as a repair with other shapes

(8) Low internal stresses, so no reinforcement required

(9) Easy to cast

(10) Easily constructed armor layer (single-unit-thickness), even in low visibility water

(11) Uses minimal casting yard or barge space

(12) Utilizes conventional construction materials and techniques

b. Example of a single-unit-thickness concrete armor layer, randomly placed:

For irregular, head-on waves, the Hudson formula has been applied, i.e.,

$$W = \frac{\rho_c \, g \, H_s^3}{K_D \left(\frac{\rho_c}{\rho_w} - 1 \right)^3 \cot \alpha}$$

as Equation VI-5-81 multiplied by the gravity constant, g, where:

H_s = the significant wave height
ρ_c = mass density of concrete
$\rho_c g$ = γ_c = unit weight concrete (assume = 140 lb/ft^3)
ρ_w = mass density of water in which unit is placed

For irregular, depth limited, breaking waves (plunging to collapsing), and the zero-damage condition (little or no rocking),

K_D = 16 is recommended for trunk section stability
K_D = 13 is recommended for head section stability

(Sheet 6 of 7)

EXAMPLE PROBLEM VI-7-15 (Concluded)

Results for the trunk section with $K_D = 16$ are:

cot α	W, lb	W, tons
1.333	10,302	5.2
1.5	9,155	4.6
2.0	6,866	3.4
2.5	5,493	2.8

and, for the head section with $K_D = 13$

cot α	W, lb	W, tons
1.333	12,680	6.3
1.5	11,268	5.6
2.0	8,450	4.2
2.5	6,760	3.4

Note that Core-Loc® units have been tested and remain stable on steeper slopes (3V:4H, cot α = 1.333) which can be very important economically for deepwater applications requiring large stone volumes.

The weight of the Core-Loc® units are roughly 3.5 to 4 times less than the rubble-mound stone armor units. But many other factors are involved in selecting the type (natural or concrete units) for repair of the north jetty at Barnegat Inlet, New Jersey. These factors include primarily:

Factor 1. Cost Data
Price of rock-fill per ton (including quarrying, transport, and positioning on structures)
Price of concrete armor unit, Core-Loc®, per cubic yard (including manufacture, handling, and positioning on structure)

and

Factor 2. Construction Method/Available Equipment
From land or using offshore equipment
Crane capacity for lifting and span

c. Other Artificial Concrete Unit Types
Many other types (dolos, tetrapod, tribar, accropode, etc.) have been successfully applied, but all have weaknesses in one or more of the areas previously cited as strengths of the Core-Loc® design. None have been specifically designed to be used as repair units. For these reasons, it is recommended to use the Core-Loc® unit.

(Sheet 7 of 7)

VI-7-5. Stability of Vertical Walled Bulkheads and Caissons.

EXAMPLE PROBLEM VI-7-16

FIND: The width of a vertical walled, caisson-type breakwater/jetty structure for the north jetty site at Barnegat Inlet, New Jersey.

GIVEN: The same information as found in Example Problem VI-7-15 summarized here (using non-SI units as given in the report):

Recurrence interval for design, T_r = 50 years.
Spectral significant wave height for design, H_{m0} = 13.8 ft.
Peak wave period, T_p = 14 s (consider also 11 s, 9 s).
Water depth near structure head, h = 24 ft (MLLW to seabed).
Storm surge for design, s = +8.1 ft (MLLW).
Structure crest elevation = +9.0 ft (MLLW).
Wave direction for design, β = 0 deg.
Seawater, γ_w = 64 lb/ft³.
Concrete, γ_c = 140 lb/ft³.
Sand, γ_s = 125 lb/ft³.

SOLUTION:
Item 1. Additional assumptions.

A suitable, rubble-mound foundation (core layer) with toe protection armor layer will support the concrete/sand-filled caisson structure. Assume the bottom of the caisson structure h^* is located at 0.85 h water depth below datum (MLLW), so take

h^* = caisson bottom depth (MLLW)
h^* = 0.85 (h) = 0.85 (24 ft) = 20.4 ft

assume

h^* = 20 ft. This leaves 4 ft for suitable rubble-foundation.

Adding the structure crest elevation above MLLW gives the total caisson structure height, h_w

h_w = 20 ft +9.0 ft = 29.0 ft

(Sheet 1 of 12)

EXAMPLE PROBLEM VI-7-16 (Continued)

The caisson concrete wall thickness, t, and concrete cap height, nt, can be varied where n is cap height multiplier of the wall thickness.

assume
$$t = 2.0 \text{ ft}$$
$$n = 2$$

The rock thickness, r, for the toe armor layer can be specified as some fraction of the core layer depth, $(h - h^*)$. Assume
$$r = 0.75 \, (h - h^*) = 0.75 \, (4) = 3.0 \text{ ft}$$

Finally, the rock berm width, B_m, can be taken as some multiplier of the rock thickness, say $2r$ (ASSUME).
$$B_m = 2r = 6.0 \text{ ft}$$

Item 2. Calculated dimensions

See the sketch in Table VI-5-53 (Goda formula for irregular waves) for additional, dimensions.

Total water depth,
$$h_s = \text{h} + \text{s}$$
$$h_s = 24.0 \text{ ft} + 8.1 \text{ ft} = 32.1 \text{ ft}$$

Structure submerged depth,
$$h' = h^* + s$$
$$h' = 20 \text{ ft} + 8.1 \text{ ft} = 28.1 \text{ ft}$$

Armor rock depth,
$$d = h' - r$$
$$d = 28.1 \text{ ft} - 3.0 \text{ ft} = 25.1 \text{ ft}$$

Freeboard
$$h_c = h_w - h' = 29.0 \text{ ft} - 28.1 \text{ ft} = 0.9 \text{ ft}$$

Item 3. Calculated coefficients and pressures

3.1. Wave height for design, H_{design}

(Sheet 2 of 12)

EXAMPLE PROBLEM VI-7-16 (Continued)

As seen in Table VI-5-53, most of the coefficients (α_i) and pressures (p) for the net wave pressure distribution against the vertical wall depend on H$_{\text{design}}$. If seaward of the surf zone, Goda (1985) recommends for practical design.

$$H_{\text{design}} = 1.8\,H_s = 1.8\,(13.8\text{ ft}) = 24.8\text{ ft}$$

This is the wave height with only 0.15 percent exceedence of Rayeigh distributed waves or $H_{1/250}$.

3.2. Coefficients

The coefficients, α_1, α_2, and α_3 depend upon the local wavelength, L, at a water depth, h_b, and at a distance of 5 H_s seaward of the front wall of the vertical caisson. This requires a further assumption of the bottom slope, m, seaward of the wall.

Take

$$m \;=\; 1/100$$

and also assume

$$H_s \;=\; H_{m0}$$

Thus
$$h_b \;=\; 5\,(H_s)\,m + \text{h}_\text{s}$$
$$h_b \;=\; 5\,(13.8\text{ ft})\,(0.01) + 32.1\text{ ft}$$
$$h_b \;=\; 32.8\text{ ft}$$

For $T_p = 14$ s and $h_b = 32.8$ ft, the local wavelength is

$$L \;=\; 439.1\text{ ft}$$

Now find the coefficients α_1, α_2, and α_3.

$$\alpha_1 = 0.6 + 0.5\left[\frac{4\pi\,h_s\,/\,L}{\sinh\left(4\pi\,h_s\,/\,L\right)}\right]^2$$

$$= 0.6 + 0.5\left[\frac{4\pi\left(32.1\text{ ft}\,/\,439.1\text{ ft}\right)}{\sinh 4\pi\left(32.1\text{ ft}\,/\,439.1\text{ ft}\right)}\right]^2$$

$$\alpha_1 = 0.980$$

(Sheet 3 of 12)

EXAMPLE PROBLEM VI-7-16 (Continued)

and

$$\alpha_2 = \text{smallest of} \left[\frac{(h_b - d)}{3h_b} \left(\frac{H_{design}}{d} \right)^2 \right] \text{ and } \frac{2d}{H_{design}}$$

Therefore,

$$\alpha_2 = \text{smallest of} \left[\frac{(32.8 \text{ ft} - 25.1 \text{ ft})}{3(32.8 \text{ ft})} \left(\frac{24.8 \text{ ft}}{25.1 \text{ ft}} \right)^2 \right] \text{ and } \frac{2(25.1 \text{ ft})}{24.8 \text{ ft}}$$

α_2 = smallest of 0.0764 and 2.024

so $\alpha_2 = 0.0764$

and

$$\alpha_3 = 1 - \frac{h_w - h_c}{h_s} \left[1 - \frac{1}{\cosh(2\pi \, h_s / L)} \right]$$

$$\alpha_3 = 1 - \frac{(29.0 \text{ ft} - 0.9 \text{ ft})}{32.1 \text{ ft}} \left[1 - \frac{1}{\cosh(2\pi \, (32.1 \text{ ft} / 439.1 \text{ ft}))} \right]$$

$$\alpha_3 = 0.9151$$

and note

$$\alpha_* = \alpha_2$$

so that

$$\alpha_* = 0.0764$$

Now, the pressures against the vertical wall are calculated.

3.3. Pressures

The theoretical, zero pressure is at a vertical distance, η^*, above the design water level. This distance and the pressures, $p_1, p_2, p_3,$ and p_u, all include a coefficient λ ($\lambda_1, \lambda_2, \lambda_3$) that are "modification" factors depending on the structure type (i.e., inclined, curved, etc.). For conventional, vertical wall structures,

$$\lambda_1 = \lambda_2 = \lambda_3 = 1$$

(Sheet 4 of 12)

EXAMPLE PROBLEM VI-7-16 (Continued)

$$\eta^* = 0.75(1+\cos\beta)\lambda_1 H_{design}$$
$$= 0.75(1+\cos 0^o)(1.0)(24.8 \text{ ft})$$
$$\eta^* = 37.2 \text{ ft}$$

and

$$p_1 = 0.5(1+\cos\beta)(\lambda_1\alpha_1 + \lambda_2\alpha_* \cos^2\beta)(\rho_w g)H_{design}$$
$$= 0.5(1+\cos 0^o)(1.0\cdot 0.980 + 1.0\cdot 0.0764 \cos^2 O^o)(64 \text{ lb/ft}^3)(24.8 \text{ ft})$$
$$p_1 = 1,677 \text{ lb/ft}^2$$

and

$$p_2 = \begin{cases} \left(1-\dfrac{h_c}{\eta^*}\right)p_1 & \text{for } \eta^* > h_c \\ \\ 0 & \text{for } \eta^* \le h_c \end{cases}$$

and clearly $\eta^* > h_c$ so that

$$p_2 = \left[1-(0.9 \text{ ft}/37.2 \text{ ft})\right](1,677 \text{ lb/ft}^2)$$
$$p_2 = 1,636 \text{ lb/ft}^2$$

and

$$p_3 = \alpha_3 p_1$$
$$= 0.9151(1,677 \text{ lb/ft}^2)$$
$$= 1,535 \text{ lb/ft}^2$$

(Sheet 5 of 12)

EXAMPLE PROBLEM VI-7-16 (Continued)

The bottom uplift pressure, p_u, is not equal to p_3 but found from

$$p_u = 0.5(1+\cos \beta)\lambda_3 \alpha_1 \alpha_3 (\rho_w g) H_{design}$$
$$= 0.5(1+\cos 0°)(1.0)(0.980)(0.9151)(64 \text{ lb/ft}^3)(24.8 \text{ lb/ft})$$
$$p_u = 1,423 \text{ lb/ft}^2$$

These pressures could be plotted to create the pressure distribution diagrams, similar to that shown in Table VI-5-53.

Item 4. Calculated forces and moments

Specifying a structure width, B, is now needed which is the horizontal dimension normal to h_w (not shown in Table VI-5-53). For illustration purposes of the calculations, assume

$B = 40$ ft

This structure width will significantly influence the structure weight (in air and water) and the uplift force on the bottom.

Table VI-5-55 presents the resulting wave induced forces and moments (including uncertainty factors) when calculated from the wave load equations by Goda (Table VI-5-53), and Takahashi, et al. (Table VI-5-54).

The formulas assume that the hydrostatic, horizontal pressures are in balance on both sides of the structure and are per running length of the structure (i.e., per unit length).

4.1.Forces

The horizontal force, F_H is:

$$F_H = U_{FH}\left[\frac{1}{2}(p_1 + p_2)h_c + \frac{1}{2}(p_1 + p_3)h'\right]$$

Use the mean values of the uncertainty and bias factors as presented in Table VI-5-56. So $U_{FH} = 0.90$ and $U_{FU} = 0.77$

$$F_H = 0.9\left[\frac{1}{2}(1,677 \text{ lb/ft}^2 + 1,636 \text{ lb/ft}^2)(0.9 \text{ ft}) + \frac{1}{2}(1,677 \text{ lb/ft}^2 + 1,535 \text{ lb/ft}^2)28.1 \text{ ft}\right]$$
$$F_H = 41,958 \text{ lb/ft}$$

(Sheet 6 of 12)

EXAMPLE PROBLEM VI-7-16 (Continued)

The uplift force, F_U, depends on caisson width, B.

$$F_U = U_{FH} \left[\frac{1}{2} p_u (B) \right]$$

$$F_U = 0.77 \left[\frac{1}{2} \left(1,423 \text{ lb/ft}^2 \right) \left(40 \text{ ft} \right) \right]$$

$$F_U = 21,914 \text{ lb/ft}$$

The structure submerged weight is more complicated. The formula in Table VI-5-55 for F_G says

$$F_G = (\rho_c g) \, B \cdot h_w - (\rho_w g) \, B \cdot h'$$

where

$\rho_c g$ = unit weight of the combined, concrete/sand structure

The first term is the total weight and the second term is the buoyant uplift force.

Because the walls are concrete ($\gamma_c = 140 \text{ lb/ft}^3$) and the caisson is filled with sand ($\gamma_s = 125 \text{ lb/ft}^3$), clearly the total weight (per lineal foot) will depend on the wall thickness, t, and the cap dimension, nt. Using the values

$t = 2.0$ ft
$n = 2$
$B = 40$ ft

gives

$W_{\text{concrete}} = 140 \text{ lb/ft}^3 \, [(40 \text{ ft})(29 \text{ ft}) - (36 \text{ ft})(23 \text{ ft})] = 46,480 \text{ lb/ft}$

$W_{\text{sand}} = 125 \text{ lb/ft}^3 \, [(36 \text{ ft})(23 \text{ ft})] = 103,500 \text{ lb/ft}$

$W_{\text{Total}} = 46,480 \text{ lb/ft} + 103,500 \text{ lb/ft} = 149,980 \text{ lb/ft}$

Therefore,

$$F_G = 149,980 \text{ lb/ft} - \left(64 \text{ lb/ft}^3 \right) \left(40 \text{ ft} \right) \left(28.1 \text{ ft} \right)$$

$$F_G = 149,980 \text{ lb/ft} - 71,936 \text{ lb/ft}$$

$$F_G = 78,044 \text{ lb/ft}$$

(Sheet 7 of 12)

EXAMPLE PROBLEM VI-7-16 (Continued)

4.2.Moments about heel of the caisson

Using the recommended uncertainty and bias factors from Table VI-5-55 (U_{MH} = 0.81 and U_{MU} = 0.72), the overturning moment due to the horizontal force is

$$M_H = U_{MH}\left[\frac{1}{6}(2p_1 + p_3)h'^2 + \frac{1}{2}(p_1 + p_2)h'h_c + \frac{1}{6}(p_1 + 2p_2)h_c^2\right]$$

$$= 0.81\begin{bmatrix}\frac{1}{6}(2(1,677 \text{ lb/ft}^2) + 1,535 \text{ lb/ft}^2)(28.1 \text{ ft})^2 + \\ \frac{1}{2}(1,677 \text{ lb/ft}^2 + 1,636 \text{ lb/ft}^2)28.1 \text{ ft} \cdot (0.9 \text{ ft}) \\ + \frac{1}{6}(1,677 \text{ lb/ft}^2 + 2(1,636 \text{ lb/ft}^2))(0.9 \text{ ft})^2\end{bmatrix}$$

$$M_H = 555,630 \text{ lb-ft/ft}$$

The overturning moment of the uplift force is given by

$$M_U = U_{MU}\frac{1}{3}p_u \cdot B^2$$

$$M_U = 0.72(1/3)(1,423 \text{ lb/ft}^2)(40 \text{ ft})^2 = 546,430 \text{ lb-ft/ft}$$

The compensating moment of the structure weight (net)

$$M_G = \frac{1}{2}B^2 g(\rho_c h_w - \rho_w h')$$

is the moment formula found in Table VI-5-55 where again, the composite structure unit weight, ($\rho_c g$) is required. However, because the caisson cross section is symmetrical, the total force, F_G, is assumed to act at a horizontal distance $B/2$ from the back side, so

$$M_G = \frac{1}{2}B \cdot F_G$$

$$= \frac{1}{2}(40 \text{ ft})(78,044 \text{ lb/ft})$$

$$M_G = 1,560,880 \text{ lb-ft/ft}$$

This completes the basic calculation of forces and moments (per unit length) due to the wave loading formulations of Goda and Takahashi. Further analysis requires basic design practices, safety factors, and codes, i.e., DESIGN CRITERIA for the engineering evaluation of the assumed structure width, B.

(Sheet 8 of 12)

EXAMPLE PROBLEM VI-7-16 (Continued)

5. Safety against sliding, overturning, and heal bearing pressures.

5.1. Design criteria
Goda (1985) gave the following coefficients, safety factors as practiced by coastal engineers in Japan. Others are possible (see Tables VI-5-62 and VI-5-63).

Sliding friction coefficient, $\mu = 0.6$
Safety factor against sliding, $SF_s \geq 1.2$
Safety factor against overturning, $SF_o \geq 1.2$
Heal bearing pressures, $\quad P_e < 8000 - 10,000 \text{ lb/ft}^2$
$\qquad\qquad\qquad\qquad P_e \max = 12,000 \text{ lb/ft}^2$

5.2. Sliding from Equation VI-5-190

$$SF_s \equiv \frac{\mu\left(F_G - F_U\right)}{F_H}$$

and is simply the ratio of frictional resistance force to the applied force in the horizontal direction.

$$SF_s = \frac{0.6\left(78,044 \text{ lb/ft} - 21,914 \text{ lb/ft}\right)}{41,958 \text{ lb/ft}}$$

$$= \frac{33,678}{41,958} \qquad \text{(WARNING: Structure unstable in sliding mode)}$$

$$SF_s = 0.80$$

The structure width with $B = 40$ ft is too narrow to prevent sliding because $SF_s < 1.2$.

5.3. Overturning from Equation VI-5-191

$$SF_O \equiv \frac{M_G}{M_U + M_H}$$

which is the ratio of resistive moment due to self-weight to overturning moments.

$$SF_O = \frac{1,560,880 \text{ lb-ft/ft}}{\left(546,430 + 555,630\right) \text{ lb-ft/ft}}$$

$$SF_O = \frac{1,560,880}{1,102,060}$$

$$SF_O = 1.42$$

(Sheet 9 of 12)

EXAMPLE PROBLEM VI-7-16 (Continued)

which is stable against overturning because SF_O exceeds the allowable safety factor, i.e., $SF_O \geq 1.2$.

5.4. Heel bearing pressure, Pe

See Goda (1985) and other marine foundation design references for eccentric, inclined loadings. Assume the seafloor is dense sand and the supporting, core layer material of good bearing capacity.

Define:

$W_e \equiv$ net vertical force

$W_e \equiv F_G - F_u = 78{,}044 \text{ lb/ft} - 21{,}914 \text{ lb/ft} = 56{,}130 \text{ lb/ft}$

and

$M_e \equiv$ net moment about heel (positive counterclockwise)

$M_e \equiv M_G - M_u - M_H = (1{,}560{,}880 - 546{,}430 - 555{,}630) \text{ lb-ft/ft} = 458{,}820 \text{ lb-ft/ft}$

Then $M_e = W_e \cong t_e$

or

$t_e \equiv M_e/W_e$

where t_e is the moment arm of the net vertical force.

so

$$t_e = \frac{458{,}820 \text{ lb-ft/ft}}{56{,}130 \text{ lb/ft}} = 8.17 \text{ ft}$$

Because $t_e \leq B/3$, assume a triangular bearing pressure distribution, with

$$P_e = \frac{2W_e}{3t_e}$$

$$P_e = \frac{2(56{,}130 \text{ lb/ft})}{3(8.17 \text{ ft})} = 4{,}580 \text{ lb/ft}^2$$

which is within safe bearing capacity.

Note: If $t_e > B/3$ a trapezodal bearing pressure distribution would be assumed with

$$P_e = \frac{2W_e}{B}\left(2 - 3\frac{t_e}{B}\right)$$

(Sheet 10 of 12)

EXAMPLE PROBLEM VI-7-16 (Continued)

5.5. Summary, $B = 40$ ft caisson width ($T_p = 14.0$ s)

For the design wave, $H_{design} = 1.8\, H_{m0}$ and structure dimensions, $h_w = 29.0$ ft, $B = 40$ ft, this vertical, caisson-type jetty is unstable against sliding and stable against overturning. Heel bearing pressures are acceptable. Several design modifications could be considered to determine a stable caisson width.

Structure width, B, is the key variable because wider structures provide more weight for stability against sliding and overturning. However, the uplift forces and moments also increase, and cost per unit structure length also increase.

For the preceding example, keeping everything constant, it turns out that B must be increased to about 60.3 ft width for

$$(SF)_s = 1.2$$

5.6. Additional considerations for increased stability.

a. Increasing the jetty crest elevation, h_c. The structure weight could be increased by increasing the crest elevation, which was previously set at +9.0 ft (MLLW). Increasing by 1.0 ft increments, +10.0 ft, 11.0 ft, etc., produces ever decreasing stable width, B in a nonlinear trend.

b. Increasing the unit weight of the fill material, γ_s. The only practical way remaining is to find a suitable caisson fill material, e.g., slag from blast furnaces that is more dense and has a higher unit weight, γ_s, than saturated sand ($\gamma_s = 125$ s/ft^3).

c. Reliability analyses. Chapter VI-6 contains tables with partial safety factors for:

Sliding failure, Table VI-6-25
Overturning failure, Table VI-6-26
Foundation failure (Heel bearing pressure)
    Table VI-6-23, sand subsoil
    Table VI-6-24, clay subsoil

The width of the foundation sublay can be extended seaward and the vertical-walled, concrete caisson protected by a rubble-mound structure, seaward. Table VI-5-58 presents changes in the $\lambda_1, \lambda_2, \lambda_3$ coefficients in the Goda formula for this case.

(Sheet 11 of 12)

EXAMPLE PROBLEM VI-7-16 (Continued)

The caisson sections exposed to wave breaking directly on the structure in the surf zone should be checked to ensure stability. Wave heights will be lower so that the same cross-sectional design may be sufficient.

6.0. Summary: In general, for relatively shallow-water depths, vertical-walled concrete caisson structures are not economically competitive with the rubbble-mound structures, but each site is different and should be checked.

(Sheet 12 of 12)

EXAMPLE PROBLEM VI-7-17

FIND: Find the total force per unit lateral length on a vertical baffle breakwater having
penetration depths of 1/3 and 2/3 of the water depth. The breakwater will protect a mooring
facility on Lake Michigan.

GIVEN: The following baffle breakwater parameters are defined in Part VI-5,
Figure VI-5-61.

Water depth, $h = 15.0$ ft.
Wall penetration relative to SWL, $w = 5.0$ ft, 10 ft.
Specific weight of fresh water, $\rho_w\, g = 62.4$ lb/ft^3.
Zeroth-moment wave height at breakwater, $H_{mo} = 3.0$ ft.
Wave period associated with the spectral peak, $T_p = 5.0$ s.

SOLUTION: The appropriate wavelength is found using linear wave theory with a depth of
$h = 15$ ft and a wave period of $T_p = 5$ s, i.e., $L_p = 96.3$ ft. The corresponding wave number is

$$k_p = \frac{2\pi}{L_p} = \frac{2\pi}{96.3\text{ft}} = 0.0652 \quad 1/\text{ft}$$

Force on a Partially Penetrating Wall:

The first step is to calculate the parameter F_o which represents the significant force per unit
length on a vertical wall extending all the way to the bottom as determined using linear wave
theory. F_o is calculated using Eq. VI-5-163.

Important Note: The F_o-factor should only be used as part of the calculation for
partially-penetrating walls (it is a normalizing factor). Estimates for forces on full-depth
walls should be calculated using the Goda method detailed in Table VI-5-53.

(Sheet 1 of 2)

EXAMPLE PROBLEM VI-7-17 (Continued)

$$F_o = \rho_w \ g \ H_{mo} \left[\frac{\sinh \ (k_p h)}{k_p \ \cosh \ (k_p h)} \right]$$

$$F_o = \left(62.4 \ \text{lb/ft}^3 \right) \ (3 \ ft) \left[\frac{\sinh \ \left[(0.0652 \ 1/\text{ft}) \ (15 \ \text{ft}) \right]}{(0.0652 \ 1/\text{ft}) \ \cosh \left[(0.0652 \ 1/\text{ft}) \ (15 \ \text{ft}) \right]} \right]$$

$$F_o = 2,159 \ \text{lb/ft}$$

Substituting given values of h and L_p and the calculated value of F_o into total force empirical formula for partially penetrating vertical walls given by Eq. VI-5-164 yields

$$F_{mo} = F_o \left(\frac{w}{h} \right)^{0.386 \left(h/L_p \right)^{-0.7}}$$

$$F_{mo} = \left(2,159 \ \text{lb/ft} \right) \left(\frac{w}{h} \right)^{0.386 \left(15 \ \text{ft}/96.3 \ \text{ft} \right)^{-0.7}}$$

$$F_{mo} = \left(2,159 \ \text{lb/ft} \right) \left(\frac{w}{h} \right)^{1.419}$$

Now substitute the different wall penetration values to find the significant total force at each depth.
$w = 5$ ft

$$F_{mo} = \left(2,159 \ \text{lb/ft} \right) \left(\frac{5 \ \text{ft}}{15 \ \text{ft}} \right)^{1.419} = 454.2 \ \text{lb/f}t$$

$w = 10$ ft

$$F_{mo} = \left(2,159 \ \text{lb/ft} \right) \left(\frac{10 \ \text{ft}}{15 \ \text{ft}} \right)^{1.419} = 1,214.4 \ \text{lb/ft}$$

Recall for final design it is recommended that the significant total force be increased by a factor of 1.8,

$$F_{design} = 1.8 \ F_{mo}$$

This increase corresponds roughly to the Goda-recommended design wave

$$H_{design} = H_{1/250} = 1.8 \ H_{mo}$$

(Sheet 2 of 2)

EXAMPLE PROBLEM VI-7-18

FIND: Find the total force and moment per unit lateral length on a vertical seawall located inside the surf zone on an ocean beach.

GIVEN: The following parameters are defined in Part VI-5, Figure VI-5-71.

Water depth at breaking, $h_b = 8.0$ ft.
Breaking wave height, $H_b = 7.0$ ft.
Still-water depth at vertical wall toe, $h_s = 5.0$ ft.
Specific weight of salt water, $\rho_w g = 64.0$ lb/ft^3.

SOLUTION: The broken wave height at the structure, as represented by the turbulent bore rushing up the beach, is estimated using Eq. VI-5-174, i.e.,

$$H_w = \left(0.2 + 0.58 \ \frac{h_s}{h_b} \right) H_b$$

$$H_w = \left[0.2 + 0.58 \ \frac{(5 \ \text{ft})}{(8 \ \text{ft})} \right] (7 \ \text{ft}) = 3.94 \ \text{ft}$$

The total horizontal force (excluding backfill soil or standing water pressure behind the wall) is the sum of the dynamic pressure force (Eq. VI-5-176) and the static pressure force (Eq. VI-5-179), i.e.,

$$R_T = \frac{\rho_w g \, h_b \, H_w}{2} + \frac{\rho_w g}{2} (h_s + H_w)^2$$

$$R_T = \frac{(64.0 \ \text{lb/ft}^3)(8 \ \text{ft})(3.94 \ \text{ft})}{2} + \frac{(64.0 \ \text{lb/ft}^3)}{2} (5 \ \text{ft} + 3.94 \ \text{ft})^2$$

$$R_T = 1,008.6 \ \text{lb/ft} + 2,557.6 \ \text{lb/ft} = 3,566 \ \text{lb/ft}$$

(Sheet 1 of 2)

EXAMPLE PROBLEM VI-7-18 (Continued)

The total moment about the toe of the wall (excluding backfill soil or standing water pressure behind the wall) is the sum of the moments due to the dynamic pressure force (Eq. VI-5-177) and the static pressure force (Eq. VI-5-180), i.e.,

$$M_T = R_d \left(h_s + \frac{H_w}{2} \right) + R_s \left(\frac{h_s + H_w}{3} \right)$$

$$M_T = \left(1{,}008.6 \ \text{lb/ft} \right) \left(5 \ \text{ft} + \frac{3.94 \ \text{ft}}{2} \right) + \left(2{,}557.6 \ \text{lb/ft} \right) \left(\frac{5 \ \text{ft} + 3.94 \ \text{ft}}{3} \right)$$

$$M_T = 7{,}030 \ \frac{\text{ft-lb}}{\text{ft}} + 7{,}622 \ \frac{\text{ft-lb}}{\text{ft}} = 14{,}652 \ \frac{\text{ft-lb}}{\text{ft}}$$

(Sheet 2 of 2)

VI-7-6. Forces on Cylindrical Piles.

EXAMPLE PROBLEM VI-7-19

FIND: Find the total force and moment on a small-diameter cylindrical vertical pile situated in the nearshore area of the Atlantic Ocean (salt water).

GIVEN:
Water depth, $d = 16.4$ ft.
Wave height, $H = 9.5$ ft (approx. ¾ H_b).
Wave period, $T = 8.0$ s
Cylindrical pile diameter, $D = 2.3$ ft.
Specific weight of salt water, $\rho_w\, g = 64.0$ lb/ft³.
Kinematic viscosity of salt water, $v = 1.076(10)^{-5}$ ft²/s

SOLUTION:
1. Estimate drag and inertia force coefficients.
An estimate of the maximum horizontal velocity under the prescribed wave is found from Stream Function theory to vary between

$u_m = 13.1$ ft/s   (at crest of wave)
$u_m = 6.1$ ft/s   (at the seabed)

The average maximum velocity of

$u_m = 9.5$ ft/s   (average)

is used for calculation of the pile Reynolds number.

From Eqn. VI-5-307 the pile Reynolds number is calculated as

$$R_e = \frac{u_m\, D}{v} = \frac{(9.5\ \text{ft/s})\ (2.3\ \text{ft})}{1.076\ (10)^{-5}\ \text{ft}^2/\text{s}} = 2.03\ (10)^6$$

Figures VI-5-139 to VI-5-143 give plots for drag and inertia coefficients for Reynolds much lower than the previous calculation. In this case, use the guidance given after Figure VI-5-143 for higher Reynolds number. Because $R_e = 2.03(10)^6$ is greater than $5(10)^5$, the drag and inertia coefficients are assumed to have constant values given by

$C_D = 0.7$
$C_M = 1.5$

(Sheet 1 of 3)

EXAMPLE PROBLEM VI-7-19 (Continued)

2. Estimate force and moment coefficients from nomograms.
The *Coastal Engineering Manual* provides two methods for estimating the maximum force
and moment acting on a cylindrical pile. One method estimates the individual maximums for
the drag and inertia forces and moments, and then sums the values. This method is
conservative because the two maximums are out of phase. The preferred method is to
estimate the maximum combined drag and inertia forces and moment forces using Equations
VI-5-300 and VI-5-301, reproduced as follows.

$$F_m = \varphi_m \ C_D \ \rho_w g \ H^2 \ D$$
$$M_m = \alpha_m \ C_D \ \rho_w g \ H^2 \ D \ d$$

Values for the nondimensional coefficients φ_m and α_m are taken from the appropriate
nomograms contained in Figures VI-5-131 to VI-5-138 based on the nondimensional
parameter W given by Eqn. VI-5-299. For the values of drag and inertia coefficients in this
example,

$$W = \frac{C_M}{C_D} \frac{D}{H} = \frac{1.5}{0.7} \frac{(2.3 \text{ ft})}{(9.5 \text{ ft})} = 0.52$$

Also needed are values of relative wave steepness and relative water depth that are the
ordinate and abscissa, respectively, of the nomograms. For this example

$$\frac{H}{gT^2} = \frac{9.5 \text{ ft}}{(32.2 \text{ ft/s}^2) (8 \text{ s})^2} = 0.0046$$

$$\frac{d}{gT^2} = \frac{16.4 \text{ ft}}{(32.2 \text{ ft/s}^2) (8 \text{ s})^2} = 0.0080$$

Figure VI-5-133 was constructed for a value of $W = 0.5$, and it is the closest to the value of
$W = 0.52$ calculated for this example. Using the values for $H/(gT^2)$ and $d/(gT^2)$, the force
coefficient in found from Figure VI-5-133 to be

$$\varphi_m = 0.32$$

Similarly, the moment coefficient for $W = 0.5$ is found from Figure VI-5-137 as

$$\alpha_m = 0.30$$

(Sheet 2 of 3)

EXAMPLE PROBLEM VI-7-19 (Continued)

3. Estimate maximum total force and moment.

Finally, the total maximum force and total maximum moment are calculated as

$$F_m = \varphi_m \, C_D \, \rho_w g \, H^2 \, D$$

$$F_m = 0.32 \; (0.7) \left(64.0 \;\; \text{lb/ft}^3\right) \left(9.5 \;\; \text{ft}\right)^2 \left(2.3 \;\; \text{ft}\right) = 2,976 \;\; \text{lb}$$

and

$$M_m = \alpha_m \, C_D \, \rho_w g \, H^2 \, D \, d$$

$$M_m = 0.30 \; (0.7) \left(64.0 \;\; \text{lb/ft}^3\right) \left(9.5 \;\; \text{ft}\right)^2 \left(2.3 \;\; \text{ft}\right) \left(16.4 \;\; \text{ft}\right) = 45,752 \;\; \text{ft-lb}$$

The corresponding moment arm is found simply as

$$\text{Moment \;\; Arm} = \frac{M_m}{F_n} = \frac{45,752 \;\; \text{ft-lb}}{2,976 \;\; lb} = 15.4 \;\; ft$$

which is one foot below the still waterline.

NOTE: In this example the calculated value of W was reasonably close to that of Figures VI-5-133 and VI-5-137. For cases when W falls between nomograms, finds values of φ_m and α_m from the bracketing nomograms and linearly interpolate.

(Sheet 3 of 3)

VI-7-7. References.

Ahrens 1987
Ahrens, J. P. 1987. "Characteristics of Reef Breakwaters," Technical Report CERC-87-17, Coastal Engineering Research Center, U.S. Army Engineer Waterways Experiment Station, Vicksburg, MS.

Brooks and Brandon 1995
Brooks, R. M., and Brandon, W. A. 1995. "Hindcast Wave Information for the U.S. Atlantic Coast: Update 1976-1993." WIS Report 33, Coastal Engineering Research Center, U.S. Army Engineer Waterways Experiment Station, Vicksburg, MS.

Carver and Heimbaugh 1989
Carver, R. D., and Heimbaugh, M. S. 1989. "Stability of Stone- and Dolos-Armored Rubble-Mound Breakwater Heads Subjected to Breaking and Nonbreaking Waves with No Overtopping," Technical Report CERC-89-4, Coastal Engineering Research Center, U.S. Army Engineer Waterways Experiment Station, Vicksburg, MS.

Chasten et al. 1993
Chasten, M. A., Rosati, J. D., McCormick, J. W., and Randall, R. E. 1993. "Engineering Design Guidance for Detached Breakwaters as Shoreline Stabilization Structure," Technical Report CERC-93-19, Coastal Engineering Researach Center, U.S. Army Engineer Waterways Experiment Station, Vicksburg, MS.

de Waal and van der Meer 1992
de Waal, J. P., and van der Meer, J. W. 1992. "Wave Run-Up and Overtopping on Coastal Structures, *Proceedings 23rd International Coastal Engineering Conference*, American Society of Civil Engineers (ASCE), Vol. 2, 1,758-1,771.

Goda 1985
Goda, Y. 1985. *Random Seas and Design of Maritime Structures*. University of Tokyo Press, Tokyo, Japan.

Hudson 1974
Hudson, R. Y. (editor). 1974. "Concrete Armor Units for Protection Against Wave Attack," Miscellaneous Paper H-74-2, U.S. Army Engineer Waterways Experiment Station, Vicksburg, MS.

Junger 1997
Junger, S. 1997. *The Perfect Storm: A True Story of Men Against the Sea*. W.W. Norton & Co., New York, 227 pp.

Leenknecht et al. 1992
Leenknecht, D. A., Szuwalski, A., and Sherlock, A. R. 1992. "Automated Coastal Engineering System: User Guide and Technical Reference, Version 1.07," U.S. Army Engineer Waterways Experiment Station, Vicksburg, MS.

Melby and Turk 1994
Melby, J. A. and Turk, G. F. 1994. "The CORE-LOC: Optimized Concrete Armor," *Proceedings 24th International Coastal Engineering Conference*, ASCE, Vol. 2, 1,426-1,438.

Nielsen 1984
Nielson, P. 1984. "Analytical Determination of Wave Height Variation," *Journal of Waterway Port, Coastal, and Ocean Engineering*, ASCE, Vol. 110, No. 2, pp. 283-287.

Owen 1980
Owen, M. W. 1980. "Design of Seawalls Allowing for Wave Overtopping," Report No. 924, Hydraulics Research Station, Wallingford, England.

Owen 1982
Owen, M. W. 1982. "The Hydraulic Design of Seawall Profiles," *Proceedings of Coastal Protection Conference*, Institution of Civil Engineers, Thomas Telford, London, 185-192.

Pitcher 1964
Pitcher, G. 1964. *The Philosophy of Wittgenstein, Part I*. Prentice-Hall, Inc., Englewood Cliffs, NJ.

Shore Protection Manual 1977
Shore protection manual. 1977. 3rd ed., 3 Vol, U.S. Army Engineer Waterways Experiment Station, U.S. Government Printing Office, Washington, DC.

Takahashi, Tanimoto, and Shimosako 1994
Takahashi, S., Tanimoto, K., and Shimosako, K. 1994. "A Proposal of Impulsive Pressure Coefficient for Design of Composite Breakwaters," *Proceedings of the International Conference on Hydro-Technical Engineering for Port and Harbor Construction*, Port and Harbour Research Institute, Yokosuka, Japan, pp. 489-504.

Shore Protection Manual 1984
Shore protection manual. 1984. 4th ed., 2 Vol, U.S. Army Engineer Waterways Experiment Station, U.S. Government Printing Office, Washington, DC.

U.S. Army Engineer District, Philadelphia 2001
U.S. Army Engineer District, Philadelphia. 2001. "Information Report on Effects of the North Jetty Weir on Navigation, Barnegat Inlet, New Jersey: Chapter 4, Hydraulic Analysis," Draft Report, February 2001.

van der Meer 1988
van der Meer, J. W. 1988. "Rock Slopes and Gravel Beaches Under Wave Attack," Ph.D. diss., Delft Hydraulics Publication No. 396, Delft University of Technology, The Netherlands.

van der Meer 1990
van der Meer, J. W. 1990. "Low-Crested and Reef Breakwaters," Report H198/Q638, Delft Hydraulics Laboratory, The Netherlands.

van der Meer 1991
van der Meer, J. W. 1991. "Stability and Transmission at Low-Crested Structures," Delft
Hydraulics Publication No. 453, Delft Hydraulics Laboratory, The Netherlands.

van der Meer and Janssen 1995
van der Meer, J. W., and Janssen, W. 1995. "Wave Run-Up and Wave Overtopping at Dikes," in
Wave Forces on Inclinded and Vertical Wall Structures. N. Kobayashi and Z. Demirbilek, ed.,
ASCE, pp. 1-27.

VI-7-8. Acknowledgements.

Authors: Dr. David R. Basco, Department of Civil Engineering, Old Dominion University,
Norfolk, VA; and Dr. Steven A. Hughes, Coastal and Hydraulics Laboratory (CHL), U.S. Army
Engineer Research and Development Center, Vicksburg, MS.

Reviewers: H. Lee Butler, CHL (retired); Dr. Lyndell Z. Hales, CHL; and Dr. Lee E. Harris,
Department of Marine and Environmental Systems, Florida Institute of Technology, Melbourne,
FL.

VI-7-9. List of Symbols.

Symbols have been defined in each example problem at the time they are first used or derived.
Also see VI-5 for symbols.

CHAPTER 8

Monitoring, Maintenance, and Repair of Coastal Projects

TABLE OF CONTENTS

List of Figures

List of Tables

CHAPTER VI-8

Monitoring, Maintenance, and Repair of Coastal Projects

VI-8-1. <u>Maintenance of Coastal Projects</u>. This chapter covers maintenance requirements of coastal engineering projects. Ongoing maintenance at some level is necessary for most existing coastal projects to assure continued acceptable project performance. Major topics included in this chapter are monitoring of projects, evaluation of project condition, repair and rehabilitation guidelines, and project modifications. Available guidance related to specific repair and rehabilitation situations is included. However, in many cases design guidance suitable for new construction is used to design repairs.

    a. Aging of coastal projects.

    (1) The U.S. Army Corps of Engineers (USACE) has responsibility for constructing and maintaining federally authorized coastal engineering projects in the United States. These include navigation channels, navigation structures, flood-control structures, and erosion control projects. Pope (1992) summarized a series of reports (see Table VI-8-10) on Corps-maintained breakwaters and jetties and noted that 77 percent of the 265 navigation projects constructed in the United States were over 50 years old. Even more revealing is the fact that about 40 percent of the breakwaters and jetties originated in the 1800s. This means that a majority of the Corps' structures were designed and built before the introduction of even rudimentary design guidance and armor stability criteria, and in many cases the structures have survived well beyond their intended service life because they have been properly maintained. Most developed countries undoubtedly have a similar situation.

    (2) Over the projected life of a project, the structural components are susceptible to damage and deterioration. Damage is usually thought of as structure degradation that occurs over a relatively short period such as a single storm event, a unique occurrence, or perhaps a winter storm season. Damage might be due to storm events that exceed design levels, impacts by vessels, seismic events, unexpected combinations of waves and currents, or some other environmental loading condition.

    (3) Deterioration is a gradual aging of the structure and/or its components over time. Deterioration can progress slowly, and often goes undetected because the project continues to function as originally intended even in its diminished condition. However, if left uncorrected, continual deterioration can lead to partial or complete failure of the structure.

    (4) Pope (1992) distinguished between two types of aging processes that occur at coastal projects. Structure aging is a change to a portion of the structure that affects its function. Examples of structure aging include: settlement or lateral displacement of the structure, loss of slope toe support, partial slope failure, loss of core or backfill material, and loss of armor units.

    (5) Unit aging is defined as deterioration of individual components which could eventually affect the structure's function. Examples of unit aging include: breakage of concrete armor units, fracturing of armor stone, below-water deterioration of wood or sheet metal pilings,

corrosion of metal supports and fittings, concrete spalling, ripping of geotextile bags, and failure of individual gabion or timber crib units.

(6) Because coastal structure aging is a slow process, and the severity of deterioration may be hidden from casual inspection, rehabilitation often is given a low priority and may be postponed if the structure is still functioning at an acceptable level. Saving money by neglecting needed repairs runs the risk of facing a far more expensive (and possibly urgent) repair later.

b. Project maintenance.

(1) Project maintenance is a continuous process spanning the life of the coastal project. The goal of maintenance is to recognize potential problems and to take appropriate actions to assure the project continues to function at an acceptable level.

(2) Maintenance consists of the following essential elements (Vrijling, Leeuwestein, and Kuiper 1995):

(a) Periodic project inspection and monitoring of environmental conditions and structure response.

(b) Evaluation of inspection and monitoring data to access the structure's physical condition and its performance relative to the design specifications.

(c) Determining an appropriate response based on evaluation results. Possible responses are

- Take no action (no problems identified or problems are minor)

- Rehabilitate all or portions of the structure

- Repair all or portions of the structure

(3) Rehabilitation is defined in the dictionary as ARestoring to good condition, operation, or capacity." This implies that steps are taken to correct problems before the structure functionality is significantly degraded. For example, replacing broken concrete armor units, filling and capping scour holes, replacing corroded steel sheet pile, or patching spalled concrete might be considered structure rehabilitation. Rehabilitation can also be thought of as preventative maintenance. There are two types of preventative maintenance: condition-based maintenance which is rehabilitation based on the observed condition of the project; and periodic maintenance which is rehabilitation that occurs after a prescribed time period or when a particular loading level is exceeded.

(4) Repair is defined in the dictionary as "Restoring to sound condition after decay, damage, or injury." The major implication in this definition for repair is that damage has occurred and structure functionality is significantly reduced. For example, rebuilding a slumped armor slope, resetting breakwater crown blocks, rebuilding damaged pier decks, repairing vertical seawall, and backfilling eroded fill could be considered structure repair. Repair can also

be thought of as corrective maintenance. Obviously there are many situations where it is difficult to distinguish between repair and rehabilitation. The concepts behind coastal structure maintenance are straightforward; the difficulties lie in determining

(a) What to monitor.

(b) How to evaluate the monitoring data.

(c) Whether or not to undertake preventative or corrective action.

(d) How to access the economic benefits of the possible responses.

(5) Because of the wide variety of coastal structures and the varied environments in which they are sited, development of a generic project maintenance plan is difficult. Perhaps the best source of guidance is past experience maintaining similar projects.

(6) In addition to repair and rehabilitation, a third response that might arise during maintenance is a decision to modify a project even if it shows no damage or deterioration. Monitoring might reveal the project is not performing as expected, or the goals of the project might have changed or expanded, necessitating structure additions or modification. Examples include raising breakwater crest elevation to reduce overtopping into a harbor, modifying jetty length to reduce downdrift erosion problems, and sand tightening jetties to block passage of sediment.

VI-8-2. Inspecting and Monitoring Coastal Structures.

a. Introduction and overview.

(1) Project monitoring is a vital part of any successful maintenance program. The complexity and scope of a monitoring effort can vary widely from simple periodic onsite visual inspections at the low end of the scale to elaborate and expensive long-term measurement programs at the other extreme. The most important aspect in any monitoring and inspection program is to determine carefully the purpose of the monitoring. Without a clear definition of the monitoring goals, resources and instruments will not be used in the most beneficial manner; and most likely, the monitoring information will be insufficient to evaluate the project and recommend appropriate maintenance responses.

(2) Project monitoring can be divided into two major categories:

(a) Project condition monitoring consists of periodic inspections and measurements conducted as part of project maintenance. Condition monitoring provides the information necessary to make an updated assessment of the structure state on a periodic basis or after extreme events.

(b) Project performance/function monitoring consists of observations and measurements aimed at evaluating the project's performance relative to the design objectives. Typically, performance monitoring is a short-duration program relative to the life of the project.

(c) There are substantial differences between monitoring plans developed for project condition monitoring and plans developed for monitoring project performance. However, when developing either type of plan, several guidelines should be followed.

- First, establish the goals of the monitoring. Once the goals are known, every component suggested for the monitoring program can be assessed in terms of how it supports the goals. If a proposed element does not support the goals, there is little justification for including it.

- Second, review the project planning and design information to identify the physical processes that affect the project. These processes are then ranked in order of importance with respect to the monitoring goals. For some situations this step will be difficult because of uncertainties about the interaction between project elements and the environmental loadings. Once the monitoring goals are determined and the principal physical processes are identified, it is then possible to proceed with developing a plan to acquire the necessary monitoring data.

- An essential component of any plan is a provision for gathering sufficient project baseline data. Baseline data provide the basis for meaningful interpretation of measurements and observations. Elements of the baseline data collection are determined directly from the monitoring plan. For example, if the cross-section profile of a rubble-mound structure is to be monitored, it is necessary to establish the profile relative to known survey monuments at the start of the monitoring period. The as-built drawings often serve as part of the baseline survey information for project condition monitoring. Note that as-built drawings based on after-construction surveys are not always prepared. Thus, original design drawings may have to serve as baseline information.

b. Project condition monitoring. Project condition monitoring and inspection are necessary only for preventative maintenance programs. Failure-based maintenance does not require a monitoring program (Vrijling, Leeuwestein, and Kuper 1995). However, even failure-based maintenance must have some means of discovering whether or not severe damage or failure has occurred. If damage is not reported, there is a risk that additional damage or complete failure may occur, resulting in more costly repairs. Choosing which aspects of the project to inspect and monitor should be based on an understanding of the potential damage and failure modes for that particular type of project. This includes understanding the failure modes and deterioration traits of individual project components, as well as the project as a whole. Some failure modes may have a higher likelihood of occurrence, but may occur gradually without immediate impact to project functionality. On the other hand, there may be other failure modes with lower probability of occurrence that cause immediate, catastrophic damage. Just as important as identifying failure modes is knowing the physical signs of impending failure associated with each particular mode. For example, loss of armor stone from a slope or armor unit breakage may be a precursor to slope failure. The monitoring plan should outline what signs to look for, and if possible, how to quantify the changes. Some identified failure modes may give no warning signs of impending doom; and in these cases, monitoring will not help. Past experience with similar projects is beneficial in establishing what aspects of the project to monitor for change. Project condition monitoring always involves at least visual inspection of the project, and in some cases the inspection is augmented with measurements meant to quantify the current structure condition relative to the baseline condition. These observations are then used to evaluate the current project condition and make decisions on the course of action. Condition monitoring should be performed when changes are most likely to occur. Most changes happen during construction and in the first year or two after a project is completed. During this period, there can be dynamic adjustments such as structure settlement, armor units nesting, and bathymetry change. After initial structure adjustment, most significant changes occur during storm events. The monitoring plan should provide enough flexibility in scheduling to accommodate the irregularity of severe storms.

(1) Periodic inspections. The Corps of Engineers' policy relative to periodic inspection of navigation and flood-control structures is as follows:

(a) "Civil Works structures, whose failure or partial failure could jeopardize the operational integrity of the project, endanger the lives and safety of the public, or cause substantial property damage shall be periodically inspected and evaluated to ensure their structural stability, safety, and operation adequacy."

(b) The major USACE District and Division commands have responsibility for establishing periodic inspection procedures, intervals, etc., for civil works projects. However, standardized inspection methodology across all USACE Field Offices is lacking due to specific guidance, credentials of the individuals performing the inspections as well as the wide diversity in projects, sites, and environmental conditions.

(c) Above-water visual inspection of structural components can be accomplished by walking on the structure, or viewing it from a boat or an airplane. The effectiveness of visual inspection depends heavily on knowing what symptoms of deterioration to look for and being able to gauge changes that have occurred since the previous inspection. For example, broken armor units and displaced stone are obvious signs of potential problems (Figure VI-8-1).

(d) Visual inspections are subjective by nature; and, as in most practical aspects of coastal engineering, experience is paramount in recognizing potential problems. Inexperienced engineers, new to the inspection process, should accompany the seasoned engineers during inspection tours so they can learn to recognize the important signs of deterioration. This also helps provide monitoring continuity over the life of the structure as senior personnel retire and younger engineers move into senior positions.

(e) When observations indicate the need to quantify the structure changes, a few simple, inexpensive techniques can be used during the onsite inspection. These measures include: counting broken armor units, spray paint marking of cracks or suspected displacements, using a tape to measure distances between established points on the structure, shooting the elevation of selected locations using a level, and repeated photo-documentation from the same vantage point (Pope 1992).

(f) Frequency of periodic walking inspections of coastal structures varies a great deal across the USACE District and Division offices and even between different structures in the same jurisdiction. Typical options are included in Table VI-8-1.

Figure VI-8-1. Dolos breakage on Crescent City, California, breakwater

Table VI-8-1
Frequency of Walking Inspections

Annual walking inspections

Annual walking inspections for recently completed structures and repairs; less frequent inspections for older structures

Walking inspections every 2 years

Walking inspections every 3 years if the structure has not changed for 4 consecutive years

Walking inspections only after major storm events

Walking inspections only when personnel are in the region for other purposes and time permits

Walking inspections only after local users report a problem

(g) In general, the frequency of inspection of a particular structure should be determined on a case-by case basis. Factors that influence inspection frequency, along with recommended general guidelines, are listed in the following paragraphs:

• Geographic location. Structures situated in exposed locations on high-energy shores (e.g., northwest coast of the U.S.) should be inspected annually. Structures in sheltered areas or on low-energy coasts can stretch the inspection interval to 2 or 3 years. If no significant damage occurs for 5 years, inspections can be less frequent. Structures that undergo seasonal ice loading and freeze/thaw conditions should be inspected at least once every other year with particular emphasis on stone integrity or armor displacement.

- Structure age. Recently constructed, rehabilitated, or repaired structures should be inspected annually. Older structures with a good stability record for at least 5 years can be inspected less often. Frequently repaired structures should have annual inspections, but chronic damage should be addressed by a more robust design.

- Storm damage. Structures should be inspected after major storm or other events that might cause damage (e.g., earthquake or ship collision). Annual inspections are warranted in cases where damaged structures would impact navigation, property, and life. Reports of damage by local users of the project should be investigated immediately.

- Available funds. Sufficient funds and available personnel dictate both the frequency and priority of periodic inspections, particularly in Districts with many coastal structures. Past experience will help establish an inspection schedule that optimizes available funds.

(h) In summary, periodic inspection methods and frequency are determined based on repair history, past experience, engineering judgment, and available funding and manpower.

(i) All inspections should be documented to provide information and guidance for future assessments, and careful consideration should be given as to how the inspection information is to be preserved. Even the most observant visual inspection has little value if others cannot review the information and understand what was observed. Cryptic shorthand notes, rough sketches, etc., should be translated and expanded shortly after the inspection.

(j) Aerial visual inspection of coastal projects by fixed-wing or rotor-wing aircraft is an option that has several advantages. Aircraft provide easier access to remote project sites and to structures that are not attached to shore. They also allow the inspector to witness structure performance during wave conditions that would be unsafe for a walking inspection. Finally, several individual projects along a stretch of coastline can be inspected from the air in a short time span. During the aerial inspection, still photographs and video can be taken to augment the inspection notes. These images can be compared to previous photographs to see if obvious changes have occurred.

(k) The major disadvantage of aerial inspections is that only obvious changes can be identified whereas subtle changes and signs of deterioration may go unnoticed, even when inspecting enlarged aerial photographs. Nevertheless, aerial inspections make economic sense for projects with good performance histories and for making quick assessments after major storms to determine which projects need closer inspection.

(l) Underwater and interior visual inspection of structure condition is difficult, if not impossible, for many projects. These inspections require professional divers who also understand the signs of damage and deterioration for the particular type of structure. Water visibility plays a big role in underwater inspection. Some inspections, such as examining the condition of piers and piles, can be performed during poor visibility. However, other visual inspections, such as assessing the integrity of armor slopes beneath the surface, require sufficient visibility to see enough of the slope to recognize missing armor and slope discontinuities. Even in the best of conditions, information from diver surveys is subjective and spatial detail is sparse. Around tidal inlets, underwater inspections can only take place during the slack water. Above all, safety for both the divers and their support crew on the surface is the most important criterion for underwater visual inspections. Figure VI-8-2 shows the interior of the concrete parapet of the Great Sodus east breakwater, Lake Ontario, New York. Notice the missing timber and interior fill. Further information on diver inspections is given in Thomas (1985).

Looking along Interior Centerline of Great Sodus East Breakwater Parapet, Lake Ontario, New York

Figure VI-8-2. Interior of Great Sodus east breakwater, Lake Ontario, New York

(m) In some circumstances, it may be possible to use video cameras lowered into the water or mounted on remotely-controlled vehicles to inspect underwater portions of a structure. Other methods for quantifying underwater portions of structures are listed in the next section.

(2) Measurements. Measurements to quantify specific aspects of a coastal project that cannot be judged from a visual inspection may be included as part of the long-term project condition monitoring. Such measurements may be acquired concurrently with the visual inspection, or they may be part of longer-duration monitoring. Generally, project condition measurements focus on physical changes of the structure and its foundation. Examples include

repeated elevation surveys of selected structure cross sections to quantify settlement or loss of armor, bathymetry soundings to document scour hole development, quantifying underwater structure profiles with sensors, and spot testing of materials undergoing deterioration, such as concrete, timber, and geosynthetics. Basically, any measurement that aids in evaluating structure condition can be considered for the condition monitoring program. Most measurements require baseline data for comparison, and sequential measurements help to assess the rate of change for the monitored property.

(a) Photogrammetry. Photogrammetry is a term applied to the technique of acquiring and analyzing aerial photography to quantify the three-dimensional (3-D) geometry of objects in the photographs relative to a fixed coordinate system. One area particularly well suited to photogrammetry techniques is profiling rubble-mound structure cross sections and monitoring movement of armor units on exposed structures based on properly acquired aerial photography (Kendell 1988; Hughes et al. 1995a,b). Traditionally, this task was accomplished by surveying targets placed on individual armor units, a difficult, expensive, and often very dangerous undertaking. Naturally, photogrammetric analysis can only be applied to that portion of the structure visible above the waterline; hence, aerial overflights are scheduled to coincide with low tide level to maximize the benefits. An example is shown in Figure VI-8-3.

• The first step in photogrammetric monitoring of a rubble-mound structure is establishing permanent benchmarks on or near the structure that can be easily recognized in the aerial photographs. The horizontal and vertical position of these benchmarks are established using conventional ground surveying techniques, and they are used in the photogrammetry analysis to correct for aircraft tilt, roll, and yaw; to determine the camera position and orientation relative to ground features; and to compensate for the earth's curvature. Next, high-quality, low-level stereo photographs of the structure are obtained using standard stereo-mapping equipment and techniques. The photographic stereo pairs are used along with the ground survey information to establish a stereo-model, which is a 3-D representation of the study area that is free of geometric distortion. Stereo-models are usually constructed using a computer. Annual flights of the same structure using the same control reference points facilitate comparisons between stereo-models to extract information such as stone movement and yearly structure profile change above water level.

• Several requirements for successful monitoring using aerial photogrammetry are as follows:

– Good quality equipment and experienced personnel must be employed. If possible the same equipment and personnel should be retained throughout the entire monitoring program.

– The pilot should be experienced in low-level, low-speed flight in order to obtain blur-free, high resolution photographic images.

– Best results come during calm weather with clear visibility and low water levels to maximize coverage of the structure. The sun should be nearly overhead to minimize shadows.

– Photograph forward overlap should be at least 60 percent.

Figure VI-8-3. Aerial photogrammetry image of Yaquina, Oregon, north jetty

 – There should be at least five or six evenly distributed control points in each photographic stereo pair in order to remove geometric distortions.

 • Additional information on photogrammetry related to rubble-mound structures can be found in Cialone (1984) and U.S. Army Engineer Waterways Experiment Station (1991). Corps monitoring of the Crescent City breakwater using aerial photogrammetry was described by Kendall (1988); monitoring of the Yaquina north breakwater was documented in Hughes et al. (1995a,b).

 (b) Underwater inspection. Quantifying underwater changes to coastal structures is difficult, but it is an important part of monitoring structure condition. Underwater problems that go undetected can lead to sudden, unexpected failures. At least four measuring systems (not including visual inspection) are available for obtaining information about the underwater condition of coastal structures.

- The most limited method for sloping-front structures is using a crane situated on the structure crest to make soundings of the underwater portions of the structure. Horizontal and vertical position of the survey lead can be established using modern Global Positioning Systems (GPS). This method is contingent on crane availability/capability and access to the structure crest. Similar techniques have been attempted using a sounding line attached to a helicopter. See McGehee (1987) for additional information.

- Side-scan sonar returns obtained by towing the instrument off a vessel running parallel to the structure can be interpreted to give general information of underwater structure condition, particularly near the seabed. The main advantage of side-scan sonar is the coverage and the speed of surveying. The disadvantage of this technology is the skill needed to interpret the record in a meaningful way. Side-scan is perhaps best used to identify structure portions that need to be examined using more sophisticated instruments. Additional information and operating rules-of-thumb are given by Kucharski and Clausner (1989, 1990) and Morang, Larson, and Gorman (1997b).

- For accurate mapping of the underwater portions of rubble-mound structures, the best solution to date is a commercial system named SeaBat®. The SeaBat is a portable, downward and side-looking single-transducer multibeam sonar system. The instrument is mounted to a vessel with the sonar head positioned to transmit on a plane perpendicular to the vessel's heading. The sonar transmits 60 sonar beams on radials spaced at 1.5 deg, giving total swath coverage of 90 deg. By tilting the sonar head, the instrument can provide data for mapping almost the entire underwater portion of a sloping rubble-mound structure from just below the sea surface to the structure toe as illustrated on Figure VI-8-4. SeaBat data must be synchronized with simultaneous readings of vessel position, heading, and motion (heave, pitch, and roll). The final analyzed product is a spatially rectified map of the structures below-water condition. Although it is difficult to identify individual armor displacement, any slope irregularities due to construction or subsequent damage are easily spotted on the map. SeaBat systems have been extensively tested by Corps Districts, and the technology is considered quite mature and highly reliable. Additional information on SeaBat multibeam sonars can be found in EM 1110-2-1003 and in Prickett (1996, 1998).

- The final method for mapping, which can be used for both the underwater and above-water portions of sloping structures, is the airborne lidar technologies as provided by the Scanning Hydrographic Operational Airborne Lidar Survey (SHOALS) system (Parsons and Lillycrop 1998). Normally a SHOALS survey is not conducted with the sole purpose of examining structures; instead the structure mapping is an added benefit that occurs during the survey of a much larger portion of the surrounding area.

- Typically, the spatial distribution of SHOALS data will not be sufficient to recognize smaller irregularities in the armor layer, such as individual movements. However, larger problems in the armor slope and details of adjacent scour holes are readily apparent in SHOALS topography/bathymetry. It is impractical to include a SHOALS component when planning structure condition monitoring unless SHOALS surveys are planned as part of the overall project monitoring.

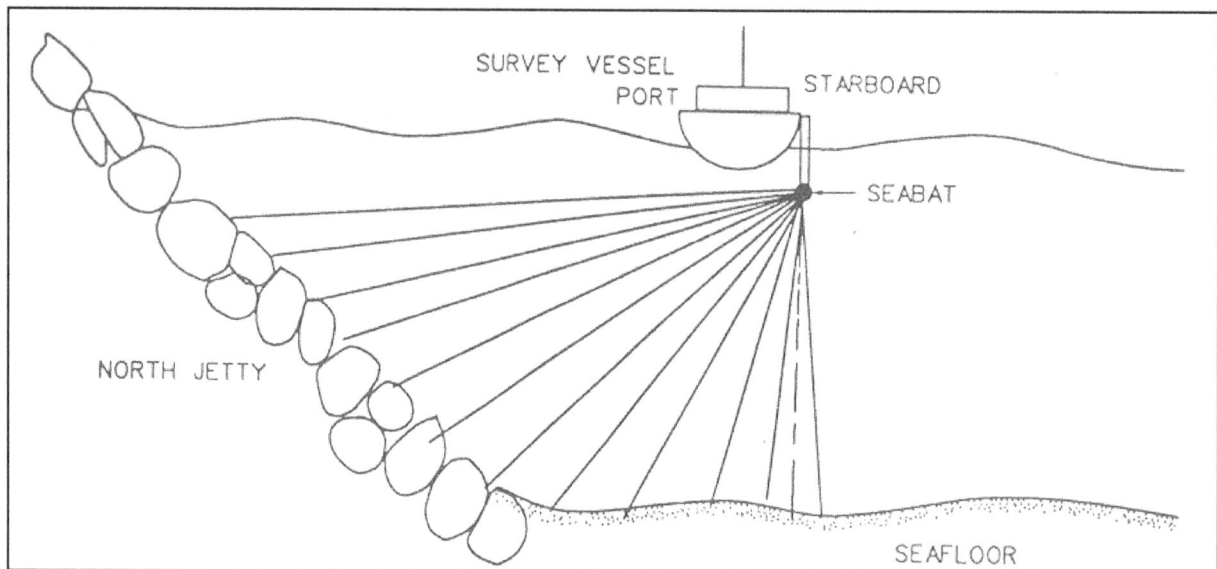

Figure VI-8-4. Multibeam scanner mounted on survey vessel

(3) Evaluating structure condition. Inspections are highly subjective, and the overall assessment of structure condition given by one observer may differ substantially from the opinion of another. Many factors can influence structure condition assessment with experience and previous visits to the site perhaps being the most important. In the 1990s, USACE developed guidelines aimed at providing a more uniform and consistent method for evaluating the physical condition and functional performance of coastal structures. These procedures, while still subjective, give a more meaningful evaluation by quantifying inspection observations in terms of uniform condition and performance criteria. The resulting numerical ratings allow better comparisons of condition and performance between similar structures, and better tracking of structure condition over time. A major benefit of a consistent condition rating system is the prospect of obtaining similar evaluations from different observers over the life of the structure. The procedure described in this section is a condition and performance rating system for rubble-mound breakwaters and jetties armored with stone or concrete armor units. Similar methods could be adopted for other coastal structures (revetments, bulkheads, monolithic structures, etc.) but have not yet been formally developed. The primary source for the information given in this section was a technical report (Oliver et al. 1998) available in Portable Document Format (PDF) from the Web site of the Repair, Evaluation, Maintenance, and Rehabilitation (REMR) Research Program (see Oliver et al. 1998 citation in this chapter=s References for Web address). A few tables from the REMR report are reproduced in this section, but application of the methodology requires obtaining a copy of the complete report, which also includes useful examples. An associated software program is also available at the REMR Web site.

(a) Condition and performance rating system for breakwaters and jetties. The condition and performance rating system is a performance-based evaluation system where emphasis is placed more on the question "how well is the structure functioning?" than on "what is the physical condition relative to the as-built structure?" This emphasis recognizes that coastal structures have some level of deterioration tolerance before significant loss of functionality, thus condition alone is not sufficient justification for rehabilitation.

• The end result of a performance evaluation is structure ratings given in terms of condition index numbers ranging between 0 and 100. Table VI-8-2 lists the general condition index range along with corresponding descriptions of the structure condition and damage levels. The different categories in Table VI-8-2 are fairly generic because the general condition index scale is intended to apply to a variety of USACE navigation and control structures, not just coastal breakwaters and jetties. For coastal structures the condition index (CI) is determined from a functional index (FI) and a structural index (SI). The FI indicates how well a structure (or a portion of the structure) is performing its intended functions, whereas the SI for a structure (or structure component) indicates the level of physical condition and structural integrity.

Table VI-8-2
General Condition Index Scale (from Oliver et al. 1998)

Observed Damage Level	Zone	Index Range	Condition Level	Description
Minor	1	85 to 100	EXCELLENT	No noticeable defects. Some aging or wear may be visible.
		70 to 84	GOOD	Only minor deterioration or defects are evident.
Moderate	2	55 to 69	FAIR	Some deterioration or defects are evident, but function is not significantly affected.
		40 to 54	MARGINAL	Moderate deterioration. Function is still adequate.
Major	3	25 to 39	POOR	Serious deterioration in at least some portions of the structure. Function is inadequate.
		10 to 24	VERY POOR	Extensive deterioration. Barely functional.
		0 to 9	FAILED	No longer functions. General failure or complete failure of a major structural component.

• The condition index is primarily a planning tool, with the index value serving as an indicator of the structure's general condition level. A series of condition index evaluations can be used to judge likely future functional performance degradation based on the trend of the condition index. For this reason, it is critical that all evaluations reflect the condition of the structure at the time of inspection, and not the condition that is expected at some future time. The condition ratings and index values are simply a numerical shorthand for describing structure physical condition and functional performance, and they represent only one part of the information required to make decisions about when, where, and how to spend maintenance dollars. Other necessary factors include knowledge of the structure's history, budget constraints, policies, etc. Furthermore, the condition index system is not intended to replace the detailed investigations which are needed to document fully structure deficiencies, to identify their causes, and to formulate plans for corrective action (see CEM, Part VI-8-2, "Project performance/ function monitoring").

(b) Initial procedures for rubble-mound breakwaters and jetties. The condition and performance rating system involves eight steps as shown in Table VI-8-3. Steps 1-5 are initial procedures that are performed once for a given structure. The only other time some or all of steps 1-5 will need to be repeated is after major rehabilitation or project alteration. Steps 6-8 are performed for each condition assessment based on structure inspection. Each of the steps in Table VI-8-3 is discussed in greater detail in the following paragraphs.

- Steps 1-2: Determining structure functions and dividing into major reaches. The functions served by a structure can vary over different portions of the structure. Division of a structure into major reaches by function is performed as an office study using authorizing documents and project history in combination with the functional descriptions given in Table VI-8-4.

Table VI-8-3
Steps in Condition and Performance Rating System (after Oliver et al. 1996)

Step	Description
Steps 1-5 are Initial (One-Time Only)	
1	Determine structure function
2	Divide structure into major reaches based on function
3	Subdivide major reaches into subreaches by structural and length criteria
4	Establish functional performance criteria
5	Establish structural requirements
Steps 6-8 are Repeated as Necessary	
6	a) Inspect structure b) Produce structural rating
7	a) Produce functional rating b) Calculate condition index
8	Review structural requirements

Table VI-8-4
Functional and Structural Rating Categories (from Oliver et al. 1996)

Functional Area	Functional Rating Categories	Structural Rating Categories
Harbor area	Harbor navigation harbor use	Breach
Navigation channel	Entrance use channel	Core-exposure
Sediment management	Ebb shoal	Armor loss
	Flood shoal	Loss of armor contact and interlock
	Harbor shoal	Armor quality defects
	Shoreline impacts	Slope defects
Structure protection	Nearby structures	
	Toe erosion[1]	
	Trunk protection[1]	

[1] Not included in condition index calculation.

- The four functional areas and associated 11 rating categories are described as follows.

- Harbor area. How well the structure controls waves and currents to allow full use of the harbor area during all conditions and for all vessels, as compared with design expectations or current requirements.

- *Harbor navigation.* Indicates how well navigable conditions are maintained within the harbor itself. Difficulty in maneuvering and restrictions on vessel drafts or lengths are indications of problems.

- *Harbor use.* Normal usage may be restricted by waves, currents, or seiches at support facilities, such as docks and mooring facilities. Problematic conditions may be seasonal. Three subcategories are considered.

  - Moored vessels. How well moored vessels are protected from damage, and the degree to which portions of the harbor are unusable during certain conditions. Functional deficiency can be measured by the frequency and degree to which vessels sustain damage from excessive waves or currents.

  - Harbor structures. How well the harbor infrastructure is kept usable and free from damage. Structures include mooring facilities, docks, slips, bulkheads, revetments, etc. Functional deficiency exists if waves or currents damage or impair use of these facilities.

  - Other facilities. This subcategory includes facilities set back from the water's edge and facilities that are part of the surrounding commercial and recreational infrastructure.

- Navigation channel. How well the structure controls waves and currents to allow full use of the navigation channel and entrance during all conditions and for all vessels.

- *Entrance use.* This category indicates the success of the structure in maintaining a safe channel by controlling waves and currents as stipulated by authorizing documents. Functional deficiencies are indicated if certain sizes or types of vessels are unable to navigate the channel safely or are delayed in entering.

- *Channels.* This category indicates how well the structure controls waves and currents in the channel through which vessels may operate without difficulty, delay, or damage. Functional deficiencies include strong cross-channel currents or crossing wave trains, channel obstructions, grounding occurrences, and vessel collisions with structures or other vessels.

- Sediment management. How well the structure controls the depth, character, and pattern of sedimentation in the navigation channel; the depth of ebb and flood shoals in tidal entrances; and the buildup or loss of sediments on nearby shorelines.

- *Ebb shoal.* This category indicates the impact of the ebb shoal on navigation depths and widths in the approach channel. Functional deficiency is indicated by vessel delays, groundings, or maneuvering difficulties. Currents and wave transformation caused by the ebb shoal are evaluated under "Entrance use."

- *Flood shoal.* This category covers the impact flood shoals may have on navigation channel depth and location. Negative impacts due to modification of hydrodynamic conditions by flood shoals are evaluated under the "Channels" category.

- *Harbor shoal.* This category applies to shoaling in mooring and maneuvering areas, and a functional rating is developed for structures designed to prevent or limit shoaling.

- *Shoreline impacts.* This category indicates the structure's impact on the adjacent shore. Functional deficiency occurs where profiles and shoreline location are not maintained within acceptable limits. Separate shoreline maintaining systems, such as bypassing plants, are not considered in this rating.

- Structure protection. How well the structure protects nearby structures, or portions of itself, from wave attack or erosion damage.

- *Nearby structures.* This category indicates the protection provided to other structures located in the lee or in the diffraction zone from damaging waves and currents. This is the only item included in the functional rating.

- *Toe erosion.* This category indicates the structure's level of resistance to toe scour and subsequent undermining of the toe by waves and currents. Toe erosion is accounted for in the structure rating.

- *Trunk protection.* This category mostly applies to structure heads, and it indicates the success of the head in preventing unraveling of the structure trunk.

- Figure VI-8-5 illustrates the division of a jetty structure into major functional regions along with identification of structure functions for each reach.

- Step 3: Subdivision of major reaches by structural and length criteria.

- Further division of a structure into more manageable lengths is based primarily on changes in construction characteristics. Examples include changes in type of construction, type or size of armor, change in cross-sectional dimensions or geometry, and rehabilitated sections. Final subdivisions are based on length where function and construction are uniform over long reaches. Generally, each final division length should be between 60 and 150 m (200 to 500 ft) with the head section always being considered as a separate reach with a length of at least 30 m (100 ft) unless construction differences dictate otherwise.

- Subdivisions within major reaches are illustrated in Figure VI-8-5. The recommended numbering scheme begins at the shoreward end and proceeds seaward with the digit representing the major functional reach and the second letter corresponding subdivisions within the reach. Both structural and functional ratings use the same demarcations. Permanent markers should be established delineating the structure reaches to assure uniformity in future evaluations.

- Step 4: Establishing functional performance criteria.

- Once structure functions have been determined for each structure reach, the next step is to determine the expected performance level for each rating category. These criteria must be based on how well the structure could perform when in perfect physical condition. (Note: Design deficiencies cannot be corrected through the maintenance and repair process, and thus, should not be considered in this analysis.) Begin by reviewing the authorizing documents and structure history. Check if the original expectations have been changed, or if they need to be changed based on past observations.

- Defining performance requirements for each structure reach should be done using the rating tables provided in the report by Oliver et al. (1998). An extracted portion from the functional rating tables for the functional subcategory of ANavigation channels@ is shown in Table VI-8-5 as an example.

- Notice in Table VI-8-5 that the functional performance descriptions are given for three different wave conditions defined as:

  – Design storm condition. The design storm is the largest storm (or most adverse combination of storm conditions) that the structure (or project) is intended to withstand without allowing disruption of navigation or harbor activities, or damage to the structure or shore facilities. Design storm conditions include: wave height, direction, and period; water level; storm duration; and combinations of these factors. The design storm is usually designated by frequency of occurrence or probability of occurrence. Authorizing documents, design notes, project history, and present-day requirements should be used to confirm the appropriate design storms for a project.

  – Intermediate storms. This level refers to storms (or combinations of adverse conditions) of intermediate intensity that occur on the order of twice as often as the design storm. This level is intended to represent a midway point between the maximum storm levels (design storm) and small or minor intensity storms that may occur more frequently, especially during certain periods of the year.

  – Low intensity storm conditions. This level refers to storms (or combinations of adverse conditions) of low intensity that may occur frequently throughout the year, and includes common rainstorms or periods of above normal winds. This level is the next stage above normal nonstorm conditions.

- Establishing the functional performance criteria essentially means determining to what extent the structure should control: waves, currents, and seiches; sediment movement; and shoreline erosion and accretion at the project.

- Factors that help decide how much control of waves, currents, and sediment movements is needed include:

  – Determining the normal dredging frequencies and sand bypassing requirements.

  – Deciding what size ships should be able to pass through the entrance and channel under normal conditions and during higher wave or storm conditions.

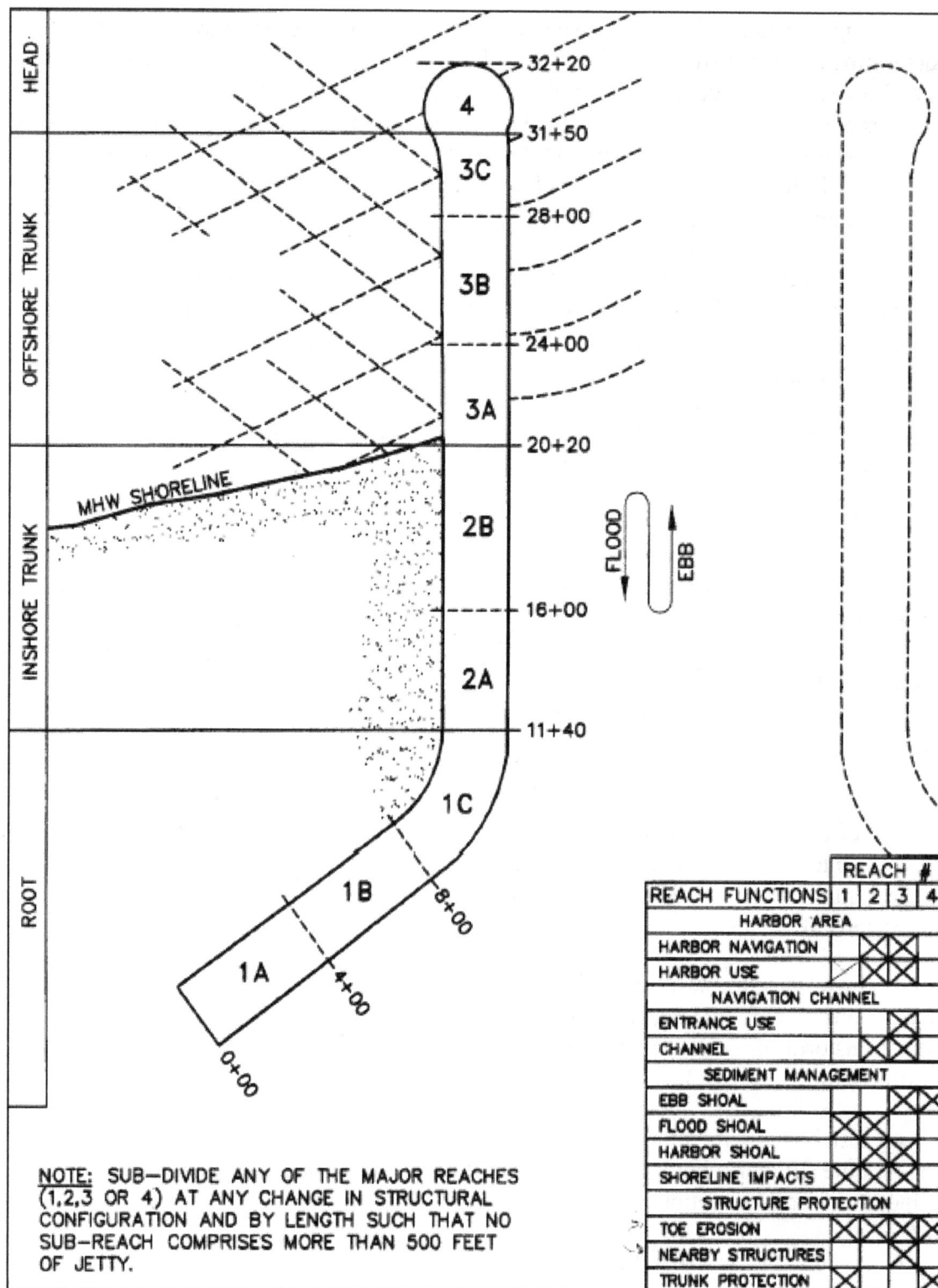

Figure VI-8-5. Major reaches and structure functions for typical jetty

Table VI-8-5
Functional Rating Guidance for Navigation Channel (partial table extract from Oliver et al. 1998)

Rating	Rating Category	Design Storm Conditions	Intermediate Storms (2X Design Storm Frequency)	Low Intensity Storm Conditions
Moderate Functional Loss				
55 to 69	Entrance Use	Vessels generally have little difficulty in the entrance when seeking shelter.	Vessels generally have no difficulty in the entrance when seeking shelter.	Vessels experience no difficulties in the entrance.
	Channel	There are generally few vessel delays in the channel within the shelter of the breakwaters or jetties, except in a few exposed locations. Some vessels using the harbor do not have enough water under the keel to go safely. Small vessels have some problems with conditions at exposed locations.	There are generally no vessel delays in the channel within the shelter of the breakwater or jetties, except at exposed locations.	There are no vessel delays in the channel within the shelter of the breakwaters or jetties. No vessels using the harbor are limited by either insufficient depth or by severe wave conditions.
40 to 54	Entrance Use	Vessels generally have some difficulty in the entrance when seeking shelter. Vessel entrance may be delayed until flood tide.	Vessels generally have no difficulty in the entrance when seeking shelter	Vessels have little or no difficulty in the entrance.
	Channel	There are vessel delays, in the channel, within the shelter of the breakwaters or jetties. In a few locations the delays can be significant for larger vessels that do not have enough water under the keel to proceed safely. Small vessels have problems with wave conditions at a number of locations.	There are some vessel delays in the channel within the shelter of the breakwaters or jetties. A few vessels that would normally use the harbor are limited by either insufficient depth or severe wave conditions.	Vessels experience little or no difficulty in the channel.

    – Determining if any flooding of shoreline facilities should be expected during storm events, and if so, to what extent.

    • Step 5: Establishing structural requirements. Structural ratings are produced by comparing the current physical condition, alignment, and cross-sectional dimensions of a structure to that of a "like new" structure built as intended, according to good practice, and with good quality materials. Because rubble-mound structures tolerate a degree of damage before loss of functionality, structural damage does not automatically equate to loss of function.

    • The structural requirements are established by determining what minimum structure cross-sectional dimensions, crest elevation, and level of structural integrity are needed to meet the functional performance requirements. Initial efforts in determining these structural dimensions can be aided by estimating the impact on project functionality if the reach under study were to be completely destroyed. Project history, authorizing documents, public input, and analysis may be required to identify these dimensions. As this is not an exact science, and some engineering judgment is necessary to produce reasonable estimates. Once established, these structural requirements are used to help identify sources of functional deficiencies in the existing structure. Structural rating categories are shown in the right-hand column of Table VI-8-4 and briefly summarized in the following section.

    (c) Recurring procedures. Once a structure has been divided into reaches, and the functional performance criteria and associated structural requirements have been established, the condition of the structure can be evaluated after each periodic inspection in a logical and consistent manner. Forms designed for use with this evaluation method are available in Oliver et al. (1998). Filling out these forms during inspection assures that key aspects of the rating system are evaluated.

    • Step 6a: Inspection process. The inspector (or inspection team) should be familiar with the structure and past inspection reports before the inspection begins. The beginning and end of each reach should also be known. A copy of the most recent inspection report should be brought to the work site to help judge changes in condition.

    – Other items to help conduct an effective inspection and document findings include: project maps and photographs, still and video cameras, tape measures, hand levels, and tide information. Ratings may be best determined by first walking the length of the structure and making notes of observed defects, their station location, and their severity. On the return walk, ratings may then be selected based on having seen the whole structure and on a second opportunity to scrutinize defects.

    – Providing thorough comments on the rating form is a very important part of the process. Comments should note the location, character, size, and actual or potential effects of structure defects. The comments serve as backup and explanation for the ratings and suggested actions chosen by the inspector. Comments also provide a good record for future reference.

    • Step 6b: Producing a structural rating. The result of the structural rating procedure is an index (SI) that can be generally related to structure condition by Table VI-8-6.

Table VI-8-6
Structural Index Scale for Coastal Structures (from Oliver et al. 1998)

Observed Damage Level	Zone	Structural Index	Condition Level	Description
Minor	1	85 to 100	Excellent	No significant defects - only slight imperfections may exist.
		70 to 84	Good	Only minor deterioration or defects are evident.
Moderate	2	55 to 69	Fair	Deterioration is clearly evident, but the structure still appears sound.
		40 to 54	Marginal	Moderate deterioration.
Major	3	25 to 39	Poor	Serious deterioration in some portions of the structure.
		10 to 24	Very poor	Extensive deterioration.
		0 to 9	Failed	General failure.

- The form for rating structures includes sections for rating the six categories described in the following paragraphs. Illustrative figures for each of the previously described categories are given in Oliver et al. (1998).

- Breach/loss of crest elevation. A breach is a depression (or gap) in the crest of a rubble-mound structure to a depth at or below the bottom of the armor layer due to armor displacement. A breach is not present unless the gap extends across the full width of the crest. Loss of crest elevation is primarily due to settlement of the structure or foundation. Both result in a reduced structure height.

- Core (or underlayer) exposure or loss. Core exposure is present when the underlayer or core stones can be readily seen through gaps between the primary armor stones. Core loss occurs when underlayer or core stone is removed from the structure by waves passing through openings or gaps in the armor layer. Movement and separation of armor often result in the exposure of the underlayer or core stone.

- Armor loss. Three cases of armor loss are considered on the inspection form:

    - *Displacement* is most likely to occur near the still-water line where dynamic wave and uplift forces are greatest. Localized loss of armor (up to 4 or 5 armor stones in length) is typically like a pocket in the armor layer at the waterline with the displaced stones having moved downslope to the toe of the structure. (If the area is longer than 4 or 5 armor stones, use the rating for "Slope defects."

    - *Settling* may occur along or transverse to the slope, and may be caused by consolidation or settlement of underlayer stones, core, or foundation soils.

- *Bridging* is a form of armor loss that may apply to the side slopes or crest of a rubble-mound structure. Bridging occurs when the underlying layers settle but the top armor layer remains in position (at or near its original elevation) by bridging over the resulting cavity much like an arch.

• Loss of armor contact or armor interlock. Armor contact is the edge-to-edge, edge-to-surface, or surface-to-surface contact between adjacent armor units, particularly large quarrystones. Armor interlock refers to the physical containment by adjacent armor units. Good contact and interlock tie adjoining units together into a larger interconnected mass. Certain types of concrete armor units are designed to permit part of one unit to nest with its neighbors. In this arrangement, one or more additional units would have to move significantly to free any given unit from the matrix. Any special armor placement should be stated in the inspection notes.

• Armor quality defects. This rating category deals with structural damage to the armor units. It is not a rating of potential armor durability, but rather a reflection of how much damage or deterioration has already occurred. Four kinds of armor quality defects are defined in the following paragraphs.

- *Rounding* of armor stones, riprap, or concrete armor units with angular edges is caused by cyclic small movements or by abrasion. The result is edges that are worn into smoother, rounded contours. This reduces the overall stability of the armor layer because edge-to-edge or edge-to-surface contact between units is less effective and movement is easier when the edges become rounded.

- *Spalling* is the loss of material from the surface of the armor unit. Spalling can be caused by mechanical impacts between units, stress concentrations at edges or points of armor units, deterioration of both rock and concrete by chemical reactions in seawater, freeze-thaw cycles, ice abrasion, or other causes.

- *Cracking* involves visible fractures in the surface of either rock or concrete armor units. Cracks may be either superficial or may penetrate deep into the body of the armor unit. Cracking is potentially most serious in slender concrete armor units such as dolosse.

- *Fracturing* occurs where cracks progress to the stage that the armor unit breaks into at least two major pieces. Fracturing has serious consequences on armor layer stability, and it brings a risk of imminent and catastrophic failure.

• Slope defects. When armor loss or settlement occurs over a large enough area that the shape or angle of the side slope is effectively changed at that section, then a slope defect exists. Slope defects occur when many adjacent armor units (or underlayer stones) appear to have settled or slid as if they were a single mass. Two forms of slope defects are described in the following paragraphs.

- *Slope Steepening* is a localized process where the surface appears to have a steeper slope than for which it was designed or constructed, and it is evidence of a failure in progress on a rubble-mound structure slope.

  - *Sliding* is a general loss of the armor layer directly down the slope. Unlike slope steepening, this problem is usually caused by more serious failures at the toe of the structure. Slope failure can be caused by severe toe scour, such as can occur at a tidal inlet with strong currents, or by failure within weak, cohesive soils when soil shear strength is exceeded.

  • Rating tables are provided in Oliver et al. (1998) for each of the six structural rating categories. These tables are structured like Table VI-8-6, but they are specific for each category to help guide the inspector while assigning an appropriate rating number. Ratings are based on a comparison of existing structure condition with the Aperfect@ condition for that particular structure. The structural rating table for armor loss is reproduced in Table VI-8-7 as an example. The other five rating tables have similar format.

Table VI-8-7
Rating Guidance for Armor Loss

Structural Rating	Description
Minor or No Damage	
85 to 100	At most, slight movement of the armor in a few isolated spots. Movement has left a depression no larger than 1/4 of one armor stone (or unit) diameter.
70 to 84	Armor movement has caused some waviness along the slope surface with depressions less than 3/4 the armor layer thickness. Any bridging is over a void less than 1/2 of the armor diameter. Underlayer may be seen in spots, but none have been lost.
Moderate Damage	
55 to 69	Some loss of armor in spots, leaving voids or depressions about the size of an armor unit. Units surrounding the void may be rocking or gradually moving out of place. Underlayer or core might be seen at these spots, but armor position still prevents loss of this material. Bridging to a diameter of an armor stone may be visible in several places.
40 to 54	Armor units have been lost or displaced in some portions of the reach length. Voids are just large enough to allow loss of underlayer.
Major Damage	
25 to 39	Armor units have been fully displaced or lost. Voids are large enough to easily allow underlayer and core loss.
10 to 24	Armor units have been fully displaced or lost. Underlayer loss is evident.
0 to 9	Armor units are gone or fully displaced. Structure is unraveling.

  • The numerical ratings are used to calculate the structural index according to the formulae given in Step 7b (Calculating the condition index).

  • Step 7a: Producing a functional rating. The structure's functional performance is the most critical portion of the condition index for coastal structures. Functional index (FI) values are expressed as numbers from 0 to 100 and have the general interpretation as shown in Table VI-8-8.

Table VI-8-8 Functional Index Scale for Coastal Structures				
Functional Loss Level	Zone	Functional Index	Condition Level	Description
Minor	1	85 to 100	EXCELLENT	Functions well, as intended. May have slight loss of function during extreme storm events.
		70 to 84	GOOD	Slight loss of function generally.
Moderate	2	55 to 69	FAIR	Noticeable loss of function, but still adequate under most conditions.
		40 to 54	MARGINAL	Function is barely adequate in general and inadequate under extreme conditions.
Major	3	25 to 39	POOR	Function is generally inadequate.
		10 to 24	VERY POOR	Barely functions.
		0 to 9	FAILED	No longer functions.

• For each designated reach of the structure, functions were assigned during step 4 from the four major functional categories containing 11 subcategories (see Table VI-8-4). The assigned functions should not change unless major changes are made to the structure or project. The functional index for each reach will then be based on the same selected functional rating categories every time a functional rating is done.

• Special forms are used for completing the functional ratings which are determined using rating tables provided in Oliver et al. (1998). Table VI-8-5 is a partial example showing the functional rating descriptions extracted from one of the 11 categories. The process of producing a functional rating for each reach will involve:

– Reviewing original authorizing documents.

– Reviewing previously established functional performance criteria.

– Examining available inspection reports, dredging records, project history, and other office records relating to project performance.

– Reviewing the structural ratings and comments.

– Reviewing the environmental setting in and around the project.

– Gathering information from vessel operators, harbor masters, the U.S. Coast Guard, etc., on any known navigation difficulties, facility damage, or other project deficiencies.

• Further guidance on determining the functional rating is provided in the report by Oliver et al. (1998) along with examples that detail how to use the rating forms and tables.

• The key point to remember is that functional ratings are made in reference to structure performance criteria. Any detected design deficiencies are not included in the rating, but should be reported for separate action. Thus, to affect the ratings, functional deficiencies must be caused by structural deterioration, or in some cases, changed requirements. In any case, situations that a structure could not reasonably correct or control should not be taken into account. Also, functional ratings must be based on the condition of the structure at the time it was inspected.

• Step 7b: Calculating the condition index. Calculation of the structural index, functional index, and overall condition index can be performed using BREAKWATER, a DOS-based computer program available for downloading (Aguirre and Plotkin 1998). The equations used by the computer program for determining the indices are listed in the following paragraphs.

• *Structure index.* For each reach of the structure a reach index is calculated by first determining individual structure indices for the crest, sea-side, and harbor-side components of the structure cross section using the same formula

$$\left.\begin{array}{c} CR \\ SE \\ CH \end{array}\right\} = R_5 + 0.3\,(R_1 - R_5)\left(\frac{R_2 + R_3 + R_4}{300}\right) \tag{VI-8-1}$$

where

CR = structural index for crest/cap

SE = structural index for sea-side slope

CH = structural index for channel/harbor-side slope

R_5 = lowest of the five ratings for the cross-section component

R_1 = highest of the five ratings for the cross-section component

R_2, R_3, R_4 = values for the second, third, and forth highest ratings

• For a reach that forms a structure head, the channel/harbor-side (*CH*) index does not apply. The individual component indices are then combined using the following equation to create a structural index for the reach

$$SI_R = I_L + 0.3\,(I_H - I_L)\frac{I_M}{100} \tag{VI-8-2}$$

where

SI_R = structural index for the reach

I_L = lowest of the three cross-sectional indices

I_H = highest of the three cross-sectional indices

I_M = middle value of the three cross-sectional indices

- For a reach that forms a structure head, there will be just two cross-sectional index values, and the term $(I_M/100)$ becomes 1.

- Finally, the structural index for the entire structure is determined by the formula

$$SI = I_L + 0.3\,(I_H - I_L)\left[\frac{\%1}{100}\left(\frac{S1}{100}\right) + \frac{\%2}{100}\left(\frac{S2}{100}\right) + \frac{\%3}{100}\left(\frac{S3}{100}\right) + \bullet\bullet\bullet\right] \qquad \text{(VI-8-3)}$$

where

SI = structural index for the structure

I_L = lowest of the reach structural indices

I_H = highest of the reach sectional indices

$\%1, \%2, \%3, \ldots$ = percentage of the structure length occupied by reaches 1, 2, 3, ...

$S1, S2, S3,\ldots$ = structural indices for reaches 1, 2, 3, ...

- *Functional index.* The functional index is calculated using rating values determined for categories within the harbor area, navigation channel, sediment management, and nearby structures functional areas. First, a functional index is calculated for each reach of the structure using the formula

$$FI_R = R_L + 0.3\,(R_H - R_L)\left[\frac{\left(R_2/100 + R_3/100 + R_4/100 + \bullet\bullet\bullet\right)}{N}\right] \qquad \text{(VI-8-4)}$$

where

FI_R = structural index for the reach

R_L = lowest of the functional ratings for the reach

R_H = highest of the functional ratings for the reach

$R_2, R_3, R_4,\ldots$ = values for the second, third, fourth, etc., highest ratings

N = number of rated functions for the reach

- Functional indices for all the reaches are combined to create the overall functional index for the entire structure using the formula

$$FI = I_L + 0.3\,(I_H - I_L)\left[\frac{\left(I_2/100 + I_3/100 + I_4/100 + \bullet\bullet\bullet\right)}{N}\right] \qquad \text{(VI-8-5)}$$

where

FI = structural index for the structure

I_L = lowest of the reach functional indices

I_H = highest of the reach functional indices

$I_2, I_3, I_4,...$ = values for the second, third, fourth, etc., highest reach indices

N = number of reaches in the structure

- The *condition index* for a reach or structure is the same as the functional index.

- Step 8: Reviewing structural requirements.

- This final step in the condition and performance rating system is to assess the structural requirements previously established for each reach in view of the structural and functional performance ratings determined after the inspection. It may be necessary to modify the structural requirements for several reaches as knowledge about the structure increases with repeated condition evaluations.

- This review can also result in recommendations related to preventative maintenance or repair. It may be possible to project any trends identified over a number of inspection periods to the future to obtain estimates of future maintenance requirements so appropriate funding can be requested.

c. Project performance/function monitoring.

(1) Project performance/function monitoring consists of measurements and observations that are used to evaluate actual project performance relative to expected design performance. Typically, performance monitoring programs are implemented early in a project's life, and monitoring duration is short (several years) relative to the project's design life.

(2) In the broadest sense, performance monitoring is making observations and acquiring measurements necessary to document the project's response to environmental forcing. Four of the most common reasons for project performance monitoring are the following:

(a) Provide a basis for improving project goal attainment. The uncertainties involved in coastal engineering design may result in a project that is not performing as well as originally anticipated. Before corrective actions can be taken, a monitoring program is needed to establish the circumstances under which the project performance is below expectation. For example, if wave action in a harbor exceeds design criteria, it is necessary to determine the incident wave conditions (forcing) and the mechanisms (refraction/diffraction/transmission) that cause unacceptable waves action (response) in the harbor.

(b) Verify and improve design procedures. Much design guidance was developed based on systematic laboratory testing combined with practical experience gained from earlier projects. However, most coastal projects have some degree of uniqueness, such as location, exposure to waves and currents, available construction materials, combined functions, and existing project features. Consequently, the generic design guidance may not be entirely applicable for a specific structure or project, and in many cases physical modeling is either too expense or not appropriate. Performance monitoring will verify whether the design is functioning as intended, and it will also provide data that can be used to improve existing design procedures or extend the design guidance over a wider range of applicability. For example, estimates can be made regarding the rate of infilling expected for a deposition basin serving a jetty weir section. Short-term monitoring will provide verification of the shoaling rate, and this could lead to cost savings associated with scheduled maintenance dredging.

(c) Validate construction and repair methods. Construction techniques for coastal projects vary depending on the specific project, availability of suitable equipment, contractor experience, environmental exposure, and whether the construction is land-based or from floating plant. Careful construction is paramount for project success. In addition, there is little guidance for designing repairs to deteriorated structures. In these cases the practical experience of the engineers can be as important as the available design guidance formulated for new construction. Performance monitoring may be needed to validate the procedures and to spot problems before serious damage can occur in these particular situations. For example, monitoring might be needed to evaluate the impacts of repairing a stone-armored rubble-mound structure with concrete armor units.

(d) Examine operation and maintenance procedures. Many coastal projects entail ongoing postconstruction operation procedures, and periodic project maintenance is usually required. Performance monitoring is useful for evaluating the efficiencies and economy associated with specific operating procedures. For example, if periodic navigation channel maintenance results in beach quality sand being placed on downdrift beaches, monitoring could be established to determine the best location for sand discharge to provide the most benefit with minimum amounts of sand re-entering the channel during episodes of littoral drift reversal.

(3) Performance/function monitoring is very project specific so there is no exact set of ingredients that constitute the perfect monitoring plan. A wide variation in monitoring plans arises from the different target goals and objectives for each project. Some performance monitoring plans are one-time, comprehensive postconstruction efforts spanning several months of continuous data collection and analyses. Other monitoring plans consist of repetitive data collection episodes spanning several years, perhaps augmented with continuous recording of environmental parameters such as wave and wind data. Comprehensive guidance for developing monitoring plans is available in EM-1110-2-1004, and several of the key points are summarized in the following paragraphs.

(4) The success of a performance monitoring program depends on developing a comprehensive and implementable plan. Several key steps for monitoring plan development are listed in the following paragraphs. (Elements contained in several of the steps are discussed in greater detail in the following section.)

(a) Identify monitoring objectives. This is the single most important step because it provides a basis for justifying every component of the monitoring plan. If there are multiple objectives, attempt to prioritize the objectives in terms of maximum benefits to the project (or to future similar projects). Whether or not a particular objective is achievable will be evident later in the plan development process.

(b) Ranking physical processes. Review the project planning and design documents carefully to identify the dominant physical processes (forcing) affecting the project. The identified physical processes should be ranked in order of importance with regard to project performance of its required functions, and each process should be linked to one or more of the monitoring objectives. Higher ranking physical processes that have significant influence on the project and adjacent shorelines are strong candidates for measurement, whereas less emphasis should be placed on lower ranked processes.

(c) Monitoring parameters. Decisions have to be made regarding which parameters of each physical process must be measured or otherwise observed. This requires knowledge of which parameters best characterize the particular aspect of the physical processes affecting the project. For instance, wave height and period might be most useful when monitoring wave runup or harbor agitation, but wave orbital bottom velocities may be most important for scour and deposition problems. A partial listing of measurable aspects of coastal projects is given in Table V-2-4, CEM Part V-2-17, "Site Characterization - Monitoring." Completing this step requires knowledge about which parameters can be reasonably estimated with instrumentation or quantified through visual or photographic observations.

(d) Scope of data collection. For each physical parameter consideration must be given to the duration and frequency of measurement. Some environmental parameters vary in time and space so that decisions must be made on where to collect information and over what duration. For example, wave parameters might be collected continuously throughout the monitoring, whereas tidal elevations need only be collected for a few months, and aerial photography may only be needed once or twice. Be sure to consider availability of skilled personnel to install and service equipment and analyze results. Fiscal constraints of the monitoring program factor heavily during this step.

(e) Instrument selection. There may be several different instruments available to accomplish measurement of a particular physical parameter. Selection of the appropriate instrument depends on factors such as accuracy, reliability, robustness, expense, availability, and installation/servicing requirements (more about this in the next section). Instruments with shore-based electronics require secure and environmentally protected locations at the monitoring site. Local troubleshooting capability is also beneficial.

(f) Implementation. Adequate funding is necessary to implement a comprehensive performance monitoring program. If funding is lower than needed for an optimal plan covering all the objectives, it is usually better to scale back the objectives rather than attempting to meet all the objectives by scaling back the measurements associated with each objective. Performance monitoring seldom last longer than 5 years except for unusual projects or situations where longer data records are needed for statistical stability. Decisions on the length and extent of a monitoring effort must be made with a realistic evaluation of the importance of the data to the objectives. No plan should be implemented until it is clear that essential data can be obtained.

(5) To be effective, performance monitoring should be implemented when changes to the project are most likely to occur. For new construction, this period is during and immediately after construction, when the project and adjacent shoreline undergoes significant adjustment. Often, this means the monitoring plan should be completed before construction starts in order to

obtain preconstruction bathymetry and measurements of physical parameters likely to be impacted by the project. The importance of preconstruction information cannot be stressed enough as it provides a basis for evaluating changes brought about by the project. After an initial adjustment of one to several years, performance monitoring can be terminated or reduced in scope and converted to condition monitoring.

(6) Performance monitoring plans implemented after problems are identified or after project repair/rehabilitation may not have good baseline data for comparison, and this must be factored into the data collection scheme. An inventory of existing data should be conducted, and data quality should be assessed to determine what baseline data are needed to meet monitoring goals.

(7) The Corps has maintained an ongoing program of monitoring completed and repaired coastal and navigation structures (Hemsley 1990) with the purpose of obtaining information that can be applied to future projects. Numerous reports have been prepared evaluating functional and structural performance for projects on the Atlantic, Pacific, Gulf, and Great Lakes coasts. The following project types and physical processes have been monitored at one or more locations:

(a) Jettied inlets (inlet hydraulics, sedimentation, and structure evaluation).

(b) Harbors (harbor waves, currents, and resonance).

(c) Breakwaters (wave and shoreline interaction and structural stability).

(d) Beach fills (longshore and cross-shore sediment motion).

(8) Lessons learned from the monitoring program, grouped by project type, are given in Camfield and Holmes (1992, 1995). The individual project reports, published by the Corps' monitoring program, are a good source of information for use when developing monitoring plans for similar projects. The bibliography section of Camfield and Holmes (1995) lists references for reports completed before 1995.

d. Monitoring plan considerations. Condition monitoring and project performance/function monitoring are distinctly different in goals and execution. However, many elements of both monitoring types have similar considerations, particularly those aspects related to measurements. This section covers those considerations most commonly encountered when developing and implementing monitoring plans. The main reference for this section is EM-1110-2-1004, "Coastal Project Monitoring."

(1) Fiscal constraints. The single most important major consideration for both condition monitoring and performance monitoring is the availability of funding because fiscal constraints impact the level of data collection. Generally, performance monitoring requires greater fiscal resources per year because of the emphasis on quantifying measurements. Condition monitoring is less expensive on a yearly basis, but often may continue for decades or longer, adding up to a substantial sum over the life of the project. Realistic funding is essential for a successful

monitoring effort, and monitoring costs should be included in the overall project budget, instead of being funded separately after project completion.

(a) Determining an appropriate funding level for project monitoring is not an easy task. A typical approach is to first determine the monitoring goals and then outline the components necessary to achieve those goals. For example, monitoring excessive wave agitation in a harbor would require one or more wave gauges outside the harbor to establish the forcing condition and several strategically-placed gauges inside the harbor to measure response. In addition, there must be analyses performed on the measurements, and there might be a numerical modeling effort or a physical modeling component that makes use of the measured wave data along with bathymetric data inside and outside the harbor. If initial evaluation indicated that wave transmission through a breakwater is contributing to the excessive harbor wave heights, then measurements are needed to quantify this aspect.

(b) The cost estimate for each major activity could vary widely, depending on factors such as the duration of data collection, instrument quality, level of analysis applied, variations in conditions examined by the models, etc. The result is a range of monitoring cost estimates from an inadequate bare-bones minimum to an expensive feature-packed plan. Available funding will tend to be much less than the highest estimate. If funding is less than the minimum estimate, then something will have to be eliminated from the plan or the monitoring goals will have to be adjusted. Attempting to accomplish all the original monitoring goals with an inadequate budget increases the likelihood of not achieving any of the goals. It is better to eliminate tasks in order to fund the remaining tasks at an appropriate level. This requires a realistic evaluation of every measurement's importance. Don't waste funding collecting data that falls into the category of Amight be useful.@ Each measurement must contribute directly to the monitoring goal, and it must have an associated analysis component. If funds appear inadequate to achieve monitoring goals, it may be wise not to attempt any monitoring because improperly collected data can lead to conclusions that are worse than those drawn when no data are available.

(2) Data considerations. There are three overriding considerations that apply to data in general: accuracy, quality, and quantity.

(a) Data accuracy is an assessment of how close the value of a recorded piece of information is to the true value that occurred at the time of observation. Data accuracy relates directly to the means of measuring or observing the physical process. As an extreme example, visual estimates of wave height and period are much less accurate than similar estimates obtained using bottom-mounted pressure transducers that infer sea surface elevation through pressure change. In turn, surface-piercing wave gauges that measure directly the change in sea surface elevation are likely to be more accurate than the pressure gauge. Generally, expect greater accuracy to carry greater cost, although this is not an absolute truth.

(b) Data quality is more encompassing than data accuracy because it includes site specific factors as well as other influences such as instrument calibration. High-quality, accurate instrumentation is necessary for quality data, but instrumentation alone does not guarantee data quality. For example, the best acoustic Doppler current meter will return poor-quality data if air bubbles pass through the sampling volume (breaking wave zone). Similarly, the methods used to analyze measurements can compromise data quality, such as using linear wave theory to interpret bottom pressures under highly nonlinear waves. Finally, data quality is jeopardized if the wrong instrument is selected, or incorrect parameters are set on an appropriate instrument. For example, sampling waves at a 1-Hz rate may not adequately resolve the shorter waves or waves with steep crests.

(c) Data quantity can greatly influence the bottom line cost for obtaining data. For some measurements, well established guidelines exist detailing necessary data quantity for success. For example, one month of tidal elevation data will suffice for many tidal circulation studies. On the other hand, greater uncertainty exists concerning the duration necessary to measure waves in order to establish reasonable wave climatology for a location. For specific measurements, rely on past experience or advice from experts familiar with measuring that particular parameter. A realistic evaluation of data quantity will need to balance multiple factors such as cost, importance of the data, instrument reliability, and natural (and seasonal) variability of the measured parameter.

• For each measurement included in a monitoring plan, the maximum range of the measured physical parameters must be estimated to assure an appropriate instrument or procedure is selected. The maximum parameter value is often most important for coastal projects because of potential adverse impacts stemming from storms. Therefore, if available measurement technology cannot span the entire range of anticipated parameter values, it may be wise to focus on the portion of the range most likely to influence project performance or cause damage.

• When feasible, data should be recorded in digital form to reduce handling and translation errors, and to simplify data reduction and analysis. Data recorded in manual or analog format should be converted to digital form (if appropriate) as soon as practical. Both the raw and transferred data should be stored together with a description of the data type, measurement location and other pertinent information (e.g., water depth), period of measurement, and a statement of data quality and completeness. A detailed description of the format of both the raw and processed data is important for future processing and analyses.

• Data reduction is a term encompassing: conversion of raw data into engineering units by applying calibration and conversion factors, identifying and correcting erroneous data (e.g., data spikes), flagging or removing obviously corrupt data that cannot be recovered, and possibility converting to digital format. Data reduction can be difficult and expensive, particularly if the raw data are noisy, erratic, incomplete, or poorly documented. Unique measurements may involve developing new data reduction techniques which may be more susceptible to errors than previously tested techniques. Of course, even using established methodology for data reduction can introduce errors if applied inappropriately.

• Data reduction costs can be significant for some types of measurements, and considerable thought should be given to data reduction during monitoring plan formulation. This includes determining the final form of the data for analysis and reporting purposes. Ideally, acquired data will be reduced and inspected for quality often, particularly near the start of data acquisition. This will catch potential measurement problems while there is still time to make corrections. The worst scenario is waiting until the end of data collection to begin data reduction and analysis. In cases where there is no possibility of interim data checking (e.g. internally recording wave gauges without shore connection), carefully select the instrument based on its reliability record.

• The complexity, and associated expense, of data analysis varies widely with the type of measurement. For example, analysis of current meter data is relatively straightforward using computer programs supplied with the instrument or programs previously developed. Conversely, photogrammetric analysis of rubble-mound structures still requires several steps, along with human intervention and interpretation of the data. Automated analysis of data has greatly increased our ability to display and interpret large quantities of data, and it has also eliminated much human error. But by automating much of the mundane data reduction, we have also lessened some of the human quality control that was present when data were reduced and manipulated by hand. Therefore, it is imperative that automatically reduced and analyzed measurements be subjected to some form of quality assurance before being used to form conclusions. Simple mistakes such as using the wrong calibration input will produce errant results. In other words, be suspect of the analyses until you have assured yourself the values are reasonable. The corollary is that unbelievable results are more likely analysis error than abnormal physical processes.

(3) Frequency of monitoring. Previous sections discussed the differences between condition monitoring and performance monitoring and the corresponding general monitoring approaches. Generally, the tasks within a condition monitoring plan tend to be somewhat evenly spaced in time over the structure service life. Some tasks may be more frequent for several years immediately after construction to assure the structure is reacting as intended. The frequency of tasks conducted as part of performance monitoring usually are more closely spaced in time in order to collect sufficient data to judge project performance and functionality. It is difficult to estimate the frequency of monitoring tasks that are conducted only after major storm events.

(a) There are few rules-of-thumb suggesting appropriate intervals between repetitive monitoring tasks, and once again past experience with similar projects is the best guideline. When monitoring project aspects where seasonal change occurs, such as beach profiles, the monitoring frequency needs to be at least semiannual to separate long-term change from seasonal variations. The data should also be collected at the same time during the year.

(b) The conduct of monitoring elements associated with potential failure modes may increase in frequency if the monitoring indicates deterioration that could put the structure at risk. Conversely, if monitoring indicates some aspect of the project is holding up better than anticipated, the interval between periodic monitoring tasks can be increased. The important point is that monitoring plans should allow flexibility in scheduling repetitive monitoring elements to react to evolving circumstances.

(c) In addition to the frequency of monitoring considerations related to the overall monitoring plan, there are also several considerations that must be determined for each type of instrument or measurement technique. These include sampling rate, sample length, sampling interval, and measurement duration.

• Sampling rate is the rate at which measurements of a specific parameter are made during a measurement interval. Most importantly, the sampling rate must be rapid enough to describe completely the physical phenomenon being measured. For example, waves must be sampled at one or more samples per second, whereas tides may be sampled at rates of several samples per hour.

- Sample length refers to the period over which samples are taken at a selected sample rate. The sample length depends on the length of time needed to determine the various components of a given phenomenon. A typical sample length for waves is 20 to 40 min in order to collect enough waves to give stable statistics, but tides are recorded for at least 28 days to cover a full lunar tidal cycle.

- Sampling interval is the time between measurement samples. This interval depends on how the measured parameter changes with time. Waves are typically sampled hourly or every 3 hr because during storms the sea condition can change fairly rapidly. Other parameters such as water temperature change at slower rates, allowing a greater sampling interval.

- Measurement duration depends on the physical process being measured and the monitoring goals. For example, it may be necessary to measure waves continuously over a 1- or 2-year period to establish average wave climatology; but if the wave measurements are to be used to validate a harbor response numerical model, the measurement duration can be shorter, provided the target condition occurs.

(d) Recommendations on appropriate sampling parameters for specific types of instruments and techniques are given in Morang, Larson, and Gorman (1997a,b); Larson, Morang, and Gorman (1997), and Gorman, Morang, and Larson (1998). Also, EM-1110-2-1004 contains specific sampling information.

(4) Instrument selection. Instrument selection can be a challenging task if several options exist for measuring the same physical parameter. In the broadest sense, instrument selection is a tradeoff between data quality/quantity and cost. In the following paragraphs are listed somewhat general considerations for instrument selection. Not every item applies to every monitoring instrument or technique, and the considerations are not in any specific order of importance.

(a) Initial instrument cost factors heavily into the aforementioned budgetary constraints of the monitoring effort if instrumentation is to be purchased or leased. Bear in mind that instruments placed in an ocean environment may be lost or damaged beyond repair.

(b) Instrument accuracy corresponds to measurement accuracy, which in turn depends on how the data will be used. Although accuracy is important, it might be overshadowed by other considerations. For example, it might be prudent to opt for an instrument giving less accuracy, but greater reliability in situations where loss of data would have a greater impact than less accurate data.

(c) Instrument reliability is usually a critical consideration for instruments that are installed at a monitoring site and left unattended for some length of time. For internally recording instruments, very high reliability is necessary to prevent loss of the entire measurement sequence. High reliability is also needed for any instruments where failure and subsequent replacement would involve substantial mobilization costs. Less reliable instruments can be used for measurements during site visits or other situations where a failure can be overcome by using a backup instrument or returning to the site later to obtain the measurement.

(d) Instrument ruggedness should be evaluated in the context of where the instrument will be deployed, and under what circumstances. Delicate instruments must be well protected during transport, and they may require special handling. Instruments mounted on the seabed must be able to resist such hazards as trawlers, anchors, etc. Instrument buoys should be able to sustain impacts from vessels or debris.

(e) Instrument stability usually refers to the capability of the instrument to give consistent output for the same input over a range of environmental conditions, such as air and water temperature fluctuations. Unexpected linear trends in data records almost always point to stability problems with electronic circuits.

(f) Type and capacity of data recording associated with an instrument will dictate data sampling rates and duration. Some instruments record data onboard for later processing, some transmit the data stream via cable or telemetry to a receiving station in either analog or digital form, and some instruments have dual capability.

(g) Instrument calibration procedures must be considered, and a clear understanding of how the calibration is applied to the recorded data is essential. The cost of calibrating instruments at special calibration facilities must be included in the budget. When practical, postcalibration of the instrument after retrieval adds veracity to collected data.

(h) Requirements for instrument installation must be known before selecting the instrument because field mobilization, installation, and retrieval can be a major portion of the cost for acquiring data with that instrument.

(i) Instruments placed in water may become fouled with marine growth. For locations where fouling is likely to occur, either avoid using instruments that suffer loss of functionality when fouled, or plan to periodically remove marine growth before data quality degrades.

(j) Some instruments, either in the water or mounted on a structure, transmit data through cables to data loggers. Consideration must be given to where and how to route the cables so they are protected from damage. Typical hazards include boat anchors, harsh environmental conditions, and vandals. Cables coming ashore through the surf zone must be buried.

(k) Decisions on instrument location are often made giving consideration to the type of instrument that will be deployed. Locations that expose the instrument to harsh conditions or other hazards require robust equipment, whereas protected locations can use less rugged instruments.

(l) Any instrument maintenance will factor into monitoring costs. For remote locations, select instruments requiring minimal maintenance to reduce site visits. Also consider the skills necessary to perform field maintenance. If special training or tools are required, maintenance costs will be higher.

(m) Security is a very important consideration for any instrument location that might be accessible by thieves and vandals. At high risk locations, choose inexpensive equipment that is easily replaced and immediately stream data to a secure location offsite. This will help minimize data loss if the equipment is stolen or damaged. Similarly, shore-based electronics for instruments placed in the water need to be in a secure, environmentally protected enclosure.

(n) If feasible, establish a local source for troubleshooting onsite instrumentation if problems develop. A local contact that has access to the instrument and can perform basic tasks will save substantial costs. Examples include reconnecting severed power lines, changing fuses and resetting breakers, periodically replacing recording media, and reporting on equipment status. A competent local contact may be able to correct a problem successfully via telephone under guidance of an expert.

(o) A strong consideration in instrument selection is the ease of data handling and analysis. Some instruments have companion analysis software that simplifies the analysis and significantly shortens the time it takes to have results for interpretation. If no such software exists, or the instrument provides data that cannot be reduced and analyzed by computer, be certain that necessary analysis procedures and equipment are available. Attempting to develop appropriate analysis techniques after the data have been acquired is not recommended.

(5) Other considerations. A concern of paramount importance is the safety and well-being of the personnel involved onsite during monitoring activities and instrument installation and retrieval. The coastal environment is harsh and deadly at times. Do not include monitoring elements that inherently require risky behavior either during installation or during routine observation. Monitoring elements that may periodically involve risk to personnel should be carefully explained to the monitoring personnel so they know when they can and cannot perform that monitoring element. Examples of unacceptable risky behavior include walking on structures when wave overtopping is occurring, being on a rubble-mound structure at night, snorkeling adjacent to a structure in adverse wave, current or surge conditions, and many more too numerous to mention. Diving operations must be conducted according to established USACE procedures and regulations.

(a) The effect of project monitoring on the environment must be considered in the development of a monitoring plan. This includes the effects of personnel, equipment, and techniques on the environment. Minimizing environmental impacts may influence sensor selection and instrument location, and it may limit site access for the monitoring personnel. The plan should incorporate methods and approaches that will reduce or eliminate adverse effects that the monitoring might have on the environment. This may result in seasonal windows for monitoring operations.

(b) The monitoring plan should consider the logistics required to install and maintain the monitoring equipment, and to perform onsite inspections. Contracting of services such as aerial photography or geophysical surveys should be carefully considered, particular for repeated work over an extended time period.

(c) Any monitoring equipment that might interfere with normal navigation, such as bottom-mounted instruments or instrument buoys, should be clearly marked with warning signs, and a description of the hazard should be published as a "Notice to Mariners." Some equipment installations may require a permit from the U.S. Coast Guard or other state or local authorities.

VI-8-3. <u>Repair and Rehabilitation of Coastal Structures</u>. In the United States and other developed countries, most of the coastal infrastructure is already in place. Thus, emphasis has shifted from developing coastal protection projects and navigation facilities to maintaining or expanding these facilities to accommodate future growth. Most of the cost associated with maintaining existing projects is either completing repairs after damage occurs or rehabilitating projects that have deteriorated. Project condition monitoring (Part VI-8-2-b) is essential for knowing when maintenance is required. The first part of this section presents general guidance useful for planning repair or rehabilitation of coastal structures. This general introduction is followed by more specific design guidance, suggested methodologies, and literature references for repairing and rehabilitating sloping-front rubble-mound structures. Finally, several case histories are presented that illustrate repairs and rehabilitations that have been successful for rubble-mound structures.

    a.   General aspects of repair and rehabilitation.

    (1) Design guidance related to structure repair and rehabilitation is not as abundant as it is for new project construction. Part of the difficulty in developing general design guidance is that often damage or deterioration is localized, and possibly quite specific to a particular project. Conducting repairs or rehabilitating a coastal structure is analogous to repairing a house damaged by a fallen tree. The success depends greatly on assessing what needs to be repaired, knowing how much needs to be removed and what can be salvaged, and mating the repaired portion with existing structure in a way that minimizes weaknesses.

    (2) Earlier in this chapter, general definitions are given for repair and rehabilitation. Without changing the general meaning, the same definitions are given as follows, tailored to coastal projects.

    (a) Repair: Fixing portions of a structure that have been damaged by waves, winds, currents, surges, impacts, or seismic activity.

    (b) Rehabilitation: Renovation of deteriorated structure components to original condition or upgrading the structure to withstand greater design loads.

    (3) One of the hardest questions to answer is, "When should a structure or coastal project be repaired or rehabilitated?" This, of course, depends on what functions are served by the project, and how critical the project is relative to other projects in need of repair. Usual indications that a project needs some type of repair or rehabilitation are listed in Table VI-8-9.

    (4) Repair and rehabilitation of coastal projects is costly, and the uncertainty in predicting when projects might need repair or rehabilitation precludes detailed advanced budget planning. Usually, critical repairs that must be completed immediately can be funded from contingency funds established for emergency operations. Less critical repairs, where loss of functionality is not great and additional damage is unlikely to occur, can be included in future budgets as a specific cost item. The same caveat applies to project rehabilitation. Once included

in a proposed budget, the repair or rehabilitation must compete with other funding priorities. Thus, those projects serving vital functions have a higher likelihood of being repaired quickly, whereas less critical ones may be left in a damaged or deteriorated condition for many years until funding is available.

Table VI-8-9
When Coastal Project Might Need Repairs or Rehabilitation

After damaging storms or other events that cause damage such as vessel impacts, earthquakes, etc.

If periodic condition inspections indicate progressive deterioration to the point where functionality is jeopardized.

If performance monitoring indicates the project is not functioning as planned.

If the project is suffering chronic damage from underestimation of design loads.

If the intended structure function is modified to provide new or enhanced service that was not originally in the design.

---

(5) There are distinct differences between designing new projects and repairing or rehabilitating existing projects of similar type. Implementation considerations relate to specific structure type, such as rubble-mound structures, monolithic concrete structures, etc., and these are discussed in the subsections related to specific types of structures. Design environment considerations encompass differences between new design and repair/rehabilitation design that include both the physical environment that provides the forcing and the societal environment in which the design occurs. Listed in the following paragraphs are several general design environment considerations that might apply to the design of a coastal project repair or rehabilitation.

(a) In general there will be no difference in the actual design parameters (waves, water levels, storm frequency) from when the project was originally constructed to the time when repairs are needed. In other words, the physical environment has not changed in time. Exceptions might occur where exposure to the wave climate has been altered (e.g., construction of an offshore breakwater) or where bathymetry has been altered (e.g., growth of an ebb-shoal bar, profile deepening/steepening, etc.).

(b) In some circumstances there may be more reliable estimates of the design parameters than were available during original construction, or during the previous repair/rehabilitation. For example, several years of wave measurements from a nearby wave gauge will provide better representations of wave climate than might have been available previously. This is particularly true if the measurements include the storm or series of storms thought to have caused the damage. In the case of general project deterioration, this will be a less important consideration.

(c) Designers of repairs and rehabilitations can draw on whatever available knowledge exists about past project performance, including performance of similar projects. Monitoring data collected before damage occurs can be crucial in understanding why the project was damaged and how to prevent a reoccurrence in the future.

(d) Depending on the age of the project, there may be new regulations or environmental restrictions that did not exist at the time of original construction. Consequently, it may not be feasible to repair or rehabilitate the project using the same construction methods or materials. Similarly, design standards may have changed or have been implemented for certain types of structures or structural components.

(e) Availability of materials and construction plant may be considerations. For example, a local quarry that produced the original armor stone may not be in operation, and there might not be any other local suppliers of adequate armor stone. This may require using concrete armor units in lieu of stone.

(f) Access to the project site for construction and staging materials could be significantly different due to surrounding development since previous work on the project. This, in turn, will influence the design by limiting construction sequence options.

(6) Every repair or rehabilitation of a coastal project will be unique. However, the general guidelines listed in the following paragraphs apply to many projects.

(a) Review the original design criteria, plans, and specifications. These documents, if available, will provide valuable insight into what the original designers held to be most important in the project design. As-built drawings are especially important because they document what was actually constructed, and thus, captured any onsite changes dictated by circumstances at the time. For older structures, the original environmental forcing design criteria become less important because the environmental forcing was probably not well characterized.

(b) Determine the cause of the problem. At times the cause will be obvious, such as in the case of a vessel impact. Other times, the cause will not be so easily determined. For example, loss of concrete armor units from a breakwater might be due to extreme wave events, breakage of units into smaller pieces, or slumping of the entire armor layer. Examine the failure modes given in CEM Part VI-2 for the structure type, and try to determine how the structure sustained damage. Bear in mind that damage or failure may have been caused by a combination of circumstances rather than to any single factor. If the true cause of damage is not identified, there is a risk of future damage caused in the same manner as previous damage.

(c) If damage can be attributed to a single storm or series of storms, estimate the severity of the events to the extent possible using available data and observations. Accurate estimates are critical in designing a repair that will withstand future events of similar strength.

(d) Investigate the present project relative to the as-built plans and locate discrepancies. This may help isolate problem areas as well as identify regions where future problems might develop.

(e) Devise a solution for the problem. If possible, propose several different solutions and develop each one to the point that reasonable cost estimates can be made. Be sure to factor in any costs associated with testing or optimizing the final design using a physical model. For large repairs or rehabilitation, physical modeling will be a small fraction of the total cost, and the modeling will more than pay for itself in cost savings. For smaller projects, potential cost savings may not justify extensive laboratory testing.

(f) Design a repair that fixes the problem without extensive modification. If a project needs extensive modifications in order to regain (or initially achieve) functionality, it may be that the project was not well designed in the first place, and the designer should reconsider the project in its entirety.

(7) An intangible factor in developing a plan for repairing or rehabilitating a structure is past history of the structure as captured through condition monitoring or through corporate memory. The quality and quantity of corporate memory varies greatly depending on how well documentation was completed originally and how well it survived through the years. Case histories of breakwater and jetty structures were reported in a series of nine reports produced by the USACE REMR Program. Table VI-8-10 lists the reports by Corps Division along with the general geographical coverage in the United States. Besides providing structure histories, in many cases these reports discuss repairs or modifications completed and the variety of repair methods that have been used.

b. Repair and rehabilitation of rubble-mound structures.

(1) General considerations.

(a) One of the advantages of rubble-mound structures is that they are relatively flexible and can tolerate slight movement caused by settlement or wave action without noticeable change in function. Damage to rubble-mound structures generally consists of armor unit wearing or breakage, dislocation of armor units, or removal of a section of the armor layer as shown on Figure VI-8-6. Generally, repairs of rubble-mound structures consist primarily of rebuilding the stone structure or replacing the stone with new material. In some cases repair can be achieved with concrete or asphalt grout.

(b) In addition to the general design environment considerations listed in the previous section, there are several implementation considerations that distinguish the design of repairs to rubble-mound structures from the design of new structures.

•   The repairs are made to an existing rubble-mound structure that may be deflated with lowered crest elevation and milder slopes than originally built.

•   Original armor may be mixed with underlayer stone.

•   Changing armor slope to suit design parameters is difficult.

•   Embedding and securing a new armor slope toe is more difficult than new construction.

•   Transitions between the repair section and the existing undamaged slope must be accomplished without creating weaknesses in the armor layer.

Table VI-8-10
Case Histories of USACE Breakwater and Jetty Structures

Corps Division	States Covered	Projects	Author(s)
New England	Maine, New Hampshire, Massachusetts, Rhode Island, Connecticut	52	Sargent and Bottin 1989a
North Atlantic	New York, Vermont, New Jersey, Pennsylvania, Delaware, Maryland, Virginia	58	Smith 1988
North Central	Minnesota, Wisconsin, Michigan, Illinois, Indiana, Ohio, Pennsylvania, New York	107	Bottin 1988b
South Atlantic	North Carolina, South Carolina, Georgia, Florida, Alabama	32	Sargent 1988
Lower Mississippi Valley	Louisiana	10	Sargent and Bottin 1989b
Southwestern	Texas	12	Sargent and Bottin 1989c
South Pacific	California	28	Bottin 1988a
North Pacific	Alaska, Washington, Oregon	48	Ward 1988
Pacific Ocean	Hawaii, Am. Samoa, Guam	14	Sargent, Markle, and Grace 1988

- Repairs to armor slopes may involve mixing of armor unit sizes and types (e.g., overlaying rock with concrete armor units, overlaying laid-up cut stone with rubble-mound, etc.).

- It may be necessary to remove part or all of a damaged armor slope in order to begin repairs. In some instances broken armor units may need to be removed. This will temporarily expose the underlayers to wave action, and it requires either removing material from the site or stockpiling it for reuse during the repair.

- Spot repairs to isolated damage on armor slopes require substantial mobilization of equipment, and thus might be postponed unless economical methods can be devised.

- Importing small quantities of armor stone is expensive, particularly if the source quarry is not local. (Stockpiling spare material onsite should be considered during initial construction.)

- There will be fewer options regarding equipment and site access for repairing rubble-mound structures. For example, the original structure might have been constructed from an access road atop the structure crest, whereas the repair might have to be accomplished from floating plant. Table VI-8-11 lists construction equipment that can be used for repair of rubble-mound structures.

Figure VI-8-6. Damage at Redondo Harbor breakwater looking from inside harbor (circa 1988)

(c) Depending on the extent of damage or deterioration on a rubble-mound structure, repair options range from minor redressing of the primary armor layer to complete replacement of the structure. Pope (1992) listed the common options for repairing rubble-mound structures shown in Table VI-8-12.

(2) Armor and underlayers. There are four general categories of armor layer repair: spot replacement of broken or dislodged armor units; overlaying existing armor layers; replacing armor layers, and rebuilding the structure. Each of these categories is discussed in the following subsections. Design guidance for armor layer repairs is sparse, and most of the following discussion consists of common sense rules of thumbs that can be applied along with consideration of those unique aspects of each particular repair. As always, past repair experience on the same or similar structures is valuable design input.

(a) Spot replacement of broken or dislodged armor units. If damage to the primary armor layer consists of displaced individual armor units, and the percentage of displaced units is less than 5 percent, the armor layer can be repaired by replacing the dislodged armor units with units of similar type and size (Groeneveld, Mol, Nieuwenhuys 1984). Reusing displaced armor units, supplemented with new units is acceptable practice provided the old armor units are still sound and have not been broken into smaller pieces.

• It is important to establish that the damage occurred as a result of forcing conditions that were similar to the design event. If damage was caused by lesser storm events, then repairing with similar units may not provide adequate long-term protection. Therefore, repair by spot replacement should be considered on a case-by-case basis.

Table VI-8-11
Construction Equipment for Repair of Rubble-Mound Structures (from CIRIA/CUR 1991)

Equipment	Handling Attachment	Comment	Access
Tracked hydraulic excavator	Bucket	Positive pickup and placement. Limited placement and movement. Drops stone.	Suitable for beaches, over crest stones
	Fixed-arm	Positive pickup and placement.	As above
	Orange peel grab	Non-positive pickup and placement. Difficult to pick up individual stones from face.	As above
Wheeled hydraulic excavator	Bucket grapple grab	As above	Only suitable for hard smooth traffic surfaces, small stones, and limited reach.
Crawler crane	Orange peel grab	As above, slower than excavator because of attachment	Suitable for remote areas of structure where hard traffic surface available adjacent to damage.
Jack-up pontoon with crane or excavator	Bucket grapple grab	As above	Suitable for non-drying sites without access from structure.
Cranes and excavator for reinforcing failed sections by other means	Buckets and skips	Applicable to areas where importing large stones is difficult.	Need good access close to structure to reinforce sections with concrete or asphalt grout
Block and tackle	Chains and lifting eyes	Suitable where rock does not have to be moved far distances.	Access for large plant materials difficult.

- Spot replacement of dislodged armor units is the least expensive of the repair options because of shorter time onsite, less new material costs, and less rehandling of existing armor. Individual armor units are generally dislodged in the vicinity of the still-water line, and repairs can be achieved using cranes perched on the structure crest if accessible. For structures where crest access in unavailable or where building temporary construction roads is too expensive, repairs can be made from floating plant.

Table VI-8-12
Options for Repairing Rubble-Mound Structures (Pope 1992)

Problem Area	Options
Slope and crest repair	Chinking, resurfacing
	Addition of dissimilar armor
	Layer reconstruction
	Crest raising
	Burial
Toe and foundation repair	Toe reconstruction
	Scour apron
	Addition of a berm or toe trench
Core repair or void sealing	Precast concrete blocks
	Filter cloth
	Grout
Replace original structure	Type of structure
	Entirely remove existing structure

• Different techniques for replacing armor stones in a damaged layer were examined using 2-D and 3-D small-scale laboratory tests (Ward and Markle 1990). Three different localized repair methods were tried using replacement stone that was similar in size to the original. For each repair method, loose armor stones (usually one to three units) were removed from the damaged area before commencing repairs.

– Voids above the still-water level (SWL) were filled with new armor stone without handling any of the surrounding undisturbed armor stones.

– Voids about SWL were filled with new armor stone, and adjacent undisturbed armor stones were reoriented to provide better seating of the new stones. Up to two stones away from the void were handled to assure the voids were completely filled and the armor was well keyed in the layer.

– Existing stones from the armor layer above the void were used to fill the void, progressively moving the void upslope to the crest where new stones were added. This method helped assure good contact between stones in the repaired area and in the slope above the repair. This method also eliminates multiple handling of armor stone and the need to stockpile existing armor stones.

• Intuitively, repair method 3 would seem better than method 2 which, in turn, would be better than method 1. However, the model tests proved inconclusive. As expected, the tests did indicate that achieving interlocking of armor stones is critical for stability of the spot repair. If an existing damaged armor layer has good interlocking in the undamaged portions, reseating or shifting those stones during spot repairs may introduce new weaknesses in the armor layer and reduce the armor stability. Similarly, new stone placed on the structure must be well seated with maximum contact with surrounding stones to achieve a stable repair.

• For interlocking concrete armor units, such as dolos or Core-LocsJ, stability is based on the strong interlocking between adjacent units. Groeneveld, Mol, and Nieuwenhuys (1984) recommended that repair of damage to concrete armor unit layers should precede by removing both damaged and undamaged units from the repair location all the way up the slope to the crest, and then replacing the entire area with undamaged units. This assures proper interlocking throughout the armor slope.

• Turk and Melby (1997) suggested two methods for repairing concrete armor slopes. The spot repair method is used to repair a small cluster of broken armor units. The broken units are removed from the slope and replaced with new units. Because there is little handling of adjacent undamaged units, care must be taken to achieve good interlocking of the new units with the existing units. The AV-notch@ method of repair is more extensive because armor is removed from the point of damage up the slope in a V-shape that widens as it approaches the structure crest. The notch is then filled in using either all new units or a combination of new and original armor units. It was noted that any spot repair is only as stable as the surrounding original armor unit matrix. The report provides guidelines for repairing dolos armor layers using the Core-Loc concrete armor unit.

(b) Overlaying damaged armor layers. If an existing structure experiences widespread armor layer damage resulting in large sections of the armor being displaced or slumped, it may be feasible to repair the structure by adding an overlay consisting of similar or dissimilar armor units. Overlays can also be used to increase crest elevation to reduce overtopping, or to decrease the armor slope for more stability. Constructing an overlay repair is expensive because of the quantity of new armor units required, but an overlay is not as expensive as replacing the armor layer completely. Degradation of the structure cross section might be caused by dislodging of armor stones, loss of underlayer materials, or toe failure and slumping of the slope armor. In some cases the remaining structure cross section has a lower crest elevation and milder side slopes than originally built. A key factor in designing the overlay is determining what caused the damage. If the damage is a result of armor instability, the overlay will need to consist of armor units capable of resisting the wave loading. For rubble-mound structures armored with stone, this implies overlying stone armor must be larger than the original armor or placed at a flatter slope. Where sufficiently large stone is unavailable, concrete armor units are the only option. There are some significant concerns related to the design and construction of rubble-mound structure overlays. These are listed as follows in no particular order of importance.

• *Single-layer stone overlays.* There are no established stability coefficients that can be used in stability formulas for single-layer rock overlays placed on existing structures. It is inappropriate to use published stability coefficients intended either for two-layer armor layers or for single-layer new construction. Physical model tests should be used to verify and optimize a stable one-layer overlay design. Wolf (1989) presented an overview of new breakwater construction in the Detroit District using one layer of armor stone. Although this is not the same as an overlay, the report provides useful guidance on armor placement and stability requirements for single layers.

• *Two-layer stone overlays.* Most stability coefficients for stone are based on two-layer design, and these coefficients should be adequate for two-layer overlays provided sufficient care is taken to interface with the underlying existing armor slope. For large projects physical model tests are warranted.

- *Overlays using dissimilar armor units.* Typically, this refers to overlaying an existing stone-armored structure with an armor layer composed of concrete armor units. Concrete armor units are used to obtain increased stability through increased interlocking due to unit shape (dolos, accropode, Core-Loc), increased mass (cubes), or a combination of both.

- Carver (1989a) surveyed existing USACE projects where dissimilar armor was used for repairs. In all cases design of the overlayer was based on:

  – Design guidance for new construction.

  – Prototype experience.

  – Engineering judgment.

  – Evaluation of model tests of similar structures.

  – Site-specific model tests.

- The total lack of appropriate design guidance for dissimilar armor overlays led to a series of laboratory model studies investigating various types of dissimilar armor overlays. Table VI-8-13 briefly summarizes the results.

Table VI-8-13
Summary of Stability Results for Dissimilar Armor Overlays

Overlay Type	Stability Coef., K_D	Comments[1]	Reference
Dolos over stone (trunk)	12	Randomly placed. Minimum stability for long waves in shallow water.	Carver and Wright (1988a)
Dolos over stone (head)	8	Slope 1:1.5	Carver (1989b)
	7	Slopes 1:2 thru 1:3.5	
	6	Slopes 1:4 thru 1:5	
Tribar over stone (trunk)	9	Randomly placed on slopes no steeper than 1:2	Carver and Wright (1988a)
	7	Uniformly placed on slopes as steep as 1:1.5	
Dolos over dolos (trunk)	15.6	Randomly placed. Minimum stability for long waves in shallow water.	Carver and Wright (1988b)
Dolos over tribars (Trunk and head)	15	Randomly placed. Exceeds stability of new construction.	Carver and Wright (1988c)

[1] Slopes are given as Avertical rise: horizontal run,@ e.g., 1:2 means A1 vertical to 2 horizontal.@

- *Armor interface with existing armor.* During placement, care is needed to maximize interlocking between the new armor layer and the existing armor stones lying beneath. Typically, the profile of the underlying armor stone will be irregular as shown in Figure VI-8-7. Thus, in some places the new armor stones may rest directly on existing armor, and in other places new underlayer stone may be needed to restore the existing slope to a uniform grade. Construction of underwater portions of armor slopes is always difficult, but this difficulty is compounded when the existing slope is irregular. Care must be taken to assure the underwater portion of the armor overlay is reasonably uniform and free of gaps.

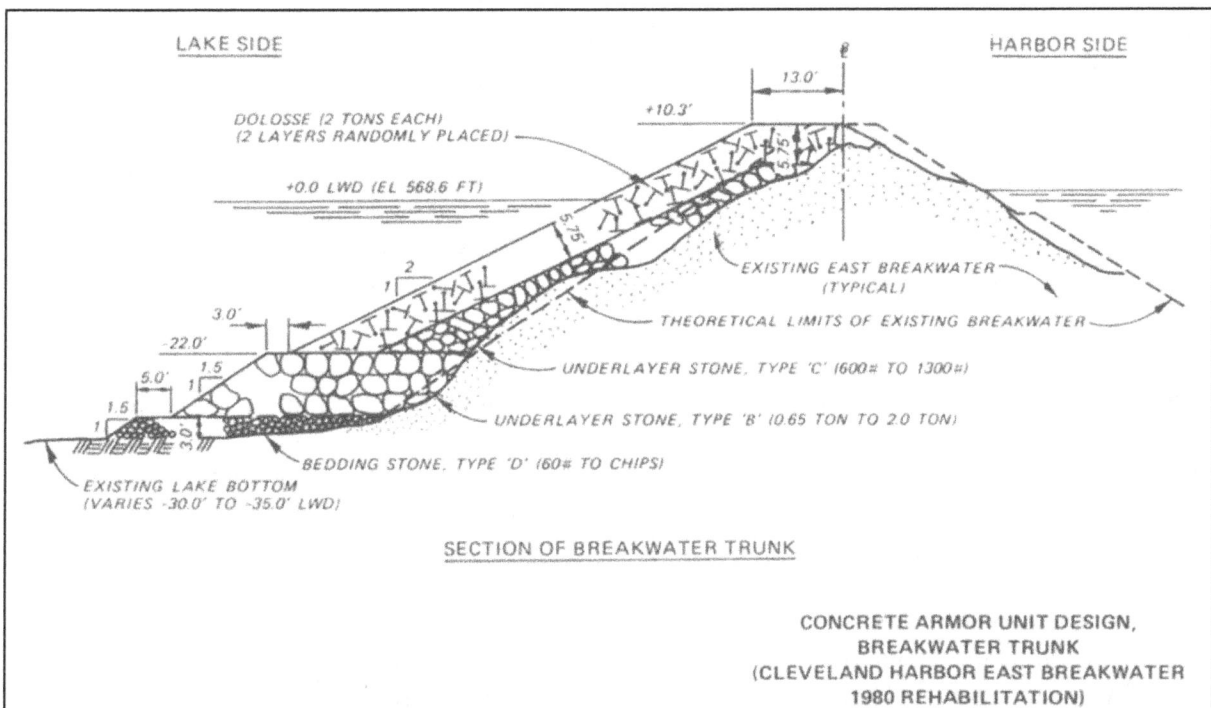

Figure VI-8-7. Cleveland Harbor breakwater rehabilitation, 1980 (from Carver 1989a)

- *Leeside crest armor units.* If heavy overtopping of the new overlay or transmission through the structure is expected, care must be taken to assure the leeside crest armor units are securely keyed into the existing structure. Otherwise, the leeward most armor units might be lost, initiating unraveling of the crest.

- *Overlay toe.* The new overlay toe must be securely positioned and adequately protected by either a toe berm or toe trench. This may range from construction of a new toe berm as illustrated on Figure VI-8-7 to excavation of a toe trench for a shallow-water structure. Difficulties arise where dislodged armor stones litter the toe region. It may be necessary to remove some displaced armor stones to clear the area for construction of the new toe.

- *Construction methods.* Constructing an overlay is quite similar to new construction in that armor placement begins at the toe and proceeds upslope. Where the existing slope is irregular, extra effort is needed to achieve good interlocking between the overlayer and existing armor. In some cases, it may be necessary to remove or relocate existing armor units. As always, past experience on similar projects is very helpful.

(c) Replacement of armor layer. A more expensive alternative than constructing an overlay is entirely replacing or rebuilding the armor layer over a portion of the structure. Replacing the armor layer might be necessary where the original armor protection has proven to be inadequate either structurally or functionally. Example situations that might warrant armor layer replacement include excessive broken armor units, undersized armor units, excessive wave overtopping, or excessive wave transmission. Rebuilding the armor layer is only advisable when it can be determined that damage was caused by something other than armor instability. Examples include faulty construction or damage due to seismic events. Although very expensive, armor layer replacement or rebuilding is justified if anticipated future maintenance costs of the existing structure (based on past performance) are projected to be higher than the replacement cost.

- Replacement armor layers should be designed using the guidance available for new construction. Be certain that the existing underlayer on which the new armor will be placed meets proper specifications to prevent loss of underlayer through voids in the primary armor. Construction of replacement armor layers requires removing all existing armor units in the section to be repaired, and replacing with new armor units or a similar or dissimilar type. Construction typically will begin at the toe and work up the slope. Disposing of the removed original armor units is a substantial expense. If possible, minimize handling costs by recycling the old armor units on a nearby project for which they are adequate. One alternative might be to place the damaged and undersized armor units at the toe of the structure to create an elevated berm that serves as a base for the new armor layer. In general, broken pieces of armor (especially rounded concrete pieces) should not be used to fill in as underlayer material.

- To rebuild a damaged armor layer, a construction plan must be devised for removing the existing stone over a section of the structure, then replacing the armor units starting with the toe and working upslope. Depending on the particular projects, efficient stockpiling of armor units and minimizing rehandling of armor can help reduce costs for rebuilding.

- Armor layer rebuilding or replacing is similar to new construction except that the core and underlayers already exist. Depending on the extent of damage, it may be necessary to replace, add to, or adjust portions of the first underlayer to accept the new armor layer. The underlayer should be checked for correct thickness and compaction. If the new armor layer consists of much larger armor units, the stone sizes in the first underlayer may have to be increased to avoid loss of material through the voids. The slope of the new primary armor layer can be decreased by placing additional underlayer stone on the existing slope. Figure VI-8-8 shows a section of Cleveland dike 14 that was typical of damage caused by severe stone cracking and deterioration. The original armor layer consisted of large blocky stones placed on a 1V-to-1.5H slope. Repair to the armor layer involved removing existing armor stones and placing them downslope (where practical) to form a new underlayer at a 1V-to-2H slope which was then covered with a new armor layer as shown in Figure VI-8-9.

(d) Reconstruction of rubble mound. Structures that sustain catastrophic damage where the integrity of the structure has been lost, or where repair can only be realized through a major redesign, will need to be entirely rebuilt. If the project function requires the structure to be rebuilt at the same location, it may be necessary to bury or completely remove the existing structure. Burial of an existing structure probably means that a larger cross section will result (Pope 1992). In cases where the same functionality can be achieved by construction of a new structure adjacent to the damaged structure, it may be possible to abandon the old structure pursuant to regulatory approvals. Design of replacement structures follows the same guidance as new construction. If the new structure is to be placed atop the remnants of the existing structure, special attention is needed preparing the existing structure to serve as the rubble base of the new construction. This may involve removing material, preparing a new toe, and laying down new bedding material. Removing materials from the old structure will be a major expense, and consideration should be given to the possible reuse of the material in the new construction. Otherwise, reconstruction will be more costly than building a new structure.

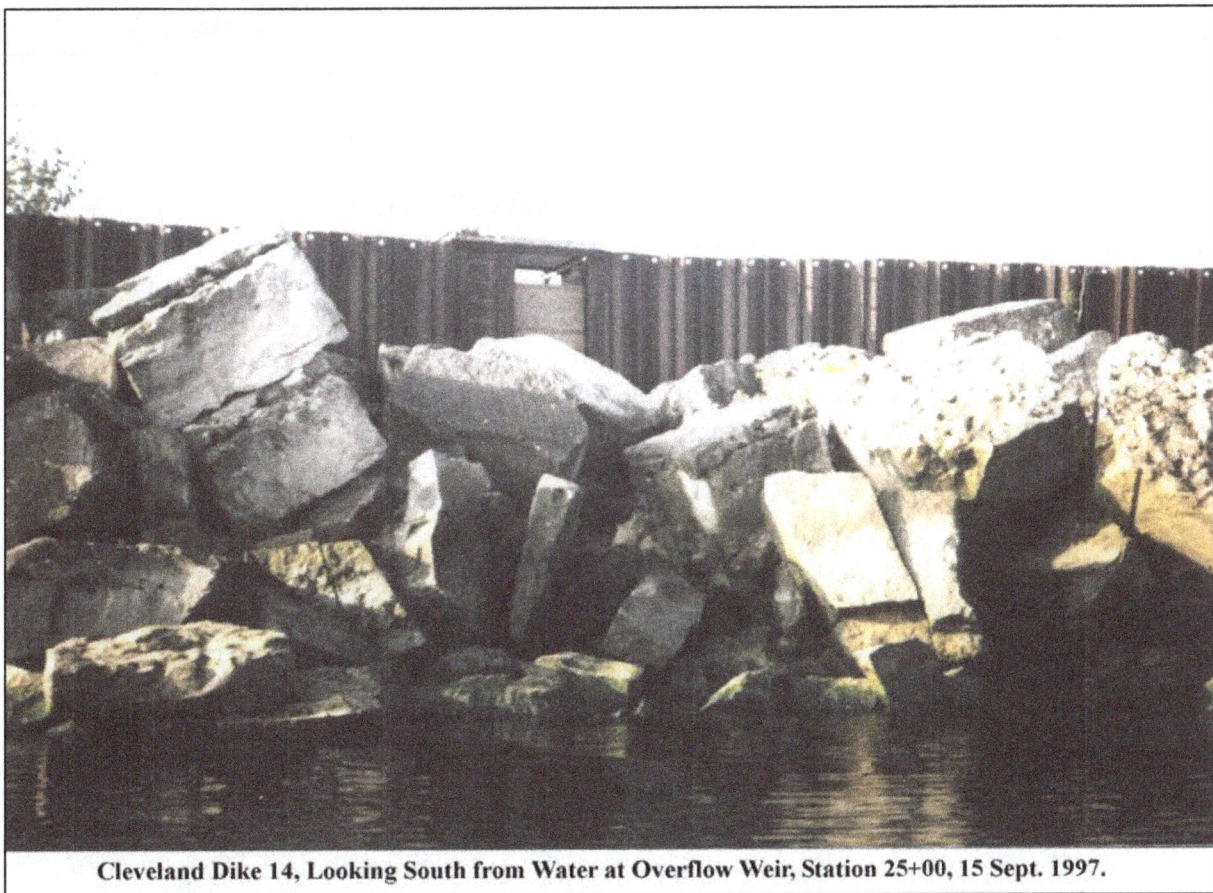

Cleveland Dike 14, Looking South from Water at Overflow Weir, Station 25+00, 15 Sept. 1997.

Figure VI-8-8. Cleveland dike 14 before rehabilitation

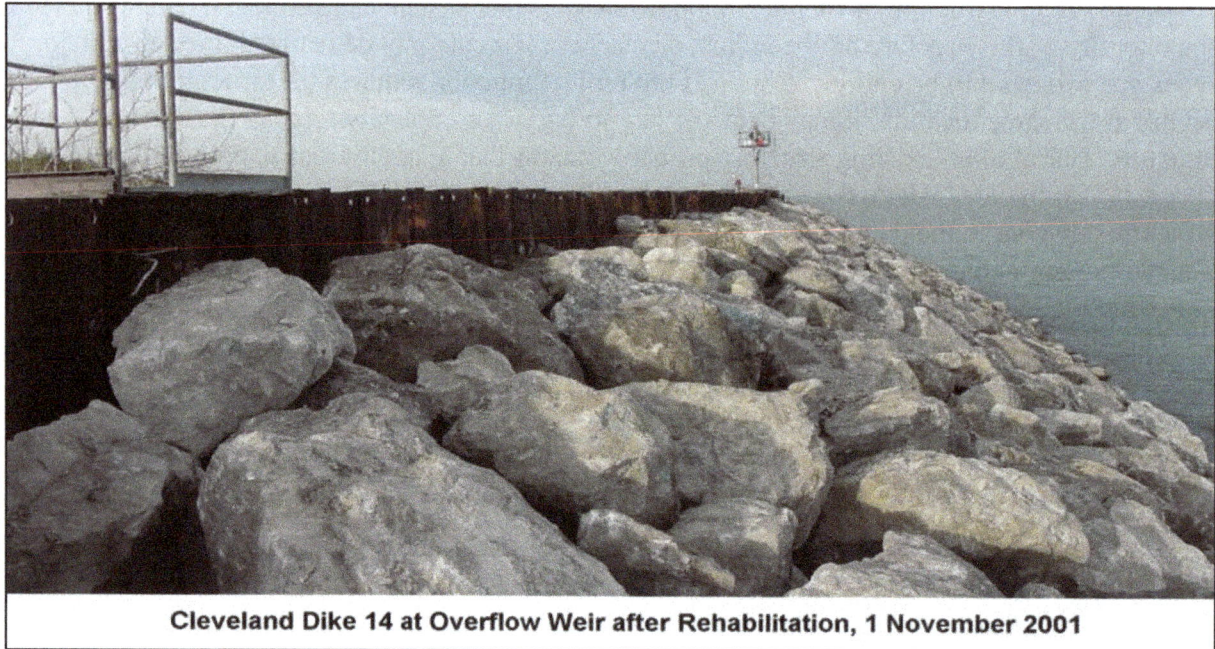

Cleveland Dike 14 at Overflow Weir after Rehabilitation, 1 November 2001

Figure VI-8-9. Cleveland dike 14 after rehabilitation

(3) Caps and crowns. Some structures have caps or crown walls placed on the crest. Usually the purpose for these crest structures is one or more of the following: to increase structure integrity of the rubble mound, reduce core loss, enhance safe public access, increase crest height to reduce overtopping, or provide harbor and ship loading facilities. Caps and crown walls come in a variety of configurations, but usually they are constructed of concrete. It is possible to cast a cap in place after construction of the structure crest, but provision must be made to prevent poured concrete from running through voids in the crest. When cured, the new cap conforms to the irregularities of the crest. Some caps are constructed of prefabricated concrete units that are placed on the rubble crest.

(a) Damage to structure caps and crown walls (excluding concrete deterioration) occurs when the concrete sections are dislodged or displaced by wave action, the supporting underlying armor stones are lost, or differential settlement distorts portions of the structure (see Figures VI-2-31 to VI-2-34, CEM Part VI-2-4, "Failure Modes of Typical Structure Types"). The displaced cap elements might remain intact, or they may break into smaller pieces depending on the cap design, the extent of movement, and the load configuration and support in the new position.

(b) Repair methods for displaced or broken caps and crown walls are not well established, and depend primarily on the specifics of the structure and damage. Minor displacements or settlements might be corrected with hydraulic jacks or cranes, or by lifting and repositioning the cap elements. In cases where the supporting armor layer is removed, it may be necessary to remove the surviving cap or crown elements, then replace them after the armor layer has been reconstructed. Caps that have been cast in place are less likely to be reused, especially if underlying crest armor has moved or has been replaced. Cap or crown elements broken during movement may not be usable, and will have to be removed from the structure and replaced with new elements. Adding a new cap or crown wall as part of a rehabilitation or enhancement of an existing rubble-mound structure will proceed much like new construction. Consideration must be given to structure access for equipment and materials required for the addition. The design should include provision for upward venting through the cap, just as in new design.

(4) Toes and berms. Damaged rubble-mound berms may still provide some degree of protection for the primary armor layer. Because berm crests are typically submerged, repairing a berm is most effectively accomplished by adding new material atop of the existing berm to restore the berm to design cross section. If damage is minor and thought to be caused by exceedence of the design condition, the new material can be the same size or slightly larger than originally placed.

(a) If berm damage is severe, it may be necessary to redesign the berm using larger armor units. It may be possible to place the new berm on the remnants of the damaged one, but in some cases it will be necessary to remove some or all of the existing berm material. Removal of scattered material may be necessary if it encroaches on the navigation channel or otherwise interferes with navigation or other activities. Existing projects can be upgraded by adding a berm to reduce wave energy and increase stability of the primary armor layer. A berm also can decrease wave runup and overtopping on projects which are not functioning as intended.

(b) Stability problems with rubble-mound breakwater and jetty toes experienced by the USACE were documented by Markle (1986). Twelve of 21 Corps Districts reported existing or past problems with rubble-mound structure toe stability, and Markle=s report includes numerous case histories. Most toe instability was related to both inadequate sizing and placing of the toe stones, or due to undermining of the toe berm by scour. Subsequent to the field survey, a small-scale laboratory test program was conducted at the U.S. Army Engineer Waterways Experiment Station (currently the U.S. Army Engineer Research and Development Center) to develop appropriate design guidance for toe berms (Markle 1989). The resulting design guidance for rubble berms and toes is given in CEM Part VI-5-3-d, "Toe Stability and Protection," and this same guidance should be used for repairing damaged toe berms if the problem appears related to toe stone stability.

(c) Toe instability caused by waves and currents scouring the seabed adjacent to the structure toe is more problematic. Additional toe material can be placed to rebuild the toe profile that has been degraded by materials falling into the scour hole, or a scour blanket can be constructed to protect the toe from damage by scour. A third, more expensive solution is to excavate and reconstruct the structure toe, but care must be taken not to initiate a slope failure of the main armor layer during repair. If warranted, it may be advantageous to reconstruct the damaged toe berm to a larger size than originally designed in order to reduce runup and overtopping and to increase armor stability.

(5) Scour holes and bed protection. Some initial designs include provisions for scour protection, usually in the form of a stone apron extending some distance from the structure toe. More often, however, scour protection is added to a project after monitoring has revealed formation of scour holes or trenches due to currents, waves, or a combination of both. For

example, in situations where scour has undermined the structure toe berm, the repair plan should include some type of scour protection to prevent future occurrences.

(a) Prediction of where scour may occur at a coastal project is rudimentary at best, with past experience being the best gauge. There is little difference between designing scour protection as part of a structure repair or rehabilitation and designing the protection for a new project. See CEM Part VI-5-6, "Scour and Scour Protection," for appropriate design guidance for estimating scour and designing scour protection. Design of scour blankets in a current field is given in CEM Part VI-5-3-f, "Blanket Stability in Current Fields."

(b) One of the main decisions facing the designer of scour protection after scour has occurred is whether or not to fill in the scour hole before placing the protective stone blanket. If the scour hole is close to the structure toe and has relatively steep side slopes, there is a risk of the toe falling into the scour hole either by armor slope failure or slip-circle failure. (See Figures VI-2-37, VI-2-64, VI-2-69, CEM Part VI-2-4, "Failure Modes of Typical Structure Types." This is a hard design decision because no guidance exists regarding armor slope instability relative to scour hole side slopes adjacent to the structure toe. In addition, deep scour holes indicate strong local currents, and filling in these scour holes may substantially increase currents and cause scour in adjacent, unprotected portions of the seabed. When judging qualitatively the risk a scour hole may have on the stability of the adjacent rubble-mound structure, be sure to examine only cross sections that have been drawn without distortion of the vertical axis which makes the scour hole side slopes appear steeper than they actually are.

(6) Void sealing. Rubble-mound breakwaters, jetties, and groins have some degree of permeability that varies significantly with cross-section design. Rubble-mound permeability helps absorb wave energy, reduce wave runup and overtopping, decrease wave reflection, and generally increase armor layer stability. However, wave transmission through the structure increases with increased permeability, and excessive wave transmission can affect vessel navigation, mooring, and harbor activities. For projects where rubble-mound structures are placed within active littoral zones, sand can flow through permeable rubble mounds and deposit in shoals on the leeside of the structure. If the shoals result in depths less than the authorized project depths, maintenance dredging is required to assure continued project functionality. Permeability of newly constructed rubble-mound structures might decrease slightly as the structure Anests@ and stones shift into the voids. Conversely, it is possible for rubble-mound permeability to increase over time if smaller core material is washed out of the structure by wave action or if portions of the armor layer suffer damage. In situations where structure permeability is causing problems, it may be possible to decrease permeability by a process called void sealing where grout or some type of sealant is injected into the structure to fill the voids. Using techniques developed in civil and mining engineering, it is possible to fill interior voids on the order of 1 m in diameter. However, the longevity of grouts and sealants placed in coastal structures is unknown due to lack of long-term field experience. The following paragraphs summarize a study conducted as part of the REMR Program that was documented in a series of four reports (Simpson 1989; Simpson and Thomas 1990; Rosati and Denes 1990; and Simpson et al. 1990). These reports should be consulted for more thorough information before undertaking feasibility studies related to breakwater or jetty void sealing.

(a) Determining need for structure sealing. Void sealing of rubble-mound structures is an alternative where project functionality is impacted by excessive wave transmission through the structure or by sand moving through the structure and infilling the navigation channel or mooring areas. Measurements should be made to establish the existing degree of wave transmission through the structure for a range of typical operational wave conditions. Also note the variation of wave transmission along the structure with a view toward minimizing the distance to be sealed.

- If sand transmission through the structure is suspected, it must be shown that deposition results from sand actually passing through the structure rather than moving over or around the structure. If dye released in the water on the sea side of a structure appears on the lee side of the rubble mound in less than 1 min, it is safe to assume that sand can easily flow through the structure and deposit in a shoal. Hydrographic surveys scheduled in concert with dredging activities or storm events will reveal the location of sand leakage and the rate of shoal growth due to sand transport through the structure. Dyed tracer sand may provide an indication of sediment pathways and the upstream source of sediment. This information can be used to assess alternate solutions that may be less costly than void sealing. For example, it may be feasible to trap some of littoral sediment before it reaches the leaky structure.

- Simpson et al. (1990) stated: "The quantity of material moving through the structure which contributes to the shoal must be of such a magnitude that the cost of its elimination is offset by the savings in mobilization, demobilization, and dredging which would otherwise be attributed to it."

- Alternately, if shoaling caused by sand movement through the structure does not affect navigation significantly between regularly scheduled channel maintenance dredging, then void sealing of the structure would not be cost-effective relative to the small additional cost of dredging the shoal as part of the regular dredging cycle. Structures that are nearly at the end of their service life or are scheduled for rehabilitation are not likely candidates for void sealing. In the case of structure rehabilitation, it may be cost-effective to include provisions for reducing sediment transmission as part of the rehabilitation. Benefits gained by reduction of future maintenance dredging may offset the additional cost of a rehabilitation that includes reduction of sand transmission.

- An important consideration in determining the viability of void sealing is the engineering practicality of sealing a particular structure. Factors such as equipment (drilling and sealing) availability and site accessibility should be investigated early in the planning process. Sealing structures requires placing drilling and other equipment on the structure. Low crested, frequently overtopped structures present a higher risk of equipment damage or loss, and this risk must be included in the economic analysis. In addition, fully successful void sealing of rubble-mound structures is difficult to achieve, and field experience with this type of remedy is sparse.

- Finally, void filling of a permeable rubble-mound structure will alter how waves interact with the structure. An assessment must be made as to whether these changes will have any impact on the overall project functionality. The primary concerns are increased wave runup and possibly overtopping, increased wave reflection from the structure, and the possibility of decreased armor unit stability.

(b) Determining extent of structure void sealing. The lateral extent proposed for void sealing should cover the area where wave and/or sand transmission is a problem. In addition, attempt to determine additional locations where transmission may become a problem after sealing the structure due to altered wave characteristics or breaker angles. The sealed structure may cause sand accretion on the seaward side that eventually could flank the sealed portion as a new equilibrium shoreline develops adjacent to the structure. For prevention of sand transmission the existing sand layer should be stabilized (sealant is injected into the sand bed, displacing the sand), and then the void sealant is injected to an elevation greater than the expected elevation of sand accretion on the seaward side of the structure. To decrease wave transmission the void sealant barrier should be placed to approximately the mean high-water level (Simpson et al. 1990).

(c) Sealing techniques. Two cementitious sealants and two sodium silicate sealants were shown to be potentially effective for permeable coastal structures. Correct mixing and handling of these grouting materials is essential for good bonding and reasonable durability. See the previous referenced reports for further details on the mixtures and correct application.

- The general procedure for grouting a permeable structure is to drill primary grouting holes along the rubble-mound structure center line at about 3-m (10-ft) spacings. The pumping nozzle is placed into the hole and an estimated volume of grout/sealant needed to create the barrier is injected starting near the bottom. After an estimated quantity of sealant has been placed, the nozzle is raised vertically a specified distance and another volume of sealant is placed. This procedure continues until the grout reaches final elevation. The quantities injected at each elevation must be determined from a few test injections or by monitoring the filling of the first few grout holes. The resulting grout structure will have a somewhat conical shape with low "valleys" between the adjacent peaks.

- After the primary holes are filled, the gaps between the injection points are filled via secondary holes that are evenly spaced between the primary holes. Again, the amount of sealant injected has to be estimated a priori. Depending on the particular application, additional holes may be needed between the secondary holes.

- Simpson et al. (1990) described four types of sealing treatment that might be considered in coastal engineering applications: barrier curtain sealing, cavity filling, soil sealing, and riprap sealing. They also discussed equipment commonly used for sealing operations such as drills, sealant pumps, concrete pumps, mixers, sealant lines, and monitoring equipment.

(d) Sealing design parameters. A preliminary field investigation is crucial for designing an appropriate sealing strategy.

- Factors to be evaluated during the investigation include the following:

- An estimate of stone size and type in the section to be sealed. Stone size provides information about the void sizes and the degree of communication with other voids. Stone type gives a preliminary indication of drilling difficulty.

- The water flow rate through the structure which factors into the type of grout or sealant suitable for the conditions. Higher flow rates dictate faster setting sealant mixtures.

- The present sand level beneath the structure should be estimated so that sealant quantities can be calculated. This may require drilling of exploratory holes through the structure. (Note: Exploratory drilling also reveals the drilling difficulty that can be expected.)

- Water chemistry (pH, salinity, and temperature) can affect the stability and durability of the sealant mixture.

• The main factors affecting a sealant mix intended for rubble-mound structure void filling are listed as follows.

- Potential for dilution and dispersion of the sealant by water movement through the structure during emplacement

- Consistency of the sealant and the homogeneity of the rubble mound

- Size of the voids within the rubble-mound structure

- Top elevation of the grout curtain

- Permanence of the grout mass

• Satisfying these criteria requires a grout or sealant mixture that is highly viscous very soon after placement, but still fluid enough under low pumping pressure to be placed at a rate much greater than the water flow rate through the cavities. The mixture must set fast and be durable and stable under cyclic wetting/drying conditions. Finally, the mixture must be economical for batching in large quantities. Sealants injected to stabilize the sand bed must penetrate the sand mass for a distance of around half a meter at low pressure.

• According to Simpson et al. (1990): "The most important part of the entire sealing operation is retaining a contractor who is competent in this type of work and conscientious enough to understand the indications of what is occurring below the surface and make necessary adjustments. The flow of sealant to adjacent holes or out the sides of a structure must be monitored to adjust the mix or injection procedures."

• Detailed guidance on void sealing field procedures, quantity estimation, contract specification, contracting procedures, and site supervision/inspection of grouting operations is included in the REMR reports cited at the beginning of this section. The summary report (Simpson et al. 1990) also includes results of a sealant durability study where specimens were exposed to the elements.

(e) Field experience. Seven completed void sealing projects, including four USACE projects, were summarized in Simpson et al. (1990). A more detailed description of void sealing at Port Everglades, Florida, was given by Rosati and Denes (1990). The Port Everglades study included pre- and postconstruction monitoring to gauge the effectiveness of the void sealing operation.

• Sodium silicate-cement and sodium silicate-diacetin chemical sealants were used to seal man-sized voids in the Port Everglades south jetty. The purpose of the grouting was to reduce sand flow through the structure. Dye injection and measurement of dye concentrations at sealed and unsealed portions of the jetty indicated that the void sealing was successful in reducing water flow transmission by a factor of 2. The unsealed jetty was estimated to be 4.0 percent transmissible whereas the sealed structure reduced this value to about 1.9 percent. It was also noted that the structure was reflecting more wave energy as indicated by a shift in the average current vector measured adjacent to the jetty. Doubts were expressed about the long-term viability of the sealants used in this project based on exposure tests conducted as part of the REMR Program.

• Sand moving through the south jetty at Palm Beach Harbor, Florida, was depositing in a channel-side shoal that interfered with navigation and required expensive dredge mobilization each year. The south jetty was sealed in 1984 by the U.S. Army Engineer District, Jacksonville, using a mixture of cement and silicate. This sealing mixture was suggested by the contractor after initial efforts using the two originally specified sealants were unsuccessful. The final hole spacing was about 1 m (3 ft), and it was during this project the alternate hole filling sequence was seen to produce better buildup of sealant. After completion of the sealing, samples of the void barrier were extracted from exploratory holes, and these samples indicated that design objectives had been met. The jetty sealing significantly reduced shoaling into the channel to the point that no dredging mobilization has been required specifically to remove the shoal.

• In 1985 the U.S. Army Engineer District, San Francisco, sealed a newly constructed groin and a new extension of an existing groin at the Buhne Point, California, Shoreline Erosion Demonstration Project. On the new groin the sealing was accomplished in three phases with the initial holes drilled at about 3-m (10-ft) centers. After filling the first set of holes, holes spaced at the midpoints between the initial holes were drilled and filled. Finally, another set of midpoint holes were drilled giving a final hole spacing of about 0.75 m (2.5 ft). The groin extension was drilled at 1.5-m (5-ft) centers. The originally specified sealant was a cementitious mixture that proved difficult to pump, and the mixture was modified, as shown in Table VI-8-14.

• Holes spaced at 3 m (10 ft) on the Buhne Point groin took about 5.4 m³ (7 yd³) on average. The holes were 4.4 m (14.5 ft) deep. The holes centered between the first set of holes had an average intake of 3.8 m³ (5 yd³). Examination during high tide and 1.2-m (4-ft) seas revealed that a third set of holes centered between existing holes were needed to prevent leaks at the +1.8-m (+6-ft) mlw elevation and above. Sealant intake for the final set of holes at 0.75-m (2.5-ft) centers was about 0.8 m³ (1 yd³).

Table VI-8-14
Buhne Point Cementitious Sealant (recipe for 1 yd^3)

Component	Specified Weight (lbs)	Modified Weight (lbs)
Coarse aggregate	1,000	1,115
Fine aggregate	1,450	1,655
Cement	705	705
Clay	305	37
Water	500	371
Calcium chloride	15	15
Air	B	0.41

- Inspection after completion of the project revealed 25 to 30 possible leaks in the total groin length of 365 m (1,200 ft). None of the leaks were thought to be below an elevation of 1.5 m (+5 ft mlw), and there did not appear to be any problems related to sand leakage through the groin. Based on experience from sealing of the Buhne Point groins it was recommended that drill holes to be at least 15 cm (6 in.) in diameter to facilitate injection operations and visual inspection.

- Several other void sealing projects are summarized in Simpson et al. (1990). Engineers involved with a project of this nature should study the REMR reports for additional insights. One very obvious conclusion from the REMR study was that much work is needed in the area of developing ways to estimate the quantity of mixture per unit length of structure.

(7) Rubble-mound structure repair case histories. Many rubble-mound structures have a history of rehabilitation, repair, and modification as witnessed by the narratives contained in the REMR reports listed in Table VI-8-10. Lessons learned from previous experience at the same or similar structures are a valuable source of ideas and methodologies that can be considered when a structure needs repair or maintenance. However, each structure is unique in its location, exposure, construction, and intended functionality, so it is important to not adopt blindly a repair procedure without first carefully evaluating all aspects of the project. With the exception of the Sines west breakwater, documented case histories of damage and repair of rubble-mound structure are not abundant in the literature. Several papers reporting on three repair case histories are summarized in the following paragraphs. Engineers with similar repair projects may avoid unforeseen situations by examining the cited literature.

(a) Port of Sines, Portugal, west breakwater. Numerous papers have documented the failure and subsequent repair of the west breakwater at the Port of Sines, Portugal. This 1.7-km-long breakwater was armored with 40-tonne dolosse, and it was nearing completion in February 1978 when the structure was hit by a moderate storm (H_s estimated to be 8.5 - 9.0 m) that caused serious damage. A year later a storm with H_s estimated near the 11-m design wave height caused near catastrophic failure of the armor slope and concrete superstructure (Weggel et al. 1994).

• Baird et al. (1980) summarized the final report of the Port Sines Investigating Panel (PSIP 1982) which examined possible causes of failure, and they presented their own hypothesis on the series of events leading to failure. Several Sines-related papers were presented in the Coastal Structures '83 conference. Toppler et al. (1983) overviewed rehabilitation studies for the west breakwater which included wave climatology, geotechnical studies, and hydraulic stability analyses. Details of study components were given in companion papers. Mynett, de Voogt, and Schmeltz (1983) presented numerical hindcast results for 20 major storms spanning a 25-year period at Sines. Barends et al. (1983) examined geotechnical stability of large breakwaters with reference to quantified results from physical model tests of the Sines west breakwater and from emergency repairs. Groeneveld, Mol, Zwetsloot (1983) compared structural strength of dolos, tetrapods, and cubes and concluded that beyond a certain size the residual static strength of dolos and tetrapods disappears. Armor stability of the emergency repair and rehabilitation of the outer breakwater portion were examined in hydraulic model tests described by Mol et al. (1983).

• Reconstruction of the Sines west breakwater was completed in December 1992, and the engineering aspects of the new breakwater design and construction were documented in the proceedings of a seminar held in September 1993 (RMSC 1994). The seminar proceedings contain 17 papers covering history, rehabilitation studies, geotechnical issues, construction, inspection, monitoring, and maintenance of the new Sines west breakwater and other port constructions. A short summary of the breakwater reconstruction was given by Weggel et al. (1994).

• The Sines west breakwater is fairly unique because of its size, exposure, and 45-m water depth at the toe. The failure and subsequent reconstruction is the most documented in history, and the lessons learned from Sines have benefited our engineering understanding of large rubble-mound structures.

(b) Jetty rehabilitation at Humboldt Bay, California. Edge et al. (1994) and Bottin (1988a) described the historical development of the two jetties protecting the entrance to Humboldt Bay on the northern California coast. Original construction began in 1888, and over the course of the next 80 years the structures have undergone several damage/repair cycles. By 1970 both jetty heads were destroyed and in need of reconstruction. A 1-to-50 scale model was used at WES (Davidson 1971) to develop a stable head section. Tests were conducted using armor slopes constructed of concrete blocks, tribars, tri-longs, tetrapods, and dolos, and the final recommended head section consisted of reinforced 42-ton dolos; the first application of this armor unit in the United States (see Figure VI-8-10). Since reconstruction of the two jetties, the heads have remained generally stable with occasional maintenance needed to repair damage caused by severe storms. Keen insight into the process of actually building the new jetty heads at Humboldt Bay was given by Hanson (1994). He was the USACE construction manager for the project, and his paper presented an interesting chronological record of the complex, and at times, adversarial relationship between the contractor and client. From bidding issues through to the end of the contract, Hanson presented an inside view of what types of issues arise, how they sometimes are resolved, and what can be learned from the experience. A focus of this paper is on the controversy surrounding the use of reinforcement in the dolos. It was decided to use the prototype construction as a testing opportunity by placing some unreinforced and fiber-reinforced dolos in the armor layer. Subsequent examination in 1993 indicated less than

1 percent dolos breakage on both jetties. No conclusions about the value of reinforcement could be drawn from this small sample other than the steel reinforcing prevented the broken pieces from becoming smaller projectiles that could impact and break intact units (Melby and Turk 1994).

(c) Breakwaters at a port in northern Spain. Groeneveld, Mol, Zwetsloot (1984) presented a case study examining options for repairing two dolos-armored breakwaters protecting an unidentified port on the northern coast of Spain. The structures were built in water depths up to 20 m. Shortly after original construction, breakage of 50-ton dolosse armor units occurred under not particularly severe wave conditions. The concrete armor units were placed on slopes of 1:1.5 and 1:2. Three years after construction 17 percent (3,000 units) of the total placed armor on one breakwater and 25 percent of the units on the other were broken. Differential settlement of the south breakwater caused large cracks in the crown wall.

- Six alternative repair options were considered for the north breakwater, and relative cost was determined in terms of the most conventional repair option, which was given as option 1. The options with the relative cost given in brackets are as follows:

  - Remove 50-ton dolosse armor layer and replace with two layers of 81,646.6-kg (90-ton) cube. New rock underlayers would also be placed.

  - Fill in existing voids with quarry-run material, then cover the existing dolos layer with a new berm-like layer consisting of 81,646.6-kg (90-ton) cubes. (140 percent of option 1).

  - Repair the damaged armor layer with 45,359.2-kg (50-ton) dolos and construct a toe berm to a level of about one-third of the water depth. New armor (unspecified) would be placed above the berm. (165 percent of option 1).

  - Install a wide berm (approximately 100-m length) to reduce wave energy. Dolos left in place, but new 60-ton cubes added to slope above the berm. (65 percent of option 1).

  - Leave breakwater unrepaired, but install a new submerged rubble-mound breakwater at some distance seaward to break up the incident waves. Waves reaching the damaged breakwater would not cause additional damage. (105 percent of option 1).

  - Same concept as option 5, with the submerged structure being concrete caissons resting on a rubble base. (200 percent of option 1).

- Costs for option 2 were increased by the need to pump gravel into the existing dolos layer to provide proper sublayer porosity for the armoring system. Higher costs for option 6 included the need to repair the head section using the option 1 technique because the berm would not protect the head adequately. The north breakwater was repaired using option 1, and the risk associated with exposing the underlayer during rehabilitation was acknowledged and accepted.

- The south breakwater needed both armor layer repair and general strengthening for geotechnical stability. The two alternatives considered were the following:

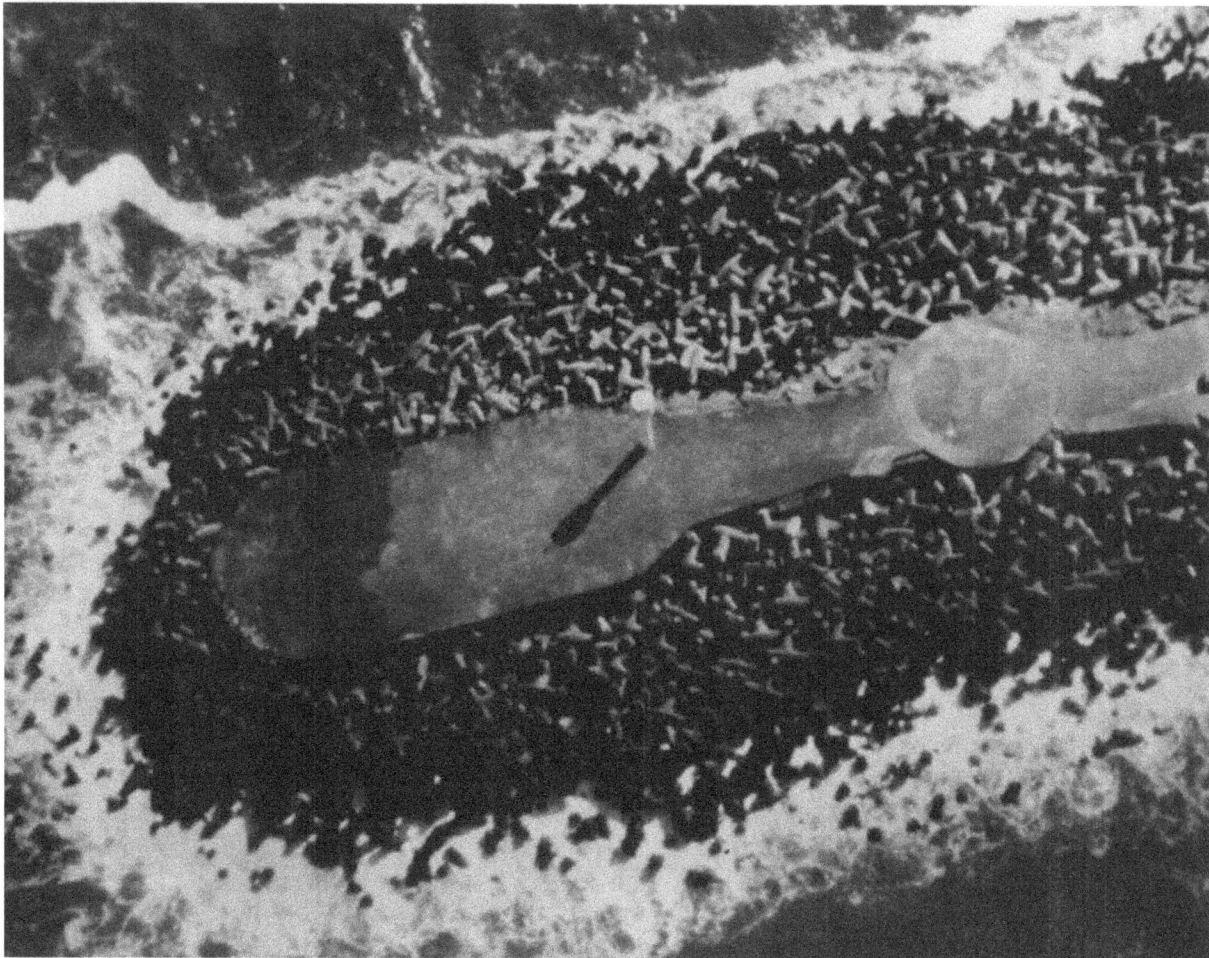

Figure VI-8-10. North jetty at entrance to Humboldt Bay, California

    – Removal of the dolosse armor layer and placement of filter material over the existing core at 1:2 slope. A new underlayer would be placed, protected by concrete cubes varying between 45,359.2 to 81,646.6 kg (50 to 90 tons).

    – Similar to option 1, but the new filter, underlayer, and armor layer would be placed at a 1:3 slope. Primary armor would be smaller than option 1, and a toe berm would be included.

    • Groeneveld, Mol, Nieuwenhuys (1984) stated that the cost of option 2 would be 50 percent higher than option 1. No details were given about repair of the concrete crown wall.

VI-8-4. References.

EM 1110-2-1003
Hydrographic Surveying

EM 1110-2-1004
Coastal Project Monitoring

Aguirre and Plotkin 1998
Aguirre, R., and Plotkin, D. 1998. "REMR Management Systems - BREAKWATER Computer
Program User Manual (Version 1.0)," Technical Report REMR-OM-20, U.S. Army Construction
Engineering Research Laboratory, Champaign, IL.
(*http://owww.cecer.army.mil/fl/remr/break.html*)

Baird et al. 1980
Baird, W. F., Caldwell, J. M., Edge, B. L., Magoon, O. T., and Treadwell, D. D. 1980. "Report
on the Damage to the Sines Breakwater, Portugal." *Proceedings of the 17th International
Coastal Engineering Conference*. American Society of Civil Engineers, Vol 3, pp 3063-3077.

Barends et al. 1983
Barends, F. B., van der Kogel, H., Uijttewaal, F. J., and Hagenaar, J. 1983. "West Breakwater -
Sines, Dynamic-Geotechnical Stability of Breakwaters." *Proceedings of Coastal Structures '83*.
ASCE, pp 31-44.

Bottin 1988a
Bottin, R. R., Jr. 1988a. "Case Histories of Corps Breakwater and Jetty Structures: Report 1,
South Pacific Division," Technical Report REMR-CO-3, U.S. Army Engineer Waterways
Experiment Station, Vicksburg, MS.

Bottin 1988b
Bottin, R. R., Jr. 1988b. "Case Histories of Corps Breakwater and Jetty Structures: Report 3,
North Central Division," Technical Report REMR-CO-3, U.S. Army Engineer Waterways
Experiment Station, Vicksburg, MS.

Camfield and Holmes 1992
Camfield, F. E., and Holmes, C. M. 1992. "Monitoring Completed Coastal Projects - Lessons
Learned I," Coastal Engineering Technical Note, CETN-III-50, U.S. Army Engineer Waterways
Experiment Station, Vicksburg, MS.

Camfield and Holmes 1995
Camfield, F. E., and Holmes, C. M. 1995. "Monitoring Completed Coastal Projects," *Journal of
Performance of Constructed Facilities*, ASCE, Vol 9, No. 3, pp 161-171.

Carver 1989a
Carver, R. D. 1989a. "Prototype Experience with the Use of Dissimilar Armor for Repair and
Rehabilitation of Rubble-Mound Coastal Structures," Technical Report REMR-CO-2, U.S.
Army Engineer Waterways Experiment Station, Vicksburg, MS.

Carver 1989b
Carver, R. D. 1989b. "Stability of Dolos Overlays for Rehabilitation of Stone-Armored
Rubble-Mound Breakwater Heads Subjected to Breaking Waves," Technical Report
REMR-CO-9, U.S. Army Engineer Waterways Experiment Station, Vicksburg, MS.

Carver and Wright 1988a
Carver, R. D., and Wright, B. J. 1988a. "Stability of Dolos and Tribar Overlays for Rehabilitation of Stone-Armored Rubble-Mound Breakwater and Jetty Trunks Subjected to Breaking Waves," Technical Report REMR-CO-4, U.S. Army Engineer Waterways Experiment Station, Vicksburg, MS.

Carver and Wright 1988b
Carver, R. D., and Wright, B. J. 1988b. "Stability of Dolos Overlays for Rehabilitation of Dolos-Armored Rubble-Mound Breakwater and Jetty Trunks Subjected to Breaking Waves," Technical Report REMR-CO-5, U.S. Army Engineer Waterways Experiment Station, Vicksburg, MS.

Carver and Wright 1988c
Carver, R. D., and Wright, B. J. 1988c. "Stability of Dolos Overlays for Rehabilitation of Tribar-Armored Rubble-Mound Breakwater and Jetty Trunks Subjected to Breaking Waves," Technical Report REMR-CO-6, U.S. Army Engineer Waterways Experiment Station, Vicksburg, MS.

Cialone 1984
Cialone, M. A. 1984. "Monitoring Rubble-Mound Coastal Structures with Photogrammetry," Coastal Engineering Technical Note, CETN III-21, U.S. Army Engineer Waterways Experiment Station, Vicksburg, MS.

CIRIA/CUR 1991
Construction Industry Research and Information Association (CIRIA) and Centre for Civil Engineering Research and Codes (CUR). 1991. "Manual on the Use of Rock in Coastal and Shoreline Engineering," CIRIA Special Publication 83/CUR Report 154, CIRIA, London and CUR, The Netherlands.

Davidson 1971
Davidson, D. D. 1971. "Proposed Jetty Head Repair Sections, Humboldt Bay, California," Technical Report H-71-8, U.S. Army Engineer Waterways Experiment Station, Vicksburg, MS.

Edge et al. 1994
Edge, B. L., Magoon, O. T., Davidson, D. D., Hanson, D., and Sloan, R. L. 1994. "The History of the Humboldt Jetties, Eureka, California, 1880 to 1994." *Proceedings of Case Histories of the Design Construction, and Maintenance of Rubble Mound Structures.* ASCE, pp 172-194.

Gorman, Morang, and Larson 1998
Gorman, L., Morang, A, and Larson, R. 1998. "Monitoring the Coastal Environment; Part III: Mapping, Shoreline Changes, and Bathymetric Analysis," *Journal of Coastal Research*, Vol 14, No. 1, pp 61-92.

Groeneveld, Mol, and Zwetsloot 1983
Groeneveld, R. L., Mol, A., and Zwetsloot, P. A. 1983. "West Breakwater - Sines, New Aspects of Armor Units." *Proceedings of Coastal Structures '83.* ASCE, pp 45-56.

Groeneveld, Mol, and Nieuwenhuys 1984
Groeneveld, R. L., Mol, A., and Nieuwenhuys, E. H. 1984. "Rehabilitation Methods for Damaged Breakwaters." *Proceedings of the 19th International Coastal Engineering Conference.* ASCE, Vol 3, pp 2467-2486.

Hanson 1994
Hanson, D. 1994. "Dolosse Repair of Humboldt Bay Breakwaters: Construction Management Aspects." *Proceedings of Case Histories of the Design Construction, and Maintenance of Rubble Mound Structures.* ASCE, pp 195-210.

Hemsley 1990
Hemsley, M. J. 1990. "Monitoring Completed Coastal Projects - Status of a Program," *Journal of Coastal Research*, Vol 6, No. 2, pp 253-263.

Hughes et al. 1995a
Hughes, S. A., Prickett, T. L., Tubman, M. W., and Corson, W. D. 1995a. "Monitoring of the Yaquina Bay Entrance North Jetty at Newport, Oregon; Summary and Results," Technical Report CERC-95-9, U.S. Army Engineer Waterways Experiment Station, Vicksburg, MS.

Hughes et al. 1995b
Hughes, S. A., Peak, R. C., Carver, R. D., Francis, J. D., and Bertrand, G. M. 1995b. "Investigation of Damage at the Yaquina Bay North Jetty." *Proceedings of Case Histories of the Design, Construction, and Maintenance of Rubble Mound Structures.* ASCE, New York, pp 227-249.

Kendall 1988
Kendall, T. R. 1988. "Analysis of 42-ton Dolos Motions at Crescent City." *Proceedings of the 21st International Coastal Engineering Conference.* ASCE, New York, Vol 3, pp 2129-2143.

Kucharski and Clausner 1989
Kucharski, W. M., and Clausner, J. E. 1989. "Side Scan Sonar for Inspection of Coastal Structures," REMR Technical Note CO-SE-1.4, U.S. Army Engineer Waterways Experiment Station, Vicksburg, MS.

Kucharski and Clausner 1990
Kucharski, W. M., and Clausner, J. E. 1990. "Underwater Inspection of Coastal Structures Using Commercially Available Sonars," REMR Technical Report CO-11, U.S. Army Engineer Waterways Experiment Station, Vicksburg, MS.

Larson, Morang, and Gorman 1997
Larson, R., Morang, A., and Gorman, L. 1997. "Monitoring the Coastal Environment; Part II: Sediment Sampling and Geotechnical Methods," *Journal of Coastal Research*, Vol 13, No. 2, pp 308-330.

Markle 1986
Markle, D. G. 1986. "Stability of Rubble-Mound Breakwater and Jetty Toes; Survey of Field Experience," Technical Report REMR-CO-1, U.S. Army Engineer Waterways Experiment Station, Vicksburg, MS.

Markle 1989
Markle, D. G. 1989. "Stability of Toe Berm Armor Stone and Toe Buttressing Stone on Rubble-Mound Breakwaters and Jetties," Technical Report REMR-CO-12, U.S. Army Engineer Waterways Experiment Station, Vicksburg, MS.

McGehee 1987
McGehee, D. D. 1987. "Crane Survey of Submerged Rubble-Mound Coastal Structures," REMR Technical Note CO-SE-1.2, U.S. Army Engineer Waterways Experiment Station, Vicksburg, MS.

Melby and Turk 1994
Melby, J. A., and Turk, G. F. 1994. "Concrete Armor Unit Performance in Light of Recent Research Results." *Proceedings of Case Histories of the Design Construction, and Maintenance of Rubble Mound Structures*. ASCE, pp 48-67.

Mol et al. 1983
Mol, A., Ligeringen, H., Groeneveld, R. L., and Pita, C. R. 1983. "West Breakwater - Sines, Study of Armour Stability." *Proceedings of Coastal Structures '83*. ASCE, pp 57-70.

Morang, Larson, and Gorman 1997a
Morang, A., Larson, R., and Gorman, L. 1997a. "Monitoring the Coastal Environment; Part I: Waves and Currents," *Journal of Coastal Research*, Vol 13, No. 1, pp 111-133.

Morang, Larson, and Gorman 1997b
Morang, A., Larson, R., and Gorman, L. 1997b. "Monitoring the Coastal Environment; Part III: Geophysical and Research Methods," *Journal of Coastal Research*, Vol 13, No. 4, pp 1064-1085.

Mynett, de Voogt, and Schemltz 1983
Mynett, A. E., de Voogt, W. J., and Schmeltz, E. J. 1983. "West Breakwater - Sines, Wave Climatology." *Proceedings of Coastal Structures '83*. ASCE, pp 17-30.

Oliver et al. 1996
Oliver, J., Plotkin, D., Lesnik, J., Pirie, D. 1996. "Condition and Performance Rating System for Breakwaters and Jetties." *Proceedings of the 25th International Coastal Engineering Conference*. ASCE, New York, Vol 2, pp 1852-1861.

Oliver et al. 1998
Oliver, J., Plotkin, D., Lesnik, J., Pirie, D. 1998. "Condition and Performance Rating Procedures for Rubble Breakwaters and Jetties," Technical Report REMR-OM-24, U.S. Army Construction Engineering Research Laboratory, Champaign, IL.
(*http://owww.cecer.army.mil/techreports/plorub/plorubb.remr.post.pdf*)

Parsons and Lillycrop 1988
Parsons, L. E., and Lillycrop, W. J. 1988. "The SHOALS system - A Comprehensive Surveying Tool," Coastal Engineering Technical Note, CETN VI-31, U.S. Army Engineer Waterways Experiment Station, Vicksburg, MS.

Pope 1992
Pope, J. 1992. "Our Ageing Coastal Infrastructure." *Proceedings of Coastal Engineering Practice >92*. ASCE, New York, pp 1055-1068.

Prickett 1996
Prickett, T. L. 1996. "Coastal Structure Underwater Inspection Technologies," Coastal Engineering Technical Note, CETN III-62, U.S. Army Engineer Waterways Experiment Station, Vicksburg, MS.

Prickett 1998
Prickett, T. L. 1998. "Coastal Structure Inspection Technologies; Investigation of Multibeam Sonars for Coastal Structure Surveys," REMR Technical Report CO-19, U.S. Army Engineer Waterways Experiment Station, Vicksburg, MS.

PSIP 1982
Port Sines Investigation Panel. 1982. *Failure of the Breakwater at Port Sines, Portugal.* ASCE, New York.

RMSC 1994
Rubble Mound Structures Committee, ASCE. 1994. Reconstruction of the West Breakwater at Port Sines, Portugal. ASCE, New York.

Rosati and Denes 1990
Rosati, J. D., and Denes, T. A. 1990. "Field Evaluation of Port Everglades, Florida, Rehabilitation of South Jetty by Void Sealing," Technical Report REMR-CO-15, U.S. Army Engineer Waterways Experiment Station, Vicksburg, MS.

Sargent 1988
Sargent, F. E. 1988. "Case Histories of Corps Breakwater and Jetty Structures: Report 2, South Atlantic Division," Technical Report REMR-CO-3, U.S. Army Engineer Waterways Experiment Station, Vicksburg, MS.

Sargent and Bottin 1989a
Sargent, F. E., and Bottin, R. R., Jr. 1989a. "Case Histories of Corps Breakwater and Jetty Structures: Report 7, New England Division," Technical Report REMR-CO-3, U.S. Army Engineer Waterways Experiment Station, Vicksburg, MS.

Sargent and Bottin 1989b
Sargent, F. E., and Bottin, R. R., Jr. 1989b. "Case Histories of Corps Breakwater and Jetty Structures: Report 8, Lower Mississippi Valley Division," Technical Report REMR-CO-3, U.S. Army Engineer Waterways Experiment Station, Vicksburg, MS.

Sargent and Bottin 1989c
Sargent, F. E., and Bottin, R. R., Jr. 1989c. "Case Histories of Corps Breakwater and Jetty Structures: Report 9, Southwestern Division," Technical Report REMR-CO-3, U.S. Army Engineer Waterways Experiment Station, Vicksburg, MS.

Sargent, Markle, and Grace 1988
Sargent, F. E., Markle, D. G., and Grace, P. J. 1988. "Case Histories of Corps Breakwater and Jetty Structures: Report 4, Pacific Ocean Division," Technical Report REMR-CO-3, U.S. Army Engineer Waterways Experiment Station, Vicksburg, MS.

Simpson 1989
Simpson, D. P. 1989. "State-of-the-Art Procedures for Sealing Coastal Structures with Grouts and Concretes," Technical Report REMR-CO-8, U.S. Army Engineer Waterways Experiment Station, Vicksburg, MS.

Simpson and Thomas 1990
Simpson, D. P., and Thomas, J. L. 1990. "Laboratory Techniques for Evaluating Effectiveness of Sealing Voids in Rubble-Mound Breakwaters and Jetties with Grouts and Concretes," Technical Report REMR-CO-13, U.S. Army Engineer Waterways Experiment Station, Vicksburg, MS.

Simpson et al. 1990
Simpson, D. P., Rosati, J. D., Hales, L. Z., Denes, T. A., and Thomas, J. L. 1990. "Rehabilitation of Permeable Breakwaters and Jetties by Void Sealing: Summary Report," Technical Report REMR-CO-16, U.S. Army Engineer Waterways Experiment Station, Vicksburg, MS.

Smith 1988
Smith, E. R. 1988. "Case Histories of Corps Breakwater and Jetty Structures: Report 5, North Atlantic Division," Technical Report REMR-CO-3, U.S. Army Engineer Waterways Experiment Station, Vicksburg, MS.

Thomas 1985
Thomas, J. 1985. "Diver Inspection of Coastal Structures," REMR Technical Note CO-SE-1.1, U.S. Army Engineer Waterways Experiment Station, Vicksburg, MS.

Toppler et al 1983
Toppler, J. F., Steenmeyer, A., Goncalves da Silva, M. A., Ligteringen, H., and Silverira Ramos, F. 1983. "West Breakwater - Sines, Overview of Rehabilitation and Synthesis of Project." *Proceedings of Coastal Structures '83*. ASCE, pp 3-16.

Turk and Melby 1997
Turk, G. F., and Melby, J. A. 1997. "Preliminary 3-D Testing of CORE-LOC^{TM} as a Repair Concrete Armor Unit for Dolos-Armored Breakwater Slopes," Technical Report REMR-CO-18, U.S. Army Engineer Waterways Experiment Station, Vicksburg, MS.

U.S. Army Engineer Waterways Experiment Station 1991
U.S. Army Engineer Waterways Experiment Station. 1991. "Surveys of Coastal Structures," Coastal Engineering Technical Note, CETN III-41, U.S. Army Engineer Waterways Experiment Station, Vicksburg, MS.

Vrijling, Leeuwestein, and Kuiper 1995
Vrijling, J. K., Leeuwestein, W., and Kuiper, H. 1995. "The Maintenance of Hydraulic Rock Structures," in *River, Coastal and Shoreline Protection; Erosion Control Using Riprap and Armourstone*, Thorne, Abt, Barends, Maynord, and Pilarczyk, ed., John Wiley & Sons, Inc., New York, pp 651-666.

Ward 1988
Ward, D. L. 1988. "Case Histories of Corps Breakwater and Jetty Structures: Report 6, North Pacific Division," Technical Report REMR-CO-3, U.S. Army Engineer Waterways Experiment Station, Vicksburg, MS.

Ward and Markle 1990
Ward, D. L., and Markle, D. G. 1984. "Repair of Localized Armor Stone Damage on Rubble-Mound Structures," Technical Report REMR-CO-14, U.S. Army Engineer Waterways Experiment Station, Vicksburg, MS.

Weggel et al. 1994
Weggel, J. R., Baird, W. F., Edge, B., Magoon, O. T., Mansard, E., Threadwell, D. D., and Whalin, R. W. 1994. "Sines Breakwater Revisited - Repair and Reconstruction." *Proceedings of Case Histories of the Design Construction, and Maintenance of Rubble Mound Structures.* ASCE, pp 272-280.

Wolf 1989
Wolf, J. R. 1989. "Study of Breakwaters Constructed with one Layer of Armor Stone; Detroit District," Technical Report REMR-CO-10, U.S. Army Engineer Waterways Experiment Station, Vicksburg, MS.

VI-8-5. Acknowledgments.

Author: Dr. Steven A. Hughes, Coastal and Hydraulics Laboratory (CHL), U.S. Army Engineer Research and Development Center, Vicksburg, MS.

Reviewers: Donald D. Davidson and Lyndell Z. Hales, Coastal and Hydraulics Laboratory (CHL), U.S. Army Engineer Research and Development Center, Vicksburg, MS.

VI-8-6. Symbols.

CH	Structural index for channel/harbor-side slope
CR	Structural index for crest or cap
FI	Structural index for the structure
FI_R	Structural index for the reach

H_s	Significant wave height (length)
I_H	Highest of the reach sectional indices
I_L	Lowest of the three cross-sectional indices
I_L	Lowest of the reach structural indices
I_M	Middle value of the three cross-sectional indices
$I_2, I_3, I_4, ...$	Values for the second, third, fourth, etc., highest reach indices
N	Number of rated functions for the reach
N	Number of reaches in the structure
R_H	Highest of the functional ratings for the reach
R_I	Highest of the five ratings for the cross-section component
R_L	Lowest of the functional ratings for the reach
$R_2, R_3, R_4, ...$	Values for the second, third, fourth, etc., highest ratings
R_5	Lowest of the five ratings for the cross-section component
SE	Structural index for sea-side slope
SI	Structural index for the structure
SI_R	Structural index for the reach

www.ingramcontent.com/pod-product-compliance
Lightning Source LLC
Chambersburg PA
CBHW081346190326
41458CB00018B/6093